ELASTIC WAVE PROPAGATION

NORTH-HOLLAND SERIES IN

APPLIED MATHEMATICS AND MECHANICS

EDITORS:

J. D. ACHENBACH
Northwestern University

B. BUDIANSKY
Harvard University

H. A. LAUWERIER
University of Amsterdam

P. G. SAFFMAN
California Institute of Technology

L. VAN WIJNGAARDEN
Twente University of Technology

J. R. WILLIS
University of Bath

VOLUME *35*

NORTH-HOLLAND-AMSTERDAM • NEW YORK • OXFORD • TOKYO

ELASTIC WAVE PROPAGATION

Proceedings of the Second I.U.T.A.M. - I.U.P.A.P. Symposium on Elastic Wave Propagation, Galway, Ireland, March 20–25, 1988

Edited by:

M. F. McCARTHY and M. A. HAYES
Department of Mathematical Physics
University College
Galway, Ireland

1989

NORTH-HOLLAND-AMSTERDAM • NEW YORK • OXFORD • TOKYO

ISBN: 0 444 87272 8

Publishers:

ELSEVIER SCIENCE PUBLISHERS B.V.
P.O. Box 1991
1000 BZ Amsterdam
The Netherlands

Sole distributors for the U.S.A. and Canada:

ELSEVIER SCIENCE PUBLISHING COMPANY. INC.
655 Avenue of the Americas
New York, N.Y. 10010
U.S.A.

Library of Congress Cataloging-in-Publication Data

Elastic wave propagation / edited by M.F. McCarthy and M.A. Hayes.
p. cm. -- (North-Holland series in applied mathematics and mechanics ; v. 35)
"Volume contains ... contributions presented at the I.U.T.A.M. -I.U.P.A.P. Symposium on Elastic Wave Propagation held in University College, Galway, Ireland from March 20-25, 1988"--Pref.
Includes bibliographies and index.
ISBN 0-444-87272-8 (U.S.)
1. Elastic waves--Congresses. 2. Wave-motion, Theory of--Congresses. I. McCarthy, M. F. (Matthew F.) II. Hayes, M. A. (Michael A.) III. I.U.T.A.M.-I.U.P.A.P. Symposium on Elastic Wave Propagation (1988 : University College, Galway) IV. International Union of Theoretical and Applied Mechanics. V. International Union of Pure and Applied Physics. VI. Series.
QA935.E415 1989
531'.1133--dc19 88-38986
CIP

PRINTED IN THE NETHERLANDS

PREFACE

This volume contains the texts of the contributions presented at the I.U.T.A.M. – I.U.T.A.P. Symposium on Elastic Wave Propagation held in University College, Galway, Ireland from March 20-25, 1988.

The first I.U.T.A.M. Symposium on Elastic Wave Propagation was held in 1977 at Northwestern University, Evanston, U.S.A. In constructing the programme for this second Symposium the International Scientific Committee was deeply conscious of the significant progress that has been made since 1977 and aimed at as broad a coverage as possible of the area. Because of the tremendous interest shown in the Symposium together with the excellent standard of the contributions submitted, a greater number of papers than is customary at such Symposia was accepted for presentation.

Seven distinguished scholars were invited to deliver sessional lectures which surveyed recent developments. A wide range of theoretical, experimental and numerical research was presented and discussed and an overview session in which an effort was made to identify new significant problems was held. There was an extremely good mix of topics presented and this volume contains topics as diverse as surface existence theory, spreading of nonlinear surface waves, and propagation through anisotropic microcracked rocks.

Financial support was provided by the International Union of Theoretical and Applied Mechanics, the International Union of Pure and Applied Physics, the Royal Irish Academy and University College, Galway. It would not, however, have been possible to organise the Symposium without the most generous support of the U.S. Office of Naval Research and the many Irish commerical organisations who made financial grants to the Symposium. The generosity of all our sponsors is greatly appreciated.

We are extremely grateful to the members of the International Scientific Committee for their advice and encouragement. The preparatory management of the Symposium and related events were the responsibility of the local organising committee and we wish to express particular thanks to Dr. P.M. O'Leary and Professor J.N. Flavin who did much to ensure the success of the meeting. We also wish to express our deepest appreciation to Ms. Noelle Scully who prepared the documentation for the Symposium and who, together with Ms. Maeve McCarthy, helped to make the Symposium successful as well as enjoyable for the participants.

We hope that our efforts in organising this Symposium and editing its proceedings will motivate researchers to further study resulting in significant advances in the field of Elastic Wave Propagation.

July, 1988

Matthew F. McCarthy
Michael A. Hayes

Opening Address of Dr. Colm O'hEocha, President, University College, Galway

It is my pleasure to welcome delegates from around the globe to University College Galway to attend a prestigious Symposium on Elastic Waves. The selection of the location is a tribute to the international standing of Galway/Irish researchers in this area of science.

The meeting is, of course, held under joint auspices with the Royal Irish Academy. The first volume of the *Transactions* of the Academy (1787) contained the following:

"The researches of the mathematician are the only sure ground on which we can reason from experiments; and how far experimental science may assist the commercial interest of a state is clearly evinced by the success of those several manufacturers in the neighbouring countries of England and France, where the hand of the artificier has taken its direction from the philosopher".

It is in this spirit that I express the hope that economic, no less than intellectual and academic benefit will accrue to Ireland as a result of this Symposium being held here.

University Presidents, as they age, become repositories of out-of-date knowledge and so are increasingly more valued for what little wisdom they possess. Therefore, I asked Professor McCarthy for advice on the historical basis for holding the meeting at Galway. We searched the list of Joseph Larmor's publications and found some references to elasticity.

Joseph Larmor was Professor of Natural Philosophy in Galway during the period 1880-85.

When he died in his native Antrim in 1942, an obituary was written by D'Arcy Wentworth Thompson, whose father was Larmor's contemporary at Galway — as Professor of Greek. He noted that the young Larmor came out of St. John's College

Cambridge as Senior Wrangler in 1880 with J.J. Thompson in second place.
"In the same year (1880) being twenty-three years old, he went to Queen's College, Galway, as Professor of Natural Philosophy ... Galway was then a nest of scholars, like one of the little old German Universities – as Larmor said, looking backward later on. My father was there, and Davies, the learned editor of the Eumenides; Allman was there, writing his small but famous book on the Greek Geometers; Rowney and King, chemist and geologist, were demolishing the old figment of *Eozoon canadense*; and Larmor came and wrote his first papers, including one on *Least Action as the Fundamental Formulation in Dynamics and Physics* (1884). It was a subject of which he never tired; it was a confession of his scientific faith. He spent five happy years in Galway, went back to Cambridge not without reluctance. There in 1903 he succeeded Stokes in the Lucasian Professorship – Isaac Newton's Chair – and held it for nearly thirty years.

It is topical to recall that, as D'Arcy Thompson says, Larmor, like many old Galway students and staff was unionist Irish. Towards the end of his days he referred to the contributions of such as Kelvin and Maxwell, and could not forgive the new generation for forgetting, as he believed they did, "that the Scoto-Irish school of Physics dominated the world in the middle of the last century, but has now vanished from the face of the earth".

In another letter on the relation of Time to Natural Theology, Larmor wrote that he hoped that his reflections might bear fruit "when the present phase has worn off, and people are willing to consider ideas not based on the negation of the great age of Mathematical Physics".

Sir D'Arcy Wentworth Thompson himself (1860-1948) held chairs of Natural History at Dundee University, and later, at St. Andrews. His great seminal work, in which he linked Mathematics and Biology, was on *Growth and Form* (1917). Sir Peter Medawer, Professor of Zoology, and Nobel Laureate in Medicine, wrote as follows of D'Arcy in *The Art of the Soluble*:

"D'Arcy Wentworth Thompson was an aristocrat of learning whose intellectual endowments are not likely ever again to be combined within one man. He was a classicist of sufficient distinction to have become President of the Classical Associations of England and Wales and of Scotland; a mathematician good enough to have had an entirely mathematical paper accepted for publication by the Royal Society; and a naturalist who held important chairs for sixty-four years, that is, for all but the length of time into which we must nowadays squeeze the whole of our lives from birth until professional retirement. He was a famous conversationalist and lecturer (the two are often thought to go together, but seldom do), and the author of a work which considered as literature, is the equal of anything of Pater's or Logan Pearsall Smith's in its complete mastery of the *bel canto* style. Add to all this that he was over six feet tall, with the build and carriage of a Viking and with the pride of bearing that comes from good looks known to be possessed".

And so, Galway has many contemporary and historic claims, both direct and indirect, to host this Symposium.

I wish to offer my congratulations to the organisers, notably Professor M.F. McCarthy.

It gives me pleasure to declare the Symposium on Elastic Waves officially open.

COMMITTEES

INTERNATIONAL SCIENTIFIC COMMITTEE

M.F. McCarthy, Ireland, Chairman
J.D. Achenbach, U.S.A.
A. Boström, Sweden
M.M. Carroll, U.S.A.
P. Chadwick, U.K.
A.T. de Hoop, The Netherlands
G.M.L. Gladwell, Canada
M.A. Hayes, Ireland
K. Kawata, Japan
A. Morro, Italy
Y.H. Pao, U.S.A.
M. Roseau, France
W. Schiehlen, Germany
H.F. Tiersten, U.S.A.
V.F. Zhuravlev, U.S.S.R.
F. Ziegler, Austria

LOCAL ORGANIZING COMMITTEE

A. Brock
J.N. Flavin
M.A. Hayes
M.F. McCarthy
J.E. Nash
P.M. O'Leary

SESSION CHAIRMEN

D.M. Barnett
A. Boström
L.M. Brock
R. Burridge
M.M. Carroll
P. Chadwick
A.T. de Hoop
G. Herrmann
K. Kawata
A. Mal
G.A. Maugin
J.J. McCoy
P.M. Naghdi
P.M. Quinlan

LIST OF PARTICIPANTS

AUSTRALIA

C.R. Rao

AUSTRIA

P. Borejko
H.P. Rossmanith
F. Ziegler

BELGIUM

Ph. Boulanger

CANADA

S. Dost
J. Haddow
J. Wegner

CHINA, Republic of

M.C. Kuo
H.J. Yang
C.S. Yeh

DENMARK

P.L. Christiansen

FRANCE

N. Daher
J. Dieulesaint
T. Ha-Duong
B. Hosten
P. Joly
G.A. Maugin
M. Osmont
M. Piau
M. Planat
J. Pouget
P. Quentin
M. Rosseau
M.J.M. Servas

GERMANY, Federal Republic of

J. Lenz
F.O. Speck

IRELAND

A. Brock
A. Cox
J.N. Flavin
M.A. Hayes
B. Lenoach
M.F. McCarthy
J.E. Nash
P. O'Leary
P.M. Quinlan
D. Ryan

ISRAEL

A.I. Beltzer
M. Ziv

POLAND

Z. Konczak
W. Kosinski
K. Majorkowska-Knap
B. Maruszewski
J.P. Nowacki

SWEDEN

A. Boström
R.T.K. Hsieh
P. Olsson
L. Petersen

SWITZERLAND

M. Sayir

CONTENTS

Preface v
Opening Address vii
Committees xi
List of Participants xiii

A. ELASTIC SURFACE WAVES

Recent Developments in the Theory of Elastic Surface and Interfacial Waves
P. Chadwick 3

Probing of Acoustic Wave Surface Displacements
E. Dieulesaint and D. Royer 17

Surface Wave Existence Theory for the Case of Zero Curvature Transonic States
D.M. Barnett and J. Lothe 33

Pulsed Cylindrical Elastic Waves in a Half-Space with a Dipping Surface Layer
P. Borejko and F. Ziegler 39

The Influence of Applied Stress on the Response of Thin Films
G.C. Johnson and Shih-Emn Chen 45

The Caustic of a Rayleigh Wave
H.P. Rossmanith 51

Ground Motions Generated by a Finite Fault
C.S. Yeh, P.L. Chen, M.K. Kuo and T.K. Wang 59

Elastic Waves in Layered Media with Interface Features
A.K. Mal and P.-C. Xu 67

Ultrasonic Propagation in Textured Polycrystalline Metals
M. Hirao and H. Fukuoka 75

Propagation of SH-Surface Waves in Elastic Layered Dielectrics
K. Majorkowska-Knap and J. Lenz 81

Attenuation of Elastic Surface Waves Propagating along Rough Surfaces
M. deBilly and G. Quentin 89

Ceramic-Ceramic Composites High Temperature Characterization
B. Hosten and S. Baste 95

Ground Motion Amplification on Alluvial Valleys
A. Boström, S.K. Datta and P. Olsson 101

Elastic Interfacial Standing Waves
J. Dunwoody 107

Ray Approximation and Caustics Formation inside an Anisotropic Elastic Half Space
M. Planat and Y. Zhang 113

B. NONLINEAR ELASTIC WAVES

Spreading of Nonlinear Surface Waves on Piezoelectric Solids
D.F. Parker and E.A. David 125

Steady, Structured Shock Waves of Arbitrary Intensity in Thermoelastic Materials
J.E. Dunn and R.L. Fosdick 133

The Evolutionary Behaviour of Plane Transverse Weak Nonlinear Shock Waves in Unstrained Incompressible Isotropic Elastic Non-conductors
Y.B. Fu and N.H. Scott 141

Nonlinear Waves of Small Amplitude in Anisotropic Elastic Solids
N. Daher and G.A. Maugin 147

Nonlinear Interaction between Optical and Acoustical Waves in Diatomic Elastic Solids
M. Teymur 155

Finite Amplitude Wave Propagation in a Stretched Elastic String
J. Wegner, J.B. Haddow and R.J. Tait 161

Solitary Waves on Nonlinear Elastic Rods
P.L. Christiansen, V. Muto and M.P. Soerensen 167

The Evolution of Resonant Acoustic Oscillations with Damping
E.A. Cox and M.P. Mortell 173

Dynamics and Stability of Twin Boundaries in Martensitic Transformations
J. Pouget 179

C. WAVE PROPAGATION IN LAYERED AND BOUNDED MEDIA

Stress Wave Propagation in a Laminated Sandwich Plate
W.A. Green 189

Beam and Mode Analysis of Weak Bonding Flaws in a Layered Aluminum Plate 195
I.T. Lu, L.B. Felsen, J.M. Klosner and C. Gabay

Wave Propagation on Thin-Walled Elastic Cylindrical Shells
A.D. Pierce 205

Nonspecular Reflection of Bounded Acoustic Beams from Layered Viscoelastic Halfspaces
F.B. Jensen and H. Schmidt 211

Scattering of Point-Source Generated Transient Elastodynamic Waves by a Semi-Infinite Void Crack in a Half-Space
B.J. Kooij 217

Analysis of Elastic Bar Method for Materials Characterization in High Velocity Tension
K. Kawata and M. Itabashi 223

Pulse Evolution in a Multimode, One Dimensional Highly Discontinuous Medium
R. Burridge and H.W. Chang 229

Wave Propagation from a Point Source in a Wedge-Shaped Layer on Top of a Half-Space
Y.-H. Pao and F. Ziegler 235

Elastic Guided Waves in Heterogeneous Media – Mathematical Analysis
A. Bamberger, Y. Dermenjian and P. Joly 241

Scattering by Fluid-Loaded Cöplanar Half-Plane Plates
P.R. Brazier-Smith 247

Wave Propagation in Thin Elastic Bodies
C. Molina and F. Pastrone 253

Group Theoretic Formulation for Elastic Properties of Media Composed of Anisotropic Layers
M. Schoenberg and F. Muir 259

Steady Propagation of Buckle in Elastic Pipelines
N. Sugimoto 265

Coupled Wave Motion of a Layered Periodic Beam
H.J. Yang 273

Love Waves in a Fluid-Saturated Porous Anisotropic Layer Resting on a Heterogeneous Elastic Half-Space
Z. Kończak 281

Coupled Thermoeleastic Waves in Periodically Laminated Plates
S.Y. Lee, L.C. Chin and H.J. Yang 287

Direct Inversion for a Layered Elastic Medium
J.A.H. Alkemade and J.W. de Maag 293

D. FLUID/SOLID WAVE INTERACTION

Wave Propagation in Submerged Layered Composites
A.M.B. Braga and G. Herrmann 301

Elastic Wave Scattering from an Interface Crack in a Layered Half Space Submerged in Water
D.B. Bogy and S.M. Gracewski 307

Vibrations of a Point-Excited Infinite Elastic Plate Interacting with a Fluid
M.J. Ghaleb and M. Piau 309

Elastodynamic Radiation from an Impulsive Point Source in a Poroelastic Fluid/Solid Medium
S.M. de Vries 315

On Variational Formulations of Boundary Value Problems for Fluid-Solid Interactions
G.C. Hsiao, R.E. Kleinman and L.S. Schuetz 321

A Simple Self-Consistent Analysis of Dispersion and Attenuation of Elastic Waves in a Porous Medium
F.J. Sabina and J.R. Willis 327

E. SCATTERING OF ELASTIC WAVES

Resonance Acoustic Scattering from Underwater Elastic Bodies
G.C. Gaunaurd 335

Transient Direct and Inverse Scattering
J. Corones 347

Propagation Modelling Based on Wave Field Factorization and Path Integration
J.J. McCoy 357

Elastodynamic Time-Domain Reciprocity Theorems for Solids with Relaxation
A.T. de Hoop and H.J. Stam 367

Elastodynamic Scattering and Matrix Factorization
E. Meister and F.O. Speck 373

Scattering of Elastic Waves by Inclusions Surrounded by Thin Interface Layers
P. Olsson, S.K. Datta and A. Boström 381

Single Integral Equations for Scattering of Elastic Waves by an Elastic Inclusion
P.A. Martin

Resonances in the Backscattered Echoes from Elastic Shells near an Interface
M.F. Werby and G.C. Gaunaurd

Dynamic Stress Intensity Factors for 3D Non-Planar Cracks
P. Olsson and A. Boström

Optimal Geometrical and Physical Bounds for Elastic Rayleigh Scattering
G. Dassios

Beyond the Born Approximation
R.T. Coates and C.H. Chapman

The Effect of Near-Tip Residual Cohesive Tractions
M.K. Kuo, J.D. Achenbach and H.J. Yang

The Dynamic Response of Random Viscoelastic Composites
A.I. Beltzer

Elastic Wave Propagation in Anisotropic Microcracked Rocks
C.M. Sayers

Elastic Wave Scattering by Two Penny-Shaped Cracks
L. Peterson

On the Boundary Integral Equation for Flat Cracks
T. Ha-Duong

Born Series Approach Applied to Three Dimensional Elastodynamic Inclusion Analysis by BIE Methods
M. Kitahara and K. Nakagawa

Elastic and Thermoelastic Scattering Problems and Non-Homogeneous Media
M. Codegone

Advanced Boundary Element Calculations for Elastic Wave Scattering
W.S. Hall and W.H. Robertson

A Practical Method to Measure the Depth of Cracks in Concrete Structures By Using Diffracted P Waves
M. Nakano 465

Time-Domain Born Approximation to the Far-Field Scattering of Plane Elastic Waves by an Elastic Heterogeneity
D. Quak 471

The Plane Plate Model Applied to the Scattering of the Ultrasonic Waves from Cylindrical Shells
G. Quentin and M. Talmant 477

A Space-Time Finite-Element Method for the Computation of Three-Dimensional Elastodynamic Wave Fields (Theory)
H.J. Stam and A.T. de Hoop 483

F. GENERAL THEORY OF ELASTIC WAVE PROPAGATION

Gaussian Wave Packets in Linear and Nonlinear Anisotropic Elastic Solids
A.N. Norris 491

An Asymptotic Description of an Elastodynamic Beam
J.G. Harris 505

Identification of Elastic Material Symmetry by Acoustic Measurement
S.C. Cowin 511

Solution to Multi-Dimensional Wave Propagation Problems in Solids by the Method of Characteristics
M. Ziv 517

Coupling Effects in Structural Wave Propagation
M. Sayir 523

Exact Solutions for Diffraction – Induced Crack Propagation and Dislocation Emission
L.M. Brock 529

Induced Discontinuities in a Class of Linear Materials with Internal State Parameters
P.J. Chen and A.C. Morro 535

Estimates for Dynamical Solutions in Finite Thermoelastodynamics
M. Aron and R.E. Craine 539

Exterior Boundary Value Problems in Elastodynamics
G.C. Hsiao and W.L. Wendland 545

Circularly Polarised Plane Waves in Transversely-Isotropic Elastic Materials
D. Ryan and M. Hayes 551

G. MAGNETO-THERMO-ELECTROMAGNETO ELASTIC WAVE PROPAGATION

Electroelastic Nonlinearities, Biasing Deformations and Piezoelectric Vibrations
H.F. Tiersten 557

Nonlinear Wave Phenomena in Magnetostrictive Elastic Bodies
G.A. Maugin 569

Acceleration Waves in Elastic Dielectrics with Polarization Inertia
S. Dost 575

Electromagnetic Wave Study of Elastic Wave Propagation
R.E. Green 581

Electromagnetically Coupled Elastic Wave Propagation Problems in Eddy Current Non Destructive Testing
R.K.T. Hsieh 587

Lower Symmetry of the Elastic and Piezoelectric Tensors from Rotational Coupling
D.F. Nelson 595

Inhomogeneous Magnetoelastic Plane Waves
Ph. Boulanger 601

Electromechanical Waves in Elastic Dielectrics
J.P. Nowacki 607

Finite Amplitude Wave Propagation in Magnetized Perfectly Electrically Conducting Elastic Materials
M.M. Carroll and M.F. McCarthy 615

The Stability of Plane Waves in Generalized Thermoelasticity
N.H. Scott 623

Elastic Waves in the Presence of a New Temperature Scale
W. Kosinski 629

Author Index 635

A
ELASTIC SURFACE WAVES

ELASTIC WAVE PROPAGATION
M.F. McCarthy, M.A. Hayes, (Editors)

RECENT DEVELOPMENTS IN THE THEORY OF ELASTIC SURFACE AND INTERFACIAL WAVES

P. CHADWICK

School of Mathematics, University of East Anglia,
Norwich, NR4 7TJ, UK.

An outline of the basic theory of elastic surface waves in the state of development reached in 1977 is followed by a survey of subsequent contributions, including advances in the related theory of interfacial waves.

1. PLANE WAVES AND SURFACE WAVES

We consider a homogeneous elastic body which is anisotropic relative to a natural reference configuration N. The material forming the body is assumed to have a positive definite strain energy, so that the fourth-order linear elasticity tensor **B** in N has the properties

$$B_{jikl} = B_{klij} = B_{ijkl} \tag{1}$$

and

$$B_{pqrs}S_{pq}S_{rs} > 0 \quad \forall \text{ non-zero real symmetric tensors } \mathbf{S}. \tag{2}$$

(All vector and tensor components in this section relate to a fixed orthonormal basis.) The constitutive equations describing the mechanical response of the material to infinitesimal deformations from N are

$$\sigma_{ij} = B_{ijrs}u_{s,r}, \tag{3}$$

where **u** denotes the displacement, σ the stress and $,i$ differentiation with respect to the component x_i of the spatial position **x**. In the absence of external body forces the displacement equations of motion are

$$B_{pirs}u_{s,pr} = \rho\partial^2 u_i/\partial t^2, \tag{4}$$

t being the time and ρ the density in N.

Our basic concern in this article is with plane-wave solutions of (4) of the form

$$\mathbf{u} = \exp\{i\kappa(\mathbf{e}_1.\mathbf{x} + p\mathbf{e}_2.\mathbf{x} - vt)\}\mathbf{a}. \tag{5}$$

The positive real quantities κ and v are respectively a wave number and a speed of propagation, and the plane R spanned by the orthogonal unit vectors $\mathbf{e}_1$ and $\mathbf{e}_2$ is the *reference plane* of the wave. The slowness is

$$\mathbf{s} = v^{-1}(\mathbf{e}_1 + p\mathbf{e}_2), \tag{6}$$

and the propagation condition resulting from the substitution of (5) into (4) can be expressed as

$$\mathbf{Q}(\mathbf{s})\mathbf{a} = \rho\mathbf{a},$$

or

$$[p^2\mathbf{Q}(\mathbf{e}_2) + p\{\mathbf{R}(\mathbf{e}_1,\mathbf{e}_2) + \mathbf{R}(\mathbf{e}_2,\mathbf{e}_1)\} + \mathbf{Q}(\mathbf{e}_1) - \rho v^2\mathbf{I}]\mathbf{a} = \mathbf{0}. \tag{7}$$

Here I is the identity tensor and Q(.) and R(.,.) are the acoustical and associated acoustical tensors defined by the component relations

$$Q_{ij}(v) = B_{pirj}v_pv_r, \quad R_{ij}(v,w) = B_{pirj}v_pw_r \quad \forall \text{ non-zero real vectors } v,w. \tag{8}$$

The traction produced by the wave on a surface element with inward unit normal e_2 is, from (3) and (5),

$$t = -\sigma^T e_2 = -i\kappa\exp\{i\kappa(e_1.x + pe_2.x - vt)\}\ell, \tag{9}$$

where

$$\ell = \{R(e_2,e_1) + pQ(e_2)\}a.$$

It will be convenient to refer to a and ℓ as the *displacement* and *traction amplitudes*.

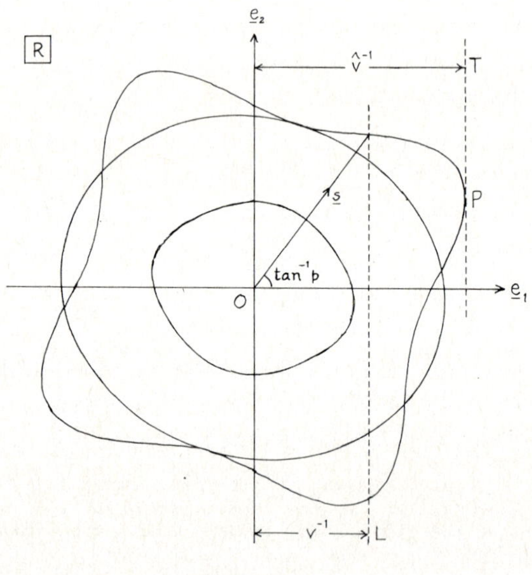

FIGURE 1
Illustration of the slowness section S∩R and the transonic state.

The eigenvalue problem (7) determines the possible values of p and, to within an arbitrary multiplier, a. The eigenvalues are the roots of the sextic equation

$$\det[p^2Q(e_2) + p\{R(e_1,e_2) + R(e_2,e_1)\} + Q(e_1) - \rho v^2 I] = 0: \tag{10}$$

in consequence of (2) they are all real when v is large enough and form three complex conjugate pairs when v is sufficiently small. This limiting behaviour of the p's can be interpreted amd amplified with reference to the slowness surface S of the material, a possible form of which is shown in Figure 1. When p is real, the representative point in slowness space of the wave (5), with position vector s given by (6) relative to the centre O, is at the intersection with S of the line L in R parallel to e_2 and at perpendicular distance 1/v from O. In Figure 1, L has four points of intersection with S∩R, indicating that (10) has four real and two conjugate complex roots for this particular value of v. The plane wave (5) is homogeneous when p is real and inhomogeneous when p is complex. When L lies to the right of the tangent T, all six eigenvalues are complex. The speed v is then in the interval $I = [0,\hat{v})$, $\hat{v}$ being the reciprocal of the perpendicular distance of O from T. T defines the *transonic state* in R relative to $\mathbf{e}_2$; the homogeneous plane wave represented by P, the point of contact of T with S, is the *limiting wave*, $\hat{v}$ the *limiting speed* and I the *subsonic interval* [1,Sect.VI]. In the transonic state illustrated in Figure 1, T touches a single sheet of S just once and the state is accordingly said to be of *type 1*. There are five other types of transonic states in which n, the number of limiting waves, is two or three [1,Sect.VI,C], but it is not necessary to particularize them here.

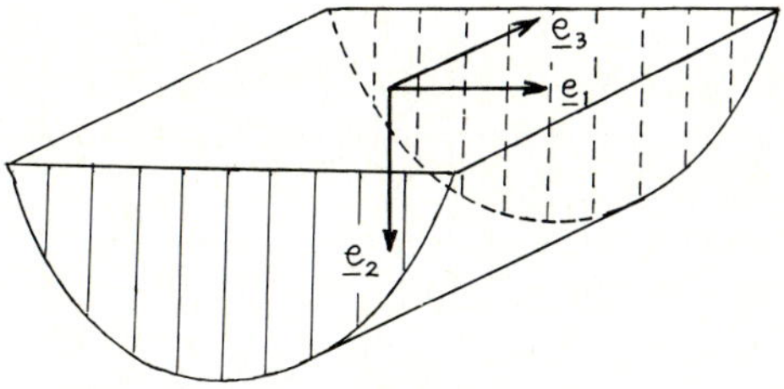

FIGURE 2
Definition of the surface-wave basis $\{\mathbf{e}_1,\mathbf{e}_2,\mathbf{e}_3\}$.

A surface wave in the semi-infinite body sketched in Figure 2, propagating in the direction of $\mathbf{e}_1$, is a linear combination of inhomogeneous plane waves of the form (5) leaving the plane boundary $\mathbf{e}_2.\mathbf{x} = 0$ traction free and satisfying the decay condition

$$\mathbf{u} \to 0 \text{ as } \mathbf{e}_2.\mathbf{x} \to \infty. \tag{11}$$

In view of (11) we can only use complex roots of (10) with positive real part in the construction of a surface wave. If p_α, $\alpha = 1,2,3$, are the roots in question and $\mathbf{a}_\alpha$ and $\boldsymbol{\ell}_\alpha$ corresponding amplitudes, the displacement-traction field is

$$\{\mathbf{u},\mathbf{t}\} = \sum_{\alpha=1}^{3} \gamma_\alpha \exp\{i\kappa(\mathbf{e}_1.\mathbf{x} + p_\alpha \mathbf{e}_2.\mathbf{x} - v_S t)\}\{\mathbf{a}_\alpha, -i\kappa\boldsymbol{\ell}_\alpha\}, \tag{12}$$

and we set

$$\{\mathbf{A},\ \mathbf{L}\} = \sum_{\alpha=1}^{3} \gamma_\alpha\{\mathbf{a}_\alpha,\ \boldsymbol{\ell}_\alpha\}. \tag{13}$$

The boundary condition then takes the form

$$\mathbf{L} = 0, \tag{14}$$

equivalent to

$$[\boldsymbol{\ell}_1,\boldsymbol{\ell}_2,\boldsymbol{\ell}_3] = 0. \tag{15}$$

This is the *secular equation*, fixing the speed of propagation v_S. When v_S is known, the coefficients γ_α are determined, up to a common multiplier, by (14).

A surface wave is *subsonic* when $0 < v_S < \hat{v}$ and *supersonic* when $v_S > \hat{v}$. Since (10) has at least two real roots when $v > \hat{v}$ (see Figure 1), a supersonic surface wave consists of only two inhomogeneous plane waves. This suggests that such waves can only exist when there is a good deal of symmetry in the orientation of the crystallographic axes of the transmitting medium relative to the surface-wave basis $\{\mathbf{e}_1,\mathbf{e}_2,\mathbf{e}_3\}$. Some examples are considered in Section 5.

U.K.

J.D. Abraham
M. Aron
P.R. Brazier-Smith
P. Chadwick
R.T. Coates
E.A. David
J. Dunwoody
N. Dunwoody
Yi-Bin Fu
W.A. Green
A. Harvey
P.A. Martin
D.F. Parker
W.H. Robertson
N. Scott
M. Teymur
G.R. Wickham

U.S.A.

A. Almgren
D.M. Barnett
D.B. Bogy
L.M. Brock
R. Burridge
M. Carroll
P.J. Chen
J. Corones
S.C. Cowin
S.K. Datta
J.E. Dunn
D. Feit
L.B. Felsen
G.C. Gaunaurd
R.E. Green
J.G. Harris
G. Herrmann
G.C. Hsiao
G.C. Johnson
J.J. McCoy
A.K. Mal
L. Mannion
P.M. Naghdi
D.F. Nelson
A.N. Norris
A. Pierce
A. Robertson
M. Schoenberg
L. Schuetz
F.O. Speck
H. Tiersten
A. Tucker

ITALY

M. Codegone
F.B. Jensen
C. Molina
A. Morro
A. Pastrone

JAPAN

M. Hirao
K. Kawata
M. Kitahara
M. Nakano
N. Sugimoto

MEXICO

F.J. Sabina

NETHERLANDS

J.A. Alkemade
A.T. de Hoop
M.V. de Hoop
J. de Maag
S.M. de Vries
B.J. Kooij
D. Quak
C.M. Sayers
S. Stam

NORWAY

S. Gundersen
J. Lothe

2. INITIAL DEVELOPMENT OF THE THEORY OF BARNETT AND LOTHE

We now proceed to outline the main features of the theory of elastic surface waves developed by Barnett and Lothe and their coworkers between 1970 and 1976. A detailed exposition, with complete literature references, can be found in the review article [1] and a contemporary survey by the original authors [2]. Against this background subsequent advances in the basic theory of surface and interfacial waves are discussed in Sections 3 to 7.

The eigenvalue problem (7) can be recast in the standard form

$$\mathbf{N}\boldsymbol{\xi} = p\boldsymbol{\xi}, \tag{16}$$

where $\mathbf{N}$ is a real 6-dimensional tensor involving $Q(.)$, $Q^{-1}(.)$ and $R(.,.)$, the unit vectors $\mathbf{e}_1$ and $\mathbf{e}_2$, and the speed v. The 6-vector $\boldsymbol{\xi}$ is formed by juxtaposing the displacement and traction amplitudes $\mathbf{a}$ and $\boldsymbol{\ell}$ introduced in (5) and (9). This is the starting point of the theory of Barnett and Lothe, but (16) is referred to the rotated vectors

$$\mathbf{m}_\varphi = \sin\varphi\mathbf{e}_1 - \cos\varphi\mathbf{e}_2, \quad \mathbf{n}_\varphi = \cos\varphi\mathbf{e}_1 + \sin\varphi\mathbf{e}_2,$$

in place of $\mathbf{e}_1$ and $\mathbf{e}_2$, and then reads

$$N(\varphi,v)\boldsymbol{\xi}(\varphi,v) = p(\varphi,v)\boldsymbol{\xi}(\varphi,v). \tag{17}$$

It is found, however, that the eigenvectors of $N(\varphi,v)$ are independent of the orientation angle φ, while the eigenvalues have the property

$$(2\pi)^{-1}\int_0^{2\pi} p(\varphi,v)d\varphi = i\,\mathrm{sgn}\,\mathrm{Im}\,p(0,v).$$

The result of averaging the two sides of (17) over the interval $0 \leqslant \varphi \leqslant 2\pi$ is therefore

$$S(v)\boldsymbol{\xi}(v) = \pm i\boldsymbol{\xi}(v). \tag{18}$$

The decomposition of equation (18) into 3-vectors and 3-tensors can be set out in matrix format as

$$\begin{bmatrix} S_1(v) & S_2(v) \\ S_3(v) & S_1^T(v) \end{bmatrix}\begin{bmatrix} \mathbf{a}(v) \\ \boldsymbol{\ell}(v) \end{bmatrix} = \pm i\begin{bmatrix} \mathbf{a}(v) \\ \boldsymbol{\ell}(v) \end{bmatrix}. \tag{19}$$

The real tensors $S_1(v)$, $S_2(v)$ and $S_3(v)$ turn out to be of fundamental significance in the analysis of steady plane motions of an anisotropic elastic body. They are defined in I by

$$S_1(v) = -\pi^{-1}\int_{-\frac{1}{2}\pi}^{\frac{1}{2}\pi} Q^{-1}(\varphi,v)R^T(\varphi,v)d\varphi,$$

$$S_2(v) = \pi^{-1}\int_{-\frac{1}{2}\pi}^{\frac{1}{2}\pi} Q^{-1}(\varphi,v)d\varphi, \tag{20}$$

$$S_3(v) = \pi^{-1}\int_{-\frac{1}{2}\pi}^{\frac{1}{2}\pi} \{R(\varphi,v)Q^{-1}(\varphi,v)R^T(\varphi,v) - Q(\varphi,v)\}d\varphi, \tag{21}$$

with

$$Q(\varphi,v) = Q(\mathbf{n}_\varphi) - \rho v^2\cos^2\varphi I, \quad R(\varphi,v) = R(\mathbf{m}_\varphi,\mathbf{n}_\varphi) - \rho v^2\sin\varphi\cos\varphi I,$$

and obey the condition

$$S_1(v)S_2(v) + S_2(v)S_1^T(v) = 0, \tag{22}$$

among others [1,Sect.IV,D].

Since the p_α's in (12) have positive imaginary part, the relations between the corresponding displacement and traction amplitudes provided by (19) are

$$S_1(v)a_\alpha + S_2(v)\ell_\alpha = ia_\alpha, \quad S_3(v)a_\alpha + S_1^T(v)\ell_\alpha = i\ell_\alpha. \tag{23}$$

On multiplying equations (23) by γ_α, summing from $\alpha = 1$ to $\alpha = 3$, and evaluating at $v = v_S$, we obtain, with the use of (13) and (14),

$$S_1(v_S)A = iA, \quad S_3(v_S)A = 0. \tag{24}$$

Equation $(24)_1$ tells us that the polarization A of the surface wave is necessarily complex and we then deduce from $(24)_2$ that, since both the real and imaginary parts of A are null vectors of $S_3(v_S)$, the surface-wave speed satisfies the real form

$$\text{rank } S_3(v) = 1 \tag{25}$$

of the secular equation (15).

Equations $(1)_2$ and $(8)_1$ show that Q(.) is a symmetric tensor and it follows from (20) and (21) that $S_2(v)$ and $S_3(v)$ are also symmetric. The derivative $S_3'(v)$ of $S_3(v)$ with respect to v is positive definite in $(0,\hat{v})$ and, subject to the positive definiteness condition (2), $S_3(0)$ is negative definite [I,Sect.VI, D]. The eigenvalues of $S_3(v)$ are therefore negative at $v = 0$ and increase monotonically in I. From (25), two eigenvalues are zero simultaneously at $v = v_S$. If there were two subsonic surface waves, at least one eigenvalue would have to vanish twice in I. But this is ruled out by the monotonicity property, so a subsonic surface wave is unique whenever it exists.

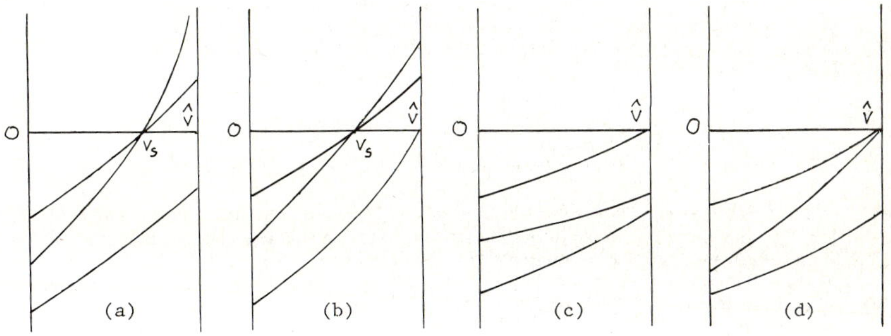

FIGURE 3
Possible patterns of behaviour of the eigenvalues of $S_3(v)$ in I.

The question of the existence of subsonic surface waves is resolved by examining the behaviour of $S_3(v)$, as given by the integral representation (21), in the limit $v \to \hat{v}$. If, for arbitrary real b, the quadratic form $b.\{S_3(v)b\}$ tends to infinity as $v \to \hat{v}$, at least one eigenvalue of $S_3(v)$ passes through zero and, as illustrated in Figure 3(a), a subsonic surface wave exists. This is found to be the case if the limiting wave depicted in Figure 1 produces non-zero traction on the planes $e_2.x$ = constant, or, when $n > 1$, if at least one limiting wave has this property [1,Sect.IX,B]. The transonic state is then said to be *normal*, and the term *exceptional* is applied to both the transonic state and the limiting wave(s) when the planes parallel to the boundary are left traction free by each limiting wave. When the transonic state is exceptional, the eigenvalues of $S_3(v)$ are bounded at $\hat{v}$ and n ($\leqslant 2$) of them vanish there [1,Sect.IX,C]. Because of the monotonicity of the eigenvalues in I, we infer that no subsonic surface wave exists when the transonic state is exceptional and not of type 1 ($\Rightarrow n = 2$) . This situation is illustrated in Figure 3(d). The possible types of behaviour of the eigenvalues when the transonic state is exceptional and of type 1 ($\Rightarrow n = 1$) are shown in Figure 3(b),(c). It is apparent from these diagrams that a subsonic surface wave does or does not exist according as the sum of the eigenvalues of $S_3(\hat{v})$ is positive or negative. We thus arrive at the existence theorem first enunciated by Barnett and Lothe [3].

(i) *If the transonic state is normal, a unique subsonic surface wave exists.*
(ii) *If the transonic state is exceptional and not of type 1, no subsonic surface wave exists.*
(iii) *If the transonic state is exceptional and of type 1, a unique subsonic surface wave exists if* $\mathrm{tr}S_3(\hat{v}) > 0$. *Otherwise no subsonic surface wave exists.*

It is also evident from Figure 3 that

> *a unique subsonic surface wave exists if and only if* $\mathrm{tr}S_3(v)$ *tends to* $+\infty$ *or a positive limit as* $v \to \hat{v}$.

Numerical calculations reported by Farnell [4] indicate that a supersonic surface wave may appear when no subsonic surface wave exists. The conclusions of the Barnett-Lothe theorem are, however, restricted to the subsonic interval I.

3. IMPROVEMENTS IN THE BASIC THEORY OF ELASTIC SURFACE WAVES

The brief but seminal study of elastic surface waves by Stroh [5,Sect.8] and the subsequent elaboration of Stroh's ideas by Barnett and Lothe *et al* relied heavily on an analogy between a surface wave and a line dislocation moving with constant velocity in the same direction in an infinite body of identical composition. No mention of dislocations has been made in Section 2, but the crucial definiteness properties of $S_3'(v)$ and $S_3(0)$ were originally derived from energy relations for a uniformly translating line dislocation [1,Sect.VI,D].

A direct proof of the positive definiteness of $S_3'(v)$ has been found by Chadwick and Jarvis [6,Sect.3(d)], and Gundersen, Barnett and Lothe [7] have recently succeeded in proving directly that $S_3(0)$ is negative definite. The first of these proofs assumes that the linear elasticity tensor B meets the strong ellipticity condition

$$B_{pqrs}v_p w_q v_r w_s > 0 \quad \forall \text{ non-zero real vectors } v,w, \tag{26}$$

but the more stringent positive definiteness condition (2) is needed in the second. The argument used by Gundersen *et al* can be rephrased neatly in terms of fourth-order tensors introduced by Walpole [8], Hill [9] and others into the analysis of interfacial discontinuities in composite bodies. An alternative proof of the negative definiteness of $S_3(0)$, constructed by Ting [10], employs a uniform strain solution of the equations of plane anisotropic elastostatics. The dependence on this particular solution can be removed, however, by the judicious use of Jacobi's theorem [11,p.25]. The proofs given in [6], [7] and

[10] effectively free the theory of Barnett and Lothe from considerations involving line dislocations.

A complicating feature of the theory, noted in passing in Section 2, is the possibility of the transonic state being of any one of six types. Barnett and Lothe [12,Sect.3] have shown that only the simplest state of type 1, exemplified in Figure 1, can be exceptional. Their proof is based on equation $(23)_1$ and makes use of the linear independence of the displacement amplitudes a_α, the boundness of $S_2^{-1}(v)$, and the vanishing of $S_2^{-1}(v)S_1(v)$ at $\hat{v}$. This result considerably simplifies the application of the theory, discussed in Section 5. It makes redundant part (ii) of the existence theorem stated in Section 2 and prohibits the mode of behaviour of the eigenvalues of $S_3(v)$ sketched in Figure 3(d). Barnett and Lothe [13] have also confirmed that parts (i) and (iii) of their existence theorem continue to hold when, in Figure 1, the section S∩R of the slowness surface has zero curvature at P.

The real secular equation (25) suffers from the limitation that $S_3(v)$ has no real continuation beyond the limiting speed $\hat{v}$. This means that no information about the possible existence of a supersonic surface wave is obtainable from (25) alone when the transonic state is exceptional and $\mathrm{tr}S_3(\hat{v}) < 0$. Another form of the secular equation to which this handicap does not apply is

$$F(v) := 1 + \tfrac{1}{2}\mathrm{tr}S_1^2(v) = 0 \tag{27}$$

[14,Sect.3.2]. An investigation of the algebraic structure of the tensors $S_1(v)$, $S_2(v)$ and $S_3(v)$ carried out by Chadwick and Ting [15] reveals that the function F defined in (27) is of fundamental significance. Connections are also established in [15] between (27), (25) and other real forms of the secular equation for elastic surface waves which have been derived (see (29) and (30) below).

Further studies of the fundamental eigenvalue problem (17) have been made by Ting [10,16] and Ting and Hwu [17]. Kirchner and Lothe [18] have devised a method of calculating $S_1(v)$ and $S_3(v)$ when $S_2(v)$ is known, based on the commutativity of the 6-tensors $N(\varphi,v)$ and $S(v)$. Lastly, in this section, we mention a systematic analysis by Chadwick [19] of subsequent transonic states, defined, with reference to Figure 1, by tangents to S∩R which are parallel and to the left of T.

4. DEVELOPMENTS BASED ON AN IMPEDANCE TENSOR

Equation $(23)_1$ can be rewritten as

$$\ell_\alpha = iZ(v)a_\alpha, \tag{28}$$

where

$$Z(v) = S_2^{-1}(v) + iS_2^{-1}(v)S_1(v).$$

Due to the symmetry of $S_2(v)$ and the skew-symmetry of $S_2^{-1}(v)S_1(v)$, implied by (22), the complex tensor $Z(v)$ is Hermitian. As signified by (28), $Z(v)$ effectively transforms a displacement amplitude into the corresponding traction amplitude and may accordingly be thought of as an impedance tensor. Combining (28) with equation (13) and evaluating at $v = v_S$ we obtain

$$Z(v_S)A = 0.$$

There follows the second alternative

$$\det Z(v) = 0 \tag{29}$$

to the secular equation (25).

The basic theory of elastic surface waves can be developed in terms of the impedance tensor $\mathbf{Z}(v)$ rather than $\mathbf{S}_3(v)$ and this approach has been favoured by Barnett and Lothe in their recent work [12,13,20,21]. They have shown in [12,Sect.4] how an investigation of the eigenvalues of $\mathbf{Z}(v)$ leads to a uniqueness theorem for subsonic surface waves, an existence theorem equivalent to parts (i) and (iii) of their original result, and a classification of possible patterns of behaviour parallel to that given in Figure 3(a)-(c). A major advantage of $\mathbf{Z}(v)$ over $\mathbf{S}_3(v)$ is that, regardless of the nature of the transonic state, it is bounded at $\hat{v}$ and may have a real continuation to supersonic values of v. There is, however, a countervailing difficulty: it has been shown by an indirect method, not involving dislocations, that $\mathbf{Z}(0)$ is positive definite [1,Sect.VII,D], but a wholly satisfactory proof of the negative definiteness of $\mathbf{Z}'(v)$ in I is still lacking. In sum, the methodology based on the impedance tensor is somewhat simpler and more natural than the formulation outlined in Section 2, but its mathematical foundations are as yet less secure.

Direct treatment of the existence of elastic surface waves

The existence theory for elastic surface waves has been approached along lines rather different from those followed above by Taylor [22]. As first pointed out by Burridge [23,Sect.5], the eigenvalues of the problem (7) may be regarded as the branches of an algebraic function $p(\lambda)$ with $\lambda = \rho v^2$. Burridge also showed how the construction of the Riemann surface of $p(\lambda)$ provides a means of analytically continuing the branches from the intervals in which they are real, and determined by the slowness surface as indicated in Figure 1, to the whole of the real line. This analysis is central to Taylor's procedure which puts equation (7) into matrix form and arrives, by matrix manipulation, at the secular equation

$$\det \mathbf{C}(\lambda) = 0, \qquad (30)$$

in which $\mathbf{C}$ is a skew-Hermitian matrix. An existence theorem is then established by studying the behaviour of the eigenvalues of $\mathbf{C}(\lambda)$ and applying a special case of Gersgorin's disc theorem [24,p.225]. An examination of the details of Taylor's paper shows that $\mathbf{C}(\lambda)$ is the matrix of components of $-i\mathbf{Z}(v)$: hence the inclusion of his contribution in this section.

The properties of surface waves find a more direct expression in Taylor's work than in the theory of Barnett and Lothe, but against this benefit must be set the necessity of analyzing in detail the complex-valued parts of branches of $p(\lambda)$. The singularities of $p(\lambda)$, in particular, are the cause of technical difficulties which are largely devoid of physical significance. Taylor has not drawn up a formal statement of his existence theorem, but the sense of his paper is that, *subject to the condition* (2), *a surface wave can propagate with a speed not exceeding* $\sqrt{(\hat{\lambda}/\rho)}$, $\hat{\lambda}$ *being the value of* λ *below which all the branches of* $p(\lambda)$ *are complex-valued*. Clearly, from Figure 1, $\hat{\lambda} = \rho\hat{v}^2$. This result is less informative than the Barnett-Lothe theorem stated in Section 2, and the two are compatible only if the limiting wave of an exceptional transonic state is admitted as a surface wave. Surprisingly, Taylor did not prove that a surface wave with speed $\leqslant \sqrt{(\hat{\lambda}/\rho)}$ is unique, but conditions sufficient for uniqueness are surmised at the end of his paper. Properties of the eigenvalues of $\mathbf{Z}(v)$ established in the later work of Barnett and Lothe [12,Sect.4] fulfil these requirements and thus supply the missing uniqueness theorem.

Interfacial waves

When the viewpoint is enlarged to include interfacial waves the balance between the versions of the basic theoretical framework described in Section 2 and the opening paragraphs of this section tilts decisively in favour of the impedance tensor approach.

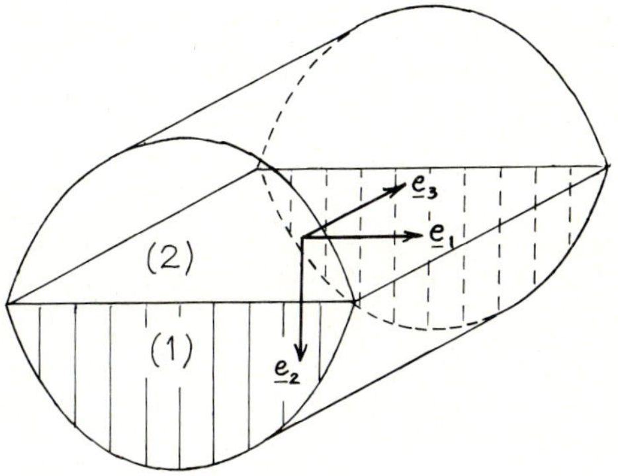

FIGURE 4
Composite body formed from semi-infinite sub-bodies (1) and (2) made of different anisotropic elastic materials or the same material in distinct orientations.

An interfacial wave in the composite body represented in Figure 4, propagating in the direction $\mathbf{e}_1$, consists of fields of the form (12) in each of the sub-bodies labelled (1) and (2), giving continuity of displacement and traction at the welded interface $\mathbf{e}_2.\mathbf{x} = 0$ and satisfying the decay condition

$$u \to 0 \text{ as } |\mathbf{e}_2.\mathbf{x}| \to \infty. \tag{31}$$

Let

$$\{u^{(r)}, -\sigma^{(r)T}\mathbf{e}_2\}, \quad r = 1,2,$$

be the fields obtained by adapting (12) to the sub-bodies. In order to comply with (31) in the limit $\mathbf{e}_2.\mathbf{x} \to -\infty$, we must replace the fields in sub-body (2) by their complex conjugates so as to reverse the signs of the imaginary parts of the p_α's. The interface conditions then require that

$$A^{(1)} = \bar{A}^{(2)}, \quad L^{(1)} = \bar{L}^{(2)}, \tag{32}$$

where an overbar denotes a conjugate and, in line with (13),

$$\{A^{(r)}, L^{(r)}\} = \sum_{\alpha=1}^{3} \gamma_\alpha^{(r)}\{a_\alpha^{(r)}, \ell_\alpha^{(r)}\}, \quad r = 1,2.$$

From (28),

$$L^{(r)} = iZ^{(r)}(v)A^{(r)}, \tag{33}$$

whence, with successive use of $(32)_2$, $(32)_1$ and the Hermitian property of $Z^{(2)}$,

$$iZ^{(1)}(v)A^{(1)} = -i\bar{Z}^{(2)}(v)\bar{A}^{(2)} = -iZ^{(2)T}A^{(1)}.$$

We deduce that

$$\Lambda(v)A^{(1)} = 0, \quad \text{with } \Lambda(v) = Z^{(1)}(v) + Z^{(2)T}(v), \tag{34}$$

and the speed of propagation v_i of the interfacial wave thus satisfies the real secular equation

$$\det\Lambda(v) = 0. \tag{35}$$

We see from $(34)_2$ that $\Lambda(v)$ is a Hermitian tensor and hence, from $(34)_1$ and (35), that there is a marked similarity between the treatments of surface and interfacial waves stemming from the impedance tensor approach. This correspondence has been exploited by Barnett, Lothe, Gavazza and Musgrave [20] to yield a tolerably complete theory of interfacial waves in a composite elastic body of general anisotropy. Defining

$$\hat{v} = \min(\hat{v}^{(1)}, \hat{v}^{(2)}),$$

and calling an interfacial wave subsonic when $0 < v_i < \hat{v}$, we can state the main results as follows.

(i) *If a subsonic interfacial wave exists it is unique and its speed of propagation is not less than the smaller of the subsonic surface-wave speeds in the individual sub-bodies. If neither sub-body, in isolation, transmits a subsonic surface wave, no subsonic interfacial wave exists.*

(ii) *A unique subsonic interfacial wave exists if and only if*

$$\{\mathrm{tr}\Lambda(\hat{v})\}^2 - \mathrm{tr}\Lambda^2(\hat{v}) < 0 \quad \text{or} \quad \det\Lambda(\hat{v}) < 0.$$

Barnett, Gavazza and Lothe [21] have gone on to consider the situation in which frictionless sliding can occur at the interface. The interfacial tractions are then purely normal, so that

$$L^{(r)} = L^{(r)} e_2, \quad r = 1,2, \tag{36}$$

and the conditions (32) are replaced by

$$A^{(1)}.e_2 = \bar{A}^{(2)}.e_2, \quad L^{(1)} = \bar{L}^{(2)}.$$

There follows, with the use of (33) and (36), the real secular equation

$$e_2.[\{Z^{(1)-1}(v) + Z^{(2)-1}(v)\}.e_2] = 0.$$

It is found that at most one slip wave exists when both sub-bodies are isotropic, but that anisotropy admits the possibility of there being a second mode of slip-wave propagation.

5. APPLICATION OF THE THEORY OF SURFACE WAVES TO MEDIA OF SPECIFIED SYMMETRY

The realization of the key role played by homogeneous plane waves in comprehending the nature of surface-wave propagation is one of the principal insights afforded by the theory of Barnett and Lothe, suggesting a systematic procedure for determining the behaviour of surface waves in elastic media with a specific type of symmetry. The programme [14,Sect.4] proceeds in three stages, aimed at, first, a sufficient understanding of the form of the slowness surface, second, a characterization of the exceptional configurations, that is the orientations of the elements of material symmetry relative to the surface-wave basis for which the transonic state is exceptional, and, third, the calculation and study of the function F, defined by (27), for the exceptional configurations. Realizations have been carried out *in extenso* for media with cubic symmetry [14] and for transversely isotropic media [25].

Cubic media

The results of [14,Sects 6-16] confirm and underpin the detailed numerical investigations reviewed by Farnell [4,Sect.V,B-E]. The exceptional configurations form three continuous families in which, modulo the overriding symmetry, unit vectors along the cube axes are as follows.

$$\begin{aligned}&1. \quad \sin\alpha e_1 + \cos\alpha e_2,\ -\cos\alpha e_1 + \sin\alpha e_2,\ e_3.\\ &2,3. \quad 2^{-\frac{1}{2}}(\sin\beta e_1 - \cos\beta e_2 - e_3),\ \cos\beta e_1 + \sin\beta e_2,\ 2^{-\frac{1}{2}}(\sin\beta e_1 - \cos\beta e_2 + e_3).\end{aligned} \tag{37}$$

For certain ranges of values of the elastic moduli the configurations 1 are exceptional for all $0 < \alpha \leqslant \frac{1}{4}\pi$, the configurations 2 for $0 < \beta < \beta_-$, and the configurations 3 for $\beta_+ < \beta \leqslant \frac{1}{2}\pi$, β_- and β_+ being defined by properties of the slowness surface [14,Sect.11].

Interest is focussed in [14,Sects 12-16] on the configurations $\beta = \frac{1}{2}\pi$, $\alpha = \frac{1}{4}\pi$, and $\beta = \cot^{-1}\sqrt{2}$ for which the boundary of the transmitting body is, in turn, a cube face, a face bisector, and an octahedral plane and the direction of surface-wave propagation is along a face diagonal in the first two cases and an octahedral diagonal (that is the intersection of two face bisectors) in the third. For each of these configurations F(v) has a real continuation beyond $\hat{v}$ and a unique supersonic surface wave appears when no subsonic wave exists. As shown by Farnell [4,Figs 4,13,20], the subsonic surface wave which can propagate in neighbouring directions on these boundaries degenerates into an exceptional homogeneous plane wave as the exceptional configuration is approached, and the supersonic wave is contiguous to a branch of pseudo, or leaky, surface waves violating the decay condition (11).

The discussion in [14,Sect.8] of the incidence of the various types of transonic states has been corrected and extended by Barnett and Lothe [26].

Transversely isotropic media

For media of this type the exceptional configurations consist of a continuous family (1) and two discrete members (2 and 3), the directions of a unit vector along the axis of rotational symmetry of the material being given by

$$\begin{aligned}&1. \quad \sin\alpha e_1 + \cos\alpha e_2 \quad \forall\ \alpha \in i,\\ &2. \quad e_3,\\ &3. \quad \cos\beta_0 e_1 + \sin\beta_0 e_3,\end{aligned} \tag{38}$$

together with their equivalents under the prevailing symmetry. Each of (38) is subject to restrictions on the elastic moduli. In 1, the interval i is $[0, \alpha_0)$, $(\alpha_0, \frac{1}{2}\pi)$ or $[0, \frac{1}{2}\pi)$, with α_0 defined by the slowness surface, and in 3, the angle β_0 is given in terms of the elastic moduli [25,Sects 5,6].

In case 1, continuous transitions between subsonic and supersonic surface-wave propagation may occur as α increases. This behaviour has not been reported in media with cubic symmetry. It would probably be found by varying the angles α and β in (37), but a definitive calculation remains to be made. When a configuration in the family 1 in (38) admitting a supersonic surface wave is approached through configurations not belonging to 1, the transition from subsonic propagation has the discontinuous character described in connection with cubic media. In case 3, the subsonic surface wave which exists in nearby configurations may degenerate into an exceptional wave without a supersonic surface wave appearing. No surface wave of any kind can then propagate and F(v) has no real continuation beyond $\hat{v}$. This effect has not been encountered in cubic media.

All the supersonic surface waves arising in cubic and transversely isotropic media are pure modes in the sense that $e_3.u$, the component of displacement normal to R, is universally zero. It appears likely that this is always the case, but a general proof has not yet been found.

6. PRESTRESSED MEDIA

Surface waves

We have seen in Section 2 that the questions of the existence and uniqueness of elastic surface waves hinge on the behaviour in I of the eigenvalues of $\mathbf{S}_3(v)$. The equations governing the propagation of a surface wave in an elastic body which has previously been placed in a uniform state of stress $\bar{\sigma}$, with $\bar{\sigma}^T\mathbf{e}_2 = 0$, by the application of a homogeneous finite deformation are formally the same as those which apply when $\bar{\sigma} = 0$. The invariance of the linear elasticities under the interchanges of suffixes $i \leftrightarrow j$, $k \leftrightarrow \ell$, implied by (1), no longer holds, however, and the property (2) must be abandoned as over-restrictive. The impact of these changes on the nature of surface waves has been examined by Chadwick and Jarvis [6].

Although the assumption that $\mathbf{B}$ is positive definite is untenable when $\bar{\sigma} \neq 0$, the retention of the strong ellipticity condition (26) has no unacceptable consequences. The monotonicity of the eigenvalues of $\mathbf{S}_3(v)$, which, as noted in Section 2, is a consequence of (26), is then preserved and, with it, the uniqueness of subsonic surface waves. We are no longer assured, however, that $\mathbf{S}_3(0)$ has negative eigenvalues and this represents the crucial difference between the characteristics of surface waves in prestressed and initially stress-free bodies.

The fact that (12) gives a stationary corrugation of the boundary when $v_s = 0$ suggests that we view the behaviour of $\mathbf{S}_3(v)$ at $v = 0$ in terms of standing surface waves. Let D denote the primary deformation received by the body prior to the appearance of a surface disturbance and ϕ the orientation of $\mathbf{e}_1$ (see Figure 2) relative to the principal axes of stretch in D. Then, corresponding to a given pair (D,ϕ), a standing surface wave exists if and only if two of the eigenvalues of $\mathbf{S}_3(v)$ vanish together at $v = 0$. In the absence of D, $\mathbf{S}_3(0)$ is negative definite and standing surface waves are entirely precluded.

These considerations indicate that the determination of the *neutral set*, consisting of all pairs (D,ϕ) for which a standing wave exists, is the central problem in delimiting the domain of existence of surface waves in a prestressed body. The condition

$$\text{rank}\mathbf{S}_3(0) = 1,$$

derived from (25), characterizes this set.

Chadwick and Jarvis [6,Sect.4] have determined the neutral set for an isotropic elastic material possessing a restricted form of the Hadamard strain-energy function when one of the principal axes in D is aligned with the boundary normal $\mathbf{e}_2$. Their findings are summarized in Figure 5 where a_1 and a_2 are the principal stretches along the axes orthogonal to $\mathbf{e}_2$. A point of the (open) domain A corresponds to a primary deformation D for which a subsonic surface wave can propagate in every direction orthogonal to $\mathbf{e}_2$. At the opposite extreme, the prestrain associated with a point of C does not permit the transmission of a surface wave in any direction. When D relates to a point of the intermediate region B, surface-wave propagation is restricted to a sector centred on the principal axis associated with the larger of a_1 and a_2. This pattern of behaviour, strikingly different from that encountered in the absence of prestress, may be taken to imply that the semi-infinite body is dynamically stable to small-amplitude surface disturbances in the equilibrium states corresponding to the points of A and unstable in the states which map into points of B∪C.

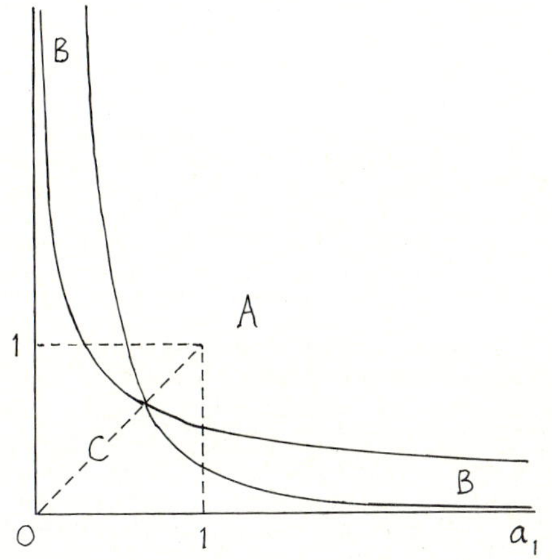

FIGURE 5
Domains of existence (A), partial existence (B) and non-existence (C) of surface waves in an isotropic elastic body subject to prestress.

Interfacial waves

The preceding argument extends to the composite infinite body shown in Figure 4 and leads to the conclusion that the primary deformations and material orientations for which standing interfacial waves exist are given by

$$\det\Lambda(0) = 0,$$

with $\Lambda(v)$ defined by $(34)_2$. Chadwick and Jarvis [27] have determined the domain of existence of interfacial waves in the case in which the sub-bodies are made of the same neo-Hookean material but have been prestrained by different amounts on common principal axes.

7. CONCLUDING REMARKS

Lack of space prevents the discussion of some further topics which deserve to be included in this article. They include surface-wave propagation in bodies subject to kinematical constraints [28,29], the formulation of efficient computational procedures for evaluating properties of surface and interfacial waves [12,Sect.5,20,Sect.5,30], the extension of the theory of Barnett and Lothe to piezoelectric media [31,32; see also 33], and the study of exceptional waves as a subject of independent interest [14,Sect.5,34,35]. Mention should also be made of the application of the sextic formalism introduced in Section 2 to other types of wave motion [36,37,38].

REFERENCES

[1] Chadwick, P. and Smith, G. D. In: Yih, C-S, (ed.) Advances in Applied Mechanics, Vol.17 (Academic Press, New York etc., 1977), p. 303.

[2] Barnett, D. M. and Lothe, J. In: Miklowitz, J. and Achenbach, J. D. (eds.) Modern Problems in Elastic Wave Propagation (Wiley, New York etc., 1978), p. 445.

[3] Barnett, D. M. and Lothe, J. J. Phys. F 4 (1974) 671.

[4] Farnell, G. W. In: Mason, W. P. and Thurston, R. N. (eds.) Physical Acoustics, Vol. VI (Academic Press, New York and London, 1970), p. 109.
[5] Stroh, A. N. J. Math. Phys. 41 (1962) 77.
[6] Chadwick, P. and Jarvis, D. A. Proc. R. Soc. Lond. A 366 (1979) 517.
[7] Gundersen, S. A., Barnett, D. M. and Lothe, J. Wave Motion 9 (1987) 319.
[8] Walpole, L. J. Math. Proc. Camb. Phil. Soc. 83 (1978) 495.
[9] Hill, R. J. Mech. Phys. Solids 31 (1983) 347.
[10] Ting, T. C. T. Some Identities and the Structure of N_i in the Stroh Formalism of Anisotropic Elasticity. Quart. Appl. Math., to appear.
[11] Mirsky, L. An Introduction to Linear Algebra (Clarendon Press, Oxford, 1955).
[12] Barnett, D. M. and Lothe, J. Proc R. Soc. Lond. A 402 (1985) 135.
[13] Barnett, D. M. and Lothe, J. Surface Wave Existence Theory for the Case of Zero Curvature Transonic States, this volume.
[14] Chadwick, P. and Smith, G. D. In: Hopkins, H. G. and Sewell, M. J. (eds.) Mechanics of Solids (Pergamon Press, Oxford, etc., 1982), p. 47.
[15] Chadwick, P. and Ting, T. C. T. Quart. Appl. Math. 45 (1987) 419.
[16] Ting, T. C. T. Int. J. Solids Structures 18 (1982) 139.
[17] Ting, T. C. T. and Hwu, C. Int. J. Solids Structures 24 (1988) 65.
[18] Kirchner, H. O. K. and Lothe, J. Phil. Mag. A 53 (1986) L7.
[19] Chadwick, P. Proc. R. Soc. Lond. A 401 (1985) 203.
[20] Barnett, D. M., Lothe, J., Gavazza, S. D. and Musgrave, M. J. P. Proc. R. Soc. Lond. A 402 (1985) 153.
[21] Barnett, D. M., Gavazza, S. D. and Lothe, J. Proc. R. Soc. Lond. A 415 (1988) 389.
[22] Taylor, D. B. Quart. J. Mech. Appl. Math. 31 (1978) 335.
[23] Burridge, R. Quart. J. Mech. Appl. Math. 23 (1970) 217.
[24] Lancaster, P. Theory of Matrices (Academic Press, New York and London, 1969).
[25] Chadwick, P. In: Parker, D. F. and Maugin, G. A. (eds.) Nonlinear and Other Non-Classical Effects in Surface Acoustic Waves (Springer-Verlag, Berlin etc., 1988).
[26] Barnett, D. M. and Lothe, J. Proc. R. Soc. Lond. A 411 (1987) 251.
[27] Chadwick, P. and Jarvis, D. A. Quart. J. Mech. Appl. Math. 32 (1979) 387, 401.
[28] Chadwick, P. and Wilson, N. J. Surface Waves in Incompressible, Transversely Isotropic Elastic Media, to appear.
[29] Captain, V. S. and Chadwick, P. Quart. J. Mech. Appl. Math. 39 (1986) 327.
[30] Gundersen, S. A. and Lothe, J. In: Parker, D. F. and Maugin, G. A. (eds.) Nonlinear and Other Non-Classical Effects in Surface Acoustic Waves (Springer-Verlag, Berlin etc. 1988).
[31] Lothe, J. and Barnett, D. M. Phys. Norv. 8 (1977) 239.
[32] Lothe, J. and Barnett, D. M. Wave Motion 1 (1979) 107.
[33] Taylor, D. B. and Crampin, S. Proc. R. Soc. Lond. A 364 (1978) 161.
[34] Al'shits, V. I. and Lothe, J. Sov. Phys. Crystallogr. 24 (1979) 644.
[35] Chadwick, P. and Whitworth, A. M. Quart. J. Mech. Appl. Math. 39 (1986) 309.
[36] Al'shits, V. I. and Lothe, J. Wave Motion 3 (1981) 297.
[37] Braga, A. M. B. and Herrman, G. Wave Propagation in Layered Submerged Composities, this volume.
[38] Ting, T. C. T. and Chadwick, P. In: Mal, A. K. and Ting, T. C. T. (eds.) Wave Propagation in Structural Composites (ASME, 1988).

ELASTIC WAVE PROPAGATION
M.F. McCarthy, M.A. Hayes, (Editors)

PROBING OF ACOUSTIC WAVE SURFACE DISPLACEMENTS

Eugène DIEULESAINT and Daniel ROYER

Laboratoire d'Acoustoélectricité
Université Pierre et Marie Curie
10, rue Vauquelin, 75231 Paris Cedex 05
France

The development of applications and correlative studies of acoustic waves in non destructive testing, medical imaging and signal processing, require the measurement of mechanical displacements whose amplitude is of the order of one Angström. In this paper, we review the suitable methods which do not necessitate any mechanical contact, namely the condenser method and the optical methods. These have the advantage of allowing local and absolute measurements. A new, compact and stable structure of a heterodyne interferometric probe having a resolution of 10^{-4} Å/$\sqrt{\text{Hz}}$ is described. Examples of local measurements of surface mechanical displacements, generated in a steady harmonic or transient regime, by piezoelectric and photothermal techniques, are presented. Detection of surface waves excited on a sphere by a laser pulse is also presented.

1. INTRODUCTION

For almost twenty years, the applications of elastic waves have developed regularly in the fields of instrumentation, communication, metallurgy and medicine. These waves have brought original and elegant solutions to various problems in signal processing, non destructive testing, photoacoustic spectroscopy, microscopy [1-5] .

This development is explained, on the one hand, by the properties of these waves to propagate at low velocity, in the range 10^3 to 10^4 m/s (a low velocity gives rise to small volume components), and to penetrate optically opaque media (materials or the human body) and on the other hand, by the fact that they are readily generated and detected in any substrate with the aid of piezoelectric plates or thin layers provided with one or several pairs of electrodes and mechanically bonded to the substrate. The excitation of surface acoustic waves (SAW) on piezoelectric substrates is particularly simple : the electrodes are made by photoetching a sputtered or vapor deposited metallic layer. This technique from microelectronics as well as the fact that SAW propagate in a material thickness of about a wavelength, that is to say, very small for normal frequency range in signal processing ($f = 100$ MHz $\rightarrow \lambda = 30$ µm), accounts for their extensive use in filters of all kinds in both professional (radar) and consumer (television) fields.

As the applications expand progressively, the performance requirements become more stringent. For instance, regarding the filters referred to, the rejection ratio has to be increased, the losses decreased, the cost reduced ; the resolution of echographs, in obstetrical examinations has to be improved. For non destructive testing, it appears necessary to measure the mechanical displacement at a distance and so on.

Accordingly, the development of applications implies new studies whose objectives, apart from theoretical modelling, include

- more precise analysis of real conditions of generation and propagation taking into account beam steering and diffraction, harmonic generation, focusing, waveguide confinement, scattering of waves by surface or bulk defects.

- substitution of (or adjunction to) piezoelectric solids by (of) non piezoelectric solids (implementation of resonators operating at higher frequencies or having greater Q-factors, composite medical transducers with better efficiency).

- excitation of acoustic waves for non destructive testing by photothermal techniques requiring no mechanical contact.

- measurement of solid elastic parameters in a hostile environment (high pressure, elevated temperature).

Most of these studies require the plotting of acoustic fields, namely the visualization and measurement of mechanical displacements whose amplitude is, at most, of the order of a few Angströms. Several methods are available to record the stationary patterns of resonators. The main purpose of these methods (optical interferometry and holography, X-ray or neutron diffraction topography, scanning electron microscope graphs) whose review is found in reference [6] is to map the mode pattern and not to measure the displacement amplitude which is naturally significant (> 100 Å), given the resonance phenomena. For transducers more or less direct techniques exist [7]. For instance, a bulk wave piezoelectric transducer, used for medical applications or non destructive testing, can be placed in à tank of water ànd its acoustic field plotted with the aid of a mobile small piezoelectric probe or spherical reflector. This technique can only be utilized at comparatively low frequencies (fmax ≈10 MHz). For it to be transposed at higher frequencies, the transducer has to be bonded to the end of a solid block, the other end, polished, playing the role of reflector ; however, the drawback of using the transducer to be tested as a receiver remains. In the restrictive case of a transducer bonded to a transparent solid, the acoustic field can be explored by acousto-optic interaction i.e. with a laser beam diffracted by the zones whose optical index varies. The surface displacements of Rayleigh or Lamb waves can be locally plotted with a piezoelectric probe, possibly provided with a cone and pressed onto the substrate [8]. However, if this kind of probe is useful to detect the instant of arrival of a wave train [9] and thus to measure the velocity of propagation or simply determine the width of an acoustic beam, it can not give a measurement of the surface displacement because the induced voltage depends on the pressure at the point of contact.

The topic of this paper is to describe the methods which do not require any mechanical contact for measuring a surface displacement and are sensitive enough to detect displacements less than one Angström. With the exception of the capacitor detection, all are optical. An example of the variable capacitor technique, which requires a very precise mounting, will be given. The main part of the paper is devoted to optical methods applied to the measurement of the displacement component perpendicular to the surface.

2. THE CAPACITIVE DETECTOR

The well known principle of operation of this detector consists in measuring the variation of the capacitance between two electrodes one of which is forced to vibrate by ultrasounds. We propose to illustrate this technique by describing the sensitive device built by Gauster and Breazeale [10] and used to study the elastic non linearity of solids [11]. The purpose of their experiments was to determine the non linearity parameters along various symmetry directions by measuring the amplitude of both the fundamental component and the generated second harmonic component of longitudinal waves launched along the corresponding directions. They used the pulse-echo technique and chose a fundamental frequency of 30 MHz. The largest ultrasonic displacement amplitude obtained at this frequency was about 10 Å. As the second harmonic amplitude was 1 % of that of the fundamental, the detector had to be able to measure displacement amplitudes of the order of 0.1 Å.

Figure 1 shows the schema of their mounting. The longitudinal wave trains are generated by a quartz transducer pressed against the sample. They are reflected at both ends of the solid, assumed to be metallic. The capacitive receiver is made up by the down end sample face acting as a mobile electrode and by a nearby fixed electrode held in place by an insulating fused-silica part placed on an outer grounded ring. Every wave train impinging on the sample face causes it to vibrate. The capacitor gap spacing is varied correspondingly and an alternating voltage is induced when a bias voltage is applied across the electrodes.

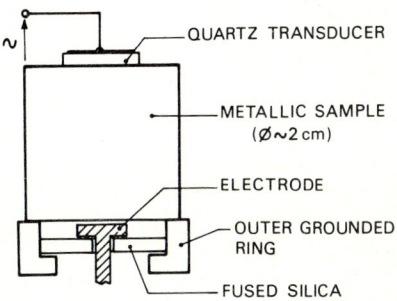

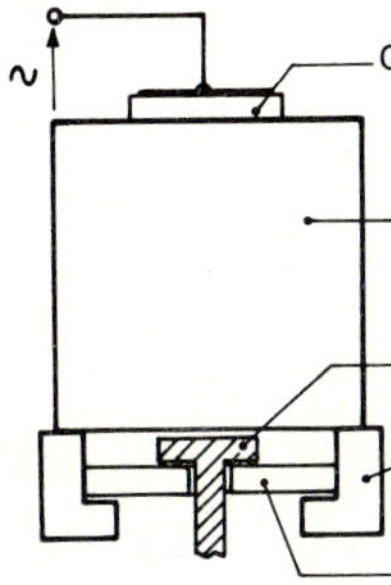

Fig.1. Capacitive detector. After (11)

If ω is the angular frequency and u the amplitude of the acoustic wave in the sample, the capacitor gap spacing d changes according to the relation

$$d = d_0 + 2u\cos(\omega t + \phi) \quad (1)$$

d_0 is the gap in the absence of wave. Factor 2 appears because the face of the sample being stress-free, its vibration amplitude is twice the wave amplitude in the sample. The capacitance of the detector of surface area S expresses as $C = \varepsilon S/d$ where ε is the permittivity of the medium (air) between the electrodes. The relative variation of the capacitance is then equal to the relative variation of the gap spacing :

$$\left|\frac{\Delta C}{C}\right| = \left|\frac{\Delta d}{d}\right| = \frac{2u}{d_0} \quad (2)$$

For $d_0 = 10$ μm, for instance, a displacement amplitude of 0.5 Å gives rise to a capacitance relative variation of 10^{-5}. It is possible to measure much smaller variations. Displacements of 10^{-4} Å which correspond to capacitance variations of 10^{-9} were detected [10] . This capacitive detector was calibrated by substituting an electrical signal of the same frequency to the acoustical signal. The RF signal from the capacitive detector was fed to either a 30 MHz or 60 MHz bandpass amplifiers. The output signals of the amplifiers were measured with a boxcar integrator.

Such resolving power requires very precise assembly. Typical figures are for capacitance : 100 pF, for surface area S = 1 cm^2, and for static gap $d_0 = 10$ μm. Consequently, the sample face, electrode surface, fused silica support and outer ring have to be flat and optically polished with an accuracy of a fraction of a wavelength. The necessary parallelism of the sample-electrode surfaces demands a fringe technique verification. Any accidental bringing together of the capacitor electrodes may generate a spark damaging for the sample since the inter-electrode electric field is very high : 10 V for a gap of 10 μm create a field of 10 kV/cm. Another drawback of this device lies in the necessity of depositing a metallic layer on the end of any non metallic sample. Despite all the difficulties, Breazeale and Philip have succeeded in using this capacitor technique, from room temperature down to 3 K, by controlling pneumatically the inter-electrode spacing. Nevertheless, this variable capacitance detector is not a true probe since it does not allow a local measurement and furthermore, the sample to be checked has to be part of the detector.

3. THE OPTICAL DETECTORS

The optical detection of mechanical displacements presents advantages : the vibrating surface can be examined at a distance without perturbing the acoustic field, the measurement is local and readily calibrated from optical wavelength if a heterodyne technique is used, the resolution is higher than that of the capacitive detection, the bandpass is wide ; furthermore, as there is no contact between the light beam and the surface, a large area can be explored by displacing either the object or the beam (scanning). On the other hand, in addition to the requirement of a good surface state, there are some drawbacks : the alignment procedure is usually critical, the measurement is sensitive to low frequency mechanical vibrations and thermal fluctuations, the resolving power is much less than that of piezoelectric transducers. The different devices of interest can be placed in three main categories as noticed by Whitman and Korpel [12]. The first category relates to the diffraction by phase surface grating of an optical beam which overlaps several acoustic wavelengths. The second category refers to the deflexion of a narrow optical beam by the local variation of the slope of the surface. These two techniques are mainly used to characterize Rayleigh wave displacements. The third category (interferometric probes) comprises the techniques based on the phase modulation of the light beam by the normal component of the displacement. For it to be exploited, this phase modulation has to be either directly converted into an amplitude modulation of the photodetector current with the aid of a homodyne interferometer (Michelson, Fabry-Pérot) or transposed into a phase modulation of the current in the RF domain by heterodyne interferometry, that is to say, with a change of optical frequency.

3.1. Diffraction by phase surface grating.

Any acoustic wave is accompanied by a refractive index wave which constitutes a mobile phase grating. The period of this grating is equal to the wavelength. Any light wave striking this grating is diffracted i.e. gives rise to waves which propagate along directions imposed by the grating period [13]. If the acoustic wave is a bulk wave propagating in the core of the solid, assumed to be optically transparent, the light waves have to pass through the acoustic beam and then the solid. If the acoustic wave propagates on the surface of the solid (case of interest for us), though the index variation is still produced in the material, it is not necessary for the light waves to penetrate the solid as the acoustic wave generates a movement of the surface. A Rayleigh wave, for example, causes a corrugation of the surface which behaves as a phase grating for light waves.

Let us now examine the conditions of diffraction. Let ω be the Rayleigh wave angular frequency, λ the wavelength in the material i.e. the grating period. The light beam, of angular frequency Ω, wave number $K = 2\pi/\Lambda$, power P_L, overlaps a zone of several acoustic wavelengths. For an angle of incidence θ_0, it is diffracted into beams of frequencies $\Omega \pm m\omega$ along directions defined by angles θ_m (figure 2a) such as [14] :

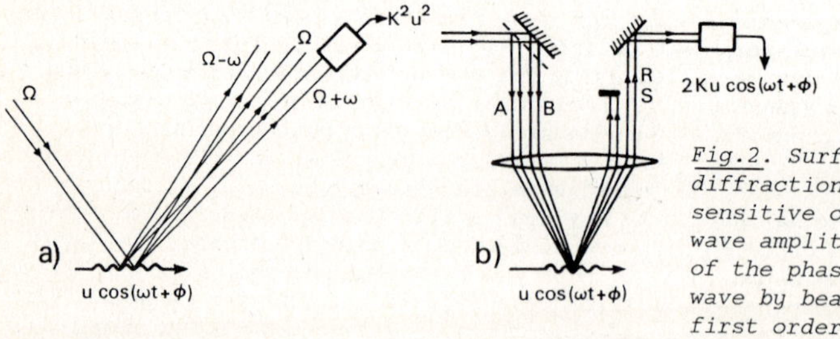

Fig.2. Surface phase grating diffraction. a) Diffraction sensitive only to the acoustic wave amplitude. b) Recovering of the phase of the acoustic wave by beating of zero and first order diffracted beams.

$$\sin\theta_m = \sin\theta_0 \pm m\Lambda/\lambda, \quad m = 0,1,\cdots \tag{3}$$

When the surface is perfectly reflective, the power of the beam of order m is given by the square of the Bessel function $J_m(\Delta\phi)$ whose argument is equal to the phase shift $\Delta\phi = 2Ku\cos\theta_0$ produced by the displacement normal to the surface of amplitude u

$$\frac{P_m}{P_L} = [J_m(\Delta\phi)]^2 \tag{4}$$

For $u \ll \Lambda/2\pi$, $\Delta\phi \ll 1$, the ratio, for m = 1, is proportional to the square of the displacement :

$$\frac{P_1}{P_L} = (Ku\cos\theta_0)^2 \tag{5}$$

Thus, the measurement of the relative intensities of the diffracted beams of order ± 1 provides the spatial power distribution of the Rayleigh wave. This simple technique is usable only when the diffracted beams are distinguishable, namely at high frequencies (f > 100 MHz), and in a steady regime. As the photodetector current is proportional to u^2, the sensitivity for small displacements is low : $Ku = 10^{-3}$ for u = 1 Å and a He-Ne laser (Λ = 6328 Å $\Rightarrow K = 2\pi/\Lambda \simeq 10^{-3}$ Å^{-1})

The optical receiver, usually a photodiode, is a quadratic detector only sensitive to light intensity. Now, the phase $(\omega t + \phi)$ of the acoustic wave is contained in the phase $(\Omega t + \omega t + \phi + \Phi_S)$ of the electric field of the diffracted beam of order 1 ($\Phi_S = 2\pi L_S/\Lambda$ is the phase shift corresponding to the optical path L_S) :

$$E_S = E_0 Ku \exp i(\Omega t + \omega t + \phi + \Phi_S), \quad (\theta_0 \sim 0) \tag{6}$$

To recover the acoustic wave phase, the field E_S must beat with that of the specularly reflected beam (i.e. of order 0) taken as reference beam ($\Phi_R = 2\pi L_R/\Lambda$)

$$E_R = E_0 \exp i(\Omega t + \Phi_R) \tag{7}$$

The resulting current from a photodiode having a response factor S :

$$I = S(E_R + E_S)(E_R + E_S)^* = I_0 + i(t) \tag{8}$$

is composed of a d.c part ($P_0 = E_0 E_0^*$) :

$$I_0 = SP_0[1 + (Ku)^2] \sim SP_0 \tag{9}$$

and of a part i(t) which contains the frequency of the acoustic wave :

$$i(t) = 2KuI_0\cos(\omega t + \phi + \Phi_S - \Phi_R) \tag{10}$$

The amplitude of i(t) is proportional to the normal displacement amplitude u. As Ku is very small with respect to unity, the sensitivity of this heterodyne method is better than that of the previous one (signal $\propto K^2u^2$). It is limited by the shot noise from the d.c part I_0 from the photodiode. For an electronic detection bandwidth B, the mean square value of the noise current intensity is given by

$$\langle i_N^2 \rangle = 2eI_0B, \quad e : \text{electron charge} \tag{11}$$

The signal to noise ratio is :

$$\frac{S}{N} = \left(\frac{\langle i^2(t)\rangle}{\langle i_N^2\rangle}\right)^{1/2} = Ku\left(\frac{I_0}{eB}\right)^{1/2} \tag{12}$$

The minimum measurable displacement u_{min} is found for S/N = 1 :

$$\boxed{u_{min} = \frac{\Lambda}{2\pi}\left(\frac{eB}{I_0}\right)^{1/2}} \qquad (13)$$

For a He-Ne laser and an optical beam power onto the photodiode $P_0 = 1$ mW (d.c intensity $I_0 = 0.3$ mA) and a detection bandwidth B = 1Hz

$$u_{min} = 2.3 \times 10^{-5}\,\text{Å}/\sqrt{\text{Hz}} \qquad (14)$$

This limit is reached only if the thermal noise in the resistor R loading the photodiode

$$\langle i_N^2 \rangle_{th} = 4kTB/R \qquad (15)$$

is less than the shot noise (11) :

$$R\,I_0 \gg 2kT/e \sim 50 \text{ mV} \quad \text{at room temperature} \qquad (16)$$

For $P_0 = 1$ mW, R has to be larger than 150 Ω. This condition is difficult to satisfy at high frequencies because of the parasitic capacitances which reduce the bandpass.

This method has been implemented by Rouvaen et al [15], according to the set-up in figure 2b, itself derived from a previous experiment with bulk acoustic waves [16]. The incident light beam is divided into two parallel beams A and B by a splitter and a mirror. The specularly reflected part R from beam A, considered as reference, beats with the order 1 diffracted part S from beam B. The signal, detected at the frequency of the acoustic wave has a relatively low sensitivity to the optical path fluctuations because of the proximity of both beams A and B.

3.2. Deflection by surface ripple.

This method, also called knife edge technique and first demonstrated by Adler et al [17] has been extensively used to visualize surface acoustic wave fields. Its interest is to permit the scanning of the surface. As shown in figure 3, the laser beam focused onto the surface is deflected by the acoustic wave ripple. The knife edge (in practice the edge of the photodiode) masking partially the oscillating beam, the diode output current is modulated.

This efficient and simple technique is exploited in an instrument (SLAM : Scanning Laser Acoustic Microscope [18]) used for non destructive testing. Its disadvantages are that it requires a good surface state and provides a signal proportional to the slope of the surface displacement (and not to its amplitude). The signal intensity depends on the wave vector orientation with respect to the knife edge. The amplitude of the angular deviation of the reflected beam is $2\alpha = 4\pi u/\lambda$. This corresponds to a beam displacement $\Delta x = 2\alpha F$ with regard to the knife edge for a lens focal length F. Assuming that half of the beam of diameter D and power P_L is masked by the knife, the photocurrent intensity variation is $i = SP_L\Delta x/D$, i.e

$$i(t) = 4\pi\frac{u}{\lambda}\frac{F}{D}SP_L\cos(\omega t + \phi) \qquad (17)$$

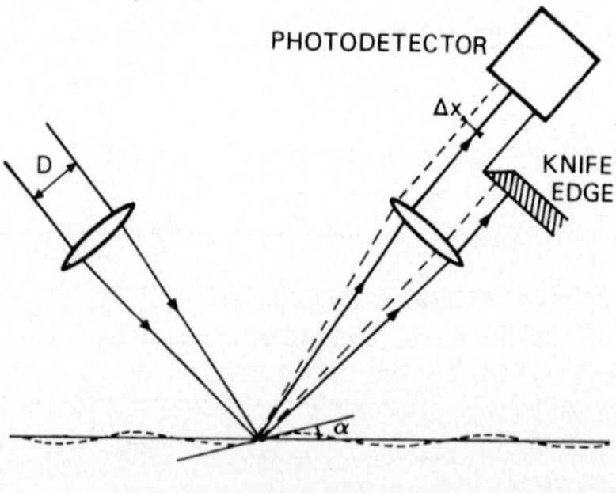

Fig.3. Knife edge technique. The ripple of the surface changes the direction of the reflected beam at the frequency of the surface acoustic wave.

The aperture D/F has to be chosen so as to optimize the diameter d of the light spot on the surface. This is presumed limited by diffraction :

$$d = \Lambda\frac{F}{D} = \frac{2\pi}{K}\cdot\frac{F}{D} \qquad (18)$$

The amplitude of the alternative current appears to be proportional to d and to the d.c component $I_0 = SP_L/2$:

$$i(t) = 4\,\frac{d}{\lambda}\,KuI_0\cos(\omega t + \phi) \qquad (19)$$

This reasoning is valid as long as $d < \lambda$. If the spot overlaps several wavelengths, the beam is diffracted (§.3.1). The signal at the acoustic wave frequency vanishes. As the optimal value is $d \cong \lambda/2$ [12], the current intensity i(t) is expressed by formula 10. The sensitivity of both methods is the same.

From the foregoing, the highest frequency is obtained when $d = \lambda$. Formula (18) gives : $f_{max} = DV/\Lambda F$, i.e for a focal length F = 25 cm, a He-Ne laser (Λ = 633 nm, D = 1 mm) and a wave velocity V = 3000 m/s, f_{max} is 20 MHz. This is a limitation for application to surface wave filters whose central frequency is more than 50 MHz. In order to increase this limit i.e to maintain at high frequencies the necessary superposition of beams of different orders, Engan [19] uses a short focal length lens and a double photodiode (figure 4). Furthermore, to avoid the detection and processing of high frequency signals, he modulates the laser intensity at the elastic wave frequency. This stroboscopic effect, previously used by Parker [20], freezes the motion of the surface. The detected signal is continuous. It is distinguished from the photodiode mean current by modulating, at a low frequency (1 kHz), the amplitude of the signal applied to the transducer and using a lock-in amplifier.

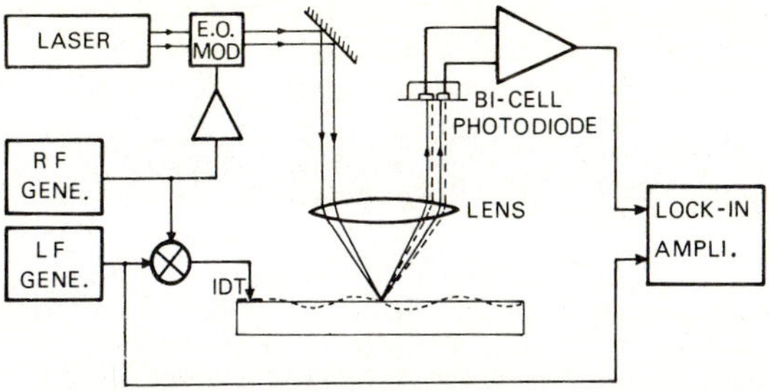

Fig.4. Diagram of the probe used by Engan to investigate surface acoustic wave propagation phenomena (f → 500 MHz). The intensity of the light beam is modulated at the frequency of the elastic wave so that the wave pattern is frozen on the substrate by stroboscopic effect.

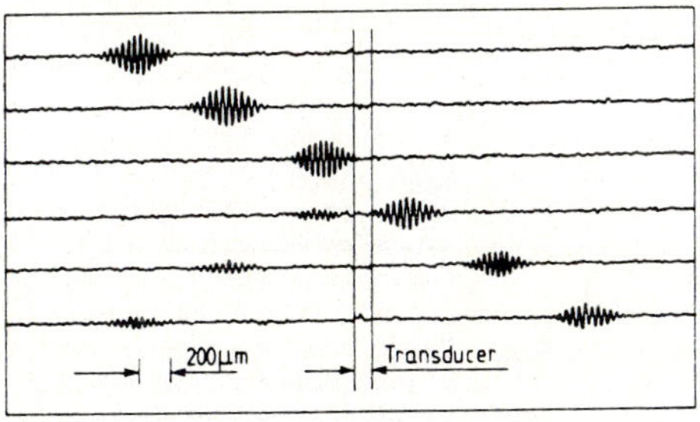

Fig.5. Reflection of a Rayleigh wave pulse by a six-finger interdigital transducer. The pulses are recorded by step by step scanning along a line. The wave trains are tracked by delaying the optical pulse (150 ns) with respect to the electrical signal applied to the transmitter (after Engan [21]).

This technique operates in a steady state regime. Engan has mapped Rayleigh wave displacements of amplitude 5 Å, frequency 100 MHz with a signal to noise ratio of 60 dB [19]. The maximum acoustic wave frequency can reach 500 MHz. The moving forward or backward of a wave train may be plotted with the aid of electronic gates (figure 5). Nevertheless, it is not possible to sense a transient response such as the impulse response of a transducer.

3.3. Interferometric probes.

The phase of a light beam back-reflected by the vibrating surface of an object is modulated by the normal surface displacement $u\cos(\omega t+\phi)$:

$$E_S = E_0 \exp i[\Omega t + \Phi_S + 2Ku\cos(\omega t + \phi)] \quad (20)$$

When the amplitude u is small with respect to the optical wavelength Λ ($Ku \ll 1$), the spectrum of the optical signal is only composed of two side components at frequencies $\Omega \pm \omega$ whose relative amplitudes are Ku with regard to that of the central line at frequency Ω; with $\phi' = \phi + \pi/2$:

$$E_S = E_0\{\exp i(\Omega t + \Phi_S) + Ku \exp i[(\Omega+\omega)t + \Phi_S + \phi'] - Ku \exp i[(\Omega-\omega)t + \Phi_S - \phi']\} \quad (21)$$

If the acoustic wave frequency is sufficiently high (a few MHz), the information on mechanical displacement can be extracted directly by optical spectroscopy with a Fabry-Pérot type interferometer used as a frequency discriminator. However, the most common methods consist in mixing the probe beam S with a reference beam R of same intensity (formula 7) coming from the same laser source. The beams S and R are superposed and directed to the photodetector. The photocurrent produced by the beating of the two waves is given by formula 8 as :

$$I = I_0\{1 + \cos[2Ku\cos(\omega t + \phi) + \Phi_S - \Phi_R]\} \quad (22)$$

Comparing this formula with (10) shows that the effect of thermal or mechanical fluctuations on optical paths L_S and L_R is much more significant in two wave interferometric devices than in the preceding techniques. Indeed, the phase variations $\Phi_S - \Phi_R = 2\pi(L_S - L_R)/\Lambda$ must be small with respect to the term Ku (equal to 10^{-3} rad. for instance, for a displacement u = 1 Å and a He-Ne laser) and not to the phase ϕ of the acoustic signal.

To reduce the effects of these fluctuations, three solutions are employed:i) the reference mirror is displaced, with the aid of a piezoelectric actuator so that the phase difference $\Phi_S - \Phi_R$ is maintained constant (stabilized Michelson interferometer), ii) the arrangement is chosen for the optical paths of reference and probe beams to be adjacent, iii) the electronic signals are appropriately processed (heterodyne interferometer with change of the frequency of one beam or both beams).

3.3.1. Fabry-Pérot cavity.

This technique based on the Doppler frequency shift of the light reflected or scattered by a vibrating surface has been studied by Monchalin [22] in order to detect mechanical displacements generated by pulsed lasers in solids at elevated temperatures and having an optically poor quality surface. Whatever the goal : measuring elastic parameters or detecting defects, the problem raised is to operate at a distance and to collect a sufficient amount of light. Monchalin used a confocal Fabry-Pérot cavity of length 40 cm (figure 6).

The optical mounting enabled the reception of light at 1.5 m from a spot of diameter 1 mm. The bandwidth of the cavity was 10 or 1.5 MHz according to the reflectivity of the mirrors. Its thickness was controlled with the aid of a piezoelectric actuator displacing a mirror so as to maintain the laser frequency near the inflexion point of a transmission peak (such a cavity operates as a frequency discriminator). So far, the experiments have been performed at room temperature by exciting ultrasounds in a machined steel plate of thickness 12.5mm. The ultrasounds were produced by a YAG laser focused on the side of the plate with a lens having a focal length of 2 m. The lowest amplitude of the displacement detected on the opposite side was found to be 6 Å ; the theoretical limit predicted was 3 Å.

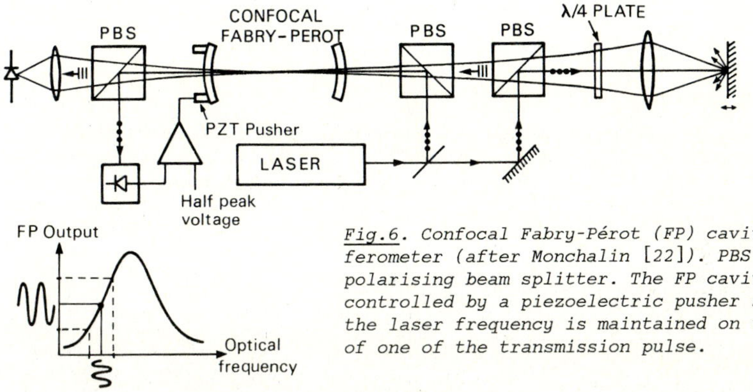

Fig.6. Confocal Fabry-Pérot (FP) cavity interferometer (after Monchalin [22]). PBS is for polarising beam splitter. The FP cavity is controlled by a piezoelectric pusher so that the laser frequency is maintained on the slope of one of the transmission pulse.

3.3.2. Homodyne technique

The interferometric two-wave settings, described under this title, do not resort to any frequency change. Two types can be considered.

• a- Stabilized-path Michelson interferometer

As in a standard Michelson interferometer, the laser power P_L is divided, by a beamsplitter, into two equal parts which reflect respectively on the object (probe beam S) and on the mirror (reference beam R) and then mix again (figure 7a). Half reaches the photodetector. The other half is prevented from entering the laser cavity and generating instabilities, by an optical isolator.

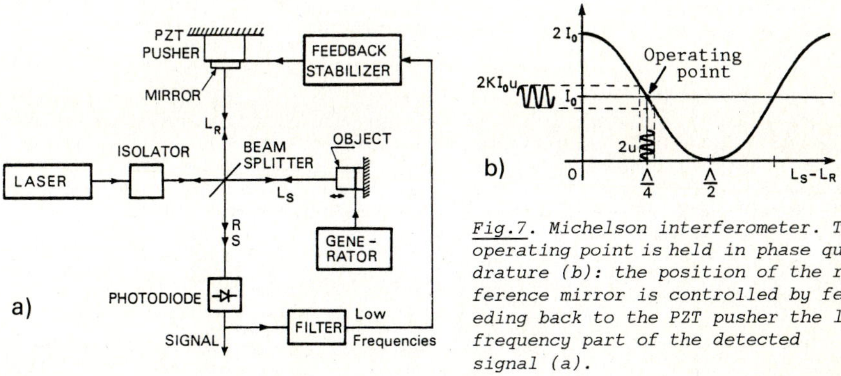

Fig.7. Michelson interferometer. The operating point is held in phase quadrature (b): the position of the reference mirror is controlled by feeding back to the PZT pusher the low frequency part of the detected signal (a).

The intensity of the photocurrent resulting from the beating of beams R and S, given by eq. 22 depends sinusoïdally on the optical path difference $L_S - L_R$ with a period Λ (figure 7b). Maximum sensitivity is obtained for

$$L_S - L_R = (n + \frac{1}{4})\Lambda \rightarrow \phi_S - \phi_R = \frac{\pi}{2} \text{ (mod } 2\pi) \quad (23)$$

In order to maintain the operating point in this phase quadrature position independently from optical path random variations due to low frequency thermal and mechanical disturbances, the position of the reference mirror has to be controlled [23] . The mirror is usually displaced by a piezoelectric plate driven by the lower frequency part ($f < 1$ kHz) of the detected signal. The limited dynamic range of the piezoelectric control necessitates periodical adjustments.

For phase quadrature operating point and small displacements ($Ku \ll 1$) the photocurrent intensity (22) is given by

$$I = I_0 [1 + 2Ku \cos(\omega t + \phi)]. \quad (24)$$

Thus, the sensitivity of homodyne interferometric techniques is the same as that of diffraction and reflection techniques. The smallest measurable displacement is given by formula 13 if I_0 is the average photodiode current : $I_0 = SP_L/2$. Using a 5 mW He-Ne laser chosen among ten for its stability, and a detection bandwidth B = 40 MHz, Hutchins et al [24] reached a limit (0.2 Å) twice the theoretical figure (0.1 Å).

• b - Differential interferometer

Here, the means for reducing unwanted long-wavelength effects consists in comparing the phase of two parallel beams which strike the vibrating surface at two points close together (figure 8). The low frequency vibrations tend to displace both spots in unison and not affect the relative phase of the two beams. These can be obtained in different ways (Bragg cell, calcite wedge, Wollaston prism...).

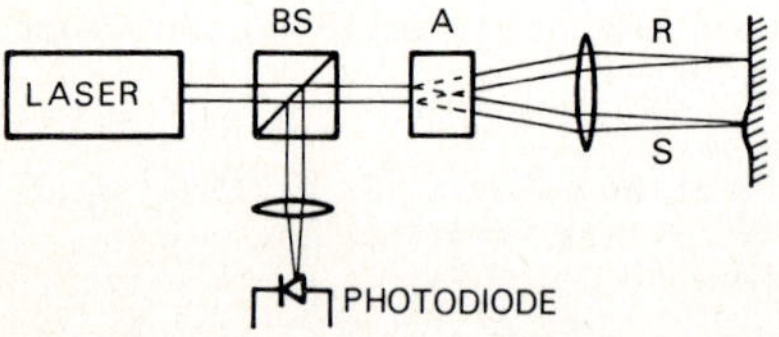

Fig. 8. Differential Interferometer. The beam coming out of the BS cube is divided into two parallel adjacent beams by part A. The low frequency vibrations displace both spots in the same way. Only the phase of the probe beam is changed by the local surface displacement. The two beams recombine and interfere on the photodetector.

In their experiments on Rayleigh waves, Palmer et al [25] chose two spots half a wavelength apart coming from a varifocal lens and achieved a sensitivity of about 10^{-3} Å/$\sqrt{\text{Hz}}$.

Recently, Fanton and Kino applied the method to the measurement of displacements generated by local heating of the sample [26]. Their design is similar to that of Heinrich et al [27] who measured free-carrier densities in silicon through their effect on the refractive index except that they used a Wollaston prism instead of a calcite one. One of the beams, R, focused onto an undisturbed portion of the sample, serves as the reference. The other beam (S) hits the point which is periodically heated. The phase modulation of this beam caused by the thermal expansion of the surface appears as intensity modulation after the two beams recombined in the Wollaston prism, and reflected by the beam splitter, have interfered on the photodetector.

In this compact device, the possible equality of optical path lengths allows the use of a small coherence length source i.e. a laser diode easy to modulate. Fanton and Kino checked the sensitivity of this interferometer by causing a half of a piezoelectric transducer to vibrate, the other half being inactive and serving as a reference. They found that a vibration of amplitude 3.4×10^{-4} Å close to the theoretical limit (2.8×10^{-4} Å) given by eq. 13, could be detected in a 100 Hz bandwidth.

3.3.3. Heterodyne techniques

The feature in these techniques is the introduction of a frequency shifter, usually an acousto-optic Bragg cell, in either (or both) arm (s) of the interferometer. Let $\pm f_B$ be the frequency shift i.e. the acoustic wave frequency (typically 100 MHz) in the cell. The electric fields E_R et E_S are given by eq. 7 and 20 in replacing Ω by Ω_R for the reference beam and Ω by Ω_S for the probe beam. In expression (22) of the photocurrent, deduced from 8, a term at the frequency difference $\Omega_S - \Omega_R = \omega_0$ (equal to ω_B or $2\omega_B$) appears in the alternative part :

$$i(t) = I_0 \cos[\omega_0 t + 2Ku\cos(\omega t + \phi) + \Phi_S - \Phi_R] \qquad (25)$$

The optical phase modulation of the probe beam by the mechanical surface displacement is transposed into the RF domain. The spectrum of the electric signal comprises a central line at ω_0 and lateral lines at $\omega_0 \pm m\omega$ whose heights are given by Bessel functions $J_m(2Ku)$. If the displacement u is small with respect to the optical wavelength ($Ku \ll 1$) i.e. less than 100 Å :

$$J_0(2Ku) \cong 1, \quad J_1(2Ku) \cong Ku, \quad J_m(2Ku) \cong 0 \quad \text{for } m \geqslant 2. \qquad (26)$$

The spectrum reduces to the carrier at ω_0 and two lateral lines at $\omega_0 \pm \omega$:

$$i(t) = I_0\{\cos(\omega_0 t + \Phi_S - \Phi_R) + Ku\cos[(\omega_0 + \omega)t + \phi + \Phi_S - \Phi_R] - Ku\cos[(\omega_0 - \omega)t - \phi + \Phi_S - \Phi_R]\} \qquad (27)$$

The ratio r of the levels of the carrier and one side component provides the absolute amplitude of the mechanical vibration in steady regime independently of the light power reflected by the sample :

$$u = \frac{1000}{r}\ \text{Å} \quad \text{with a He-Ne laser} \qquad (28)$$

a - Coherent electronic detection

The random phase fluctuations $\Phi_S - \Phi_R$ affect the carrier and the sidebands in the same way. Their effect can be cancelled or strongly reduced by coherent electronic detection [28] which restores the "acoustic" phase $\omega t + \phi$, for example, from the difference of the phase of one side component and the central component. Several designs can be devised ; we will describe two. The first one has a narrow bandwidth detection suitable for harmonic regime, the second one, a broad bandwidth suitable for transient regime.

Narrow bandpass.

The spectrum of the amplitude photocurrent is transposed [29] by mixing with a signal from a local oscillator whose frequency ω_{LO} is chosen so that one of the side components is extracted with a narrow bandpass filter (a few kHz) and mixed with the carrier (figure 9). The resulting signal is the product :

$$s(t) \propto Ku\cos[(\omega_0 - \omega_{LO})t + \Phi_S - \Phi_R] \times \cos[(\omega_0 - \omega_{LO} + \omega)t + \phi + \Phi_S - \Phi_R] \qquad (29)$$

After eliminating the part sum at high frequency $(2\omega_0 - 2\omega_{LO} + \omega)$ the mechanical displacement is recovered independently of the fluctuations $\Phi_S - \Phi_R$:

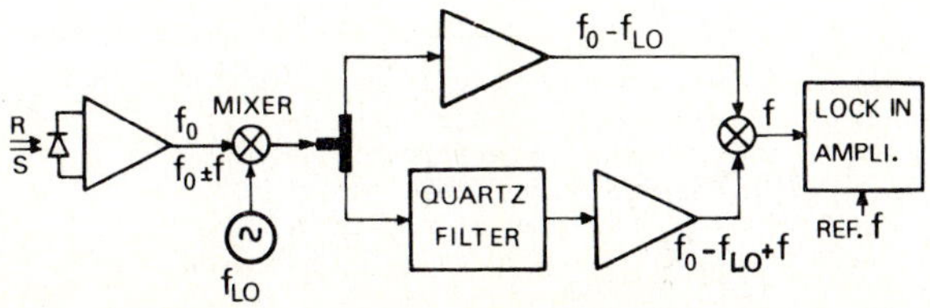

Fig. 9. Narrow bandpass coherent detection diagram.

$$s(t) \propto Ku\cos(\omega t + \phi) \quad (30)$$

The resolution is increased when the signal is detected in a narrow bandwidth by a lock-in amplifier (B = 1Hz for instance).

Broad bandpass.

The photocurrent (25) is amplified and then split into two parts (figure 10.a). One of these passes through a narrow band filter centered at frequency f_0 which eliminates the acoustic term and is then $\pi/2$ phase-shifted. The mixing of this signal

$$i_F(t) \propto \cos(\omega_0 t + \frac{\pi}{2} + \Phi_S - \Phi_R) \quad (31)$$

with the other part of the photocurrent i(t) gives a signal

$$s(t) \propto \sin[2Ku\cos(\omega t + \phi)] \cong 2Ku\cos(\omega t + \phi) \quad (32)$$

proportional to the mechanical displacement when Ku << 1 and a signal at the frequency $2\omega_0$. A low-pass filter of bandwidth B eliminates this high frequency term [30]. The amplitude of the signal is twice that of the previous one (formula 30) because the two spectral sidebands are taken into account. The bandwidth Δf of the filter centered at the carrier frequency f_0 has to be larger than the spectrum width of the phase fluctuations $\Phi_S - \Phi_R$, i.e. a few kHz for mechanical vibrations, for them to be eliminated by difference (figure 10.b). A quartz filter is perfectly appropriate. The electronic detection bandpass is $[\Delta f/2, B]$. The sensitivity is still given by formula 13.

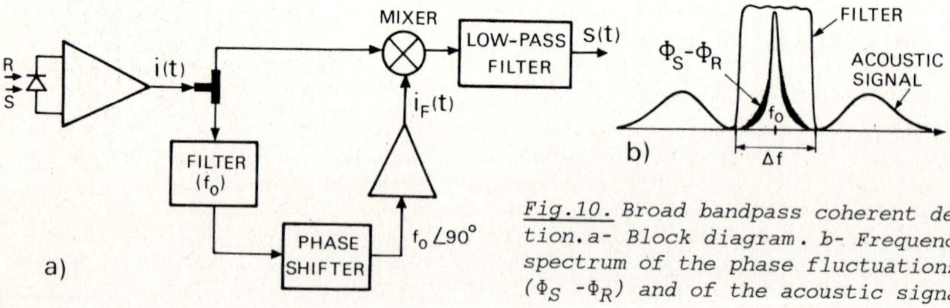

Fig.10. Broad bandpass coherent detection. a- Block diagram. b- Frequency spectrum of the phase fluctuations ($\Phi_S - \Phi_R$) and of the acoustic signal.

In order to measure surface displacements at very low frequencies (f < 1 kHz) a detection circuit based on a phase locked loop has been implemented and applied to photothermal microscopy [31].

b - Optical configurations

The devices constructed so far [12,28] are modified Michelson interferometers in which the beam splitter is an acousto-optic Bragg cell driven at frequency f_B (figure 11). The frequency of the reference beam (R) is upshifted by f_B,

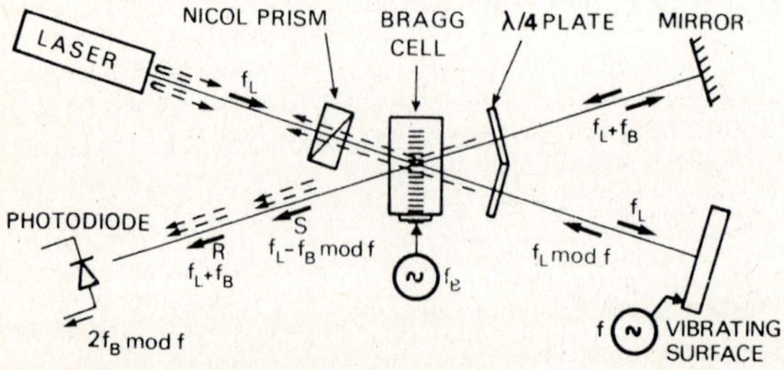

Fig.11. Symmetrical heterodyne configuration. Dashed arrows represent spurious beams that cause instabilities.

that of the probe beam (S) is downshifted by f_B. The carrier frequency of the photocurrent is $f_0 = 2f_B$. This configuration shows some drawbacks : critical alignment procedure, instabilities. The cause of instability is partially to be found in the symmetry of the structure because, on the one hand it gives rise to long optical paths since the angle between beams coming out of the Bragg cell is small (15 mrad for 100 MHz), on the other hand, it causes spurious beams to return toward the laser source. The output signal exhibits slow temporal variations. The variations of amplitude may exceed 10 %,those of phase 10 degrees. Accordingly dynamic range and practical sensitivity are reduced [32].

We will describe an asymmetrical configuration that we have implemented and which reduces the drawbacks mentioned. As shown in figure 12a, the two beams are extracted by the cube beam splitter BS from the laser source. The horizontally polarized reference beam R is directed through a prism towards the photodiode. The probe beam S whose frequency is shifted by a colinear Bragg cell (f_B) is reflected by the vibrating sample (f). After passing (twice) through the quarter wavelength plate this beam is vertically polarized and, accordingly, reflected at $\pi/2$ by the polarizing beam splitter PBS, along the direction of the reference beam R. The two beams, respectively at frequencies f_L and $f_L + f_B$, pass through the analyser, oriented at $\pi/4$ relative to their polarizations, and interfere on the photodetector. The photocurrent at frequency f_B is phase modulated by the vibration of the sample surface at the frequency f (figure 12b).

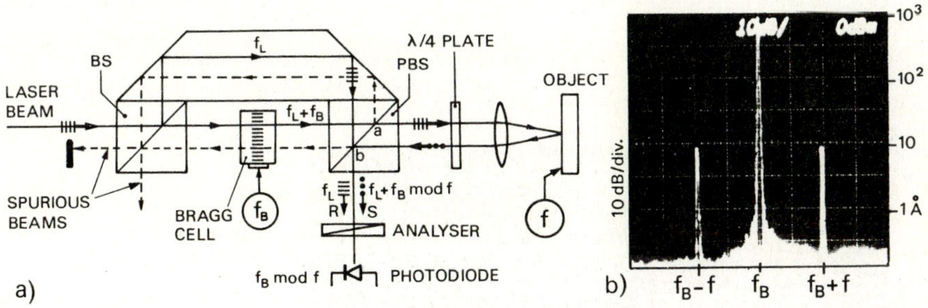

Fig.12. Asymmetrical heterodyne probe. a) Optical part. b) Spectrum of the photodiode signal. The amplitude of the surface displacement at frequency f is given by the level of the side component compared to that of the central peak (28).

The focusing of the probe beam onto the object is useful because the spot is more precisely defined, a larger amount of reflected light is collected and the probe becomes less sensitive to the tilt of the sample when this is displaced laterally. The off-centering, with respect to the axis of the two cubes, eliminates the spurious signals which come from the reflexions of the probe beam (point a) and of the reference beam (point b) on the splitter interface of the PBS cube.

The advantages of the asymmetrical probe compared to the symmetrical one derive from the fact that only the probe beam passes (once) through the acousto-optic cell and comes out without change of direction. The carrier frequency of the photocurrent is then f_B instead of $2f_B$. The optical part is compact (less than 40 cm long, laser included) and can operate in a vertical or horizontal position. The stability is improved : the amplitude and phase of the output signal variations are respectively about 0.2 % and 0.2° [32] .

c - Experimental results

We will present here a few results to illustrate the possibilities of heterodyne optical probes. Figure 13 is the image of a crack in a block of glass recorded by probing (with a symmetrical probe) the displacement generated by periodically

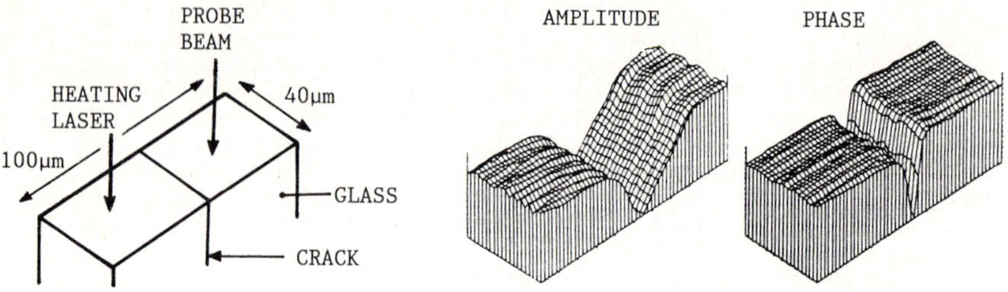

Fig.13. Image of a closed crack in a block of glass. After Martin and Ash [33].

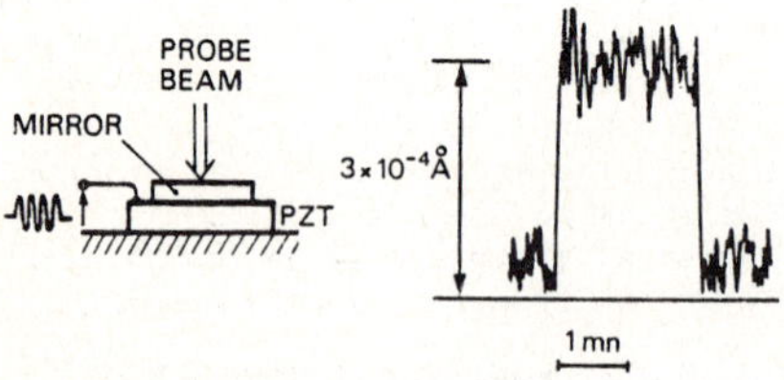

Fig.14. Harmonic vibration amplitude of a piezoelectric (PZT) plate showing the sensitivity of the asymetrical probe (B = 1 Hz).

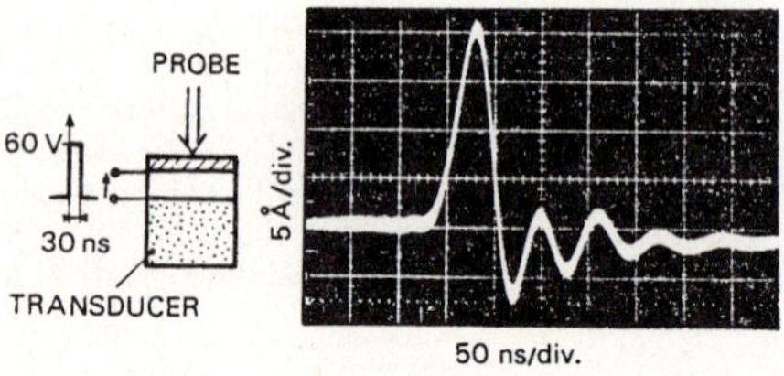

Fig.15. Impulse response of a Panametrics transducer at center (B = 20 MHz).

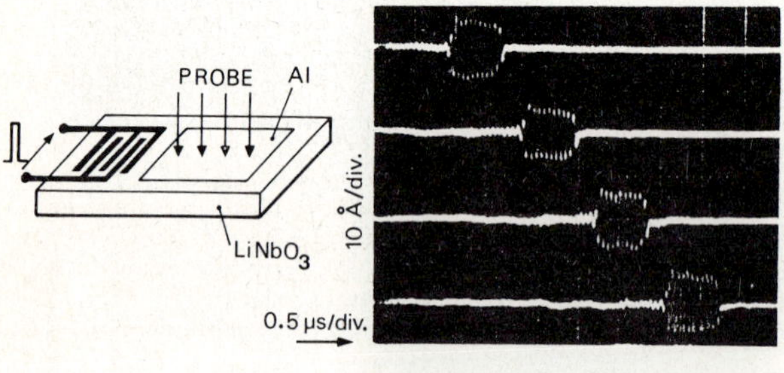

Fig.16. Response of an interdigital Rayleigh wave transducer on lithium niobate substrate to a pulse (60 V, 20 ns) probed at points 3 mm apart (B = 50 MHz).

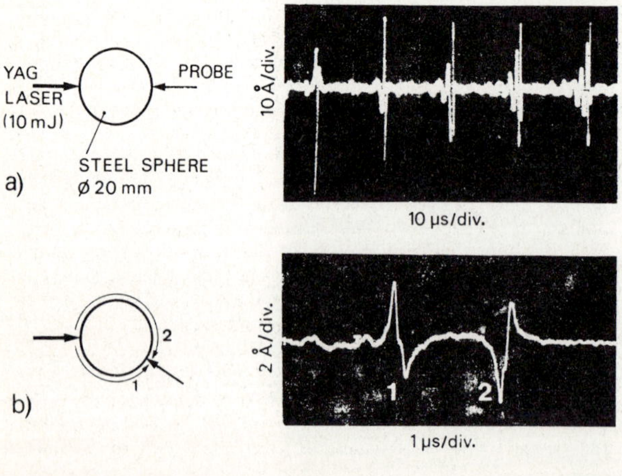

Fig.17. Spherical surface acoustic waves excited by a YAG laser pulse (10 mJ, 100 ns).
a) Wave train launched by a pulse and observed at the pole. The spreading caused by dispersion is noticeable as the number of turns increase.
b) First pulse going to the pole (1) leaving the pole (2). The passge through the pole produces a π phase-shift.

heating the surface with a laser diode [33]. All the following results were obtained with the asymmetrical probe. The sensitivity of this probe provided with a narrow bandpass electronic detection is emphasized in figure 14 which is the recording of the harmonic vibration of the surface of a piezoelectric plate (f = 160 kHz). The step of 3×10^{-4} Å shows that 10^{-4} Å is detectable i.e. that the theoretical limit corresponding to the conditions of experiment (2 mW He-Ne laser, B = 1 Hz) is nearly reached. Figure 15 is the impulse response of a Panametrics transducer ($\phi \cong 5$ mm) recorded with a broad bandpass circuit (B = 20 MHz). The peak indicates a maximum displacement of 20 Å. Of course, a complete map of the peak displacement can be obtained by translating the transducer along two orthogonal directions [30]. Figure 16 illustrates the propagation of Rayleigh waves on lithium niobate substrate. The wave trains are the impulse response of an interdigital transducer (10 finger pairs, center frequency : 17 MHz) observed at points separated by 3 mm (B = 50 MHz).

The ultrasounds in these three foregoing experiments were excited by electrical signals. In another set of experiments we have generated surface acoustic waves by YAG laser pulses (10 mJ). Figure 17.a shows wave trains propagating on a steel sphere ($\phi = 20$ mm), launched by a single laser pulse. The observation at a distance (5 mm) away (b) from the pole shows the wave converging (1) toward the pole and then diverging (2) from the pole. Pulse 2 undergoes a π-phase shift on passing through the pole [34].

4. CONCLUSION

The purpose of this paper was to examine the means of detecting surface displacements less than one Angström not requiring any mechanical contact. This restriction rules out piezoelectric effect (which is extremely sensitive since it is able to detect displacements of 10^{-8} Å $/\sqrt{\text{Hz}}$).

The capacitive detection which can achieve a sensitivity of 10^{-4} Å is apparently a good technique. However, we have described only one example of application because this technique presents, in practice, several shortcomings - the surface has to be metallic - the capacitive receiver generally collects parasitic or direct feedthrough electromagnetic signals which may saturate the amplifier associated with the receiver. - exploration of a surface of area a few square centimeters, point by point, is difficult because the electrode gap has to be maintained constant. To summarize, this capacitive method may be useful to detect the frequency [35] rather than the amplitude of a displacement.

This remark also partly applies to a few of the optical methods that we have described, for instance to the detection by surface ripple deflexion, however the optical techniques remain unrivalled. Indeed, they lend themselves to a local examination of a sample. The area of the vibrating surface may be as small as a few square micrometers (light spot size). Zones of area several square centimeters, not necessarily flat, can be explored by displacing either the object or the light beam (which can periodically scan the surface if an acousto-optic or electro-optic deflector is used). The theoretical sensitivity, whatever the method, is better than 10^{-4} Å. Heterodyne techniques, which have low sensitivity to environmental noise, are easy to calibrate. They are most efficient if they are provided with electronic coherent detection circuits. The drawbacks, often so far mentioned with regard to these techniques and referring to their bulkiness, instabilities and their requirement of a polished surface, are no longer totally justified [36]. Compact and stable configurations have been implemented which are now being developed in the form of instruments.

ACKNOWLEDGEMENTS

The authors would like to acknowledge the support given by DRET (Paris) and to thank H. Duchaussoy (DRET) for his helpful remarks.

REFERENCES

1- Dieulesaint,E.,and Royer,D.,"Elastic Waves in Solids" (Wiley, Chichester, 1980)

2- Morgan,D.P.,"Surface Wave Devices for Signal Processing" (Elsevier, 1985)

3- Ristic,V.M.,"Principles of Acoustic Devices" (Wiley, New-York, 1983)

4- Hutchins, D.A.,and TAM, A.C.,"Special Issue on Photoacoustics" IEEE Trans. Ultrason. Ferroelectrics and Freq. Control, UFFC-33, 427 (1986)

5- Lemons, R.A.,and Quate, C.F.,"Acoustic Microscopy" In Physical Acoustics (W.P. Mason and R.N. Thurston Eds), Vol.14,chap.1(Academic Press,N.Y.,1979)

6- Bahadur,H.,and Parshad,R., "Acoustic Vibrational Modes in Quartz Crystals : Their Frequency, Amplitude and Shape Determination" In Physical Acoustics (W.P. Mason and R.N. Thurston Eds),Vol.16, Chap.2 (Academic Press, N.Y.,1982)

7- Sachse, W.,and Hsu,N.,"Ultrasonic Transducers for Materials Testing and their Characterisation"In Physical Acoustics (W.P. Mason and R.N. Thurston Eds) Vol. 14, Chap.4 (Academic Press, New-York, 1979)

8- Ishii,A., and Hashimoto,S., IEEE Ultrason. Symp. Proc., 167 (1981)

9- Dieulesaint,E.,Royer,D.,Chaabi,A.,and Formery,B.,Electronics Letters,23,982 (1987)

10- Gauster,W.B.,and Breazeale,M.A.,Rev. Sci. Instrum.,37, 1544 (1966)

11- Breazeale,M.A.,and Philip,J.,"Determination of Third-Order Constants from Ultrasonic Harmonic Generation Measurements" In Physical Acoustics (W.P. Mason and R.N. Thurston Eds), Vol.17 Chap.1. (Academic Press, New-York ,1984)

12- Whitman,R.L.,and Korpel,A.,Applied Optics,8, 1567 (1969)

13- Ref. 1, Chap. 8.

14- Stegeman,G.I.,IEEE Trans. Son. Ultrason.,SU-23, 33 (1976)

15- Rouvaen,J.M.,et al, Electronics Letters, 10, 297 (1974)

16- Ref.1, p. 434

17- Adler,R.,Korpel,A.,and Desmares,P.,IEEE Trans.Son. Ultrason.SU-15, 157 (1968)

18- Kessler,L.W.,and Yuhas,D.E.,Proc. IEEE,67, 526 (1979)

19- Engan,H.,IEEE Trans. Son. Ultrason.,SU-25,372 (1978)

20- Parker,T.E.,IEEE Ultrason. Symp. Proc., 365 (1974)

21- Engan,H.,IEEE Trans. Son. Ultrason.,SU-29, 281 (1982)

22- Monchalin,J.P., Appl. Phys. Lett.,47, 14 (1985)

23- Kroll,M.,and Djordjevic,B.B.,IEEE Ultrason. Symp. Proc.,864 (1982)

24- Hutchins,D.A.,and Nadeau,F.,IEEE Ultrason. Symp. Proc.,1175 (1983)

25- Palmer,C.H., Richard,O.C., and Fick,S.E., Applied Optics,16, 1849 (1977)

26- Fanton,J.T.,and Kino,G.S.,Appl. Phys. Lett.,51, 66 (1987)

27- Heinrich,H.K.,Hemenway,B.R.,McGroddy,K.A.,and Bloom,D.M.,Electron. Letters, 22, 650 (1986)

28- De La Rue,R.M., Humphryes,R.F.,Mason,I.M.,and Ash,E.A.Proc.IEE,119,117(1972)

29- Martin,Y.,and Ash,E.A., Electron. Lett.,18, 763 (1982)

30- Royer,D.,and Dieulesaint,E.,IEEE Ultrason. Symp. Proc.,527 (1986)

31- Cretin,B.,and Hauden,D.,IEEE Ultrason. Symp. Proc., 656 (1984)

32- Royer,D.,Dieulesaint,E. and Martin,Y.,IEEE Ultrason. Symp. Proc., 432(1985)

33- Martin,Y. and Ash,E.A. IEEE Ultrason. Symp. Proc.,647 (1984)

34- Royer,D.,Dieulesaint,E.,Jia,X.,and Shui,Y.,Appl. Phys. Lett.,to be published

35- Dieulesaint,E.,Mazerolle,D., and Royer,D.,Electron. Lett., 23, 581 (1987)

36- Monchalin,J.P., IEEE Trans. Ultrason. Ferro. and Freq. Control, 33,485(1986)

ELASTIC WAVE PROPAGATION
M.F. McCarthy, M.A. Hayes, (Editors)

SURFACE WAVE EXISTENCE THEORY FOR THE CASE OF ZERO CURVATURE TRANSONIC STATES

David M. BARNETT

Department of Materials Science and Engineering, Stanford University, Stanford, California 94305-2205, USA.

Jens LOTHE

Department of Physics, University of Oslo, P.O.Box 1048 Blindern, 0316 Oslo 3, Norway.

The existence theory is extended to take zero curvature transonic states into account. A subsonic free surface wave will always exist.

1. INTRODUCTION

Consider surface waves along $\vec{m}$ in a surface whose unit normal is $\vec{n}$. The inverse limiting velocity $\hat{v}^{-1}$ is found from the slowness curve as shown in Fig. 1. There are six types of transonic states differing by the number of tangency points and coincidences [1,2]. A tangency point represents a limiting bulk wave. A limiting wave which leaves the plane traction free is an *exceptional* limiting wave. It has been proved [1,3,4] that *unless the transonic state is of type 1 and with an exceptional limiting wave, a subsonic surface wave exists.* This conclusion rests on the assumption that the slowness curve has non-zero curvature at a tangency point, or equivalently that at a tangency point coalescence of only one pair of roots has occurred. A zero curvature tangency point signifies coalescence of two pairs. Since there are only six roots, we are assured that types 3,5 and 6 are ordinary transonic states. The new possibilities are zero curvature variants of types 1, 4 and 2 which we call *extraordinary* transonic states E1, E2 and E3 respectively. See Fig. 2. E1 is known to occur in Zn. Here we develop general theory for E1 states.

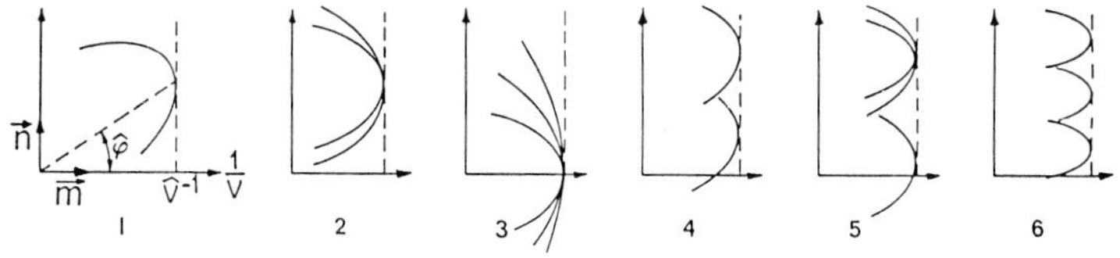

Figure 1: The six types of transonic states.

2. GENERAL FORMALISM

$$\vec{u} = \vec{A} \exp ik(\vec{m} \cdot \vec{x} + p\vec{n} \cdot \vec{x} - vt), \quad C_{ijkl} \frac{\partial^2 u_k}{\partial x_i \partial x_l} = \rho \frac{\partial^2 u_j}{\partial t^2} \tag{2.1}$$

$$\text{i.e.} \quad \{(mm) - \rho v^2 I + p((mn) + (nm)) + p^2(nn)\}\vec{A} = 0 \tag{2.2}$$

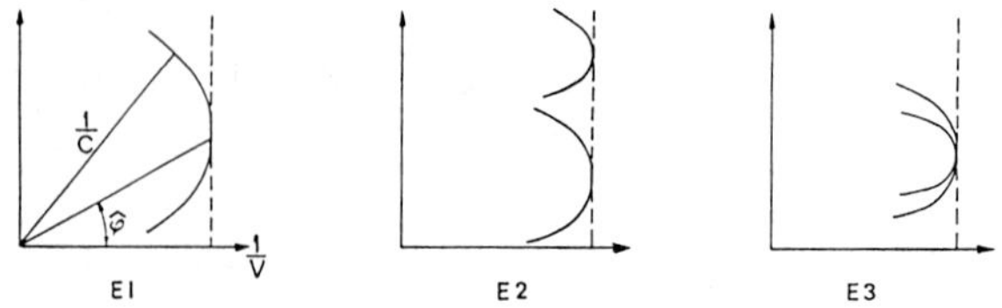

Figure 2: Extraordinary transonic states.

The matrices (mm) etc. are defined by the convention that $(ab)_{jk} = a_i C_{ijkl} b_l$. When $0 \leq v < \hat{v}$ solutions $\vec{A}_\alpha$ exist for six complex p_α in pairs of complex conjugates. For $\alpha = 1,2,3$ $\mathrm{im}\, p_\alpha > 0$. As $v \to \hat{v}$, one or more pairs coalesce to real roots corresponding to limiting bulk waves. Introduce [5] $\vec{L}$ vectors and sixdimensional ξ vectors,

$$\vec{L}_\alpha = -((nm) + p_\alpha (nn))\vec{A}_\alpha, \quad \xi_\alpha = (\vec{A}_\alpha, \vec{L}_\alpha). \tag{2.3}$$

$$N\xi_\alpha = p_\alpha \xi_\alpha, \tag{2.4}$$

$$\text{where } N = -\begin{pmatrix} (nn)^{-1}(nm) & (nn)^{-1} \\ (mn)(nn)^{-1}(nm) - (mm) + \rho v^2 I & (mn)(nn)^{-1} \end{pmatrix}. \tag{2.5}$$

give a reformulation of (2.2). N has such symmetry that $N^T = T \cdot N \cdot T$, where T as a superscript means tranposition and as a sixdimensional matrix means

$$T = \begin{pmatrix} 0 & I \\ I & 0 \end{pmatrix}. \tag{2.6}$$

Stroh [5] derives the orthogonality relation

$$\xi_\alpha \cdot T \cdot \xi_\beta = 0 \text{ when } p_\alpha \neq p_\beta. \tag{2.7}$$

When no degeneracies are present, we may normalize so that

$$\vec{A}_\alpha \cdot \vec{L}_\beta + \vec{A}_\beta \cdot \vec{L}_\alpha = \delta_{\alpha\beta}. \tag{2.8}$$

$\vec{L}_\alpha$ is related to the amplitude $\vec{T}_\alpha$ of surface traction needed to support the partial wave

$$\vec{T}_\alpha = -ik\vec{L}_\alpha. \tag{2.9}$$

Thus a subsonic s.w. exists if at some $v < \hat{v}$ the $\vec{L}_\alpha$, $\alpha = 1,2,3$, are linearly dependent,

$$\| L_{i\alpha} \| = 0. \tag{2.10}$$

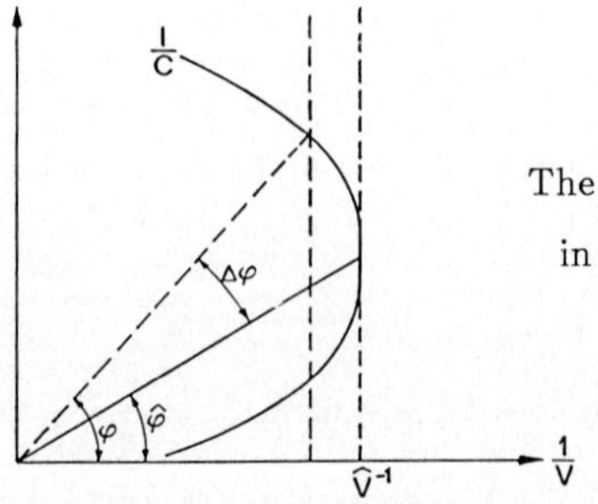

Figure 3:
The relation between φ and v in the supersonic region.

The surface impedance Z [3,6,7] relates $\vec{L}_\alpha$ to $\vec{A}_\alpha$ by the relation

$$\vec{L}_\alpha = -iZ \cdot \vec{A}_\alpha. \tag{2.11}$$

In terms of the eigenvectors $(\vec{A}_\alpha, \vec{L}_\alpha)$ Z is

$$Z = i \sum_{\alpha \neq \beta \neq \gamma} \frac{\vec{L}_\alpha \otimes (\vec{A}_\beta \times \vec{A}_\gamma)}{2[\vec{A}_\alpha, \vec{A}_\beta, \vec{A}_\gamma]}, \quad \alpha,\ \beta,\ \gamma = 1, 2, 3 \tag{2.12}$$

where $[\vec{A}_\alpha, \vec{A}_\beta, \vec{A}_\gamma]$ is the triple scalar product. $\otimes$ denotes dyadic multiplication. The three $\vec{A}_\alpha$ are independent in the subsonic interval [5,8]. Thus Z is well defined and condition (2.10) reduces to

$$\| Z \| = 0. \tag{2.13}$$

Z is positive definite at $v = 0$ and its eigenvalues decrease monotonically with increasing velocity [2,3,7]. Surface wave existence theory thus reduces to the question of whether or not Z gets a negative eigenvalue as $v \to \hat{v}$.

3. GENERALIZED EIGENVECTORS

Consider an E1 transonic state. $\hat{N}$, (2.5) at $v = \hat{v}$, has then a fourfold degenerate root $\hat{p} = \tan\hat{\varphi}$, but there is only one eigenvector, the limiting vector $\hat{\xi} = (\hat{\vec{A}}, \hat{\vec{L}})$ which satisfies

$$\hat{N}\hat{\xi} = \tan\hat{\varphi}\,\hat{\xi}. \tag{3.1}$$

In addition there is a complex pair p_c and p_c^* with corresponding eigenvectors ξ_c and ξ_c^*. A complete set can be obtained by construction of additional generalized eigenvectors by a ladder scheme [9,10]. N can be written as

$$N(v) = \hat{N} - \rho(v^2 - \hat{v}^2)M, \text{ where } M = \begin{pmatrix} 0 & 0 \\ I & 0 \end{pmatrix}. \tag{3.2}$$

In the supersonic region we can parameterize v by the angle φ defined from the intersection between the zero curvature slowness curve and a vertical line through $\frac{1}{v}$; see Fig. 3. Thus

$$v^2 = \hat{v}^2 + \alpha(\varphi - \hat{\varphi})^4 + \beta(\varphi - \hat{\varphi})^5 + \gamma(\varphi - \hat{\varphi})^6 + \dots, \tag{3.3}$$

$$N(\varphi) = \hat{N} - \rho(\alpha(\varphi - \hat{\varphi})^4 + \beta(\varphi - \hat{\varphi})^5 + \dots)M, \tag{3.4}$$

$$N(\varphi)\xi(\varphi) = \tan\varphi\,\xi(\varphi). \tag{3.5}$$

$\xi(\varphi)$ can be taken to be real. Letting $\varphi \to \hat{\varphi}$ in (3.5) we regain (3.1). Differentiating (3.5) with respeat to φ and letting $\varphi \to \hat{\varphi}$, we obtain

$$(\hat{N} - \tan\hat{\varphi})(\frac{d\xi}{d\varphi})_{\hat{\varphi}} = \tan'\hat{\varphi} \cdot \hat{\xi}, \tag{3.6}$$

$$(\hat{N} - \tan\hat{\varphi})(\frac{d^2\xi}{d\varphi^2})_{\hat{\varphi}} = 2\tan'\hat{\varphi}(\frac{d\xi}{d\varphi})_{\hat{\varphi}} + \tan''\hat{\varphi} \cdot \hat{\xi}, \tag{3.7}$$

$$(\hat{N} - \tan\hat{\varphi})(\frac{d^3\xi}{d\varphi^3})_{\hat{\varphi}} = 3\tan'\hat{\varphi}(\frac{d^2\xi}{d\varphi^2})_{\hat{\varphi}} + 3\tan''\hat{\varphi}(\frac{d\xi}{d\varphi})_{\hat{\varphi}} + \tan'''\hat{\varphi} \cdot \hat{\xi}. \tag{3.8}$$

Since (3.1) and (3.6)-(3.8) form a ladder similar to the standard ladder [9,10]

the eigenvectors and $(\frac{d\xi}{d\varphi})_{\hat{\varphi}}$, $(\frac{d^2\xi}{d\varphi^2})_{\hat{\varphi}}$ and $(\frac{d^3\xi}{d\varphi^3})_{\hat{\varphi}}$ form a complete set. (3.9)

These generalized eigenvectors have a simple interpretation in terms of plane waves belonging to the slowness curve. A number of orthogonality relations follow:

$$\hat{\xi} \cdot T \cdot \hat{\xi} = 0, \qquad \hat{\xi} \cdot T \cdot (\frac{d\xi}{d\varphi})_{\hat{\varphi}} = 0$$

$$\hat{\xi} \cdot T \cdot (\frac{d^2\xi}{d\varphi^2})_{\hat{\varphi}} = 0, \qquad (\frac{d\xi}{d\varphi})_{\hat{\varphi}} \cdot T \cdot (\frac{d\xi}{d\varphi})_{\hat{\varphi}} = 0. \tag{3.10}$$

These relations together with

$$\hat{\xi} \cdot T \cdot \xi_c = 0; \quad (\frac{d^k\xi}{d\varphi^k})_{\hat{\varphi}} \cdot T \cdot \xi_c = 0, \quad k = 1, 2, 3, ; \quad \xi_c^* \cdot T \cdot \xi_c = 0 \tag{3.11}$$

will suffice for our purposes.

4. LINEAR INDEPENDENCE CONSIDERATIONS

$\hat{\xi}$, $(\frac{d\xi}{d\varphi})_{\hat{\varphi}}$ and ξ_c are independent, but is it true that $\hat{\vec{A}}$, $(\frac{d\vec{A}}{d\varphi})_{\hat{\varphi}}$ and $\vec{A}_c$ are independent? Our analysis of this question will also rely on a corollary to (3.9), namely that

since ξ_c is an eigenvector for a nondegenerate root p_c, $\xi_c \cdot T \cdot \xi_c \neq 0$. (4.1)

$$\text{i.e.} \quad \vec{A}_c \cdot \vec{L}_c = \frac{1}{2} \tag{4.2}$$

is possible ((2.8)). Consider $\hat{\vec{A}}$ and $(\frac{d\vec{A}}{d\varphi})_{\hat{\varphi}}$. With $\vec{A}(\varphi) \cdot \vec{A}(\varphi) = 1$ we have

$$\hat{\vec{A}} \cdot (\frac{d\vec{A}}{d\varphi})_{\hat{\varphi}} = 0. \tag{4.3}$$

Either a) $(\frac{d\vec{A}}{d\varphi})_{\hat{\varphi}}$ is nonzero and orthogonal to $\hat{\vec{A}}$, or b)

$$(\frac{d\vec{A}}{d\varphi})_{\hat{\varphi}} = 0. \tag{4.4}$$

According to (2.3)

$$\vec{L}(\varphi) = -((nm) + \tan\varphi(nn))\vec{A}(\varphi), \tag{4.5}$$

$$\text{i.e.} \quad (\frac{d\vec{L}}{d\varphi})_{\hat{\varphi}} = -\frac{1}{\cos^2\hat{\varphi}}(nn)\hat{\vec{A}} \tag{4.6}$$

when (4.4) is true. Then, from (4.6),

$$\hat{\vec{A}} \cdot (\frac{d\vec{L}}{d\varphi})_{\hat{\varphi}} < 0 \tag{4.7}$$

since (nn) is positive definite. One of eqs. (3.10) states that

$$\hat{\vec{A}} \cdot (\frac{d\vec{L}}{d\varphi})_{\hat{\varphi}} + \hat{\vec{L}} \cdot (\frac{d\vec{A}}{d\varphi})_{\hat{\varphi}} = 0 \tag{4.8}$$

which when (4.4) is true, reduces to

$$\hat{\vec{A}} \cdot (\frac{d\vec{L}}{d\varphi})_{\hat{\varphi}} = 0, \tag{4.9}$$

in contradiction with (4.7). (4.4) is untrue. $\hat{\vec{A}}$ and $(\frac{d\vec{A}}{d\varphi})_{\hat{\varphi}}$ are independent. But could

$$\vec{A}_c + \gamma_1 \hat{\vec{A}} + \gamma_2 (\frac{d\vec{A}}{d\varphi})_{\hat{\varphi}} = 0 \tag{4.10}$$

be true? According to (3.11),

$$(\frac{d\vec{A}}{d\varphi})_{\hat{\varphi}} \cdot \vec{L}_c + (\frac{d\vec{L}}{d\varphi})_{\hat{\varphi}} \cdot \vec{A}_c = 0 \text{ and } \hat{\vec{A}} \cdot \vec{L}_c + \hat{\vec{L}} \cdot \vec{A}_c = 0. \tag{4.11}$$

If (4.10) is true, it follows from (4.2) and (4.11) that

$$-[\gamma_1 \hat{\vec{L}} + \gamma_2 (\frac{d\vec{L}}{d\varphi})_{\hat{\varphi}}] \cdot [\gamma_1 \hat{\vec{A}} + \gamma_2 (\frac{d\vec{A}}{d\varphi})_{\hat{\varphi}}] = \frac{1}{2}. \tag{4.12}$$

Multiplying out (4.12) and using (3.10) we derive $0 = \frac{1}{2}$ which is a contradiction. Eq. (4.10) cannot be true. In conclusion

For an E1 transonic state $\hat{\vec{A}}$, $(\frac{d\vec{A}}{d\varphi})_{\hat{\varphi}}$ and $\vec{A}_c$ are linearly independent. (4.13)

5. THE LIMITING IMPEDANCE

Consider the operator $N(v)$, (3.2), for subsonic velocities near $\hat{v}$. In terms of the complete set (3.9) and to lowest order in $(\hat{v}^2 - v^2)^{\frac{1}{4}}$ the three eigenvectors for which $\operatorname{im} p > 0$ are

$$\xi_{1,2} = \hat{\xi} \pm \delta(1 \pm i)(\frac{d\xi}{d\varphi})_{\hat{\varphi}}, \quad \xi_3 = \xi_c \tag{5.1}$$

where $\delta = \frac{1}{2}\sqrt{2}\alpha^{-\frac{1}{4}}(\hat{v}^2 - v^2)^{\frac{1}{4}}$. Inserting (5.1) into (2.12) and letting $\delta \to 0$,

$$\begin{aligned} \hat{Z} &= i[\hat{\vec{A}}, (\frac{d\vec{A}}{d\varphi})_{\hat{\varphi}}, \vec{A}_c]^{-1}[\hat{\vec{L}} \otimes ((\frac{d\vec{A}}{d\varphi})_{\hat{\varphi}} \times \vec{A}_c) \\ &+ (\frac{d\vec{L}}{d\varphi})_{\hat{\varphi}} \otimes (\vec{A}_c \times \hat{\vec{A}}) + \vec{L}_c \otimes (\hat{\vec{A}} \times (\frac{d\vec{A}}{d\varphi})_{\hat{\varphi}})]. \end{aligned} \tag{5.2}$$

Eq. (4.13) guarantees that (5.2) is well defined. $\hat{Z}$ behaves relative to $\hat{\xi}$ and $(\frac{d\xi}{d\varphi})_{\hat{\varphi}}$ in the usual way, (2.11)

$$\hat{\vec{L}} = -i\hat{Z} \cdot \hat{\vec{A}}; \quad (\frac{d\vec{L}}{d\varphi})_{\hat{\varphi}} = -i\hat{Z} \cdot (\frac{d\vec{A}}{d\varphi})_{\hat{\varphi}}. \tag{5.3}$$

6. SURFACE WAVE EXISTENCE CONSIDERATIONS

Considering ordinary transonic states we [6] showed that from the facts that $\hat{Z}$ is hermitian, that $\vec{V} \cdot \hat{Z} \cdot \vec{V} \geq 0$ for all real vectors $\vec{V}$, that a real limiting eigenvector $\hat{\xi}$ (a limiting wave) obeys $\hat{\vec{L}} = -i\hat{Z} \cdot \hat{\vec{A}}$, and that a limiting wave $\hat{\xi}$ is selforthogonal, $\hat{\vec{A}} \cdot \hat{\vec{L}} = 0$, it follows that unless all limiting $\hat{\vec{L}}$ are zero, $\hat{Z}$ has a negative eigenvalue and a subsonic surface wave solution exists. By the same reasoning, from (5.3) and the selforthogonalities in (3.10), we conclude that

unless *both* $\hat{\vec{L}}$ and $(\frac{d\vec{L}}{d\varphi})_{\hat{\varphi}}$ are zero, there is a negative eigenvalue in $\hat{Z}$ for an E1 transonic state and a subsonic s.w. solution exists. (6.1)

We may go further. Eq. (3.16) in ref. [3] can be written as

$$(\vec{m} + \tan\varphi\,\vec{n})\cdot\vec{L} = -\rho\Big(\frac{c(\varphi)}{\cos\varphi}\Big)^2\vec{n}\cdot\vec{A} \tag{6.2}$$

since $v^{-1} = \cos\varphi\cdot c^{-1}$; see fig.3. Taking (6.2) to the limit $\hat{\varphi}$, we obtain

$$(\vec{m} + \tan\hat{\varphi}\,\vec{n})\cdot\hat{\vec{L}} = -\rho\hat{v}^2\vec{n}\cdot\hat{\vec{A}}. \tag{6.3}$$

Differentiating (6.2) with respect to φ and then going to the limit,

$$(\vec{m} + \tan\hat{\varphi}\,\vec{n})\Big(\frac{d\vec{L}}{d\varphi}\Big)_{\hat{\varphi}} + \frac{1}{\cos^2\hat{\varphi}}\vec{n}\cdot\hat{\vec{L}} = -\rho\hat{v}^2\vec{n}\cdot\Big(\frac{d\vec{A}}{d\varphi}\Big)_{\hat{\varphi}} \tag{6.4}$$

since $\frac{d}{d\varphi}(\cos\varphi\, c^{-1}(\varphi))_{\hat{\varphi}} = 0$. It follows from (6.3) and (6.4) that

when *both* $\hat{\vec{L}}$ and $(\frac{d\vec{L}}{d\varphi})_{\hat{\varphi}}$ are zero then $\vec{n}\cdot\hat{\vec{A}}$ and $\vec{n}\cdot(\frac{d\vec{A}}{d\varphi})_{\hat{\varphi}}$ are zero. (6.5)

From analogy with our discussion for type 4 transonic states [3], with the first generalized eigenvector playing the rôle of the second limiting eigenvector in that discussion, we conclude that a situation such as in (6.5) contradicts $\operatorname{im} p_c > 0$. Therefore $\hat{\vec{L}}$ and $(\frac{d\vec{L}}{d\varphi})_{\hat{\varphi}}$ cannot both be zero. Thus from (6.1)

for an E1 transonic state a subsonic surface wave solution always exists. (6.6)

ACKNOWLEDGEMENT

D. M. B. & J. L. acknowledge support from U.S. Dept. of Energy, Office of Basic Sci.,Div. of Eng., Math., and Geosci., through a contract with Stanford University, and from STATOIL, respectively.

REFERENCES

[1] D. M. Barnett and J. Lothe, *J.Phys.F* **4**, 671 (1974).

[2] P. Chadwick and G. D. Smith, *Adv.appl.Mech.* **17**, 303 (1977).

[3] D. M. Barnett and J. Lothe, *Proc.Roy.Soc.Lond.A* **402**, 135 (1985).

[4] P. Chadwick, *Proc.Roy.Soc.Lond.A* **401**, 203 (1985).

[5] A. N. Stroh, *J.Math.& Phys.* **41**, 77 (1962).

[6] J. Lothe and D. M. Barnett, *J.Appl.Phys.* **47**, 428 (1976).

[7] K. A. Ingebrigtsen and A. Tonning, *Phys.Rev.* **184**, 942 (1969).

[8] P. K. Currie, *Q.J.Mech.Appl.Math.* **27**, 489 (1974).

[9] E. D. Nering, Linear Algebra and Matrix Theory, (Wiley, London, 1963).

[10] M. C. Pease III, Methods of Matrix Algebra, (Academic Press, New York, 1972).

ELASTIC WAVE PROPAGATION
M.F. McCarthy, M.A. Hayes, (Editors)
© Elsevier Science Publishers B.V. (North-Holland), 1989

PULSED CYLINDRICAL ELASTIC WAVES IN A HALF-SPACE WITH A DIPPING SURFACE LAYER

Piotr BOREJKO* and Franz ZIEGLER*

Civil Engineering Department,
Technical University of Vienna,
Karlsplatz 13-201, A-1040 Vienna, Austria

The theory of generalized ray integrals is applied to analyzing transient P- and SV-waves subjected to multiple reflection and refraction when propagating in a dipping layer from a line source to a receiving point fixed in space. Ray integrals are sorted by arrival times and their time signatures are presented.

1. INTRODUCTION

The generalized ray theory is well established for wave propagation in parallel linear elastic and homogeneous layers; for a reviewing article see e.g., Pao, Gajewski [1]. SH-waves in a dipping layer were studied by Ishii, Ellis [2,3]. Hong, Helmberger [4] developed the generalized ray theory for dipping structures. Using the Cagniard transformation for inversion of the Laplace transforms in connection with a constructive way of assigning proper phase functions to the ray integrals resulted in a semi-numerical method discussed in a series of papers by Pao, Ziegler (et al) [5-10]. Since all the reflection and transmission coefficients of plane waves depend on local wave slowness, the transformation properties during rotations of coordinate systems through the dipping angle α are crucial. The concept of apparent source ray introduced in [6] and [7] and further applied to wave mode conversion in [9] determines the fundamental steps of successive transformations up to the source ray slowness which is the integration variable. The line source is located at tip distance x_0 in a depth $z_0 = h/2$ where h denotes the layer thickness at source location. Receiver is put on the traction free surface of the half-space at various epicentral distances (x,0). According to Fig. 1 a coordinate system (x',z') rotated against (x,z) by angle α is considered. Interface (2) is a perfect bond between the top layer and the underlying fast bedrock.

The outgoing cylindrical P- wave from a line source has the Laplace transformed displacement potential in Weyl-Sommerfeld representation

$$\bar{\phi}_0(s) = \bar{F}(s) \int_{-\infty}^{\infty} S(\xi)e^{sg_0(\xi)}d\xi = \bar{F}(s) \int_{-\infty}^{\infty} S'(\xi')e^{sg_0'(\xi')}d\xi' , \quad (1)$$

in (x,z) or (x',z') rotated coordinates. The emittance function for an explosive source is simply

----

* Sponsored by the Austrian FWF, grant S30-01, 1986-1988

$$S(\xi) = \eta^{-1}, \quad \eta = (a_2^2+\xi^2)^{1/2}, \quad a_2 = c_1^{-1}, \quad S'(\xi') = (\eta')^{-1}. \tag{2}$$

The phase functions of the "plane" waves are

$$g_o(\xi) = i\xi x - \eta|z-z_o| = g_o'(\xi') = i\xi'(x'-z_o \sin\alpha) - \eta'|z'-z_o \cos\alpha| \tag{3}$$

due to the invariance condition under coordinate rotation through the dipping angle α, further $Sd\xi = S'd\xi'$. Hence, the divergence relation of slowness of one and the same plane wave must hold

$$i\xi' = i\,\xi\cos\alpha - \eta\sin\alpha, \quad \eta' = (a_2^2+\xi'^2)^{1/2} = i\,\xi\sin\alpha + \eta\cos\alpha, \tag{4}$$

$$D = \begin{Bmatrix} \cos\alpha & \sin\alpha \\ -\sin\alpha & \cos\alpha \end{Bmatrix} \tag{5}$$

which in matrix notation becomes the linear vector transformation

$$\underline{\xi}'^T = \underline{\xi}^T.D \quad , \quad \underline{\xi}^T = (i\xi, \eta), \tag{6}$$

and, with direction factor $\varepsilon = \pm 1$,

$$g_o = \underline{\xi}^T(\underline{x}-\underline{x}_o), \quad \underline{x}^T = (x+x_o, -\varepsilon z), \quad \underline{x}^T_{\mp o} = (x_o, \pm z_o) \tag{7}$$

Reflection at the free surface renders two apparent source rays with respect to further considerations. The displacement potentials are given in classical form by

$$\text{Pp} \ldots \quad \bar{\phi}_{-o}(s) = \bar{F}(s) \int_{-\infty}^{\infty} [S(\xi)R^{pp}(\xi)] e^{s g_{-o}(\xi)} d\xi \tag{8}$$

$$\text{Ps} \ldots \quad \bar{\psi}_{-o}(s) = \bar{F}(s) \int_{-\infty}^{\infty} [S(\xi)R^{ps}(\xi)] e^{s h_{-o}(\xi)} d\xi \tag{9}$$

with the phase functions

$$g_{-o} = -\eta z_o + i\xi x - \eta z, \quad h_{-o} = -\eta z_o + i\xi x - \zeta z, \quad \zeta = (a_1^2+\xi^2)^{1/2}, \quad a_1 = C_1^{-1}. \tag{10}$$

The general expression of reflection coefficients $R(\xi)$ of plane waves at plane interfaces are listed e.g. in [1].

First reflection at the dipping interface of a source ray pointing downward renders a classical form of the ray integrals in primed coordinates:

$$\begin{aligned} \text{pP} \ldots \quad \bar{\phi}_1(s) &= \bar{F}(s) \int_{-\infty}^{\infty} S'(\xi')R_{pp}(\xi') e^{s g_1'(\xi')} d\xi' \\ \text{pS} \ldots \quad \bar{\psi}_{01}(s) &= \bar{F}(s) \int_{-\infty}^{\infty} S'(\xi')R_{ps}(\xi') e^{s h_{01}'(\xi')} d\xi' \end{aligned} \tag{11}$$

with the phase functions, note $(z'-z_1')<0$, $z_1' = h\cos\alpha$, unspecified coordinate of point of reflection x_1' cancels,

$$\begin{aligned} g_1'(\xi') &= g_o'(x_1', z_1', \xi') + i\xi_1'(x'-x_1') + \eta_1'(z'-z_1'), \quad \xi_1' = \xi', \; \eta_1' = \eta' \\ h_1'(\xi') &= g_o'(x_1', z_1', \xi') + i\xi_1'(x'-x_1') + \zeta_{01}'(z'-z_1'), \quad \zeta_{01}' = (a_1+\xi_1'^2)^{1/2} \end{aligned} \tag{12}$$

The Cagniard path is somewhat simplified in the primed source ray slowness plane ξ' for downward pointing source rays.

Analogously, the generalized ray integrals of waves reflected twice become, $\xi_1' = \xi'(\xi)$ of Eq. (4)

$$\text{PpP} \ldots \quad \bar{\phi}_{-1}(s) = \bar{F}(s) \int_{-\infty}^{\infty} S(\xi) R^{pp}(\xi) R_{pp}(\xi_1') e^{sg_{-1}(\xi)} d\xi$$

$$\text{PpS} \ldots \quad \bar{\psi}_{-01}(s) = \bar{F}(s) \int_{-\infty}^{\infty} S(\xi) R^{pp}(\xi) R_{ps}(\xi_1') e^{sh_{-01}(\xi)} d\xi \qquad (13)$$

with phase functions in final form transformed to slowness ξ :

$$g_{\mp 1} = -\underline{\xi}^T \underline{x}_{\mp o} + \underline{\xi}^T D^2 \underline{x} \qquad (14)$$

$$h_{\mp 01} = -\underline{\xi}^T \underline{x}_{\mp o} + \underline{\xi}_{01}'^T D \underline{x} \quad , \qquad (15)$$

$i\xi_{01}' = i\xi_1' = i\xi \cos\alpha - \eta \sin\alpha$, and $\zeta_{01}' = (a_1^2 + \xi_{01}'^2)^{1/2}$ are the components of $\underline{\xi}_{01}'^T = (i\xi_{01}', \zeta_{01}')$ and are substituted above.

In addition to the first group of 2x2 rays, # ∓1, as discussed above, a second group of 8x2 rays, # ∓3 is received at the surface. Exemplarily the pSpS ray is listed:

$$\bar{\psi}_{0111}(s) = \bar{F}(s) \int_{-\infty}^{\infty} S(\xi) R_{ps}(\xi_1') R^{sp}(\xi_{11}) R_{ps}(\xi_{0111}') e^{sh_{0111}(\xi)} d\xi \qquad (16)$$

where the phase function is manipulated to become finally

$$h_{0111}(\xi) = -\underline{\xi}^T \underline{x}_o + \underline{\xi}_{0111}'^T D\underline{x} \ , \ \xi_{0111}'^T = (i\xi_{0111}', \zeta_{0111}') \ , \qquad (17)$$

and $i\xi_{01}' = i\xi_1' = i\xi \cos\alpha - \eta \sin\alpha$, $\zeta_{01}' = (a_1^2 + \xi_{01}'^2)^{1/2}$, $\eta_{011} = (a_2^2 + \xi_{011}^2)^{1/2}$, $i\xi_{011} = i\xi_{01} = i\xi_{01}' \cos\alpha - \zeta_1' \sin\alpha$, $\zeta_{0111}' = (a_1^2 + \xi_{0111}'^2)^{1/2}$, $i\xi_{0111}' = i\xi_{011}' = i\xi_{011} \cos\alpha - \eta_{011} \sin\alpha$ are the corresponding slowness of the apparent source rays to be substituted.

A table of the phase functions including a third group of 32x2 rays, # ∓5, is included in [10].

2. CAGNIARD TRANSFORM

The phase function $g(\xi)$, of a particular ray with last segment P (or $h(\xi)$ when last segment is in S-mode) is changed to time t by the complex mapping

$$t = -g(\xi) \qquad (18)$$

Assuming real time t renders the Cagniard path Γ of integration by the inverse mapping (the nonlinear equation is solved by Müller's method)

$$\xi = g^{-1}(t) \qquad (19)$$

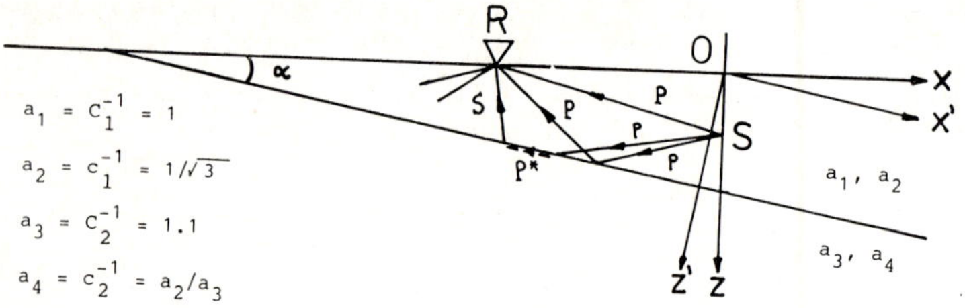

FIGURE 1
Source ray P, refracted ray pP*S,
direct ray pP from S to R

see Fig. 2. A portion of Γ, AM is along the imaginary ξ-axis and M denotes the stationary point at $\xi = \xi_M = ib_M$ which is the root of

$$dt/d\xi = 0 \quad . \tag{20}$$

The arrival time t_A of a direct ray, when none of the relevant branch points (ia_3, ia_4) falls below ξ_M, is then given by

$$t_M = -g(\xi_M) \; . \tag{21}$$

From the point M, the contour extends into the complex plane, observation time $t \geq t_M$. In case of a fast bottom a local slowness $\xi_k' = \pm ia_3$ or $\xi_k' = \pm ia_4$ may correspond to a source ray slowness of a branch point $\xi_E = ib_E$ falling below ξ_M and indicating the refracted ray ... P*. or ... S*. associated to the arrival of a head wave (receiver outside of the corresponding critical epicentral distance)

$$t_E = -g(\xi_E), \qquad t \geq t_E. \tag{22}$$

A part of Γ in that case is EM before branching out into the ξ-plane. The Rayleigh pole is located above M on the imaginary ξ-axis. Thus, the original Cagniard method of inversion of the Laplace integral renders a finite integration in the ξ-plane and a convolution according to the time signature f(t) of the source

$$\begin{aligned} u_x(x,0,t) &= 2AH(t-t_A) \int_{t_A}^{t} \ddot{f}(t-\tau) \left[\mathrm{Im} \int_\Gamma \xi E_x(\xi) d\xi \right] d\tau \\ u_z(x,0,t) &= 2AH(t-t_A) \int_{t_A}^{t} \ddot{f}(t-\tau) \left[\mathrm{Re} \int_\Gamma E_z(\xi) d\xi \right] d\tau \end{aligned} \tag{23}$$

where $E(\xi)$ is the even function given by the product of reflection coefficients times the emittance function times the receiver function of Table 1. Fig. 2 shows the integration path for numerical integration by Gauss-quadrature.

Table 1. Receiver functions of displacements in ray integral #k

Displacement	s^q	Last segment		At free Surface	
	q	P	S	P	S
$\bar{u}_x(s)$	1	$i\xi_k$	$-\zeta_k$	$i\xi_k + i\xi_k R^{pp} + \zeta_{k+1} R^{ps}$	$-\zeta_k + i\xi_k R^{sp} + \zeta_k R^{ss}$
$\bar{u}_z(s)$	1	η_k	ξ_k	$\eta_k - \eta_k R^{pp} + i\xi_k R^{ps}$	$i\xi_k - \eta_{k+1} R^{sp} + i\xi_k R^{ss}$

3. NUMERICAL RESULTS

An explosive line source located at (0,1/2) is considered with a parabolic ramp function with rounded shoulders, raise time 2Δ, dimensionless time step $\Delta = 0.1$. The signals received in an observation point on the free surface with epicentral distance $x = \pm 2h$ are shown in Figs. 3-5. A fast bottom is assumed $C_2 = 1.1C$; $C=1$, $c_2 = 1.1c$, $c = C\sqrt{3}$: $a_4 = c_2^{-1} < a_2 = c^{-1} < a_3 = C_2^{-1} < a_1 = C^{-1} = 1$.

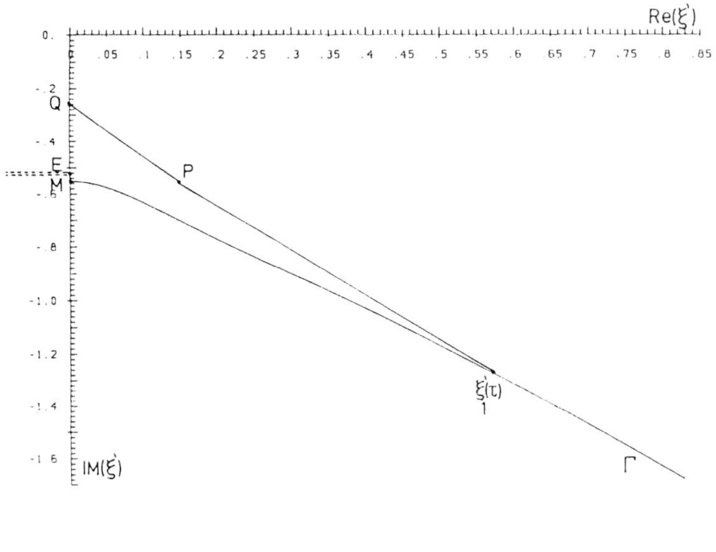

FIGURE 2
Cagniard path Γ, pP*S-ray integral (S-R of Figure 1)

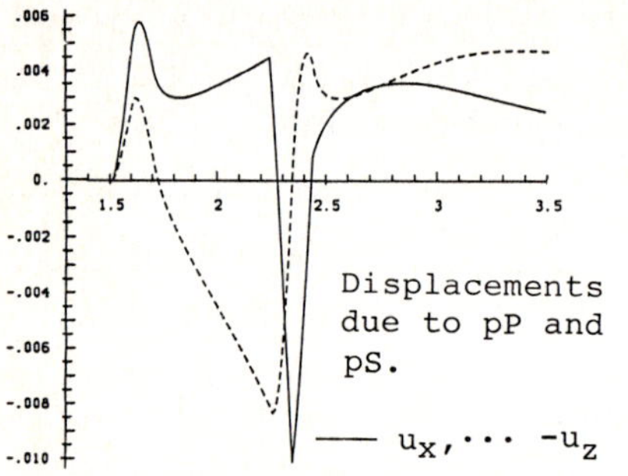

FIGURE 3
Dipping angle $\alpha = 10°$. Receiver at (2h,0)

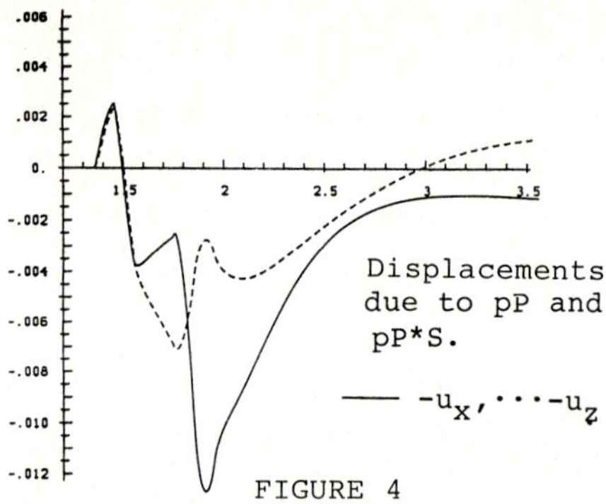

FIGURE 4
Dipping angle $\alpha = 10°$. Receiver at (-2h,0)

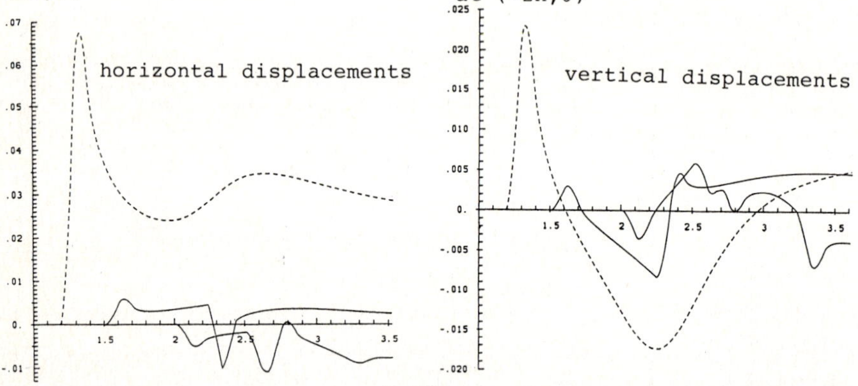

FIGURE 5
Dipping angle $\alpha = 10°$. Receiver at(2h,0).Total displacement is the sum of the partial contributions shown: P+(pP,pS)+(PpP,PpS,PsP,PsP*S)

REFERENCES

1 Pao, Y.-H. and Gajewski, R.R., The Generalized Ray Theory and Transient Responses of Layered Elastic Solids, in Mason, P. and Thurston, R.N., (eds.), Physical Acoustics 13 (Academic Press, New York, 1977) 183-265.

2 Ishii, H. and Ellis, R.M., J. Phys. Earth 18 (1970) 1-17.

3 Ishii, H. and Ellis, R.M., Bull. Seismolog. Soc.America 60 (1970) 15-28.

4 Hong, T.L. and Helmberger, D.V., Bull. Seismolog. Soc. America 67 (1977) 995-1008.

5 Pao, Y.-H. and Ziegler, F., Geophys. J.R. astr. Soc. 71 (1982) 57-77.

6 Ziegler, F. and Pao, Y.-H., Wave Motion 7 (1985) 1-24.

7 Ziegler, F., Pao, Y.-H. and Wang, Y.-S., J. Geophysics 57 (1985) 23-32.

8 Ziegler, F., Pao, Y.-H. and Wang, Y.-S., Acta Mechanica 56 (1985) 1-15.

9 Ziegler, F. and Pao, Y.-H., Acta Mechanica 52 (1984) 133-163.

10 Ziegler, F. and Pao, Y.-H., Sitzungsbericht, Abt. II, Austrian Academy of Sciences 193 (1984) 501-512.

11 Borejko, P. ZAMM 68 (1988) T161 - T162.

ELASTIC WAVE PROPAGATION
M.F. McCarthy, M.A. Hayes, (Editors)

THE INFLUENCE OF APPLIED STRESS ON THE RESPONSE OF THIN FILMS

George C. Johnson and Shih-Emn Chen

Department of Mechanical Engineering
University of California
Berkeley, CA 94720 USA

The effect of an applied in-plane stress on the Rayleigh - Lamb spectrum of a plate is considered in the context of acousto-elasticity. The basic equations of acoustoelasticity are presented and several of the standard solutions are reviewed. It is recalled that in these solutions the magnitude of the stress-induced variation in wave speed is rather small. The influence of an in-plane stress on the spectrum curves for various modes in a flat plate is then considered. It is shown that the applied stress causes a large change in the the phase velocity of the lowest order flexural mode at long wavelengths where the application of the stress causes the membrane effects to dominate the plate response.

1. INTRODUCTION

The deposition of a thin film on a substrate often leads to large in-plane stresses in the film. As free-standing structures (elements of micro-mechanical sensors and actuators, for example [1]) are made with these stressed films, the dynamic characteristics of the structures can be substantially affected by the stress. This paper considers the effect that an in-plane stress state has on the Rayleigh - Lamb spectrum of a free plate. The analysis is carried out in the framework of acoustoelasticity by considering the strain energy as a function which is cubic in the strain.

We begin by reviewing the basic equations which govern the acoustoelastic problem. A summary of some of the standard acoustoelastic results for bulk waves travelling in a body subject to plane stress is also presented. The governing displacement equations of motion have been solved directly (numerically) for the symmetric and antisymmetric plate modes with and without an applied in-plane stress. We show that the effect of the stress is most pronounced in the behavior of the lowest order antisymmetric (flexural) mode. The results presented here are in agreement with plate analyses [2-4] which consider the long wavelength limit of this investigation.

2. BACKGROUND EQUATIONS

The acoustoelastic analysis is based on the superposition of an infinitesimal wave on an elastic medium which is subjected to a finite deformation. The equations of motion for the superposed disturbance are first expressed in terms of Cauchy stress and are then linearized in the infinitesimal displacement from the initially deformed state. This linearized displacement equation is then solved subject to the boundary conditions appropriate to the problem.

Let vectors with cartesian components X_A, x_i, and x^*_i denote the position of a material point X in the reference, statically deformed, and current configurations, respectively. The reference configuration is taken to be the initial, unstressed configuration; the statically deformed configuration is the body's configuration prior to the application of the infinitesimal disturbance; the current configuration includes both the static deformation and the superposed disturbance. Associated with the two deformed configurations are deformation gradients F_{iA} and F^*_{iA}, which are related as

$$F^*_{iA} = (\delta_{ij} + u_{i,j})\, F_{jA} \tag{1}$$

where δ_{ij} is the Kronecker delta and u_i is the displacement of the superposed disturbance from the statically deformed configuration,

$$u_i = x^*_i - x_i \ . \tag{2}$$

Lagrangian strains E_{AB} and E^*_{AB} are defined in the usual way as

$$E_{AB} = \tfrac{1}{2}(F_{iA}F_{iB} - \delta_{AB}), \quad E^*_{AB} = \tfrac{1}{2}(F^*_{iA}F^*_{iB} - \delta_{AB}) = E_{AB} + F_{iA}F_{jB}e_{ij} \tag{3}$$

where e_{ij} is the strain due to u_i referred to the statically deformed state. Assuming that u_i is infinitesimal, e_{ij} may be expressed as

$$e_{ij} = \tfrac{1}{2}(u_{i,j} + u_{j,i}). \tag{4}$$

The constitutive equations relating stress and strain are obtained through the introduction of a strain energy function W. Specifically, let T_{ij} be the Cauchy stress in the statically deformed configuration and let W be a function of E_{AB}. The constitutive relation for T_{ij} may be expressed as

$$T_{ij} = \frac{1}{J} F_{iA}F_{jB}\frac{\partial}{\partial E_{AB}}(\rho_o W) \tag{5}$$

where J=det $F_{iA} > 0$, and ρ_o is the mass density in the initial configuration. A similar expression may be written for the Cauchy stress T^*_{ij} associated with the current configuration if all quantities (except for ρ_o) are replaced by "starred" quantites associated with the current configuration.

The equations of motion governing the response of the body in the current configuration may be written in the absence of body forces as

$$\frac{\partial T^*_{ij}}{\partial x^*_i} = \rho^* \ddot{u}_j \ . \tag{6}$$

Equation (6) may be linearized in u_i by expanding T^*_{ij}, ρ^*, and the derivative with respect to x^*_i about the associated quantities in the statically deformed state. Assuming that the material properties and static stress are homogeneous, the linearized displacement equations are

$$(C_{ijkl} + T_{il}\delta_{jk})u_{k,il} = \rho\ddot{u}_j, \tag{7}$$

where

$$C_{ijkl} = \frac{1}{J} F_{iA}F_{jB}F_{kC}F_{lD}\frac{\partial^2 (\rho_o W)}{\partial E_{AB}\partial E_{CD}} \tag{8}$$

is the stiffness tensor in the statically deformed configuration.

The development to this point is valid for homogeneous elastic materials with arbitrary nonlinearity. It is customary in the acoustoelastic analysis to let the strain energy be cubic in the strain,

$$\rho_o W = \frac{1}{2} C_{ABCD}E_{AB}E_{CD} + \frac{1}{6} C_{ABCDEF}E_{AB}E_{CD}E_{EF}, \tag{9}$$

where C_{ABCD} and C_{ABCDEF} are constants referred to as the second-order elastic coefficients (SOEC) and third-order elastic coefficients (TOEC), respectively.

Let us now consider an isotropic material subject to a state of plane stress, with the axes x_α, α=1,2, aligned with the principal in-plane directions of the stress, and with x_3 along the stress-free direction. Let V_{ij} denote the phase velocity of a plane wave propagating along x_i and polarized along x_j. Further, let V_{lo} and V_{so} denote the bulk longitudinal and shear wave speeds in the unstressed material. For waves propagating along x_3, we find that the stress-induced changes in velocity may be written

$$\frac{V_{33} - V_{lo}}{V_{lo}} = A_1(T_{11} + T_{22}),$$
$$\frac{V_{31} - V_{32}}{V_{so}} = A_2(T_{11} - T_{22}), \tag{10}$$

where A_1 and A_2 are acoustoelastic constants which are particular combinations of the SOEC and TOEC. For most structural materials, A_1 and A_2 are rather small - on the order of 10 to 50 TPa^{-1}. (This magnitude indicates a relative velocity change of 10 to 50 parts per million per MPa of applied stress.)

Another interesting case is that of SH_0 waves propagating in the plane of stress. In this case,

$$V_{12}^2 - V_{21}^2 = \rho_o(T_{11} - T_{22}). \tag{11}$$

Equation (11) can be linearized in the velocity difference in a form similar to Eq. $(10)_2$, yielding

$$\frac{V_{12} - V_{21}}{V_{so}} = \frac{1}{2\mu}(T_{11} - T_{22}), \tag{12}$$

where μ is the shear modulus in the reference configuration. Again, the coefficient multiplying the stress difference is on the order of 50 TPa^{-1}.

Most acoustoelastic work to date has been based on Eqs. (10) and (11), with the attendant requirements on the precision with which the velocities (or in some cases times-of-flight) must be measured [5]. It is shown in the next section that certain plate mode speeds are more sensitive to the presence of stress than are the bulk wave speeds.

3. PLATE MODES

We consider now an infinite plate of thickness 2h subject to in-plane (principal) stresses T_{11} and T_{22}. Let a straight crested harmonic wave travel in the x_1 direction. We assume, then, that

$$u_\alpha = U_\alpha(x_3)\exp\{j(kx_1-\omega t), \quad \alpha=1,3$$
$$u_2 = 0, \tag{13}$$

where $j=\sqrt{-1}$, k is the wave number, and ω is the frequency. Using Eq. (13) in the displacement equations, Eq. (7), and applying the stress free boundary conditions at $x_3=\pm h$ allows us to obtain the frequency spectrum (ω as a function of k).

We have numerically evaluated this spectrum for the cases of an unloaded plate (the usual Rayleigh - Lamb spectrum) and a plate loaded in uniaxial tension T_{11} (with T_{22} = 0). The properties of this plate are typical of those for aluminum. In presenting the results, we use the dimensionless wave number ξ and frequency Ω, where

$$\xi = \frac{2hk}{\pi}, \qquad \Omega = \frac{2h\omega}{\pi V_{so}}. \tag{14}$$

Figure 1 shows the spectra for the first three symmetric (a) and flexural (b) modes in the unloaded plate.

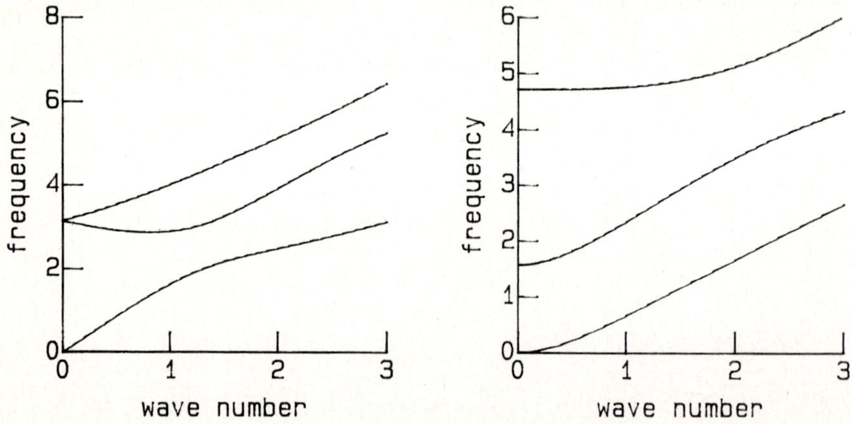

Figure 1. Frequency spectra for symmetric (a) and anitsymmetric (b) modes in an unloaded plate.

Figure 2a shows the first flexural mode for the unloaded plate (solid line) and the loaded plate (dashed line). At the scale of this figure, it is clear that the application of the stress has increased the frequency associated with this mode, but the extent of the increase is not clear. Figure 2b focusses on the spectrum in the long wavelength region (ξ small). At this scale, the effect of the loading is much more clear. Of particular interest is the fact that the spectrum curve in the loaded case approaches the origin with nonzero slope, while in the unloaded case it approaches the origin with zero slope. This effect may also be seen by calculating the phase velocity V_p,

$$V_p = \frac{\Omega}{\xi}. \tag{15}$$

Figure 3 shows the variation of phase velocity with wave number for this lowest flexural mode. The major effect of the applied tension is to shift the long wavelength behavior from that of a plate dominated by flexural stiffness to a membrane dominated by the in-plane stress. The relative change in the phase velocity from the unloaded case is infinite at $\xi = 0$, and remains substantial up to relatively short wavelengths. This is in sharp contrast to the small differences experienced by the bulk wave speeds V_{ij}.

An examination of the higher modes in flexure or any of the symmetric modes indicates again that the influence of the loading is minor. Examples showing the spectra of the second flexural and the first symmetric modes are given in Fig. 4.

REFERENCES

[1] Transducers '87, Proceedings of the 4th International Conference on Solid-State Sensors and Actuators, IEEE and Electrochemical Society, Tokyo, Japan (1987).
[2] Mindlin, R.D., and Medick, M.A., J. Appl. Mech., 26 (1959) 561.
[3] Tiersten, H.F., Linear piezoelectric plate vibrations (Plenum, New York, 1969)
[4] Lee, P.C.Y., Wang, Y.S., and Markenscoff, X., J. Acoust. Soc. Am., 57 (1975) 95.
[5] Pao, Y.H., Sachse, W., and Fukuoka, H., Physical Acoustics, 17 (1983) 61.

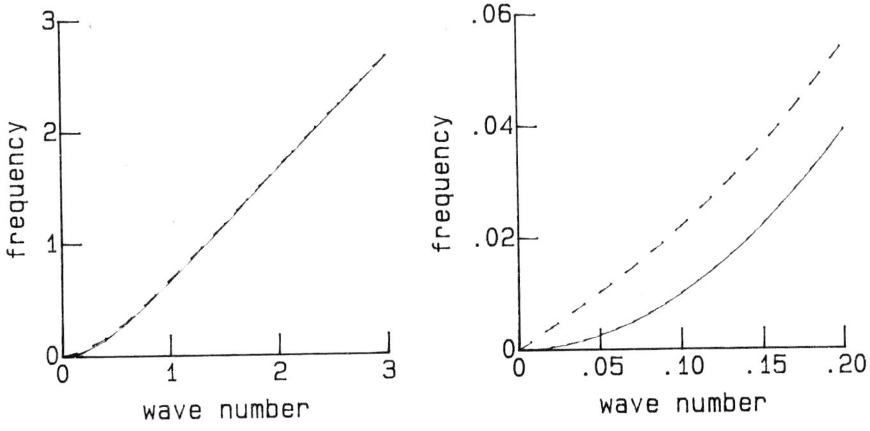

Figure 2. Frequency spectra for the lowest flexural mode in the loaded plate (solid line) and the loaded plate (dashed line).

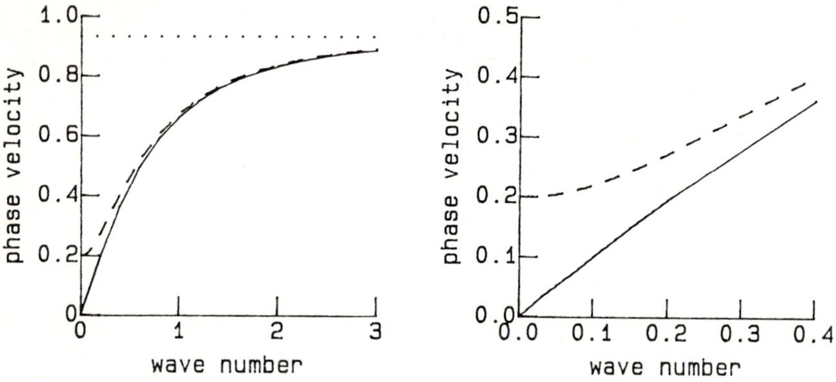

Figure 3. Phase velocity versus wave number for the unloaded plate (solid line) and the loaded plate (dashed line).

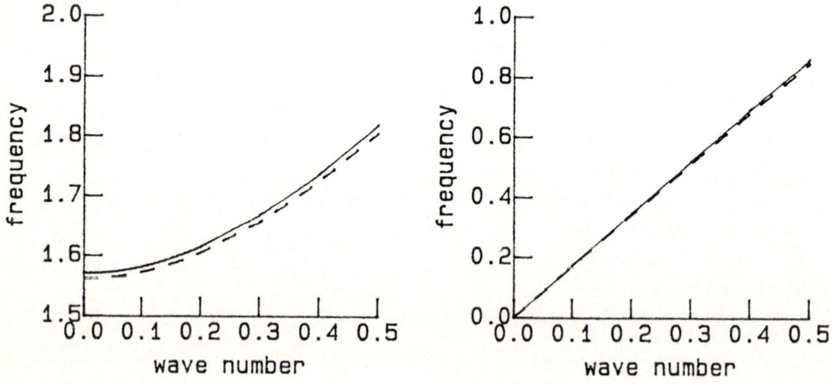

Figure 4. Frequency spectra for the second antisymmetric mode (a) and first symmetric mode (b). The solid lines are for the unloaded plate; the dashed lines are for the loaded plate.

ELASTIC WAVE PROPAGATION
M.F. McCarthy, M.A. Hayes, (Editors)

THE CAUSTIC OF A RAYLEIGH-WAVE

Hans Peter Rossmanith

Institute of Mechanics
Technical University Vienna
Vienna, Austria

The method of caustic or shadow-spot technique is utilized to determine the caustic of a Rayleigh-wave. The size and shape of the caustic allow for characterization of the source of the Rayleigh-wave disturbance.

1. Introduction

Since the pioneering paper by Manogg [1] on light deflection due to stress intensification at structural discontinuities the method of caustics or shadow spot technique has become a powerful tool in experimental mechanics. The method originally introduced by Manogg for transparent materials has been adapted by Theocaris and his co-workers [2,3] for nontransparent materials. The general equations of caustics for plane static and dynamic elasticity theory may be found in the review article by Kalthoff and Beinert [4] and Rossmanith [5]. In this contribution the method of caustics in conjunction with high-speed photography is employed to correlate the size and shape of the shadow area associated with a travelling Rayleigh surface wave to the nature of its generating source of disturbance. In particular, the Rayleigh surface wave emerging from the scattered wave field upon impact carries a moving caustic which can be observed in the recordings. Therefrom characteristics of the source of the disturbance may be identified and series of sequentially recorded shadow patterns provide pertinent information about the dynamic event.

2. The Method of Caustics

The physical principle of the method of caustics is the inhomogeneous deflection of parallel light rays during their passage through a plate specimen due to two effects: the reduction of the thickness of the specimen and the change of the refractive index of the material as a consequence of stress intensification. The transmitted and/or reflected light rays form a shadow space and the intersection of this shadow space with a screen produces a shadow area (shadow spot) surrounded by a bright curve (caustic). The formation of this shadow space and the shadow region on the image-plane (screen) at a distance behind the specimen for a half-plane subjected to a concentrated edge-load is schematically shown in Fig. 1 for the transmitted light method.

For nontransparent materials with a mirrored surface the reflection-light method is utilized where qualitatively similar shadow patterns can be observed. In caustic analysis the cases of plane stress and plane strain, transmitted light and reflected light turn out to be basically similar and differ only by the values of the elasto-optical parameters. In a third method, where part of the light is reflected at the front surface and part of the light is reflected at the back surface the resulting caustic is a combination of the caustics obtained by transmission and simple reflection.

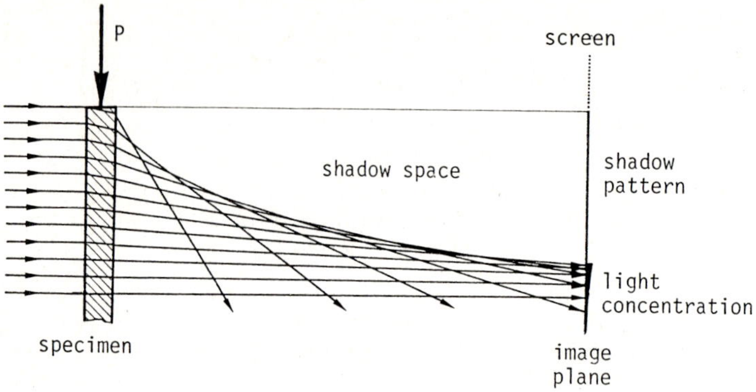

Figure 1: The transmission method of caustics applied to contact problems

3. Analysis

When a normally impinging light beam traverses the unloaded specimen at the Point $P_0(r,\theta)$ in the object plane, its image on the shadow image plane is defined by the vector $\vec{r}$ (Fig. 2). Load application induces deflection of the beam to the point $P'(x',y')$. With the deflection $\vec{w} = \vec{w}(r,\theta)$ the vector of the image point P' is given by

$$\vec{W} = \vec{r} + \vec{w}. \tag{1}$$

This deflection gives rise to the formation of a shadow space formed by the deflected light rays upon passage of the object. The caustic is a singular curve of the image equ. (1) and is generated by the intersection of the image plane with the shadow space and ray field. A necessary condition for the existence of the singular caustic curve is that the Jacobian function determinant, $\mathbf{J}$, of equ. (1) vanishes:

$$\mathbf{J} = \frac{\partial x'}{\partial r}\frac{\partial y'}{\partial \theta} - \frac{\partial x'}{\partial \theta}\frac{\partial y'}{\partial r} \tag{2}$$

for all points $P_c'(r,\theta)$ of the caustic. Hence, the fundamental problem of the method of caustic is to solve equ. (2) under certain conditions imposed by the real physical problem. An elasto-optical analysis yields a relation between light deflection, the stress field parameters, the elasto-optical material parameters and the geometry of the experimental set-up:

$$\vec{w} = -z_0 h\kappa\, grad[(\sigma_1 + \sigma_2) \pm \varepsilon(\sigma_1 - \sigma_2)] \tag{3}$$

where h is the effective model thickness, $\sigma_{1,2}$ denote the principal normal stresses, z_0 is the distance between model and screen, κ is an elasto-optical parameter and ε accounts for the optical anisotropy of the material. For optically isotropic or inert material (e.g. Plexiglass) $\varepsilon = 0$ and one obtains one single caustic. Optically anisotropic or birefringent materials (e.g. plate glass) give rise to a double-caustic. In addition, the mapping of the deformed boundaries of the elastic bodies in contact produces socalled *'pseudo-caustics'* [7].

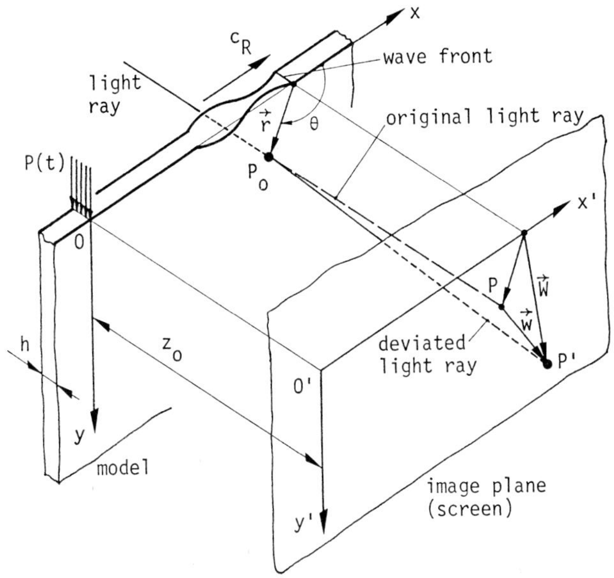

Figure 2: Optical arrangement, light ray deflection and coordinate systems for the Rayleigh-wave caustic problem

4. The Caustic of a Rayleigh-wave

In Lamb's classical paper *"On the propagation of tremors over the surface of an elastic solid"* [8] the propagation of vibrations along the surface of an elastic isotropic half-space which are caused by an arbitrary application of a line load is examined. Thereby special attention is given to the Rayleigh wave field produced by the line load. The part of the stress field with the Rayleigh-wave is given by

$$\begin{aligned}\sigma_x &= \alpha Re\Phi''(\zeta_P) + \beta Re\Phi''(\zeta_S)\\ \sigma_y &= \beta\{Re\Phi''(\zeta_P) - Re\Phi''(\zeta_S)\}\\ \sigma_{xy} &= 2\gamma\{Im\Phi''(\zeta_P) - Im\Phi''(\zeta_S)\},\end{aligned} \tag{4}$$

where the coefficients α, β and γ depend on the wave propagation speeds

$$\begin{aligned}\alpha = \mu(s^2 - 2r^2 - k^2), \quad \beta = \mu(s^2 + k^2), \quad \gamma = \mu r k\\ s^2 = c_R^{-2} - c_S^{-2}, \qquad r^2 = c_R^{-2} - c_P^{-2}, \qquad k = c_R/c_S\end{aligned} \tag{5}$$

and c_R = Rayleigh wave speed, c_P = longitudinal wave speed, c_S = transverse wave speed and μ = Poisson's ratio. Complex variables ζ_j with respect to a constant speed moving coordinate system

$$\zeta_j = \xi + i\eta_j = (\tau - \frac{x}{c_R}) + iy\sqrt{c_R^{-2} - c_j^{-2}} \qquad (j = P, S) \tag{6}$$

have been introduced. Following Dally and Thau [10] the stress functions $\Phi'(\zeta_P)$ can be related to the loading function $P(\tau)$

$$\Phi'(\zeta_P) = \frac{i}{6\mu\pi c_R(s^2 - r^2)} \int_{-\infty}^{+\infty} \frac{P(\tau)d(\tau)}{\zeta_P - \tau} \tag{7}$$

and the caustic image equations given by equs. (1) and (3) take the form

$$\begin{aligned} x' &= x + C_1 Re\Phi''' = x + C_1' \int_{-\infty}^{+\infty} P''(\tau)\frac{\eta_P}{(\xi-\tau)^2 + \eta_P^2} d\tau \\ y' &= y + C_2 Im\Phi''' = y + C_2' \int_{-\infty}^{+\infty} P''(\tau)\frac{\xi - \tau}{(\xi-\tau)^2 + \eta_P^2} d\tau \end{aligned} \tag{8}$$

where $P''(\tau)$ is the second time-derivative of the loading function and

$$C_1 = 2\mu z_0 h\kappa(s^2 - r^2)/c_R, \quad C_2 = 2\mu z_0 h\kappa(s^2 - r^2)r, \quad C_1' = \frac{z_0 h\kappa}{3\pi c_R^2}, \quad C_2' = \frac{z_0 h\kappa r}{3\pi c_R}. \tag{9}$$

The pseudo-caustic represents the mapping of the boundary $y = 0$ of the half-plane i.e. $\eta_P = 0$ and is associated with the image equation by

$$x' = x + \pi C_1' P''(\tau), \qquad y' = C_2' \int_{-\infty}^{+\infty} P''(\tau)\frac{d\tau}{\xi - \tau}. \tag{10}$$

Because of $P''(\tau) = 0$ for $\tau < 0$ and $\tau > T$, the upper and lower limits of the Cauchy-integral in equ. (10) may be replaced by T and 0 respectively. The initial curve for the caustic is given by the solution of the Jacobian,

$$\mathbf{J} = \frac{\partial(x', y')}{\partial(x, y)} = 1 - \varepsilon Re\Phi^{IV} - \omega|\Phi^{IV}|^2 = 0, \tag{11}$$

where the constants ε and ω depend on the optical and elastical parameters:

$$\varepsilon = C_1/c_R - C_2 r, \qquad \omega = C_1 C_2 r/c_R. \tag{12}$$

5. Example

Numerical calculations have been performed on the basis of dynamic caustic experiments with PMMA-models. Optical, geometrical and material data are as follows: model thickness $h = 10\,mm$, Young's modulus $E = 3.24\ 10^9\,Pa$, density $\varrho = 1260\,kg/m^3$, Poisson's ratio $\mu = 0.35$, elastooptic constant $\kappa = -1.08\ 10^{-10} m^2/N$, distance $z_0 = -1.5\,m$, wave speeds: $c_P = 1712\,m/s$, $c_S = 976\,m/s$, and $c_R = 899\,m/s$. Consider e.g. the example loading function

$$P(\tau) = \alpha(\tau - T/2)^4 - \beta(\tau - T/2)^2 = \frac{64P_0}{T^4}[\frac{\tau^4}{4} - \frac{T\tau^3}{2} + \frac{T^2\tau^2}{4}] \tag{13}$$

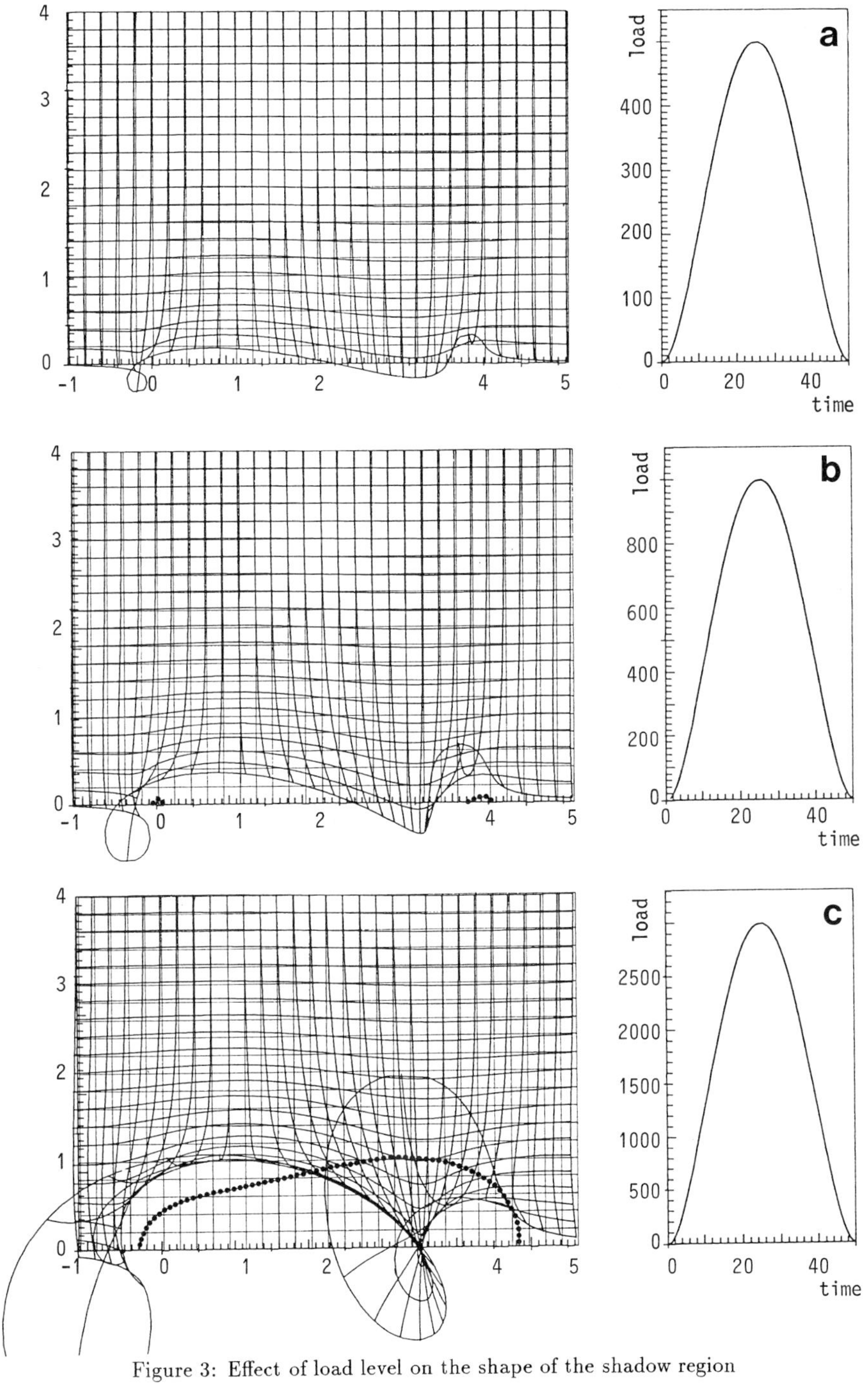

Figure 3: Effect of load level on the shape of the shadow region

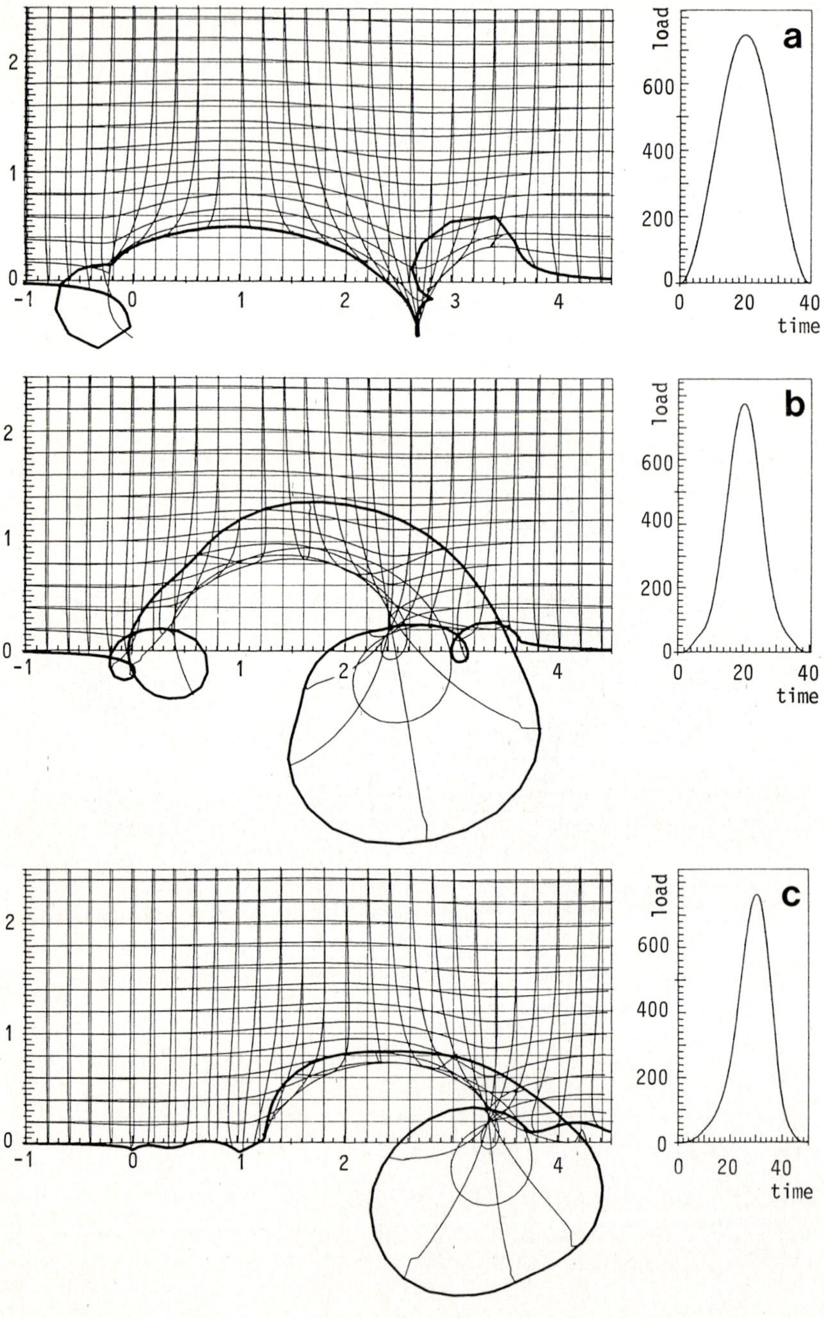

Figure 4: Effect of loading pulse width on the shape of the shadow region

where α and β are real coefficients and T is the loading interval. The corresponding load-time relationship is shown in Figs. 3a, 3b and 3c and the initial curve, equ. (11), is represented by the dotted line in the left part of Fig. 3c. The associated caustic has a rather complex shape and divides and engulfes two separate shadow regions. The size of the caustic is directly proportional to the load magnitude and inversely proportional to the fourth-power of the pulse width in the central peak region. The influence of the load level on the size and shape of the shadow area is shown in Figs. 3a, 3b and 3c where the load increases from the level 500 to 1000 and 3000. The pulse width has an even more pronounced effect on the shape of the shadow region. Broad pulses (Fig. 4a) are associated with spread-out caustics and a reduction of the pulse width is accompanied by a shape transformation towards the shape of a semi-ellipse (Figs. 4b and 4c).

The semi-elliptical shadow-region as shown in Fig. 4c is in good agreement with caustics observed in impact experiments as shown in Fig. 5. In these experiments a centrally explosively loaded circular disc is in dynamic edge contact with a half-plane. In Fig. 5 the four semi-elliptical shadow spots correspond to the four Rayleigh-waves generated upon impact and emerging from the dynamic contact site. From an experimental caustic recording it is possible to indirectly determine the load-time trace of the input load function. Direct inversion of the caustic equations is not possible because of the numerical evaluation of the initial curve. This is a major difference from the application of the method of caustics for the determination of stress intensity factors in fracture mechanics.

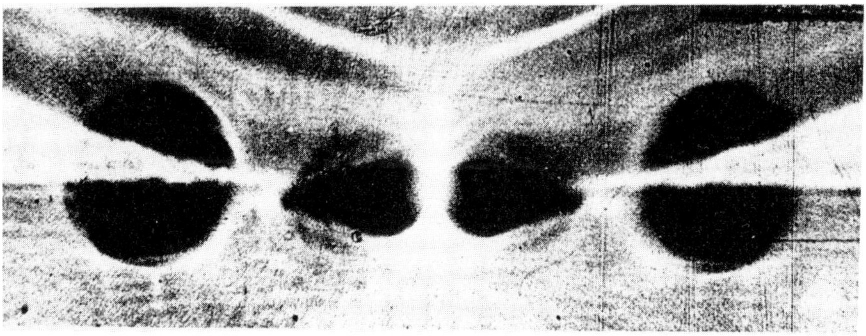

Figure 5: Rayleigh-wave caustics and dynamic caustic for an explosively excited disc impacting on a half-plane

Acknowledgement

The author would like to acknowledge the financial support granted by the *Austrian Science Foundation* under project number # 5814.

References

[1] P. Manogg: *Die Lichtablenkung durch eine elastisch beanspruchte Platte und die Schattenfiguren von Kreis- und Rißkerbe.* Glastechnische Berichte 39, 323-329, 1966.

[2] P.S. Theocaris and E. Gdoutos: *An optical method for determining opening-mode and edge-sliding-mode stress intensity factors.* J. Appl. Mechanics 39, 91-97, 1972.

[3] P.S. Theocaris: *The method of caustics applied to elasticity problems.* In: Development in Stress Analysis (G. Hollister, Ed.), 27-63, 1979.

[4] J. Beinert and J.F. Kalthoff: *Experimental determination of dynamic stress intensity factors by the method of shadow patterns.* In: Mechanics of Fracture, Vol. VII (Ed. G.C. Sih), 1979.

[5] H.P. Rossmanith: *The method of caustics for plane elasticity problems.* J. of Elasticity 12, 193-200, 1982.

[6] H.P. Rossmanith,R.E. Knasmillner and A. Shukla: *Experimental investigation of dynamic contact problems by means of the method of caustics.* Proc. of SESA Conference, 407-416, 1985.

[7] P.S. Theocaris and C. Razem: *Deformed boundaries determined by the method of caustics.* J. Strain Analysis 12, 223-232, 1977.

[8] H. Lamb: *On the propagation of tremors over the surface of an elastic solid.* Phil. Trans. Vol. CCIII-A 359, B1-B42, 1903.

[9] J.F. Cardenas: *On Rayleigh waves and Rayleigh wave extension of surface micro-cracks.* PhD Thesis, University of Maryland, MD, USA, 1983.

[10] J.W. Dally and S.A. Thau: *Observations of stress wave propagation in a half-plane with boundary loading.* Int. J. Solids Structures 3, 293-308, 1967.

ELASTIC WAVE PROPAGATION
M.F. McCarthy, M.A. Hayes, (Editors)

GROUND MOTIONS GENERATED BY A FINITE FAULT

C. S. Yeh[1], P. L. Chen[2], M. K. Kuo[3] and T. K. Wang[3]

Department of Civil Engineering and Institute of Applied Mechanics, National Taiwan University, Taipei, Taiwan 10764, R. O. C.
[2]Department of Civil Engineering, National Cheng–Kung University, Tainan, Taiwan 70101, R. O. C.
[3]Institute of Applied Mechanics, National Taiwan University, Taipei, Taiwan 10764, R. O. C.

Transient response on the surface of an elastic half–space due to a finite rectangular fault is investigated. In this paper, the case of shear fault motions in a horizontal plane is considered.

In the analysis, the solution due to a quadrantal fault is first derived. The superposition of four quadrantal faults then leads to a desired finite rectangular fault. The crucial steps of the analysis are the direct application of the one–sided Laplace transform and the double two–sided Laplace transform over the time variable t and spatial variables x and y, respectively. Then the generalized ray theory is employed to take into account the free surface effect. The extended Cagniard–de Hoop method is subsequently applied to invert the integral transforms. The final solution consists of a number of finite range integration plus some contributions from poles. Numerical results of the ground acceleration due to several different fault motions are presented and compared. The procedure can easily be extended to faults with arbitrary dip.

INTRODUCTION

Since the theory of elasticity for dislocations was introduced by Steketee [1,2], there have been many investigators trying to determine the seismic source parameters from the analysis of observed results. It is recognized that the near–field data may be more useful than the far–field data for the study of the focal process [3]. For the case of dipolar point source, Haskell [4] obtained the solutions of ground displacements in an elastic half–space. Haskell [5], Harkrider [6] and Ben–Menahem and Singh [7] derived the ground displacements of layered media in frequency domain. Müller [8,9] applied generalized ray [10–13] theory combining with a modified Cagniard method to analyze the ground responses of layered media and expressed the solutions in time domain.

In order to study the effects of fault size and displacement pattern of fault on ground motions, Yeh and Lin [14] derived a solution for ground displacements of a layered medium generated by an infinite strike–slip fault by using ray theory. The common approach used in the study of finite fault always involves a space integral of the point source solution over fault plane either directly in the time domain (e.g. Kawasaki [15] and Hartzell et al [16]) or in the frequency domain then followed by a Fourier synthesis (e.g. Levy and Mal [17]). Extensive literature reviews can be found in Rice [18] and Luco and Anderson [19]. In this paper an entirely different approach is taken by utilizing the Cagniard–de Hoop technique to obtain the solution for a uniform dislocation over a horizontal rectangular fault. A similar approach was reported by Madariaga [20] to study a similar problem except in an unbounded medium. The approach can easily be extended to more general faults with arbitrary dip.

PROBLEM STATEMENT

The equations of motion in an elastic medium can be reduced to wave equations as

$$c_p^2\nabla^2\Phi = \ddot{\Phi} \quad ; \quad c_s^2\nabla^2\Psi = \ddot{\Psi} \quad ; \quad c_s^2\nabla^2\chi = \ddot{\chi} \tag{1}$$

and

$$c_p^2 = (\lambda + 2\mu)/\rho \quad ; \quad c_s^2 = \mu/\rho \tag{2}$$

where ∇^2 is the Laplacian operator, c_p and c_s are the pressure and shear wave speeds, respectively, $(\cdot)$ denotes the derivative with respect to time t, λ and μ are Lamé constants and ρ is the mass density of solid. The Lamé potential Φ, χ and Ψ are related to displacements by a tensor form [21] as

$$u_i = \Phi_{,i} + e_{ijz}\chi_{,j} + \Psi_{,zi} - \delta_{iz}\Psi_{,jj} \quad , \ i,j = x,y \tag{3}$$

Here (,) denotes the derivative with respect to the spatial variables, and δ_{ij} and e_{ijk} are the Kronecker delta and the permutation symbol, respectively. By employing Hooke's law, the stresses τ_{ij} can be expressed by

$$\tau_{ij} = \lambda u_{k,k}\delta_{ij} + \mu(u_{i,j} + u_{j,i}) \quad , \ i,j,k = x,y,z \tag{4}$$

Consider a finite rectangular shear fault with length L and width W embeded in an elastic half–space and the fault plane parallel to the free surface. The rupture initiates at an edge and propagates in a unidirection at a uniform velocity until it covers the whole rectangular fault. The prescribed displacements inside the fault region can be written, along $0 < x < L$, $0 < y < W$, $z = z_0$, as

$$[u_x](x, y, z_0, t) = H(t)f(t - x/v_r) \tag{5}$$

$$[u_j](x, y, z_0, t) = 0 \qquad j = y, z \tag{6}$$

$$[u_i](x,y,z_0,t) = 0 \quad , \ i = x, y, z, \ \text{outside the fault region} \tag{7}$$

where $[\cdot]$ denotes the jump of a quantity from face $z = z_0 - 0$ to face $z = z_0 + 0$, and v_r is the rupture velocity.

METHOD OF SOLUTION

The problem formulated by eqns (5)–(7) can be solved by considering the equivalent body force to the shear fault, then applying the integral transform and obtaining the solution in the transformed domain satisfying the corresponding radiation condition at infinity and boundary condition at the free surface. Alternatively, this transformed solution can be constructed rather easily by generalized ray theory. A detailed review of the generalized ray theory can be found in [13]. In this method, the solution is expressed as the sum of several terms. Each term represents the contribution from a particular ray and it contains only the product of a source function, a receiver function and a phase term. The source functions are actually related to the transformed responses of Lamé potentials for an unbounded medium. The receiver functions relate the Lamé potentials to the desired displacements and/or stress components and account for the boundary effects. This method can easily be extended to the case of faults with arbitrary orientation or even in a layered medium. The results from these two approaches are exactly the same. In this paper, we present only the latter approach in detail.

SOURCE FUNCTION

The responses due to a finite shear fault defined as eqns (5)–(7) in an unbounded elastic medium have been studied by Madariaga [20]. The transformed responses of Lamé potentials in an infinite space are related to the source functions used in the generalized ray theory. Due to the property of anti–symmetry with respect to the fault plane, the solution of the infinite space problem can be obtained by solving an equivalent mathematical problem on a half–space as discussed in [20]. The crucial step of the analysis is the direct application of the one–sided Laplace transform and the double two–sided Laplace transform over the time variable t and spatial variables x and y with transformed parameters s, $(s/c_P)\xi$ and $(s/c_P)\eta$, respectively. The solution due to a quadrantal fault is first derived. The superposition of four quadrantal faults will lead to a desired finite rectangular fault. After inverting the double two–sided Laplace transform, the Laplace transformed solution of Lamé potentials due to a quadrantal fault is written as

$$\tilde{\Phi}(x,y,z,s;z_0) = -\frac{\tilde{f}(s)c_P}{4\pi^2 s}\int_{L_\xi}\int_{L_\eta} S_p(\xi,\eta)e^{\frac{s}{c_p}(\xi x + \eta y - \zeta_p|z-z_0|)}\,d\xi d\eta \ , \tag{8}$$

$$\tilde{\Psi}(x,y,z,s;z_0) = -\frac{\tilde{f}(s)c_P{}^2}{4\pi^2 s^2}\int_{L_\xi}\int_{L_\eta} S_v(\xi,\eta)e^{\frac{s}{c_p}(\xi x + \eta y - \zeta_s|z-z_0|)}\,d\xi d\eta \ , \tag{9}$$

$$\tilde{\chi}(x,y,z,s;z_0) = -\frac{\tilde{f}(s)c_P}{4\pi^2 s}\int_{L_\xi}\int_{L_\eta} S_h(\xi,\eta)e^{\frac{s}{c_p}(\xi x + \eta y - \zeta_s|z-z_0|)}d\xi d\eta \quad , \tag{10}$$

where

$$\zeta_\alpha^2 = \kappa_\alpha^2 - \xi^2 - \eta^2 \quad , \alpha = p,s \quad , \tag{11a}$$

and

$$\kappa_p = 1 \quad , \kappa_s = \frac{c_p}{c_s} \quad , \tag{11b}$$

and source functions S_p, S_v and S_h are defined as

$$S_p(\xi,\eta) = \frac{\xi}{\kappa_s^2\eta(\xi+\gamma)} \quad , \tag{12}$$

$$S_v(\xi,\eta) = \frac{-\xi(\zeta_s^2-\xi^2-\eta^2)}{2\kappa_s^2\zeta_s\eta(\xi^2+\eta^2)(\xi+\gamma)}\,\mathrm{sgn}(z-z_0) \quad , \tag{13}$$

$$S_h(\xi,\eta) = \frac{\eta}{2\eta(\xi^2+\eta^2)(\xi+\gamma)} \quad , \tag{14}$$

where γ is nondimensional rupture slowness

$$\gamma = \frac{c_p}{v_r} \quad , \tag{15}$$

and $\tilde{f}(s)$ is the Laplace transform of function $f(t)$. L_ξ and L_η denote the Bromwich contours in the complex ξ–plane and η–plane, respectively.

RECEIVER FUNCTIONS

Once the source functions $\tilde{\Phi}$, $\tilde{\Psi}$ and $\tilde{\chi}$ are constructed, the displacements and stresses can be determined according to Eqs. (3) and (4), respectively. The displacements of the infinite space problem in one–sided Laplace transformed domain may be written as follows,

$$\tilde{u}(x,y,z,s;z_0) = -\frac{\tilde{f}(s)}{4\pi^2}\left\{\int_{L_\xi}\int_{L_\eta} D_p S_p \exp[\tfrac{s}{c_P}(\xi x + \eta y - \zeta_p|z-z_0|)]d\xi d\eta \right.$$
$$\left. + \int_{L_\xi}\int_{L_\eta}(D_v S_v + D_h S_h)\exp[\tfrac{s}{c_P}(\xi x + \eta y - \zeta_s|z-z_0|)]d\xi d\eta\right\} \quad , \tag{16}$$

where D_p, D_v, D_h are the receiver functions of the displacements in the infinite space generated by P–wave, SV–wave and SH–wave sources, respectively.

For a half–space problem, a free surface presents itself. If a receiver is at the free surface of the half–space, the source P ray, reflected P–P and P–SV rays coalesce into one P ray [13]. Thus the surface receiver function for a source P ray, D_p^*, is modified to

$$D_p^* = D_p + R^{pp} D_p + R^{pv} D_v \quad . \tag{17}$$

Similarly, the surface receiver functions for the source S rays, D_v^* and D_h^*, are modified to

$$D_v^* = D_v + R^{vv} D_v + R^{vp} D_p \quad , \tag{18}$$

$$D_h^* = D_h + R^{hh} D_h \quad , \tag{19}$$

where R^{pp}, R^{pv}, R^{vp}, R^{vv}, R^{hh} are reflection coefficients at the free surface. The first superscript denotes the mode of incident plane wave and the second one denotes the mode of reflected plane wave. For example, R^{pv} represents the ratio of the amplitude of reflected plane SV–wave to the amplitude of incident plane P–wave. The explicit expressions of reflection coefficients are

$$R^{pp} = R^{vv} = [4\zeta_p\zeta_s(\xi^2 + \eta^2) - (\zeta_s^{\,2} - \xi^2 - \eta^2)^2] \,/\, \Delta_R \quad , \tag{20}$$

$$R^{vp} = 4\zeta_s(\xi^2 + \eta^2)(\zeta_s^{\,2} - \xi^2 - \eta^2) \,/\, \Delta_R \quad , \tag{21}$$

$$R^{pv} = -4\zeta_p(\zeta_s^{\,2} - \xi^2 - \eta^2) \,/\Delta_R \quad , \tag{22}$$

$$R^{hh} = 1 \quad , \tag{23}$$

where

$$\Delta_R = 4\zeta_p\zeta_s(\xi^2 + \eta^2) + (\zeta_s^{\,2} - \eta^2 - \eta^2)^2 \tag{24}$$

These surface receiver functions for the corresponding displacement components are listed in Table 1 for reference. Notice that it can be shown very easily that the reflection coefficients vanish at the faces of prescribed displacement discontinuities. Hence it eludes the possibility of multiple reflections for the half–space problem with fault.

Cagniard–de Hoop method

For simplicity, let us consider only the responses at the ground (z = 0), and assume the source time function, f, to be a step function. For a more general source time function, the results involve only one additional convolution integral. The one–sided Laplace transformed ground velocities due to a quadrantal fault, from eqns (16)–(24), are expressed as

$$\tilde{\dot{u}}(x,y,0,s;z_0) = -\frac{1}{4\pi^2}\left\{\int_{L_\xi}\int_{L_\eta} \overset{*}{D}_p S_p \exp[\tfrac{s}{c_p}(\xi x + \eta y - \zeta_p z_0)]d\xi d\eta \right.$$

$$\left. + \int_{L_\xi}\int_{L_\eta} (\overset{*}{D}_v S_v + \overset{*}{D}_h S_h)\exp[\tfrac{s}{c_p}(\xi x + \eta y - \zeta_s z_0)]d\xi d\eta \right\} \tag{25}$$

The double integral in(25) can be evaluated and be transformed back to the time domain at the same time by employing the extended Cagniard–de Hoop method [20, 22, 23]. The modified de Hoop transformation is adopted first. Let

$$\xi = g\cos\phi - iw\sin\phi \quad , \quad \eta = g\sin\phi + iw\cos\phi \quad . \tag{26}$$

Equation (25) can then be written in the form as

$$\tilde{\dot{u}}_i(x,y,0,s;z_0) = \tilde{\dot{u}}_i^p(x,y,0,s;z_0) + \tilde{\dot{u}}_i^s(x,y,0,s;z_0), \;\; i = x,y,z \tag{27}$$

where

$$\tilde{\dot{u}}_i^\alpha(x,y,0,s;z_0) = -\frac{1}{4\pi^2}\int_{-\infty}^{\infty} dw \int_{L_g} \frac{A_i^\alpha \exp[\frac{s}{c_p}(gr - \zeta_\alpha z_0)]dg}{(\xi + \gamma)\,\eta\,\Delta_R} , \quad \alpha = p,s \tag{28}$$

and the expressions A_i^α, i = x,y,z; α = p,s are defined as

$$A_x^p = 4\zeta_p\zeta_s\xi^2 \tag{29a}$$

$$A_y^p = 4\zeta_p\zeta_s\eta\xi \tag{29b}$$

$$A_z^p = 2\zeta_p\xi(\zeta_s^{\,2} - \xi^2 - \eta^2) \tag{29c}$$

$$A_x^s = (\zeta_s^{\,2} - \xi^2 - \eta^2)^2 + 4\zeta_p\zeta_s\eta^2 \tag{30a}$$

$$A_y^s = -4\zeta_p\zeta_s\xi\eta \tag{30b}$$

$$A_z^s = -2\zeta_p\xi(\zeta_s^{\,2} - \xi^2 - \eta^2) \tag{30c}$$

In eqns (26)–(28), r and ϕ are defined as $r^2 = x^2 + y^2$, $\phi = \tan^{-1}(y/x)$ and L_g is a Bromwich contour in the complex g–plane.

The integration over g can be deformed into Cagniard's contour plus contributions from poles and a branch line integral. Then for each of them we employ the Cagniard–de Hoop method once again to evaluate the integration over w and transform back to the time domain at the same step. The analysis is very similar to that of [20] except some extra contributions from branch line integrals (head waves). The detailed analysis can be found in

[22, 24] and will appear in a different paper.

Contribution from the pole at $\xi = -\gamma$ *in the g–plane* (spherical waves)

$$\eta = -\frac{\nu_y}{1-\nu_x^2}(\tau - \gamma\nu_x) + i\frac{\nu_z}{1-\nu_x^2}[(\tau - \nu_x\gamma)^2 + (1-\nu_x^2)(\gamma^2 - \kappa_\alpha^2)]^{\frac{1}{2}} \tag{31}$$

and

$$\dot{u}_i^\alpha(x,y,0,t) = \frac{c_p H(x) H[\tau - \kappa_\alpha(1-\nu_x^2)^{\frac{1}{2}}] H(\tau - \gamma/\nu_x)}{\pi R[(\tau - \nu_x\gamma)^2 + (1-\nu_x^2)(\gamma^2 - \kappa_\alpha^2)]^{\frac{1}{2}}} \mathrm{Re}\left\{\frac{A_i^\alpha \zeta_\alpha}{\eta \Delta_R}\right\}, \quad \alpha = p,s\,. \tag{32}$$

Contribution from the pole at $\eta = 0$ *in the g–plane*

(i) Cylindrical waves

$$\xi = -\frac{\nu_x}{1-\nu_y^2}\tau + i\frac{\nu_z}{1-\nu_y^2}[\tau^2 - (1-\nu_y^2)\kappa_\alpha^2]^{\frac{1}{2}}\quad, \tag{33}$$

$$\dot{u}_i^\alpha(x,y,0,t) = \frac{c_p H(y) H[\tau - \kappa_\alpha(1-\nu_y^2)^{\frac{1}{2}}]}{\pi R[\tau^2 - \kappa_\alpha^2(1-\nu_y^2)]^{\frac{1}{2}}} \mathrm{Re}\left\{\frac{A_i^\alpha \zeta_\alpha}{(\xi + \gamma)\Delta_R}\right\}\quad, \quad \alpha = p,s\,. \tag{34}$$

(ii) head waves

$$\xi = -\frac{\nu_x}{1-\nu_y^2}\tau + \mathrm{sgn}(x)\frac{\nu_z[(1-\nu_y^2)\kappa_s^2 - \tau^2]^{\frac{1}{2}}}{1-\nu_y^2}\quad, \tag{35a}$$

$$\zeta_p = -i\,\mathrm{sgn}(x)(\xi^2 + \eta^2 - 1)^{\frac{1}{2}}\quad, \tag{35b}$$

$$\dot{u}_i^s(x,y,0,t) = \frac{c_p H(y)\,\mathrm{sgn}(x)}{\pi R[(1-\nu_y^2)\kappa_s^2 - \tau^2]^{\frac{1}{2}}} \mathrm{Im}\left\{\frac{A_i^s \zeta_s}{(\xi + \gamma)\Delta_R}\right\}\cdot$$

$$H[(1-\nu_y^2)^{\frac{1}{2}}\kappa_s - \tau]\, H[\tau - |\nu_x| - (\kappa_s^2 - 1)^{\frac{1}{2}}\nu_z] \cdot H[\kappa_s|\nu_x| - (1-\nu_y^2)^{\frac{1}{2}}]\,. \tag{36}$$

Contribution from the Cagniard's contour in the g–plane (spherical waves)

$$\xi = -\tau\nu_x - iw\sin\phi + i(\tau^2 - \kappa_\alpha^2 - w^2)^{\frac{1}{2}}\nu_z\cos\phi\quad, \tag{37a}$$

$$\eta = -\tau\nu_y + iw\cos\phi + i(\tau^2 - \kappa_\alpha^2 - w^2)^{\frac{1}{2}}\nu_z\sin\phi\quad, \tag{37b}$$

$$\dot{u}_i^\alpha(x,y,0,t) = \frac{c_p}{2\pi^2 R} J_i\, H(\tau - \kappa_\alpha)\ , \ \alpha = p,s\quad, \tag{38a}$$

$$J_i = \mathrm{Re}\int_{-(\tau^2-\kappa_\alpha^2)^{\frac{1}{2}}}^{(\tau^2-\kappa_\alpha^2)^{\frac{1}{2}}} \left[\frac{A_i^\alpha \zeta_\alpha}{(\xi + \gamma)\eta\Delta_R(\tau^2 - \kappa_\alpha^2 - w^2)^{\frac{1}{2}}}\right] dw\quad. \tag{38b}$$

Contribution from the branch line integral in the g–plane (head waves)

$$\xi = -\tau\nu_x + (w^2 - \tau^2 + \kappa_s^2)^{\frac{1}{2}}\nu_z\cos\phi - iw\sin\phi\quad, \tag{39a}$$

$$\eta = -\tau\nu_y + (w^2 - \tau^2 + \kappa_s^2)^{\frac{1}{2}}\nu_z\sin\phi + iw\cos\phi\quad, \tag{39b}$$

$$\zeta_p = i(\xi^2 + \eta^2 - 1)^{\frac{1}{2}}\quad, \tag{39c}$$

$$\dot{u}_i^s(x,y,0,t) = \frac{-c_p}{2\pi^2 R} H(\tau - \tau_{sp}) H(\kappa_s - \tau) H[(1-\nu_z^2)^{\frac{1}{2}} - 1/\kappa_s]\, M_i(-w_p, w_p)$$

$$- \frac{c_p}{2\pi^2 R} H(\tau_{sc} - \tau) H(\tau - \kappa_s) \cdot H[(1-\nu_z^2)^{\frac{1}{2}} - 1/\kappa_s]$$

$$\cdot\,[M_i(-w_p, -(\tau^2 - \kappa_s^2)^{\frac{1}{2}}) + M_i((\tau^2 - \kappa_s^2)^{\frac{1}{2}}, w_p)]\quad, \tag{40a}$$

where

$$M_i(a,b) = \mathrm{Im}\int_a^b \left[\frac{A_i^s \zeta_s}{(\xi + \gamma)\eta\Delta_R(w^2 - \tau^2 + \kappa_s{}^2)^{\frac{1}{2}}} \right] dw \quad , \tag{40b}$$

$$\tau_{sp} = (1-\nu_z^2)^{\frac{1}{2}} + \nu_z(\kappa_s^2-1)^{\frac{1}{2}} \quad , \quad \tau_{sc} = (\kappa_s^2-1)^{\frac{1}{2}}/\nu_z \quad , \tag{41}$$

$$w_p = \{[\tau-(\kappa_s^2-1)^{\frac{1}{2}}\nu_z]^2 - (1-\nu_z^2)\}^{\frac{1}{2}}/(1-\nu_z^2)^{\frac{1}{2}} \quad . \tag{42}$$

In eqns(31)–(42), R, ν_x, ν_y, ν_z and τ are defined as

$$R^2 = x^2 + y^2 + z_0{}^2 \quad , \tag{43}$$

$$\nu_x = x/R \quad , \quad \nu_y = y/R \quad , \quad \nu_z = z_0/R \quad , \tag{44}$$

$$\tau = c_p t/R \quad . \tag{45}$$

In summary, the ground velocities due to a quadrantal fault in an elastic half–space are the sum of terms listed in equations (32), (34), (36), (38) and (40). Moreover, the superscript α runs from p to s. In addition, by the use of a superposition argument, the field due to a desired finite fault described as eqns (5)–(7) can be expressed as the superposition of the corresponding fields due to four identical quadrantal faults but shifted in space and time. Consequently, the ground velocities due to the finite fault in a half space are written as

$$\dot{u}_i^f(x,y,0,t;z_0) = \dot{u}_i(x,y,0,t;z_0) - \dot{u}_i(x,y-W,0,t;z_0) - H(t- L/v_r)\dot{u}_i(x-L,y,0,t-L/v_r;z_0)$$
$$+ H(t- L/v_r)\dot{u}_i(x-L,y-W,0,t-L/v_r;z_0) \quad , \tag{46}$$

where the superscript f indicates the response due to the finite fault.

NUMERICAL RESULTS

Ground accelerations due to a finite fault with unidirectional propagation in x direction are (case 1) presented. The specific rupture mode is chosen as

$$[u_x](x,y,z_0,t) = H(t)H(t-x/v_r) \quad , \ 0<x<L, \ 0<y<W \quad , \tag{47a}$$

$$[u_x](x,y,z_0,t) = 0 \quad , \ \text{otherwise} \quad . \tag{47b}$$

The numerical results for a uniform dislocation (spontaneous rupture velocity) with rise time over a rectangular fault (case 2) are also presented for comparison. The rupture form of this mode is chosen as

$$[u_x](x,y,z_0,t) = (t/t_0)H(t)H(t_0-t) + H(t-t_0) \quad , \ 0<x<L, \ 0<y<W \quad , \tag{48a}$$

$$[u_x](x,y,z_0,t) = 0 \quad , \ \text{otherwise} \quad . \tag{48b}$$

To make the comparison meaningful, the rise time t_0 is chosen as $t_0 = L/v_r$.

As in the case 1, the field due to the desired finite fault described as eq. (48) can be constructed from the superposition of the corresponding fields of quadrantal faults but shifted in space,

$$\ddot{u}_i^f(x,y,0,t;z_0) = \ddot{u}_i(x,y,0,t;z_0) - \ddot{u}_i(x,y-W,0,t;z_0)$$
$$- \ddot{u}_i(x-L,y,0,t;z_0) + \ddot{u}_i(x-L,y-W,0,t;z_0) \quad . \tag{49}$$

The solution for the ground accelerations due to a quadrantal fault of this case can be obtained simply of the form $(t/t_0)H(t)H(x)H(y)$ with the corresponding results due to another rupture of the form $[-(t-t_0)/t_0]H(t-t_0)H(x)H(y)$. Notice that the results of these two specific ruptures differ from each other only by a time shift t_0 and the signs on the amplitudes. The ground acceleration due a rupture of the form $tH(t)H(x)H(y)$ is simply the direct summation of terms of the right hand sides of equations (34),(36),(38) and (40) with $\gamma = 0$. In addition, there is a term representing the plane wave,

$$\ddot{u}_i^\alpha(x,y,0,t) = \frac{A_i^\alpha c_p}{\kappa_s^4 R} H(x)H(y)\delta(\tau - \kappa_\alpha \nu_z) \tag{56}$$

and one more set of contributions, which are exactly the same as the right hand sides of equations (34) and (36) except with $\gamma = 0$ and x and y interchanged.

For the numerical examples, we set $L=W=z_0$ and the Lamé constants of the half–space are chosen as $\lambda=\mu$ (i.e. Poisson's ratio = 0.25). Nondimensional x–component of ground acceleration $a_x = \dot{u}_x z_0/c_s^2$ versus nondimensional time $c_s t/z_0$ are plotted for the case of $v_r/c_s = 0.9$. Figs 1(a)–1(b) are the responses at $\phi = \pi/6$, $r/z_0 = 5, 20$, for case 1 and case 2 respectively. Similar results for $\phi = \pi/3$ are shown in Figs 2(a)–2(b), respectively.

Table 1: free surface receiver functions D_p^*, D_v^* and D_h^* for displacement components

	D_p^*	D_v^*	D_h^*
$\tilde{u}_x$	$\xi(1+R^{pp}-\zeta_s R^{pv})$	$\zeta_s\xi(1-R^{vv})+\xi R^{vp}$	2η
$\tilde{u}_y$	$\eta(1+R^{pp}-\zeta_s R^{pv})$	$\zeta_s\eta(1-R^{vv})+\eta R^{vp}$	-2ξ
$\tilde{u}_z$	$\zeta_p(1-R^{pp})-(\xi^2+\eta^2)R^{pv}$	$-(\xi^2+\eta^2)(1+R^{vv})-\zeta_p R^{vp}$	0

REFERENCES

1. Steketee, J.A., Can. J. Phys., Vol. 36, (1958), p. 192.
2. Steketee, J.A., Can. J. Phys., Vol. 36, (1958), p. 1168.
3. Aki, K. and Richards, P., Quantitative Seismology: Theory and Methods, Freeman and Co. (1980), Chap. 14.
4. Haskell, N.A., Bull. Seism. Soc. Am., Vol. 53, (1963), p. 619.
5. Haskell, N.A., Bull. Seism. Soc. Am., Vol. 54, (1964), p. 377.
6. Harkrider, D.G., Bull. Seism. Soc. Am., Vol. 54, (1964), p. 627.
7. Ben–Menahem, A. and Singh, S.J., Bull. Seism. Soc. Am., Vol. 58, (1968), p. 1519.
8. Müller, G., Z. Geophys., Vol. 34, (1968), p. 147.
9. Müller, G., Z. Geophys., Vol. 35, (1969), p. 347.
10. Spencer, T.W., Geophys., Vol. 25, (1960), p. 625.
11. Spencer, T.W., Geophys., Vol. 30, (1965), p. 363.
12. Spencer, T.W., Geophys., Vol. 30, (1965), p. 369.
13. Pao, Y.H. and Gajewski, R.R., "The generalized ray theory and transient responses of layered elastic solids," Physical Acoustics, Vol. 13, (1977), p. 183.
14. Yeh, C.S. and Lin, T.W., "Ground motions generated by a strike–slip fault in a layered medium," Proc. Ist Conf. on Theo. and Appl. Mech., R.O.C., (1977), p. 9.
15. Kawasaki, I., "J. Phys. Earth, Vol. 23, (1975), p.127.
16. Hartzell, S. H., Frazier, G.A., and Brune, J.N., Bull. Seism. Soc. Am., Vol. 68, (1978), p.301.
17. Levy, N.A. and Mal, A.K., Bull. Seism. Soc. Am., Vol. 66, (1976), p.405.
18. Rice, J. R.,"The mechanics of earthquake rupture," in Physics of the Earth Interior, A. M. Dziewonski and Boshi (ed.), North–Holland Pub. Co., (1980), p.555.
19. Luco, J.E., and Anderson, J.G., Bull. Seism. Soc. Am., Vol. 73, (1983), P. 1.
20. Madariaga, R., Bull. Seism. Soc. Am., Vol. 68, (1978), P. 869.
21. Eringen, A.C., and Suhubi, E.S., Elastodynamics, Vol. II, Academic, (1975), P. 717.
22. Chen, P.L., "Ground response due to fault motion," Ph.D. Dissertation, National Taiwan University, Taipei, R.O.C. (1981).
23. Miklowitz, J., The Theory of Elastic Wave and Waveguides, North–Holland Pub. Comp., New York, (1978).
24. Wang, T.K. "On Haskell's rectangular fault with arbitrary dip in an eastic half–space," Masters thesis, National Taiwan University, Taipei, R.O.C. (1988) (in preparation)

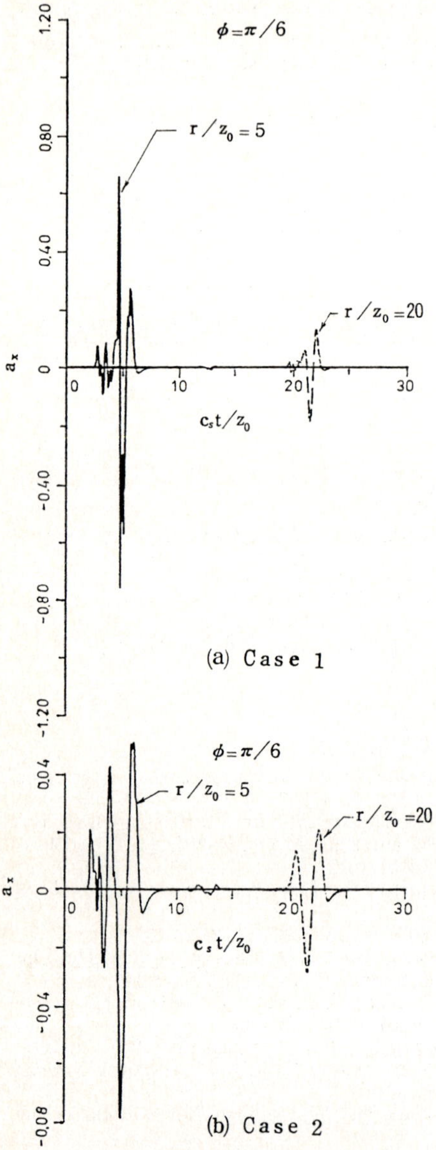

Fig1. Nondimensional x–component of ground accelation a_x versus nondimensional time $c_s t/z_0$, for $v_r/c_s = 0.9$ at $\phi = \pi/6$.

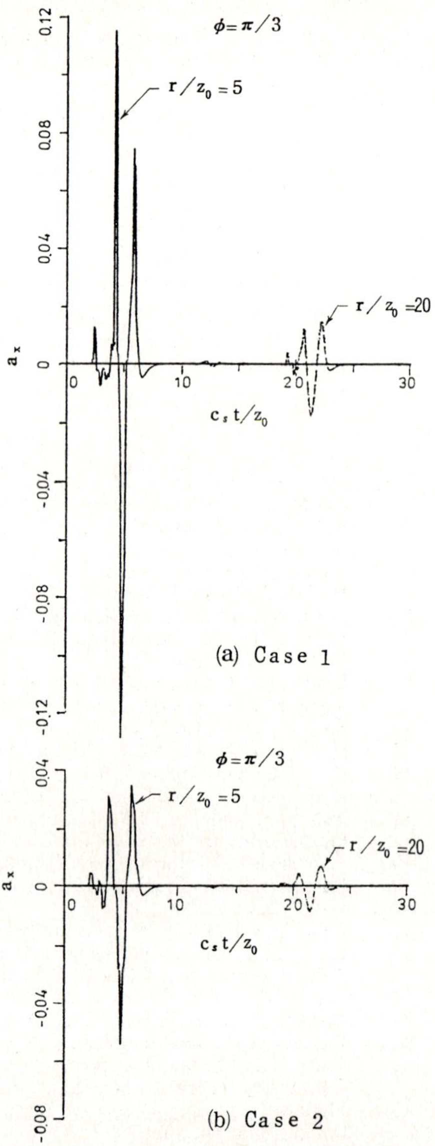

Fig2. Nondimensional x–component of ground accelation a_x versus nondimensional time $c_s t/z_0$, for $v_r/c_s = 0.9$ at $\phi = \pi/3$.

ELASTIC WAVE PROPAGATION
M.F. McCarthy, M.A. Hayes, (Editors)

ELASTIC WAVES IN LAYERED MEDIA WITH INTERFACE FEATURES

A.K. Mal and P.-C. Xu

Mechanical, Aerospace and Nuclear Engineering Department
University of California
Los Angeles, California 90024-1597, U.S.A.

A simple and efficient matrix method is applied to analyze the reflected waves from a layered plate with certain interface features. It is shown that a thin, adhesive layer can be modeled as an equivalent elastic interface with zero thickness characterized by two adjustable parameters. The accuracy of the model is examined through comparison with calculations from the exact case in which the adhesive layer is explicitly modeled. The influence of the imperfect interface on the velocity of layer guided waves is investigated.

1. INTRODUCTION

Bonded solids are being used with increasing frequency in modern aerospace and other structures. The quality of the bond is clearly of great importance in controlling the failure properties of the bonded components and therefore in the performance of the structure, both during its design and service phases. Thus it is necessary to have nondestructive evaluation (NDE) methods available to assess the quality of bonds. Ultrasonic methods can, in principle be used for this purpose. However, their present capability is limited to the detection of total unbonds and other gross defects in the bond region. Results of current research appear to indicate that the capability of ultrasonic methods can be significantly improved through careful modeling of the interfacial zone followed by theoretical analysis [1,2]. This paper is concerned with the NDE of interfaces by means of one of the recently developed methods based on leaky guided waves.

The theory of guided wave propagation in bonded, layered solids is well established. The bulk of the research on the subject can be found in the seismological literature where the the layers are usually assumed to be perfectly bonded at their common interfaces [3]. For engineering applications it is necessary to model the inteface more precisely, including the adhesive layer, and the possible presence of imperfect bonding resulting in partial slippage between the materials across the interface. The influence of these additional factors on the velocity and other properties of the guided waves in the solid needs to be determined theoretically before a reliable NDE technique can be developed. A theory of guided wave propagation in multilayered solids containing interface imperfections has been developed by Mal [4]. In this paper the theory is used to determine the influence of certain interface features on leaky Lamb waves from a bonded two layered plate immersed in water. In particular, the influence of the adhesive properties on the velocity of the Lamb type waves is investigated.

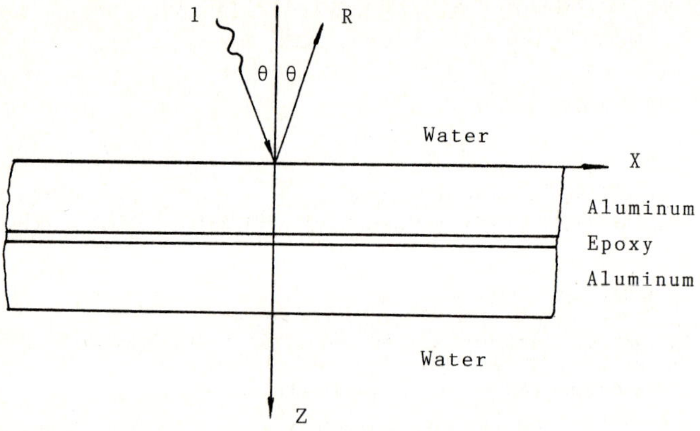

Figure 1: Geometry of the Problem.

2. THEORETICAL MODEL

The problem under consideration is sketched in Fig. 1. The two plates are separated by a bonding layer of small thickness. The material properties of the plates are denoted by $\rho(m)$ (density), $\alpha(m)$ (P-wave speed), $\beta(m)$ (S-wave speed) and $h(m)$ (thickness), $m = 2,3$, while those of the adhesive by the same letters without an argument. The bonded plate is fully immersed in water and is insonified by a plane, harmonic acoustic wave of frequency f(MHz) and unit amplitude. The objective is to calculate the reflected acoustic field for arbitrary angle and frequency of the incident waves.

Two models of the interface have been proposed in Mal [4]. In one, the adhesive is modeled as an additional layer which is perfectly bonded to the adherents at the two interfaces. In the second method, the effect of the thin layer of adhesive material is represented by a displacement jump across the interface with zero thickness. This jump is assumed to be proportional to the interfacial traction. It should be noted that the second model contains adjustable "interface parameters", which may be related to the "quality" of the interface.

It is assumed that the incident ray is on the xz plane and the relevant displacement and stress components at any point in the medium are u, w, τ_{xz} and σ_{zz} with the stipulation that τ_{xz} vanishes in water. Following the usual procedure [4,5] a four dimensional vector is defined through the equation

$$\{S(z)\} = \{U(z)\ T(z)\} \tag{1}$$

where

$$\{U(z)\} = \{u\ w\}e^{-i(kx-\omega t)} \qquad \{T(z)\} = \{\tau_{xz}\ \sigma_{zz}\}e^{-i(kx-\omega t)}$$

$$k = k_1(1)\sin\theta, \qquad k_1(1) = \omega/\alpha(1), \qquad \omega = 2\pi f$$

and $\alpha(1)$ is the sound wave speed in water.

In general, the displacement-stress vector $\{S(z)\}$ in the mth layer ($z_{m-1}<z<z_m$) can be expressed in the form

$$\{S(z)\} = \begin{bmatrix} Q_a(m) & Q_b(M) \\ Q_c(m) & Q_d(m) \end{bmatrix} [E(z,m)] \begin{Bmatrix} C^+(m) \\ C^-(m) \end{Bmatrix} \tag{2}$$

where $\{C^+\}$ and $\{C^-\}$ are two-dimensional constant vectors corresponding to the downgoing and upgoing waves, respectively, and the matrices $[Q(m)]$ and $[E(z,m)]$ have been defined in [4].

It follows that the jump in the displacement-stress vector at the interface between the mth and $(m+1)^{th}$ layers is given by

$$\begin{bmatrix} Q_a(m+1) & Q_b(m+1)E(m+1) \\ Q_c(m+1) & Q_d(m+1)E(m+1) \end{bmatrix} \begin{Bmatrix} C^+(m+1) \\ C^-(m+1) \end{Bmatrix} - \begin{bmatrix} Q_a(m)E(m) & Q_b(m) \\ Q_c(m)E(m) & Q_d(m) \end{bmatrix} \begin{Bmatrix} C^+(m) \\ C^-(m) \end{Bmatrix} = \begin{Bmatrix} \{U\}^+_- \\ \{T\}^+_- \end{Bmatrix} \tag{3}$$

where

$$\left[E(m)\right] = \text{Diag}\left[e^{i\eta_1(m)h(m)} \quad e^{i\eta_2(m)h(m)}\right] \tag{4}$$

and $\eta_1(m)$, $\eta_2(m)$ are the vertical wavenumbers of the P and S waves in the mth layer with

$$\text{Im}\ \eta_\alpha(m) > 0.$$

The above matrix equation is a general form of the displacement-stress discontinuity conditions across an interface. The following special cases are of practical interest,

1. $\{U\}^+_- = \{T\}^+_- = \{0\}$: perfect bonding.

2. $\{U\}^+_- = \{0\}$, $\{T\}^+_- \neq \{0\}$: perfect bonding, presence of interfacial force.

3. $\{U\}^+_- \neq \{0\}$, $\{T\}^+_- = \{0\}$: perfect bonding, presence of interfacial dislocation.

4. $\{U\}^+_- = [F]\{T\}$, $\{T\}^+_- = \{0\}$: elastic bonding. (5)

where $[F]$ is in general, a frequency dependent matrix and will be called the "flexibility matrix" for the interface. The last case is of interest here.

For an interface with elastic bonding, an asymptotic expression for the flexibility matrix $[F]$ for small $h|\eta_\alpha|$, $\alpha = 1,2$ can be derived from the adhesive layer model. To this end, the exponential matrix $[E]$ is expanded in the form,

$$[E] = [I] - h\ \text{Diag}[\eta_1 \quad \eta_2] + \text{Diag}[0(h\eta_1)^2 \quad 0(h\eta_2)^2] \tag{6}$$

where $[I]$ is the 2x2 unity matrix. Assuming that $T^+ = T^- = T$, Eqs. (3), (5) and (6) lead to

$$[\tilde{F}] \simeq \text{Diag}\ \left[\frac{h}{\mu}\gamma_1(\theta) \quad \frac{h}{\lambda+2\mu}\gamma_2(\theta)\right] \tag{7}$$

where

$$\gamma_1(\theta) = 1 - \beta^2\sin^2\theta/\alpha^2(1)$$

$$\gamma_2(\theta) = \left(\alpha(1)^2 - \alpha^2\sin^2\theta\right)/\left(\alpha^2(1) - 2\beta^2\sin^2\theta(1 - \beta^2/\alpha^2)\right) \qquad (8)$$

It can be seen that the matrix [F] depends only on the material properties of the adhesive layer and the incident angle so that it is independent of frequency. The parameters γ_1 and γ_2 are inversely proportional to the shear and longitudinal stiffness of the adhesive layer, and since the matrix [F] is diagonal, the two effects are decoupled in this model. It should be noted Eq. (7) is meaningful only for wavelengths long compared to the adhesive layer thickness, and as $\theta \to 0$,

$$[\tilde{F}] \to \mathrm{Diag}\left[\frac{h}{\mu} \;\; \frac{h}{\lambda+2\mu}\right]$$

More generally, it is reasonable to assume that the flexibility matrix [F] takes the form

$$[F] = \mathrm{Diag}[p_1\gamma_1 \;\; p_2\gamma_2] \qquad (9)$$

where p_1 and p_2 are dimensionless but in general, frequency dependent parameters.

Thus, for elastic bonding, the interfacial condition (3) may be expressed in the general form,

$$\begin{bmatrix} Q_a(m+1) & Q_b(m+1)E(m+1) \\ Q_c(m+1) & Q_d(m+1)E(m+1) \end{bmatrix} \begin{Bmatrix} C^+(m+1) \\ C^-(m+1) \end{Bmatrix} - \begin{bmatrix} \tilde{Q}_a(m)E(m) & \tilde{Q}_b(m) \\ Q_c(m)E(m) & Q_d(m) \end{bmatrix} \begin{Bmatrix} C^+(m) \\ C^-(m) \end{Bmatrix} = \begin{Bmatrix} 0 \\ 0 \end{Bmatrix} \qquad (10)$$

where

$$[\tilde{Q}_a] = [Q_a] + [F][Q_c] \qquad [\tilde{Q}_b] = [Q_b] + [F][Q_d] \qquad (11)$$

Finally, based on the above mentioned considerations, the problem illustrated in Fig. 1 leads to the linear system of order 12,

$$[Q]\{C\} = \{B\} \qquad (12)$$

where

$$[Q] = \begin{bmatrix} -Q_b(1) & Q_a(2) & Q_b(2)E(2) & 0 & 0 & 0 \\ -Q_d(1) & Q_c(2) & Q_d(2)E(2) & 0 & 0 & 0 \\ 0 & -\tilde{Q}_a(2)E(2) & -\tilde{Q}_b(2) & Q_a(3) & Q_b(3)E(3) & 0 \\ 0 & -Q_c(2)E(2) & -Q_d(2) & Q_c(3) & Q_d(3)E(3) & 0 \\ 0 & 0 & 0 & -Q_a(3)E(3) & -Q_b(3) & Q_a(4) \\ 0 & 0 & 0 & -Q_c(3)E(3) & -Q_d(3) & Q_c(4) \end{bmatrix}$$

$$\{C\} = \{C^-(1)\;\; C^+(2)\;\; C^-(2)\;\; C^+(3)\;\; C^-(3)\;\; C^+(4)\}$$

and

$$\{B\} = \{Q_a(1)C^+(1)\;\; Q_c(1)C^+(1)\;\; 0\;\; 0\;\; 0\;\; 0\} \qquad (15)$$

In the above, $\{C^+(1)\} = \{1\ 0\}$, $\{C^-(1)\} = \{R\ 0\}$ and $\{C^+(4)\} = \{T\ 0\}$ where R is the reflection coefficient and T is the transmission coefficient, for the given problem. The system of equations (12) can be easily solved to calculate R and T.

3. NUMERICAL RESULTS AND CONCLUSIONS

Numerical results are presented for the exact three layered plate containing the adhesive layer as well as the approximate model discussed above. Figure 2 compares the amplitudes of reflection coefficients calculated by means of these two different models. It can be seen that for smaller values of fh sinθ, the results for the elastic bonding model agree quite well with those for the exact model and as expected, the agreement is not as good at higher frequencies. If the adhesive layer thickness is increased, the agreement deteriorates. The details of these results could not be presented here due to space limitations.

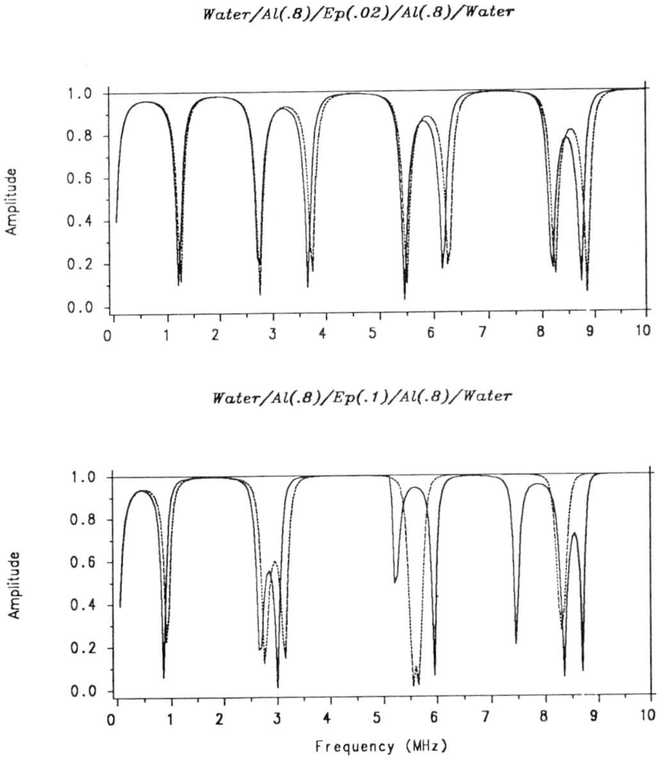

Figure 2: Comparison of the amplitude spectra of the reflection coefficient calculated by the adhesive layer model (solid line) and by the elastic interface model (dashed line). The incident angle is 20 degrees, and the exact model is water/Aℓ(.8 mm)/Ep(.02 and .1 mm)/ Aℓ(.8 mm)/water.

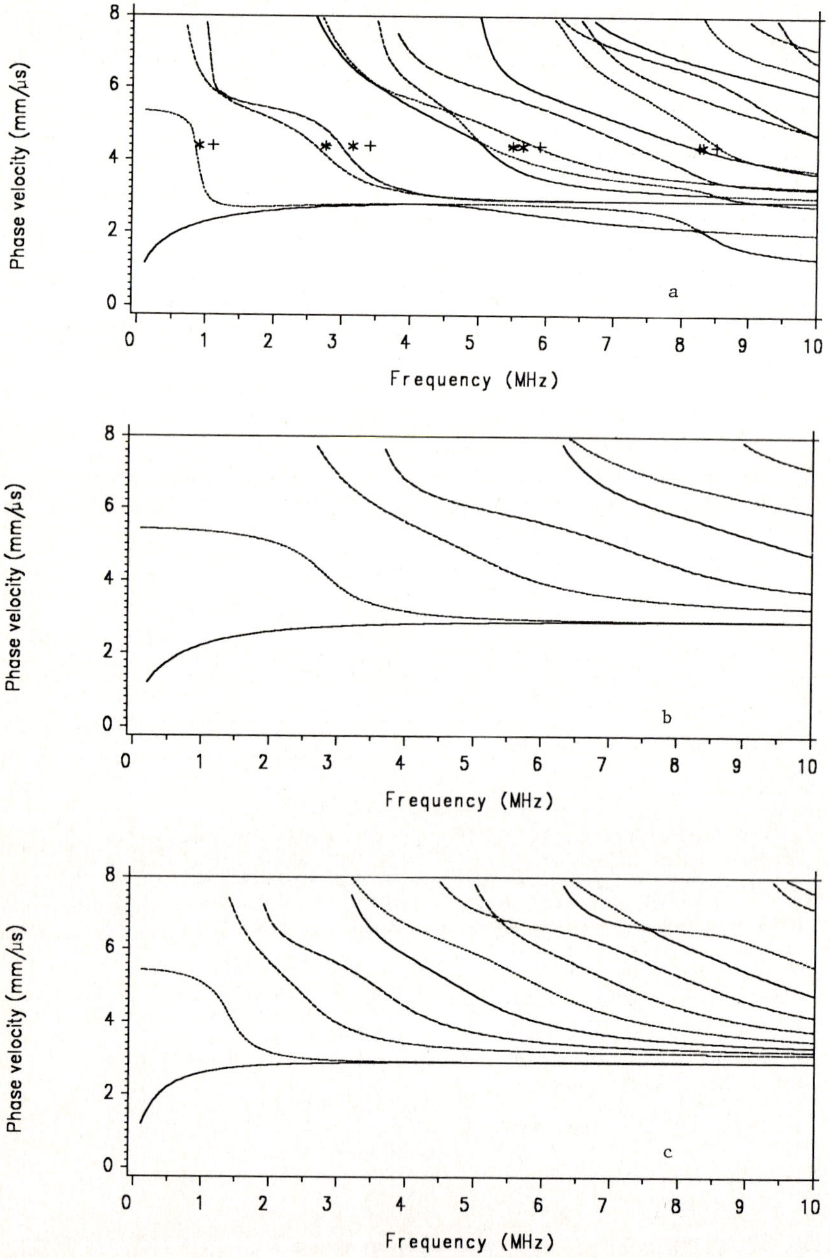

Figure 3: Dispersion curves for (a) the bonded plate Aℓ(0.81 mm)/Ep(0.10 mm)/Aℓ(0.81 mm); (b) an aluminum plate of thickness 0.81 mm; (c) an aluminum plate of thickness 1.62 mm. The "*" in (a) are corresponding to $p_1 = 1$, $p_2 = 1$ while the "+" are $p_1 = 1$, $p_2 = 0.5$.

Figures 3(a), (b), (c) give the dispersion curves for the three layered plate for three possible bonding conditions. As can be expected, the changes introduced in the dispersion curves by the epoxy layer are significant.

The objective of introducing the elastic interface model is not only to replace the effect of the adhesive layer by equivalent interface conditions, but to represent certain features of an imperfect interface which can not be modeled by such a layer. If the adhesive layer thickness is not very small, the parameters p_1, p_2 can be adjusted to improve the agreement with the exact result. As an example, the Lamb wave phase velocity-frequency relation obtained from the reflection coefficients for two sets of p_1, p_2 is shown in Fig. 3(a). Obviously, the adjustment $p_2 = 0.5$ does not give good results except at about 5.9 MHz. This clearly suggests that p_1 and p_2 should be frequency dependent. The influence of these two parameters on the reflection coefficient is under current investigation.

In conclusion, adhesive layers in plates can be modeled as an elastic interface of zero thickness. The computer implementation of the solution for this model is very convenient and efficient under the matrix formulation of Mal [4]. This model gives results close to those produced by the model containing the adhesive layer only for small thicknesses of the adhesive layer. The elastic interface model can be used to represent certain types of imperfect interfaces by suitably selecting the two adjustable parameters.

4. ACKNOWLEDGMENT

This research was supported by the Office of Naval Research under contract N00014-87-0351. We wish to thank Dr. Y. Rajapakse for helpful discussions during the course of this work.

5. REFERENCES

1. Mal, A.K. and Weglein, R., Characterization of Film Adhesion by Acoustic Microscopy, in: Thompson, D.O. and Chimenti, D.E. (Eds.), Quantitative Nondestructuve Evaluation, 7 (1987, in press).

2. Mal, A.K., Yin, C.-C. and Bar-Cohen, Y., The Influence of Attenuation and Bonding Layers on Leaky Waves, ibid.

3. Aki, K. and Richards, P.G., Quantitative Seismology, Theory and Methods (Freeman, San Francisco, 1980).

4. Mal, A.K., Guided Wave in Layered Solids with Interface Zones, Int'l. J. Eng. Sci. (1988, in press).

5. Xu, P.-C and Mal, A.K., Bull. Seism. Soc. Am. 77, (1987) 1823-1837.

ELASTIC WAVE PROPAGATION
M.F. McCarthy, M.A. Hayes, (Editors)

ULTRASONIC PROPAGATION IN TEXTURED POLYCRYSTALLINE METALS

Masahiko Hirao and Hidekazu Fukuoka

Faculty of Engineering Science
Osaka University
Toyonaka, Osaka 560, Japan

A generalized dispersion relation is presented for the guided modes in polycrystalline plates with orthorhombic texture. The texture is brought into the analysis through the crystallites' orientation distribution coefficients (ODCs). Its presence gives rise to the anisotropy of the elastic response and then to the elastic wave velocity. Ultrasonic experiments were made for the angular variations of propagation velocities of Rayleigh surface wave in aluminum alloy plates and the lowest symmetric plate mode (S_o) in the cold-rolled steel sheets. The velocity measurements allow the determination of the ODCs and well characterize the texture as illustrated by a favorable comparison with the conventional testing techniques for texture.

1. INTRODUCTION

The polycrystalline metals usually possess nonrandom distribution of crystallite orientation, which is developed by and sensitive to the plastic flow and heat treatment underwent during the fabrication process. These materials are said to have a texture or preferred grain orientation. Texture is the principal source of directional material properties, which may be advantageous and disadvantageous. Examples of advantageous texture-induced anisotropy include the high permeability of ferromagnetic sheets used for the electrical transformer cores and the plastic formability of steel sheets pressed to form the automobile bodies, etc. The yield stress and creep strength also vary with direction. Accordingly, there has been a requirement to monitor, in process if possible, the texture for the materials characterization. The elastic anisotropy itself is of less industrial interest than other aspects of macroscopic anisotopy. But, the ultrasonics, whose velocities detect the elastic behavior, are now recognized to be important for the texture measurement because of its nondestructive nature.

In this paper the dispersion relations are presented for the P-SV and SH elastic wave modes guided in textured plates. Approaching to the high- and low-frequency limits, the dispersion relation reduces to the velocity equations for the Rayleigh surface wave and the lowest symmetric plate mode (S_o mode), respectively. These formulae are useful to interpret the experimentally observed angular variations of ultrasonic velocities and to characterize the texture in aluminum alloy plates and cold-rolled steel sheets. It is found that the ultrasonics provide an attractive way to extract the quantitative textural information.

2. ELASTIC CONSTANTS OF POLYCRYSTALS WITH ORTHORHOMBIC TEXTURE

The elastic constants of textured polycrystalline materials are calculated from the single-crystal elastic constants using the orientation distribution function (ODF) as a weighting function in the Voigt-Reuss-Hill (VRH) average [1,2]. We use the Hill average throughout this paper. The ODF is a function of Euler angles relating the orthogonal axes in a crystallite and the sample axes and gives the probability for a specific orientation. It can be series-expanded in terms of the generalized spherical harmonics with expansion coefficients W_{lmn} called orientation distribution coefficients (ODCs) [3]. The number of independent ODCs decreases with the crystallite and sample symmetry properties.

For the polycrystalline cubic materials with orthorhombic texture, only three independent ODCs of the fourth order (l=4) determine the anisotropy of the (second-order) elastic constants, [C]. The elastic constants are expressed in the linear combinations of W_{4m0} (m=0,2,4) as

$$\begin{aligned}
C_{11} &= \lambda + 2\mu + (12\sqrt{2}c\pi^2/35)[W_{400} - (2\sqrt{10}/3)W_{420} + (\sqrt{70}/3)W_{440}],\\
C_{22} &= \lambda + 2\mu + (12\sqrt{2}c\pi^2/35)[W_{400} + (2\sqrt{10}/3)W_{420} + (\sqrt{70}/3)W_{440}],\\
C_{33} &= \lambda + 2\mu + (32\sqrt{2}c\pi^2/35)W_{400},\\
C_{44} &= \mu - (16\sqrt{2}c\pi^2/35)(W_{400} + \sqrt{5/2}\, W_{420}),\\
C_{55} &= \mu - (16\sqrt{2}c\pi^2/35)(W_{400} - \sqrt{5/2}\, W_{420}),\\
C_{66} &= \mu + (4\sqrt{2}c\pi^2/35)(W_{400} - \sqrt{70}\, W_{440}),\\
C_{23} &= \lambda - (16\sqrt{2}c\pi^2/35)(W_{400} + \sqrt{5/2}\, W_{420}),\\
C_{31} &= \lambda - (16\sqrt{2}c\pi^2/35)(W_{400} - \sqrt{5/2}\, W_{420}),\\
C_{12} &= \lambda + (4\sqrt{2}c\pi^2/35)(W_{400} - \sqrt{70}\, W_{440}),
\end{aligned} \tag{1}$$

where we adopted Voigt's two-suffix notation to write [C] and the Cartesian coordinate system in which x_1=rolling direction (RD), x_2 =transverse direction (TD), and x_3=normal direction to the sheet surfaces (ND). Each element of [C] consists of the isotropic part of λ and μ and the anisotropic part of c times W_{4m0}. The anisotropic factor c measures the inherent strength of elastic anisotropy of crystals. The tensor [C] shows an orthorhombic symmetry, but independent constants are six at most, not nine. In the derivation, an approximation was made by retaining only the first-order terms in W_{4m0}. The typical order of magnitude of W_{4m0}, say ε, is $O(10^{-2}\sim10^{-3})$ or less even for completely aligned aggregates (single crystals).

For the preparation to derive the dispersion relation for the arbitrary propagation direction in a plate plane, we clockwise rotate the coordinate system around its x_3 axis by an angle θ. With respect to the new coordinate system $O-X_1X_2X_3$, the elastic constant tensor, denoted by $[\bar{C}]$, is modified by multiplying the W_{420} terms in Eqs. (1) by cos2θ and the W_{440} terms by cos4θ: $X_3=x_3$ and $\bar{C}_{33} = C_{33}$. Besides that, the following additional elements occur:

$$\begin{aligned}
\bar{C}_{16} &= (8\sqrt{2}c\pi^2/35)(\sqrt{5/2}W_{420}\sin2\theta - \sqrt{70}/2\, W_{440}\sin4\theta),\\
\bar{C}_{26} &= -(8\sqrt{2}c\pi^2/35)(\sqrt{5/2}W_{420}\sin2\theta + \sqrt{70}/2\, W_{440}\sin4\theta),\\
\bar{C}_{36} &= \bar{C}_{45} = -(16\sqrt{5}c\pi^2/35)W_{420}\, \sin2\theta.
\end{aligned} \tag{2}$$

The Bond transformation procedure [4] is convenient to calculate $[\bar{C}]$.

3. DISPERSION RELATION OF PLATE MODES [5]

We consider the plate wave propagation along the X_1 direction, being independent of X_2 coordinate, in a textured plate of thickness 2d. For this, we shall solve the linear elastodynamic equation for the displacement $\underset{\sim}{u}(X_1, X_3, t)$

$$\rho\ddot{u}_k = \bar{C}_{klmn}(\partial^2 u_m/\partial X_l \partial X_n), \tag{3}$$

(k=1,2,3), with the boundary condition for traction-free surfaces at $X_3 = \pm d$. Because of the lower class of symmetry in $[\bar{C}]$, the P-SV motion (u_1 and u_3) and the SH motion (u_2) are coupled to each other in Eq. (3). It is found that this coupling occurs through the elastic constants of $O(\varepsilon)$ shown in Eqs. (2). These elastic constants are seen to be zero for traveling along the principal texture axes (RD and TD) and two types of motion are then decoupled.

As far as the first-order approximation in ε is used, however, we can neglect these coupling terms and treat the P-SV and SH motions separately for any θ. Trying to solve Eq. (3) in a perturbation framework, we may prescribe the form

$$\underset{\sim}{u} = \underset{\sim}{U}(X_1, X_3, t) + \varepsilon\underset{\sim}{U}'(X_1, X_3, t; \theta) + O(\varepsilon^2). \tag{4}$$

The classical solution for an isotropic plate is denoted by $\underset{\sim}{U}$ and the small deviation from it due to the texture presence is by $\varepsilon\underset{\sim}{U}'$. We can start with $U_2 = 0$ to solve them for P-SV motion. Putting this form into Eq. (3) together with the boundary condition and collecting terms of $O(\varepsilon)$, we have a set of equations to be solved for the perturbed field $\underset{\sim}{U}'$. They imply that the P-SV motion is now not "pure" except θ=0 and 90°, in the sense that $U_2' \neq 0$. But, the displacement $\varepsilon U_2'$ (= u_2 now) has nothing to do with the solutions for U_1' and U_3': it follows U_1 and U_3. Consequently, we can assume $u_2 = 0$ from the very beginning when deriving the dispersion relation of the P-SV modes. The reversed situation occurs in case of solving for the SH motion. The coupling terms become effective in the second-order approximation.

The harmonic solution has the phase velocity V determined by the transcendental function of kd = wavenumber x (half thickness) as well as W_{4m0} (m=0,2,4) and θ:

$$\tan(m_a kd)/\tan(m_b kd) = [G_b H_a/G_a H_b]^{\pm 1}, \tag{5}$$

with

$$\begin{aligned} G_i &= \bar{C}_{11} - \bar{C}_{13} m_i{}^2 - z, \\ H_i &= [\bar{C}_{13}(\bar{C}_{13}+\bar{C}_{55}) - \bar{C}_{33}(\bar{C}_{11}+\bar{C}_{55} m_i{}^2 - z)] m_i, \end{aligned} \tag{6}$$

in the case of the P-SV modes, where $z = \rho V^2$ and $m_i{}^2$ (i=a,b) are the roots of bi-quadratic characteristic equation

$$\bar{C}_{33}\bar{C}_{55}m^4 - [\bar{C}_{33}(z-\bar{C}_{11}) + \bar{C}_{55}(z-\bar{C}_{55}) + (\bar{C}_{13}+\bar{C}_{55})^2]m^2 + (z-\bar{C}_{11})(z-\bar{C}_{55}) = 0. \tag{7}$$

The (+) and (-) refer, respectively, to the symmetric and anti-symmetric modes of vibration about the central plane (X_3=0). Equation (5) is the Rayleigh-Lamb frequency equation generalized for the plates with orthorhombic texture. When W_{4m0} are known, the dispersion curves are numerically calculated for any θ with the use of Eq. (5).

For the SH modes, the dispersion relation is explicitly obtained as

$$\rho V_{SH}{}^2 = \bar{C}_{66} + \bar{C}_{44}(n\pi/2kd)^2, \tag{8}$$

where n is an integer: even n corresponds to symmetric mode and odd n to anti-symmetric mode. The phase velocity of the fundamental SH_0 mode (n=0) is exceptionally independent of kd and W_{420}, showing only the $\cos 4\theta$ variation [6]. Other SH modes will display the velocity anisotropy, depending on all of W_{4m0}.

4. VELOCITIES IN HIGH- AND LOW-FREQUENCY LIMITS

We are interested in inferring the ODCs from the observed angular variation of ultrasonic velocities. The dispersion relations obtained above are not in a convenient form for a practical use. It is then worth considering the limiting cases at the high and low frequencies (or equivalently thick and thin plates).

When kd increases to infinity, the left-hand side of Eq. (5) is asymptotic to unity and we have the velocity equation for the Rayleigh surface wave

$$\bar{C}_{33}\bar{C}_{55}z^2(z-\bar{C}_{11}) = (z-\bar{C}_{55})[\bar{C}_{33}(z-\bar{C}_{11})+\bar{C}_{13}{}^2]^2, \quad (9)$$

or

$$V_r(\theta) = V_{ro} + (c/\rho V_{ro})(r_0W_{400}+r_2W_{420}\cos2\theta+r_4W_{440}\cos4\theta), \quad (10)$$

which has been derived by Sayers [7] and Delsanto and Clark [8]. It shows that V_r changes with a period of 90°. The coefficients r_i (i=0,2,4) are functions of Poisson's ratio. For most materials, r_2 is much larger than r_4, which means that the cos2θ dependence is dominant in the V_r angular variation.

On the other hand, using the Taylor expansion $\tan x = x + x^3/3$ for small x and characteristic equation Eq. (7) to eliminate $m_a{}^2+m_b{}^2$ and $m_a{}^2m_b{}^2$, we find from Eq. (5) of the (+) sign

$$\rho V_{SO}{}^2 = (\bar{C}_{11}-\bar{C}_{13}{}^2/\bar{C}_{33})[1-(kd)^2\bar{C}_{13}{}^2/3\bar{C}_{33}{}^2], \quad (11)$$

or

$$V_{SO}(\theta) = V_o\sqrt{1-\delta} + (c/\rho V_o)(s_0W_{400}+s_2W_{420}\cos2\theta+s_4W_{440}\cos4\theta), \quad (12)$$

where $\delta = [\lambda/(\lambda+2\mu)]^2(kd)^2/3$, $V_o = \sqrt{4\mu(\lambda+\mu)/\rho(\lambda+2\mu)}$ is the isotropic velocity at kd=0, and $k=2\pi f/V_o$, f being the wave frequency. The coefficients s' can be expressed in terms of Poisson's ratio and δ: s_4 is alone independent of δ. This is the dispersion relation for the lowest symmetric (S_o) mode for small values of kd, probably less than 1. The above relation includes the effects of texture and dispersion both to the first order. The dispersion changes the isotropic velocity from V_o to $V_o\sqrt{1-\delta}$. The magnitude of this shift may be comparable to or larger than that of W_{400}. If it is neglected, then the resultant W_{400} would be totally useless. The δ-dependence of s_i (i=0,2), however, has a minor influence on the evaluation of W_{400} and W_{420}. Equating kd to zero, we reproduce the solution in the low-frequency limit [9].

5. ULTRASONIC EXPERIMENTS

Two ultrasonic experiments were carried out corresponding to the theoretical results. A sing-around technique allows the velocity measurement with an accuracy of ±2 m/sec. In the first experiment, V_r was measured for the surface texture in 4 rolled plates of Al 7075-T651 with different rolling reductions. The Rayleigh wavelength is small enough (∿0.6 mm) to consider the plates as

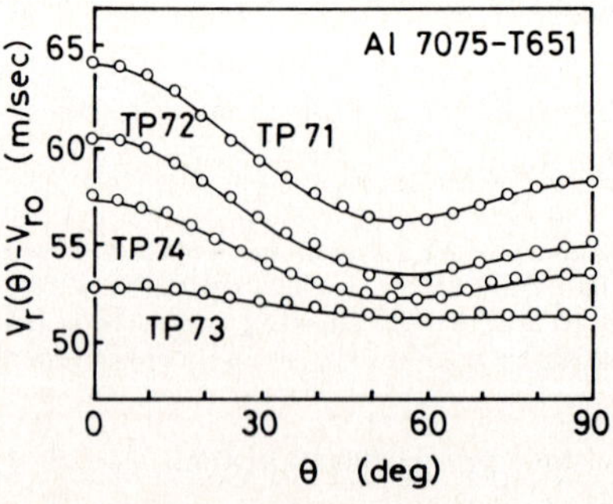

FIG. 1 Experimental plots and fitted curves for angular dependence of Rayleigh wave velocity for rolled aluminum alloy 7075-T651.

semi-infinite bodies. The variation of V_r with θ is shown in Figure 1. The least-square fitted curves shown are made up with constant, $\cos 2\theta$, and $\cos 4\theta$ components. The corresponding Fourier coefficients are used to determine W_{4m0} and then to draw the "ultrasonic pole figures" [10], which are compared with x-ray diffraction pole figures in Figure 2. Both represent the single texture, (001)[110].

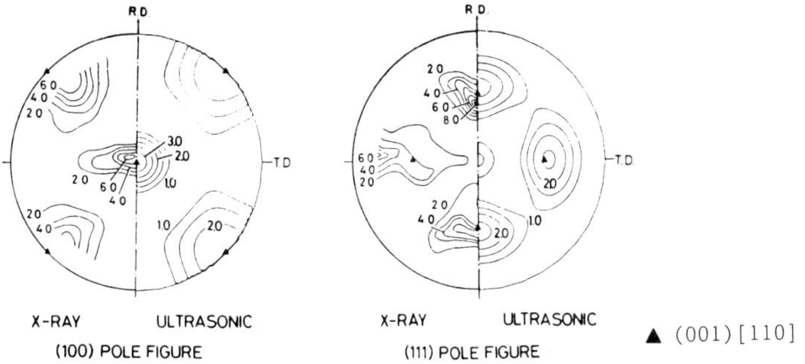

FIG. 2 Comparison of ultrasonic pole figures to x-ray pole figures (TP 71).

The second experiment was done with the S_o and SH_o modes using 11 cold-rolled steel sheets of different deformation and thermal histories. They are 0.677 mm to 0.902 mm thick. These sheets contain various intensity of rolling recrystallization texture centered around {111} component. That is, the {111} crystallographic planes tend to be aligned parallel to the sheet surfaces. Figure 3 presents the typical results for the S_o and SH_o modes obtained using 1 MHz and 2 MHz PZT transducers, respectively. The major effect of dispersion, δV_o^2, has been compensated. It is found again that the two or three Fourier components well interpret the observed velocity anisotropy as predicted by the preceding analysis. The values of W_{440} measured through V_{SO} and V_{SHO} differ by less than 0.0003. These observations lead to a conclusion that the directional velocity variations are mainly induced by the texture presence.

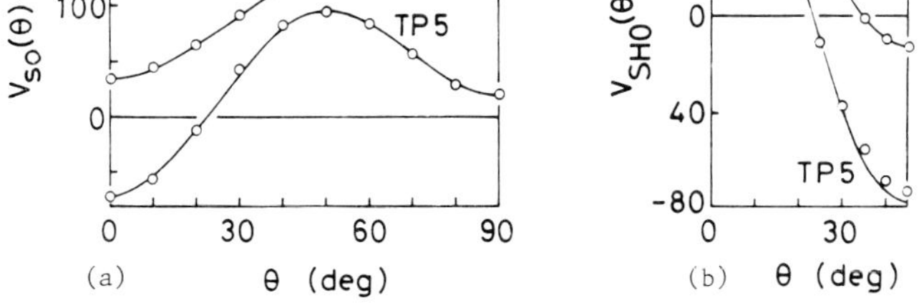

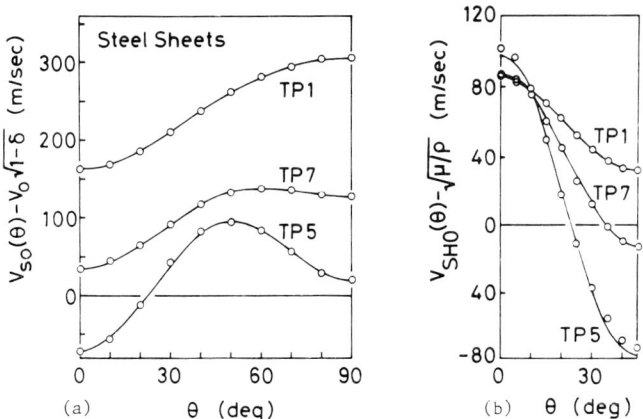

FIG. 3 Experimental plots and fitted curves for angular dependence of phase velocities of (a) S_o and (b) SH_o plate modes in cold-rolled steel sheets. From $V_{SO}(\theta)$, W_{400}=-0.01312, W_{420}=-0.00279, W_{440}=0.00032 for TP 1, W_{400}=-0.00241, W_{420}=-0.00149, W_{440}=0.00274 for TP 5, and W_{400}=-0.00621, W_{420}=-0.00139, W_{440}=0.00139 for TP 7.

Special attention was given to the overall shift in the V_{SO} variation, $\bar{V}_{SO}$, off the isotropic background, from which W_{400} can be evaluated. Figure 4 compares thus deduced W_{400} against $\bar{r} = (r_0 + 2r_{45} + r_{90})/4$, the planer average of the r-values. The r-value or the plastic strain ratio is defined as the ratio of strains in thickness to width directions in tensile specimen elongated to 15∿20 %: r_α denotes the strain ratio at an angle α from the rolling direction. The value of $\bar{r}$ is important in industry, because it measures the press-formability of metal sheets. The correlation shown in Figure 4 implies that $\bar{V}_{SO}$ and $\bar{r}$ view essentially the same aspect of texture. A crystal-plasticity discussion was made by Hirao et al [11] as for their close correlation. We expect a practical use of the S_o-mode velocity measurement for nondestructive evaluation of $\bar{r}$.

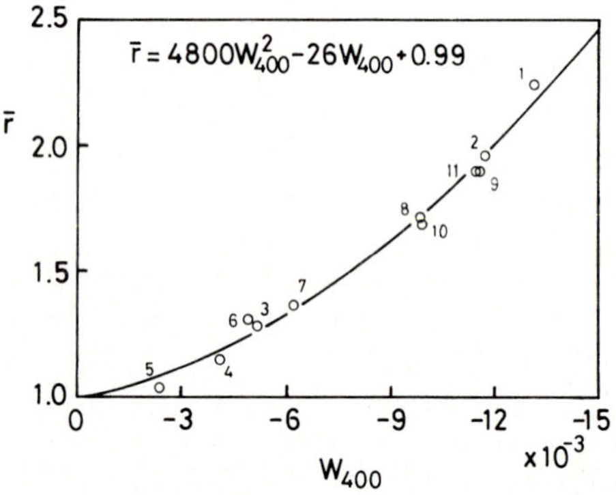

FIG. 4 Relation between $\bar{r}$ and W_{400} deduced form the S_o-mode experiment on cold-rolled steel sheets. Small numbers near the plots indicate TP numbers.

6. CONCLUSIONS

The dispersion relation of the plate modes was extended to include the orthorhombic elastic anisotropy due to the texture in polycrystalline materials. Dispersion has a large influence on W_{400} reduced from the S_o-mode experiments operated at finite frequencies. Although the texture modifies the phase velocities as shown, many basic features of the plate mode propagation in the isotropic case remain unchanged because of the relative weakness of the texture parameters, W_{4m0}. That is, the dispersion relation can be split into the P-SV and SH motions, each of which has an infinite number of symmetric and anti-symmetric normal modes. The experiments support the theoretical results, demonstrating the usefulness of ultrasonics for the texture characterization.

REFERENCES

[1] Sayers, C.M., J. Phys. D15 (1982) 2157.
[2] Hirao, M., Aoki, K., and Fukuoka, H., J. Acoust. Soc. Am. 81 (1987) 1434.
[3] Roe, R.-J., J. Appl. Phys. 36 (1965) 2024.
[4] Auld, B.A., Acoustic Fields and Waves in Solids (Wiley-Interscience, New York, 1973), Chap. 3.
[5] Hirao, M. and Fukuoka, H., submitted.
[6] Allen, D.R., Langman, R., and Sayers, C.M., Ultrasonics 23 (1985) 215.
[7] Sayers, C.M., Proc. R. Soc. London Ser.A400 (1985) 175.
[8] Delsanto, P.P. and Clark, Jr., A.V., J. Acoust. Soc. Am. 81 (1987) 952.
[9] Thompson, R.B., Lee, S.S., and Smith, J.B., Ultrasonics 25 (1987) 133.
[10] Hirao, M. and Hara, N., Appl. Phys. Lett. 50 (1987) 1411.
[11] Hirao, M., Hara, N., Fukuoka, H., and Fujisawa, K., submitted.

ELASTIC WAVE PROPAGATION
M.F. McCarthy, M.A. Hayes, (Editors)

PROPAGATION OF SH-SURFACE WAVES IN ELASTIC LAYERED DIELECTRICS

Krystyna MAJORKOWSKA-KNAP and Jürgen LENZ

Institute of Mechanics of Engineering Constructions, Technical University of Warsaw, Płock, Poland

Institut für Theoretische Mechanik, Universität Karlsruhe, Karlsruhe, Fed. Rep. of Germany

On the basis of Mindlin's model of piezoelectricity which takes into account the effect of the polarization gradient, the propagation of electromechanically coupled Stoneley waves is studied, representing transverse interface modes with horizontal polarization in a medium consisting of a centrosymmetric, isotropic, dielectric, elastic layer between two isotropic, purely elastic half-spaces.

1. INTRODUCTION

It is well-known that, due to the piezoelectric effect, the propagation of waves in elastic dielectrics leads in general to a coupling of mechanical deformation and electric fields [1,2]. These surface and interface acoustic waves have been studied intensely in recent years in view of their applicability to signal processing in electronic systems, seismology, nondestructive evaluation, flaw detection etc.

This contribution deals with the propagation of piezoelectric SH-interface waves (generalized Stoneley waves) along a medium consisting of a centrosymmetric, isotropic, elastic, dielectric layer between two isotropic, (in general materially distinct) purely elastic half-spaces. The dielectric material is idealized according to Mindlin's gradient theory of piezoelectricity [1,3]. This model takes into consideration the influence of the polarization gradient and thus accounts more precisely for the microscopic aspects of structure and interatomic interactions. Furthermore, it accommodates — in contrast to the classical theory of piezoelectricity [4,5] — an electromechanical interaction also in centrosymmetric (including isotropic) media.

The problem in question is considered as an idealization of a special real wave delay line which is composed of a thin dielectric, elastic stratum attached to comparatively thick elastic substrates on both sides. Such structures might be important in practice with regard to the excitation, processing and amplification of interface waves.

2. FIELD EQUATIONS

As presented in [3] and briefly outlined in [6] (where also a survey of the related literature can be found), Mindlin's theory of elastic, dielectric continua leads for a centrosymmetric, isotropic material (in the quasistatic approximation with regard to the electromagnetic field) in the absence of body forces and external electric fields to the *field equations*

$$\left.\begin{aligned} c_{44}\Delta\mathbf{u} + (c_{12}+c_{44})\,\mathrm{grad\,div}\,\mathbf{u} + d_{44}\Delta\mathbf{P} + \qquad & \\ +(d_{12}+d_{44})\,\mathrm{grad\,div}\,\mathbf{P} &= \varrho\ddot{\mathbf{u}} \\ d_{44}\Delta\mathbf{u} + (d_{12}+d_{44})\,\mathrm{grad\,div}\,\mathbf{u} + (b_{44}+b_{77})\Delta\mathbf{P} + \qquad & \\ +(b_{12}+b_{44}-b_{77})\,\mathrm{grad\,div}\,\mathbf{P} - a\mathbf{P} - \mathrm{grad}\,\varphi &= 0 \\ -\varepsilon_0\Delta\varphi + \mathrm{div}\,\mathbf{P} &= 0 \end{aligned}\right\}\ \text{in } V \qquad (1,2,3)$$

and *boundary conditions*

$$\boldsymbol{\sigma}\mathbf{n} = \mathbf{t} \quad , \quad \mathbf{E}^L\mathbf{n} = \mathbf{0} \quad , \quad \mathbf{n}\circ\{\mathbf{P} - \varepsilon_0[\![\mathrm{grad}\,\varphi]\!]\} = 0 \qquad \text{on } S \qquad (4)$$

(V: volume of the body, S: surface separating V from vacuum; $[\![\ ]\!]$ denotes the jump across S), where

$\mathbf{u}$	: displacement vector ,	$\boldsymbol{\sigma}$	: stress tensor ,	$\mathbf{P}$	: polarization vector ,
$\boldsymbol{E}^L$	: local electric tensor ,	$\mathbf{t}$	: surface traction ,	$\mathbf{n}$	: outer unit normal vector ,
φ	: potential of the Maxwell self field ,	ϱ	: mass density ,	ε_0	: permittivity of vacuum .

Hereby c_{12} and c_{44} are the elastic constants; $a, d_{12}, d_{44}, b_{12}, b_{44}, b_{77}$ are additional material coefficients [6].

It is immediately clear from (1)–(3) that for a centrosymmetric, isotropic medium the classical Toupin-Voigt theory [4,5] (i.e. $d_{ij} = 0$ and $b_{ij} = 0$) cannot account for electromechanical coupling effects, as accomplished by Mindlin's model.

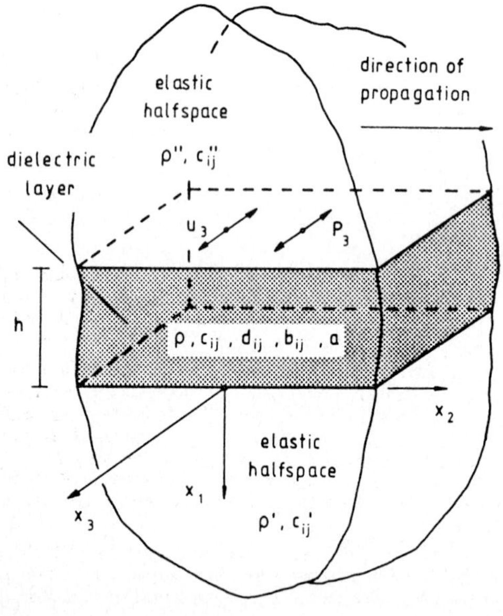

FIGURE 1
Illustration of problem.

Furthermore, it is well-known from the classical theory that the effect of piezoelectricity is to *increase* the effective elastic constants, and therefore piezoelectricity is said to "*stiffen*" the solid, thus giving rise to a fractional *increase* in the phase velocities of waves [7]. Mindlin's gradient theory, however, delivers the inverse, i.e. piezoelectrically "*softening*" effect for *centrosymmetric, isotropic* media. This may easily be shown by determining the phase velocities of plane bulk waves by means of (1)–(3). For sodium chloride as dielectric (cf. last section), the maximum "*softening*" effect in the phase velocity, as compared with the purely elastic continuum, amounts to $\sim 1.7\,‰$ and $\sim 0.35\,‰$ for transverse and longitudinal plane bulk waves, respectively.

3. FORMULATION OF PROBLEM

As pictured in Fig. 1, we consider an infinite continuum consisting of the layer $-h \leq x_1 \leq 0$ of a centrosymmetric, isotropic, elastic dielectric, attached to two centrosymmetric, isotropic, purely elastic half-spaces $x_1 \geq 0$ and $x_1 \leq -h$, respectively, which in general may be materially distinct. The material coefficients of the dielectric layer (of thickness h) are given by $\varrho, c_{ij}, d_{ij}, b_{ij}, a$; those of the elastic half-spaces by ϱ', c'_{ij} and ϱ'', c''_{ij}, respectively.

In the absence of surface tractions ($\mathbf{t} = \mathbf{0}$), according to (1)–(3) the *field equations* are given by

$$\left.\begin{aligned} c_{44}\Delta\mathbf{u} + (c_{12}+c_{44})\operatorname{grad}\operatorname{div}\mathbf{u} + d_{44}\Delta\mathbf{P} + \\ +(d_{12}+d_{44})\operatorname{grad}\operatorname{div}\mathbf{P} &= \varrho\ddot{\mathbf{u}} \\ d_{44}\Delta\mathbf{u} + (d_{12}+d_{44})\operatorname{grad}\operatorname{div}\mathbf{u} + (b_{44}+b_{77})\Delta\mathbf{P} + \\ +(b_{12}+b_{44}-b_{77})\operatorname{grad}\operatorname{div}\mathbf{P} - a\mathbf{P} - \operatorname{grad}\varphi &= \mathbf{0} \\ -\varepsilon_0\Delta\varphi + \operatorname{div}\mathbf{P} &= 0 \end{aligned}\right\} \quad \text{for } -h \leq x_1 \leq 0 \tag{5}$$

$$c'_{44}\Delta\mathbf{u}' + (c'_{12}+c'_{44})\operatorname{grad}\operatorname{div}\mathbf{u}' = \varrho'\ddot{\mathbf{u}}' \qquad \text{for } x_1 \geq 0 \tag{6}$$

$$c''_{44}\Delta\mathbf{u}'' + (c''_{12}+c''_{44})\operatorname{grad}\operatorname{div}\mathbf{u}'' = \varrho''\ddot{\mathbf{u}}'' \qquad \text{for } x_1 \leq -h \tag{7}$$

which, according to (4), have to be augmented by the *continuity conditions*

$$\mathbf{u} = \mathbf{u}' \;,\; \boldsymbol{\sigma}\mathbf{e}_1 = -\boldsymbol{\sigma}'(-\mathbf{e}_1) \;,\; \boldsymbol{E}^L\mathbf{e}_1 = \mathbf{0} \;,\; \mathbf{e}_1 \circ \{\mathbf{P} - \varepsilon_0[\![\operatorname{grad}\varphi]\!]\} = 0 \qquad \text{on } x_1 = 0 \tag{8}$$

and

$$\mathbf{u} = \mathbf{u}'' \ , \ \boldsymbol{\sigma}(-\mathbf{e}_1) = -\boldsymbol{\sigma}''\mathbf{e}_1 \ , \ \mathbf{E}^L(-\mathbf{e}_1) = \mathbf{0} \ , \ -\mathbf{e}_1 \circ \{\mathbf{P} - \varepsilon_0[\![\operatorname{grad}\varphi]\!]\} = 0 \qquad \text{on } x_1 = -h \tag{9}$$

and *regularity conditions* at infinity.

Being interested in the propagation of *plane* Stoneley waves with constant phase velocity p, we are looking for solutions of (5)–(9) of the form

$$\begin{aligned} &\mathbf{u} = u_3(x_1, x_2, t)\mathbf{e}_3 \quad , \quad \mathbf{P} = P_3(x_1, x_2, t)\mathbf{e}_3 \quad , \quad \varphi = \varphi(x_1, x_2, t) \ ; \\ &\mathbf{u}' = u_3'(x_1, x_2, t)\mathbf{e}_3 \quad , \quad \mathbf{u}'' = u_3''(x_1, x_2, t)\mathbf{e}_3 \end{aligned} \tag{10}$$

4. ANALYTICAL SOLUTION

Under the assumptions (10) the system (5)–(9) reduces to the set

$$\left.\begin{aligned} c_{44}\left(\Delta - \frac{\varrho}{c_{44}}\partial_t^2\right)u_3 + d_{44}\Delta P_3 = 0 \\ d_{44}\Delta u_3 + \{(b_{44} + b_{77})\Delta - a\}P_3 = 0 \end{aligned}\right\} \quad \text{for } -h \le x_1 \le 0 \tag{11,12}$$

$$c'_{44}\left(\Delta - \frac{\varrho'}{c'_{44}}\partial_t^2\right)u'_3 = 0 \quad \text{for } x_1 \ge 0 \quad , \quad c''_{44}\left(\Delta - \frac{\varrho''}{c''_{44}}\partial_t^2\right)u''_3 = 0 \quad \text{for } x_1 \le -h \tag{13,14}$$

and

$$\begin{aligned} u_3(0, x_2, t) &= u'_3(0, x_2, t) && (15) \ , & u_3(-h, x_2, t) &= u''_3(-h, x_2, t) && (18) \\ \sigma_{13}(0, x_2, t) &= \sigma'_{13}(0, x_2, t) && (16) \ , & \sigma_{13}(-h, x_2, t) &= \sigma''_{13}(-h, x_2, t) && (19) \\ E_{13}(0, x_2, t) &= 0 && (17) \ , & E_{13}(-h, x_2, t) &= 0 && (20) \end{aligned}$$

We seek solutions of (11)–(20) in the form of *plane harmonic* waves:

$$\begin{aligned} &u_3(x_1, x_2, t) = u_3^0(x_1)\exp\{iK(x_2 - pt)\} \quad , \quad P_3(x_1, x_2, t) = P_3^0(x_1)\exp\{iK(x_2 - pt)\} \ , \\ &u'_3(x_1, x_2, t) = u_3^{0'}(x_1)\exp\{iK(x_2 - pt)\} \quad , \quad u''_3(x_1, x_2, t) = u_3^{0''}(x_1)\exp\{iK(x_2 - pt)\} \end{aligned} \tag{21}$$

and express the phase velocity $p = \omega/K$ (ω: angular frequency, K: wave number) as

$$p = c\sqrt{c_{44}/\varrho} = c'\sqrt{c'_{44}/\varrho'} = c''\sqrt{c''_{44}/\varrho''} \ . \tag{22}$$

c, c' and c'' denote relative phase velocities, and the roots characterize the propagation velocities of transverse bulk waves in the layer and in the half-spaces.

The regularity conditions at infinity demand that the solutions of (13) and (14) have the form

$$u_3^{0'}(x_1) = A'\exp(-Ks'x_1) \quad , \quad u_3^{0''}(x_1) = A''\exp(Ks''x_1) \tag{23,24}$$

where

$$s' = \sqrt{1 - c'^2} \quad , \quad s''_2 = \sqrt{1 - c''^2} \ . \tag{25}$$

The general solution of (11), (12) is given by [6]

$$u_3^0(x_1) = \bar{A}\sin(s_1x_1) + \bar{B}\cos(s_1x_1) + \bar{C}\sinh(s_2x_1) + \bar{D}\cosh(s_2x_1) \tag{26}$$

$$P_3^0(x_1) = \kappa_1\left[\bar{A}\sin(s_1x_1) + \bar{B}\cos(s_1x_1)\right] + \kappa_2\left[\bar{C}\sinh(s_2x_1) + \bar{D}\cosh(s_2x_1)\right] \tag{27}$$

where the abbreviations [6]

$$s_1 = \sqrt{\frac{-B + \sqrt{B^2 - 4AD}}{2A}} \quad , \quad s_2 = \sqrt{\frac{B + \sqrt{B^2 - 4AD}}{2A}} \tag{28}$$

with

$$A = \frac{c_{44}\hat{b}_{44} - d_{44}^2}{a c_{44}} > 0\,, \qquad \hat{b}_{44} = b_{44} + b_{77}\,,$$

$$B = 1 + \frac{\kappa^2}{h^2}\left\{2A - \frac{\hat{b}_{44}}{a}c^2\right\}\,, \qquad D = \frac{\kappa^2}{h^2}\left\{1 - c^2 + \frac{\kappa^2}{h^2}\left(A - \frac{\hat{b}_{44}}{a}c^2\right)\right\}\,, \tag{29}$$

$$\kappa_1 = -\frac{d_{44}(h^2 s_1^2 + \kappa^2)}{\hat{b}_{44}(h^2 s_1^2 + \kappa^2) + h^2 a}\,, \qquad \kappa_2 = -\frac{d_{44}(h^2 s_2^2 - \kappa^2)}{\hat{b}_{44}(h^2 s_2^2 - \kappa^2) - h^2 a}$$

have been used, and the dimensionless wave number

$$\kappa = Kh \tag{30}$$

has been introduced.

From (25) and the derivation of (26) and (27) it can be shown that the phase velocity $p = p(\kappa)$ must satisfy the inequality

$$\left.\begin{aligned} &\frac{c_{44}}{\varrho}\left[1 - \frac{d_{44}^2\kappa^2}{c_{44}(h^2 a + \hat{b}_{44}\kappa^2)}\right] < p^2 < \frac{c'_{44}}{\varrho'} < \frac{c''_{44}}{\varrho''} \\ \text{or}\quad & \\ &\frac{c_{44}}{\varrho}\left[1 - \frac{d_{44}^2\kappa^2}{c_{44}(h^2 a + \hat{b}_{44}\kappa^2)}\right] < p^2 < \frac{c''_{44}}{\varrho''} < \frac{c'_{44}}{\varrho'} \end{aligned}\right\} \tag{31}$$

These restrictions allow for the following *existence conditions* (cf. [6]):

(A) *Symmetric problem* ($c''_{44} = c'_{44}$, $\varrho'' = \varrho'$):

(I) "Classical" case: $\dfrac{c_{44}}{\varrho} < \dfrac{c'_{44}}{\varrho'}$ for every ω; (32)

(II) "Non-classical" case: $\dfrac{c_{44}}{\varrho} > \dfrac{c'_{44}}{\varrho'}$ (33)

for *cut-off frequencies*:
$$\omega^2 > \frac{a\dfrac{c'_{44}}{\varrho'}\left(\dfrac{c_{44}}{\varrho} - \dfrac{c'_{44}}{\varrho'}\right)}{\dfrac{d_{44}^2}{\varrho} + \hat{b}_{44}\left(\dfrac{c'_{44}}{\varrho'} - \dfrac{c_{44}}{\varrho}\right)}\,. \tag{34}$$

(B) *Non-symmetric problem* ($c''_{44} \neq c'_{44}$, $\varrho'' \neq \varrho'$): In the

(I) "classical" case: $\dfrac{c_{44}}{\varrho} < \dfrac{c'_{44}}{\varrho'} < \dfrac{c''_{44}}{\varrho''}$ or $\dfrac{c_{44}}{\varrho} < \dfrac{c''_{44}}{\varrho''} < \dfrac{c'_{44}}{\varrho'}$ (35)

as well as in the

(II) "non-classical" case: $\dfrac{c_{44}}{\varrho} > \dfrac{c'_{44}}{\varrho'} > \dfrac{c''_{44}}{\varrho''}$ or $\dfrac{c_{44}}{\varrho} > \dfrac{c''_{44}}{\varrho''} > \dfrac{c'_{44}}{\varrho'}$ (36)

there exist, as in the totally elastic medium, *cut-off frequencies* which might in principle be determined by further analysing the dispersion relation (40). In any case, however, (36) demands additionally the *cut-off frequencies*

$$\omega^2 > \max\left\{\frac{a\dfrac{c'_{44}}{\varrho'}\left(\dfrac{c_{44}}{\varrho} - \dfrac{c'_{44}}{\varrho'}\right)}{\dfrac{d_{44}^2}{\varrho} + \hat{b}_{44}\left(\dfrac{c'_{44}}{\varrho'} - \dfrac{c_{44}}{\varrho}\right)}\,,\ \frac{a\dfrac{c''_{44}}{\varrho''}\left(\dfrac{c_{44}}{\varrho} - \dfrac{c''_{44}}{\varrho''}\right)}{\dfrac{d_{44}^2}{\varrho} + \hat{b}_{44}\left(\dfrac{c''_{44}}{\varrho''} - \dfrac{c_{44}}{\varrho}\right)}\right\}\,. \tag{37}$$

In the order (19),(18),(20),(16),(17),(15),the continuity conditions lead to the following homogeneous system of equations for the amplitude factors $\bar{A}, \bar{B}, \bar{C}, \bar{D}, A'$ and A'':

$$\begin{bmatrix} A_{11} & A_{12} & A_{13} & A_{14} & 0 & A_{16} \\ A_{21} & A_{22} & A_{23} & A_{24} & 0 & A_{26} \\ A_{31} & A_{32} & A_{33} & A_{34} & 0 & 0 \\ A_{41} & 0 & A_{43} & 0 & A_{45} & 0 \\ A_{51} & 0 & A_{53} & 0 & 0 & 0 \\ 0 & 1 & 0 & 1 & -1 & 0 \end{bmatrix} \begin{bmatrix} \bar{A} \\ \bar{B} \\ \bar{C} \\ \bar{D} \\ A' \\ A'' \end{bmatrix} = \begin{bmatrix} 0 \\ 0 \\ 0 \\ 0 \\ 0 \\ 0 \end{bmatrix} \tag{38}$$

where

$$\begin{aligned} A_{11} &= s_1(c_{44} + d_{44}\kappa_1) \cdot \cos(s_1 h) & A_{31} &= s_1(d_{44} + \hat{b}_{44}\kappa_1) \cdot \cos(s_1 h) \\ A_{12} &= s_1(c_{44} + d_{44}\kappa_1) \cdot \sin(s_1 h) & A_{32} &= s_1(d_{44} + \hat{b}_{44}\kappa_1) \cdot \sin(s_1 h) \\ A_{13} &= s_2(c_{44} + d_{44}\kappa_2) \cdot \cosh(s_2 h) & A_{33} &= s_2(d_{44} + \hat{b}_{44}\kappa_2) \cdot \cosh(s_2 h) \\ A_{14} &= -s_2(c_{44} + d_{44}\kappa_2) \cdot \sinh(s_2 h) & A_{34} &= -s_2(d_{44} + \hat{b}_{44}\kappa_2) \cdot \sinh(s_2 h) \\ A_{16} &= -c''_{44} s''(\kappa/h) \cdot \exp(-s''\kappa) & A_{41} &= s_1(c_{44} + d_{44}\kappa_1) \\ A_{21} &= -\sin(s_1 h) & A_{43} &= s_2(c_{44} + d_{44}\kappa_2) \\ A_{22} &= \cos(s_1 h) & A_{45} &= c'_{44} s'(\kappa/h) \\ A_{23} &= -\sinh(s_2 h) & A_{51} &= s_1(d_{44} + \hat{b}_{44}\kappa_1) \\ A_{24} &= \cosh(s_2 h) & A_{53} &= s_2(d_{44} + \hat{b}_{44}\kappa_2) \\ A_{26} &= -\exp(-s''\kappa) \end{aligned} \tag{39}$$

Non-trivial solutions of (38) exist, if the determinant of the matrix of coefficients vanishes. After some lengthy and tedious calculations, which we omit, this matrix can be transformed into triangular form with the aid of the Gaussian method of elimination. Thus we arrive at the following analytical expression for the *dispersion relation*:

$$\begin{aligned} &[(B_{33}A_{45} - B_{43}A_{32})B_{24} - (B_{23}A_{45} - B_{43}A_{22})B_{34}]\,A_{16} + \\ &+ [(B_{13}A_{45} - B_{43}A_{12})B_{34} - (B_{33}A_{45} - B_{43}A_{32})B_{14}]\,A_{26} = 0 \end{aligned} \tag{40}$$

where the abbreviations

$$\begin{aligned} B_{13} &:= A_{13}A_{51} - A_{11}A_{53} \;,\; B_{23} := A_{23}A_{51} - A_{21}A_{53} \;,\; B_{33} := A_{33}A_{51} - A_{31}A_{53} \\ B_{43} &:= A_{43}A_{51} - A_{41}A_{53} \;,\; B_{14} := A_{14} - A_{12} \;,\; B_{24} := A_{24} - A_{22} \;,\; B_{34} := A_{34} - A_{32} \end{aligned} \tag{41}$$

have been used. The *amplitude factors* in (23),(24),(26) and (27) are finally computed as

$$\begin{aligned} &\bar{B} = \frac{B_{43}A_{34} - B_{33}A_{45}}{B_{34}A_{45}A_{53}}\bar{A} \;,\; \bar{C} = -\frac{A_{51}}{A_{53}}\bar{A} \;,\; \bar{D} = \frac{B_{43}(B_{34} - A_{34}) + B_{33}A_{45}}{B_{34}A_{45}A_{53}}\bar{A} \;, \\ &A' = \frac{B_{43}}{A_{45}A_{53}}\bar{A} \;,\; A'' = \frac{(B_{13}A_{45} - B_{43}A_{14})B_{34} - (B_{33}A_{45} - B_{43}A_{34})B_{14}}{B_{34}A_{45}A_{53}A_{16}}\bar{A} \;. \end{aligned} \tag{42}$$

5. RESULTS

The presented results were obtained numerically on a SIEMENS 7780-computer of the "Rechenzentrum der Universität Karlsruhe".

The thickness of the dielectric layer was assumed as $h = 10^{-3}$ cm. As dielectric material *sodium chloride (NaCl)* was chosen throughout with [8] $c_{44} = 1.266 \cdot 10^6$ N/cm^2; $\varrho = 2.164$ g/cm^3; $b_{44} = b_{77} = 3.44 \cdot 10^{-2}$ Ncm^4/C^2; $d_{44} = -1.7 \cdot 10$ Ncm/C; $a = 1.74 \cdot 10^4$ Ncm^2/C^2. For the purely elastic substrates the calculations were carried through for the following materials [9] (piezoelectric properties neglected): *silicon* ($c'_{44} = 7.96 \cdot 10^6$ N/cm^2; $\varrho' = 2.33$ g/cm^3), *fused quartz* ($c''_{44} =$

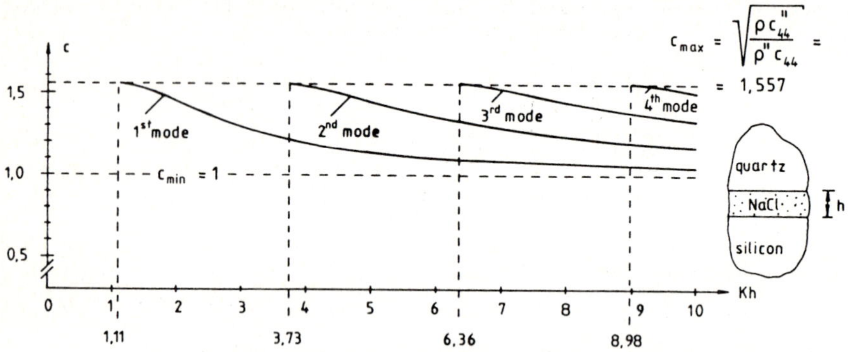

FIGURE 2
Dispersion curves in the non-symmetric, "classical" case.

$3.12 \cdot 10^6$ N/cm^2; $\varrho'' = 2.2$ g/cm^3) and a *fictitous material* (cf. [6]) with $c'''_{44} = 1.168 \cdot 10^6$ N/cm^2; $\varrho''' = 2$ g/cm^3. Whereas fused quartz is isotropic, NaCl and silicon both belong to the cubic crystal system (class m3m). It can be shown, however, that for such materials the field equations, presented above, remain unchanged.
Fig. 2 shows the dispersion curves for the combination silicon-NaCl-fused quartz (non-symmetric, "classical" case). The cut-off value $\kappa = Kh = 1.1097$ which corresponds to the angular frequency $\omega = 4.179 \cdot 10^8$ s^{-1}, is identical with the value computed for the totally elastic medium.

The dispersion curve of the first mode for the combination fictitous continuum-NaCl-fictitous continuum (symmetric, "non-classical" case) is pictured in Fig. 3. The cut-off value $\kappa = Kh = 0.533$ corresponds to $\omega = 1.17 \cdot 10^8$ s^{-1}. The accompanying displacement and polarization profiles prove that this wave resembles much a transverse *bulk* wave (very triflingly decreasing amplitudes into the half-spaces).

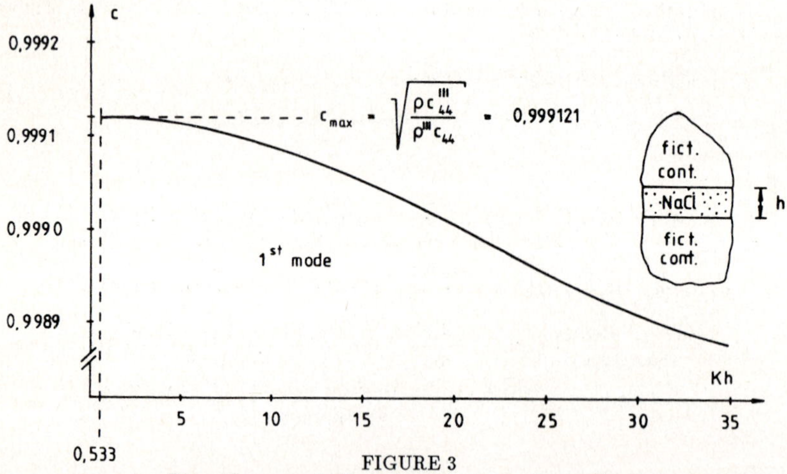

FIGURE 3
Dispersion curve in the symmetric, "non-classical" case.

The displacement and polarization profiles which belong to the first three modes of the non-symmetric, "classical" case (Fig. 2) for the value $\kappa = Kh = 10$, are presented in Fig. 4 ($\hat{u}_3$ and $\hat{P}_3$ denote normalized deflexions). For a given value of κ the wave penetrates the deeper into the half-spaces the higher the mode-number, whereby the penetration into the stiffer medium is less pronounced.

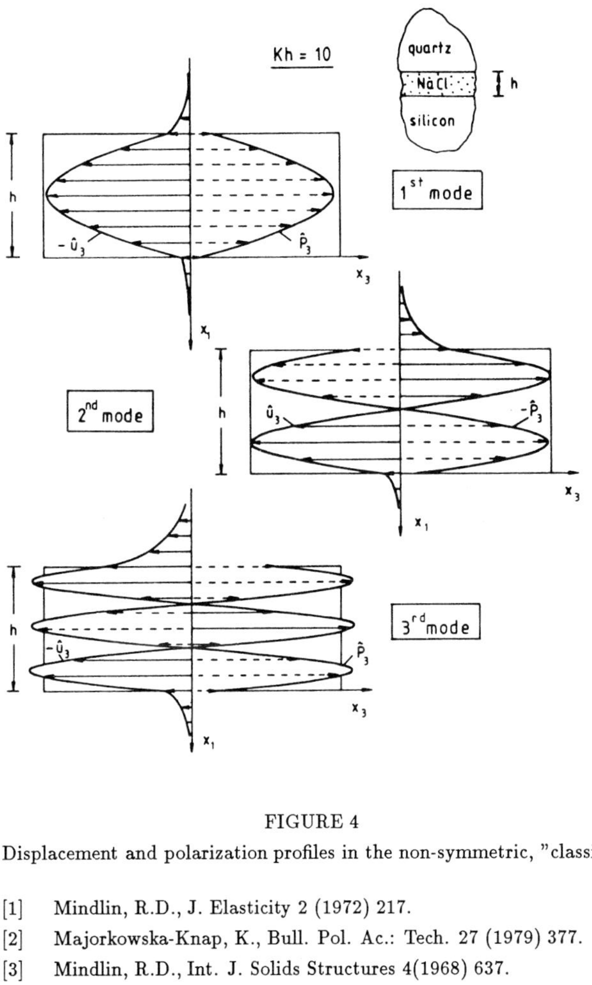

FIGURE 4

Displacement and polarization profiles in the non-symmetric, "classical" case.

A numerical analysis shows that the phase velocity depends only very weakly on the thickness of the layer (in the symmetrical, "classical" case the variation of c is less than 1‰ in the wide range $10^{-5} \leq h[\text{cm}] \leq 10^{-1}$). This also holds for the dependency of c on d_{44} for smaller, and on $\hat{b}_{44}$ for higher than the values given above.

The authors wish to thank S. Schade and H. Göhr for their assistance in the numerical work and in the preparation of the figures.

[1] Mindlin, R.D., J. Elasticity 2 (1972) 217.

[2] Majorkowska-Knap, K., Bull. Pol. Ac.: Tech. 27 (1979) 377.

[3] Mindlin, R.D., Int. J. Solids Structures 4(1968) 637.

[4] Voigt, W., Lehrbuch der Kristallphysik (Johnson Reprint Co., New York; Teubner, Stuttgart, 1966).

[5] Toupin, R.A., J. Rat. Mech. Anal. 5 (1956) 849.

[6] Majorkowska-Knap, K. and Lenz, J., Springer Series on Wave Phenomena, in print.

[7] Farnell, G.W., Types and Properties of Surface Waves, in: Oliner, A.A. (ed.), Acoustic Surface Waves (Springer, Berlin, New York, 1978) pp. 13–60.

[8] Askar, A., Lee, P.C.Y. and Cakmak, A.S., Int. J. Solids Structures 7 (1971) 523.

[9] Slobodnik, A.J., Conway, E.D. and Delmonico, R.T., Microwave Acoustics Handbook, Vol. 1A (Air Force Cambridge Res. Lab., Hanscom AFB, Mass., 1973).

ELASTIC WAVE PROPAGATION
M.F. McCarthy, M.A. Hayes, (Editors)
Elsevier Science Publishers B.V. (North-Holland), 1989

ATTENUATION OF ELASTIC SURFACE WAVES PROPAGATING ALONG ROUGH SURFACES

M. de BILLY and G. QUENTIN

Groupe de Physique des Solides - Université Paris 7 - Tour 23 -
2 place Jussieu 75251 Paris Cedex 05 - France

The attenuation of elastic surface waves is investigated on one and two dimensional rough surfaces whose the root mean square - or roughness - h varies from 8.5 to 90 μm. These values satisfy the condition of validity of the perturbation theory model $\varepsilon = h/\lambda_R << 1$ developed by A.G. Eguiluz and A.A. Maradudin [Phys. Rev. B - 28, 728-747 (1983)] The experimental results and the theoretical calculations show a reasonably good agreement and confirms the frequency dependence of the attenuation for low values of ka (a represents the autocorrelation length).

1. INTRODUCTION

The attenuation of surface acoustic waves propagating across randomly rough surfaces has been studied theoretically assuming that the surface is one or two-dimensionally rough [1-5]. If the wavelength is longer than the transverse autocorrelation length, the attenuation varies proportionnaly to the fifth or fourth power of the frequency according to the configuration of the non uniform boundary surfaces. The principal mechanism which is responsible of this attenuation is due to the scattering into bulk elastic waves.
In this paper we study the attenuation of Rayleigh surface waves which propagate on different surfaces : polished plane surfaces, uneven surfaces on which random gratings were grooved (one-dimensional rough surfaces) and randomly rough surfaces (two dimensional rough surfaces). The experimental results that we report here have been initiated at the laboratory by Baron [6, 7]. The data are compared with the theoretical calculations obtained by Maradudin et al [1-3] in the limit that the incident wavelength is larger than the transverse correlation length. The agreement is reasonably good if we consider that the results depend on the actual statistical properties of the investigated samples which are difficult to evaluate.

2. EXPERIMENTAL PROCEDURE

A classical technique equivalent to a liquid-wedge method was used for the generation and the detection of the surface acoustic wave (SAW). Two small tanks filled up with water are attached with a very thin coupling film to the semi-infinite elastic solid on which measurements are done. This technique insures rather good reproducible results. The intermediate zone between the two tanks is in contact with air and the SAW will be only attenuated by the interactions of the wave with internal discontinuities in the solid.
The angles of incidence of the two transducers (emitter and receiver) satisfy the relationship $\sin \theta_i = C_{water}/C_{Rayleigh}$ (where C_{water} = 1480 m/s is the wave velocity in water and $C_{Rayleigh}$ the Rayleigh wave velocity). The experimental procedure for attenuation measurements by uneven surfaces consists of calculating the quantity $\alpha = 20/L \log A_t/A_o$ (in dB/cm). L is the length of the corrugated surface crossed by the SAW ; A_t and A_o are respectively the amplitudes of the transmitted signal with and without irregularities for any frequency with the same value of D and the same experimental set-up. The

amplitude A_o will be used as reference level. This normalization eliminates the absorption and diffraction effects. The measurements were taken at discrete frequencies F (2 < F < 10 MHz) and the pulse length was long enough (8 μs) to consider the excitation of the transducer as monofrequential. The transmitted signal is detected, amplified and visualized on the screen of an oscilloscope for recording the amplitude variations. Typical signal are given in figure 1 in case of a duraluminum sample with a rough surface the roughness of which is h = 43 μs.

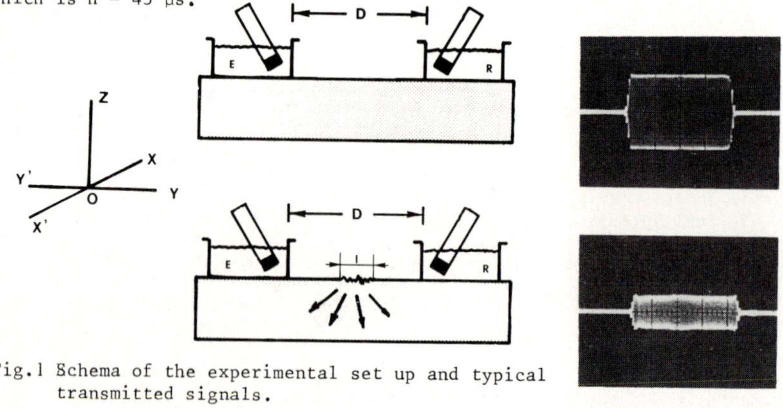

Fig.1 Schema of the experimental set up and typical transmitted signals.

To get the influence of the roughness on the variations versus the frequency, of the attenuation α_o of the surface acoustic wave in case of a plane surface, we recorded the amplitudes of the received signals observed for different distances D and D' between the transducers. The attenuation is then given by the formula :

$$\alpha_o = \frac{20}{D'-D} \log \frac{A'_o}{A_o} \text{ (dB/cm).}$$

3. EXPERIMENTAL RESULTS

In this paper we investigate the attenuation of SAW by plane and partially one or two-dimensional rough surfaces. Various samples with different values of the correlation length L and roughness h were prepared.

3. 1. Attenuation by plane surfaces

Experimental measurements were achieved on duraluminum samples. The results are given in figure 2 where are also indicated the variations of α_o measured by Viktorov [1] and Urazakov [5] versus the incident frequency. The agreement is reasonnably good. Our results do not take into account the divergence of the acoustic beam which can be important at low frequency. These data show that the behavior of the attenuation (expressed in dB/cm) for a plane plate can be described by a linear function (continuous line in figure 2). The differences of the amplitudes are very small, so we needed to use a quite long distance between the two small tanks.

3.2. Attenuation by rough surfaces

The measurements reported here were obtained by recording the ratio of the amplitudes of the received signals as the incident wave propagates over the rough and the plane surface with the same experimental conditions. This normalization eliminates the beam diffraction effects and the attenuation by the

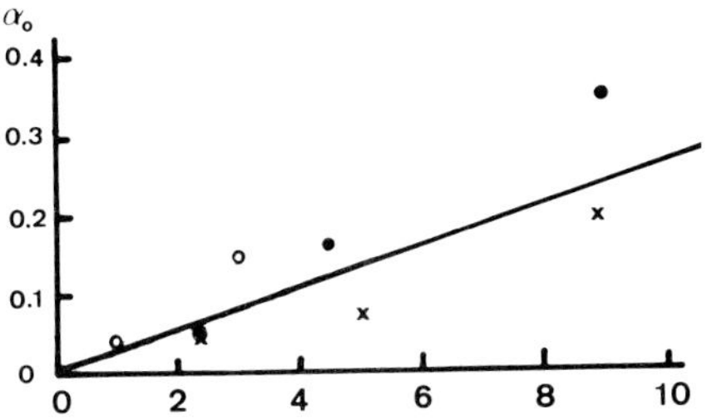

Fig.2 Variation of the attenuation versus the frequency(in MHz).Case of flat surface.

elastic medium. The uneven part of the samples was one or two dimensional randomly roughened. The one-dimensional samples were prepared by engraving grooves with random depths and periodicities but without any variation in the direction perpendicular to the sagittal plane. The two-dimensional rough samples were obtained by hitting the middle of the investigated surface whith the corner of a hammer.

3.2.1. Statistical properties of the samples

From discrete records of the profiles of rough samples we calculated the actual height distribution , the root mean square value (h or roughness) and the auto-correlation length (a). The roughness ranges from 8 to 90 μm and the mean slope (h/a) varies from 2.6 to 14.1.The statistical parameters which measure respectively the symmetry of the profile about its mean (skewness) and the sharpness of the amplitude density function (kurtosis) confirm that the height distribution are not perfectly Gaussian.The statistical properties of the investigated samples are given in table 1.

TABLE 1. Statistical properties of the samples

Sample (identification)	h(μm)	a(μm)	a/h
Duraluminum (M_2)	18	100	5
Duraluminum (M_3)	43	120	2.8
Duraluminum (M_4)	85	220	2.6
Duraluminum (D_0)	28	180	6.4
Duraluminum (D_1)	54	200	3.7
Duraluminum (D_2)	9	125	14.1
Duraluminum (D_3)	20	185	9.2
Duraluminum (D_4)	85	240	2.8
Titanium T_1	$\simeq 0$	-	-
Titanium T_2	8.5	120	14
Titanium T_3	17	220	13
Titanium T_4	45	280	6.2

3.2.2. Results for one-dimensional rough surfaces

The measurements were done on duraluminum samples (M_i - see table 1) at frequencies ranged in the 2-8 MHz interval. The experimental data are given in figure 3 where are plotted the variation of w_2 (see § 4) versus the normalized correlation length a $q_{//}$.

3.2.3. Results for two - dimensional rough surfaces

In figures 4 and 5 are given the experimental curves obtained respectively for duraluminum and titanium samples.They show the variations of α(in dB/cm) versus the frequency for different values of the roughness h. Notice that the attenuation increases drastically with the frequency F for both samples. A brief analysis of the transmitted beam profiles in the X'OX and Y'OY directions (see figu-

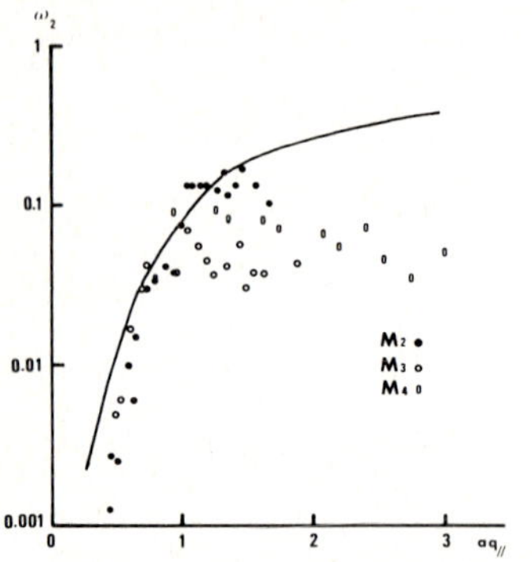

Fig.3 Variations of the function w_2 versus the normalized correlation length.

re 1) confirms that, as the wavelength of the incident waves has value closed to the roughness, the scattering predominates and the profiles are less and less symmetric. This is illustrated in figure 6.

4. THEORETICAL APPROACH and DISCUSSION

The first theoretical formulation of the attenuation phenomon were proposed by Urazakov and Fal'kovskii [5] and by Maradudin and Mills [1]. More recent papers devoted to this problem were carried out by Equiluz and Maradudin [2] for a two-dimensional problem and by Huang and Maradudin [3]for a one-dimensional roughness.

If the roughness h is small in comparaison with the wavelength, the authors show that we may write $w_R(q_{//}) = w_0(q_{//}) + \Delta w(q_{//})$ were Δw ($\Delta w = \nu_1 - i\nu_2$) is a pertubation term the imaginary part of which can be related to the attenuation length l of the surface wave by the equation :

$$l^{-1} = \frac{2\nu_2}{C_R} = 2\left(\frac{h}{a}\right)^2 q_{//}\, w_2(aq_{//}) \simeq 2\left(\frac{h}{a}\right)^2 q_{//}\, w_2(aq_{//})$$

l is defined as the distance at which the energy of the Rayleigh wave falls to $^1/e$ of its initial value).

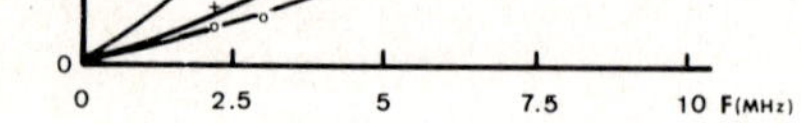

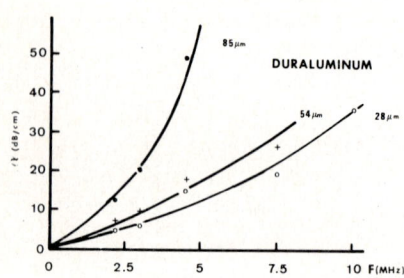

Fig.4 Variations of the attenuation versus the frequency for different values of the roughness. Case of a duraluminum sample.

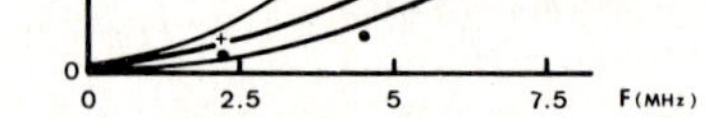

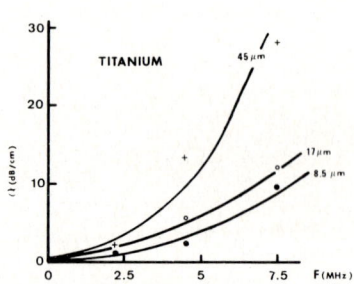

Fig.5 Variations of the attenuation versus the frequency for different values of the roughness. Case of a titanium sample.

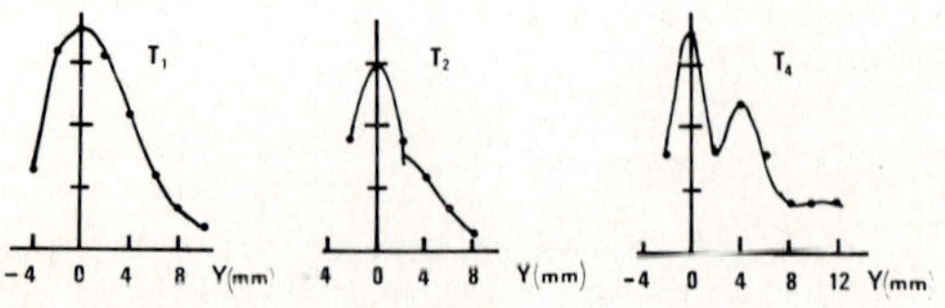

Fig.6 Transmitted beam profiles observed in Y'OY direction as function of the roughness: $h_1 = 0$ μm, $h_2 = 8.5$ μm and $h_4 = 45$ μm.

For small values of $aq_{//}$ the authors pointed out that the frequency dependance of w_2 varies according to the dimension of the uneven surface. The behaviour of the function w_2 is proportional to F^4 and F^5 respectively for one and two dimensional rough surface. The predominant factor due to the attenuation rate comes from its scattering into bulk waves.

In figure 3 we have plotted (continuous line) the theoretical varations of w_2 versus the dimensionless number $aq_{//}$ [3]. For small values of $aq_{//}$ the theoretical plot seems to be shifted in comparaison with experimental data. For higher $aq_{//}$ values the disagreement is important because probably the theoretical approach is only valid for values of $aq_{//} \lessapprox 1$. From these results a limit of the ratio $\varepsilon = h/\lambda_R$ is pointed out for agreement between theory and experiences : $\varepsilon \simeq 0.4$.

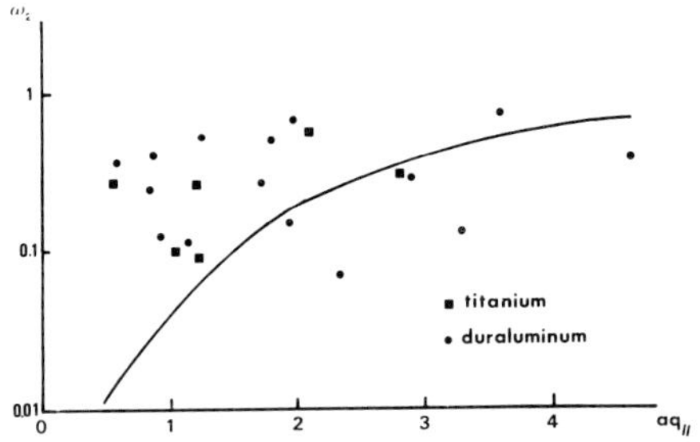

Fig.7 Variations of the function w_2 versus the normalized correlation length.(from Ref 7)

In figure 7 are plotted the variations of a two-dimensional rough surfaces (continuous line). The experimental results (points) predicts an attenuation approximately ten times higher than the two dimensional modes predictions. These results indicate that the damping of SAW by randomly rough surface is much larger for two-dimensional roughness than for surfaces exhibiting one-dimensional irregularities. The discrepancies noted for $aq_{//} < 1$ in figure 7 between theoretical and experimental results can be explained by the non verification of the theoretical hypothesis (gaussian height dis-distribution and autocorrelation functions).

5 CONCLUSION

This experimental study confirms the great dependance of the attenuation of SAW by irregularities on flat surfaces. The results are characterized by a good agreement with theory for small values of $aq_{//}$ and offer a large dipersion of the data for higher values in both configuration of the rough surface (one and two-dimensional). This experimental technique should be extended to samples with cracks into randomly rough surface.

REFERENCES

[1] A.A. Maradudin and D.L. Mills. "Attenuation of Rayleigh surface waves by surface roughness". Ann. Phys. 100, 262 - 309 (1976)

[2] A.G. Eguiluz and A.A. Maradudin. "Frequency shift and attenuation length of a Rayleigh wave. Phys. Rev B - 28, 728 -747 (1983)

[3] X. Huang and A.A. Maradudin. "Propagation of surface acoustic waves across random gratings". Submitted for publication to Phys. Rev B.

[4] I.A. Viktorov. "Damping of surface and spatial ultrasonic waves" Sov. Phys. Acoust. 10, 91 - 92 (1964)

[5] E.I. Urazakov and L.A. Fal'kovskii "Propagation of a rayleigh wave along a rough surface". Sov. Phys. JETP. 36, 1214 - 1216 (1973)

[6] E. Baron. Etude de l'interaction d'une onde de Rayleigh avec des discontinuités superficielles. Thèse de 3ème Cycle - Paris 1986.

[7] M. de Billy, G. Quentin and E. Baron. "Attenuation measurements of an ultrasonic Rayleigh wave propagating along rough surfaces. J. Appl, Phys. 61, 2140 - 2145 (1987)

ACKNOWLEDGMENTS

This study has been supported by the D.R.E.T. (contract 83 - 070)

ELASTIC WAVE PROPAGATION
M.F. McCarthy, M.A. Hayes, (Editors)

Ceramic-Ceramic Composites High Temperature Characterization

Bernard Hosten and Stéphane Baste*

Laboratoire de Mécanique Physique
Université de Bordeaux I U.A. C.N.R.S.867
351 cours de la Libération 33405 Talence France.

* Société Européenne de Propulsion
Les 5 Chemins Le Haillan B.P.37 33165 St Médard En Jalle France.

Abstract.

Ceramic/Ceramic composites like carbon/carbon, silicon-carbide/ silicon-carbide (Sic/Sic), ..., are made to work at very high temperatures (1000°-2000° C), for which elastic characterization is not easy to obtain with classical mechanical tests.

An ultrasonic method is under way to measure phase velocities of longitudinal and transverse waves. Waves are launched in sample, through a thermal buffer which let us to use commercial, room temperature transducers. Time delays, between interface echos and free surface echos are precisely measured by means of appropriate signal processing. Preliminary measurements of C_{11} elastic coefficients in the range 100°C/700°C will be given for two dimensional Sic/Sic composites (CeraSep^{R}) produced by S.E.P.

I Introduction.

Measurements of high temperatures elastic moduli in solids are easily obtained by means of ultrasonic method using a shouldered region [1,2]. Other methods use a waveguide with ka small (a : section of sample and k: wave number) in order to measure Young's modulus [3].

In this paper, we describe an ultrasonic method of which the long term purpose is the measurements of stiffness coefficients of composites in the range 1000°C/2000°C. Because samples are thin plates (few mm thick), the shoulder method does not fit and in order to measure separatly the evolutions of each stiffness coefficient, it is necessary to use high ka.

We present preliminary results for Silicon Carbide/Silicon Carbide in the range 100°/700°C of an ultrasonic method based on a graphite buffer.

II Silicon Carbide-Silicon Carbide (Sic/Sic) Composites.

The method is tested with 2D Sic/Sic composites produced by Société Européenne de Propulsion (France). The composite is processed by superimposing clothes made from woven silicon carbide nicalon fibers, densified with a silicon carbide matrix. They are prepared by low pressure chemical vapor infiltration technology (CVI) .

III Thermal Buffer.

The buffer method is based on times of flight (TOF) measurement for buffer/sample interface echo 1 and free surface echo 2 [Fig.1].

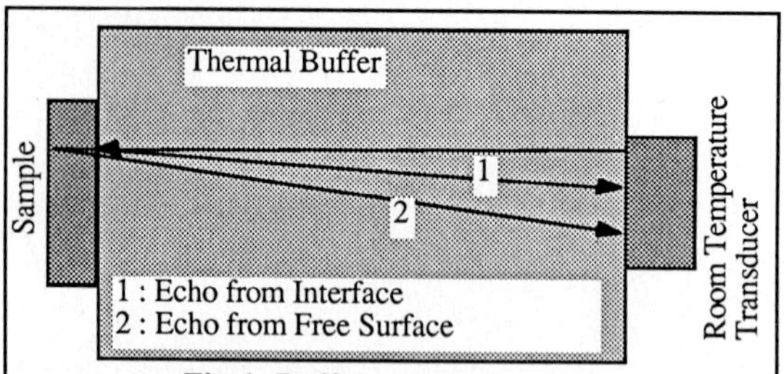

Fig.1 Buffer method.

Table 1 : Graphite acoustical properties.

Density :	**1.783**	
Longitudinal velocity :	**3080 ± 5**	**M/S**
Shear velocity :	**1680 ± 10**	**M/S**

The main problems of this method are :

1- to find a buffer material of which the impedance gives an appropriate energy repartition between the two echos and have stable properties at high temperatures.

2- to use a non-porous, high temperature binder between sample and buffer.

In vacuum, Graphite can be easily used in the range 1000/ 2000°C.

The acustical properties of the graphite buffer we used, were measured and are given in table 1. These values give a good ratio between graphite and Sic/Sic impedances. (see § V).

The buffer is a cylindrical rod, the length of which is 296 mm and the smaller diameter is d = 50 mm. At 1 Mhz, the product kd (= 186) for shear waves, is very large in buffer and the unlimited medium hypothesis is very well verified.

IV Room temperature elastic constants of SIC-SIC.

A Sic/Sic sample was previously characterized at room temperature by means of an immersion ultrasonic test bed [4] for composites. Results are given in table 2, for which 1 means the normal sample direction (weak direction). The 2 and 3 directions define the clothes plane, with identical elastic properties.

<u>Table 2 : Elastic coefficients of SIC-SIC sample.</u>

Dimensions 5.30 x 50 x 50 mm Density = 2.61

C_{11} = 61 GPa	C_{22} = 206 GPa
C_{12} = 19 GPa	C_{66} = 39 GPa
v_{l11} = 4834 m/s	v_{t11} = 3865 m/s

Fig.2 Stress-Strain Curve in the direction 2.

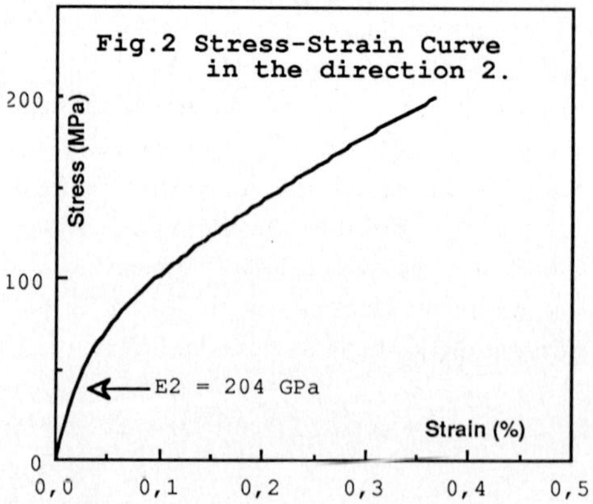

Sic/Sic are built from brittle components and have the typical stress-strain curves [5] showed in figure 2. A shear wave velocity close to the longitudinal wave velocity is a characteristic of ceramic-ceramic composites [6].

V Graphite/Sic-sic Interface.

With a perfect adhesion, the stress reflection coefficient is computed by the formula $R_{GS} = (Z_S - Z_G) /(Z_S + Z_G)$ where Z_S and Z_G are acoustical impedances of sample and graphite. With data given in tables 1 and 2 :

$R_{GSl} =$ 0.41 for a longitudinal wave $R_{GSt} =$ 0.55 for a shear wave.

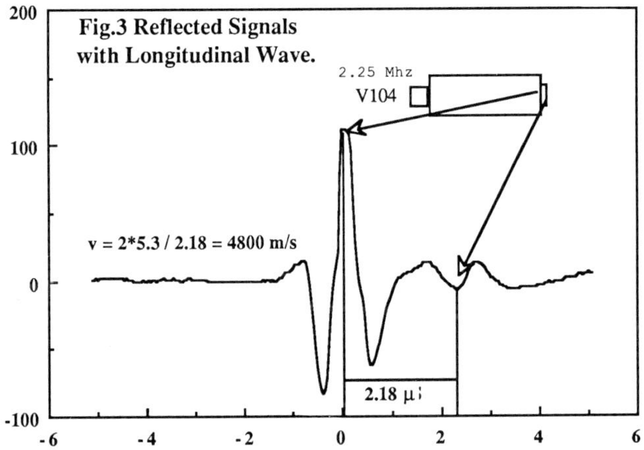

Fig.3 Reflected Signals with Longitudinal Wave.

With longitudinal transducer (Panametrix V104 2.25 Mhz) and low temperature glue, separated, quasi-equal amplitudes echos are produced.

The high temperature glue must maintain these values to give a good energy repartition. Many ceramic high temperature glues were tested. They became very porous during the cure. At this time, only one, based on Zirconia, gives satisfaction although the transmission is still low. The signals acquired in these conditions are shown in figure 3.

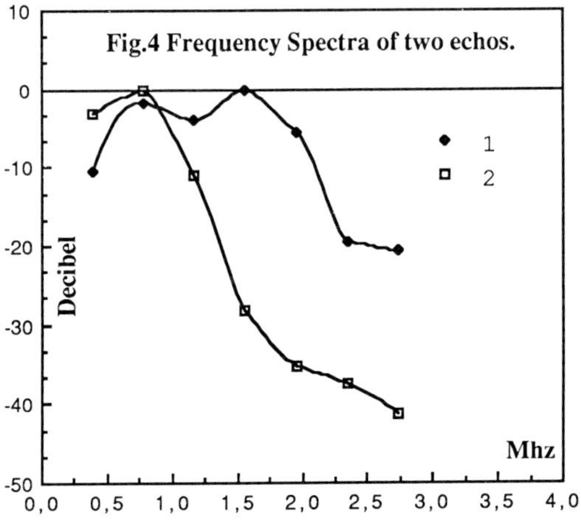

Fig.4 Frequency Spectra of two echos.

The frequency spectra [Fig.4] of echos are different. In graphite, echo 1 has a spectrum not very different than spectrum of input signal. But echo 2, after round-trip in sample, has a cut off frequency around 1 Mhz.

Hence TOF are not precisely measured by level method. So with appropriate temporal windows, signals are separated and acquired in micro-computer. Because of free surface reflection coefficient, the echo 2 is multiplied by -1 and the TOF difference is computed from intercorrelation and filtering around 1 Mhz frequency.

At room temperature, the measured velocity differs less than 1% of the value measured by immersion method.

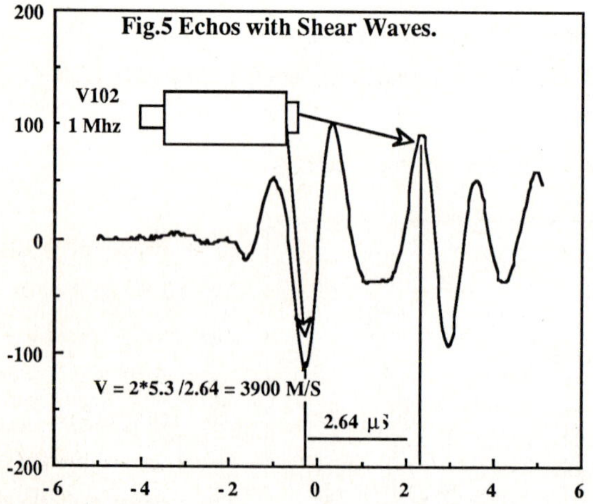

Fig.5 Echos with Shear Waves.

Figure 5 shows the two echos obtained with room temperature glue and shear wave at 1 Mhz. For high temperatures, we tried ceramics glues with little success until now.

VI Measurements set up and results.

To avoid oxydation of graphite, furnace and buffer are placed in a vacuum chamber [Fig.6].

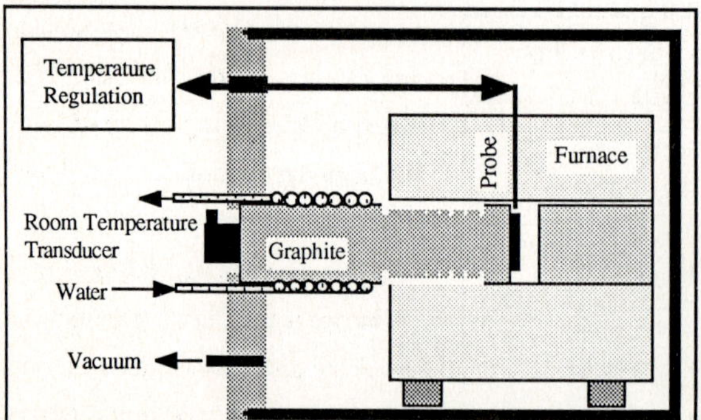

Fig.6 Experimental arrangement

The cooling system imposes room temperature to the ultrasonic transducer. An electronic device and a probe (Pt-Rh thermocouple) placed very close to the sample, regulate high temperatures in a small volume. The graphite buffer-rod is threaded to let a better contact between cooling tube and to avoid spurious signals by modes conversion.

VII Preliminary results.

Figures 7 and 8 show the evolution of longitudinal velocity and stiffness coefficient in the weak direction.

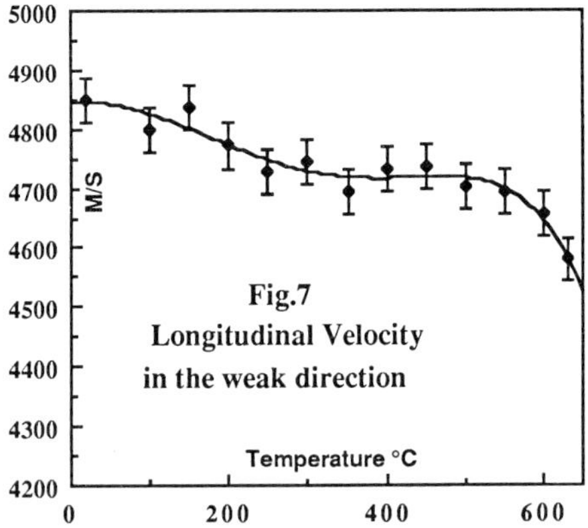

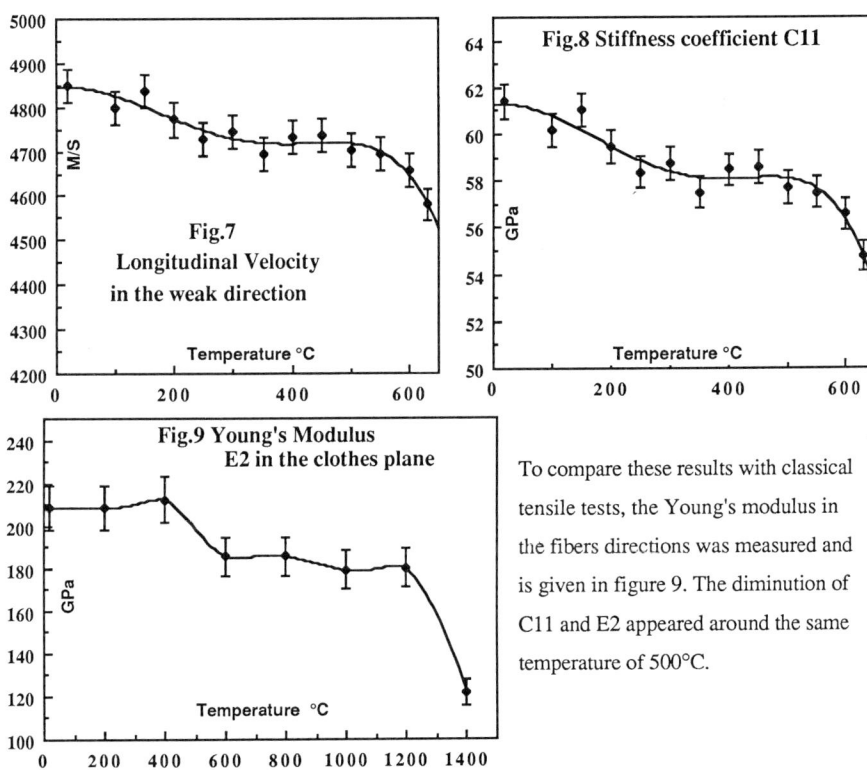

Fig.7 Longitudinal Velocity in the weak direction

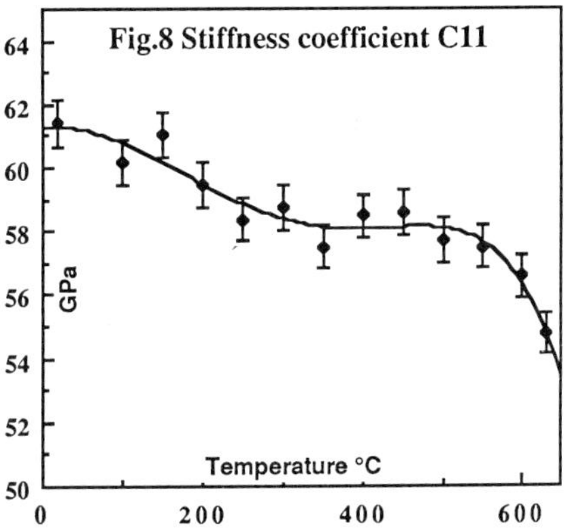

Fig.8 Stiffness coefficient C11

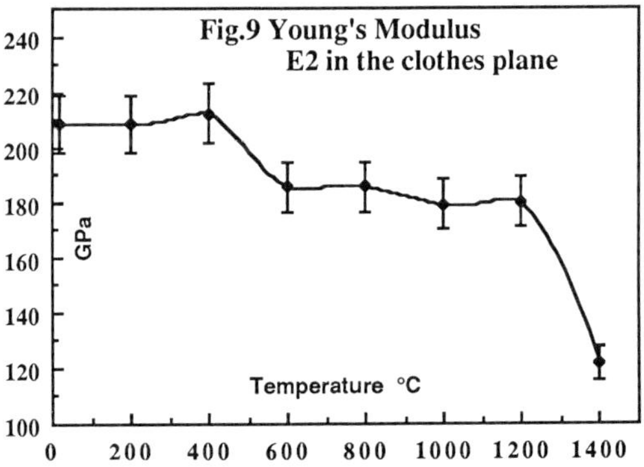

Fig.9 Young's Modulus E2 in the clothes plane

To compare these results with classical tensile tests, the Young's modulus in the fibers directions was measured and is given in figure 9. The diminution of C11 and E2 appeared around the same temperature of 500°C.

VII Conclusions.

The elastic constants in the out-plane direction are very difficult to measure by classical mechanical tests. The ultrasonic system presented in this paper can measure these constants in front of temperature and we presented results for thin Sic/Sic samples.

The temperature range (100°/700°C) of this system is now only limited by the furnace and the cooling system we used and will be enhanced soon.

More effort would be necessary for shear wave transmission.

References :

[1] "Elastic moduli of solids - a method suitable for high temperature measurements." M W Grigg,T J Davis, A.G.Cimmino, A.G.Klein and G.I.Opat J.Phys.E:Sci.Instrum.19(1986).

[2] "Longitudinal Ultrasonic Wave Velocity and Attenuation Dependence on temperature Between 20° and 1374 °C" R.Fenn and J.J.Wooton Nondes.Testing Communications, 1986, Vol.2 pp 115-126.

[3] "Improved ultrasonic measurement of Young's modulus at high temperatures in composites ceramics:application to refractory concretes" P.Lamidieu and C.Gault Mat.Sci.and Engi.,77(1986) L11-L15.

[4] "Inhomogeneous wave generation and propagation in lossy anisotropic solids. Application to the viscoelastic characterization of composite materials". B.Hosten, M.Deschamps and B.R.Tittmann. J.Acoust.Soc.Am.,Vol.82, N.5,November 1987.

[5]"Fiabilité des composites céramiques-céramiques" G.Bernhart, P.Lamicq, J.Mace L'industrie céramique, n° 790,1/85.

[6] "Elastic anisotropy of carbon_carbon composites during the fabrication process". B.Hosten and B.R.Tittmann. Proceeding of Ultrasonics Symposium (IEEE 1986) Williamsburg Nov.17-19 1986.

ELASTIC WAVE PROPAGATION
M.F. McCarthy, M.A. Hayes, (Editors)

GROUND MOTION AMPLIFICATION ON ALLUVIAL VALLEYS

Anders BOSTRÖM[a†], Subhendu K. DATTA[b*], and Peter OLSSON[a†]

[a]Division of Mechanics
Chalmers University of Technology
S- 412 96 GÖTEBORG
Sweden

[b]Department of Mechanical Engineering and CIRES
University of Colorado
Boulder, CO 80309-0427
USA

Seismic ground motion amplification on alluvial valleys is studied in this paper. The alluvial valley is modelled as an elastic inhomogeneity in an elastic half-space. To solve for the ground motion amplification on such an inhomogeneity the null field approach is adopted. This requires modifications of the null field approach for an inhomogeneity inside a half-space and various ways of doing this are discussed.

1. INTRODUCTION

Ground motion amplification due to local topographic and geologic features has been studied extensively in recent years. It is well documented that significant dynamic amplification can occur on alluvial valleys and canyons. Estimates of maximum amplitudes of ground displacements are important for seismic resistant design of structures. A review of the many recent studies of this problem can be found in [1].

Because of the complexity of the problem most of the theoretical studies have been confined to two-dimensional cases, although some three-dimensional cases have also been investigated [2,3]. In both of these studies displacements were expressed in power series or in terms of spherical wave functions appropriate for an infinite medium and the stress free boundary conditions on the surface of the half-space were satisfied approximately.

In the present paper the null field approach (T matrix method) that has previously been used for buried inhomogeneities [4,5,6] is extended to the problem of a three-dimensional alluvial valley.

†Sponsored by the National Swedish Board for Technical Development (STU)

*Supported in part by grants from the National Science Foundation (INT-8610487 and ECE-8518604)

To our knowledge this is perhaps the only complete solution to this difficult problem. As a way of illustration results for an incident Rayleigh wave on a soft semi-spheroidal shaped valley are presented. The cases of more generally shaped valleys and other incident wave types will be reported in a subsequent publication.

2. THE NULL FIELD APPROACH

Consider an elastic half-space with an alluvial valley as depicted in Fig. 1. The interface between the half-space and the valley is denoted S_1, the surface of the valley is S_{01}, and the rest of the planar surface of the half-space is S_0. The outward-pointing normals are shown in Fig. 1. An origin is chosen inside the valley and both spherical and rectangular coordinates are used. The material parameters and the wavenumbers in the half-space are ρ, λ, μ, k_p, and k_s and in the valley the corresponding quantities are ρ_1, λ_1, μ_1, k_{p1}, and k_{s1}.

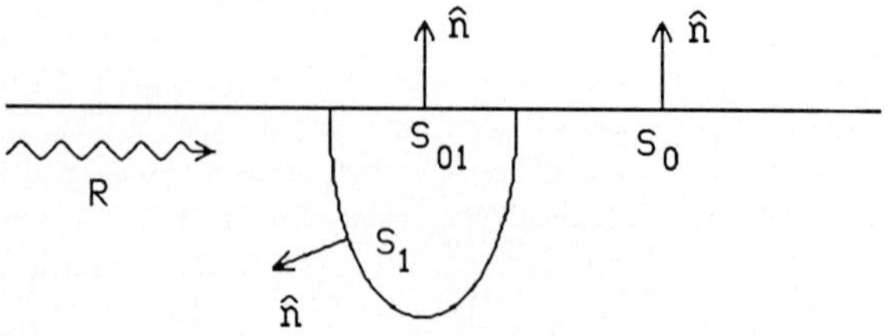

FIGURE 1
The geometry.

In the null field approach the starting point is the following integral representation

$$\mathbf{u}^i + k_s/\mu \int_{S_1 - S_0} [\mathbf{u} \cdot (\hat{n} \cdot \Sigma) - \mathbf{t} \cdot G]\, dS = \begin{cases} 0 & \text{above } S_0 \text{ and } S_1 \\ \mathbf{u} & \text{below } S_0 \text{ and } S_1 \end{cases} \tag{1}$$

where **u** is the displacement field, **t** the traction, G the Green tensor, and $\hat{n} \cdot \Sigma$ the surface traction Green tensor. The boundary conditions that the traction is zero on S_0 and that the displacement and traction are continuous on S_1 are immediatly used in (1).

All fields including the Green tensors are to be expanded in spherical and/or plane vector waves. The notations and conventions from [5] are used. Thus ψ_n and $\mathrm{Re}\psi_n$ are outgoing and regular spherical waves where n is a quadruple index and $\varphi_\uparrow$ and $\varphi_\downarrow$ are plane waves where the arrows stand for one mode index and two continuous parameters giving the direction of propagation upwards or downwards (including evanescent waves). The corresponding waves in the valley are ψ_n^1, $\mathrm{Re}\psi_n^1$, $\varphi_\uparrow^1$, and $\varphi_\downarrow^1$. The summation (and integration) convention that a repeated n (arrow) should be summed (integrated) over is used. Thus the transformations between the plane and spherical waves

$$\varphi_{\uparrow} = \mathrm{Re}\psi_n B^{\dagger}_{n\uparrow} \tag{2}$$

$$\psi_n = 2\varphi_{\uparrow} B_{\uparrow n} \tag{3}$$

contain a quadruple sum and a double integration and mode sum, respectively. The transformation functions $B^{\dagger}_{n\uparrow}$ and $B_{\uparrow n}$ are essentially components of spherical vector harmonics, see [5].

For the prescribed incoming displacement field two expansions are used:

$$\mathbf{u}^i = \varphi_{\uparrow} a_{\uparrow} \qquad \text{above } S_0 + S_{01} \tag{4}$$

$$\mathbf{u}^i = \mathrm{Re}\psi_n a_n \quad \text{inside } S_1 + S_{01} \tag{5}$$

where the coefficients $a_{\uparrow}$ and a_n are known. The surface fields are expanded as

$$\mathbf{u} = \varphi_{\uparrow} \alpha_{\uparrow} \qquad \text{on } S_0 + S_{01} \tag{6}$$

$$\mathbf{u} = \mathrm{Re}\psi^1_n \beta_n \quad \text{on } S_1 + S_{01} \tag{7}$$

Using the inner integral representation for the valley it is then easily shown that the surface traction on the valley is [7]

$$\mathbf{t} = \mathbf{t}^1(\mathrm{Re}\psi^1_n)\ \beta_n \quad \text{on } S_1 + S_{01} \tag{8}$$

where the same coefficients as in (7) appear. $\mathbf{t}^1(\cdot)$ is the traction operator in the valley [5]. Note that the expansions in (6)-(8) are over a plane and a closed surface, respectively, so as to avoid problems with incomplete or linearly dependent expansion systems.

Expanding also the Green tensors in (1) the same procedure as in [4,5] gives

$$a_{\uparrow} = Q^{S_0}_{\uparrow\uparrow} \alpha_{\uparrow} - 2B_{\uparrow n}\, \mathrm{Re}Q^{S_1}_{nn'} \beta_{n'} \tag{9}$$

$$a_n = B^{\dagger}_{n\downarrow}\, Q^{S_0}_{\downarrow\uparrow} \alpha_{\uparrow} - Q^{S_1}_{nn'} \beta_{n'} \tag{10}$$

where

$$Q^{S}_{\downarrow\uparrow} = 2ik_s/\mu \int_S \mathbf{t}(\varphi^{\dagger}_{\downarrow}) \cdot \varphi_{\uparrow} dS \tag{11}$$

$$Q^{S}_{nn'} = ik_s/\mu \int_S [t(\psi_n) \cdot Re\psi^1_{n'} - \psi_n \cdot t^1(Re\psi^1_{n'})]\, dS \tag{12}$$

and $Q^{S_0}_{\uparrow\uparrow}$ contains upgoing waves in both positions and $ReQ^{S_1}_{nn'}$ contains regular waves in both positions . The problem with (9) and (10) is that the integrals over S_0 are not over a complete plane so that $Q^{S_0}_{\uparrow\uparrow}$ and $Q^{S_0}_{\downarrow\uparrow}$ are "nondiagonal" and (9) and (10) are integral equations for $\alpha_\uparrow$.

To save the situation the so far unused boundary conditions on S_{01} are now used:

$$\varphi_\uparrow \alpha_\uparrow = Re\psi^1_{n'}\, \beta_{n'} \tag{13}$$

$$t^1(Re\psi^1_{n'})\beta_{n'} = 0 \tag{14}$$

To make matrix equations out of these, project (13) onto $2t(\varphi^\dagger_\uparrow) = 2B_{\uparrow n}t(Re\psi_n)$ and (14) onto $2B_{\uparrow n}\, Re\psi_n$ and add:

$$Q^{S_{01}}_{\uparrow\uparrow}\, \alpha_\uparrow = 2B_{\uparrow n}\, ReQ^{S_{01}}_{nn'}\, \beta_{n'} \tag{15}$$

Likewise, project (13) onto $t(\psi_n) = 2B^\dagger_{n\downarrow}t(\varphi^\dagger_\downarrow)$ and (14) onto ψ_n and add:

$$B^\dagger_{n\downarrow}Q^{S_{01}}_{\downarrow\uparrow}\, \alpha_\uparrow = Q^{S_{01}}_{nn'}\, \beta_{n'} \tag{16}$$

If (15) and (16) are added to (9) and (10) exactly the same formalism as for an inhomogeneity inside a half-space [4] is obtained. If $\alpha_\uparrow$ is eliminated from the equations there results:

$$[Q^{S_1+S_{01}}_{nn'} - 2B^\dagger_{n\downarrow}Q^{S_0+S_{01}}_{\downarrow\uparrow}\,(Q^{S_0+S_{01}}_{\uparrow\uparrow})^{-1}\, B_{\uparrow n}\, ReQ^{S_1+S_{01}}_{nn'}]\beta_{n'}$$

$$= -\, a_n + B^\dagger_{n\downarrow}Q^{S_0+S_{01}}_{\downarrow\uparrow}\,(Q^{S_0+S_{01}}_{\uparrow\uparrow})^{-1}\, a_\uparrow \tag{17}$$

which can be solved for β_n and thereby for the displacement field on the alluvial valley. It is also possible to solve for $\alpha_\uparrow$ and then, from the integral representation (1) for the scattered field.

There are possible modifications of the above procedure, but if integral equations involving $\alpha_\uparrow$ and double integrals are to be avoided, the freedom is really very limited. To avoid integral equations it seems that the only possible way is to use the sum of (9) and (15) to solve for $\alpha_\uparrow$ ($Q^{S_0+S_{01}}_{\uparrow\uparrow}$ and $Q^{S_0+S_{01}}_{\downarrow\uparrow}$ are "diagonal"). To avoid double integrals (10) and (16) can not be used separately but only in summed form as in the derivation of (17) above. The only simple extension that avoids the problems is to use the traction-free boundary condition (14) projected on some suitable system, $Re\psi^1_n$ say, to obtain

$$P^{S_{01}}_{nn'}\, \beta_{n'} = 0 \tag{18}$$

$$P^{S_{01}}_{nn'} = \int_{S_{01}} Re\psi^1_n \cdot t^1(Re\psi^1_{n'})dS \tag{19}$$

One procedure would then be to use (17) and (18), suitably truncated, to solve for β_n. This could be done in a least-square sense with more equations then unknowns or in the ordinary sense, and the number of equations taken from (17) and (18) could be varied. Some numerical experimentation indicates that it is best to skip (18) altogether, but this is presently being more fully explored. The numerical examples in the next section use only (17).

3. NUMERICAL RESULTS

For the numerical computations the incoming field is chosen as a Rayleigh wave travelling in the positive x direction with unit normal amplitude. The shape of the alluvial valley is chosen as half a prolate spheriod with axis a in the free surface and the other axis 2a.

The valley is softer than the surroundings with $\rho_1/\rho = 0.8$ and $\mu_1/\mu = 0.45$ and the Poisson ratios are $\nu = \nu_1 = 0.3$. In the figures the absolute value of the vertical displacement u_z and the horizontal displacement u_x are shown as functions of the normalized frequency $k_{s1}a$ at some locations on the valleys.

Figures 2 and 3 show $|u_z|$ and $|u_x|$, respectively, at the three points $x/a = 0.5$, $y/a = 0$, $x/a = -0.5$, $y/a = 0$, and $x/a = 0$, $y/a = 0.5$. At low frequencies the curves go to the values $|u_z| = 1$ and $|u_x| = 0.68$ of the undisturbed Rayleigh wave. With increasing frequency the displacements then generally grow and at the upper end of the frequency band there is some sort of resonance with large displacements. It should be noted that the numerical accuracy is not too good at the highest frequencies; as judged from changing the truncations the errors could be a few per cent.

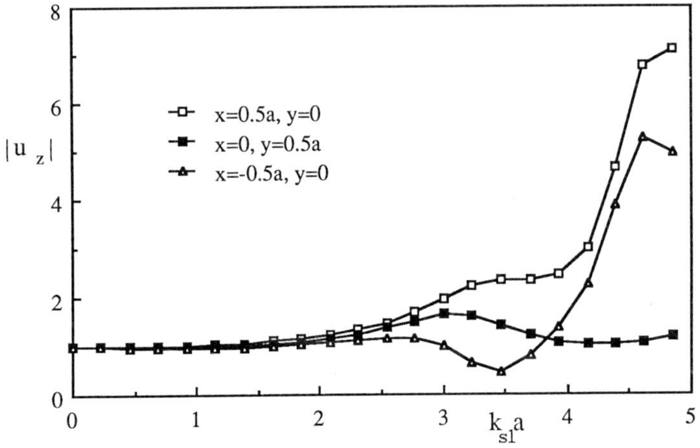

FIGURE 2

$|u_z|$ as a function of frequency at the three points $x/a = 0.5$, $y/a = 0$, $x/a = -0.5$, $y/a = 0$, and $x/a = 0$, $y/a = 0.5$.

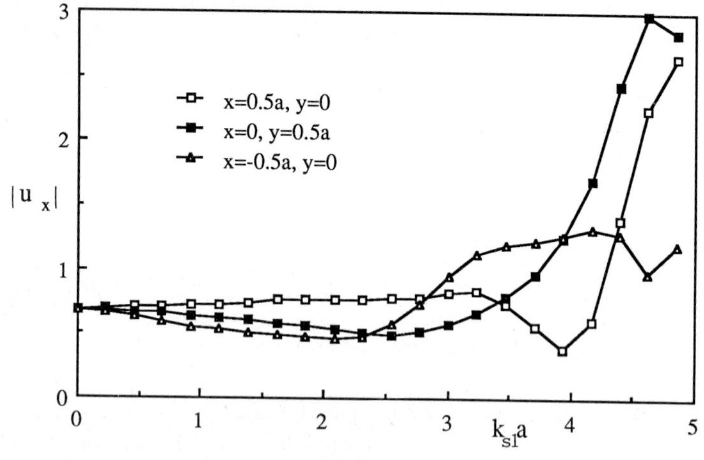

FIGURE 3

$|u_x|$ as a function of frequency at the three points x/a = 0.5, y/a = 0, x/a = − 0.5, y/a = 0, and x/a = 0, y/a =0.5.

The computations shown so far are only to be regarded as preliminary, as there are good reasons to perform further calculations with other incoming waves and other shapes of the valley. Combining the present method with earlier results [5,6] it is also possible to consider an alluvial valley in a layered half-space (but the layering must be below the valley).

ACKNOWLEDGEMENT

S.K.D. gratefully acknowledges the hospitality extended to him at Chalmers University of Technology during his visit there when this work was initiated.

REFERENCES

[1] Sanchez-Sesma, F.J., Soil Dynamics and Earthquake Eng. 6 (1987) 124.
[2] Lee, V.W., Soil Dynamics and Earthquake Eng. 1 (1982) 122.
[3] Sanchez-Sesma, F.J., Bull. Seism. Soc. Am. 73 (1983) 1621.
[4] Boström, A. and Kristensson, G., Wave Motion 2 (1980) 335.
[5] Boström, A. and Karlsson, A., Geophys. J. R. astr. Soc. 79 (1984) 835.
[6] Boström, A. and Karlsson, A., Geophys. J. R. astr. Soc. 89 (1987) 527.
[7] Waterman, P.C., J. Acoust. Soc. Am. 60 (1976) 567.

ELASTIC WAVE PROPAGATION
M.F. McCarthy, M.A. Hayes, (Editors)

ELASTIC INTERFACIAL STANDING WAVES

by

J DUNWOODY*+

1. THE PROBLEM

Within the theory of infinitesimal deformations imposed on the finite we pose this problem:
When two hyperelastic half-spaces, uniformly prestressed into states of finite deformation are brought into contact at their plane surfaces and bonded there, so that the principal axes of pre-stress in each space are mutual with two of them parallel to the bonded surface, what should be the outcome if instantaneously they adjust to one another so as to remain in mechanical equilibrium without loss or gain of energy?
To avoid ambiguity arising from local non-uniqueness of the initial states of the separated half-spaces, we insist that both half-spaces in their separated pre-stressed states have elastic response which is 'strongly elliptic'.

The presentation here is an extension of an analysis by Dunwoody and Villaggio (1). Complementary results for incompressible neo-Hookean materials are contained in an analysis of interfacial, Stoneley type waves by Chadwick and Jarvis (2); that is, they found *standing wave solutions*. The results we present are for non-isochoric deformations in compressible general hyperelastic materials.

2. PRELIMINARIES

The initial Cauchy stress tensors $\overset{(m)}{\boldsymbol{\sigma}_o}$, m = (1,2) denoting upper and lower, are diagonal in either half-space with respect to a co-ordinate frame whose x_1x_3 plane is parallel to the bounding surfaces, i.e.

$$\overset{(m)}{\boldsymbol{\sigma}_o} = dg(\overset{(m)}{p_1}, \overset{(m)}{p_2}, \overset{(m)}{p_3}), \tag{2.1}$$

Additional plane displacements $\overset{(m)}{u} = \varepsilon \overset{(m)}{v}(x_1, x_2)$, $0 < \varepsilon << 1$, of any point $\mathbf{x}$ in the total space formed by the bonded system give rise to infinitesimal strain gradients

$$\overset{(m)}{H} = \nabla \overset{(m)}{u}(x) \tag{2.2}$$

* Department of Engineering Mathematics, The Queen's University of Belfast, Ashby Building, Stranmillis Road, Belfast BT9 5AH
+ The author acknowledges support received from CNR, Italy

and hence the perturbation stresses

$$\overset{(m)}{\boldsymbol{\sigma}} = (-\mathrm{tr}\,\overset{(m)}{\mathbf{H}})\,\overset{(m)}{\boldsymbol{\sigma}_o} + \overset{(m)}{\boldsymbol{\sigma}_o} \cdot \overset{(m)}{\mathbf{H}}^T + \overset{(m)}{\mathbf{H}} \cdot \overset{(m)}{\boldsymbol{\sigma}_o} + \overset{(m)}{\mathbb{E}} : \mathbf{H} + O(\varepsilon^2). \qquad (2.3)$$

In (2.3) the elasticity tensors $\overset{(m)}{\mathbb{E}}$ are derived from a strain energy function $\overset{(m)}{W}$ (cf. Truesdell and Noll (3)). Here our interest reduces to the four independent components in which 3 does not appear as a subscript, viz. $\overset{(m)}{\mathbb{E}}_{1111}$, $\overset{(m)}{\mathbb{E}}_{2222}$, $\overset{(m)}{\mathbb{E}}_{1122}$, $\overset{(m)}{\mathbb{E}}_{2211}$ and $\overset{(m)}{\mathbb{E}}_{1212}$.

There are but two sub-cases of initial deformation to be considered in each half-space: (i) $\overset{(m)}{p}_1 \neq \overset{(m)}{p}_2$, (ii) $\overset{(m)}{p}_1 = \overset{(m)}{p}_2 = \overset{(m)}{p}$. In all instances we assume

$$||[\boldsymbol{\sigma}_o].\mathbf{i}|| = O(\varepsilon) \quad \text{and} \quad ||[\boldsymbol{\sigma}_o].\mathbf{j}|| = O(\varepsilon), \qquad (2.4)$$

where $[\boldsymbol{\sigma}_o] = \overset{(1)}{\boldsymbol{\sigma}_o} - \overset{(2)}{\boldsymbol{\sigma}_o}$ and $\mathbf{i}$, $\mathbf{j}$ are the co-ordinate base vectors in the x_1x_2-plane; but, it is essential to allow the differences to be non-zero. For $\overset{(m)}{p}_1 = \overset{(m)}{p}_2 = \overset{(m)}{p}$ the elasticities are determined by just two constants λ and μ and $\overset{(m)}{\mathbb{E}}_{ijkl} = \overset{(m)}{\lambda}\,\delta_{ij}\,\delta_{kl} + \overset{(m)}{\mu}(\delta_{ik}\,\delta_{jl} + \delta_{il}\,\delta_{jk})$, each of $(i,j,k,l) = (1,2)$.

The following constraints on the elasticities $\overset{(m)}{\mathbb{E}}_{ijkl}$ sufficient to guarantee satisfaction of the SE condition (cf. Truesdell and Noll (3))

$$(\mathbf{a} \otimes \mathbf{b}) : \overset{(m)}{\mathbb{E}} : (\mathbf{a} \otimes \mathbf{b}) + \overset{(m)}{\boldsymbol{\sigma}_o} : (\mathbf{b} \otimes \mathbf{a}).(\mathbf{a} \otimes \mathbf{b}) > 0 \qquad (2.5)$$

for all $\mathbf{a} = (a_1, a_2, 0) \neq 0$ and $\mathbf{b} = (b_1, b_2, 0) \neq 0$ are adopted:

$$\text{(i)}\ (\overset{(m)}{\mathbb{E}}_{nnnn} + \overset{(m)}{p}_n) > 0, \quad \text{(ii)}\ \overset{(m)}{\mathbb{E}}_{1212} + \overset{(m)}{p}_n > 0, \quad \text{(iii)}\ \overset{(m)}{p}_n + \overset{(m)}{\mathbb{E}}_{1122} + 2\overset{(m)}{\mathbb{E}}_{1212} > 0,$$

$$\text{(iv)}\ \overset{(m)}{p}_n + \overset{(m)}{\mathbb{E}}_{nnnn} - (\overset{(m)}{\mathbb{E}}_{1122} + \overset{(m)}{\mathbb{E}}_{1212}) > 0 \qquad (2.6)$$

for $n = (1,2)$. They reduce to just two when $\overset{(m)}{p}_1 = \overset{(m)}{p}_2$, viz.

$$\overset{(m)}{p} + \overset{(m)}{\lambda} + 2\overset{(m)}{\mu} > 0 \quad \text{and} \quad \overset{(m)}{p} + \overset{(m)}{\mu} > 0, \qquad (2.7)$$

and are then also necessary for SE for all plane $\mathbf{a}$ and $\mathbf{b}$.

3. INTERFACE CONDITIONS

That the common plane surface of the half-spaces is bonded requires

$$\overset{(1)}{u}(x,o) = \overset{(2)}{u}(x,o) = h(x), \quad -\infty < x < \infty, \qquad (3.1)$$

at the resulting interface denoted by $y = 0$, where the notation (x,y) replaces (x_1,x_2) for the co-ordinates in the plane of the deformations $\overset{(m)}{u}$. The normal n_b to the deformed interface is

$$n_b = -h_{2,x}\,\mathbf{i} + \mathbf{j} + O(\varepsilon^2). \qquad (3.2)$$

We consider periodic deformations with unspecified frequency α, i.e.

$$\overset{(m)}{u}(x + \frac{2\pi}{\alpha}, y) = \overset{(m)}{u}(x,y), \tag{3.3}$$

which are either uniform with $||\varepsilon^{-1}\nabla\, u^{(m)}|| = ||dg(c_1, c_2)||$ bounded, or non-uniform with $||\varepsilon^{-1}\, u^{(m)}||$ and $||\varepsilon^{-1}\nabla\, u^{(m)}||$ bounded.

On account of (3.3) and the restrictions on $u^{(m)}$ enunciated following it, the further conditions which must be satisfied by the deformations may be expressed on any periodic cell

$$\overset{(1)}{S} \cup \overset{(2)}{S} = \{(x,y): x_o < x < x_o + \frac{2\pi}{\alpha}, -\infty < y < \infty\}. \tag{3.4}$$

They are

$$\nabla \cdot \overset{(m)}{\boldsymbol{\sigma}} = 0, \qquad y \neq 0, \tag{3.5}$$

$$[\boldsymbol{\sigma}_o].\mathbf{j} + [\boldsymbol{\sigma}].\mathbf{j} = 0, \qquad \text{for all } x \text{ on } y = 0, \tag{SSi}$$

$$\lim_{y\to\infty} (\overset{(1)}{\boldsymbol{\sigma}}_o + \overset{(1)}{\boldsymbol{\sigma}}).\mathbf{j} - \lim_{y\to-\infty} (\overset{(2)}{\boldsymbol{\sigma}}_o + \overset{(2)}{\boldsymbol{\sigma}}).\mathbf{j} = 0 \text{ for all } x, \tag{SSii}$$

and

$$\lim_{y\to\infty} \overset{(1)}{\mathbf{u}} \cdot \overset{(1)}{\boldsymbol{\sigma}}_o \cdot \mathbf{j} - \lim_{y\to-\infty} \overset{(2)}{\mathbf{u}} \cdot \overset{(2)}{\boldsymbol{\sigma}}_o \cdot \mathbf{j} = 0 \text{ for all } x \tag{SSiii}$$

The last three are the *conditions of self-straining* and of them (SSiii) guarantees that the total work done in the deformation is zero.

Since (3.1), (3.5) and (SSiii) are homogeneous in $u^{(m)}$, we consider (SSi) and (SSii) in their homogeneous and non-homogeneous forms:

$$[\boldsymbol{\sigma}_o].\mathbf{j} + [\boldsymbol{\sigma}].\mathbf{j} = 0, \qquad y = 0, \tag{SSia}$$

$$\lim_{y\to\infty} (\overset{(1)}{\boldsymbol{\sigma}}_o + \overset{(1)}{\boldsymbol{\sigma}}).\mathbf{j} - \lim_{y\to-\infty} (\overset{(2)}{\boldsymbol{\sigma}}_o + \overset{(2)}{\boldsymbol{\sigma}}).\mathbf{j} = 0 \tag{SSiia}$$

and

$$[\boldsymbol{\sigma}].\mathbf{j} = 0, \quad y = 0, \quad \text{(SSib)}, \quad \lim_{y\to\infty} \overset{(1)}{\boldsymbol{\sigma}} \cdot \mathbf{j} - \lim_{y\to-\infty} \overset{(2)}{\boldsymbol{\sigma}} \cdot \mathbf{j} = 0. \tag{SSiib}$$

Equations (3.5) and the interface conditions (3.1), (SSia), (SSiia) and (SSiii) are satisfied by particular homogeneous deformations $u^{(m)}$ as described by Dunwoody and Villaggio (1). Therefore, appealing to the superposition principle, in the sequel we concentrate on the conditions (3.1), (SSib), (SSiib) and (SSiii). Also, $[\boldsymbol{\sigma}_o].\mathbf{i} \neq 0$, although it does not feature, may also be redressed by suitable homogeneous deformations $u^{(m)}$ in order to ensure continuous conditions in the system at infinity. When $\boldsymbol{\sigma}_o^{(m)} = p^{(m)}I$ in both half-spaces one set of homogeneous deformations $u^{(m)}$ suffices.

4. INTERFACIAL STANDING WAVES

Every two dimensional perturbation $\overset{(m)}{u}(x,y)$ which decays sufficiently fast to zero at infinity has the representation

$$\overset{(m)}{u} = \{\overset{(m)}{\Phi}_{,x},\ \overset{(m)}{\Phi}_{,y}\} + \{\overset{(m)}{F}_{,y},\ -\overset{(m)}{F}_{,x}\} \tag{4.1}$$

In addition it is assumed that $\overset{(m)}{\Phi}$ and $\overset{(m)}{F}$ have the separable forms, consistent with (3.1),

$$\overset{(m)}{\Phi}(x,y) = e^{i\alpha x}\overset{(m)}{\phi}(y), \qquad \overset{(m)}{F}(x,y) = e^{i\alpha x}\overset{(m)}{f}(y). \tag{4.2}$$

On substituting from (4.1) and (4.2) in (2.3) and then in (3.5) we obtain, using the notation $D = \frac{d}{dy}$,

$$\begin{aligned} &i\alpha\{(\overset{(m)}{p}_2 + \overset{(m)}{\mathbb{E}}_{1122} + 2\overset{(m)}{\mathbb{E}}_{1212})D^2 - \alpha^2(\overset{(m)}{p}_1 + \overset{(m)}{\mathbb{E}}_{1111})\}\overset{(m)}{\phi} \\ &+ [(\overset{(m)}{p}_2 + \overset{(m)}{\mathbb{E}}_{1212})D^2 - \alpha^2\{\overset{(m)}{p}_1 + \overset{(m)}{\mathbb{E}}_{1111} - (\overset{(m)}{\mathbb{E}}_{1122} + \overset{(m)}{\mathbb{E}}_{1212})\}]D\overset{(m)}{f} = 0 \end{aligned} \tag{4.3}$$

and

$$\begin{aligned} &\{\overset{(m)}{p}_2 + \overset{(m)}{\mathbb{E}}_{2222})D^2 - \alpha^2(\overset{(m)}{p}_1 + \overset{(m)}{\mathbb{E}}_{1122} + 2\overset{(m)}{\mathbb{E}}_{1212})\}D\overset{(m)}{\phi} \\ &- i\alpha[\{\overset{(m)}{p}_2 + \overset{(m)}{\mathbb{E}}_{2222} - (\overset{(m)}{\mathbb{E}}_{1122} + \overset{(m)}{\mathbb{E}}_{1212})\}D^2 - \alpha^2(\overset{(m)}{p}_1 + \overset{(m)}{\mathbb{E}}_{1212})]\overset{(m)}{f} = 0. \end{aligned} \tag{4.4}$$

By elimination the following operator upon either $\overset{(m)}{\phi}$ or $\overset{(m)}{f}$ yields a single ordinary differential equation of order six for either function:

$$\{(D^2 - \alpha^2\,\overset{(m)}{c}{}_1^2)(D^2 - \alpha^2\,\overset{(m)}{c}{}_2^2)D^2 - \alpha^2\,\overset{(m)}{b}{}^2(D^2 - \alpha^2\,\overset{(m)}{c}{}_3^2)(D^2 - \alpha^2\,\overset{(m)}{c}{}_4^2)\} \tag{4.5}$$

where, from (2.6),

$$\overset{(m)}{c}{}_1^2 = (\overset{(m)}{\tilde{\lambda}}_1 + 2\overset{(m)}{\tilde{\mu}}_1)\,\overset{(m)}{\nu}{}_2^{-1} > 0, \qquad \overset{(m)}{c}{}_2^2 = \{\overset{(m)}{\nu}_1 - (\overset{(m)}{\tilde{\lambda}}_1 + \overset{(m)}{\tilde{\mu}}_1\}\,\overset{(m)}{\tilde{\mu}}{}_2^{-1} > 0,$$

$$\overset{(m)}{c}{}_3^2 = \{\overset{(m)}{\nu}_2 - (\overset{(m)}{\tilde{\lambda}}_2 + \overset{(m)}{\tilde{\mu}}_2)\}^{-1}\,\overset{(m)}{\tilde{\mu}}_1 > 0, \qquad \overset{(m)}{c}{}_4^2 = (\overset{(m)}{\tilde{\lambda}}_2 + 2\overset{(m)}{\tilde{\mu}}_2)^{-1}\,\overset{(m)}{\nu}_1 > 0,$$

$$\overset{(m)}{b}{}^2 = (c_3\,c_4)^{-2}\,\overset{(m)}{\tilde{\mu}}_1\,\overset{(m)}{\nu}_1(\overset{(m)}{\tilde{\mu}}_2\,\overset{(m)}{\nu}_2)^{-1}, \tag{4.6}$$

and by definition, for $n = (1,2)$,

$$\overset{(m)}{\tilde{\mu}}_n = \overset{(m)}{p}_n + \overset{(m)}{\mathbb{E}}_{1212}, \quad \overset{(m)}{\tilde{\lambda}}_n = \overset{(m)}{\mathbb{E}}_{1122} - \overset{(m)}{p}_n, \quad \overset{(m)}{\nu}_n = \overset{(m)}{p}_n + \overset{(m)}{\mathbb{E}}_{nnnn}. \tag{4.7}$$

In (4.7) only $\overset{(m)}{\tilde{\lambda}}_n$, $n = (1,2)$, may be non-positive and in agreement with (2.6).

It is clear that the differential operator coefficient (4.5) of either $\overset{(m)}{\phi}$ or $\overset{(m)}{f}$ is a polynomial $\overset{(m)}{p}(D)$ in D^2 with constant coefficients, and hence that the auxiliary equation $\overset{(m)}{p}(r) = 0$ has roots r^2 either real or complex conjugates in pairs. But, $r = \pm i\,\overset{(m)}{\gamma}\,\alpha$, $\overset{(m)}{\gamma} > 0$, are not possibilities, since $\overset{(m)}{p}(\pm i\,\overset{(m)}{\gamma}\,\alpha) < 0$. Hence, *for each* $\overset{(m)}{\phi}$ *and* $\overset{(m)}{f}$

there are just three independent solutions to (4.5) *which when substituted in* (4.1) *and* (4.2) *satisfy the condition* $||u^{(m)}|| = O(\varepsilon)$.[§] In general these solutions correspond to those roots r which have $\mathrm{Re}(r) < 0$ and multiplicity k, $1 \leq k \leq 3$. The three possibilities are exemplified in three non-exhaustive sub-cases, namely

$$\text{sub-case I } (k=3): \quad \overset{(m)}{c_1^2} = \overset{(m)}{c_2^2} = \overset{(m)}{c_3^2} = \overset{(m)}{c_4^2} = \overset{(m)}{b^2} = \overset{(m)}{c^2},$$

$$\text{sub-case II } (k=2): \quad \overset{(m)}{c_1^2} = \overset{(m)}{c_2^2} = \overset{(m)}{c_3^2} = \overset{(m)}{c_4^2} \neq \overset{(m)}{b^2},$$

$$\text{sub-case III } (k=1): \quad \overset{(m)}{c_1^2} = \overset{(m)}{c_3^2} \neq \overset{(m)}{b^2}, \quad \overset{(m)}{c_2^2} = \overset{(m)}{c_4^2} \neq \overset{(m)}{b^2}.$$

Here we treat only the most simple, sub-case I (which includes the interesting example $p_1^{(m)} = p_2^{(m)} = p^{(m)}$ implying $c^{(m)2} = 1$), when the conditions for it pertain in both half-spaces.

Sub-case I. If we factor y and α by c^{-1} and c respectively, the solutions to (4.5) have the forms

$$\overset{(m)}{\phi} = (\overset{(m)}{A_\phi} + \overset{(m)}{B_\phi}|y| + \overset{(m)}{C_\phi} y^2)\exp(-\alpha|y|)$$

$$\overset{(m)}{f} = (\overset{(m)}{A_f} + \overset{(m)}{B_f}|y| + \overset{(m)}{C_f} y^2)\exp(-\alpha|y|). \tag{4.8}$$

Substituting (4.8) into (4.3) and (4.4) we find the resulting equations consistent if and only if $C_\phi^{(m)} = C_f^{(m)} = 0$ and then

$$(-1)^{m-1}\overset{(m)}{B_\phi} = \overset{(m)}{b_1}\overset{(m)}{B_f} \iff \overset{(m)}{b_2}\overset{(m)}{B_\phi} = (-1)^{m-1}\overset{(m)}{B_f} \tag{4.9}$$

where

$$\overset{(m)}{b_1} = -i\,\overset{(m)}{c_3^{-2}}\,\overset{(m)}{\nu_2^{-1}}\,\overset{(m)}{\tilde{\mu}_1}, \quad \overset{(m)}{b_2} = i\,\overset{(m)}{c_4^{-2}}\,\overset{(m)}{\tilde{\mu}_2^{-1}}\,\overset{(m)}{\nu_1}, \quad b_1 b_2 = b^2 = 1. \tag{4.10}$$

Hence, we reduce to just three constants describing $u^{(m)}(x,y)$ in either half-space; and, *if we were to apply the condition* $u^{(m)}(x,o) = 0$ *to each, the null solution* $\mathbf{u}(x,y) \equiv 0$ *would result.* But, our study calls for (3.1) and (SSib) - (SSiib) and (SSiii) being redundant due to the exponential decay - which yield a set of four algebraic equations

$$M \cdot 1 = 0, \tag{4.11}$$

for

$$1 = (\overset{(1)}{B_\phi}, \overset{(2)}{B_\phi}, \overset{(1)}{A_\phi} + i\overset{(1)}{A_f}, \overset{(2)}{A_\phi} - i\overset{(2)}{A_f}), \tag{4.12}$$

$B_f^{(m)}$, $m = (1,2)$, having been eliminated through (4.9). *Equations* (4.11) *have a non-null solution* - $(A_\phi^{(1)} + iA_f^{(1)}$ *and* $A_\phi^{(2)} - iA_f^{(2)}$ *both zero leads to*

---

§ This theorem still applies when conditions (2.6) are replaced by the weaker set quoted in (1), though the demonstration is not so obvious.

$u^{(m)}(x,y) \equiv 0$ *for* m = (1,2)) - *if and only if* $|\mathbf{M}| = 0$; that is,

$$(i\,b_2^{(1)} - 1)[2i\,b_2^{(2)}\tilde{\mu}_1^{(2)}(w^{(1)} + w^{(2)}) - \{4\tilde{\mu}_1^{(1)} + (\tilde{\mu}_1^{(2)} + \tilde{\mu}_2^{(2)}) - (\tilde{\mu}_1^{(1)} + \tilde{\mu}_2^{(1)})\}\{\nu_2^{(2)} + (i\,b_2^{(2)} - 1)w^{(2)}\}]$$

$$+ (i\,b_2^{(2)} - 1)[2i\,b_2^{(1)}\tilde{\mu}_1^{(1)}(w^{(1)} + w^{(2)}) - \{4\tilde{\mu}_1^{(2)} + (\tilde{\mu}_1^{(1)} + \tilde{\mu}_2^{(1)}) - (\tilde{\mu}_1^{(2)} + \tilde{\mu}_2^{(2)})\}\{\nu_2^{(1)} + (i\,b_2^{(1)} - 1)w^{(1)}\}]$$

$$-4[\tilde{\mu}_1^{(1)}\{\nu_2^{(2)} + (i\,b_2^{(2)} - 1)w^{(2)}\} + \tilde{\mu}_1^{(2)}\{\nu_2^{(1)} + (i\,b_2^{(1)} - 1)w^{(1)}\}] = 0 \qquad (4.13)$$

in which $2w^{(m)} = (\nu_2^{(m)} - \tilde{\lambda}_2^{(m)}) > \tilde{\mu}_2^{(m)} > 0$, m = (1,2), by (2.6) ((iii), (v)). Also, $i\,b_2^{(m)} < 0$, $\tilde{\mu}_1^{(m)} < 0$ and $\tilde{\mu}_2^{(m)} < 0$, are fixed in sign, while $\{\nu_2^{(m)} + (i\,b_2^{(m)} - 1)w^{(m)}\}$ may take either sign if $p_1^{(m)} \neq p_2^{(m)}$, m = (1,2). Therefore, *it is always in theory possible to arrange cancellation of the positive and negative terms on the left hand side of* (4.13) *without contravening the SE condition.*

For plane hydrostatic initial states, when $p_1^{(m)} = p_2^{(m)} = p^{(m)}$, (4.13) has a particularly simple form due to the reductions

$$\tilde{\mu}_1^{(m)} = \tilde{\mu}_2^{(m)} = \tilde{w}^{(m)} = \tilde{\mu}^{(m)} = p^{(m)} + \mu, \quad \tilde{\lambda}_1^{(m)} = \tilde{\lambda}_2^{(m)} = \lambda^{(m)} - p^{(m)}$$

$$\nu_2^{(m)} + (i\,b_2^{(m)} - 1)w^{(m)} = -w^{(m)} = -\tilde{\mu}^{(m)}. \qquad (4.14)$$

Even though $\{\nu_2^{(m)} + i(b_2^{(m)} - 1)w^{(m)}\} < 0$ takes on a definite sign for this simplification, (4.13) which may be replaced by its equivalent

$$\{(\tilde{\mu}^{(1)} + \tilde{\mu}^{(2)})(\lambda^{(1)} + \mu^{(1)}) + 2\tilde{\mu}^{(1)}\tilde{\mu}^{(2)}\}\{(\mu^{(1)} + \mu^{(2)})(\lambda^{(2)} + \mu^{(2)}) + 2\tilde{\mu}^{(1)}\tilde{\mu}^{(2)}\} = 0 \qquad (4.15)$$

may still be satisfied. Necessary conditions for the satisfaction of (4.15) are

$$\{(\lambda^{(1)} + \mu^{(1)}) < 0\} \vee \{(\lambda^{(2)} + \mu^{(2)}) < 0\} \text{ and } \tilde{\mu}^{(1)} \neq \tilde{\mu}^{(2)}. \qquad (4.16)$$

The first is an inclusive disjunctive and is satisfied if the initial plane hydrostatic pressure in either half-space brings the material into an unstable region of its plane pressure-volume response curve (cf. Dunwoody and Villaggio (1)). *The second of* (4.16) *is satisfied if the materials are different, even though* $[p] = 0$, *or if the materials are the same but* $[p] = O(\varepsilon)$.

REFERENCES

1. J. Dunwoody and P. Villaggio, *Q. Jl. Mech. Appl. Math.* in proof (1988).
2. P. Chadwick and D.A. Jarvis, *Q. Jl. Mech. Appl. Math.* 32, 387-418 (1979).
3. C. Truesdell and W. Noll, *Handbuch der Physik*, Vol. III/3, Springer, Berlin (1965).

ELASTIC WAVE PROPAGATION
M.F. McCarthy, M.A. Hayes, (Editors)

RAY APPROXIMATION AND CAUSTICS FORMATION INSIDE AN ANISOTROPIC ELASTIC HALF SPACE

Michel PLANAT and Yuwen ZHANG

*Laboratoire de Physique et Métrologie des Oscillateurs du CNRS
associé à l'Université de Franche-Comté-Besançon
32, avenue de l'Observatoire - 25000 Besançon - France

We study the response of an elastic anisotropic half space to an electric excitation by a metallic interdigital transducer. The Fourier integral for the mechanical displacement is first computed at the geometrical approximation ; waves inside and near the boundary are studied for singly rotated quartz crystal cuts. This ray approximation fails when the dispersion curve of any propagation mode exhibits an inflexion point. Caustics formation is demonstrated and the field on both sides of the caustic is computed by means of the Airy integral. The theoretical analysis is illustrated by new experiments performed on Y-cut quartz crystals with transducers on both sides of the plate.

1. INTRODUCTION

The problem of finding the field excited from an arbitrary source into an isotropic half space was initiated by Lamb for elastic waves and by Sommerfeld for electromagnetic waves. A renewal of interest was associated with the work on piezoelectrics as quartz and lithium niobate which are widely used in oscillators and signal processing devices. In these cases, the classical approach is complicated by the anisotropic nature of the propagation substrate and by the finite extent of the source. In most applications, the excitation of waves is by a metallic interdigital transducer located on the boundary of the piezoelectric half space. It was used first to generate Rayleigh surface waves ; the simultaneous excitation of a continuum of bulk dispersive modes was recognized only recently [1,2].

In this paper an integral representation of the displacement at any point of the semi-infinite solid is obtained and asymptotic formulae are given for the far field.

We first apply the analysis for the case of an horizontally polarized acoustic beam in quartz (§ 3) and show that the orientation and width of the emission lobe is very sensitive to frequency and crystal orientation [3,4]. For more general cases, the integrand can become double valued for a point near the surface. The branch point contribution is computed systematically for singly rotated quartz crystal cuts (§ 4).

A particular feature of anisotropic crystals is the existence of inflexion points in dispersion curves, which lead to caustics in the radiation pattern. A preliminary investigation of this effect is presented (§ 5) both theoretically and experimentally.

2. GENERAL FORMULATION OF THE PROBLEM

We consider the excitation by a very wide interdigital transducer located on a low piezoelectric crystal. We solve the problem by taking the Fourier transform of fields along the direction Ox , and choosing exponential dependence along the depth Oy (Fig. 1).

Three components (r = 1-3) are obtained of wave numbers k_1 and $k_2^{(r)}$ along respectively Ox and Oy axis, corresponding to wave motions carrying energy away from the surface. In polar coordinates R, θ_o the displacement takes the following form

$$u_j(R,\theta_o) = \int_{-\infty}^{+\infty} \sum_{r=1}^{3} [\hat{\phi}_o A_j^{(r)}](k_1)\, e^{Rf^{(r)}(k_1)}\, dk_1 , \tag{1}$$

with

$$f^{(r)}(k_1) = - i \left[k_1 \cos\theta_o + k_2^{(r)}(k_1) \sin\theta_o\right] \quad \text{and} \quad A_j^{(r)} = \frac{B_r a_j^{(r)}}{\Delta} .$$

In these relations $k_2^{(r)}$ (r = 1-3) are the roots of the Christoffel equation, $a_j^{(r)}$ are the corresponding eigenvectors, Δ is the determinant of the boundary condition and B_r the weight of each mode r. The symbol $\hat{\phi}_o$ denotes the Fourier transform (expressed at the surface) of the applied voltage ϕ. Dispersion relations $k_2^{(r)}(k_1)$ are represented on Fig. 2 for the case of a Y-cut quartz crystal.

We evaluate the integral for each mode r by using the classical steepest descent path approximation (§ 3). In this method the integration path is deformed from the real axis into the complex plane in order to obtain the main contribution in integral (1) spread around the so-called saddle point $k_1 = k_{1c}^{(r)}$ defined by the zero first-order derivative of the phase factor $f^{(r)}(k_1)$. When the observation point (R, θ_o) is near the surface ($\theta_o \simeq 0$), the path may include a pole singularity, which leads to the familiar Rayleigh wave [4]. Also the saddle point can approach a branch point and the steepest descent approximation becomes invalid (§ 4). Finally the saddle point approximation breaks down near an inflexion point of the dispersion curve (§ 5).

3 - CONTRIBUTION OF THE SADDLE POINT. APPLICATION TO SURFACE SKIMMING BULK WAVES IN SINGLY ROTATED QUARTZ CRYSTAL CUTS WITH PROPAGATION NORMAL TO X-AXIS

For sufficiently deep observation angles θ_o, the contribution of the saddle point is dominant and takes the following form [3,4]

$$u_j(R,\theta_o) = \sum_{r=1}^{3} \left\{ \frac{\exp\left[R\, f^{(r)} + i(\pi/4)\, \mathrm{Sgn}\, f_2^{(r)}\right]}{\left[2\pi R\, \left|f_2^{(r)}\right|\right]^{1/2}} \cdot \frac{B_r\, a_j^{(r)}}{\Delta}\, \hat{\phi}_o \right\}(k_{1c}^{(r)}), \quad (2)$$

where $f_2^{(r)}$ is the second order derivative of the function $f^{(r)}$ and the expression inside the brackets has to be evaluated at the saddle point $k_{1c}^{(r)}$.

In equation (2) the saddle point $k_{1c}^{(r)}$ happens when the first order derivative of phase vanishes, i.e. when

$$f'\ (k_{1c}^{(r)}) = -\ \mathrm{cotg}\ \theta_o = -\ x/y\ . \quad (3)$$

On the other hand, since the wave length along direction Ox is equal to twice the transducer periodicity, it is apparent from Fig. 1 that waves interfere constructively for an excitation frequency f given by

$$f = V_c\ /\ (\lambda_o\ \cos\ \theta_c)\ , \quad (4)$$

where θ_c is the propagation angle for the main excited wave and V_c is the corresponding phase velocity. As shown on Fig. 2, the angle θ_o is defined from the normal to the dispersion curve and represents the power flow direction.

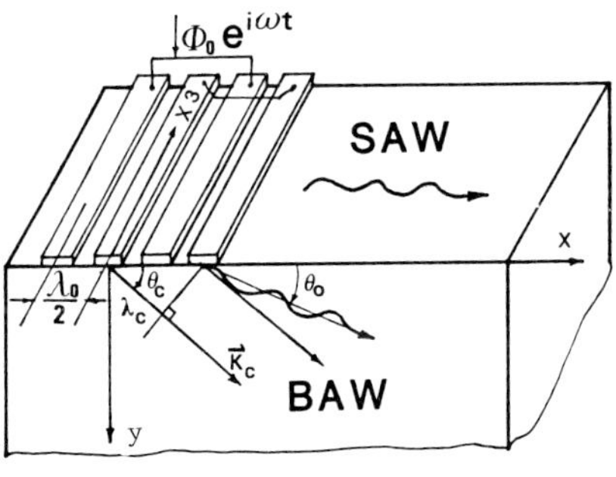

Fig. 1 :
Schematic of the excitation source

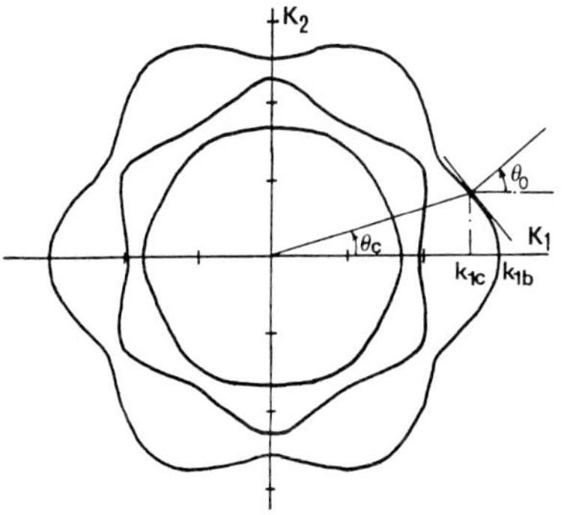

Fig. 2
Dispersion curves for a Y-cut, X prop. quartz crystal

From eq. (3) we see that the saddle point location depends on the observation angle θ_o. For different values of θ_o the propagation velocity of the wave that contributes is different and the main direction of emission is scanned by the applied frequency (eq. (4)). Moreover the total displacement at angle θ_o is the summation of all modes r.

Here we restrict ourselves for the case of singly rotated cuts of quartz with propagation normal to X-axis. This simple class of cuts is convenient for supporting the so-called surface skimming bulk waves (SSBW), because only horizontal shear motion is coupled to piezoelectric excitation [1-4].

Main characteristics of SSBW for this class of cuts were studied recently [3]. A complete analytic treatment is possible and yields an expression of power flow density in the radial direction given by

$$P_r (R, \theta_o) = \alpha_f \alpha \, \phi_o^2 / R \,, \tag{5}$$

where α_f and α are respectively transducer and material factors.

For operation at the surface, the cut with a rotation angle $\theta = 31.5°$ exhibits the highest velocity corresponding to the highest possible frequency. This cut is near the well known AT-cut $\theta = 35°$ which has been used for oscillators with frequencies up to 3.4 GHz [8]. The lowest velocity is obtained for $\theta = 121.5°$. Both cuts have zero power flow angle.

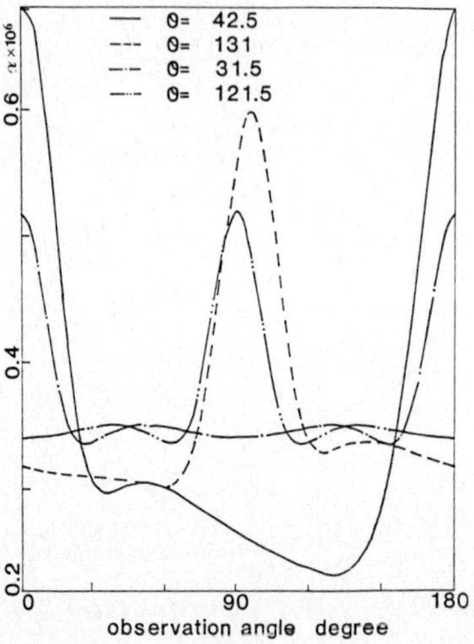

Fig. 3
Material efficiency of bulk radiation for shear horizontal waves

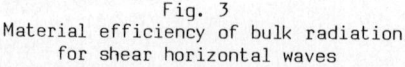

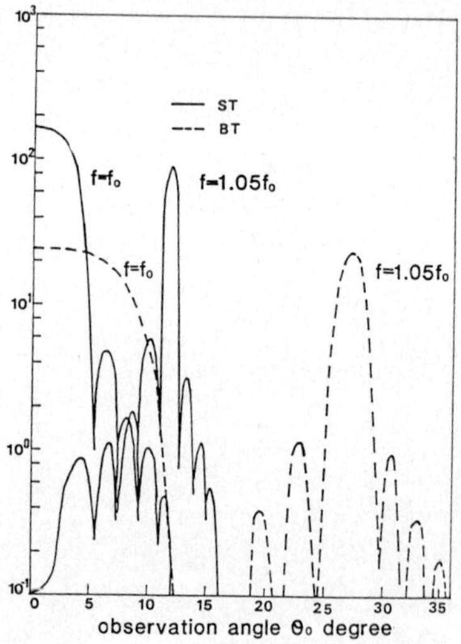

Fig. 4
Anisotropy dependence of radiation diagram

Efficiency was studied as a function of θ. At shallow angles, the best efficiency ($180\ W/m^2$) is obtained near the ST-cut ($\theta = 42^\circ$) and remains large ($100\ W/m^2$) for the $\theta = 31.5^\circ$ cut (Fig. 3). On the contrary, material efficiency α remains almost constant at shallow angles for BT-cut ($\theta = 131^\circ$) and $\theta = 121.5^\circ$ cut, and exhibits a peak for observation angles near 90°. These latter cuts may be suitable for vertical radiation application.

Plots of power density P_r versus observation angle θ_o are given on Fig. 4 for the two extreme cases of the highest velocity cut $\theta = 31.5^\circ$ and the lowest velocity cut $\theta = 121.5^\circ$. It is clearly observed that the width and height of the main lobe at the cut-off frequency f_o are sensitive to the crystalline configuration. Moreover, it is apparent that the main lobe direction depends on the operating frequency more strongly for the lowest velocity cut.

4 - CONTRIBUTION OF BRANCH POINTS : APPLICATION TO SINGLY ROTATED CUTS OF QUARTZ WITH PROPAGATION ALONG X-AXIS

By looking at Fig. 2, we see that there exist a double root for the function $k_2(k_1)$ at points k_{1b} where the tangent to the dispersion curve is vertical. These so-called branch points, where the integrand in (1) becomes double valued, have to be excluded from the closed contour by using an extra loop in the complex k_1-plane.

Branch points are found in the deformation of contour only at shallow angles. Consequently the steepest descent paths encircling branch points are vertical [5]. After lengthy calculations it can be shown that a branch point contribution takes the form

$$u_j = i\, k_o\, \hat{\phi}_o(k_{1b})\, \exp(-ik_b x)(c_j/2\pi) \int_0^\infty \frac{\eta^{1/2}}{a + b\eta}\, e^{-R\eta}\, d\eta \ , \qquad (6)$$

where $a = \Delta^2/2$, b and c_j are material coefficients and Δ is the determi nant of boundary condition and has been computed at the branch point.

Three distinct cases can be considered :

a) when $a = 0$, i.e. $\Delta = 0$, The field exactly satisfies boundary condition. We obtain $u_j \propto R^{-1/2}$. This corresponds to the case of SSBW waves considered in § 3.

b) when $a \neq 0$ and $|a| \gg |b|\eta$, The integral in (6) decays as $R^{-3/2}$.

c) in the general case, the integral can be expressed in terms of gamma and Whittaker functions [5].

The analysis was applied to singly rotated cuts of quartz with propagation along X-axis.

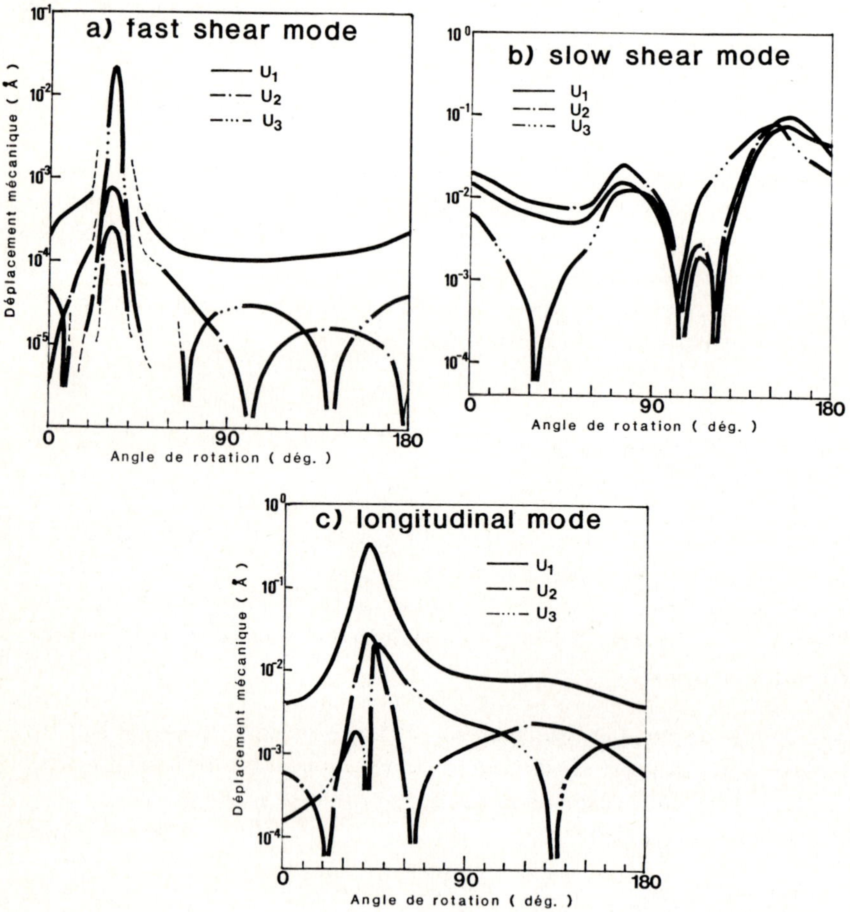

Fig. 5 : Branch point contribution for a rotated Y-X cut of quartz

Mechanical displacements corresponding respectively to fast shear, slow shear and longitudinal branches are represented on Fig. 5, as a function of rotation angle θ. The results are for a propagation distance $R = 100\ \lambda_o$ and an excitation voltage $\Phi_o = 1$ V.

For the fast shear mode and cuts located near the 32 ° cut (Fig. 5a), we have $\Delta = 0$. From (7) the decay is proportional to $R^{-1/2}$. The displacement is approximately shear horizontal and consequently the wave is a quasi-SSBW. For other cuts the decay is proportional to $R^{-3/2}$ and the contribution is lower.

For the slow shear mode (Fig. 5b), the main contribution again comes from the $R^{-1/2}$ term although the determinant Δ nearly vanishes only at $\theta = 120^\circ$. Main contributions happen at $\theta = 0$, $\theta = 80^\circ$ and $\theta = 150^\circ$. The SSBW for this mode exists in the vicinity of $\theta = 120^\circ$, since only the u_3 is dominant.

Finally for the longitudinal mode (Fig. 5c), the decay is proportional to $R^{-3/2}$. The wave is strong only around the ST-cut ($\theta = 41^\circ$) and its polarization is longitudinal.

5 - INVALIDITY OF THE RAY APPROXIMATION AT AN INFLEXION POINT OF THE DISPERSION CURVE : APPLICATION TO TRAPPED WAVES IN SINGLY ROTATED QUARTZ PLATES

The approximation considered in § 3 fails when the dispersion curve of any propagation mode exhibits an inflexion point $k_1 = k_{1I}$. Around this point, the field is considerably enhanced since the second order derivative of the phase vanishes (see eq. (2)). Such a situation can be analyzed by introducing the concept of Airy integral to heal the singularity [7]. Expanding the phase around the inflexion point, the condition of stationarity of the phase allows us to replace the equation of rays (3) by the following expression

$$x/y = - k_2'(k_{1I}) - 1/2\ (k_1 - k_{1I})^2\ k_2'''(k_{1I}) \tag{8}$$

From this relation, we see that the ray of equation $x = -y\ k_2'\ (k_{1I})$ separates a region of the half space without rays from a region twice covered by rays (Fig. 6). This ray is called a caustic and the key concept to compute the field on both sides of the caustic is the Airy integral.

After standard manipulations [7], the mechanical displacement takes the form

$$u_j(x,y) = [\hat{\phi}_o\ A_j]\ (k_{1I})\ [\tfrac{1}{2}\ y\ k_2'''(k_{1I})]^{-1/3}\ A_i\ \left\{- \frac{[x + y\ k_2'(k_{1I})]}{[\tfrac{1}{2}\ y\ k_2'''(k_{1I})]^{1/3}}\right\}$$
$$\times \exp\ \{-i\ [k_{1I}x + k_2(k_{1I})y]\}\ . \tag{9}$$

Waves are significant on the side of the caustic where the argument of the Airy integral is negative and the wave amplitude falls away exponentially on the other side.

We here consider the case of an AT cut quartz crystal ($\theta \simeq 38^\circ$) with dispersion curves given in Fig. 6a. The shear horizontal mode exhibits inflexion points at P_1 and P_2, leading to caustics C_1 and C_2.

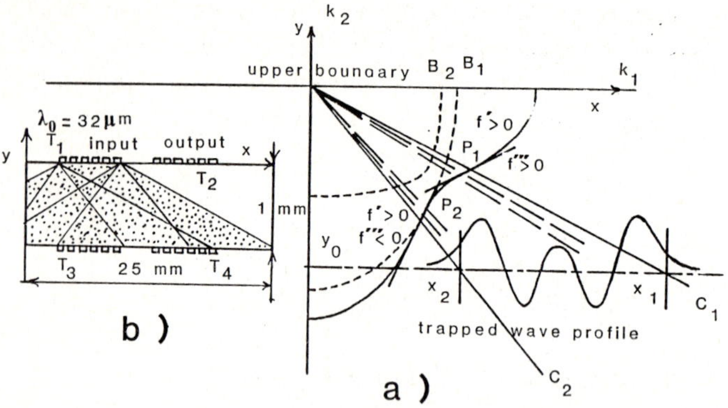

Fig. 6

a) Schematic of the caustics problem for waves in an AT cut quartz half space
b) Experimental arrangement for trapped waves in finite thickness plates

Due to the sign of third order derivatives at these points waves are significant only inside the angular region limited by caustics.

At constant depth, the amplitude of waves emanating from caustics C_I (I = 1,2) takes the following form

$$u_j = C_{Ij}\ A_i\ [\alpha_I(x_I - x] \quad ; \quad I = 1,2 \tag{10}$$

where $\alpha_I = [\frac{1}{2}\, y_0\, f'''(k_{1I})]^{-1/3}$ and $x_I = -y_0\, f'(k_{1I})$ are constants, $\alpha_1 < 0$ and $\alpha_2 > 0$. Following the approach given in [7], the two fields given by (10) will interfere constructively only at particular frequencies.

Then at constant depth and for these frequencies, waves are trapped between caustics C_1 and C_2 (Fig. 6a). However since y_0 takes a continuous set of values, the corresponding spectrum is continuous as well. A discrete spectrum is obtained for the case of plates of finite thickness (Fig. 6b).

This is illustrated by experiments performed on an AT cut quartz crystal plate with two pairs of transducers deposited on both sides of a well polished plate (Fig. 7).

Transducers on opposite sides of the plate were carefully set straight above each other to provide good matching of phases for waves emanating from upper and lower boundaries.

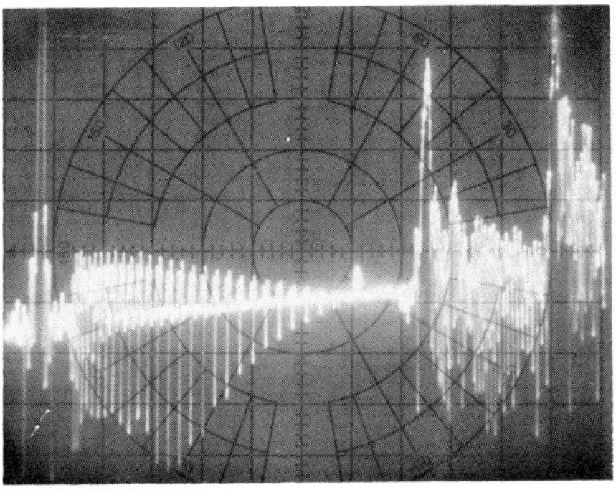

Fig. 7 : Transmission response for AT cut quartz crystal plates

We observe a main clear peak at 91 MHz corresponding to Rayleigh waves, and two secondary peaks at 149 and 167 MHz due to branch points B_1 and B_2 of fast shear and longitudinal branches (Fig. 6a). In addition, numerous sharp stop bands exist in transmission curves. This is an interesting new feature of such experiments which are believed to correspond to trapped waves considered at the beginning of this section.

The comparison of theoretical and experimental curves will be given in a future work.

ACKNOWLEDGEMENTS

Special thanks to R. Coquerel and C. Bonjour for processing of the device used in experiments.

REFERENCES

[1] Lewis, M.F., Proceedings of the Ultrasonics Symposium, de Klerk J. and McAvoy B.R. ed. (IEEE, New York, 1977) pp. 744-752.

[2] Goodberlet, M.A. and Lee D.L., IEEE Trans. Sonics and Ultrason., SU-31 (1984) 67.

[3] Zhang, Y.W. and Planat, M.P., Appl. Phys. Lett., 52 (1988) 456.

[4] Zhang, Y.W. and Planat, M.P., Elect. Lett., 23 (1987) 68.

[5] Yashiro, K. and Goto N., IEEE Trans. Sonics and Ultrason., SU-25 (1978) 146.

[6] Lau, K.F., Yen, K.H., Kagiwada, R.S. and Kong, A.M., Proceedings of the Ultrasonics Symposium, Mc Avoy B.R. ed. (IEEE, New York, 1980) pp. 240-244.

[7] Lighthill, J., Waves in fluids (Cambridge University Press, 1978), pp. 385-399.

B
NONLINEAR ELASTIC WAVES

ELASTIC WAVE PROPAGATION
M.F. McCarthy, M.A. Hayes, (Editors)

SPREADING OF NONLINEAR SURFACE WAVES ON PIEZOELECTRIC SOLIDS

David F. Parker and Easter A. David

Department of Theoretical Mechanics
University of Nottingham
Nottingham, NG7 2RD, U.K.

A theory for the effect of lateral spreading of a beam of nonlinear surface acoustic waves travelling across the surface of a homogeneous piezoelectric solid is presented. It combines procedures used previously in deriving the nonlinear evolution equation governing the Fourier transform of the surface elevation of a unidirectional piezoelectric signal, and in analysing spreading effects in elastic surface waves. The resulting equation contains a quadratic integral of the convolution type found by Lardner for elastic media and a linear diffusion operator (of Schrödinger type).

Three different electric boundary conditions are considered and in each case the coefficients in the resulting evolution equation are related to the corresponding slowness curve for linearized waves.

1. INTRODUCTION

The authors [1,2] have recently shown that the procedure described by Parker [3] for analysing the evolution of nonlinear elastic surface waves may be adapted to describe electro-elastic waves in piezoelectric media. In that work, as in other theoretical treatments of nonlinear surface waves, all disturbances are independent of one Lagrangian coordinate and so involve generalized plane strain. By suitable scaling of coordinates an evolution equation is obtained, equating the derivative of the Fourier transform of the surface elevation to a quadratic integral of evolution type.

It has recently been found [4] that, for elastic materials, the procedure for derivation of the evolution equation may be extended to allow for gradual variation in signal strength transverse to the propagation direction. The analysis merely adds to the evolution equation a term involving second derivatives transverse to the propagation rays. The effect is simply to replace a first derivative by a diffusion operator of Schrödinger type. In the present paper, these procedures are outlined and combined, in order to describe a beam of surface acoustic waves on an electro-elastic material.

Three different electrical boundary conditions on the traction-free boundary of the material halfspace are considered. It is found that, whether the surface is electrically earthed, adjoins free space or is treated by an 'open circuit' model, the resulting evolution equation has the same form as in the elastic case. Moreover, the coefficient of the 'diffusion' term is simply related to the shape of the slowness curve for the corresponding linearized waves.

2. THE NONLINEAR TREATMENT OF UNI-DIRECTIONAL WAVES

Following the treatment in [1] and [2] , a homogeneous, anisotropic, electro-elastic halfspace is taken to occupy the region $X_2 \leqslant 0$ of Lagrangian coordinates

(X_1,X_2,X_3). Small displacements have components $\varepsilon u_j(X_1,X_2,X_3,t)$ and the electrostatic approximation is assumed, with potential $\varepsilon\phi(X_1,X_2,X_3,t)$. The momentum equation and Gauss equation in the absence of free charge may then be combined into the single equation

$$T_{Lj,L} = \varepsilon\rho\eta_{jm}\ddot{u}_m \qquad \text{in } X_2 < 0 \;, \tag{1}$$

where $_{,L}$ denotes $\partial/\partial X_L$, dots denote time derivatives, upper case letters range over the values 1,2,3 while lower case letters range over the values,1,2,3,4 and summation convention is used, with $\eta_{jm} = \delta_{jm}-\delta_{j4}\delta_{m4}$. For j=1,2,3, T_{Lj} denotes the components of Piola-Kirchhoff stress, where $T_{L4}= D_L$ are the components of the material electric field while $u_4 \equiv \phi$ and ρ is the reference density. In this notation, the constitutive laws correct to $O(\varepsilon^2)$ become

$$T_{Lj} = \varepsilon c_{jLmM}\, u_{m,M} + \varepsilon^2 d_{jLmMnN}\, u_{m,M}\, u_{n,N} \quad , \tag{2}$$

where the coefficients, which are symmetric with respect to interchanges of the pairs of indices (jL), (mM) and (nN), are related to the elastic, dielectric, piezoelectric, electroelastic and electrostriction coefficients as in [1] and [2].

Three different boundary conditions on the material surface $X_2 = 0$ will be considered. The simplest,corresponding to an earthed, traction-free boundary (case E) is

$$T_{2m}\eta_{jm} = 0 \quad , \quad u_4 = 0 \qquad \text{on } X_2 = 0 \;. \tag{3}$$

For a surface adjoining free space,it is necessary to consider the potential and related Maxwell stress in the region $X_2 > 0$, so that the boundary conditions are replaced by interface conditions (case F):

$$[\![T_{2j}]\!] = 0 \quad , \quad [\![u_4]\!] = 0 \qquad \text{on } X_2 = 0 \;. \tag{4}$$

The third condition considered is the 'open circuit' limit in which $D_2 = 0$ at $X_2 = 0$ (case O), so that

$$T_{2j} = 0 \;, \qquad \text{on } X_2 = 0 \;. \tag{5}$$

For each choice of boundary condition, linear theory predicts that surface waves propagating in a chosen direction have phase speed independent of wavelength. If the X_1-axis is chosen parallel to the wave normal and c denotes the corresponding phase speed, we introduce a phase variable $\theta \equiv X_1-ct$. Then, observing that nonlinearity should cause a waveform to evolve only over the scale of $X \equiv \varepsilon X_1$, we convert to coordinates (θ,X_2,X) and deduce that, to leading order, the equations are satisfied by

$$u_j = \int_{-\infty}^{\infty} C(k,X)A_j(kX_2)e^{ik\theta}\, dk \equiv U_j(\theta,X_2,X) \quad . \tag{6}$$

Here, $A_j(kX_2)$ describes the depth-dependence of the displacements and potential in a harmonic wave of linear theory, with wavenumber k and normalized so that $A_2(0) = 1$. Then $C(k,X)$ is the Fourier transform of the (vertical) displacement normal to the surface.

The equation governing the evolution of C with range X is determined as a compatibility condition for the existence of bounded solutions to the equation arising from the $O(\varepsilon^2)$ terms in (1) and its associated boundary conditions. For both elastic waves [3] and piezoelectric waves [1,2] the compatibility condition yields the evolution equation (of a form first derived by Lardner [5] for isotropic elasticity)

$$J\frac{\partial C}{\partial X}(k,X) = i\int_0^{\infty} K\left(\frac{\ell}{k}\right)(k-\ell)\ell\; C(k-\ell,X)C(\ell,X)\, d\ell \quad , \tag{7}$$

where $C(-k,X) = C^*(k,X)$ and * denotes a complex conjugate. The (real) coefficient J and complex kernel K(.) depend on the material coefficients in (2) both directly and through integrals involving the depth dependence $A_j(kX_2)$. Details may be found in the references cited.

3. ANALYSIS OF LATERAL SPREADING

If the disturbance has gradual transverse variations, it is appropriate to allow displacements to depend also on $Z \equiv \nu(X_3 - gX_1)$, where $dX_3/dX_1 = g$ describes the 'surface rays' and ν is some $O(1)$ parameter. In order that spreading effects are comparable in magnitude to the nonlinear effects, we set $\nu = \varepsilon^{\frac{1}{2}}$. Then, after conversion to independent variables

$$\theta \equiv X_1 - ct \ , \quad X_2 \ , \quad Z \equiv \varepsilon^{\frac{1}{2}}(X_3 - gX_1) \ , \quad X = \varepsilon X_1 \ ,$$

the constitutive law (2) becomes

$$T_{Lj} = \varepsilon c_{jLm\alpha} u_{m,\alpha} + \varepsilon^{\frac{3}{2}}(c_{jLm_3} - gc_{jLm_1}) u_{m,Z} + \varepsilon^2 (c_{jLm_1} u_{m,X} + d_{jLm\alpha n\beta} u_{m,\alpha} u_{n,\beta}) + o(\varepsilon^2), \tag{8}$$

where Greek indices range only over the values 1, 2 and where $_{,1}$ now denotes $\partial/\partial\theta$. Similarly the field equation (1) becomes

$$\begin{aligned} 0 &= T_{\alpha j,\alpha} + \varepsilon^{\frac{1}{2}} g_N T_{Nj,Z} + \varepsilon T_{1j,X} - \varepsilon\rho c^2 \eta_{jm} u_{m,11} \\ &= \varepsilon\{c_{j\alpha m\beta} - \rho c^2 \eta_{jm} \delta_{\alpha 1} \delta_{\beta 1}\} u_{m,\alpha\beta} + \varepsilon^{\frac{3}{2}} g_N (c_{j\alpha mN} + c_{m\alpha jN}) u_{m,\alpha Z} \\ &\quad + \varepsilon^2 g_N g_L c_{jNmL} u_{m,ZZ} + \varepsilon^2 (c_{j\alpha m_1} + c_{m\alpha j_1}) u_{m,\alpha X} \\ &\quad + \varepsilon^2 d_{j\gamma m\alpha n\beta} (u_{m,\alpha} u_{n,\beta})_{,\gamma} + o(\varepsilon^2) \ , \end{aligned} \tag{9}$$

where, to simplify subsequent analysis, we have introduced the surface vector with components $g_1 = -g$, $g_2 = 0$, $g_3 = 1$. The boundary conditions are determined from the appropriate one of equations (3)-(5) by using expression (8) (in case F, it is necessary to introduce pseudo-displacements in $X_2 > 0$, as in [2]).

From the above equations, it is clear that, to leading order, the displacements and potential satisfy the boundary value problem describing the corresponding linear piezoelectric surface waves which depend only on X_2 and $\theta = X_1 - ct$. Thus, the leading order approximation to u_j is given by expression (6), with C and U_j now allowed to depend on Z. Here the depth dependence $A_j(kX_2)$ is governed in $X_2 < 0$ by the equation

$$\mathcal{L}_{jm}A_m \equiv k^2(c_{j_1m_1} - \rho c^2 \eta_{jm})A_m - ik(c_{j_1m_2} + c_{j_2m_1})\frac{dA_m}{dX_2} - c_{j_2m_2}\frac{d^2A_m}{dX_2^2} = 0 \tag{10}$$

for $k > 0$, and so has the form

$$A_j(kX_2) = \sum_{p=1}^{4} B^{(p)} a_j^{(p)} \exp iks^{(p)}X_2 \qquad k > 0 \tag{11}$$

where $s = s^{(p)}$ $(p=1,2,3,4)$ are the roots having $\operatorname{Im} s^{(p)} < 0$ and $a_j = a_j^{(p)}$ are the corresponding vectors satisfying

$$\{c_{j_1m_1} - \rho c^2 \eta_{jm} + s(c_{j_1m_2} + c_{j_2m_1}) + s^2 c_{j_2m_2}\} a_m = 0 \ .$$

For $k < 0$, A_j is defined by $A_j(kX_2) = A_j^*(-kX_2)$ while the boundary behaviour is denoted by

$$A_{0j}(k) = \sum_{p=1}^{4} B^{(p)} a_j^{(p)} \qquad \text{all } k > 0 \ , \qquad A_{0j}(-k) = A_{0j}^*(k) \ .$$

The coefficients $B^{(p)}$ of the partial waves are determined by the boundary conditions at $X_2 = 0$. For the earthed case (Case E) these give

$$\sum_{p=1}^{4} \eta_{j\ell}(c_{\ell 2m1} + s^{(p)} c_{\ell 2m2})\, a_m^{(p)}\, B^{(p)} = 0 \;, \qquad \sum_{p=1}^{4} a_4^{(p)}\, B^{(p)} = 0 \;,$$

which has the form $\sum M_{jp} B^{(p)} = 0$. The free-space case (case F) requires matching to the free-space potential $\varepsilon\Phi(\theta, X_2, X, Z)$, where

$$\Phi = \int_{-\infty}^{\infty} C(k,X,Z) A_{04}(k)\, e^{-|k|X_2}\, e^{ik\theta}\, dk \quad \text{in} \quad X_2 > 0 \;. \tag{12}$$

This gives rise to the boundary conditions

$$\sum_{p=1}^{4} \hat{M}_{jp} B^{(p)} \equiv \sum_{p=1}^{4} \{c_{j2m1} + s^{(p)}\, c_{j2m2} + i\,\epsilon_0 \delta_{j4} \delta_{m4}\} a_m^{(p)}\, B^{(p)} = 0 \;, \tag{13}$$

where ϵ_0 is the permittivity of free space. The open circuit case is similar. Indeed, the boundary conditions in cases E and 0 correspond to the mathematical limits $\epsilon_0 \to \infty$ and $\epsilon_0 \to 0$, respectively, of (13).

In all cases, the functions $A_m(kX_2)$ are normalized so that $A_{02}(k) = 1$, while the vanishing of the appropriate determinant det $\hat{M}_{jp}$ provides the secular equation which determines the phase speed c.

Seeking solutions to (9) correct to $0(\varepsilon^{\frac{3}{2}})$ in the form

$u_j = U_j(\theta, X_2, X, Z) + \varepsilon^{\frac{1}{2}} u_j^1(\theta, X_2, X, Z)$ in $X_2 \leqslant 0$ leads to the system

$$\{\rho c^2 \eta_{jm} \delta_{\alpha 1} \delta_{\beta 1} - c_{j\alpha m\beta}\} u^1_{m,\alpha\beta}$$

$$= g_N (c_{j\alpha mN} + c_{m\alpha jN}) U_{m,\alpha Z} \equiv P_j^1(\theta, X_2, X, Z) \;.$$

Similarly, the boundary conditions for the appropriate case are found from (3), (4) or (5) to be a linear, inhomogeneous system in u_j^1 (and, in case F, ϕ^1 also, where $\hat{\phi}$ is written as $\hat{\phi} = \Phi + \varepsilon^{\frac{1}{2}}\phi^1$ in $X_2 \geqslant 0$ and $\phi^1_{,\alpha\alpha} = 2g\Phi_{,1Z} \equiv P_5^1$ in $X_2 > 0$. Thus, the boundary-value problem for u_m^1 (and ϕ^1) is an inhomogeneous version of the problem defining U_m (and Φ). The compatibility condition is determined by taking Fourier transforms (denoted by overbars) with respect to θ. This yields

$$\mathcal{L}_{jm}\, \overline{u_m^1} = \overline{P_j^1}\,(k, X_2, X, Z) \qquad \text{in } X_2 < 0$$

with boundary conditions at $X_2 = 0$ of the form

$$\eta_{j\ell}\, \mathcal{B}_{\ell m}\, \overline{u_m^1} = -\eta_{j\ell}\, \overline{Q_\ell^1}(k,X,Z) \;, \quad \overline{u_4^1} = 0 \qquad \text{(case E)},$$

$$\mathcal{B}_{jm}\, \overline{u_m^1} = -\overline{Q_j^1}(k,X,Z) \qquad \text{(case 0)},$$

where $\mathcal{B}_{jm} \equiv ikc_{j2m1} + c_{j2m2}\, \partial/\partial X_2$ and $Q_j^1 = g_N c_{j2mN} U_{m,Z}$. It is then found (as in [1]-[3]) that the compatibility condition takes the simple form

$$\int_{-\infty}^{0} A_j^*(kX_2)\, \overline{P_j^1}\,(k, X_2, X, Z)\, dX_2 = A_{0j}^*(k)\, \overline{Q_j^1}\,(k,X,Z) \;. \tag{14}$$

After making the substitution $\overline{U_m^1} = A_m(kX_2) C_{,Z}$, it is found that $C_{,Z}$ cancels from equation (14) leaving an equation which may be rearranged as

$$\int_{-\infty}^{0} \{c_{j1m3}(A_j^{*} A_m + A_j A_m^{*}) + ic_{j2m3}(A_j^{*\prime} A_m - A_j^{\prime} A_m^{*})\}\, d\xi$$

$$= g \int_{-\infty}^{0} \{c_{j1m1}(A_j^{*} A_m + A_j A_m^{*}) + ic_{j2m1}(A_j^{*\prime} A_m - A_j^{\prime} A_m^{*})\}\, d\xi \;, \tag{15}$$

where a prime denotes differentiation with respect to the variable $\xi \equiv kX_2$. (In case F, equation (15) is unchanged except for the inclusion on the right hand side of a term $-g\epsilon_0 A_{04}A_{04}^*$ arising from matching of fields in $X_2 < 0$ to $\bar{\phi}^1$ in $X_2 > 0$). Equation (15) is the condition which defines the (real) ray direction $dX_3/dX_1 = g$.

When condition (15) is satisfied, the equations governing $\overline{u_m^1}$ are found to have solutions of the form

$$\overline{u_m^1} = ik^{-1}\, b_m(\xi)\, C_{,Z}(k,X,Z)\ .$$

The equations for $b_m(kX_2)$ are then found to specify $\mathcal{L}_{jm}b_m$ in $X_2 < 0$ and $\mathcal{B}_{jm}b_m$ (or $n_{j\ell}\mathcal{B}_{\ell m}b_m$ and b_4) at $X_2 = 0$ linearly in terms of $A_m(kX_2)$. Together with the conditions $b_m \to 0$ as $X_2 \to -\infty$ and the subsidiary condition $b_2(0) = 0$ (related to the normalization of $A_m(\xi)$), these equations determine $b_m(\xi)$ uniquely. Indeed, an explicit formula is given in Section 4.

We now repeat the process, analysing the $O(\varepsilon^2)$ terms in (9). Writing

$$u_j = U_j + \varepsilon^{\frac{1}{2}}u_j^1 + \varepsilon u_j^2(\theta,X_2,X,Z)$$

gives rise to an inhomogeneous boundary value problem involving the equation

$$\mathcal{L}_{jm}\,\overline{u_m^2} = \overline{P_j^2}(k,X_2,X,Z) \qquad \text{in } X_2 < 0$$

and appropriate transformed boundary conditions (and an equation $\mathcal{L}\overline{\phi^2} = \overline{P_5^2}$ in $X_2 > 0$, case F). Here, $\overline{P_j^2}$ contains terms linear in $\overline{u^1_{m,Z}}$, $\overline{U}_{m,ZZ}$ and $\overline{U}_{m,X}$ together with terms involving the transform of $(U_{m,\alpha}U_{n,\beta})_{,\gamma}$. Thus, the terms involve either $C_{,ZZ}$, $C_{,X}$ or a convolution product of C. A similar deduction applies to the inhomogeneous terms appearing in the boundary conditions. Consequently, when the compatibility condition analogous to (14) is analysed, it is found to have the form

$$k^{-1}M\,\frac{\partial^2 C}{\partial Z^2} + iJ\,\frac{\partial C}{\partial X} + \int_0^\infty K\!\left(\frac{\ell}{k}\right)(k-\ell)\ell\, C(k-\ell,X,Z)\,C(\ell,X,Z)\,d\ell = 0, \tag{16}$$

where the coefficient J and kernel K(.) are identical to those in (7).

Explicitly, it is found that for both cases E and O,

$$J = \int_{-\infty}^{0} \{2c_{j1m1}A_j^*A_m + ic_{j2m1}(A_j^{*\prime}A_m - A_j^{\prime}A_m^*)\}\,d\xi \tag{17}$$

$$M = \int_{-\infty}^{0} g_L g_N c_{jLmN}A_j^*A_m\,d\xi$$

$$- \int_{-\infty}^{0} \{c_{j2m2}b_j^{*\prime}b_m^{\prime} + c_{j2m1}(b_j^{*\prime}b_m - b_j^{\prime}b_m^*) + (c_{j1m1} - \rho c^2 n_{jm})b_j^*b_m\}\,d\xi, \tag{18}$$

while the kernel $K(k^{-1}\ell)$ is given in terms of $s^{(p)}$, $a_j^{(p)}$ and $B^{(p)}$ by (A.2) of [3], with the summations extended over the range 1,...,4. The difference between the two cases lies in the way $A_m(\xi)$, $b_m(\xi)$ and c are affected by the boundary conditions.

In case F, the definitions are slightly more complicated, M and J being replaced by two other *real* parameters $M-\epsilon_0A_{04}^*(gb_4(0) + \frac{1}{2}A_{04})$ and $J-\epsilon_0A_{04}^*A_{04}$, respectively. The formula for the kernel $K(k^{-1}\ell)$ is unchanged in $\ell > k$, but altered by the addition of a constant $\epsilon_0A_{04}^2(1-iA_{01}^*)$ for $0 < \ell < k$.

Equation (16) is the evolution equation for nonlinear surface acoustic waves subject to weak lateral spreading. In addition to the material function $J^{-1}K(\ell/k)$ which occurs in the evolution equation (7) for unidirectional waves, it involves only one other material constant, the ratio M/J.

4. THE DISPERSION COEFFICIENT

For all of the boundary conditions considered, the coefficients M and J involve complicated quadratic integrals of the depth dependence $A_j(\xi)$ of linear theory. However, the observation in [4] that M, J and g can be expressed in terms of the dispersion relation for linearized waves carries over to the piezoelectric case.

The procedure is to analyse harmonic surface waves

$$u_m \equiv V_m(X_2;\underset{\sim}{k}) \exp i\,(k_N X_N - \omega t) \ , \qquad k_2 \equiv 0$$

with (surface) wave vector $\underset{\sim}{k} = (k_1, 0, k_3)$ and depth variation governed by the generalization of (10) in $X_2 < 0$ to

$$\mathcal{L}_{jm}V_m \equiv (k_L k_M c_{jLmM} - \rho\omega^2 \eta_{jm})V_m - ik_M(c_{j2mM} + c_{m2jM})\frac{dV_m}{dX_2} - c_{j2m2}\frac{d^2V_m}{dX_2^2} = 0 \ , \qquad (19)$$

together with appropriate boundary conditions involving the modified operator $\mathcal{B}_{jm} \equiv ik_N c_{j2mN} + c_{j2m2}\, d/dX_2$. The condition that this system possesses solutions which decay as $X_2 \to -\infty$ imposes a dispersion relation $D(\omega, k_1, k_3) = 0$ of the form

$$\omega = kc(\hat{\underset{\sim}{k}}) = \Omega(\underset{\sim}{k}), \text{ where } \hat{\underset{\sim}{k}} = \underset{\sim}{k}/k, \quad k = |\underset{\sim}{k}| \ .$$

A convenient choice for D, utilizing solutions V_m normalized by the condition $V_2(0;\underset{\sim}{k}) = 1$, is

$$D(\omega,\underset{\sim}{k}) \equiv \int_{-\infty}^{0} V_j^* \mathcal{L}_{jm} V_m \, dX_2 = \int_{-\infty}^{0} V_m \mathcal{L}_{mj}^* V_j \, dX_2 \ , \qquad (20)$$

(except in case F, when an additional contribution arising from Φ in $X_2 > 0$ is needed). After differentiation of (20) with respect to k_N and use of integration by parts, boundary conditions and a number of identities derivable from (19), it is found that

$$D_{,N} \equiv \frac{\partial D}{\partial k_N} = \int_{-\infty}^{0} \left\{ k_M c_{mMjN}(V_m V_j^* + V_m^* V_j) + ic_{j2mN}\left(V_m \frac{dV_j^*}{dX_2} - V_m^* \frac{dV_j}{dX_2} \right) \right\} dX_2 \ . \qquad (21)$$

By specializing to the case $k_3 = 0$, $k_1 = k$ and writing $V_m(X_2;k,0) = A_m(\xi)$ $\xi \equiv kX_2$, it is then seen that equation (15) is exactly in the form

$$D_{,3}(\omega,k,0) = gD_{,1}(\omega,k,0). \qquad (22)$$

This is no more than the statement that the ray direction $dX_3/dX_1 = g$ is given by the normal to the 'slowness curve' $D(\omega,k_1,k_3) = 0$ at (arbitrary) fixed ω.

Generally, the ray direction is not parallel to the wavenormal. However, if the sagittal plane OX_1X_2 is a plane of reflectional symmetry, it is found that $g = 0$. A further observation, from (17) is that

$$J = D_{,1}(\omega,k,0) \ . \qquad (23)$$

Moreover, it may also be shown that the function $b_m(\xi)$ required in the $O(\varepsilon^{\frac{1}{2}})$ terms for u are given by

$$b_m(\xi) = -kg_N \frac{\partial V_m}{\partial k_N}(X_2;k,0) \ . \qquad (24)$$

This success in relating features of linear wave theory to $D(\omega,k_1,k_3)$ suggests that M might be related to $D_{,LN}(\omega,k,0)$. Indeed, by partial differentiation of (21) with respect to k_L, it is found that

$$M = \tfrac{1}{2}kg_L g_N\, D_{,LN}(\omega,k,0) = \tfrac{1}{2}k(D_{,33} - 2g\,D_{,13} + g^2 D_{,11}) \ . \qquad (25)$$

Thus, the parameters g and M/J which characterize the energy propagation direction and the transverse spreading are simply related to the geometry of the dispersion curve $D(\omega,k_1,k_3) = 0$, ω fixed. (The kernel $J^{-1}K(k^{-1}\ell)$, which is the sole constitutive function influencing the nonlinear term in the evolution equation (16), has been computed for some propagation directions on Y-cut $LiNbO_3$. Generally it is complex. See [1],[2] for details.)

The geometric significance of the ray direction $dX_3/dX_1 = g$ and of the ratio M/J are shown in Figure 1. The slowness curve is determined by $D(1,k_1,k_3) = 0$. If the angles ψ and ϕ denote the inclinations of the wave normal $\underset{\sim}{k}$ and ray direction to the X_1-axis, then $k_3 = k_1 \tan\psi$ and $g = \tan\phi = D_{,3}/D_{,1}$. Since it can be shown that (25) reduces to

$$M = \tfrac{1}{2}kD_{,1}\left(\frac{\partial g}{\partial k_3} - g\frac{\partial g}{\partial k_1}\right) \quad ,$$

it is found that at $\psi = 0$ $(k_3 = 0)$

$$\frac{M}{J} = \tfrac{1}{2}\sec^2\phi\,\frac{d\phi}{d\psi} = \tfrac{1}{2}(1+g^2)\,\frac{d\phi}{d\psi} \quad . \qquad (26)$$

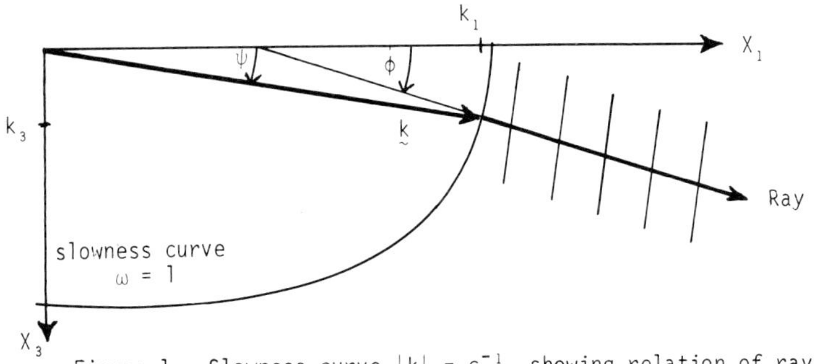

Figure 1. Slowness curve $|\underset{\sim}{k}| = c^{-1}$, showing relation of ray angle ϕ to the angle ψ of the wave normal.

Thus, M/J depends on the convexity or concavity of the dispersion curve in the vicinity of the X_1-axis. The regime $M/J > 0$ characterizes spreading, with $M/J = \tfrac{1}{2}$ corresponding to isotropic behaviour for which $\phi = \psi$. The situation $M/J < 0$ arises only if the curve is concave near the X_1-axis, so that waves tend to focus rather than to spread.

REFERENCES

[1] David, E.A. and Parker, D.F., Nonlinear Evolution of Piezoelectric SAWs, in: Parker, D.F. and Maugin, G.A. (eds) Recent Developments in Surface Acoustic Waves, Springer Ser. Wave Phen. (Springer, Berlin, Heidelberg, 1988)

[2] Parker, D.F. and David, E.A., Nonlinear Piezoelectric Surface Waves, in preparation.

[3] Parker, D.F., Int. J. Engng Sci. 26 (1988) 59.

[4] Kalyanasundaram, N., Parker, D.F. and David, E.A., The Spreading of Nonlinear Elastic Surface Waves, in preparation.

[5] Lardner, R.W., Int. J. Engng Sci. 21 (1983) 1331.

ELASTIC WAVE PROPAGATION
M.F. McCarthy, M.A. Hayes, (Editors)
Elsevier Science Publishers B.V. (North-Holland), 1989

STEADY, STRUCTURED SHOCK WAVES OF ARBITRARY INTENSITY IN THERMOELASTIC MATERIALS*

J. E. Dunn and R. L. Fosdick

Solid Dynamics Department
Sandia National Laboratories
Albuquerque, New Mexico

Department of Aerospace Engineering
and Mechanics
University of Minnesota
Minneapolis, Minnesota

1. PRELIMINARY IDEAS

If $\mathcal{B}$ is a one dimensional body with particles X, then we let $x = \chi(X,t)$ denote a *motion* of $\mathcal{B}$. The *velocity*, $\mathrm{v} = \mathrm{v}(X,t)$, and the *deformation gradient*, $F = F(X,t)$, associated with $\chi(\cdot,\cdot)$ are then given by the respective partial derivatives $\dot{\chi}(X,t)$ and $\chi,_X(X,t)$. To ensure the invertibility of $\chi(\cdot,t)$, we assume $F > 0$. For smooth enough motions we have the *condition of kinematic compatibility*:

$$\mathrm{v},_X = \dot{F}. \tag{1}$$

Governing the motion of $\mathcal{B}$ we have momentum balance, energy balance, and entropy inbalance. Here these have the respective forms

$$\sigma,_X = \rho_R \dot{\mathrm{v}}, \tag{2}$$

$$\rho_R \dot{e} = \sigma \mathrm{v},_X - q,_X, \tag{3}$$

$$\rho_R \dot{\eta} \geq -\left(\frac{q}{\theta}\right),_X, \tag{4}$$

where $\sigma = \sigma(X,t)$ is the *axial stress*, $e = e(X,t)$ is the specific *internal energy*, $q = q(X,t)$ is the *heat flux*, $\eta = \eta(X,t)$ is the specific *entropy*, $\theta = \theta(X,t)$ is the (positive) *absolute temperature*, and where ρ_R, the *referential mass density*, is a constant, independent of X. We wish to study *steady, structured shock wave* solutions of the system (1)-(4). These are smooth 7-tuples of functions

$$\pi = \pi(X,t) = (F, \mathrm{v}, \sigma, e, q, \eta, \theta)(X,t)$$

satisfying (1)-(4) and such that

$$\pi = \pi(X,t) = \hat{\pi}(\xi) = \hat{\pi}(X - \mathcal{V}t), \tag{5}$$

where $\xi = X - \mathcal{V}t \in (-\infty, \infty)$ is the *travelling wave variable* and where the number $\mathcal{V}$ is the referential *shock speed* or *shock velocity*. For convenience, we assume $\mathcal{V} > 0$; thus, the material at $+\infty$ $(-\infty)$ is *ahead* (*in back*) of the wave. We say the wave is *compressive* (*expansive*) if $F^- < F^+$ $(F^- > F^+)$, where $F^\pm \equiv \hat{F}(\pm\infty)$. The steady, structured shock

*This work performed at Sandia National Laboratories supported by the U.S. DOE under contract #DE-AC04-76DP00789.

wave hypothesis (5) allows us to integrate the system (1)-(4) one time. From (1) and (2) we have

$$\hat{v}(\xi) - v^+ = -\mathcal{V}(\hat{F}(\xi) - F^+), \tag{1'}$$

$$\hat{\sigma}(\xi) - \sigma^+ = \rho_R \mathcal{V}^2(\hat{F}(\xi) - F^+). \tag{2'}$$

In addition, the energy equation may now be put in the form

$$\hat{q}(\xi) = \mathcal{V}\left\{\rho_R(\hat{e}(\xi) - e^+) - \frac{1}{2}(\hat{\sigma}(\xi) + \sigma^+)(\hat{F}(\xi) - F^+)\right\}, \tag{3'}$$

while the entropy inequality becomes

$$\rho_R \mathcal{V}(\hat{\eta}(\xi) - \eta^+) \geq \frac{\hat{q}(\xi)}{\hat{\theta}(\xi)}. \tag{4'}$$

We, of course, assume the existence of the limits $F^\pm, v^\pm, \sigma^\pm$, etc. We also assume $q^\pm = 0$. Thus, the front and back states are taken to be in "equilibrium".

2. THERMOELASTIC MATERIALS

We assume the material composing $\mathcal{B}$ is *thermoelastic*. This means that there are two *response functions* $\overline{e}(\cdot,\cdot)$ and $\overline{q}(\cdot,\cdot,\cdot)$ for the energy and heat flux, respectively, such that

$$e = \overline{e}(\eta, F) \quad \text{and} \quad q = \overline{q}(\eta, F, G),$$

where $G \equiv \theta,_X$ is the material temperature gradient, and where $\overline{e}(\cdot,\cdot)$ serves as a potential for the temperature and the stress in the sense that

$$\begin{aligned} \theta &= \overline{\theta}(\eta, F) = \overline{e}_\eta(\eta, F), \\ \sigma &= \overline{\sigma}(\eta, F) = \rho_R \overline{e}_F(\eta, F). \end{aligned}$$

Moreover, thermodynamic considerations require that $\overline{q}(\cdot,\cdot,\cdot)$ satisfy $\overline{q}(\eta, F, 0) \equiv 0$.

For thermoelastic materials (3′) gives that at $\xi = -\infty$

$$\mathcal{H}(\eta^-, F^-) = 0,$$

where we have introduced the *Hugoniot function* based at (η^+, F^+)

$$\mathcal{H}(\eta, F) \equiv \rho_R(\overline{e}(\eta, F) - e^+) - \frac{1}{2}(\overline{\sigma}(\eta, F) + \sigma^+)(F - F^+).$$

The zero level set of $\mathcal{H}(\cdot,\cdot)$, *i.e.*

$${}^0\mathcal{H} \equiv \{(\eta, F) \mid \mathcal{H}(\eta, F) = 0\},$$

is called the *Hugoniot set* based at (η^+, F^+). For a given front state (η^+, F^+), the only possible back states (η^-, F^-) are those that lie in ${}^0\mathcal{H}$.

Momentum balance (2′) and energy balance (3′) now take the respective forms

$$\overline{\sigma}(\hat{\eta}, \hat{F}) - \sigma^+ = \rho_R \mathcal{V}^2(\hat{F} - F^+), \tag{2''}$$

$$\overline{q}(\hat{\eta}, \hat{F}, \hat{\theta}') = \mathcal{V}\mathcal{H}(\hat{\eta}, \hat{F}), \tag{3''}$$

where $(\hat{\eta}, \hat{F}) \equiv (\hat{\eta}, \hat{F})(\xi)$. From (2″) we see that $(\hat{\eta}, \hat{F})(\xi) \in \mathcal{R}(\mathcal{V})$ where

$$\mathcal{R}(\mathcal{V}) \equiv \left\{(\eta, F) \mid \bar{\sigma}(\eta, F) - \sigma^+ = \rho_R \mathcal{V}^2 (F - F^+)\right\}$$

is the **Rayleigh curve** based at (η^+, F^+). Thus, the **Rayleigh trajectory** $(\hat{\eta}, \hat{F})(\cdot)$, which solves the differential equation (3″), must lie on the Rayleigh curve $\mathcal{R}(\mathcal{V})$. For a thermoelastic material based on aluminum, the relevant portions of ${}^0\mathcal{H}$ and of four Rayleigh curves $\mathcal{R}(\mathcal{V})$, all based at $(\eta^+, F^+) = (0.0, 0.995)$, are shown in Figure 1.

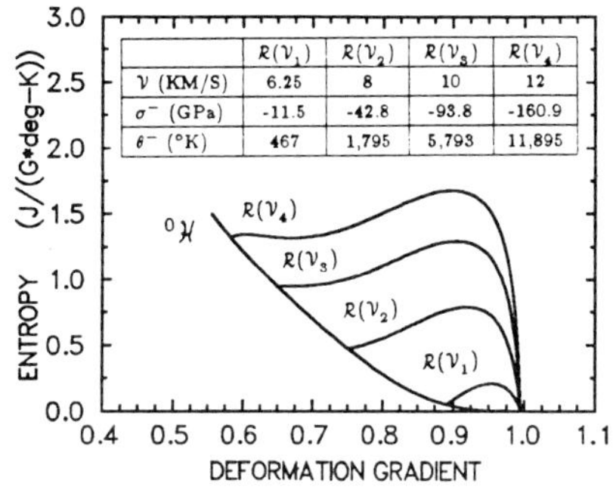

Figure 1. Portions of the Hugoniot ${}^0\mathcal{H}$ and four Rayleigh curves $\mathcal{R}(\mathcal{V})$, all based at the front state $(\eta^+, F^+) = (0.0, 0.995)$. The front state stress σ^+ is -0.526 GPa; the front state temperature θ^+ is 298° K.

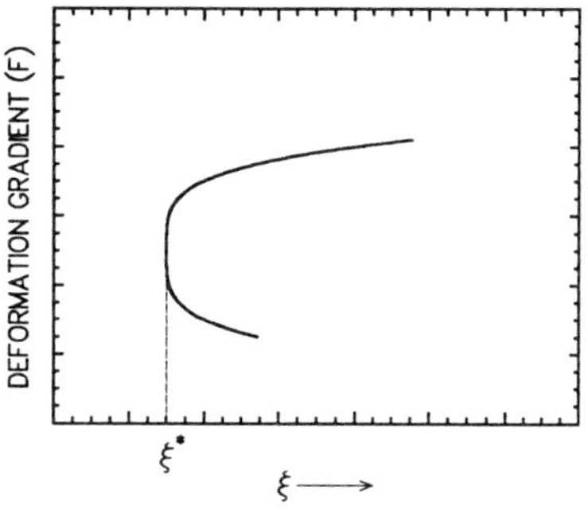

Figure 2. A portion of the graph of the deformation gradient $\hat{F}(\cdot)$ as determined by the solution to the differential equation (3.3) when $I - \rho_R \mathcal{V}^2 k$ changes sign along $\mathcal{R}(\mathcal{V})$.

3. RAYLEIGH'S NON-EXISTENCE RESULT

For perfect gases with constant specific heats that obey Fourier's law of heat conduction, the system (2″)-(3″) was, in essence, written down first by RANKINE [1] in 1869. Forty-one years later, in (1910), RAYLEIGH [2] revisted RANKINE'S analysis and made two fundamental observations. First, RAYLEIGH noted that the condition of JOUGUET and ZEMPLÉN must hold, *i.e.*, only that part of the Hugoniot ${}^0\mathcal{H}$ lying in the *upper half space* $[\eta^+, \infty) \times (0, \infty)$ of the η-F plane is available for back states. This, of course, follows at once from (4′) if we set $\xi = -\infty$ and is evidenced in Figure 1 where each η_i^-, $i =$ 1, 2, 3 , or 4, is indeed larger than η^+. (Figure 1 also makes clear the phenomenon of entropy overshoot: in a steady, structured shock wave in a thermoelastic material there are places in the center of the wave that experience a much higher entropy than the back state entropy η^-.)

RAYLEIGH'S second observation seems to have been much more crucial for structured shocks in thermoelastic materials. Indeed, at first glance it seems to be fatal for the entire theory. Essentially, RAYLEIGH noted that, for RANKINE'S materials, the differential equation (3″) delivers a single-valued solution for, say, $\hat{\sigma}(\xi)$ if and only if

$$\frac{\sigma^-}{\sigma^+} \leq \frac{1 + \mathfrak{R}}{3 - \mathfrak{R}}, \tag{6}$$

where $\mathfrak{R}$ is the *ratio of the specific heats*. Since $\mathfrak{R} \approx 1.5$ for gases, the estimate of RAYLEIGH gives $\dfrac{\sigma^-}{\sigma^+} \leq 1.66$ and we conclude, as did RAYLEIGH, that only relatively weak shocks can be supported by heat conduction alone. But, RAYLEIGH'S *estimate is clearly incomplete* since it gives a nonsensical result if $\Re > 3$, a value which is too big for a realistic gas but shows RAYLEIGH'S estimate is model-dependent. Indeed, recall that the (dimensionless) *Grüneisen coefficient* is given by

$$\gamma = \overline{\gamma}(\eta, F) \equiv \frac{-F\overline{\sigma}_\eta}{\rho_R \theta} = \frac{-F}{\theta}\overline{\theta}_F.$$

For a perfect gas, whether the specific heats be constant or not, $\Re = 1+\gamma$. Thus, in terms of the Grüneisen coefficient, RAYLEIGH'S estimate is

$$\frac{\sigma^-}{\sigma^+} \leq \frac{2+\gamma}{2-\gamma}. \tag{6'}$$

But, values of $\gamma > 2$ are common for solids. Even $\gamma < 0$ is possible.

4. RAYLEIGH'S RESULT GENERALIZED

Although he did not frame it in these terms, the key question behind RAYLEIGH'S estimate (6) is: Under what conditions are momentum balance, (2″), and energy balance, (3″), compatible ? As we have noted, (2″) means that the $(\hat{\eta}, \hat{F})$ pairs experienced in a steady, structured shock wave must lie on the *algebraically* determined Rayleigh curve $\mathcal{R}(\mathcal{V})$. On the other hand, (3″) implies that the Rayleigh trajectory $(\hat{\eta}, \hat{F})(\cdot)$ must be such that $\hat{q}(\xi)$ vanishes only on the Hugoniot ${}^0\mathcal{H}$. Let us first off dispose of a rather trivial incompatibility between (2″) and (3″). This occurs if $\overline{q}(\cdot,\cdot,\cdot)$ vanishes over set $\mathcal{Q}$ in the η-F plane (*i.e.*, $\overline{q}(\eta, F, \cdot) = 0$ for $(\eta, F) \in \mathcal{Q}$). Clearly, the placement of $\mathcal{Q}$ would have to be carefully arranged so as to prevent it from being crossed by Rayleigh curves in their passage from (η^+, F^+) to their respective back states $(\eta^-, F^-) \in {}^0\mathcal{H}$. Indeed, no Rayleigh trajectory could cross $\mathcal{Q}$ before it reaches ${}^0\mathcal{H}$ and also be consistent with (3″). To rule this out, we strengthen our earlier condition, $\overline{q}(\eta, F, 0) = 0$, by requiring that $\overline{q}(\cdot,\cdot,\cdot)$ be *strictly conductive, i.e.*, $\overline{q}(\eta, F, G) = 0 \Longleftrightarrow G = 0$.

A more serious and subtle potential incompatibility between (2″) and (3″) still remains, however. To describe it let us suppose that $\mathcal{R}(\mathcal{V})$ is *regular, i.e.*, $\overline{\sigma}_\eta$ and $\overline{\sigma}_F - \rho_R \mathcal{V}^2$ do not simultaneously vanish on $\mathcal{R}(\mathcal{V})$. Thus, $\mathcal{R}(\mathcal{V})$ is locally curve-like. Let us also suppose that the function $\overline{\theta}(\cdot, F)$ may be inverted to give us $\eta = \tilde{\eta}(\theta, F)$, and that $c_d \equiv \theta\tilde{\eta}_\theta(\theta, F) = \theta/\overline{\theta}_\eta$, the *specific heat at fixed deformation*, is positive and finite on $\mathcal{R}(\mathcal{V})$.

Now, (3″) allows $\hat{q}(\xi)$ (and so $\hat{\theta}'(\xi)$) to vanish only on ${}^0\mathcal{H}$. But, by (2″), we have that $\overline{\sigma}_\eta \hat{\eta}' + (\overline{\sigma}_F - \rho_R \mathcal{V}^2)\hat{F}' = 0$. Thus, along the Rayleigh curve $\mathcal{R}(\mathcal{V})$, we have

$$\hat{\theta}' = \overline{\theta}_\eta \hat{\eta}' + \overline{\theta}_F \hat{F}' = \overline{\theta}_\eta \left(\frac{I - \rho_R \mathcal{V}^2}{A - \rho_R \mathcal{V}^2} \right) \hat{\eta}' = \frac{-\overline{\theta}_\eta}{\overline{\sigma}_\eta} \left(I - \rho_R \mathcal{V}^2 \right) \hat{F}', \tag{7}_{1,2,3}$$

where $(7)_2$ holds anywhere on $\mathcal{R}(\mathcal{V})$ where $A - \rho_R \mathcal{V}^2 \neq 0$, where $(7)_3$ holds anywhere on $\mathcal{R}(\mathcal{V})$ where $\overline{\sigma}_\eta \neq 0$, and where $A = \overline{A}(\eta, F) \equiv \overline{\sigma}_F(\eta, F)$ is the *adiabatic modulus* and, with $\tilde{\sigma}(\theta, F) \equiv \overline{\sigma}(\tilde{\eta}(\theta, F), F)$, the quantity $I = I(\eta, F) \equiv \tilde{\sigma}_F(\theta, F)|_{\theta=\overline{\theta}(\eta,F)}$ is the *isothermal modulus*.

Next, it is not difficult to show that $I \leq A$ wherever $c_d \geq 0$. Further,

$$I - \rho_R \mathcal{V}^2 = 0 \Longleftrightarrow A - \rho_R \mathcal{V}^2 = \frac{c_d}{\rho_R \theta} \overline{\sigma}_\eta^2.$$

Therefore, *if there exists* ξ^* *such that* $(\hat{\eta}, \hat{F})(\xi^*) \notin {}^0\mathcal{H}$ *and* $\hat{I}(\xi^*) - \rho_R \mathcal{V}^2 = 0$ then, by the regularity of $\mathcal{R}(\mathcal{V})$, *neither* σ^*_η *or* $A^* - \rho_R \mathcal{V}^2$ *can vanish.* Consequently, by $(7)_{2,3}$,

$$|\hat{\eta}'(\xi^*)| = \infty \quad and \quad |\hat{F}'(\xi^*)| = \infty.$$

In addition, *if* $\hat{I}(\cdot) - \rho_R \mathcal{V}^2$ *also changes sign at* ξ^*, *then, by* $(7)_2$ *and* $(7)_3$, *so too must* $\hat{\eta}'(\cdot)$ *and* $\hat{F}'(\cdot)$ *i.e.*, as shown in Figure 2, **the "solutions"** $(\hat{\eta}, \hat{F})(\cdot)$ **are multi-valued.**

Thus, for a steady, structured shock wave of speed $\mathcal{V}$ to exist $I(\eta, F) - \rho_R \mathcal{V}^2$ *cannot vanish and change sign on that part of* $\mathcal{R}(\mathcal{V})$ *connecting* (η^+, F^+) *to* (η^-, F^-). Equivalently, on that part of $\mathcal{R}(\mathcal{V})$ connecting (η^+, F^+) to (η^-, F^-), the temperature $\bar{\theta}(\eta, F)$ must vary **monotonically.**

Recall now the *supersonic–subsonic condition*

$$A^+ < \rho_R \mathcal{V}^2 < A^-, \tag{8}$$

which, as is well-known, can be shown to follow either from certain *a priori* inequalities or from conditions of shock stability.† Clearly, the supersonic-subsonic condition implies that the difference $A - \rho_R \mathcal{V}^2$ **must** change sign on $\mathcal{R}(\mathcal{V})$, while our analysis above has revealed that the difference $I - \rho_R \mathcal{V}^2$ **cannot** change sign on $\mathcal{R}(\mathcal{V})$. Further, since $c_d > 0$, we have that $I^+ \leq A^+$. Thus, $\rho_R \mathcal{V}^2 > A^+$ means we must have $\rho_R \mathcal{V}^2 \geq I$ everywhere on that part of $\mathcal{R}(\mathcal{V})$ connecting (η^+, F^+) to (η^-, F^-). This turns out to have several additional implications (see [3] for a more complete discussion):

$\left\{\begin{array}{l}\text{Let a steady, structured shock}\\ \text{wave satisfy } \rho_R \mathcal{V}^2 > A^+.\\ \text{Also, let } \gamma \text{ and } c_d \text{ be } \textbf{positive}\\ \text{everywhere along } \mathcal{R}(\mathcal{V}).\end{array}\right\}$ Then, $\left\{\begin{array}{l}\text{the deformation gradient } \hat{F}(\cdot)\\ \text{is monotone, and the passage of}\\ \text{such a compressive (expansive)}\\ \text{shock always heats (cools) the}\\ \text{material behind it.}\end{array}\right\}$

Less common is the phenomenon of *compressive shock cooling* (found in fluid nitrogen at high densities and temperatures):

$\left\{\begin{array}{l}\text{Let a steady, structured shock}\\ \text{wave satisfy } \rho_R \mathcal{V}^2 > A^+. \text{ Also,}\\ \text{let } c_d \text{ be positive and let } \gamma\\ \text{be } \textbf{negative} \text{ everywhere on } \mathcal{R}(\mathcal{V}).\end{array}\right\}$ Then, $\left\{\begin{array}{l}\text{the deformation gradient } \hat{F}(\cdot)\\ \text{is monotone, and the passage of}\\ \text{such a compressive (expansive)}\\ \text{shock always cools (heats) the}\\ \text{material behind it.}\end{array}\right\}$

If the Grüneisen coefficient changes sign on $\mathcal{R}(\mathcal{V})$, it is no longer straightforward to tell whether the body warms or cools due to the passage of the shock; however,

$\left\{\begin{array}{l}\text{Let a steady, structured shock}\\ \text{wave have a specific heat } c_d\\ \text{of fixed sign throughout its}\\ \text{structure. Assume, however}\\ \text{that } \gamma \text{ changes sign on } \mathcal{R}(\mathcal{V}).\end{array}\right\}$ Then, $\left\{\begin{array}{l}\text{the deformation gradient}\\ \hat{F}(\cdot) \text{ is } \textbf{not} \text{ monotone}\\ \text{through the shock and is}\\ \text{locally } \textbf{flat}\text{, i.e., } \hat{F}' = 0, \text{ at}\\ \text{each place where } \gamma \text{ vanishes.}\end{array}\right\}$

†Recently, DUNN & FOSDICK have found that more refined forms of (8) follow from the dissipation inherent in the entropy inequality. See Section 3 of [3].

The phenomenon of entropy overshoot noted earlier implies that the entropy $\hat{\eta}(\cdot)$ is not monotone through the shock. The above result gives conditions when the same is true of the deformation gradient $\hat{F}(\cdot)$. That $\hat{F}(\cdot)$ itself can suffer overshoot or undershoot follows from:

$$\left\{\begin{array}{l}\text{Additionally, if } I^{\pm} - \rho_R \mathcal{V}^2 \text{ does} \\ \text{not vanish and if } \gamma^{-}\gamma^{+} < 0,\end{array}\right\} \text{ then, } \left\{\begin{array}{l}\text{besides failing to be monotone,} \\ \hat{F}(\cdot) \text{ is not everywhere} \\ \text{between } F^{+} \text{ and } F^{-}\text{, i.e.,} \\ \text{there is strain } \textbf{undershoot} \\ \text{or strain } \textbf{overshoot} \text{ at either} \\ \text{the front or the back of the wave.}\end{array}\right\}$$

To address the possible non-monotone growth of $\overline{\theta}(\cdot,\cdot)$ on $\mathcal{R}(\mathcal{V})$, consider now the function

$$\mathcal{T} = \mathcal{T}(\eta, F) \equiv \overline{\sigma}(\eta, F) - \sigma^{+} - I(\eta, F)(F - F^{+}).$$

The zero level set of $\mathcal{T}(\cdot,\cdot)$, *i.e.*,

$${}^{0}\mathcal{T} \equiv \{(\eta, F) |\ \mathcal{T}(\eta, F) = 0\}$$

is called the **curve of thermal extremes** based at (η^{+}, F^{+}). When a Rayleigh curve $\mathcal{R}(\mathcal{V})$ crosses ${}^{0}\mathcal{T}$ the difference $I - \rho_R \mathcal{V}^2$ will change sign, *i.e.*, the temperature on $\mathcal{R}(\mathcal{V})$ will experience a strict local minimum or maximum. Accordingly, *a steady, structured shock wave of speed $\mathcal{V}$ does not exist if $\mathcal{R}(\mathcal{V})$ is blocked by ${}^{0}\mathcal{T}$ in its transit to ${}^{0}\mathcal{H}$*. We close with a few results culled from [3] that both extend RAYLEIGH'S analysis and reveal that perfect gases are rather atypical in the simplicity of their Hugoniot and curve of thermal extremes behavior. Our first theorem gives fairly mild conditions for certain Rayleigh curves to be unblocked by ${}^{0}\mathcal{T}$ at least in a neighborhood of (η^{+}, F^{+}).

Theorem 1. *For any thermoelastic material, consider front states (η^{+}, F^{+}) with $\gamma^{+} \neq 0$ and $c_{d}^{+} > 0$. Let $\mathcal{V}$ be any shock speed such that*

$$A^{+} < \rho_R \mathcal{V}^2.$$

Then, near enough to (η^{+}, F^{+}), the curves $\mathcal{R}(\mathcal{V})$ and ${}^{0}\mathcal{T}$ lie on **opposite sides** *of the Hugoniot ${}^{0}\mathcal{H}$.*

Corollary. *For sufficiently weak shocks, i.e., for shocks that complete their transit from (η^{+}, F^{+}) to (η^{-}, F^{-}) in a small enough neighborhood of (η^{+}, F^{+}), the Rayleigh curve $\mathcal{R}(\mathcal{V})$ will* **not** *be blocked by ${}^{0}\mathcal{T}$.*

For perfect gases much more can be said about the intersection of ${}^{0}\mathcal{T}$ and Rayleigh curves $\mathcal{R}(\mathcal{V})$. The first half of our next result reestablishes RAYLEIGH'S estimate (6′) in terms of such intersections; the second half addresses the situation when RAYLEIGH'S estimate is nonsensical.

Theorem 2. *For the thermoelastic materials of* RANKINE *and* RAYLEIGH, *assume first that $\gamma \in (0,2)$. Then the only Rayleigh curves $\mathcal{R}(\mathcal{V})$ that are not blocked by the curve of thermal extremes ${}^{0}\mathcal{T}$ in their transit from (η^{+}, F^{+}) to (η^{-}, F^{-}), $\eta^{-} > \eta^{+}$, are those meeting*

$$\sqrt{\frac{A^{+}}{\rho_R}} < \mathcal{V} \leq \sqrt{\frac{2+3\gamma}{(2-\gamma)(1+\gamma)}}\sqrt{\frac{A^{+}}{\rho_R}},$$

i.e.. those meeting

$$1 < \frac{\sigma^-}{\sigma^+} \leq \frac{2+\gamma}{2-\gamma}.$$

On the other hand, if $\gamma \in [2, \infty)$, *then* **every** *Rayleigh curve* $\mathcal{R}(\mathcal{V})$, $\mathcal{V} > \sqrt{A^+/\rho_R}$, *is unblocked in its transit from* (η^+, F^+) *to* (η^-, F^-) *by* ${}^0\mathcal{T}$. *Steady, structured shocks with arbitrarily high* $|\sigma^-|$ *and* $\mathcal{V}$ *can be supported by heat conduction alone.*

More complicated behavior of Rayleigh curves and the curve of thermal extremes is possible. Consider just the special class of *Mie-Grüneisen materials*. These are thermoelastic materials for which

$$\gamma = \overline{\gamma}(F) \Longleftrightarrow c_d = \overline{c}_d(\eta) \Longleftrightarrow e = \alpha(F)N(\eta) + \beta(F).$$

For the same front state (η^+, F^+), Figure 3 presents the Hugoniot set ${}^0\mathcal{H}$ (solid line) and the curve of thermal extremes ${}^0\mathcal{T}$ (dashed line) for four different Mie-Grüneisen materials based on aluminum. With the exception of (d), all the materials have very similar looking Hugoniots but quite different curves of thermal extremes.

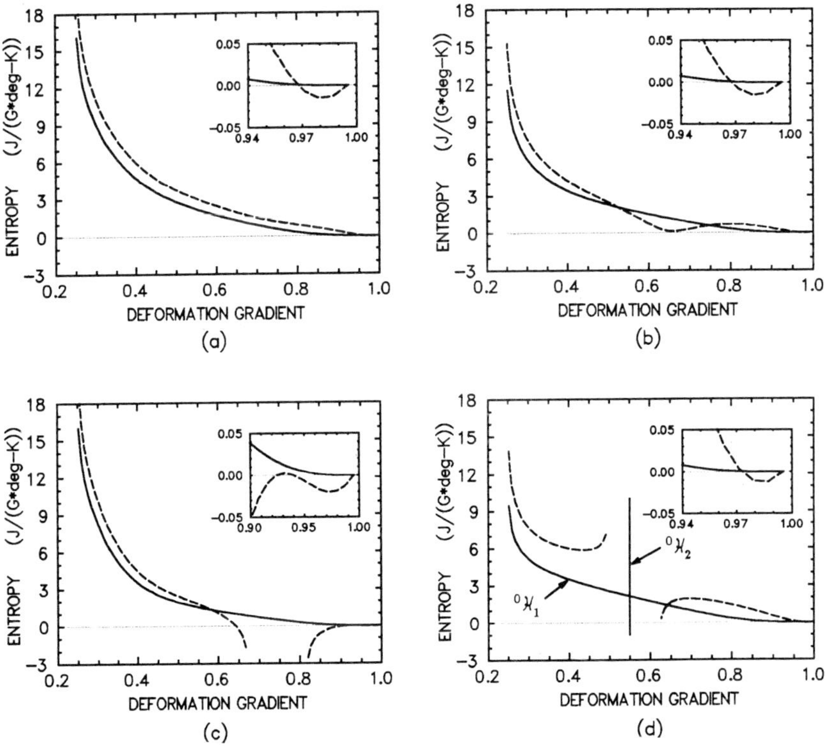

Figure 3. The Hugoniot (solid line) and the curve of thermal extremes (dashed line) for four different Mie-Grüneisen materials.

Although they are not shown, the Rayleigh curves based at (η^+, F^+) in each of the materials of Figure 3, like the Rayleigh curves of Figure 1, sweep upwards and to the left from (η^+, F^+) and so lie *above* ${}^0\mathcal{H}$ in their transit from (η^+, F^+) to (η^-, F^-). The insets to each of the four graphs thus illustrate Theorem 1 by revealing that near enough to $(\eta^+, F^+) = (0.0, 0.995)$ the curve of thermal extremes lies *under* ${}^0\mathcal{H}$. Figure 3(a) shows behavior like RAYLEIGH found for perfect gases: only very weak steady, structured shocks are not blocked by ${}^0\mathcal{T}$ in their passage from the front to the back state. Figure 3(b) presents us with a material for which any back state deformation in the *disjoint* union $[0.523, 0.745] \cup [0.968, 0.995)$ will correspond to structured shocks that are unblocked by ${}^0\mathcal{T}$. In Figure 3(c) the curve of thermal extremes "tries" to block ${}^0\mathcal{H}$ (at about $F = 0.935$) but fails. Accordingly, a wide, continuous range ($F^- \in [0.58, 0.995)$) of steady, structured shocks are supportable by heat conduction alone in this material. Finally, Figure 3(d) presents us with a case in which the Hugoniot set is *bifurcated*. Here ${}^0\mathcal{H}$ is the union of two smooth arcs which intersect at a nonzero angle. One arc, ${}^0\mathcal{H}_1$, is very much like the Hugoniots of Figures 3(a,b,c), the other arc is a vertical ray, ${}^0\mathcal{H}_2$, at $F = 0.55$. We remark, however, that the subsonic condition $\rho_R \mathcal{V}^2 < A^-$ is satisfied only on the right half of ${}^0\mathcal{H}_1$ and only on the upper half of ${}^0\mathcal{H}_2$. Only states on this kinked subset of ${}^0\mathcal{H}$ are thus available as back states in steady, structured shock waves. For a general treatment of bifurcated Hugoniots, as well as a discussion of how readily they can occur in Mie-Grüneisen materials, the reader should see Section 6 of [3].

REFERENCES

1. RANKINE, W. J. M., On the thermodynamic theory of waves of finite longitudinal disturbance. *Trans. Royal Soc. London* **160**, 277-288 (1870).

2. RAYLEIGH, J., Aerial plane waves of finite amplitude. *Proc. Royal Soc. London* Series A, **84**, 247-284 (1910).

3. DUNN, J. E., & R. FOSDICK, Steady, Structured Shock Waves. Part 1: Thermoelastic Materials. To appear in *Archive for Rational Mechanics and Analysis* (1988).

ELASTIC WAVE PROPAGATION
M.F. McCarthy, M.A. Hayes, (Editors)

The Evolutionary Behaviour of Plane Transverse Weak Nonlinear Shock Waves In Unstrained Incompressible Isotropic Elastic Non-conductors

Y.B. Fu[1] and N.H. Scott[1]

Abstract

Asymptotic evolution laws for the amplitudes of a plane transverse shock wave of the type alluded to in the title and its accompanying second order discontinuity are derived in this paper using two different approximation methods. The first method is a combination of singular surface theory and the method of multiple scales. Singular surface theory yields a set of coupled evolution equations for the shock amplitude and the amplitudes of higher order discontinuities which accompany the shock. These equations are then solved using the method of multiple scales to obtain uniformly valid solutions. The other method is the shock-fitting method based on simple wave theory. Results obtained by using these two methods are compared. We deduce from the evolution law for the shock amplitude that the effects of nonlinearity on the evolution of the transverse shock waves are cumulative, becoming most pronounced for distances of travel of order $(\zeta_T hk)^{-1}$, where h and k are respectively the initial amplitudes of the shock and the accompanying second order discontinuity, ζ_T being a dimensionless material constant characterizing the degree of nonlinearity of the material. This conclusion is in contrast with that which we obtained before for dilatational shock waves where the effects of nonlinearity are stronger, becoming most pronounced for the shorter distances of travel of order $(\zeta_D k)^{-1}$, where ζ_D corresponds to ζ_T for dilatational shock waves.

1 Introduction

Weak nonlinear shock waves in elastic non-conductors may be classified into two categories, the first consisting of those shock waves across which the entropy jump is of third order in the shock amplitude, and the second consisting of those across which it is of fourth order. Shock waves in the first category are much more strongly affected by material nonlinearity. Also, there is a correspondence between the existence of rarefactional (or condensational) shock waves in the first category and the growth to infinity in finite time of rarefactional (or condensational) acceleration waves. This is not the case for shock waves in the second category, which may exist and propagate with changing amplitude whilst the corresponding acceleration wave has constant amplitude.

In a previous paper [1], we have derived an evolution law for one dimensional decaying dilatational shock waves and have made a detailed study of the evolutionary behaviour of shock waves belonging to the first category. In the present paper, it is the second category which concerns us. We consider the propagation of plane transverse weak nonlinear shock waves in unstrained incompressible isotropic elastic non-conductors, since these can be taken to be the prototype of shock waves falling into the second category. The incompressibility assumption is generally a necessary condition for the existence of purely transverse shock waves, whilst

[1]School of Mathematics, University of East Anglia, Norwich, NR4 7TJ, U.K.

the assumption of isotropy is made only to simplify the analysis. Materials of more general constitution, such as unstrained incompressible transversely isotropic elastic non-conductors, may transmit simultaneously shock waves of both categories.

There have been many contributions to the understanding of the general properties and the instantaneous growth and decay behaviour of shock waves in elastic non-conductors, see, for example, Chen [2], Nunziato and Herrmann [3], Chen and Gurtin [4]. Reviews have been written by Eringen and Suhubi [5] and McCarthy [6]. However, little work has been done on the global evolutionary behaviour of shock waves, mainly because, unlike acceleration waves, nonlinear shock waves do not obey a single exact evolution equation; their evolutionary behaviour is always coupled with that of the higher order discontinuities which accompany them. The works most closely related to the present discussion are those by Reddy and Achenbach [7], Collins [8], Bailey and Chen [9], and Anile [10]. In the first two papers, the shock-fitting method, which was first proposed by Friedrichs [11] in the context of gas dynamics, is employed to determine both the shock amplitude and the shock path in parametric forms. Using singular surface theory, Bailey and Chen [9] have recently studied the evolutionary behaviour of a one dimensional shock wave and its accompanying second order discontinuity, deriving an asymptotic evolution law which is valid only for small distances of travel. In Anile's paper, the singular surface theory is combined with a straightforward expansion procedure to derive a leading order evolution law for a shock wave which is governed by a general quasi-linear hyperbolic system of conservation laws, but questions of the convergence of the asymptotic solutions are not studied there. In fact, his approach, when specialised to the present problem, gives merely the same results as those given by the linear theory, which we can show not to be uniformly valid with respect to the distance of travel.

Since it is not possible to enter into details, we merely explain briefly the general line of development of the two methods which we use in deriving the evolution laws, and present the main results in the following three sections.

2 Basic equations

Consider the plane transverse motion

$$x_1 = X_1,\ x_2 = X_2 + u_2(X_1, t),\ x_3 = X_3, \tag{1}$$

where x_i is the position at time t of a particle which is at position X_i in the undeformed reference configuration. If we denote the strain u_{2,X_1} and the velocity $u_{2,t}$ by m and v respectively, then, after making the isentropic assumption, the basic equations consist merely of the equation of motion and the compatibility relation:

$$c^2(m)m_{,X} = v_{,t}, \text{ and } m_{,t} = v_{,X}, \tag{2}$$

where $c^2(m) \stackrel{\text{def}}{=} \rho^{-1}\partial\pi_{21}/\partial m$, and where π_{21} and ρ are respectively the nominal stress and the constant density. We have written X for X_1, a comma signifying partial differentiation.

Across a shock wave, the following jump conditions which correspond to (2) must be satisfied:

$$[\pi_{21}] = -\rho U_N[v],\ [v] = -U_N[m], \tag{3}$$

Here U_N is the shock velocity in the reference configuration, and for any quantity f, its jump across the wavefront is defined by $[f] = f^- - f^+$, superscripts "+" and "−" signifying evaluation just ahead of the shock and immediately behind the shock, respectively.

The constitutive equation for an unstrained isotropic elastic non-conductor can be shown to be of the form

$$\pi_{21} = \rho(\xi_1 m + \xi_2 m^3) + O(m^5), \tag{4}$$

where ξ_1 and ξ_2 are material constants. Then the expression for the shock velocity in terms of the shock amplitude $[m]$ is obtained by combining (3a) with (3b) and is given by

$$U_N^2 = [\pi_{21}]/(\rho[m]) = \xi_1 + \xi_2[m]^2 + O([m]^4), \tag{5}$$

and the definition of $c(m)$ together with (4) gives

$$c^2(m) = \xi_1 + 3\xi_2 m^2 + O(m^4). \tag{6}$$

3 Singular surface theory and the perturbation method.

In general, if equation (2a) is differentiated with respect to X a total of $n-1$ times and jumps taken of each, a total of n equations (including the one obtained by taking the jump of (2a)) are obtained for the $n+1$ unknown functions $[m], [m_{,X}], \ldots, [\partial^n m/\partial X^n]$, provided that jumps in time derivatives of m are eliminated by means of higher order compatibility conditions. A tractable system of equations may be obtained by making assumptions on $[\partial^n m/\partial X^n]$ sufficient to justify neglecting the term containing it. In the sequel, we shall take $n = 2$. With the use of (5) and(6), the two exact evolution equations obtained by applying the procedure explained above reduce to

$$\frac{d[m]}{dX} + \zeta_T[m]^2[m_{,X}] = O([m]^4)[m_{,X}], \tag{7}$$

$$\frac{d[m_{,X}]}{dX} + 3\zeta_T[m][m_{,X}]^2 = max\{O([m]^3)[m_{,X}]^2, O([m]^2)[m_{,XX}]\}, \tag{8}$$

where ζ_T, defined by $\zeta_T = \xi_2/\xi_1$, is a dimensionless constant characterizing the degree of nonlinearity of the material. We note from (4) that when $\zeta_T = 0$, the material reduces to a linear material within the order of approximation considered.

We now solve (7) and (8) using the perturbation method. Let us assume that the initial conditions are prescribed as

$$[m]|_{X=0} = h, \quad [m_{,X}]|_{X=0} = k. \tag{9}$$

Since h is dimensionless, we can take $|h|$, which is assumed to be small, as our perturbation parameter and we denote it by ϵ. As we have shown before in a previous paper[1], the initial amplitude of the accompanying second order discontinuity, when appropriately scaled, can be of order either one or ϵ. It can be shown that in both cases the straightforward perturbation solutions of (7) and (8) contain secular terms and are therefore not uniformly valid. To obtain solutions which converge uniformly with respect to the distance of travel, we use the method of multiple scales [12,p.231]. Omitting the details, we summarize the results here. In the two cases considered, the "far distance variables" are different and are respectively ϵX and $\epsilon^2 X$, but the solutions have the following single representation:

$$[m] = h(1 + 4\zeta_T khX)^{-\frac{1}{4}} + \begin{cases} O(h^3) & \text{if } [m_{,XX}] = O(k^2h^p),\ p \geq 1, \\ O(h^2) & \text{if } [m_{,XX}] = O(k^2), \end{cases} \tag{10}$$

$$[m_{,X}] = k(1 + 4\zeta_T khX)^{-\frac{3}{4}} + \begin{cases} O(kh^2) & \text{if } [m_{,XX}] = O(k^2h^p),\ p \geq 1, \\ O(kh) & \text{if } [m_{,XX}] = O(k^2). \end{cases} \tag{11}$$

It is seen from the above solutions that the assumptions which justify neglecting the effects of $[m_{,XX}]$ depend on the order of k, and also that the orders of error of the leading order evolution laws can be different depending on these assumptions. It can be shown that the entropy jump being non-negative requires $\zeta_T \geq 0$, which can be taken to be the existence condition for weak shock waves. Then it is clear by inspecting (10) and (11) that the shock is decaying if $kh > 0$.

4 The shock-fitting method.

According to the simple wave theory, equations (2) admits two simple wave solutions. One simple wave is outgoing, the other being incoming. In the outgoing simple wave region, the Riemann invariant J_1, defined by $J_1 = v + \int_0^m c(m)dm$, is constant. It can be shown that the jump suffered by J_1 across the shock wave under consideration is of fifth order in $[m]$. If we neglect this fifth order jump, we can take $J_1^- = 0$ since $J_1^+ = 0$. Therefore, an outgoing simple wave can follow the shock. In the simple wave region behind the shock, the constant value of J_1 is $J_1^-(=0)$, and thus $v + \int_0^m c(m)dm = 0$. On differentiating this relation, we obtain

$$m_{,t} + c(m)m_{,X} = 0, \tag{12}$$

We now consider the evolution of a decaying transverse shock wave which is initiated at the accessible boundary of an unstrained elastic half-space $X \geq 0$. That is, we assume that $m(X,t)$ is prescribed at $X = 0$ with $m(0,0^+) = h > 0$. Equation (12) implies that

$$m(X,t) = m(0,\tau) \text{ along } t = \tau + \frac{X}{c(m(0,\tau))}. \tag{13}$$

That is, m is a constant on the wavelet which leaves the boundary at time τ. Equation (13) admits the following geometrical interpretation. For fixed X, the graph for $m(X,t)$ versus t can be obtained by simply translating and stretching the graph of $m(0,\tau)$ (see Fig.1) to the right by a distance given by $X/c(m(0,\tau))$ (see Fig.2). In Fig.2, the graph of $m(X,t)$ is multivalued when $X/c(h) \leq t \leq X/c(0)$. This is the case when $\zeta_T \geq 0$. When $\zeta_T < 0$, no shock may exist for $X > 0$, a case which does not concern us here.

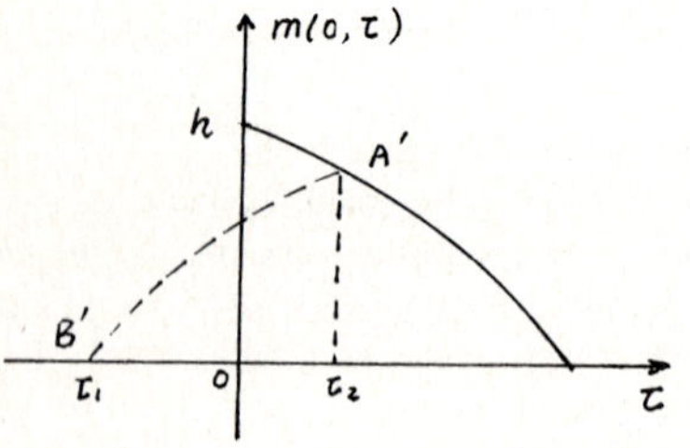

Fig. 1

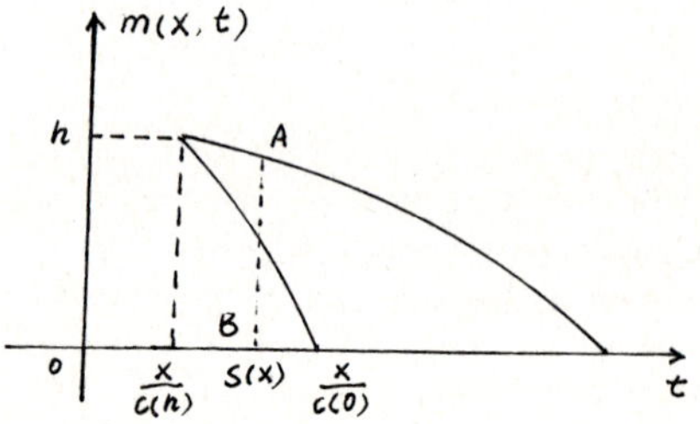

Fig. 2

To obtain a single-valued solution for $m(X,t)$, we fit a shock into the multivalued region. Suppose that the shock position $t = s(X)$ is somewhere between the two ends of the multivalued region (see Fig.2). Then from (13) we have

$$s(X) = \tau_2 + \frac{X}{c(m(0,\tau_2))}, \quad [m] = m(0,\tau_2). \tag{14}$$

On differentiating (14a,b) with respect to X and then eliminating $d\tau_2/dX$ from the resulting equations, we obtain, noting that $s'(X) = 1/U_N$,

$$\left(\frac{X}{c^2([m])}c'([m]) - \frac{1}{m'(0,\tau_2)}\right)\frac{d[m]}{dX} + \frac{1}{U_N} - \frac{1}{c([m])} = 0. \tag{15}$$

If $m(0,t)$ is monotonically decreasing, then equation (14b) can be inverted to give an expression for τ_2 in terms of $[m]$, and therefore equation (15) is an ordinary first order differential equation which can be solved to give an evolution law for the shock amplitude $[m]$. To obtain more explicit results, we assume that

$$m(0,t) = \begin{cases} (h^{\frac{1}{n}} - At)^n, & 0^+ \le t \le h^{\frac{1}{n}}/A, \\ 0, & \text{otherwise}, \end{cases} \tag{16}$$

where the parameter n may take any positive value and A is chosen as $A = c(h)kh^{1/n-1}/n$, so that the initial jumps in $[m]$ and $[m_{,X}]$ are independent of n and are the same as those given by (9). On substituting (5),(6) and (16) into (15) and integrating, we arrive at

$$\left(\frac{[m]}{h}\right)^{1+1/n} + \frac{1+n}{n\sqrt{\xi_1}}\zeta_T khc(h)X\left(\frac{[m]}{h}\right)^3 = 1 + max\{O(R_1), O(R_2)\}, \tag{17}$$

where

$$R_1 = \frac{k}{h^2}\int_0^X [m]^5 dX, \quad R_2 = \frac{k}{h^2}\int_0^X [m]^4 \frac{d[m]}{dX} X dX. \tag{18}$$

For arbitrary n, the leading order solution of $[m]$ is determined from (17) by neglecting the second term on the right hand side. Then the error term in (17) can be estimated with the aid of this leading order solution. The evolution laws for the accompanying second and third order discontinuities can be shown to have the form

$$[m_{,X}] = k\left(\frac{[m]}{h}\right)^{(1-1/n)}\left(\frac{1+O(h^2)}{1+3\zeta_T kX[m](\frac{[m]}{h})^{1-1/n}}\right), \tag{19}$$

$$[m_{,XX}] = \left(-3\zeta_T[m_{,X}]^2\frac{\frac{2n-1}{n}kX(\frac{[m]}{h})^{1-1/n} + 2[m]}{1+3\zeta_T kX[m](\frac{[m]}{h})^{1-1/n}} + \frac{n-1}{n}\frac{[m_{,X}]^2}{[m]}\right)(1+O(h^2)). \tag{20}$$

5 Discussion

In the previous two sections we have presented two different approximation methods for determining the evolution laws. It is plain that the evolution law determined from (17) is different from that given by (10) for all n. The reason for the disagreement is that the assumptions stated in (10) concerning the order of the accompanying third order discontinuity which ensure the validity of (10) are not satisfied by the specific loading programme (16) which we have considered in explaining the shock-fitting method. It is seen from (20) that $[m_{,XX}]$ is of order k^2/h if $n \neq 1$. In the case $n = 1$, equation (20) reduces to

$$[m_{,XX}] = -3\zeta_T[m_{,X}]^2\frac{kX + 2[m]}{1+3\zeta_T kX[m]}(1+O(h^2)). \tag{21}$$

It is clear by inspection that the initial order of $[m_{,XX}]$ is k^2h, which satisfies the assumptions made about $[m_{,XX}]$. But as X increases, $[m_{,XX}]$ increases in magnitude until it is $O(k^2/h)$ when $X = O((kh)^{-1})$. Therefore, the order of $[m_{,XX}]$ is not uniform with respect to the distance of travel. However, for small $\zeta_T khX$, say $O(h)$, the assumption $[m_{,XX}] = O(k^2)$ is satisfied and so the solution given by (17) should agree with that given by (10). This is indeed the case, as can be shown by looking for series solutions of (10) and (17) in terms of $\zeta_T khX$.

It can be deduced either from (10) or from (17) that the effects of nonlinearity operate through the product $\zeta_T kh$ of ζ_T characterizing the material nonlinearity with k and h, the initial amplitudes of the second order discontinuity and of the shock, respectively. This is in contrast to the corresponding solutions for dilatational shocks [1,(4.32) and (4.33)] in which the factor h does not occur. Since $| h | \ll 1$, the effects of nonlinearity are therefore much weaker for the transverse shocks considered here, becoming most pronounced for distances of travel of order $(\zeta_T k h)^{-1}$. In the limits $k \to 0$ or $\zeta_T \to 0$, the shock is of constant amplitude, as predicted by the linear theory.

In conclusion, we remark that as far as one dimensional wave propagation in an elastic non-conductor is concerned, the shock-fitting method gives much more complete results than does the singular surface method. For example, the higher order discontinuities need not be restricted and the strain and velocity fields behind the shock may also be computed using the shock-fitting method. As we shall demonstrate in future work, however, the situation is not so clear cut for three dimensional wave propagation since further approximations are needed before the shock-fitting method may be applied, whilst the singular surface theory is well-adapted to the three dimensional case. Shock waves in other kinds of materials, such as heat-conducting materials and simple materials with memory, can not be treated by the shock-fitting method without further modifications, but the singular surface method can be applied with no extra difficulties.

References

[1] Y.B.Fu and N.H.Scott, The evolution law of one dimensional weak nonlinear shock waves in elastic non-conductors, submitted to *Quart. Jl Mech. Appl. Math.*.

[2] P.J.Chen, One dimensional shock waves in elastic non-conductors, *Arch. Ratl Mech. Anal.*, **43**, 1971, 350–362.

[3] J.W.Nunziato and W.Herrmann, The general theory of shock waves in elastic non-conductors, *Arch. Ratl Mech. Anal.*, **47**, 1972, 272–287.

[4] P.J.Chen and M.E.Gurtin, The growth of one-dimensional shock waves in elastic non-conductors, *Int. J. Solids Structures*, **7**, 1971, 5–10.

[5] A.C.Eringen and E.S.Suhubi, *Elastodynamics Vol.1: Finite Motions*, Academic Press, New York and London, 1974.

[6] M.F.McCarthy, Singular surfaces and waves, *in* A.C.Eringen (ed.), *Continuum Physics Vol.II*, Academic Press, New York etc., 1975, 449–521.

[7] D.P.Reddy and J.D.Achenbach, Simple waves and shock waves in a thin prestressed elastic rod, *ZAMP*, **19**, 1968, 473–485.

[8] W.D.Collins, The propagation and interaction of one-dimensional non-linear waves in an incompressible isotropic elastic half-space, *Quart. Jl. Mech. Appl. Math.*, **20**, 1967, 429–452.

[9] P.B.Bailey and P.J.Chen, Evolutionary behaviour of induced discontinuities behind one-dimensional shock waves in non-linear elastic materials, *J. Elasticity*, 1985, **15**, 257–269.

[10] A.M.Anile, Propagation of weak shock waves, *Wave Motion*, **6**, 1984, 571–578.

[11] K.O.Friedrichs, Formation and decay of shock waves, *Comm. Pure Appl. Math.*, **1**, 1948, 211–246.

[12] A.H.Nayfeh, *Perturbation Methods*, John Wiley and Sons, New York, 1973.

ELASTIC WAVE PROPAGATION
M.F. McCarthy, M.A. Hayes, (Editors)

NONLINEAR WAVES OF SMALL AMPLITUDE IN ANISOTROPIC ELASTIC SOLIDS

N. DAHER* and G.A. MAUGIN**

*Laboratoire de Physique et Métrologie des Oscillateurs du CNRS associé à l'Université de Franche-Comté-Besançon
32, avenue de l'Observatoire - 25000 Besançon - France

**Université Pierre et Marie Curie, Laboratoire de Mécanique Théorique associé au CNRS, Tour 66, 4 place Jussieu - 75230 Paris Cedex 05 - France

Nonlinear waves of small amplitude are discussed in elastic and piezoelectric media on the basis of an asymptotic iterative procedure known as Poincaré's méthod. The controversy surrounding the elastic solution derived through different methods is resolved and the resulting expressions are compared to recent experimental verifications. After reconciliating the various theoretical and experimental results, the new version of Poincaré's method is extended to electroelastic solids.

1. INTRODUCTION

Nonlinear wave propagation has been studied extensively from both theoretical and experimental points of view. In particular, perturbation methods showed their efficiency in the treatment of the subject matter, especially for harmonic generation at different orders of nonlinearity. Several authors [1,2,3] pointed out that some problems, such as the evaluation of anisochronism or amplitude-frequency effect, require two orders of nonlinearity. Consequently, unlike previous works [2,4] dealing with piezoelectric media through quadratic equations in the small field variables the present equations deduced from Refs [5,6] are cubic. Although many authors have dealt with purely elastic waves with two orders of nonlinearity, these works are based on one-dimensional models and a disagreement is observed between the different theoretical approaches [1,3,7,8]. More precisely, the approximate solution deduced from the exact method of characteristics of hyperbolic systems -to which one applies an expansion in series because of the smallness of the amplitudes- is not consistent with the one derived through a direct iterative asymptotic procedure known as Poincaré's method. In order to reconciliate the different results, a new version of the iterative procedure is presented in the purely elastic case and the solution is compared to recent experimental verifications. Then electroelastic solids are considered.

2. NONLINEAR PROPAGATION EQUATIONS FOR PIEZOELECTRIC DIELECTRICS

The differential equations which govern the quasi-electrostatic wave propagation in an elastic piezoelectric solid are deduced from Refs [5,6] as follows

Equation of motion

$$\rho_o \frac{\partial^2 U_K}{\partial t^2} = {}_o\mathcal{T}^T_{KL,L} + \{C^T_{KLMN}\, U_{M,N} + B^T_{NKL}\, \bar{\Phi}_{,N} + \frac{1}{2} C^T_{KLMNPQ}\, (U_{M,N}\, U_{P,Q})\}_{,L}$$
$$+ \{-\frac{1}{2} E^T_{NQKL}\, (\bar{\Phi}_{,N}\, \bar{\Phi}_{,Q}) + H^T_{QKLMN}\, (\bar{\Phi}_{,Q}\, U_{M,N})\}_{,L} \qquad (2.1)$$
$$+ \{\frac{1}{3!} C^T_{LMNPQRS}\, (U_{M,N}\, U_{P,Q}\, U_{R,S}) + \frac{1}{3!} E_{NQSKL}\, (\bar{\Phi}_{,N}\, \bar{\Phi}_{,Q}\, \bar{\Phi}_{,Q})\}_{,L}$$
$$+ \{\frac{1}{2} H^T_{SKLMNPQ}\, (\bar{\Phi}_{,S}\, U_{M,N}\, U_{P,Q}) - \frac{1}{2} F^T_{QSKLMN}\, (\bar{\Phi}_{,Q}\, \bar{\Phi}_{,S}\, U_{M,N})\}_{,L}$$

Gauss' equation

$${}_oD_{R,R} = \{\epsilon_{LN}\, \bar{\Phi}_{,N} - B^T_{LMN}\, U_{M,N} - \frac{1}{2} \chi_{LNQ}\, (\bar{\Phi}_{,N}\, \bar{\Phi}_{,Q})\}_{,L}$$
$$+ \{-\frac{1}{2} H^t_{LMNPQ}\, (U_{M,N}\, U_{P,Q}) + E^t_{LQMN}\, (\bar{\Phi}_{,Q}\, U_{M,N})\}_{,L}$$
$$+ \{\frac{1}{3!} \chi_{LNQS}\, (\bar{\Phi}_{,N}\, \bar{\Phi}_{,Q}\, \bar{\Phi}_{,S}) - \frac{1}{3!} H^t_{LMNPQRS}\, (U_{M,N}\, U_{P,Q}\, U_{R,S})\}_{,L}$$
$$+ \{-\frac{1}{2} E_{LQSMN}\, (\bar{\Phi}_{,Q}\, \bar{\Phi}_{,Q}\, U_{M,N}) + \frac{1}{2} F^t_{LSMNPQ}\, (\bar{\Phi}_{,S}\, U_{M,N}\, U_{P,Q})\}_{,L} \qquad (2.2)$$

with

$${}_o\mathcal{T}^T_{KL} = {}_oT^E_{KL} + {}_oT^F_{KL} \quad , \quad {}_oD_R = {}_oW_R + {}_o\mathcal{P}_R \quad , \quad {}_oW_R = -\, {}_o\Phi_{,R} \qquad (2.3)$$

such that

$${}_oT^E_{KL} = {}_o(\frac{\partial \chi}{\partial E_{KL}}) \quad , \quad {}_o\mathcal{P}_R = -\, {}_o(\frac{\partial \chi}{\partial W_R}) \quad , \quad {}_oT^F_{KL} = {}_oW_K\, {}_oW_L - \frac{1}{2}\, {}_o\tilde{W}^2\, \delta_{KL} \qquad (2.4)$$

where χ is the free energy per unit volume, ${}_o\tilde{W}$, ${}_o\Phi$, ${}_oT^E_{KL}$, ${}_oT^F_{KL}$ and ${}_o\mathcal{P}_R$ denote respectively the electric field, the electric potential, the elastic stress tensor, the Maxwell tensor relative to electromagnetic contributions and the polarization. All these quantities are evaluated at the initial state. ρ_o is the mass density and U_K is the displacement vector. C_{KLMN}, B_{NKL}, ϵ_{LN}, C_{KLMNPQ}, H_{QKLMN}, E_{NQKL}, χ_{LNQ}, $C_{KLMNPQRS}$, $H_{SKLMNPQ}$, F_{QSKLMN}, E_{NQSKL} and χ_{LNQS} are linear and nonlinear effective coefficients accounting for elasticity, piezoelectricity, dielectricity, electrostriction and other electroelastic effects. The upperscripts "t" and "T" denote the deviation of a quantity from its thermodynamical definition. This deviation is due either to convection terms or to the presence of an initial state, or to dynamical electromagnetic contributions (see Ref [5] for details).

On applying a constant initial state $(\underset{o\sim}{W}, \underset{o\sim}{\mathcal{P}}, \underset{o\sim}{T}^E)$ the zeroth order system in eqs. (2.1), (2.2) vanishes identically. As to the linear system, we look for a solution of the form $(U_K, \bar{\Phi}) = (d_K, \phi)\, f(t - s/V)$ where $s = \underset{\sim}{b}.\underset{\sim}{X}$, $\underset{\sim}{b}$ is the direction of propagation and V is the phase velocity. Carrying this solution into the linear contribution of (2.1) and (2.2) and eliminating ϕ between the equations we obtain an eigen value problem for $\tilde{\Gamma}_{KL} = \tilde{\Gamma}_{LK}$ as follows

$$(\rho_o V^2 \delta_{KL} - \tilde{\Gamma}_{KL}) d_L = 0 \Leftrightarrow \det\,(\rho_o V^2 \delta_{KL} - \tilde{\Gamma}_{KL}) = 0\ ,\ \tilde{\Gamma}_{KL} = \Gamma_{KL} + \frac{B_K B_L}{\varepsilon} \quad (2.5)$$

where the eigenvectors d_L relative to the eigenvalues $\lambda = \rho_o V^2$ are orthogonal to one another and we have set

$$\Gamma_{KL} = C^T_{KLMN}\, b_M\, b_N \quad , \quad B_K = B^T_{MNK}\, b_M\, b_N \quad , \quad \varepsilon = \varepsilon_{LN}\, b_L\, b_N \quad (2.6)$$

3. NONLINEAR PURELY ELASTIC CASE

Before dealing with electroelastic phenomena, we first discuss the controversy surrounding the purely elastic case. To this end, the problem will be tackled from various sides with different conditions and interpretations.

The orthogonality of the eigen vectors d_L in eq. (2.5) allows one to form a new Cartesian basis on which the motion can be described. After some developments the purely elastic part of eq. (2.1) transforms to

$$\rho_o \frac{\partial^2 U^r}{\partial t^2} = \lambda^r U^r_{,ss} + \sum_{k,\ell} \Gamma^{rk\ell} (U^k_{,s} U^\ell_{,s})_{,s} + \sum_{k,\ell,m} \Delta^{rk\ell m} (U^k_{,s} U^\ell_{,s} U^m_{,s})_{,s} \quad (3.1)$$

where λ^r, $\Gamma^{rk\ell}$ and $\Delta^{rk\ell m}$ are effective elastic constants in the new Cartesian basis (see Refs [3,5,7,8,9] for details). In Refs [3,7,8] the mutual couplings between the different eigenmodes are neglected by the assumption of the "monomode hypothesis" ($r=k=\ell=m$). These couplings are taken into account in Refs [5,9] and the domain of validity of the monomode hypothesis is discussed. In particular, it is shown that in many realistic situations the monomode hypothesis holds. Here the attention is focused on the elastic case in its simplest form. On seeking an expansion as : $U = \varepsilon\overset{1}{U} + \varepsilon^2\overset{2}{U} + \varepsilon^3\overset{3}{U}$ where ε is infinitesimally small and identifying with respect to ε, eq. (3.1) with ($r=k=\ell=m$) yields

$$\rho_o \frac{\partial^2 \overset{1}{U}}{\partial t^2} - \lambda \overset{1}{U}_{,ss} = 0 \quad , \quad \rho_o \frac{\partial^2 \overset{2}{U}}{\partial t^2} - \lambda \overset{2}{U}_{,ss} = \Gamma\, (\overset{1}{U}_{,s})^2 \quad (3.2)$$

$$\rho_o \frac{\partial^2 \overset{3}{U}}{\partial t^2} - \lambda \overset{3}{U}_{,ss} = 2\Gamma\, (\overset{1}{U}_{,s}\, \overset{2}{U}_{,s})_{,s} + \Delta\, (\overset{1}{U}_{,s})^3_{,s} \quad (3.3)$$

In Refs [1,3,7,8] the problem is solved by the use of an asymptotic iterative procedure subject to a steady forcing assumption such as :

$$U = \overset{0}{U} \cos\omega t \qquad \text{at} \quad s = 0 \tag{3.4}$$

The linear differential eq. (3.2), leads to

$$\overset{1}{U} = \overset{0}{U} \cos\Psi \quad , \quad \Psi = \omega t - ks \quad , \quad k = \omega/V \tag{3.5}$$

where ω and k are respectively the angular frequency and the wave number.

On substituting the linear solution (3.5) into the quadratic eq. $(3.2)_2$ we obtain the following second order solution

$$\overset{2}{U} = \overset{+}{A} \cos 2\Psi + \bar{A} \tag{3.6}$$

where $\overset{+}{A}$ and $\bar{A}$ must satisfy the following relation

$$4(k\overset{+}{A}_{,s} - \frac{\Gamma}{4\lambda} \overset{0}{U}{}^2 k^2)\, k \sin 2\Psi + (\overset{+}{A}_{,s} - \frac{\Gamma}{2\lambda} k^2 \overset{0}{U}{}^2)_{,s} \cos 2\Psi$$
$$+ (\bar{A}_{,s} + \frac{\Gamma}{2\lambda} k^2 \overset{0}{U}{}^2)_{,s} = 0 \tag{3.7}$$

On account of the boundary condition (3.4) eq. (3.7) is satisfied if

$$\overset{+}{A} = \frac{\Gamma}{4\lambda} k^2 \overset{0}{U}{}^2 s \qquad \bar{A} = (a - \frac{\Gamma}{2\lambda} k^2 \overset{0}{U}{}^2)\, s \tag{3.8}$$

where a remains indetermined. In Refs [1,3,7,8] a is omitted by assuming a vanishing radiation stress associated with the steady state solution leading thus to $\overset{2}{U} = \overset{+}{A}\,[\cos 2\Psi - 2]$. However, according to Ref [10] presenting experimental verifications and dealing with acoustic radiation-induced static strains in solids the above-mentioned omission is not realistic. Here, we derive a new version of the iterative procedure starting with two different points of view but leading to equivalent solutions consistent with experimental results, and in full agreement with the solution deduced from the method of characteristics [3,8] where $\overset{2}{U} = \overset{+}{A}\,[\cos 2\Psi - 1]$.

A - Polytunal source and continuity argument

On assuming a polytunal source ($\omega \to \omega^\alpha$, $\alpha = 1,2, \ldots$) eqs. (3.4)-(3.7) transform to

$$U = \sum_\alpha \overset{0}{U}{}^\alpha \cos\omega^\alpha t \quad \text{at} \quad s = 0 \ , \qquad \overset{1}{U} = \sum_\alpha \overset{0}{U}{}^\alpha \cos\Psi^\alpha \tag{3.9}$$

and

$$\overset{2}{U} = \sum_{\alpha,\beta} [\overset{+}{A}{}^{\alpha\beta} \cos(\Psi^\alpha + \Psi^\beta) + \bar{A}^{\alpha\beta} \cos(\Psi^\alpha - \Psi^\beta)] \tag{3.10}$$

with the following relation

$$2\Big[\overset{\pm}{A}{}^{\alpha\beta}_{,s} \mp \frac{\Gamma}{4\lambda} k^{\alpha} k^{\beta} \overset{0}{U}{}^{\alpha} \overset{0}{U}{}^{\beta}\Big](k^{\alpha} \pm k^{\beta}) \sin(\Psi^{\alpha} \pm \Psi^{\beta}) + \overset{\pm}{A}{}^{\alpha\beta}_{,ss} \cos(\Psi^{\alpha} \pm \Psi^{\beta}) = 0 \quad (3.11)$$

so that we have necessarily

$$\overset{+}{A}{}^{\alpha\beta} = \frac{\Gamma}{4\lambda} k^{\alpha} k^{\beta} \overset{0}{U}{}^{\alpha} \overset{0}{U}{}^{\beta} s \quad , \quad \forall \alpha, \beta \quad , \quad \overset{-}{A}{}^{\alpha\beta} = -\overset{+}{A}{}^{\alpha\beta} \quad , \quad \forall \alpha, \beta - \{\alpha \equiv \beta\} \qquad (3.12)$$

By continuity, we assume that eq. $(3.12)_2$ holds for $\alpha \equiv \beta$. Thus we get

$$\overset{2}{U} = \sum_{\alpha} \overset{+}{A}{}^{\alpha\alpha} \left[\cos 2\Psi^{\alpha} - 1\right] + \sum_{\substack{\alpha,\beta \\ \alpha \neq \beta}} \overset{+}{A}{}^{\alpha\beta} \left[\cos(\Psi^{\alpha} + \Psi^{\beta}) - \cos(\Psi^{\alpha} - \Psi^{\beta})\right] \qquad (3.13)$$

In addition to the fact that the present procedure allows for the evaluation of intermodulation process it leads to the result deduced through the method of characteristics when a monotunal source is considered ($\alpha \equiv \beta$).

B - Additional boundary and initial condition

In order to obtain directly a unique solution without neglecting a in eq. $(3.8)_2$ and in agreement with experiment, it is necessary to assume an additional boundary and initial condition as follows

$$\frac{\partial U}{\partial s} = 0 \quad \text{at} \quad t = s = 0 \qquad (3.14)$$

In this case $a = \frac{\Gamma}{4\lambda} k^2 \overset{0}{U}{}^2$ does not vanish so that the so-called radiation-induced static strain reads $\overset{-}{A}_{,s} = -\overset{+}{A}_{,s}$ instead of $\overset{-}{A}_{,s} = -2\overset{+}{A}_{,s}$ as in Refs [3,7,8].

As to the third order or cubic solution needed for the evaluation of anisochronism or amplitude-frequency effect, no ambiguity appears in the identification procedure and we are led to

$$\overset{3}{U} = D\left[\cos 3\Psi - \cos\Psi\right] + E\left[\sin 3\Psi - 3\sin\Psi\right] \qquad (3.15)$$

with

$$D = \frac{1}{8}\frac{\Gamma^2}{\lambda^2} k^4 \overset{0}{U}{}^3 s^2 \qquad E = -\frac{1}{6}\left(\frac{\Gamma^2}{\lambda^2} - \frac{3}{4}\frac{\Delta}{\lambda}\right) k^3 \overset{0}{U}{}^3 s \qquad (3.16)$$

We note that the difference encountered between previous works [3,7,8] and eq. (3.15) is a direct consequence of the disagreement in the second order solution. Moreover, eq. (3.15) contains third harmonics as well as contributions at the fundamental frequency which alter the amplitude and the velocity.

In particular, it is shown that the relative velocity change reads

$$\frac{\Delta V}{V} = \frac{1}{2}\frac{\omega^2}{V^2}\overset{0}{U}{}^2\left[\left(\frac{\Gamma}{\lambda}\right)^2 - \frac{3}{4}\frac{\Delta}{\lambda}\right] \tag{3.17}$$

On comparing this result with the one deduced in Refs [3,8] we note that the relative difference between the two solutions may reach 50 %. This justifies the need for the present approach. Before ending this Section it is useful to recall that the interest in the "Poincaré method" lies in the fact that it may be applied to involved problems for which a solution by the method of characteristics is not presently available.

4. NONLINEAR ELECTROELASTIC CASE

The study of nonlinear wave propagation in complex elastic solids such as piezoelectric semiconductors [11] does not allow for the diagonalization procedure discussed in Section 1. This is due to dissipative processes through conduction and diffusion which alter the symmetry of the Christoffel acoustic tensor. More precisely, instead of $C_{KLMN} \Leftrightarrow C_{\alpha\beta}$ (in Voigt's notation) we shall have $C^{eff}_{KLMN} \Leftrightarrow C^{eff}_{\alpha\beta}$ which is no longer symmetric positive definite. In order to prepare the way for such complex descriptions we apply the successive approximation procedure directly to eqs. (2.1) and (2.2). After lengthy calculations similar to those presented in the purely elastic case the relative velocity change in a nonlinear piezoelectric medium reads

$$\frac{\Delta V}{V} = \frac{1}{8}\frac{\omega^2}{V^2}\overset{0}{U}{}^2 \, \ell_R \, (\underset{\sim}{\Gamma}^T)^{-1}_{RK} \left\{ \overset{3}{P}_K - \frac{1}{2}\overset{4}{P}_K \right\} \tag{4.1}$$

with

$$\overset{\alpha}{P}_K = \overset{\alpha}{P}_{KQXY}\, \ell_Q\, \ell_X\, \ell_Y \qquad \alpha = 3,4 \qquad \ell_K = \overset{0}{U}_K/\overset{0}{U} \quad , \quad \overset{0}{U} = \sqrt{\overset{0}{U}_K\overset{0}{U}_K}$$

where $\overset{\alpha}{P}_{KQXY}$ denote effective contributions of the elastic, electroelastic and electric coefficients present in eqs (2.1)-(2.2).

In the purely elastic case eq. (4.1) reduces to

$$\frac{\Delta V}{V} = \frac{1}{2}\frac{\omega^2}{V^2}\overset{0}{U}{}^2\, \ell_R\, \Gamma^{-1}_{RK}\left\{\Gamma_{KQP}\,\Gamma^{-1}_{PS}\,\Gamma_{STL} - \frac{3}{4}\Delta_{KQTL}\right\}\ell_Q\,\ell_T\,\ell_L \tag{4.3}$$

with

$$\Gamma_{KMP} = \frac{1}{2}C_{KLMNPQ}b_L b_N b_Q \qquad \Delta_{KMPR} = \frac{1}{3!}C_{KLMNPQRS}b_L b_N b_Q b_S \tag{4.4}$$

Notice the similar structure of eq. (4.3) and (3.17).

REFERENCES

[1] Thompson, R.B. and Tiersten, H.F., Harmonic generation of longitudinal elastic waves, J. Acoust. Soc. Am., 62(1) (1977) 33-37.

[2] Tiersten, H.F. and Baumhauer, J.C., Second harmonic generation and parametric excitation of surface waves in elastic and piezoelectric solids, J. Appl. Phys., 45(10) (1974) 4272-4287.

[3] Planat, M., Propagation non linéaire des ondes acoustiques dans les solides, Thèse d'Etat, Université de Franche-Comté (1984).

[4] Nelson, D.F., Acoustic wave mixing in piezoelectric crystals, J. Acoust. Soc. Am., 64(2) (1978) 652-657.

[5] Daher, N., Principe des puissances virtuelles étendu aux discontinuités et interfaces. Application à l'acousto-électronique, Thèse d'Etat, Université de Franche-Comté (1987).

[6] Daher, N. and Maugin, G.A., Nonlinear electroacoustic equations in semiconductors with interfaces", Int. J. Engng. Sci., 26(1) (1988) 37-58.

[7] Planat, M., Théobald, G. et Gagnepain, J.J., Propagation non linéaire d'ondes élastiques dans un solide anisotrope. I - Ondes de volume, L'Onde Electrique, 60(11) (1980) 33-40.

[8] Maugin, G.A., Nonlinear electromechanical effects and applications - A - Series of Lectures, World Scientific, Singapore (1985).

[9] Daher, N., Intermodulation and harmonic generation in electroelastic anisotropic solids, Revue d'Acoustique, (to be published in 1988).

[10] Cantrell, H.H., Yost, W.T. and Li, P., Acoustic radiation-induced static strains in solids, Phys. Rev. B, 35(18) (1987) 9780-9782.

[11] Daher, N. and Maugin, G.A., Deformable semiconductors with interfaces, Int. J. Engng. Sci., 25 (1987) 1093-1129.

ELASTIC WAVE PROPAGATION
M.F. McCarthy, M.A. Hayes, (Editors)

NONLINEAR INTERACTION BETWEEN OPTICAL AND ACOUSTICAL WAVES IN DIATOMIC ELASTIC SOLIDS

Mevlut TEYMUR

Department of Mathematics and Computer Science
The University, Dundee, DD1 4HN, Scotland, U.K.

The interaction between modulated optical and long acoustical waves in a diatomic elastic solid is considered. These waves can exchange energy in a resonant manner if the group velocity of the optical wave is sufficiently close to the phase velocity of the acoustical wave. Coupled nonlinear partial differential equations which describe the phenomenon, are obtained by an asymptotic method and the soliton solutions of these equations are given.

1) INTRODUCTION

According to the classical theory of linear elasticity, in an infinite medium , plane harmonic waves propagate with constant phase velocity; that is the angular frequency of the motion depends linearly on wave number. The waves are non-dispersive. As it is well known, there arc two kinds of elastic waves called longitudinal and transverse. The frequency-wave number (dispersion) relation, consists of two acoustic branches corresponding to these waves. There is no difference between the dispersion relation of a monoatomic solid and that of a polyatomic solid, as far as the qualitative characters of the dispersion relations are concerned.

On the other hand lattice dynamical approaches yield dispersive acoustical branches for monoatomic solids and dispersive acoustical and optical branches for polyatomic solids. For example, for the longitudinal plane harmonic waves propagating in a diatomic solid, the dispersion relation has two branches; one is acoustical and the other is optical while in a triatomic solid it has one acoustical and two optical branches .These facts have also been verified by experimental techniques such as slow neutron spectroscopy , the diffuse scattering of X-rays, infrared and Raman spectroscopy[1].

Recently , in [2], the idea of a deformable shell model is generalised to elastic solid continua with diatomic structure. The balance and constitutive laws are derived. It is shown that, in contrast to the result of classical theory of linear elasticity, this theory, in the linear approximation , gives dispersive optical branches as well as dispersive acoustical ones for time harmonic waves in unbounded media as in the lattice dynamical approaches. The acoustical ones have weakly dispersive regions when the wave number is small. Except for this region both the acoustical and optical modes are strongly dispersive. Employing this continuum approach the propagation of plane longitudinal weak nonlinear waves have been investigated in [3].It is shown that when the nonlinearity and dispersion are weak, i.e. in the weakly dispersive region, the wave propagation is governed by the Korteweg-de Vries (KdV) equation. On the other hand, in the strongly dispersive region the coupling of weak nonlinearity and strong dispersion yields the nonlinear Schrodinger (NLS) equation for the amplitude modulation of waves. As a consequence of these results it is concluded that solitons and envelope solitons which are not encountered in the classical nonlinear elasticity theory , may exist in unbounded diatomic elastic solids. These results are also in agreement with those obtained via the lattice dynamical approach to the same problem by employing the cubic or both cubic and quartic interaction potentials [4,5].

As it is noted in [3], in two exceptional cases the governing equations cannot be reduced asymptotically to the NLS equation. The first one is the second harmonic resonance in which the phase velocity of the second harmonic component coincides with that of the fundamental mode at certain critical wave number whence nonlinear resonant interaction between these two modes does occur. The other one, which is examined in this paper, is the nonlinear interaction between modulated optical and long acoustical waves in which the group velocity of the optical wave coincides with the phase velocity of the acoustical wave. By employing the derivative expansion method we first consider the case when the amplitude of the acoustical wave is small by comparison to that of the optical wave. Two coupled nonlinear partial differantial equations which describe the resonant wave interaction phenomenon asymptotically, are obtained. From the steady state solutions of these equations, they are in soliton and envelope soliton forms ; it is observed that when the propagation velocity of these solitons is close to the phase velocity of the acoustical wave, the amplitude of this wave becomes larger than that of the optical wave contrary to the assumption. Consequently the perturbation expansion ceases to be uniformly valid. Therefore it is modified by taking this fact into consideration and a new pair of coupled equations is obtained. The steady state solutions of these equations are also given and they are in soliton and envelope soliton forms as in the previous case.

2)BASIC EQUATIONS

An elastic medium consisting of diatomic molecules may be considered as two interacting media coincident at an initial configuration. For a plane longitudinal motion if X is the common material coordinate of such a generalised particle located at P_0 , upon deformation of the body each particle in the cell occupies a different space point p^α (α=1,2) whose spatial coordinates are x^α [2]. Then this motion is described by

$$x^\alpha = X + u^\alpha(X,t) , \tag{2.1}$$

where u^α is the displacement of the αth component and the direction of wave propagation is X. The equations of motion , in the reference state with vanishing body forces, are given as follows

$$\frac{\partial T^\alpha}{\partial X} - (-1)^\alpha R = \rho_\alpha^0 \frac{\partial v^\alpha}{\partial t} \quad , \quad \alpha = 1,2, \tag{2.2}$$

where T^α are the partial Piola-Kirchoff stresses , R is the rate of linear momentum transfer among the constituent bodies , ρ_α^0 is the undeformed partial mass density and v^α is the velocity of the αth costituent. It is assumed that such a homogeneous diatomic elastic continuum possesses a strain energy function of the form $\Sigma = \Sigma(p,q,w)$, where $p = \partial u^1 / \partial X$ and $q = \partial u^2 / \partial X$ are displacement gradients , $w = u^2 - u^1$ is the displacement of the second continuum relative to the first one. Then the constitutive equations for T^α and R are found to be

$$T^1 = \frac{\partial \Sigma}{\partial p} \quad , T^2 = \frac{\partial \Sigma}{\partial q} \quad , R = \frac{\partial \Sigma}{\partial w} . \tag{2.3}$$

For isotropic diatomic quadratic materials T^α and R take the forms, [3],

$$\begin{aligned} T^1 &= a_1 p + b_1 p^2 + 2b_3 pq + a_4 q + b_4 q^2 + b_5 w^2 , \\ T^2 &= a_2 q + b_2 q^2 + 2b_4 pq + a_4 p + b_3 p^2 + b_6 w^2 , \\ R &= a_3 w + 2b_5 pw + 2b_6 qw , \end{aligned} \tag{2.4}$$

where $a_1,...,a_4$ are linear and $b_1,...,b_6$ are second order nonlinear material constants. Now using (2.3), (2.2) can be rewritten explicitely as

$$a_1\frac{\partial p}{\partial X} + a_4\frac{\partial q}{\partial X} + 2b_1 p\frac{\partial p}{\partial X} + 2b_3\frac{\partial}{\partial X}(pq) + 2b_4 q\frac{\partial q}{\partial X} + a_3 w + c_1 qw - \rho_1^0\frac{\partial r}{\partial t} = 0, \tag{2.5}$$

$$a_2\frac{\partial q}{\partial X}+a_4\frac{\partial p}{\partial X}+2b_2 q\frac{\partial q}{\partial X}+2b_3 p\frac{\partial p}{\partial X}+2b_4\frac{\partial}{\partial X}(pq)-a_3 w-c_1 pw-\rho_2^0\frac{\partial s}{\partial t}=0, \quad (2.6)$$

where

$$r = v^1 \quad , s = v^2 \quad , c_1 = 2\,(b_5+b_6). \quad (2.7)$$

The following compatibility equations are also valid

$$\frac{\partial p}{\partial t}-\frac{\partial r}{\partial X}=0 \;,\; \frac{\partial q}{\partial t}-\frac{\partial s}{\partial X}=0 \;,\; \frac{\partial w}{\partial t}+r-s=0 \;,\; \frac{\partial w}{\partial X}+p-q=0. \quad (2.8)$$

The linear dispersion relation for harmonic longitudinal waves is found to be

$$\omega^4-[\omega_0^2+(v_a^2+v_o^2)\,k^2]\,\omega^2+(\omega_0^2 c_a^2)\,k^2+v_o^2 v_a^2 k^4=0, \quad (2.9)$$

where

$$\omega_0^2=a_3\,(\rho_1^0+\rho_2^0)\,/\,\rho_1^0\rho_2^0 \quad , c_a^2=(a_1+a_2+2a_4)\,/\,(\rho_1^0+\rho_2^0), \quad (2.10)$$

$$(v_a^2,v_o^2)=[a_1\rho_2^0+a_2\rho_1^0+[(a_1\rho_2^0-a_2\rho_1^0)^2+4\rho_1^0\rho_2^0 a_4^2]^{1/2}/2\rho_1^0\rho_2^0. \quad (2.11)$$

Note that the dispersion relation (2.9) has two branches, the so-called acoustical branch which involves low frequencies and the optical branch which involves a range of higher frequencies. Denoting them respectively by ω_- and ω_+ , from (2.10) we write

$$2\omega_{\mp}^2=\omega_0^2+(v_a^2+v_0^2)k^2\mp[[\omega_0^2+(v_a^2+v_o^2)k^2]^2-4[(\omega_0^2 c_a^2)k^2+v_o^2 v_a^2 k^4]]^{1/2}, \quad (2.12)$$

When $k=0$, (2.9) gives

$$\omega_+^2=\omega_0^2 \quad , \quad \omega_-^2=0, \quad (2.13)$$

where ω_0 is the cutoff frequency. The constants c_a and v_a defined above are the phase velocities of acoustical waves in the limits $k\to 0$ and $k\to\infty$, respectively , and v_o is the phase velocity of optical waves in the limit $k\to\infty$.

3) RESONANT INTERACTION OF WAVES

We now consider the nonlinear resonant interaction between acoustical and optical waves. This phenomenon occurs when the group velocity of the optical wave is sufficiently close to the phase velocity of the acoustical wave at a critical wave number ,that is if $c_a^2-(d\omega/dk)^2=O(\varepsilon^n)$, $n\geq 1$, where ε is a small positive perturbation parameter which measures the strength of the nonlinearity and the narrowness of the band-width of the optical wave centered around the critical wave number. We assume that both waves are weakly nonlinear and investigate the problem by employing an asymptotic perturbation method. Here the derivative expansion method based on multiple scales will be used [6].

Following the usual procedure of the method p,q,r,s and w are expanded in asymptotic power series in ε as

$$f(X,t)=\sum_{n=1}^{\infty}\varepsilon^{n+\alpha}\bar{f}_n(x_0,x_1,x_2,\ldots,t_0,t_1,t_2,\ldots)+\sum_{n=1}^{\infty}\varepsilon^{n+\beta}\hat{f}_n(x_1,x_2,\;\ldots\;,t_1,t_2,\ldots) \quad (3.1)$$

where the first and the second expansions on the right hand side stand for modulated and long waves, respectively, and

$$x_i=\varepsilon^i X \quad , t_i=\varepsilon^i t \quad , \quad i=0,1,2,\ldots, \quad (3.2)$$

are multiple scales. That is the asymptotic expansion is chosen as the superposition of the interacting waves. It should be noted here that, as shown in [3], in the absence of the optical wave, the long acoustical wave is asymptotically governed by the KdV equation while the modulated optical wave, in the absence of the acoustical wave, obeys the NLS equation.

Depending on α and β , various situations may occur in the problem. Let us presume that the acoustical wave is weaker than the optical one and take $\alpha=0$ and $\beta=1$. Then substituting (3.1) into (2.5)-(2.8) and collecting the terms of like powers in ε , we obtain a hierarcy of equations from which it is possible to determine p_n, q_n etc. succesively.Then the first order equations are found to be

$$a_1\frac{\partial\bar{p}_1}{\partial x_0}+a_4\frac{\partial\bar{q}_1}{\partial x_0}+a_3\bar{w}_1-\rho_1^0\frac{\partial\bar{r}_1}{\partial t_0}=a_2\frac{\partial\bar{q}_1}{\partial x_0}+a_4\frac{\partial\bar{p}_1}{\partial x_0}-a_3\bar{w}_1-\rho_2^0\frac{\partial\bar{s}_1}{\partial t_0}=0,$$

$$\frac{\partial\bar{p}_1}{\partial t_0}-\frac{\partial\bar{r}_1}{\partial x_0}=\frac{\partial\bar{q}_1}{\partial t_0}-\frac{\partial\bar{s}_1}{\partial x_0}=\frac{\partial\bar{w}_1}{\partial t_0}+\bar{r}_1-\bar{s}_1=\frac{\partial\bar{w}_1}{\partial x_0}+\bar{p}_1-\bar{q}_1=0. \tag{3.3}$$

We now seek the solutions of these equations in the following form

$$\bar{f}_1=\bar{f}_1^{(l)}e^{il\phi}+c.c. \qquad l=1,2,3,..., \tag{3.4}$$

where $\phi=k\,x_0-\omega t_0$, k is the wave number and ω is the angular velocity and c.c denotes the complex conjugate to the preceding terms. Then the substituion of (3.4) in (3.3) yields

$$\mathbf{W}_l\,\bar{\mathbf{U}}_1^{(l)}=0, \tag{3.5}$$

where

$$\mathbf{W}_l=\begin{bmatrix} ilka_1 & ilka_4 & a_3 & il\omega\rho_1^0 & 0\\ ilka_4 & ilka_2 & -a_3 & 0 & il\omega\rho_2^0\\ -il\omega & 0 & 0 & -ilk & 0\\ 0 & -il\omega & 0 & 0 & -ilk\\ 0 & 0 & -il\omega & 1 & -1\end{bmatrix},\quad \bar{\mathbf{U}}_1^{(l)}=\begin{bmatrix}\bar{p}_1^{(l)}\\ \bar{q}_1^{(l)}\\ \bar{w}_1^{(l)}\\ \bar{r}_1^{(l)}\\ \bar{s}_1^{(l)}\end{bmatrix}. \tag{3.6}$$

Note that $\det\mathbf{W}_1=0$ gives the linear dispersion relation (2.9). Since, here, we are only interested in the interaction phenomenon between modulated and long waves, to exclude the harmonic resonance it is assumed that $\det\mathbf{W}_l\neq0$ for $l\geq2$. Thus a group of optical wave centered around the critical wave number k and the corresponding critical frequency ω is considered. Then the solutions of the homogeneous algebraic equations (3.5) are found to be

$$\bar{\mathbf{U}}_1^{(1)}=A_1\mathbf{R},\qquad \bar{\mathbf{U}}_1^{(l)}=0,\qquad \text{for } l\neq1, \tag{3.7}$$

where A_1 is a complex function representing the first order slowly varying amplitude of the optical wave and it is to be determined in higher order perturbation problems; and $\mathbf{R}$ is a column vector satisfying $\mathbf{W}_1\mathbf{R}=0$. Substituting the first order solutions into the second order equations and taking the solutions for $\bar{p}_2,\bar{q}_2$, etc. in the form (3.4), as in the previous order, after some algebra we deduce

$$\mathbf{W}_l\bar{\mathbf{U}}_2^{(l)}=\mathbf{b}_2^{(l)},\quad \hat{w}_1=0,\quad \hat{r}_1-\hat{s}_1=0,\quad \hat{p}_1-\hat{q}_1=0, \tag{3.8}$$

where $\mathbf{b}_2^{(l)}\equiv0$ for $l\neq1,2$ and $\mathbf{b}_2^{(1)}$ and $\mathbf{b}_2^{(2)}$ are defined as

$$\mathbf{b}_2^{(1)}=-i\,(\frac{\partial A_1}{\partial t_1}\frac{\partial\mathbf{W}_1}{\partial\omega}-\frac{\partial A_1}{\partial x_1}\frac{\partial\mathbf{W}_1}{\partial k})\mathbf{R},\quad \mathbf{b}_2^{(2)}=\mathbf{B}A_1^2,\quad \mathbf{B}^T=(B_1,B_2,0,0,0),$$

$$B_1=-ik[2b_1+4b_3K+2b_4K^2+c_1K(1-K)/k^2],\ B_2=-ik[2b_2K^2+2b_3+4b_4K-c_1(1-K)/k^2],$$

$$K=(a_1k^2-\rho_1^0\omega^2+a_3)/(a_3-k^2a_4).$$

Since $\det\mathbf{W}_1=0$ and $\mathbf{b}_2^{(1)}\neq0$, in order that $(3.8)_1$ is algebraically solvable for $\bar{\mathbf{U}}_2^{(1)}$, the compatibility condition $\mathbf{L}.\mathbf{b}_2^{(1)}=0$ must be satisfied, where $\mathbf{L}$ is a row vector defined by $\mathbf{L}\mathbf{W}_1=0$. This condition leads to the result

$$\frac{\partial A_1}{\partial t}+V_g\frac{\partial A_1}{\partial x_1}=0,\quad V_g=\frac{d\omega}{dk}=-\frac{\mathbf{L}(\partial\mathbf{W}_1/\partial k)\mathbf{R}}{\mathbf{L}(\partial\mathbf{W}_1/\partial\omega)\mathbf{R}}, \tag{3.9}$$

which implies that the amplitude of the optical wave remains constant in a frame of reference moving with the group velocity V_g. That is, A_1 depends on x_1 and t_1 only through the combination $x_1 - V_g t_1$. Then $\overline{\mathbf{U}}_2^{(1)}$ is found to be

$$\overline{\mathbf{U}}_2^{(1)} = A_2 \mathbf{R} - i \frac{\partial A_1}{\partial x_1} \left(\frac{\partial \mathbf{R}}{\partial k} + V_g \frac{\partial \mathbf{R}}{\partial \omega} \right), \tag{3.10}$$

where A_2 is a complex function representing the second order slowly varying amplitude of the optical wave and it can be determined from the higher order perturbation problems. For $l=2$, since $\det \mathbf{W}_2 \neq 0$ by assumption, then we may write $\overline{\mathbf{U}}_2^{(2)} = A_1^2 \mathbf{W}_2^{-1} \mathbf{B}$. On the other hand for $l>2$, $\overline{\mathbf{U}}_2^{(l)} \equiv 0$ since $\det \mathbf{W}_l \neq 0$ and $\mathbf{b}_2^{(l)} \equiv 0$. Then taking $\bar{p}_3, \bar{q}_3$, etc. as in the form (2.4), and using the first and second order solutions in the third order perturbation equations we get

$$\mathbf{W}_l \overline{\mathbf{U}}_3^{(l)} = \mathbf{b}_3^{(l)}, \quad a_3 \hat{w}_2 = -(a_1 + a_4) \frac{\partial \hat{p}_1}{\partial x_1} + \rho_1^0 \frac{\partial \hat{r}_1}{\partial t_1} - \eta_1 \frac{\partial}{\partial x_1} |A_1|^2, \tag{3.11}$$

$$-a_3 \hat{w}_2 = -(a_2 + a_4) \frac{\partial \hat{p}_1}{\partial x_1} + \rho_2^0 \frac{\partial \hat{r}_1}{\partial t_1} - \eta_2 \frac{\partial}{\partial x_1} |A_1|^2, \quad \frac{\partial \hat{p}_1}{\partial t_1} - \frac{\partial \hat{r}_1}{\partial x_1} = \hat{r}_2 - \hat{s}_2 = \hat{p}_2 - \hat{q}_2 = 0,$$

where $\mathbf{b}_3^{(l)} \equiv 0$ for all $l \neq 1,2,3$ and $\mathbf{b}_3^{(1)}$, η_1 and η_2 are defined as

$$\mathbf{b}_3^{(1)} = [-i \left(\frac{\partial \mathbf{W}_1}{\partial \omega} \frac{\partial A_2}{\partial t_1} - \frac{\partial \mathbf{W}_1}{\partial k} \frac{\partial A_2}{\partial x_1} \right) - i \left(\frac{\partial \mathbf{W}_1}{\partial \omega} \frac{\partial A_1}{\partial t_2} - \frac{\partial \mathbf{W}_1}{\partial k} \frac{\partial A_1}{\partial x_2} \right)$$

$$+ \frac{1}{2} \left(\frac{\partial^2 \mathbf{W}_1}{\partial \omega^2} \frac{\partial^2 A_1}{\partial t_1^2} - 2 \frac{\partial^2 \mathbf{W}_1}{\partial \omega \partial k} \frac{\partial^2 A_1}{\partial x_1 \partial t_1} + \frac{\partial^2 \mathbf{W}_1}{\partial k^2} \frac{\partial^2 A_1}{\partial x_1^2} \right)] \mathbf{R}$$

$$+ \left(\frac{\partial \mathbf{W}_1}{\partial k} \frac{\partial^2 A_1}{\partial x_1^2} - \frac{\partial \mathbf{W}_1}{\partial \omega} \frac{\partial^2 A_1}{\partial x_1 \partial t_1} \right) \left(\frac{\partial \mathbf{R}}{\partial k} + V_g \frac{\partial \mathbf{R}}{\partial \omega} \right) + \mathbf{C} |A_1|^2 A_1 + \mathbf{D} \hat{p}_1 A_1, \tag{3.12}$$

$$\mathbf{C} = -(\mathbf{H}.\mathbf{R}^*) \mathbf{W}_2^{-1} \mathbf{B}, \qquad \mathbf{D}^T = (D_1, D_2, 0, 0, 0),$$

$$D_1 = -ik [2b_1 + 2b_3(1+K) + b_4 K + c_1(1-K)/k^2], \qquad \eta_1 = 2b_1 + 4b_3 K + 2b_4 K^2 - c_1 K(1-K)/k^2,$$

$$D_2 = -ik [2b_2 K + 2b_3 + 2b_4(1+K) - c_1(1-K)/k^2], \qquad \eta_2 = 2b_2 K^2 + 2b_3 + 4b_4 K + c_1(1-K)/k^2,$$

and $\mathbf{H}$ denotes a system of quantities having three indices with the components

$$H_{111} = 2ikb_1, \quad H_{222} = 2ikb_2, \quad H_{112} = H_{121} = H_{211} = 2ikb_3, \quad H_{122} = H_{221} = H_{212} = 2ikb_4,$$

$$H_{123} = H_{133} = -H_{213} = -H_{233} = c_1, \quad \text{all other} \quad H_{ijk} = 0, \quad i,j,k = 1,2,3,4,5.$$

Here the dot product represents the operation $H_{ijk} R_k = H_{ij}$. For $l=1$, in order that $(3.11)_1$, is algebraically solvable for $\overline{\mathbf{U}}_3^{(1)}$, the compatibility condition $\mathbf{L}.\mathbf{b}_3^{(1)} = 0$ must be satisfied. If we assume that A_2 depends on x_1 and t_1 through the combination $x_1 - V_g t_1$ as A_1, then this condition yields

$$i \left(\frac{\partial A_1}{\partial t_2} + V_g \frac{\partial A_1}{\partial x_2} \right) + \Gamma \frac{\partial^2 A_1}{\partial x_1^2} + \Delta |A_1|^2 A_1 + \gamma \hat{p}_1 A_1 = 0. \tag{3.13}$$

Also from the equations (3.11) we deduce

$$\frac{\partial^2 \hat{p}_1}{\partial t_1^2} - c_a^2 \frac{\partial^2 \hat{p}_1}{\partial x_1^2} - \delta \frac{\partial^2}{\partial x_1^2} |A_1|^2 = 0, \tag{3.14}$$

where

$$\Gamma = \frac{1}{2} \frac{d^2 \omega}{dk^2}, \quad \Delta = -i \frac{\mathbf{LC}}{\mathbf{L}(\partial \mathbf{W}_1 / \partial \omega) \mathbf{R}}, \quad \gamma = i \frac{\mathbf{LD}}{\mathbf{L}(\partial \mathbf{W}_1 / \partial \omega) \mathbf{R}}, \quad \delta = \frac{\eta_1 + \eta_2}{\rho_1^0 + \rho_2^0}.$$

These equations comprise a coupled set of the NLS type of equation and a semi-linear equation. Let us now consider travelling wave solutions to (3.13) and (3.14) of the following forms

$$A_1=A(a\zeta)e^{ib\xi}, \quad \hat{p}_1=B(a\zeta), \quad \zeta=x_1-\lambda t_1, \xi=x_2-\kappa t_2, \tag{3.15}$$

where a, b, λ, κ are real constants and, A and B are real functions of ζ. In general, the solutions for A and B can be expressed in terms of Jacobian elliptic functions. But here we only give the solutions when $A, B \to 0$ as $\zeta \to \infty$. They are

$$A_1=(2\Gamma)^{1/2}a[\Delta+\gamma B_0]^{-1/2}sech(a\zeta)\exp[ib[x_2-(V_g-\Gamma a^2/b)t_2]], \tag{3.16}$$

$$\hat{p}_1=B_0 sech^2[a\zeta], \quad B_0=[\delta/(\lambda^2-c_a^2)], \tag{3.17}$$

provided only that $\Delta+\gamma B_0>0$ and $\Gamma>0$. It is seen that for the optical wave the solution is an envelope solitary wave while for the acoustical wave it is a solitary wave.

Note that when $\lambda \to c_a$, then $A_1 \to 0$, hence the amplitude of the acoustical wave becomes comparatively large contrary to the assumption. Therefore, when this is the case the asymptotic expansion must be modified. By taking $\alpha=1$, $\beta=0$, then proceeding as before, a new pair of equations is obtained

$$i\left(\frac{\partial A_1}{\partial t_2}+V_g\frac{\partial A_1}{\partial x_2}\right)+\Gamma\frac{\partial^2 A_1}{\partial x_1^2}+\gamma\hat{p}_1A_1=0, \tag{3.18}$$

$$\frac{\partial \hat{p}_1}{\partial t_3}+c_a\frac{\partial \hat{p}_1}{\partial x_3}+\theta\hat{p}_1\frac{\partial \hat{p}_1}{\partial x_1}+\mu\frac{\partial^3\hat{p}_1}{\partial x_1^3}+\nu\frac{\partial}{\partial x_1}|A_1|^2=0. \tag{3.19}$$

They comprise a coupled set of the NLS type and KdV type of equations. Similar equations have also been obtained for plasma and water waves [6]. Travelling wave solutions to (3.18) and (3.19) of the forms given in (3.15) can be expressed , in general, in terms of Jacobian elliptic functions. But, again, we only give the solutions when $A, B \to 0$ as $\zeta \to 0$. They are

$$A_1=a(2\Gamma/\gamma B_0)^{1/2}sech(a\zeta)\exp[ib(x_2-(V_g-\Gamma a^2/b)t_2)], \tag{3.20}$$

$$\hat{p}_1=B_0 sech^2(a\zeta), \qquad B_0=\nu\varepsilon^2/[(\lambda-c_a)-4a^2\varepsilon^2\mu], \tag{3.21}$$

provided only that $(2\Gamma/\gamma B_0)>0$ and $\theta\Gamma=6\mu$.

4) CONCLUSIONS

Amplitude modulation of a plane longitudinal optical wave propagating in a diatomic elastic solid is asymptotically governed by the NLS equation while a long acoustical wave obeys the KdV equation. When the group velocity of the optical wave is close to the phase velocity of the acoustical wave at a critical wave number, these waves can exchange energy in a resonant manner, and they are asymptotically governed by a pair of coupled nonlinear equations. This also indicates that, even if the acoustical wave does not exist initially, it can be generated later if the resonance phenomenon occurs. This type of wave interaction may also exist between a longitudinal and a transverse plane wave in certain diatomic solids. Further investigation of this topic will appear elsewhere.

REFERENCES

[1] Dean, P.,J. Inst. Maths. Applics 3 (1967) 98.

[2] Demiray, H., Int. J. Engn. Sci. 11 (1973) 1237.

[3] Teymur, M., Int. J. Engn. Sci. 24 (1986) 883.

[4] Dash, P.C. and Patnaik, K., Prog. Theor. Phys.65 (1981) 1526.

[5] Pnevmatikos, S.,Remoissenet, M. and Flytzanis, N.,J.Phys.C 16 (1983) L305

[6] Jeffrey, A. and Kawahara, T., Asymptotic Methods in Nonlinear Wave Theory (Pitman, Boston, 1981)

ELASTIC WAVE PROPAGATION
M.F. McCarthy, M.A. Hayes, (Editors)

FINITE AMPLITUDE WAVE PROPAGATION IN A STRETCHED ELASTIC STRING

J. Wegner*, J.B. Haddow and R.J. Tait
University of Alberta
Edmonton, Alberta

1. INTRODUCTION

In this paper we consider finite deformation plane motion of a hyperelastic string which is assumed to be perfectly flexible. We obtain the governing equations in conservation form and note that these equations are of the same form as those obtained by Collins [1] for non-linear wave propagation in an incompressible elastic half space. An equivalent system of governing equations has been derived by Beatty and Haddow [2], this system, however, is not easily expressible in conservation form. The system of equations which we obtain is strictly valid for a non heat conducting solid. For a heat conducting solid the system is an approximation and if we also neglect the effect of the entropy jump, across a shock, on the constitutive relation we have the isentropic approximation, which is adopted in this paper. We apply the governing equations and corresponding jump relations to the symmetrical plane motion of a stretched string subjected to a transverse impact. Other problems, such as the unloading waves in a plucked string can also be solved by the techniques described. Similarity solutions are outlined for two different strain energy functions, particular cases of the Mooney-Rivlin and a three term function proposed by Ogden [3]. These solutions are valid until the first reflection occurs at a fixed end. In order to extend a solution beyond the time of the first reflection a finite difference scheme due to Godunov and described by Sod [4] was used. For this scheme a system of equations in conservation form is essential.

2. GOVERNING EQUATIONS

We consider a perfectly flexible uniform hyperelastic string. The unstressed reference configuration at temperature T_o occupies the interval $[-L,L]$ of the x_1 axis of a rectangular Cartesian coordinate system and the x_1 coordinate of a particle in the reference configuration is $X\epsilon[-L,L]$. We consider plane motion in the Ox_1x_2 plane and at time t the particle is at place $\underset{\sim}{x} = \underset{\sim}{x}(X,t)$ with coordinates $x_i = x_i(X,t)$, $i=1,2$.
If $s(\underset{\sim}{x},t)$ denotes the arc length, measured from $\underset{\sim}{x} = \underset{\sim}{x}(0,t)$, the stretch λ is given by

$$\lambda(X,t) = \frac{\partial s}{\partial X}, \qquad (1)$$

and the compatibility relations,

$$\frac{\partial(\lambda\cos\theta)}{\partial t} = \frac{\partial \dot{x}_1}{\partial X}, \quad \frac{\partial(\lambda\sin\theta)}{\partial t} = \frac{\partial \dot{x}_2}{\partial X} \qquad (2)$$

follow, where θ is the angle the tangent to the string makes with the x_1 axis as indicated in Fig. 2. Since the string is perfectly flexible the direction

*University of Alberta, Edmonton, Alberta T6G 2G8, Canada.

of the string tension is tangential to the string so that the Lagrangian equations of motion are:

$$\frac{\partial(P\cos\theta/\rho_o)}{\partial X} = \frac{\partial \dot{x}_1}{\partial t}, \quad \frac{\partial(P\sin\theta/\rho_o)}{\partial X} = \frac{\partial \dot{x}_2}{\partial t}, \tag{3}$$

where P is the nominal tensile stress and ρ_o is the density in the reference configuration. The system of equations (2) and (3) is in the conservation form

$$\frac{\partial \underset{\sim}{G}}{\partial t} + \frac{\partial \underset{\sim}{H}(\underset{\sim}{G})}{\partial X} = \underset{\sim}{0}, \tag{4}$$

where $\underset{\sim}{G} = (\lambda\cos\theta, \lambda\sin\theta, \dot{x}_1, \dot{x}_2)^T$, $\underset{\sim}{H} = (\dot{x}_1, \dot{x}_2, P\cos\theta/\rho_o, P\sin\theta/\rho_o)^T$ and a superposed T denotes the transpose. System (4) can be put in the non conservation form

$$\frac{\partial \underset{\sim}{G}}{\partial t} + \underset{\sim}{B}\frac{\partial \underset{\sim}{G}}{\partial X} = 0. \tag{5}$$

If the isentropic approximation is adopted, and $P = P(\lambda)$ is the adiabatic nominal stress-stretch relation, the matrix $\underset{\sim}{B}$ is given by

$$\{\underset{\sim}{B}\} = \left[\begin{array}{cc|cc} 0 & 0 & -1 & 0 \\ 0 & 0 & 0 & -1 \\ \hline -B_{11} & -B_{12} & 0 & 0 \\ -B_{21} & -B_{22} & 0 & 0 \end{array}\right]$$

where $B_{11} = (C_L^2\cos^2\theta + C_T^2\sin^2\theta)$, $B_{12} = B_{21} = (C_L^2 - C_T^2)\sin\theta\cos\theta$ $B_{22} = (C_L^2\sin^2\theta + C_T^2\cos^2\theta)$, $C_L^2 = (dP/d\lambda)/\rho_o$ and $C_T^2 = P/\rho_o\lambda$. The eigenvalues of $\underset{\sim}{B}$ are $\pm C_L$ and $\pm C_T$ and the system is strictly hyperbolic if $C_L^2 > 0$, $C_T^2 > 0$ and $C_T^2 \neq C_L^2$. For the strain energy functions considered these inequalities are satisfied for $\lambda > 1$ except for isolated values of λ for which $C_L = C_T$. It follows that there are 4 families of characteristics with slopes $\pm C_T$ and $\pm C_L$ in the X,t plane and that λ and θ are propagated with Lagrangian wave speeds C_L and C_T, respectively. A system of ordinary differential equations,

$$(\underset{\sim}{B} - Z\underset{\sim}{I})\frac{d\underset{\sim}{G}}{dZ} = 0, \tag{6}$$

is obtained from system (5) by introducing the similarity variable $Z = X/t$. A non-trivial solution to (6) exists if, and only if, $Z = \pm C_L$ or $Z = \pm C_T$, and the similarity solutions consist of centered simple waves and/or shocks.

Jump relations obtained from the system (4) are

$$V[\underset{\sim}{G}] = [\underset{\sim}{H}], \tag{7}$$

where V is Langrangian shock velocity and [ ] denotes the jump in the enclosed quantity. It may be deduced from (7) that there are three possibilities, $[\theta] = 0$ and $[\lambda] \neq 0$, $[\theta] = \pm\pi$ and $[\lambda] \neq 0$, $[\theta] \neq 0$ and $[\lambda] = 0$. Consequently a discontinuity with $[\lambda] \neq 0$ and $[\theta] \neq 0$ is not possible, since $[\theta] = \pm \pi$ is not physically admissible, and we have the shock velocities,

$$V_L = \pm \left\{ \frac{[P]}{\rho_o[\lambda]} \right\}^{1/2} \quad \text{and} \quad V_T = \pm \left\{ \frac{P}{\rho_o \lambda} \right\}^{1/2} , \tag{8}$$

which are velocities of propagation of $[\lambda]$ and $[\theta]$, respectively. It follows that (7) consists of two sets of jump relationships,

$$V_L[\lambda] \begin{Bmatrix} \cos\theta \\ \sin\theta \end{Bmatrix} = - \begin{bmatrix} \dot{x}_1 \\ \dot{x}_2 \end{bmatrix} , \; V_L \begin{bmatrix} \dot{x}_1 \\ \dot{x}_2 \end{bmatrix} = - \frac{[P]}{\rho_o} \begin{Bmatrix} \cos\theta \\ \sin\theta \end{Bmatrix} , \tag{9}$$

and

$$V_T\lambda \begin{bmatrix} \cos\theta \\ \sin\theta \end{bmatrix} = - \begin{bmatrix} \dot{x}_1 \\ \dot{x} \end{bmatrix} , \; V_T \begin{bmatrix} \dot{x}_1 \\ \dot{x}_2 \end{bmatrix} = - \frac{P}{\rho_o} \begin{bmatrix} \cos\theta \\ \sin\theta \end{bmatrix} . \tag{10}$$

Since $V_T = \pm C_T$ a discontinuity of θ is propagated along a characteristic.

The systems (4) and (5) are of the same form as those obtained by Collins [1] for propagation of waves in an incompressible hyperelastic half space in which a pair of transverse simple waves or shocks and a pair of circular waves can propagate. These waves are analogous to the λ and θ waves, respectively, of the string problem but the two problems are not completely analogous.

3. CONSTITUTIVE RELATIONS

The simple tension form of Ogden's three term strain energy function [3] is,

$$W(\lambda) = \sum_{i=1}^{3} \frac{\mu_i}{a_i} (\lambda^{a_i} + 2\lambda^{-a_i/2} - 3) , \tag{11}$$

which gives the isothermal nominal stress-stretch relation

$$\hat{P} = \frac{\partial W}{\partial \lambda} = \sum_{i=3}^{3} \mu_i (\lambda^{a_i - 1} - \lambda^{-a_i/2 - 1}) , \tag{12}$$

where μ is the infinitesimal shear modulus and $\sum_{i=1}^{3} \mu_i a_i = 2\mu$. Eq. (12) is in close agreement with experimental data for certain rubbers for stretches up to 6 when the μ_i and a_i take the values

$$\begin{aligned} &\mu_1/\mu = 1.491, \; \mu_2/\mu = 0.003, \; \mu_3/\mu = -0.0237, \\ &a_1 = 1.3 , \quad a_2 = 5.0 , \quad a_3 = -2.0 . \end{aligned} \tag{13}$$

With these parameters (12) gives an "S" shaped curve as indicated in Fig. 1. The Mooney-Rivlin s.e.f. is a special case of (11) with

$$\begin{aligned} &\mu_1/\mu = \alpha , \quad \mu_2/\mu = -(1-\alpha) , \quad \mu_3 = 0 , \\ &a_1 = 2.0 , \quad a_2 = -2 , \quad 0 \le \alpha \le 1 , \end{aligned} \tag{14}$$

and with $\alpha = 0.6$, is in good agreement with experimental data for stretches up to 3.5. If the strictly entropic elasticity model for an elastic solid is adopted the adiabatic P,λ relation, for deformation from the reference configuration is

$$P(\lambda) = \frac{dW}{d\lambda} \exp\left(\beta \frac{W}{\mu}\right), \qquad (15)$$

where $\beta = \mu/\rho C T_o$ and C is the specific heat, assumed to be constant. A typical value of the non-dimensional quantity β, for rubberlike materials at $T_o \simeq 290°K$, is $\beta = 10^{-3}$ and for $\lambda < 6$ the term $\exp(\beta W/\mu)$ differs neglibly from unity. For the Mooney-Rivlin string, $C_L > C_T$ if $\lambda < \lambda_{c1}$ and $C_L < C_T$ if $\lambda > \lambda_{C1}$, where $\lambda_{c1} = 2.4733$ if $\alpha = 0.6$. For the three term s.e.f. string, $C_L > C_T$ if $\lambda < \lambda_{c1}$ or $\lambda > \lambda_{c2}$ and $C_L < C_T$ if $\lambda_{c1} < \lambda < \lambda_{c2}$ where $\lambda_{c1} = 2.1674$ and $\lambda_{c2} = 3.1674$. When $\lambda = \lambda_{c1}$ or $\lambda = \lambda_{c2}$, $C_L = C_T$. The P,λ relations are shown in Fig. 1.

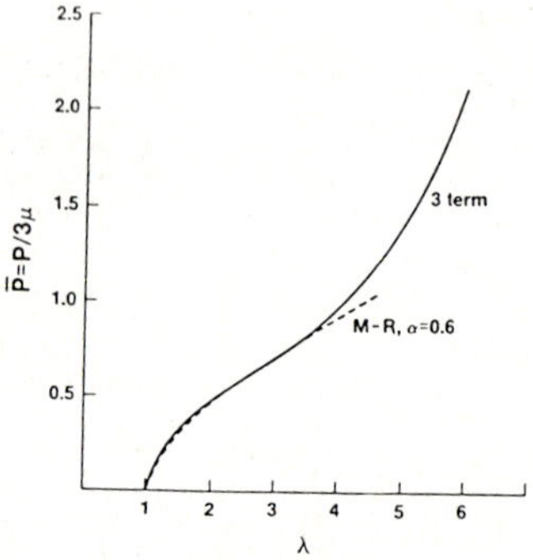

FIGURE 1

5. SIMILARITY SOLUTIONS

The ends of the string are fixed at $x_1 = \pm \ell_o$, and $\lambda_o = \ell_o/L \geq 1$. The initial conditions are $\lambda = \lambda_o$, $\dot{x}_1 = \dot{x}_2 = 0$, and at time $t = 0$ the particle $X = 0$ is given a constant normal velocity in the $-x_2$ direction so that if we consider the part $X \epsilon [0,L]$ the boundary condition is

$$\dot{x}_1(0,t) = 0\ , \quad \dot{x}_2(0,t) = -\,qH(t),$$

where q is constant and H(t) is the unit step function. First we present a solution which is valid for both s.e.f.s when $\lambda_f < \lambda_{c1}$, where λ_f is the maximum stretch after impact. Solutions for other cases are then indicated.

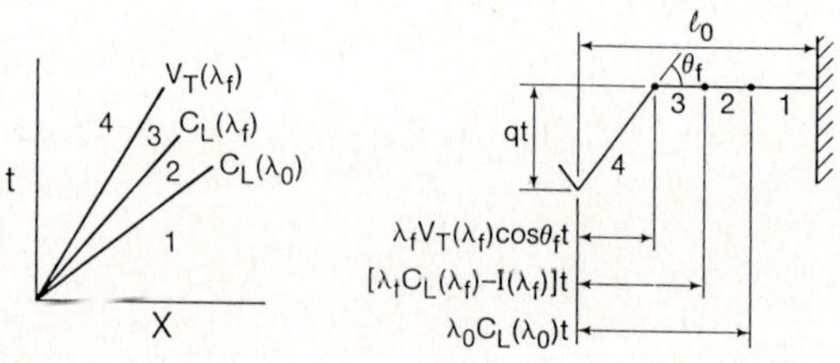

FIGURE 2

Referring to the X,t plane shown Fig. 2, the similarity solution is:
Region 1: $X/t \geq C_L(\lambda_o)$; $\lambda = \lambda_o$, $\theta = \dot{x}_1 = \dot{x}_2 = 0$.
Region 2: $C_L(\lambda_o) \geq X/T \geq C_L(\lambda_f)$; $X/t = C_L(\lambda)$, $\theta = \dot{x}_2 = 0$, $\dot{x}_1 = -I(\lambda)$.
Region 3: $C_L(\lambda_f) \geq X/t > C_T(\lambda_f)$, $\lambda = \lambda_f$, $\theta = \dot{x}_2 = 0$, $\dot{x}_1 = -I(\lambda_f)$.
Region 4: $V_T(\lambda_f) > X/t \geq 0$, $\lambda = \lambda_f$, $\theta = \theta_f$, $\dot{x}_1 = 0$, $\dot{x}_2 = -q$,

where

$$I(\lambda) = \int_{\lambda_o}^{\lambda} C_L(\eta)d\eta .$$

This solution satisfies (6), the jump relations (9) and (10) and the entropy condition across the shocks. It may be deduced from the jump relations that

$$q^2 = I(\lambda_f)\{2\lambda_f C_T(\lambda_f) - I(\lambda_f)\}, \quad \sin\theta_f = q/(\lambda_f V_T(\lambda_f)), \tag{16}$$

and it follows that $\lambda_f < \lambda_{c1}$ if

$$q^2 < I(\lambda_{c1})\{2\lambda_{c1}C_T(\lambda_{c1}) - I(\lambda_{c1})\}.$$

If λ_o and q are given, λ_f and θ_f can be obtained from (16) and the solution completed. The solution for the Mooney-Rivlin string is indicated in Fig. 3 for $\lambda_o > \lambda_{c1}$

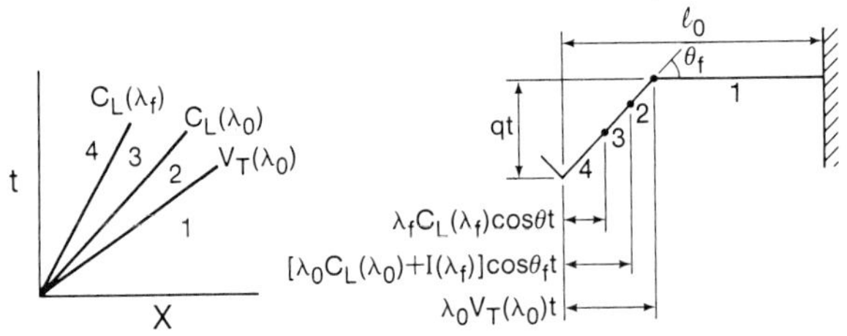

FIGURE 3

and in Fig. 4 for $\lambda_o < \lambda_{c1}$ and $\lambda_f > \lambda_{c1}$. These solutions are also valid for the 3 term s.e.f. if $\lambda_f < \lambda_i$ where λ_i is the stretch at the inflection point.

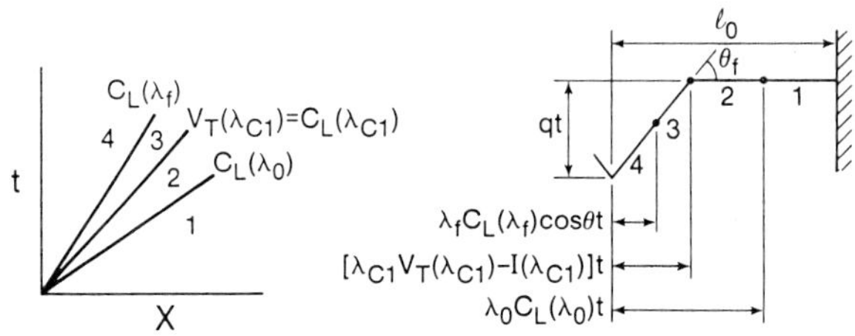

FIGURE 4

In Fig. 3, regions 1,2 and 4 and in Fig. 4 regions 1 and 4 are constant state regions, the other regions represent centered simple waves. The solution for the 3 term s.e.f. is indicated in Fig. 5 for $\lambda_o > \lambda_{c2}$. This solution consists of constant state regions separated by shocks.

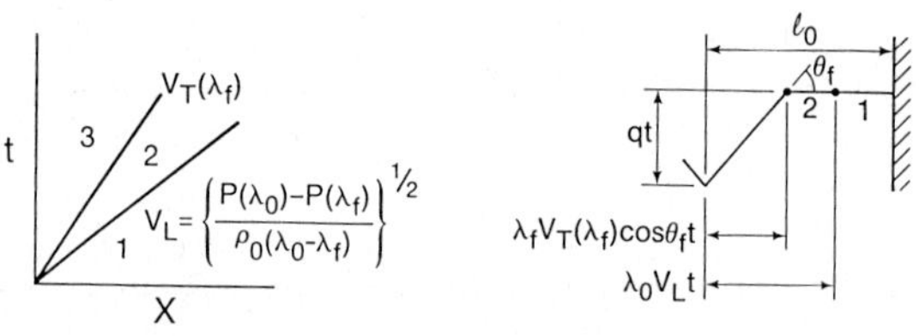

FIGURE 5

6. EXTENSION OF SOLUTIONS BEYOND FIRST REFLECTION

The Godunov finite difference scheme, which is based on the exact solution of a sequence of Riemann problems, was applied to the conservation system (4) in order to extend the similarity solutions between the time of the first reflection and the time the reflected wave front reaches $X = 0$. The results of the similarity solution are used as initial conditions. In order to apply the scheme λ, θ, $\dot{x}_1$ and $\dot{x}_2$ must be known at $X = 0$ and $X = L$. At $X = 0$ these are the same as for the similarity solution until the reflected wave reaches $X = 0$ and at $X = L$, $\dot{x}_1 = \dot{x}_2 = 0$ and λ and θ are found from the method of images.

If $t < t_r$, where t_r is the time the reflected wave reaches $X = 0$ the solution for $\lambda(X,t)$ for the transverse impact problem is the same as that for the problem of simple tension of a string initially at rest under a nominal stress $P(\lambda_o)$ which is suddenly increased to $P(\lambda_f)$. This is evident for the similarity solution and facilitates the application of the Godunov scheme for $t_i < t < t_r$, when t_i is the time the incident wave reaches $X = L$.

REFERENCES

[1] Collins, W.D., Qu. J. Appl. Math. and Mech., 14 (1966) 259.

[2] Beatty, M.F., and Haddow, J.B., J. Appl. Mech., 52 (1985) 137.

[3] Ogden, R.W., Proc. Roy. Soc. Lond. A. 236 (1972) 565.

[4] Sod, G., Numerical Methods in Fluid Dynamics, (Cambridge, 1985).

ELASTIC WAVE PROPAGATION
M.F. McCarthy, M.A. Hayes, (Editors)

SOLITARY WAVES ON NONLINEAR ELASTIC RODS

P.L. Christiansen, V. Muto, and M.P. Soerensen,
Laboratory of Applied Mathematical Physics,
The Technical University of Denmark,
DK-2800 Lyngby, Denmark.

Acoustic waves on elastic rods with circular cross section are governed by so-called improved Boussinesq equations when transverse motion and nonlinearity in the medium are taken into account. Solitary waves to these equations are shown to possess soliton-like properties in agreement with the fact that the improved Boussinesq equations are nearly integrable.
Numerical investigations of blow up (in finite time), reflection and fission of solitary waves are presented in the cases of ending rods and rods with varying cross-section.
The results are applied in a model for DNA-molecules in aqueous solution which is capable to predict the influence of anharmonicity on the spectral properties of this important molecule.

1. INTRODUCTION

In the paper the so-called improved Boussinesq equations for circular-cylindrical elastic rods with a physical nonlinearity due to a nonlinear constitutive relation for the elastic medium are given. Since the equations resemble the integrable Boussinesq equation they are nearly integrable in the sense that they support solitary waves which may interact almost like solitons. Therefore, the solitary waves are also referred to as solitons. In contrast to the Boussinesq equation, which is unstable for short wavelengths, the improved Boussinesq is numerically stable. Therefore, this latter equation is well suited for mathematical modelling work. In the case of a quadratic nonlinearity blow-up occurs in a finite time. The effect of continuously and discontinuously varying properties of the rod may lead to fission of the solitary wave. The reflection of solitary waves at the end of the rod is also investigated. Finally, a closed nonlinear elastic rod with periodic boundary conditions is used as a model for the DNA molecule. The effect of anharmonicity is predicted.

The present contribution summarizes work performed in [1-3]. These papers contain relevant further references.

2. THE BOUSSINESQ EQUATIONS AND THEIR SOLUTIONS

The Lagrangian density for an isotropic circular-cylindrical rod is

$$\mathcal{L} = \frac{1}{2} S\rho[W_T^{\,2} + (S\sigma^2/2\pi)W_{XT}^{\,2}] - \frac{1}{2} SE_2W_X^{\,2} - \frac{1}{6} SE_3W_X^{\,3} - \frac{1}{24} SE_4W_X^{\,4} . \tag{1}$$

Here W(X,T) is the longitudinal displacement component of a plane cross-section along the rod as a function of X, the undisturbed position of the cross-section, and time T. S is the cross-sectional area, ρ is the density of the elastic material, σ is the Poisson ratio, and E_2, E_3 and E_4 are the Young's modulus and higher order elastic coefficients [1].

The Euler equation for this Lagrangian density becomes, in the case of a homogeneous rod,

$$u_{xx} - u_{tt} + \frac{1}{2}(u^2)_{xx} + u_{xxtt} = 0 \tag{2a}$$

for material with quadratic Hooke's law ($E_3 \neq 0$, $E_4 = 0$) and

$$u_{xx} - u_{tt} \pm \frac{1}{3}(u^3)_{xx} + u_{xxtt} = 0 \tag{2b}$$

for material with cubic Hooke's law ($E_3 = 0$, $E_4 \gtrless 0$). In Eq. (2) dimensionless variables have been introduced as $x = X/(\sigma\sqrt{S/2\pi})$, $t = T\sqrt{E_2}/(\sigma\sqrt{S\rho/2\pi})$, and u is the dimensionless strain $u = w_x$, where $w = WE_3/(E_2\sigma\sqrt{S/2\pi})$ in (2a) and $w = W\sqrt{|E_4|}/(\sigma\sqrt{SE_2/2\pi})$ in (2b). Eq. (2a) is called the improved Boussinesq equation (IBE) and Eq. (2b) is called the modified improved Boussinesq equation (MIBE).

Solitary wave solutions to (2a) and (2b) are given by

$$u = 3(c^2-1)\operatorname{sech}^2[(\sqrt{c^2-1}/2c)(x-x_0-ct)] , \tag{3a}$$

in the case of Eq. (2a), and by

$$u = \pm \sqrt{6}\,\sqrt{c^2-1}\,\operatorname{sech}^2[(\sqrt{c^2-1}/c)(x-x_0-ct)] , \tag{3b}$$

in the case of Eq. (2b) with $+1/3(u^3)_{xx}$. Here the velocity c must satisfy $|c| > 1$, and $x = x_0$ is the position of the wave at $t = 0$.

It is easily shown [1] that energy, momentum and total displacement (defined as $\int_{-\infty}^{\infty} w_x dx$) are conserved quantities in both IBE and MIBE. There does not appear to be further conserved quantities for these two nearly integrable equations (in contrast to the Boussinesq equation, which has infinitely many conserved quantities).

Figures 1-4 show some numerical solutions of IBE and MIBE. In the latter case, Eq. (2b), solitary and antisolitary solutions exist because the nonlinearity is an odd function of u. The solitary and antisolitary waves may be bound together in breather waves. The interactions between the waves in Figures 1-4 are seen not to be perfect soliton interactions since small amounts of radiation (i.e. little ripples) are created. As a consequence, some of the energy and momentum, originally present in the solitary waves, is lost into the radiation.

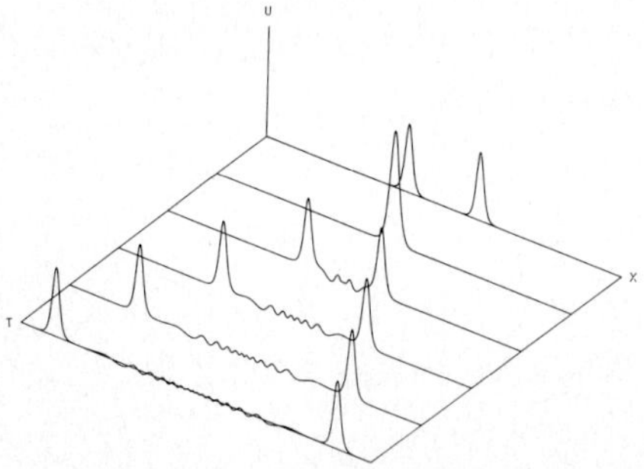

Fig.1. Head-on collision between two solitary wave solutions to Eq.(2a) [1].

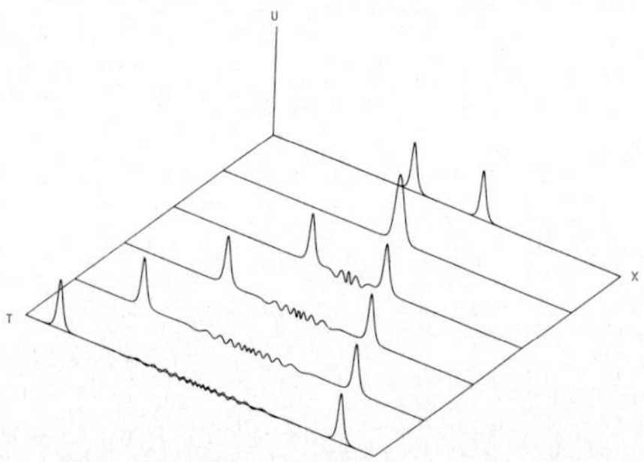

Fig.2. Head-on collision between two solitary wave solutions to Eq.(2b) [1].

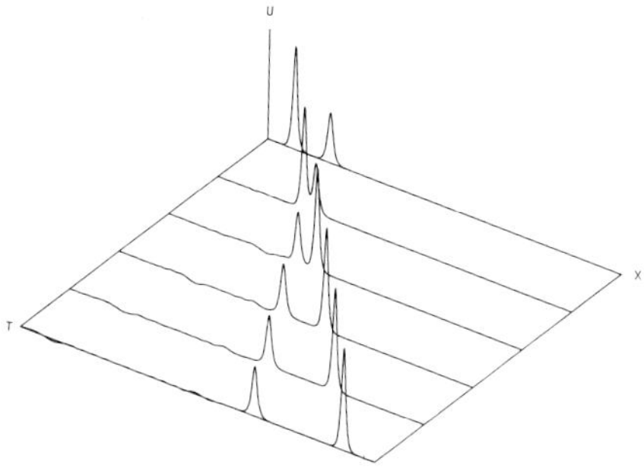

Fig.3. Takeover of two solitary wave solutions to Eq.(2b) [1].

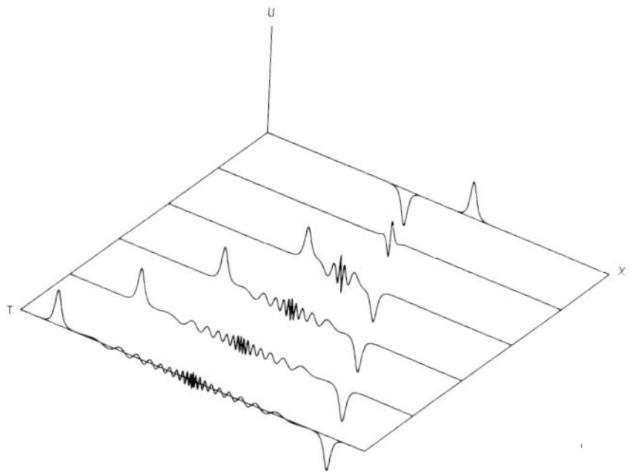

Fig.4. Head-on collision between solitary and antisolitary solution to Eq. (2b) [1].

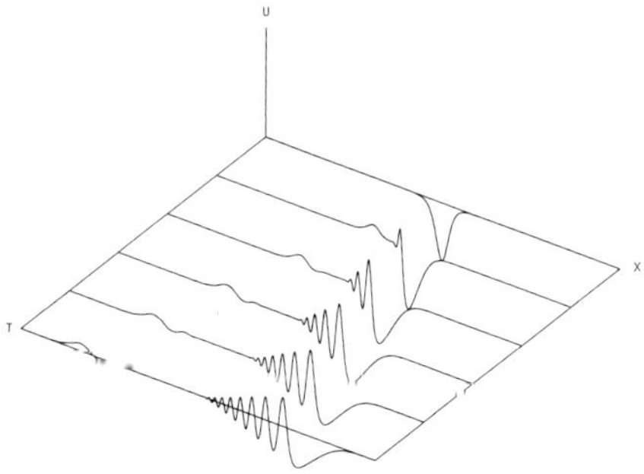

Fig.5. Time evolution of single static negative solitary wave in Eq.(2a). Oscillatory solutions result [1].

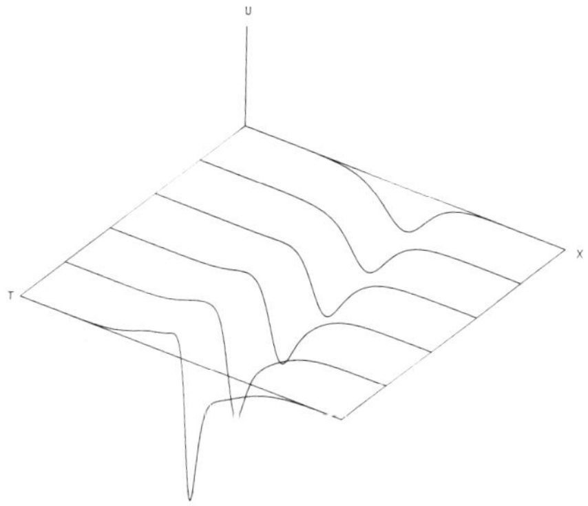

Fig.6. Blowup of single static negative solitary wave in Eq.(2a) [1].

In Eq. (2a) negative initial waves either disperse (as seen in Figure 5) or blow up in finite time (as seen in Figure 6). In Figure 7 we have studied the dependence on parameters A and B in the initial conditions

$$u(x,0) = A \operatorname{sech} B(x-x_0) \tag{4a}$$

$$u_t(x,0) = 0 \; . \tag{4b}$$

The area of the initial pulse is proportional to A/B, while the curvature at the extremum is proportional to AB^2. In AB-parameter space (Figure 7) the boundary between the blowup region and the dispersion region appears to be a tilted straight line. This indicates that the area of the initial pulse (rather than the extremum or the curvature at the extremum) decides whether blowup or dispersion occurs or not.

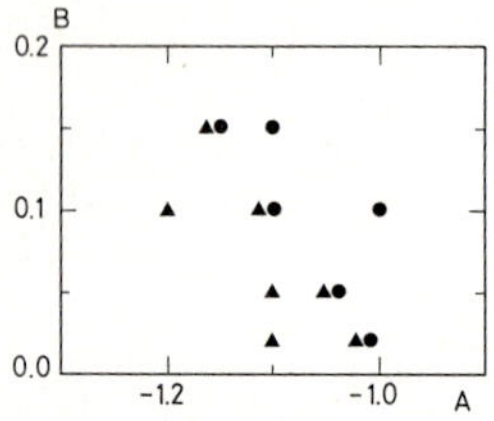

Fig.7. Numerical solution of Eq.(2a) with initial conditions (4). Blowup: ▲. Dispersion: ● [2].

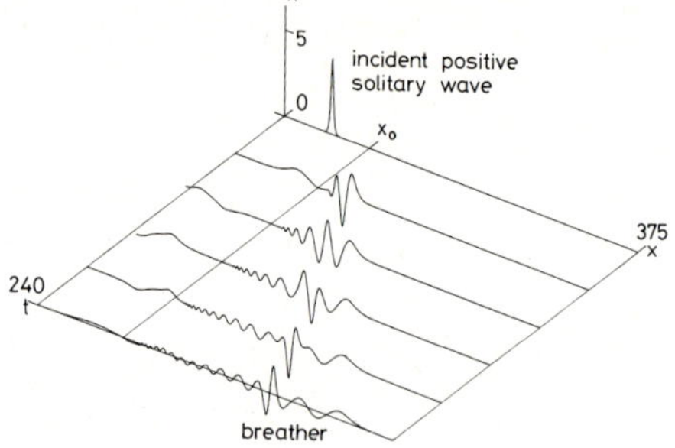

Fig.8. MIBE-solitary wave becomes a breather at $x = x_0$ where rod cross-section increases continuously [2].

3. VARYING PROPERTIES OF THE ROD AND REFLECTION AT END

Boussinesq equations with varying coefficients are easily derived in the cases where the properties (cross-section, density, elastic moduli, Poisson ratio) of the elastic rod vary. Figure 8 shows how an incident MIBE solitary wave is transmitted as a breather across a continuously increasing cross-section around $x = x_0$. Figure 9 shows the fission of an incident MIBE solitary wave into two transmitted solitary waves, a breather, and a reflected antisolitary wave at continuously decreasing cross-section aroung $x = x_0$. Even more dramatic results are observed in the case of discontinuously decreasing cross-section (Figure 10).

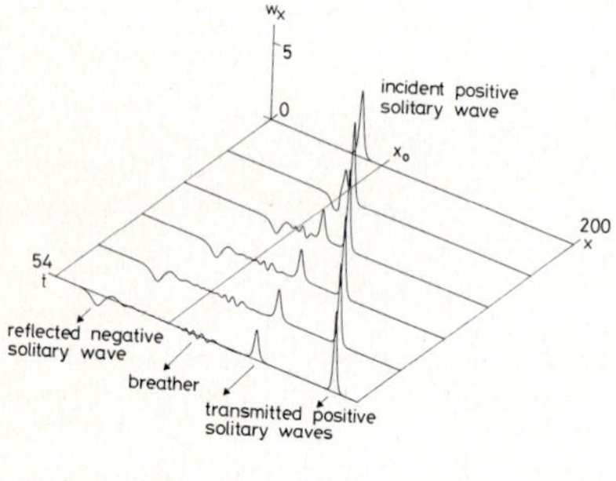

Fig.9. Fission of MIBE solitary wave at $x = x_0$ where rod cross-section decreases condinuously [2].

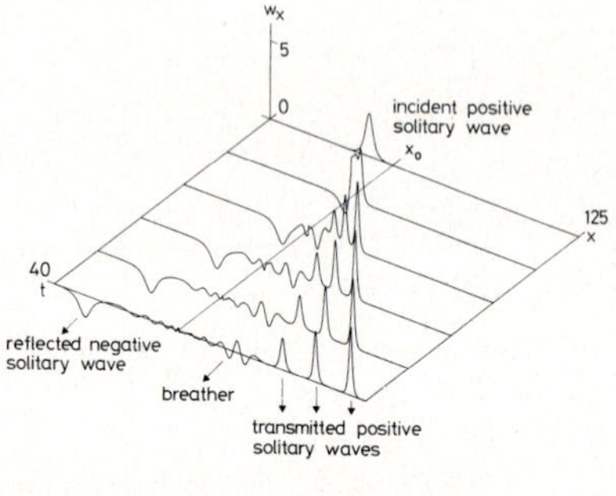

Fig.10. Fission of MIBE solitary wave at $x = x_0$ where cross-section decreases discontinuously [2].

The nonlinear discontinuity conditions are easily derived from the Lagrangian density (1) by means of Hamilton's principle [2]. The conditions can also be applied to the case of an ending rod. Figures 11 and 12 show the reflection of an incident MIBE solitary wave into an antisolitary wave at a weakly loaded end and into a solitary wave and an antisolitary wave at a heavily loaded end. In both cases a substantial amount of radiation is created.

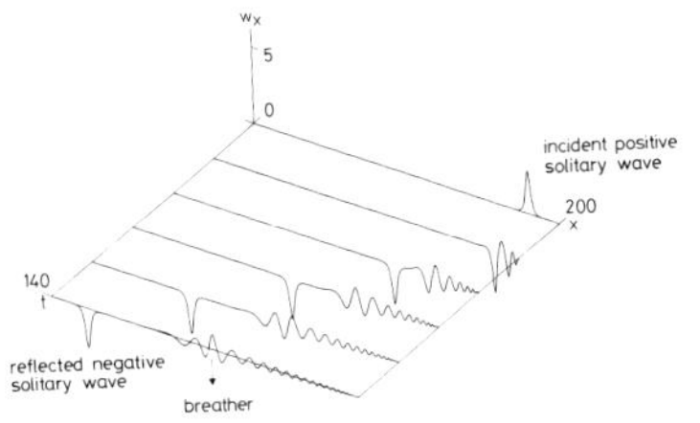

Fig.11. Reflection of MIBE solitary wave at weakly loaded end [2].

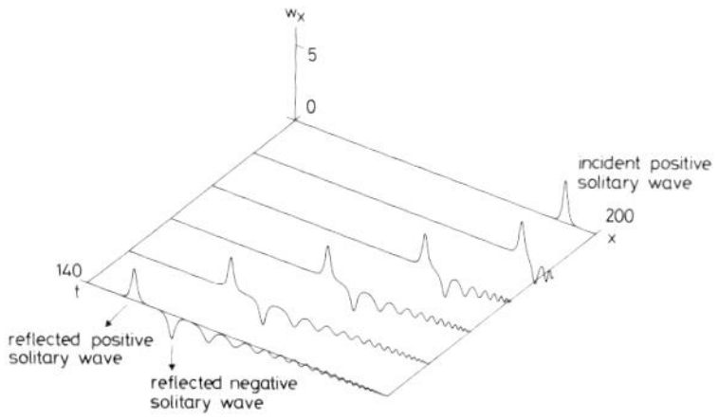

Fig.12. Reflection of MIBE solitary wave at heavily loaded end [2].

4. APPLICATION OF THE BOUSSINESQ MODEL TO DNA MOLECULES

In order to model the interaction between DNA molecules in aqueous solution and electromagnetic radiation we use the perturbed equation [3]

$$w_{xx} - w_{tt} - (w_x{}^2)_x + w_{xxtt} - \alpha w_t + f(x)\cos\omega t = 0 \qquad (5)$$

for the longitudinal displacement, w, with a different normalization of the nonlinearity. The terms $-\alpha w_t$ and $f(x)\cos\omega t$ represent viscous damping and driving with normalized frequency ω by the electromagnetic field. The spatial dependence, f(x), of the driver follows from the shape of the closed DNA-molecule. We consider $f(x) = \Gamma\delta(x)$ for pointwise interaction and $f(x) = \Gamma\cos(2\pi x/\ell)$, where Γ is constant and ℓ is the normalized circumference of the closed molecule, also entering the periodicity condition

$$w(x,t) = w(x+\ell, t) \ . \qquad (6)$$

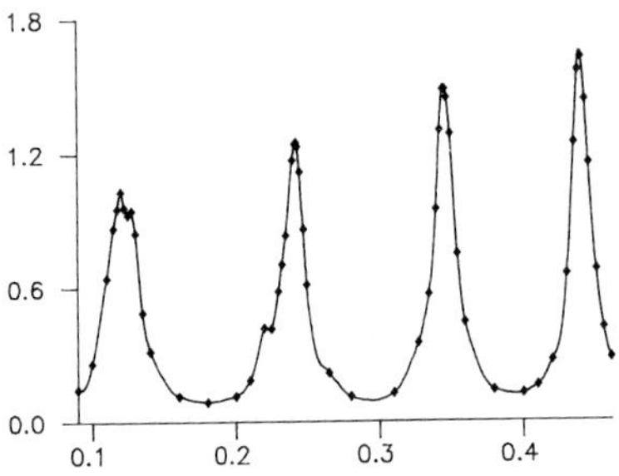

Fig.13. Total energy vs. frequency ω for solution of Eq. (5) [3].

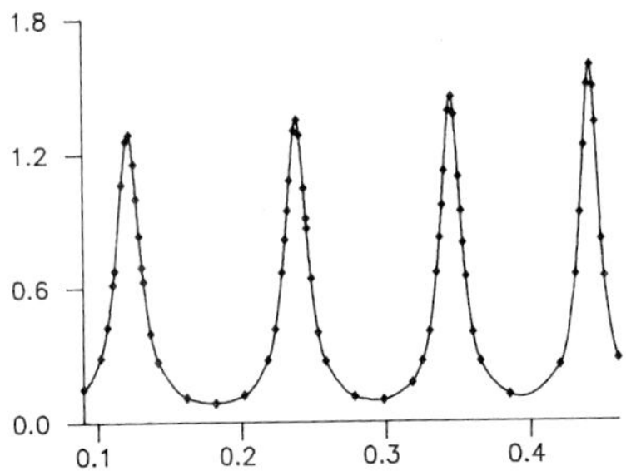

Fig.14. Total energy vs. frequency ω for solution of linearized Eq. (5) [3].

Figure 13 shows the energy content versus the frequency a of the solution of Eq. (5) with $f(x) = \Gamma\delta(x)$. To illustrate the influence of anharmonicity Figure 14 shows analogous energy dependence on frequency in the case without the nonlinear term $-(w_x{}^2)_x$. The first two resonances exhibit a fine-structure in the nonlinear case (Figure 13), but not in the linearized case (Figure 14). A detailed analysis of the structure of the first resonance is shown in Figure 15 for the case $f(x) = \Gamma\cos 2\pi x/\ell$. The very different periodic soliton behaviour of the solution corresponding to peak values for ω are computed in Figures 16-18. The fact that we only observe structure in the two first resonances (Figure 13) indicates that the anharmonicity becomes more important the longer the molecule is. This result demonstrates that the anharmonic effect is present for physically realistic lengths of the DNA molecule.

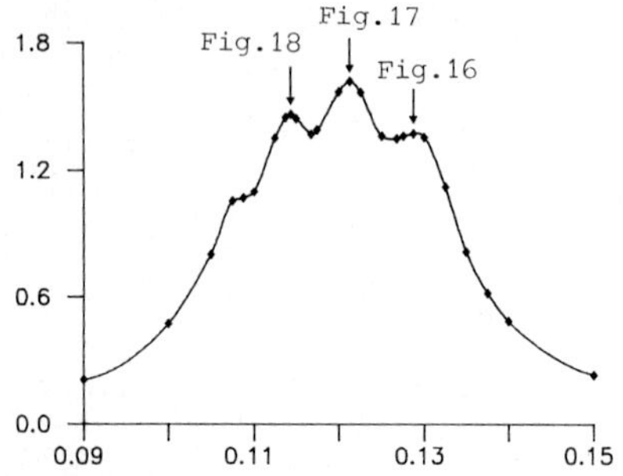

Fig.15. Total energy vs. frequency ω First resonance exhibiting structure due to different soliton behaviour [3].

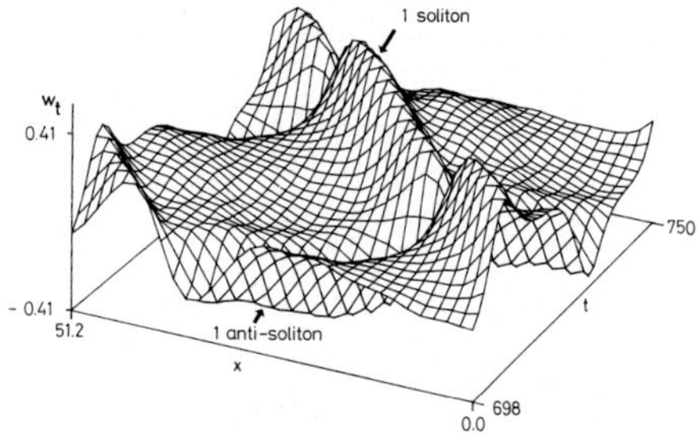

Fig.16. Soliton and antisoliton travelling in opposite directions for ω = 0.12875 [3].

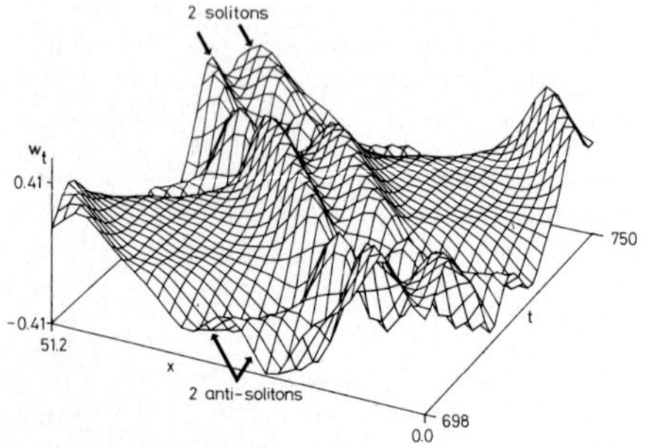

Fig.17. Two solitons and two antisolitons travelling in opposite directions for ω = 0.12125 [3].

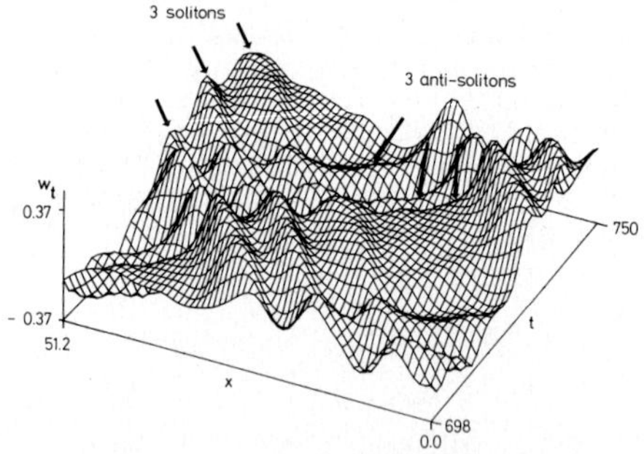

Fig.18. Three solitons and three antisolitons travelling in opposite directions for ω = 0.114375 [3].

ACKNOWLEDGEMENTS

The financial support of the Danish Council for Scientific and Industrial Research, Nato Scientific Affairs Division through grant RG 85/0396, the European Research Office of the United States Army through contract DAJA-45-85-C-0042, and of the Consiglio Nazionale delle Ricerche, Rome, Italy, is gratefully acknowledged.

REFERENCES

[1] Soerensen, M.P., Christiansen, P.L., and Lomdahl, P.S., J. Acoust. Soc. Am. 76 (1984) 871.
[2] Soerensen, M.P., Christiansen, P.L., Lomdahl, P.S., and Skovgaard, O., J. Acoust. Soc. Am. 81 (1987) 1718.
[3] Muto, V., Halding, J., Christiansen, P.L., and Scott, A.C., J. Biomole. Struct. Dyn. 5 (1988) (to appear).

ELASTIC WAVE PROPAGATION
M.F. McCarthy, M.A. Hayes, (Editors)

THE EVOLUTION OF RESONANT ACOUSTIC OSCILLATIONS WITH DAMPING

EDWARD A. COX and MICHAEL P. MORTELL

Department of Mathematical Physics, University College, Dublin

Registrar's Office, University College, Cork

ABSTRACT

This paper is concerned with the evolution of small amplitude resonant, and near resonant, oscillations of an inviscid gas in a closed tube in the presence of damping. The damping is introduced by allowing energy to radiate through the closed end. The evolution of the gas motion, from the initial rest state to the final periodic state, is examined. The motion is shown to consist of non-interacting oppositely travelling linear waves where the signal carried is determined by a nonlinear partial differential equation. There is a critical level of damping which prevents the formation of shocks in the flow, and then linear theory is an adequate discription. An unusual feature is that when shocks occur outside the resonant band, they decay during the evolution to leave the continuous periodic profile predicted by linear theory, while shocks that occur within the resonant band are maintained in the periodic state.

1. INTRODUCTION

This paper is concerned with the evolution of small amplitude resonant and near resonant oscillations of a gas in a tube of finite length that is driven by a piston at one end while energy is allowed to radiate through the other end. The motion is assumed to be one dimensional and inviscid.

The periodic motion of the gas in a closed tube has been extensively investigated both experimentally and theoretically and shocks are a feature of the motion, see Seymour and Mortell [1] for references. The influence of damping on the periodic state was treated in Seymour and Mortell [2]. The nature of the periodic state was seen to depend on the competing phenomena of nonlinear amplitude dispersion and linear damping. The theoretical analysis showed that radiative damping can prevent the occurrence of shocks provided that the damping is above some critical level. The prediction is confirmed by experiment, Sturtevant [3].

The literature on the evolution of resonantly forced periodic oscillations in a finite domain is quite sparse. The mathematical problem involves the solution of a system of nonlinear partial differential equations on a semi-infinite strip, subject to appropriate initial and boundary conditions. Cox and Mortell [4], [5] have considered the small-rate evolution of resonant oscillations in closed tubes and of resonant small-amplitude, long-wavelength oscillations of a liquid in a tank of finite length. In both papers a nonlinear differential-difference equation was derived which described the evolution of a linear Riemann invariant on one boundary. The differential-difference equations were then reduced to partial differential equations - a periodically forced simple wave equation, and a periodically forced K de V equation - using the method of multiple scales. Seymour and Mortell [6] treated the finite rate evolution of resonant oscillations in a closed tube by a direct analysis of the "standard mapping".

In this paper we show that the problem of describing the evolution of damped resonant oscillations in a closed tube can be reduced to the analysis of a periodically forced damped simple wave equation viz.,

$$\varepsilon \frac{\partial F}{\partial \tau} + F \frac{\partial F}{\partial t} + jF = j\bar{\Delta} + A \sin \pi t.$$

A phase plane analysis elucidates qualitative features of the flow and quantitative results are found numerically. It is possible to ensure a shockless evolving signal for all piston frequencies by introducing radiation damping of sufficient strength. The critical level of damping required agrees with that obtained in Seymour and Mortell [2] to ensure a shockless periodic profile. The presence of fixed points in the appropriate phase plane defines resonance. When shocks occur within the resonant band they are sustained and constitute part of the periodic state. In contrast, when shocks occur outside the resonant band they decay to zero strength during evolution to leave a continuous steady periodic profile.

2. FORMULATION

A tube of length L is closed at one end and contains a gas driven by an oscillating piston at the other end which is operating at or near a resonant frequency. Energy is allowed to radiate from the closed end of the tube. The pressure and density of the gas are measured from the undisturbed reference state p_o, ρ_o. The motion of the gas is expressed in terms of the nondimensional variables $a_o u$, $\rho_o \rho$, Lx, $\frac{L}{2\omega a_o} t$, where u, ρ and a denote the velocity, density and sound speed at particle x in the gas at time t, and ω^{-1} is the nondimensional period of the piston. The equations governing the isentropic flow of an ideal gas in Riemann Invariant form are

$$(2\omega \frac{\partial}{\omega t} + a^{\frac{\gamma+1}{\gamma-1}} \frac{\partial}{\partial x}) (u + \frac{2}{\gamma-1} a) = 0, \tag{2.1}$$

$$(2\omega \frac{\partial}{\partial t} - a^{\frac{\gamma+1}{\gamma-1}} \frac{\partial}{\partial x}) (u - \frac{2}{\gamma-1} a) = 0, \tag{2.2}$$

where γ is the gas constant.

The end x = 0 is closed but allows outward radiation of energy. The requirement of continuity of pressure and velocity across x = 0 results in the boundary condition

$$u(0, t) = \frac{-j}{\gamma} (a^{\frac{2\gamma}{\gamma-1}} - 1), \tag{2.3}$$

where $\frac{j}{\gamma} \geq 0$ is the impedance of the interface, see Mortell and Varley [7]. It has been shown in Seymour and Mortell [2] that the equation resulting from the boundary condition (2.3) can be interpreted to include the effects of internal dissipation (e.g. vibrational excitation or molecular dissociation) of the gas or the effect of boundary-layer friction when introduced through a body-force term proportional to the gas velocity.

At x = 1 the piston imparts a small amplitude sinusoidal velocity to the gas

$$u(1, t) = 2\pi\varepsilon\omega \sin \pi t, \tag{2.4}$$

where $\varepsilon = \frac{\ell}{L} << 1$ is the ratio of maximum piston displacement to the tube length.

The gas is initially undisturbed i.e.,

$$u(x,t) = 0, \ a(x,t) = 1, \quad 0 \leqq x \leqq 1, \ t \leqq 0. \tag{2.5}$$

The problem posed is to follow the evolution of the gas motion, governed by (2.1), (2.2) and the boundary conditions (2.3) and (2.4), from the initial undisturbed state (2.5) to the final periodic state.

We adopt the relationship, see Seymour and Mortell [2],

$$j = \varepsilon^{\frac{1}{2}}\lambda, \ \lambda = 0(1), \tag{2.6}$$

and we introduce the detuning parameter

$$\bar{\Delta} = \varepsilon^{\frac{1}{2}}\Delta = 2\omega - 1, \ \Delta = 0(1). \tag{2.7}$$

We expand the gas velocity and sound speed as

$$u(x,t) = \varepsilon u_1 + \varepsilon^{\frac{3}{2}} u_2 + \cdots . \tag{2.8}$$

$$a(x,t) = 1 + \varepsilon a_1 + \varepsilon^{\frac{3}{2}} a_2 + \cdots . \tag{2.9}$$

The boundary conditions on $x = 0$ and $x = 1$ are expanded in a similar manner.

The basic approximation is

$$u_1 = f(t+x-1) - f(t-x-1). \tag{2.10}$$

Following Cox and Mortell [4], [5] the representation for u from which the evolution equation will be derived is

$$u = \varepsilon u_1 + \varepsilon^{\frac{3}{2}} u_2 + \varepsilon^2 u_3. \tag{2.11}$$

The boundary condition at $x = 0$ is satisfied at each order and when the condition at $x = 1$ is applied to (2.11), the result is

$$f(t) - f(t-2) + \varepsilon^{\frac{1}{2}} \Delta[f'(t-2) + f'(t)] + \varepsilon^{\frac{1}{2}}\lambda\,[f(t) + (t-2)]$$
$$+ \frac{\varepsilon(\gamma+1)}{2}[f(t-2)\,f'(t-2) + f(t)\,f'(t)]$$
$$+ \frac{\varepsilon(\gamma+1)}{2}[f'(t)\int^{t-2} f(s)ds - f'(t-2)\int^{t} f(s)ds] = 2\pi\omega \sin \pi t. \tag{2.12}$$

Equation (2.12) is a nonlinear functional-differential equation. Terms independent of ε correspond to undamped acoustic theory; the $\varepsilon^{\frac{1}{2}}\Delta$ terms represent the off resonance effects; the $\varepsilon^{\frac{1}{2}}\lambda$ terms represent the effect of damping, while the ε terms represent nonlinear amplitude dispersion and interaction of oppositely travelling waves.

3. GOVERNING PARTIAL DIFFERENTIAL EQUATION

We assume an expansion of the form

$$f(t;\varepsilon) = f_o(t, \tau) + \varepsilon f_1(t, \tau) + \cdots , \tag{3.1}$$

where $\tau = \varepsilon t$ is the slow variable, and seek solutions which are periodic on the fast time scale with the same period as the driver, and are slowly modulated on the long time scale.

We define

$$F(t, \tau) = M\varepsilon f_o(t, \tau) + \bar{\Delta}, \quad M = \frac{\gamma+1}{2}, \tag{3.2}$$

and then the final equation is

$$\varepsilon \frac{\partial F}{\partial \tau} + F \frac{\partial F}{\partial t} + jF = j\bar{\Delta} + A \sin \pi t, \tag{3.3}$$

where $F(t+2, \tau) = F(t, \tau)$. The similarity parameter is $A = \pi \omega \varepsilon M << 1$, see Mortell and Seymour [8].

The initial condition for the partial differential equation (3.3) is

$$F(t, 0) = \bar{\Delta}. \tag{3.4}$$

The solution of the nonlinear acoustic problem on the semi-infinite strip $0 \leqq x \leqq 1$, $t \geqq 0$ defined by equations (2.1) - (2.5) has been reduced in the small amplitude, small rate limit to the linear acoustic representation (2.10), where the signal f carried by a wave is determined by the first order quasilinear partial differential equation (3.3) subject to the initial condition (3.4) and the condition that f has period 2 in the fast variable. In this formulation the signal f carried by a wave evolves on the boundary $x = 1$ according to a quasilinear partial differential equation, and this signal is then propagated into the tube according to linear theory.

The equation (3.3) is amenable to a phase plane analysis similar to that in Cox and Mortell [4].

In Figure 1 the damping is such that shocks cannot form. The periodic solution is influenced directly by the initial conditions through the characteristic curves. The rapid amplification of the signal in the early stages should be noted. Figure 2 shows the evolution of the signal slightly off resonance, and there is a shock in the final periodic state. Figure 3 shows the evolution of the signal outside the resonant band, where a shock which forms cannot be sustained and eventually a continuous periodic signal remains.

ACKNOWLEDGEMENT

One of the authors (E.A.C.) was partly supported by the U.S. Army Research Office through the Mathematical Sciences Institute of Cornell University.

REFERENCES

[1] Seymour, B.R. and Mortell, M.P. 1980. A finite rate theory of resonance in a closed tube: discontinuous solutions of a functional equation. J. Fluid Mech. 99. 365 - 382.

[2] Seymour, B.R. and Mortell, M.P. 1973. Resonant acoustic oscillations with damping: small rate theory. J. Fluid Mech. 58, 353 -373.

[3] Sturtevant, B. 1974. Nonlinear gas oscillations in pipes. Part 2. Experiment. J. Fluid Mech. 63, 97-120.

[4] Cox, E.A. and Mortell, M.P. 1983. The evolution of resonant oscillations in closed tubes. Z. angew. Math. Phys. 34, 845 - 866.

[5] Cox, E.A. and Mortell, M.P. 1986. The evolution of resonant water-wave oscillations. J. Fluid Mech., 162, 99 - 116.

[6] Seymour, B.R. and Mortell, M. P. 1985. The evolution of a finite rate oscillation. Wave Motion 7, 399 - 409.

[7] Mortell, M.P. and Varley, E. 1970. Finite amplitude waves in bounded media: free vibrations of an elastic panel. Proc. Roy. Soc. A 318, 169 - 196.

[8] Mortell, M.P. and Seymour, B.R. 1979. Nonlinear forced oscillations in a closed tube: continuous solutions of a functional equation. Proc. Roy. Soc. A 367, 253 - 270.

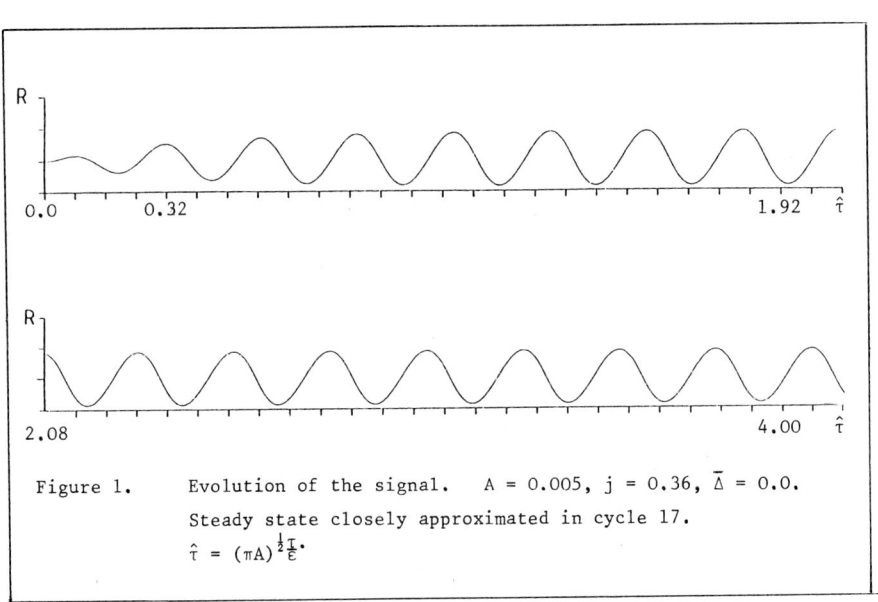

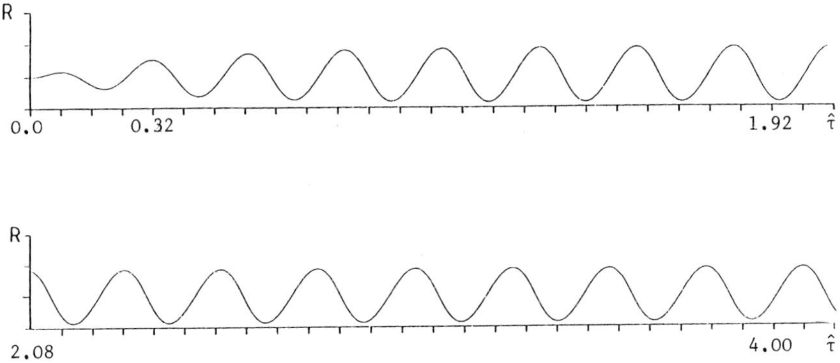

Figure 1. Evolution of the signal. $A = 0.005$, $j = 0.36$, $\bar{\Delta} = 0.0$.
Steady state closely approximated in cycle 17.
$\hat{\tau} = (\pi A)^{\frac{1}{2}} \frac{T}{\varepsilon}$.

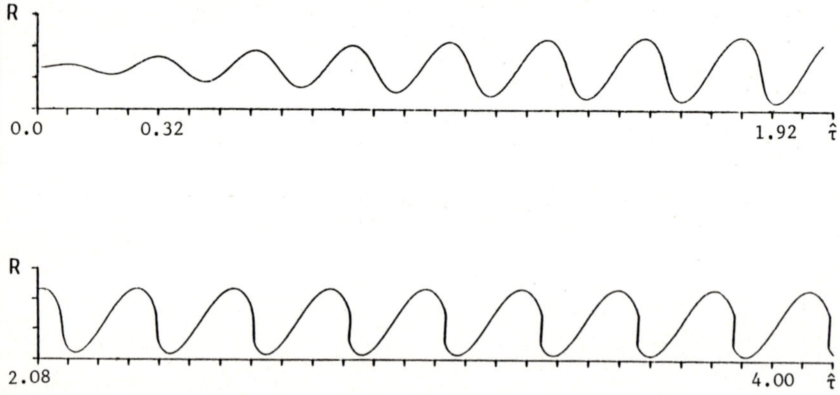

Figure 2. Evolution of the signal. $A = 0.005$, $j = 0.12$, $\bar{\Delta} = 0.1$. Steady state closely approximated in cycle 17.

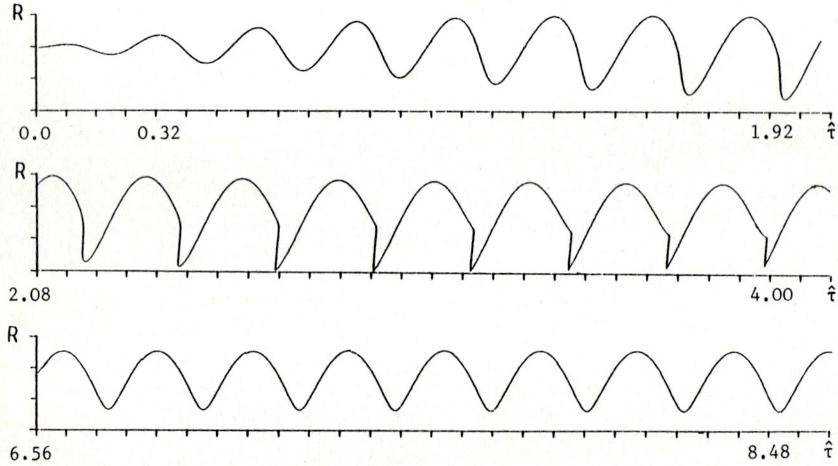

Figure 3. Evolution of the signal. $A = 0.005$, $j = 0.05$, $\bar{\Delta} = 0.055$. Steady state closely approximated in cycle 36.

ELASTIC WAVE PROPAGATION
M.F. McCarthy, M.A. Hayes, (Editors)
Elsevier Science Publishers B.V. (North-Holland), 1989

DYNAMICS AND STABILITY OF TWIN BOUNDARIES IN MARTENSITIC TRANSFORMATIONS

Joël POUGET

Laboratoire de Modélisation en Mécanique
Université Pierre et Marie Curie - Tour 66
4 place Jussieu, 75252 Paris Cedex 05, France

On the basis of a lattice model for martensitic transformations the nonlinear wave propagation of elastic strains is examined. The model accounts for both strongly nonlinear and competing interactions which allow for, in some situations, the propagation of nonlinear excitations. The phonon dispersion of the transverse acoustic wave is discussed and the results show first a phonon softening at a nonzero wave-number and next an upwards convexity of the dispersion branch near the long-wavelength limit. Then, the nonlinear problem is presented and three main classes of solutions are found : **i)** quasi-harmonic solutions corresponding to periodically modulated strain structures in space, **ii)** arrays of solitons made of periodic arrangement of elastic domains and **iii)** a moving martensitic soliton. All significant results are numerically illustrated by means of the microscopic model.

1. INTRODUCTION

The paper deals with the *nonlinear dynamics of microtexture patterns* made of elastic domains relating to *martensitic transformations*. With this in view, a two-dimensional lattice model and its continuum version are built up. In the physical point of view, martensitic transformations have been defined as a subset of diffusionless displacive phase transition involving rather large lattice distortive deformations, i.e. elastic strains, that means the transformation is mostly dominated by strain energy [1]. As a consequence, the transition is *first order* where the spontaneous strain is the order parameter in the Landau theory of phase transitions and the *ferroelastic* transition is called proper [2,3] (for instance In-Te, Ni-Ti, V_3Si provide good prototype of these). Crystals undergoing such transformations exhibit particularly interesting effects : *elasticity, pseudo-elasticity* and *ferroelasticity* which are intimately connected with phase transition phenomena as well as the *shape memory effects* of which the technological applications are especially promissing. Because of the phase transition, the crystal splits into domains of different strain strengths. The above-mentionned effects are mostly caused by the rearrangement of the domain structure which first develops a periodic arrangement of martensitic domains (deformed state) within the parent phase (undeformed state) [1,4] and a domain growths at the expense of the other one.

In the mechanical point of view, we are concerned with the competition between the *nonlocal elasticity* (inhomogeneous deformations) and a *local nonlinear elastic* energy (homogeneous deformations), which plays a crutial role in the motion of coherent elastic domain structures. We point out the interest of the lattice model, since the latter possesses the underlying physical ingredient which are the basis of the phenomenon. Particularly, there exist *competing* and *nonlinear interactions* which allow for the propagation of *nonlinear waves*. As a matter of fact, the combination of both nonlinear phenomena and competing interactions triggers the propagation of coherent structures that we can describe as *solitons*. In the framework of the present study we extend the ideas underlying the nonlinear soliton models developed in other contexts [5,6] to the long-wavelength elastic description of twin interfaces in martensitic transformations.

Starting with a two dimensional lattice model we introduce particular interatomic interactions **(i)** between the next nearest neighbours and **(ii)** of the three-body type (between three particles involving non-central interactions). Moreover, the strong anisotropy of the lattice allows to place in evidence the anomalously low, anisotropic *shear modulus softening* (transverse acoustic branch) which is a precursor effect of the lattice instability and can be seen as a pre-transitional phenomenon [7,8]. That means the domains structure nucleation can be considered as a modulated structure with periodic deformation patterns which are usually not commensurate with the lattice spacing [9]. The origin of such patterns emerges from competing interactions (involving of course next nearest neighbour and non-central interactions) but also the nonlinear elastic potential having stable and metastable minima. Accordingly, a rather fine scale of the lattice description seems to be necessary. We address a particular attention on the transition to the continuum model and quasi-continuum notion is therefore deduced which leads to a local nonlinear elastic energy (usually non-convex) and weakly nonlocal strain energy (strain gradient elasticity). Next, a special attention is devoted to the shear strain transformations (no dilatation) describing the deformation of a square into a rectangle with area preserving and the model reduces thus to the shearing motion of the close-packed atomic planes. Then, the motion can be characterized by a strongly localized deformation (soliton) which propagates in the staking direction.

2. THE MICROSCOPIC MODEL

We now consider an atomic plane extracted from a cubic lattice, moreover we assume that deformations are homogeneous in the direction perpendicular to the plane. The *geometry* of the plane consists in squares parallel to the x and y directions corresponding to crystallographic directions (see Fig.1.a). A particle of the plane is located by (i,j) in the co-ordinate system (x,y) or by (I,J) in the system (X,Y) deduced from the former by a rotation 45° clockwise. It may be more convenient to use the second coordinate system. The lattice spacing is denoted by a. Then, the absolute position of a particle at (i,j) is given by

$$\vec{x}_{i,j} = (ia + u_{i,j})\vec{i} + (ja + v_{i,j})\vec{j} \tag{1}$$

where $u_{i,j}$ and $v_{i,j}$ are the longitudinal and transverse displacements, respectively. The corresponding displacements in the system (X,Y) are

$$U(I,J) = (u_{i,j} + v_{i,j})/\sqrt{2}\ , \quad V(I,J) = (u_{i,j} - v_{i,j})/\sqrt{2}, \tag{2a}$$

$$I = i + j \quad \text{and} \quad J = i - j. \tag{2b}$$

Insofar as the interatomic interactions are concerned, we assume that the particle at (i,j) interacts with the first height nearest neighbours surrounding it. We consider two kinds of interactions leading to potentials which are functions of *particle pairs* between the nearest particles in the x,y,X and Y directions and potential considered as *three-body interactions* in the same directions [10]. We account for the lattice anisotropy by assu-

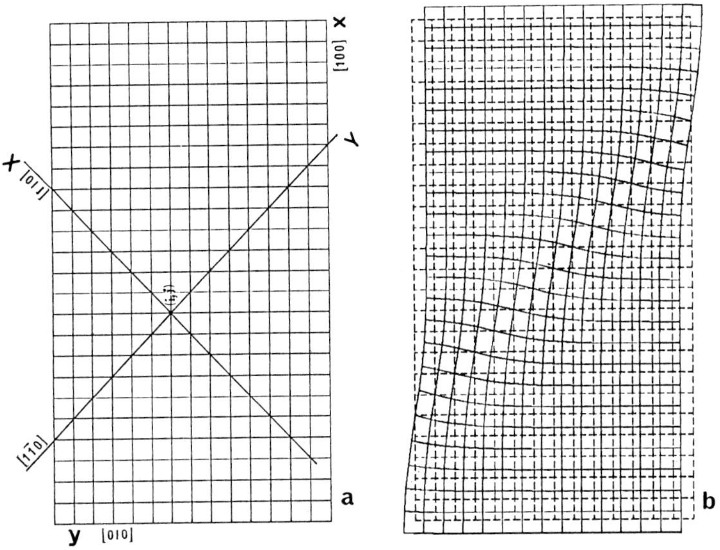

FIGURE 1 : Geometry of the plane lattice,
a) undeformed lattice, b) deformed lattice : martensitic soliton

ming potentials of different strengths. Since the lattice energy must be translationally and rotationally invariant, it depends only on the discrete Lagrangian strain tensor including geometric nonlinearities. Because of the potentials of the three body type, the energy depends also on the first order finite difference of the discrete strains and the lattice energy can be written as

$$\begin{aligned} V = \sum_{i,j} \Phi \{ & e_1(i,j), e_2(i,j), e_3(i,j), e_1(i+1,j)-e_1(i,j), e_1(i,j+1)-e_1(i,j), \\ & e_2(i+1,j)-e_2(i,j), e_2(i,j+1)-e_2(i,j), \\ & e_3(i+1,j)-e_3(i,j), e_3(i,j+1)-e_3(i,j)\} \end{aligned} \tag{3}$$

where we have set :

$$\begin{aligned} e_1(i,j) &= (\varepsilon_{11}(i,j) + \varepsilon_{12}(i,j))/\sqrt{2}, \\ e_2(i,j) &= (\varepsilon_{11}(i,j) - \varepsilon_{12}(i,j))/\sqrt{2}, \\ e_3(i,j) &= \varepsilon_{12}(i,j), \end{aligned} \tag{4}$$

and where ε_{11}, ε_{22} and ε_{12} are the discrete components of the Lagrangian strain tensor as usually defined in continuum mechanics [11]. The components e_1, e_2 and e_3 are those of the dilatational and deviatoric parts corresponding to the strain tensor. It is noteworthly that we have one-to-one mapping between the discrete functions and the corresponding counparts obtained in the continuum approximation involving weakly nonlocal behaviour.

At this stage the problem is somewhat complicated and simplifications are in order. The main feature of the model can be examined in a simple case. We then consider transformations without dilatation ($e_1 \equiv 0$) and shear ($e_3 \equiv 0$). Moreover, we assume that the remaining deformation $e_2(i,j) = S(i,j)$ is homogeneous in the Y direction. That means that S does not depend on J index. It follows that the lattice energy is now function only of the deformation S(I) and takes on the form

$$V = \sum_I \{\Phi_L(S_I) + \Phi_A(S_{I+1}+S_I) + \psi_1(S_{I+1}-S_I) + \psi_2(S_{I+2}-S_I)\} \tag{5}$$

where we have set

$$\Phi_L(x) = \frac{1}{2}\alpha x^2 - \frac{1}{3}x^3 + \frac{1}{4}x^4, \tag{6a}$$

$$\Phi_A(x) = \frac{1}{2}\beta x^2, \quad \psi_1(x) = \frac{1}{2}\delta x^2, \quad \psi_2(x) = \frac{1}{2}\eta x^2. \tag{6b}$$

On the other hand, the strain S is connected with the displacement V in the Y direction by

$$S(I) = 2(V(I) - V(I+1))/a. \tag{7}$$

In eqs. (5-6) Φ_L is the nonlinear elastic potential, Φ_A describes the interactions between the second nearest neighbours, ψ_1 and ψ_2 emerge from the three body interactions. The coefficients α, β, δ and η are functions of the interatomic potentials and their derivatives at equilibrium. Now the physical meaning of S is clear, it represents the relative shear displacement of the close-packed atomic layers <1$\bar{1}$0> (Y direction) in the stacking direction [110] (X direction).

3. THE CONTINUUM MODEL

We now consider the continum approximation of the microscopic model. Then, we must suppose dynamic processes with a characteristic length much larger than the lattice spacing a. That means that the discrete shear deformation S(I) is slowly varying in space and time. After some classical manipulations the Hamiltonian of the continuum system can be written as (in non-dimensional notation)

$$H = \int_{-\infty}^{+\infty} \left\{\frac{1}{2}\dot{V}^2 + \psi(S, S_X)\right\} dX, \tag{8}$$

where $S(x,t) = V_{,X}$ is now a continuous function of space. On minimizing the Hamiltonian (8) we arrive at the equation of motion for the transverse displacement

$$\ddot{V} = \Sigma_{,X} \tag{9}$$

Σ is then the shear stress. The latter can be decomposed into

$$\Sigma = \sigma - \chi_{,X} \tag{10}$$

with

$$\sigma = \partial\psi/\partial S \text{ and } \chi = \partial\psi/\partial S_{,x}. \tag{11}$$

In eq. (11) σ is the *macroscopic shear stress* (classical elasticity) and χ represents the *microstress* due to the lattice curvature $S_{,x}$ (arising from the non-central interactions of three-body type). The microstress notion can be introduced from a more general theory of strain gradient elasticity [12]. In agreement with the lattice model, the elastic potential takes on the form

$$\psi(S,S_{,x}) = \psi_1(S) + \psi_2(S_{,x}), \tag{12}$$

with

$$\psi_1(S) = \frac{1}{2}AS^2 - \frac{1}{3}S^3 + \frac{1}{4}S^4, \quad \psi_2(S_{,x}) = \frac{1}{2}\gamma(S_{,x})^2. \tag{13}$$

In eq. (13) $A = \alpha + 4\beta$ represents the elastic modulus $(C_{11} - C_{12})/2$ satisfying the Curie-Weiss temperature law and we note $\gamma = \delta + 4\eta - A/12$. The potential (13) is the simplest free energy which fulfils all the physical requirements of the Ginzburg-Landau theory of the first-order phase transition [7]. The elastic energy ψ_1 in eq. (13) is sketched in Fig. 2 for different values of A. For A rather high S = 0 is the single stable minimum. For intermediate values, the potential has a stable minimum for S = 0 and metastable minimum $S \neq 0$ and finally when A is small S = 0 is now metastable whereas $S \neq 0$ is stable and in the latter case the lattice is homogeneously deformed and we have the martensitic phase.

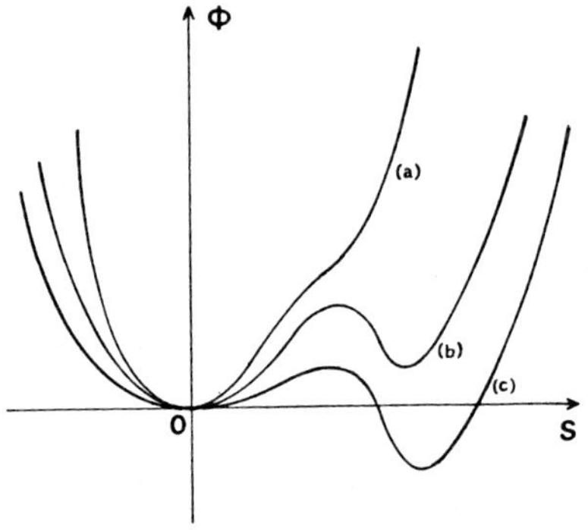

FIGURE 2

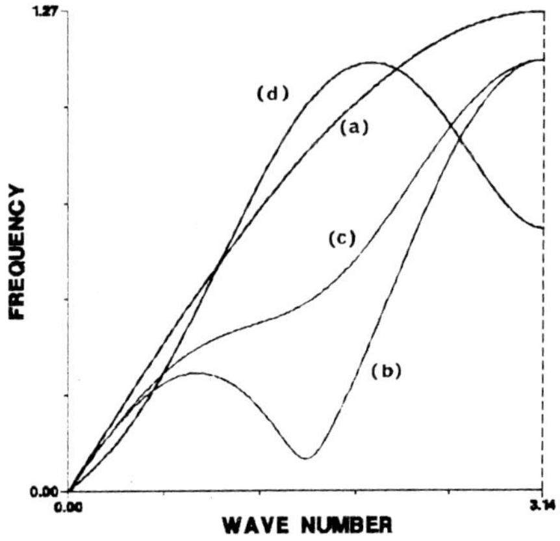

FIGURE 3

FIG. 2 : Elastic energy as function of homogeneous strain : a) high temperature phase, b) intermediate temperature and c) low temperature phase.

FIG. 3 : Phonon dispersion of the transverse acoustic waves. a) the curve differs slightly from the classical case, b) the curve exhibits a softening at a nonzero wave-number, c) the curve has a slight softening and d) the curve has an upwards convexity at a long wavelength region.

4. NONLINEAR DYNAMICS

In order to investigate dynamic processes of the lattice we can start with either the discrete equations or the equations obtained in the continuum approximation. Nevertheless, in a general situation the set of discrete coupled nonlinear equations is not tractable except for the linearized case. The strongly nonlinear feature of the equations can be however achieved by means of numerical simulations.

Insofar as the linear problem is concerned, we place the importance of the competing interactions in evidence, which gives instructive information about the nonlinear case. We then obtain the dispersion branches of the transverse acoustic waves travelling in the X direction. The dispersion curves are given in Fig. 3 for various values of the parameter A, δ and η. We can observe : (a) the curve differs slightly from the classical case, (b) the curve undergoes a softening at a nonzero wavenumber, (c) is a case where the curve has a small bend at a nonzero wave-number and (d) the curve has a positive convexity near the long-wavelength limit, which is a precursor of possible nonlinear excitations propagations. These effects are usually observed at the pre-martensitic transition [7,8].

The nonlinear problem is more complicated and an alternative situation happens in the case of the continuum approximation. Then, the equation of motion for the continuum model (see eq. (9)) can be written as

$$\ddot{S} = (AS - S^2 + S^3)_{,XX} - \gamma S_{,4X}. \qquad (14)$$

Now it is easy to look for solution of eq. (14) in the form travelling waves form. Nevertheless, we must introduce appropriate boundary conditions. We choose periodic conditions at each end of the chain, these conditions are particularly useful for the numerical simulations of the discrete equations. In the mechanical point of view the boundary conditions involve first the macroscopic stress and next the micro-stress eq. (11). We impose the equilibrium of the whole atomic chain and vanishing micro-stress at each end of the chain [6]. These conditions are fulfilled if we consider vanishing strain for X large enough. Fig. 4 collects the most significant cases which depend mainly on the total energy of the system and strain amplitude of the martensitic zone. Fig.4.a gives the strain profile corresponding to a modulated structure with a rather small amplitude which can be considered as a *precursor* feature of the transformation [4]. For larger amplitude the solution consists in an *array of solitons* describing a spatially arranged structure made of periodic martensitic and austenitic buffers (see Fig. 4.b) [13]. Finally, the limit case shown in Fig.4.c represents a martensitic soliton moving in an austenitic matrix with the corresponding distorted lattice plotted in Fig. 1.b. The fully dynamic problem of the nonlinear excitations of soliton type is given in Fig. 5 in space-time perspective representation. The problem of the interaction of a moving martensitic soliton with a defect is illustrated FiG. 6. The instability of the strain soliton is caused by the defect and the soliton splits out forming thus a smaller strain soliton travelling in the opposite direction.

5. CONCLUSION

The lattice model thus constructed on the basis of a two-dimensional lattice possesses all the needed ingredients which allow for the nonlinear excitation propagation. The model involves two important notions, first a nonlinear lattice energy is considered, next interactions of the three-body type are accounted for which fulfil all the physical requirements involved

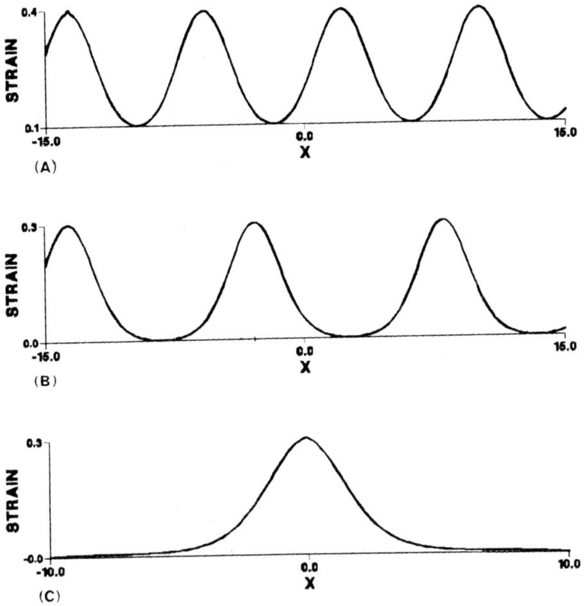

FIGURE 4 : Nonlinear excitations : a) quasi-harmonic solution, b) array of solitons and c) martensitic soliton.

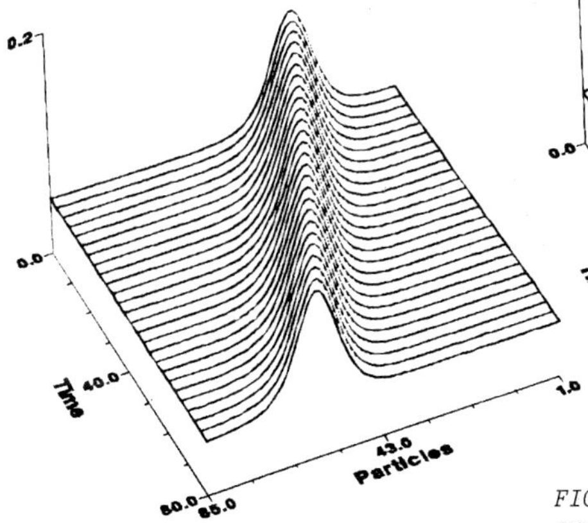

FIGURE 5 : Nonlinear dynamics of a martensitic soliton moving in an austenitic matrix.

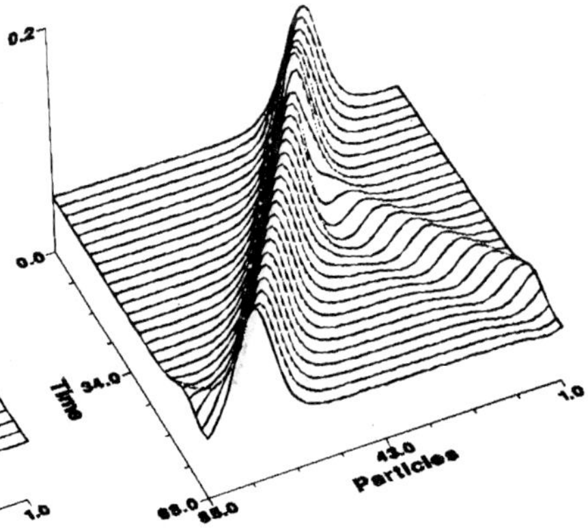

FIGURE 6 : The strain soliton splits out into two solitons travelling in opposite directions.

in martensitic transformations. The pertinant results are **i)** the partial softening of the transverse acoustic branch at a nonzero wave-number, **ii)** the upwards convexity of the dispersion branch near the long-wavelength limit and **iii)** the propagation of nonlinear excitations : array of solitons, martensitic soliton which describe the sweeping out of the crystal by phase boundaries (domain walls). These effects described here by this model are usually observed for various alloys such as Ni-Ti, Ti-Mn,In-Tℓ, etc ... just to quote some of them [4,14]. This one-dimensional version of a 2D-lattice seems to be necessary to tackle more complicated problems. We can then study, on the basis of the present model, the transverse instabilities of nonlinear excitations and what are the nature of the transverse structures developing in the Y direction. At length, since we have at hand the microscopic model, we can investigate the discreteness effects in the case of rather small domains (e.g. pinning effects, phonon radiations, etc ...).

REFERENCES

[1] Delaey, L., Krishnan, R.V., Tas, H. and Warlimont, H., J. Mat. Science, 9, (1974), 1521.

[2] Landau, L.D. and Lifschitz, E., Statistical Physics (Pergamon Press, Oxford, 1980).

[3] Tolédano, J.C. and Tolédano, P., The Landau theory of phase transitions (World Scientific, Singapore, 1987).

[4] Tanner, L.E., Pelton, A.R. and Gronsky, R., J. de Phys., 43, Colloque, (1982) C4-169.

[5] Bishop, A.R., Krumhansl, J.A. and Trullinger, S.E., Physica D, 1 (1980), 1.

[6] Pouget, J., p. 359 in NATO-ASI, series C, 225 "Physical Properties and Thermodynamic Behaviour of Minerals", ed. E.K.H. Salje (P. Reidel, Dordrecht, 1988).

[7] Mori, M., Yamada, Y. and Shirane, G., Solid State Comm., 17, (1975) 127.

[8] Satija, S.K., Shapiro, M., Salamon, M.B. and Wayman, C.M., Phys. Rev. B, 29, (1984), 6031.

[9] Aubry, S., In : Solitons and Condensed Matter Physics, ed. Bishop, A.R. and Schneider, T., (Springer-Verlag, Berlin, 1978), p. 264.

[10] Krumhansl, J.A., In : Lattice Dynamics, ed. Wallis, R.F., (Pergamon, Oxford, 1963).

[11] Eringen, A.C., Nonlinear Theory of Continuous Media, (Mc Graw-Hill, New York, 1962).

[12] Toupin, R.A., Arch. Rational Mech. Anal., 17 (1964), 85.

[13] Barsch, G.R. and Krumhansl, J.A., Met. Trans. A, 18 (1987).

[14] Knowles, K.M., Christian, J.M. and Smith, D.A., J. de Phys., 43, Colloque, (1982) C4-185.

C
WAVE PROPAGATION IN LAYERED AND BOUNDED MEDIA

ELASTIC WAVE PROPAGATION
M.F. McCarthy, M.A. Hayes, (Editors)

STRESS-WAVE PROPAGATION IN A LAMINATED SANDWICH PLATE

W. Anthony Green

Department of Theoretical Mechanics,
The University of Nottingham,
Nottingham, NG7 2RD, U.K.

The propagation of stress waves through an infinite periodically laminated medium is governed by Floquet's Theory for periodic equations. The solutions exhibit the phenomena of passing and stopping bands, whereby waves are transmitted in some ranges of frequency and wavelength and attenuated in others. The dispersion equation for harmonic waves in a plate of the same laminated material gives a relation between frequency and wavelength which intersects both the passing and the stopping bands. Thus for some of the waves travelling in the plate the core is transparent whereas for others it is opaque. The paper derives this dispersion equation and examines the parameters which determine the regions giving rise to transparency and opacity of the core.

1. INTRODUCTION

The study of elastic wave propagation in periodically layered infinite media gives rise to systems of ordinary differential equations with periodic coefficients whose solutions are governed by Floquet's Theory. Among the early applications in this field are the papers by Sun Achenbach & Herrmann [1] and Lee [2], which refer to waves propagating normal to the layering and the papers by Auld, Beaupre & Herrmann [3],[4], which are concerned with waves travelling parallel to the layering. One of the significant features of these Floquet waves is the existence of passing and stopping bands. The passing bands are regions of the wave number/frequency space in which the disturbance varies periodically in the direction normal to the layering, without attenuation. The stopping bands correspond to combinations of wave number and frequency such that the wave motion decays along the direction normal to the layering.

In this paper we are concerned with propagation in a bounded wave guide, consisting of a laminated plate of infinite lateral extent and finite depth. The plate is fabricated from layers of a fibre reinforced composite in which the reinforcement in each layer consists of a family of straight parallel fibres lying in the plane of the layer. Each of the layers is of uniform thickness 2h and of identical constitution and the laminate is formed by assembling 2n layers in a cross-ply arrangement so that the fibres in adjacent layers are orthogonal to each other. Each layer of the fibre reinforced composite is modelled as a homogeneous continuum of transversely elastic material which is inextensible in the direction of transverse isotropy (the fibre direction). The dispersion equations for waves propagating parallel to each of the fibre directions in such a laminate have been obtained by Green & Baylis [5] and our object here is to extend these results to waves travelling in the plane of the laminae at an arbitrary angle to the fibre directions. In particular we obtain the dispersion equation for this bounded plate in terms of the Floquet wavenumber for the corresponding infinite medium and we examine the conditions under which the plate is transparent and opaque to the propagating disturbance.

2. PROPAGATOR SOLUTION FOR UNIT CELL

We choose a cartesian coordinate system of axes $Ox_1x_2x_3$ with some suitable origin and we let Ox_1 be along the normal to the laminae, Ox_2 parallel to the fibre direction in one set of layers and Ox_3 parallel to the fibre direction in the other set. Despite the fact that the individual plies are all of the same fibre reinforced composite it is convenient to refer to layers with the fibre direction parallel to Ox_3 as material 1 and layers with fibre direction parallel to Ox_2 as material 2. The elastic response of the composite is completely characterised by the density ρ and three independent wave speeds c_1, c_2, c_3, together with the inextensibility constraint.

We consider displacement functions $u_i(x_1,x_2,x_3,t)$, $(i=1,2,3)$ in the form

$$u_1 = U(x_1)e^{i(\omega t-kx)}, \quad u_2 = V(x_1)e^{i(\omega t-kx)}, \quad u_3 = W(x_1)e^{i(\omega t-kx)}, \tag{1}$$

where t is the time, $x = x_3c + x_2s$ and $c = \cos\gamma$, $s = \sin\gamma$. These represent plane waves of circular frequency ω and wave number k travelling in the plane of the laminate at an angle $(\frac{\pi}{2}-\gamma)$ to the x_2-axis. The stress components $t_{ij}(x_k,t)$ $(i,j,k=1,2,3)$ then have the form

$$t_{ij}(x_k,t) = T_{ij}(x_1)e^{i(\omega t-kx)} \tag{2}$$

and we define $T(x_1)$, $S(x_1)$, $R(x_1)$ by the relations

$$\rho c_2^2 T(x_1) = T_{11}(x_1), \quad \rho c_2^2 S(x_1) = T_{12}(x_1), \quad \rho c_2^2 R(x_1) = T_{13}(x_1). \tag{3}$$

Using subscripts $_1$ and $_2$ for displacement and stresses in materials 1 and 2 respectively, the inextensibility constraint in material 1 gives $W_1(x_1) = 0$ and the solution of the equations of motions may be written in the form

$$\underset{\sim}{Y}(x_1) = \underset{\sim}{P}(x_1-\bar{x}_1)\underset{\sim}{Y}(\bar{x}_1) , \tag{4}$$

where $\underset{\sim}{Y}(x_1) = (T_1(x_1)\ S_1(x_1)\ U_1(x_1)\ V_1(x_1))^T$ and T denotes the transpose. Equation (4) expresses the vector $\underset{\sim}{Y}$ at any level x_1 in a layer of material 1 in terms of the same vector evaluated at some specified level $\bar{x}_1$ in the same layer, through the propagator matrix $\underset{\sim}{P}(x_1-\bar{x}_1)$. The elements p_{ij} of $\underset{\sim}{P}(h)$ corresponding to $x_1-\bar{x}_1 = h$ are to be found in the paper by Green & Baylis [6].

In material 2, the inextensibility condition yields $V_2(x_1) = 0$ and the solution of the equations of motion is

$$\underset{\sim}{Z}(x_1) = \underset{\sim}{Q}(x_1-\hat{x}_1)\underset{\sim}{Z}(\hat{x}_1) , \tag{5}$$

where $\underset{\sim}{Z}(x_1) = (T_2(x_1)\ R_2(x_1)\ U_2(x_1)\ W_2(x_1))^T$. Here, $\hat{x}_1$ represents some specified level in a layer of material 2, x_1 any other level in the same layer and the elements q_{rs} of the propagator matrix $\underset{\sim}{Q}(h)$ are obtained from those of $\underset{\sim}{P}(h)$ by the transformations indicated in [6].

Equations (4) and (5) yield the solution of the governing equations in individual layers of material with fibres parallel to the x_3-axis and the x_2-axis respectively. In order to obtain the solution in a multi-layered laminate it is necessary to satisfy the appropriate conditions at the upper and lower surfaces as well as continuity conditions at the interfaces. In general, the conditions at the interface between two dissimilar elastic materials which are bonded together requires continuity of all three displacement components and the three components of traction across the interface. These continuity conditions are

$$U_1 = U_2, \quad V_1 = V_2, \quad W_1 = W_2, \quad T_1 = T_2, \quad S_1 = S_2, \quad R_1 = R_2. \tag{6}$$

For the idealized inextensible material, however, there exists the possibility of a discontinuity in the tangential component of stress parallel to the fibre direction across any surface in the material and in particular this allows a discontinuity in R_1 in material 1 and in S_2 in material 2 at any interface or free boundary in these materials. Thus at an interface between materials 1 and 2 conditions (6) reduce to

$$U_1 = U_2, \quad T_1 = T_2, \quad V_1 = 0, \quad W_2 = 0 \ , \tag{7}$$

the last two conditions being a consequence of the inextensibility constraints applied to the displacement continuity conditions.

Consider now a layer of thickness 2h of material 1 sandwiched between two layers of material 2, Letting $\underset{\sim}{Y}^L$ and $\underset{\sim}{Y}^u$ denote the value of the vector $\underset{\sim}{Y}$ at the bottom and top of this layer respectively, we have, from equation (4) that

$$\underset{\sim}{Y}^u = \underset{\sim}{P}(2h)\underset{\sim}{Y}^L = \underset{\sim}{P}(h).\underset{\sim}{P}(h)\underset{\sim}{Y}^L \ , \tag{8}$$

where we have made use of the property that for any propagator matrix $\underset{\sim}{P}(a+b) = \underset{\sim}{P}(a)\underset{\sim}{P}(b)$. If we now impose the condition that $V_1^u = 0$, $V_1^L = 0$, arising from the interface conditions (7), it is possible to eliminate the (discontinuous) stresses S_1^u, S_1^L from equations (8) to obtain

$$\underset{\sim}{X}^u = \frac{\underset{\sim}{R}\,\hat{\underset{\sim}{R}}}{\det \underset{\sim}{R}}\,\underset{\sim}{X}^L \ , \tag{9}$$

where

$$\underset{\sim}{X}_1 = \begin{bmatrix} T_1 \\ U_1 \end{bmatrix}, \quad \underset{\sim}{R} = \begin{bmatrix} r_{11} & r_{12} \\ r_{21} & r_{22} \end{bmatrix}, \quad \hat{\underset{\sim}{R}} = \begin{bmatrix} r_{22} & r_{12} \\ r_{21} & r_{11} \end{bmatrix} \tag{10}$$

and the components r_{ij} of $\underset{\sim}{R}$ are given in terms of the components of p_{ij} of $\underset{\sim}{P}(h)$ by the expressions

$$\begin{aligned} r_{11} &= p_{11} - \frac{p_{14}p_{41}}{p_{44}}, \qquad & r_{12} &= p_{13} - \frac{p_{12}p_{43}}{p_{42}}, \\ r_{21} &= p_{31} - \frac{p_{34}p_{41}}{p_{44}}, \qquad & r_{22} &= p_{33} - \frac{p_{32}p_{43}}{p_{42}}. \end{aligned} \tag{11}$$

In the same way, if we consider a layer of thickness 2h of material 2 sandwiched between layers of material 1 we may express the value of $\underset{\sim}{X}_2$ at the top of the layer in terms of the value at the bottom in the form

$$\underset{\sim}{X}_2^u = \frac{\underset{\sim}{F}\,\hat{\underset{\sim}{F}}}{\det \underset{\sim}{F}}\,\underset{\sim}{X}_2^L \tag{12}$$

where $X_2 = (T_2\ U_2)^T$, $\hat{\underset{\sim}{F}}$ bears the same relation to $\underset{\sim}{F}$ as $\hat{\underset{\sim}{R}}$ does to $\underset{\sim}{R}$ and the components f_{ij} of $\underset{\sim}{F}$ are related to the components $q_{\ell m}$ of $\underset{\sim}{Q}(h)$ by equations analogous to (11) above. Using the first two of the continuity conditions (7) to relate X_2 at the top of one layer to $\underset{\sim}{X}_1$ at the bottom of the next and applying (12) and (9) gives the relation

$$\underset{\sim}{X}^{(\alpha+1)} = \frac{(\underset{\sim}{R}\,\hat{\underset{\sim}{R}})}{\det \underset{\sim}{R}}\,\frac{(\underset{\sim}{F}\,\hat{\underset{\sim}{F}})}{\det \underset{\sim}{F}}\,\underset{\sim}{X}^{(\alpha)} \ . \tag{13}$$

This gives the vector $\underset{\sim}{X}^{(\alpha+1)}$ at the top of unit cell $(\alpha+1)$ in terms of the vector $\underset{\sim}{X}^{\alpha}$ at the top of unit cell α where the unit cell is a layer of depth 2h of material 1 on top of a layer of depth 2h of material 2 embedded in a repeating pattern. It is possible to show that (13) is equivalent to the relation

$$\underset{\sim}{X}^{(\beta+1)} = \frac{\hat{\underset{\sim}{F}}\,\underset{\sim}{R}\,\hat{\underset{\sim}{R}}\,\underset{\sim}{F}}{\det \underset{\sim}{R}\,\det \underset{\sim}{F}}\,\underset{\sim}{X}^{(\beta)} = \frac{\underset{\sim}{C}\,\hat{\underset{\sim}{C}}}{\det \underset{\sim}{C}}\,\underset{\sim}{X}^{(\beta)} = \underset{\sim}{D}\,\underset{\sim}{X}^{(\beta)} \ , \tag{14}$$

where $\underset{\sim}{C} = \hat{\underset{\sim}{F}}\,\underset{\sim}{R}$, $\hat{\underset{\sim}{C}} = \hat{\underset{\sim}{R}}\,\underset{\sim}{F}$ and $\underset{\sim}{D} = \underset{\sim}{C}\,\hat{\underset{\sim}{C}}/\det \underset{\sim}{C}$.

Equation (14) is appropriate to a unit cell consisting of a layer 2h of material 1 sandwiched above and below by a layer of depth h of material 2 and embedded in a repeating pattern. It is this latter version that we shall adopt as our unit cell and it is straightforward to show that the eigenvalues λ_1, λ_2 of the transfer matrix $\underset{\sim}{D}$ for this cell may be written as $\lambda_1 = \exp(2d)$, $\lambda_2 = \exp(-2d)$ where d is defined in terms of the elements c_{ij} of $\underset{\sim}{C}$ by the expression

$$\tanh d = (c_{12}c_{21}/c_{11}c_{22})^{\frac{1}{2}} . \tag{15}$$

The eigenvectors $\underset{\sim}{L}_1$ and $\underset{\sim}{L}_2$ of $\underset{\sim}{D}$ associated with eigenvalues λ_1 and λ_2 respectively are

$$\underset{\sim}{L}_1 = \left(\sqrt{c_{11}c_{12}}, \sqrt{c_{21}c_{22}}\right)^T , \quad \underset{\sim}{L}_2 = \left(\sqrt{c_{11}c_{12}}, -\sqrt{c_{21}c_{22}}\right)^T . \tag{16}$$

If the unit cell is embedded in an infinite laminated medium then by the Bloch theorem on periodic differential equations the solution vector $\underset{\sim}{X}^{(\beta+1)}$ is related to $\underset{\sim}{X}^{(\beta)}$ by the expression

$$\underset{\sim}{X}^{(\beta+1)} = e^{ir\ell}\, \underset{\sim}{X}^{\beta} \tag{17}$$

where r is the Floquet wavenumber and $\ell = 4h$ is the wavelength. Comparing (17) with the last expression in (14) we see that r must be such that $e^{ir\ell}$ are the eigenvectors of $\underset{\sim}{D}$ and are therefore given by

$$r\ell = \pm 2id + 2s\pi, \qquad s = 0, \ \pm 1, \pm 2, \ \ldots \ . \tag{18}$$

Real values of r correspond to the passing bands and since all the elements of $\underset{\sim}{C}$ are real it may be seen from (15) and (18) that this requires

$$c_{11}c_{22}c_{12}c_{21} \leqslant 0 . \tag{19}$$

The stopping bands are given by complex values of r, which occur when the product in equation (19) is positive.

3. DISPERSION EQUATION

We now consider a laminate of finite depth $n\ell$, consisting of a set of (n-1) unit cells each in the configuration represented by equation (14), bonded together to form an inner core and bonded above and below to two half cells. Each of the half cells comprises a layer of thickness h of material 2 bonded to the core and to an outer layer of thickness h of material 1. Letting $\underset{\sim}{X}^{(n-\frac{1}{2})}$ and $\underset{\sim}{X}^{(\frac{1}{2})}$ denote the vector $\underset{\sim}{X}$ at the top and bottom of the core respectively, the repeated application of equation (14) leads to the result

$$\underset{\sim}{X}^{(n-\frac{1}{2})} = (\underset{\sim}{D})^{n-1} . \underset{\sim}{X}^{(\frac{1}{2})} . \tag{20}$$

In the two boundary half cells the inner layers of material 2 are still subject to the inextensibility constraint imposed by the outer layers of material 1, so that we may express the vector $\underset{\sim}{X}^{(n-\frac{1}{4})}$ at the mid-plane of the top half layer in terms of $\underset{\sim}{X}^{(\frac{1}{4})}$ at the mid-plane of the bottom half layer by

$$\underset{\sim}{X}^{(n-\frac{1}{4})} = \frac{\underset{\sim}{F}(\underset{\sim}{D})^{n-1}\hat{\underset{\sim}{F}}}{\det \underset{\sim}{F}} . \underset{\sim}{X}^{(\frac{1}{4})} \ . \tag{21}$$

Each of the outer layers (of thickness h) of material 1 at the top and bottom of the complete plate must be treated separately. Each is subject to the constraint $V_1 = 0$ at the interface with material 2 and we will make use of the fact that the tangential component of traction S_1 vanishes at the outer surface. Applying equation (5) to the layer on the upper surface of the plate gives $\underset{\sim}{Y}(h) = \underset{\sim}{P}(h)\underset{\sim}{Y}(0)$, and making use of the conditions $S_1(h) = 0$ and $V_1(0) = 0$ allows the values of $T_1(h)$, $U_1(h)$ to be given in terms of $T_1(0)$, $U_1(0)$ by the expression

$$\underset{\sim}{X}_1(h) = \underset{\sim}{S}\underset{\sim}{X}_1(0) , \tag{22}$$

where the elements of the 2×2 matrix $\underset{\sim}{S}$ are given in terms of the elements p_{ij} of $\underset{\sim}{P}(h)$ by

$$s_{11} = p_{11} - \frac{p_{12}p_{21}}{p_{22}} , \quad s_{12} = p_{13} - \frac{p_{12}p_{23}}{p_{22}} , \tag{23}$$

$$s_{21} = p_{31} - \frac{p_{32}p_{21}}{p_{22}} , \quad s_{22} = p_{33} - \frac{p_{32}p_{23}}{p_{22}} . \tag{23}$$

Applying the same procedure to the layer on the bottom surface of the plate leads to the expression

$$\underset{\sim}{X}_1(h) = \hat{\underset{\sim}{S}} \underset{\sim}{X}_1(0) . \tag{24}$$

where $\hat{\underset{\sim}{S}}$ is related to $\underset{\sim}{S}$ in the same way as $\hat{\underset{\sim}{R}}$ to $\underset{\sim}{R}$, and the expression in equation (24) uses the result that det $\underset{\sim}{S}$ = 1. Combining (21) with (22) and (24) gives for the complete laminate

$$\underset{\sim}{X}^{(n)} = \underset{\sim}{S} \underset{\sim}{F}(\underset{\sim}{D})^{n-1} \frac{\hat{\underset{\sim}{F}} \hat{\underset{\sim}{S}}}{\det \underset{\sim}{F}} \underset{\sim}{X}^{(0)} = \underset{\sim}{M} \underset{\sim}{X}^{(0)} , \tag{25}$$

where $\underset{\sim}{X}^{(0)}$ and $\underset{\sim}{X}^{(n)}$ denote the value of $\underset{\sim}{X}$ at the bottom surface and the top surface respectively and $\underset{\sim}{M}$ is the overall transmission matrix for the plate. Setting the normal component of traction at the lower surface to zero gives $\underset{\sim}{X}^{(0)} = (0 \ U^0)^T$ and the normal component of traction on the upper surface is then $T^{(n)} = m_{12} \ U^{(0)}$

The traction component will vanish provided m_{12} is zero and this is the condition that yields the dispersion equation for the plate. Using the Cayley Hamilton theorem to reduce the term $(\underset{\sim}{D})^{n-1}$ in equation (25) into a linear combination of $\underset{\sim}{D}$ and $\underset{\sim}{I}$ the dispersion equation may be written as

$$\sinh 2\left[(n-1)d + c\right] = 0 , \tag{26}$$

where c is defined by

$$\tanh c = v_{12} \sqrt{c_{21}c_{22}} \Big/ v_{11} \sqrt{c_{11}c_{12}} , \tag{27}$$

and v_{ij} are the components of the matrix $\underset{\sim}{V} = \underset{\sim}{S} \underset{\sim}{F}$. The elements v_{ij} are all real and it follows from (27) and (15) that d and c are both real in the stopping bands and both imaginary in the passing bands. The solution of equation (26) is given in the stopping bands by

$$c = -(n-1)d , \tag{28}$$

and in the passing bands by

$$c = i(m+\tfrac{1}{4})\pi - (n-1)d \qquad m = 0, \pm 1, \ldots . \tag{29}$$

Regions of the ω/k plane corresponding to the stopping bands are shown hatched in Figures 1, 2 and 3 for angles of propagation $\gamma = 0^0$ and 90^0, $\gamma = 30^0$ and 60^0 and $\gamma = 45^0$ respectively. In Figures 1 and 2 the boundaries between passing and stopping bands are given by some harmonic of one of the equations

$$c_{11} = 0, \quad c_{22} = 0, \quad c_{12} = 0, \quad c_{21} = 0 . \tag{30}$$

At $\gamma = 45^0$ the functions c_{11} and c_{22} are identical and the boundaries are then given by the harmonics of the last two equations in (30).

REFERENCES

[1] Sun, C.T., Achenbach, J.D. and Herrmann, G., J. Appl. Mech., 35 (1968),467.
[2] Lee, E.H., Wave Propagation in Composites with Periodic Structures, in Proceedings Fifth Canadian Congress of Appl. Mech., (Fredericton, 1975) G49-59.
[3] Auld, B.A., Beaupre, G.S. and Herrmann, G., Electronics Letters,13(1977) 525.
[4] Auld, B.A., Beaupre, G.S. and Herrmann, G., Mechanics Today, 5 (1980) 83.
[5] Green, W.A. and Baylis, E. Rhian, Archives of Mechanics, 38 (1986) 301.
[6] Green, W.A. and Baylis, E. Rhian, The propagation of impact stress waves in anisotropic fibre reinforced laminates, in Proc. Symp. on Wave Propagation in Structural Composites, A.S.M.E. (1988) in print.

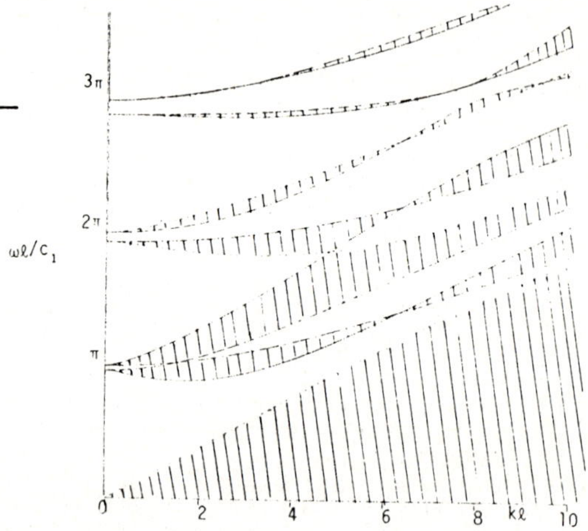

Figure 1 $\gamma = 0^0$ and 90^0

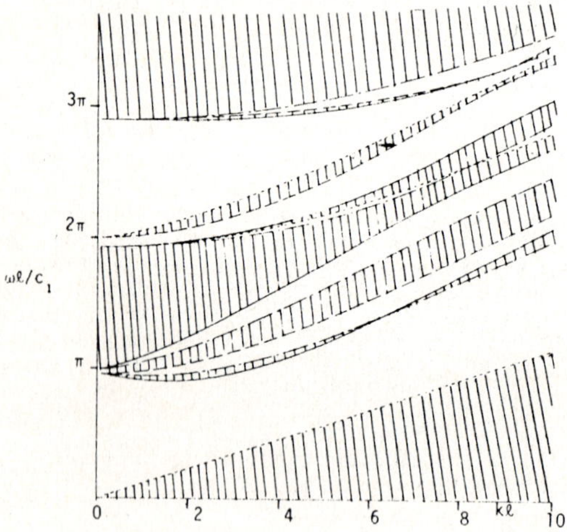

Figure 2 $\gamma = 30^0$ and 60^0

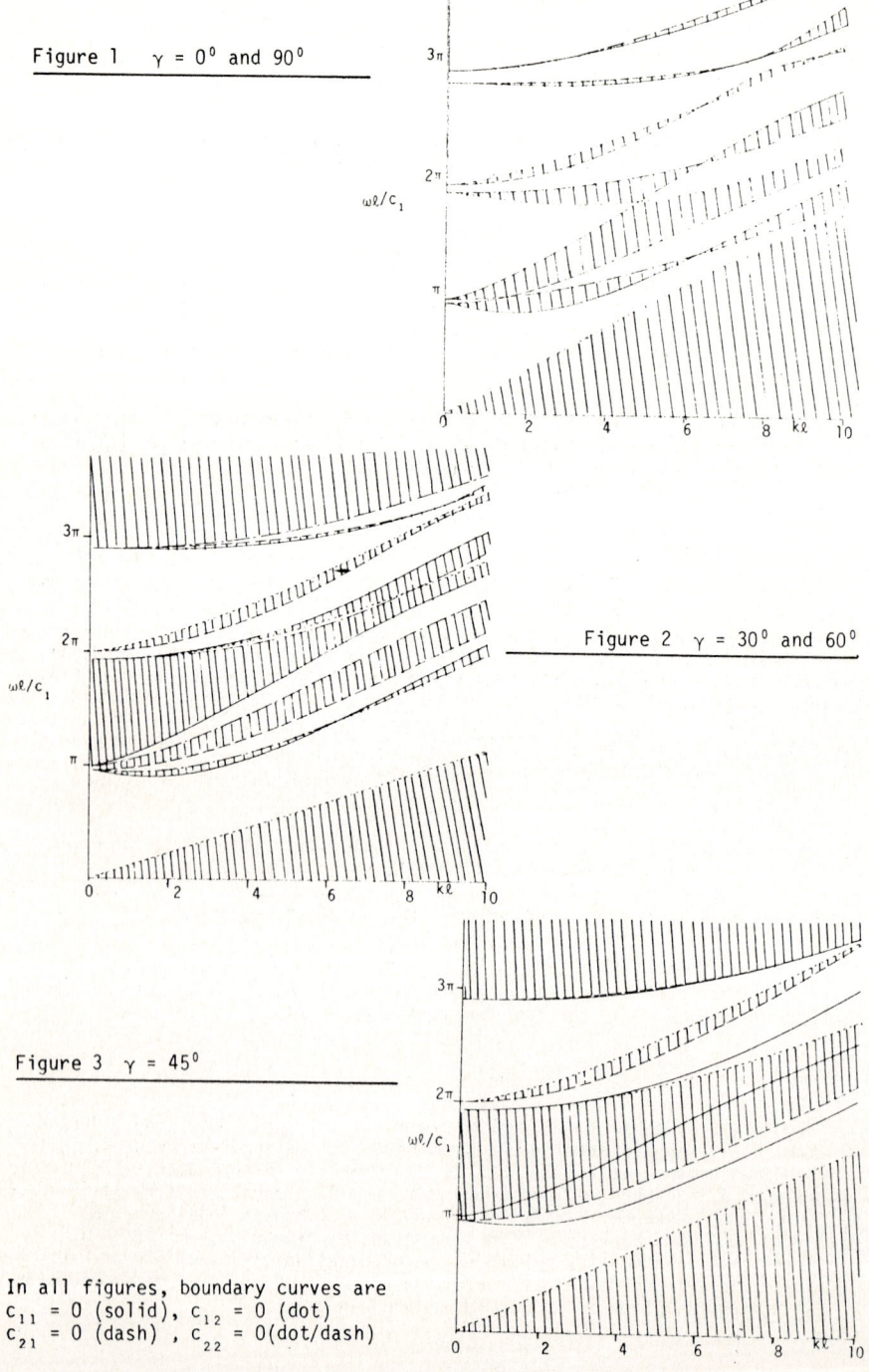

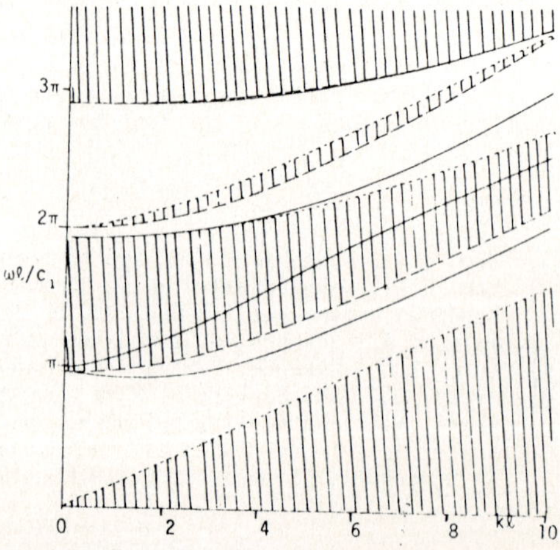

Figure 3 $\gamma = 45^0$

In all figures, boundary curves are
$c_{11} = 0$ (solid), $c_{12} = 0$ (dot)
$c_{21} = 0$ (dash), $c_{22} = 0$ (dot/dash)

ments that describe compactly the wave phenomena associated with one of the constituents are likely to be awkward and less physical for the other constituents. One attempt to cope with these difficulties is via self-consistent hybrid combinations of different basis elements, chosen so as to model the various observables in those parameter ranges where they are most pronounced.

In a previous and the present investigation, an effort has been made to address these concerns systematically. The overall problem has been decomposed by considering first the fields excited by the input beam in the unflawed environment, to be followed by the interaction of these fields with the flaw, and the consequent scattered fields produced thereby at the receiver. For simplicity, the problem is taken to be two-dimensional, and the unflawed material is a perfectly bonded layered aluminum plate (which therefore acts like a single thick layer) in vacuum. The input beam is assumed to be compressional (P-wave) and to have a Gaussian profile so that it can be modeled compactly by a line source at a complex coordinate location (complex source point (CSP) method) [1]. This permits the beam input response to be computed from the line force input response by analytic continuation of the source coordinates into the complex domain. Depending on whether the observer is near, or far from, the source region, the field in the plate has a beam-like or mode-like character, undergoing rapid conversion from the former to the latter due to multiple reflections, with consequent compressional-shear (P-SV) coupling, at the plate boundaries. A hybrid ray-mode form can describe this phenomenology. This first (unflawed) phase of the problem has already been treated in detail [2], and the results from that investigation provide the input for the flawed environment here.

To test the proposed problem strategy without undue analytic complication, an imperfect bond of finite extent aligned parallel to the plate boundaries has been postulated. The imperfection pertains only to reduced shear resistance, and thus, since the adhesive layer is thin in comparison to the wavelengths, the bond can be represented by distributed shear springs. For weak imperfections, the equivalent surface fields induced on the flaw can be approximated by the incident field as modified by the stiffness profile (Born approximation); because the flaw responds to tangential shear but not to compression, the incident field at the flaw site should have a strong longitudinal shear component. The scattering produced in the plate by the resulting equivalent source distribution can be evaluated by surface integration over the flaw contour, using the line force plate Green's function. The detector is assumed to be located on the plate boundary; therefore, primary attention is placed on determining the physical observables (i.e., the horizontal and vertical displacements) there.

The presentation below begins with a summary description of beam excited fields that have favorable properties for interaction with the flaw in the thick plate, and of "good" parametrizations of these fields. The surface integral for the scattered field is formulated next, with expressions given for the scattering induced displacements on the plate surface. The detection strategy emphasizes beam-like phenomena. Qualitative measures of "good observables" are introduced, which permit drawing inferences about the flaw size and location from the detected data. Each of these phases is accompanied by numerical calculations that demonstrate the detailed field behavior. The paper ends with conclusions based on these results.

2. FORMULATION

2.1 General Concepts

The physical configuration, schematized in Fig. 1, involves a bonded plate of thickness a, constitutive parameters (Lamé constants) λ, μ and (density) ρ, with excitation provided by a two-dimensional ultrasonic harmonic Gaussian P-beam input whose waist is centered at $(x,z) = (0,z')$ and whose inclination with respect to the x-axis is α. A longitudinally oriented thin boundary flaw of length 2ℓ is centered at (x_f, z_f). As noted in Section 1, the bond imperfection is only with respect to its shear resistance, while its full strength is retained in the normal direction. The rigidity of the weakened bond is represented by a distribution of equivalent springs along the flaw contour, the equivalent spring stiffness being defined as

$$K = \tau_{zx}/\Delta u \tag{1}$$

where τ_{zx} is the shearing stress at the joint, and $\Delta u = u^{+} - u^{-}$ denotes the x-displacement jump discontinuity across the bond line.

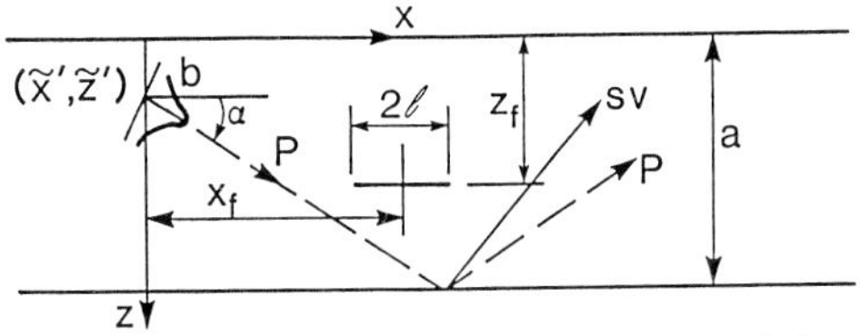

Fig. 1. Gaussian P-beam input to elastic plate with bonding flaw of length 2ℓ centered at (x_f,z_f). Incident and reflected P-beam axes (dashed) and reflected SV-beam axis (solid) are included. Numerical data in Figs. 2-6 are for an aluminum plate of thickness a = 0.3 in. and bond parameters x_f= 0.165, 0.135, 0.09, 0.045, z_f=0.15 in., 2ℓ = 0.090 in. The beam is modeled by the complex source point method, with complex source point location $\tilde{x}' = i\, b \cos\alpha$, $\tilde{z}' = z' + i\, b \sin\alpha$, z' = 0.06 in. , $\alpha=45^\circ$, b = 0.3 in. The corresponding beam waist is $w^e = (2b/k_p)^{1/2}$ = 0.0335 in.

The objective is the calculation of the horizontal and vertical displacements $u(x,z)$ and $w(x,z)$, respectively, which constitute the physical observables generated by input forcing functions. Because the problem is composite in the sense described in Section 1, various analytical building blocks are employed to characterize the relevant wave phenomena in a manner amenable to good interpretation. This may require use of different coordinate systems "matched" to particular phenomena (for example, beam centered coordinates instead of the plate coordinates (x,z)). For the force excited displacements, it may be convenient to use the scalar potentials $\hat{\Phi}(x,z)$ and $\hat{\Psi}(x,z)$ pertaining to compressional (P) and vertically polarized (SV) motion, respectively, with propagation speeds $v_p = [(\lambda+\mu)/\rho]^{1/2}$ and $v_s = [\mu/\rho]^{1/2}$. The displacements are derived from the potentials via the differentiations

$$u = \frac{\partial\hat{\Phi}}{\partial x} - \frac{\partial\hat{\Psi}}{\partial z} \quad , \quad w = \frac{\partial\hat{\Phi}}{\partial z} + \frac{\partial\hat{\Psi}}{\partial x} \tag{2}$$

Wave phenomena in the composite environment of Fig. 1 can be synthesized by spectral decomposition and recombination with respect to relevant "preferred" coordinates. For each individual beam, whether incident or multiply reflected, this is best done in the plane perpendicular to the beam axis. For waves "tied" to the plate, the spectral decomposition is along x or z. Decomposition along x organizes the response in terms of a plane wave spectral continuum with horizontal wavenumbers k,

$$\hat{A}(x,z) = \frac{1}{2\pi} \int_{-\infty}^{\infty} A(z,\kappa)\, e^{ikx}\, dx \tag{3}$$

where $A(z,\kappa)$, with $\kappa = \kappa(k)$ representing the vertical wavenumber, is the z-dependent spectral wave amplitude corresponding to the physical quantity $\hat{A}(x,z)$.

Decomposition along z organizes the response in terms of a discrete spectrum of guided modes with vertical wavenumber κ_m,

$$\hat{A}(x,z) = \sum_m \overline{A}_m(z,\kappa_m)\, e^{ik_m(\kappa_m)x}, \quad x>0 \tag{4}$$

where m is the mode index and $\overline{A}_m(z,\kappa_m)$ is the mode function. The wavenumbers k and κ are related by the dispersion equations

$$k^2 + \kappa_{p,s}^2 = (\omega/v_{p,s})^2 \qquad \text{for P, SV waves,} \tag{5}$$

with $\mathrm{Im}(k \text{ or } \kappa) \geq 0$ to ensure convergence when the wavenumber is nonreal. The continuous wave functions A in (3) or each mode function $\overline{A}_m$ in (4) can be expressed either in oscillatory form or in terms of upgoing and downgoing plane wave congruences with dependence $\exp(\pm i\kappa z)$ or $\exp(\pm i\kappa_m z)$. The latter, more general, format identifies plane waves as the fundamental spectral building blocks. It should be noted that P-SV coupling at the layer boundaries implies that both wave types have the same horizontal wavenumber k.

The spectral considerations for the beam-excited layer have been described in detail in a separate publication [2]. The phenomenological summary above has been presented to provide justification for subsequently used qualitative estimates that explain the numerically generated displacement observables. For example, the displacements in (2) involve contributions from the differentiated compressional and shear potentials simultaneously, thereby mitigating against the possibility of unscrambling their separate impact in the observed waveforms. However, in either the x-based or z-based spectral domains, it follows for each spectral plane wave element $\sim \exp(i\kappa z + ikx)$ that the derivative operations imply

$$\frac{\partial}{\partial z} \to i\kappa \quad , \quad \frac{\partial}{\partial x} = ik \tag{6}$$

Thus, depending on the ratio κ/k (i.e., the spectral plane wave propagation angle), one may encounter parameter requirements where either Φ or Ψ predominates. The observed displacements can then be interpreted from the form of the appropriate dominant potential. Conversely, this guideline has been employed by us in the selection of input beams that favor the excitation of strong compression or strong shear.

2.2 Implementation

2.2.1 *Beam excited layer without flaw*

The solution strategy for charting the evolution of an incident Gaussian P-wave beam in the perfectly bonded layer has been treated analytically and numerically in reference [2]. The beam input has been generated from the line force input by analytic continuation of the source point coordinates from real to complex values, and the potentials, stresses and displacements have been calculated in various vertical cross sections and also along the surface. Guided by the spectral concepts discussed in Section 2.1, it has been possible to explain the observed waveforms in terms of multiply reflected P and SV beams near the input, and in terms of the mode "most favored" by the incident beam in the intermediate and more distant regions. It has also been possible, based on these spectral considerations, to "tune" the input beam parameters so as to generate a desired response, for example, strong horizontal shear in the longitudinal plane that is to contain the flaw. A typical set of response curves is shown in Fig. 2.

2.2.2 *Excitation of the flaw*

With the flaw in place, as shown in Fig. 1, and knowing the incident field at the flaw from the results in Section 2.2.1, one may formulate the scattering problem. In general, this requires solution of a surface integral equation for the unknown induced sources on the flaw. However, in view of the assumed weakness of the imperfection, the induced sources can be approximated by those corresponding to the incident field on the flaw (Born approximation). Accordingly, from (1), the displacement discontinuity across the weakened zone becomes

$$\Delta u^{sc} = \frac{\tau^{i}_{zx}}{K} \tag{7}$$

where superscripts "i" and "sc" refer to incident and scattered quantities, respectively.

3. SCATTERED FIELDS

The displacements u^{sc} and w^{sc} generated by the initial conditions in (7) can be evaluated by use of Green's function techniques [3],

$$u^{sc}(x,z) = \int_{x_f-\ell}^{x_f+\ell} \Sigma_{zxx}(x',z_f;x,z)\Delta u^{sc}(x',z_f)\, dx'$$

$$w^{sc}(x,z) = \int_{x_f-\ell}^{x_f+\ell} \Sigma_{zxz}(x',z_f;x,z)\Delta u^{sc}(x',z_f)\, dx' \tag{8}$$

The Green's stress functions Σ_{zxx} and Σ_{zxz} in (8) permit calculation of the shearing stress $\tau_{zx}(x,z_f)$ in a perfectly bonded layer due to a horizontally or vertically polarized input force, respectively. They can be represented in any of the alternative forms discussed in reference [4,5]. In the calculations that follow, the modal expansion form was chosen for convenient reference, but the *interpretation* of the results has been carried out by recourse to spectral concepts tied to "good observables," as described in Section 2.1.

4. NUMERICAL RESULTS

An extensive sequence of numerical calculations has been carried out to gain understanding of the fields generated in the composite geometry of Fig. 1, in particular, the displacements on the upper boundary of the layer, where the detector is assumed to be located. To this end, the displacements generated by the input beam without the flaw, by the scattering in (8), and by the combination of both, have been calculated and plotted individually. An effort has been made to choose the input beam so as to give rise in the total displacements to *distinct features* that can be tied to flaw location and strength. This has led to oblique input beams that generate *resolvable* multiple excursions over the extent of the flaw. The scattered shear waves then become clearly visible on the surface even in the *total* response. Deconvolution techniques for separate extraction of the scattered field displacements can furnish an even more useful data base.

Figures 2-6 contain the numerical data which have been selected for presentation here. The parameters used pertain to an aluminum plate (see Fig. 1):

$$v_p = 2.36\text{x}10^5 \text{ in./sec.}, \quad v_s = 1.209\text{x}10^5 \text{ in./sec.}$$
$$\rho = 2.53\text{x}10^{-4} \text{ lb.-sec.}^2/\text{in.}^4, \quad a = 0.3 \text{ in.} \tag{9}$$

The flaw length is $2\ell = 0.090$ in., and the shear spring stiffness distribution is assumed to be Gaussian

$$K = K_o e^{[(x-x_f)/\ell]^2}, \quad K_o = 5\text{x}10^9 \text{ lb./in.}^3 . \tag{10}$$

The bond location at $z_f = 0.150$in. is midway between the plate boundaries, and the center of the flaw for the four cases considered is located at $x_f = 0.165$, 0.135, 0.090, 0.045 in.

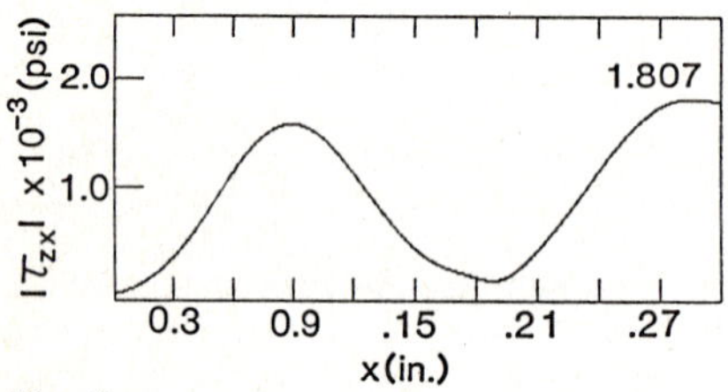

Fig. 2.
Shearing stress generated along bond line of unflawed plate. Note that the first peak corresponds to the incident P-beam and the second peak to the converted SV-beam (see Fig. 1). A number next to a peak denotes the peak amplitude.

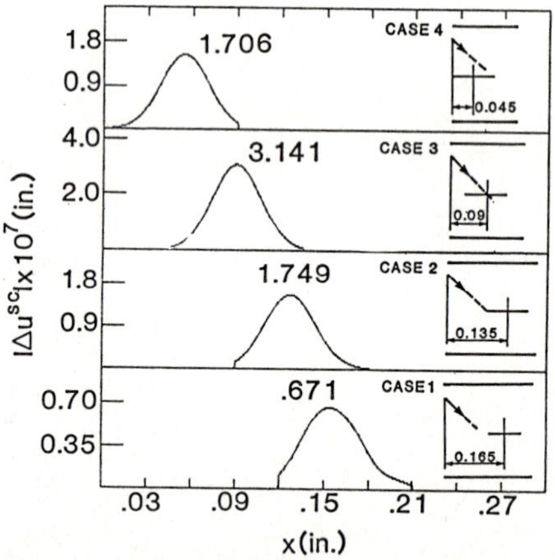

Fig. 3.
Induced source distribution for four flaw locations. Gaussian spring stiffness distribution $K = K_o \exp[(x-x_f)/\ell]^2$, $K_o = 5\text{x}10^9$ lb./in^3. Asymmetries in the distribution for Cases 1, 2, 4 arise because the incident beam illumination over the flaw is asymmetrical.

The waist of the 20MHz P-source beam has been placed at $x' = 0.0$ in., $z' = 0.06$in. with beam width parameter $b = 0.3$ in. and corresponding beam waist $w^e = (2b/k_p)^{1/2} = 0.0335$ in. The beam axis inclination angle $\alpha = 45°$ accommodates the resolvability requirement mentioned above, and it also leads to excitation of strong horizontal shear at the flaw site. For these incident beam conditions, there is a cluster of modes (with P-wave congruences surrounding the 45° axis) which contributes appreciably to the shear inducing field on the flaw; only antisymmetric modes need to be included for the calculations because symmetric modes have no horizontal shear at that location. Out of the contributing spectrum of 64 modes (see [2]), a total of 20 (filling the angular spectrum interval $58° \leq \theta \leq 30°$) was retained for the calculations; the error produced by omitting the others is too small to be discerned on the plots. The P-wave angular spectrum spread is compatible with the important spectral content of the incident beam. Spectral filtering tests have been performed to assess the loss of definition when the mode spectrum is truncated further. For example, retaining only 7 modes in the cluster, which fill the angular spectrum interval $38° < \theta < 52°$, one finds a maximum amplitude error of 1%, but more significant error in the phase. By choosing a larger beam incidence angle, one may excite a *particular* mode in purer form because the mode separation increases under these conditions. However, this is accompanied by a reduction of the inducing shear, and also by poorer resolvability of the multiple reflected beams. Such a strategy may be useful for longitudinally extended bonds with strong imperfections which can appreciably affect the mode structure in the flawed interval, thereby making the perturbed mode a good observable.

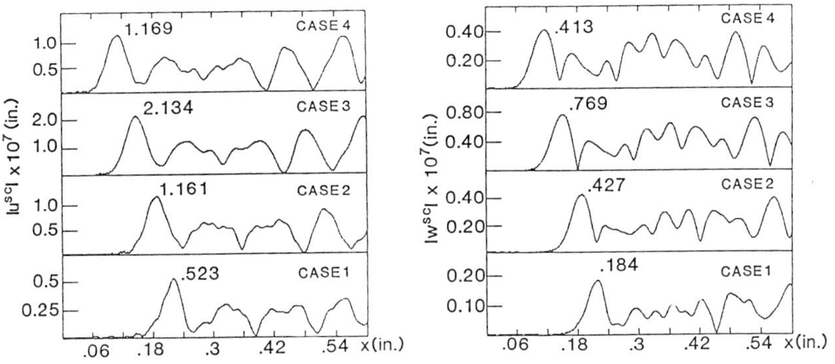

Fig. 4. Scattered vertical and horizontal displacement fields on upper surface of plate. First peaks correspond to intersection of direct scattered SV-beam with the surface.

All numerical calculations were performed on a PC/AT compatible Everex 10MHz System 1800, having a FAST 10MHz 80287 Math Compressor. The scattered displacement fields were determined by evaluating the integrals in (8) numerically, using 30 nodal points (5 to 6 nodes per wavelength) along the flaw. The shearing stress generated by the input beam along the bond line of the unflawed plate is shown in Fig. 2. The axes of the incident beam, the reflected P-beam and the converted SV-beam indicate that the first peak corresponds to the incident P-beam, while the second peak is associated with the converted SV-beam. For a steeper beam angle α, the horizontal shearing stress due to the P-beam is decreased, rendering the reflected SV contribution dominant. The induced sources along the flaw, computed from (7), are given in Fig. 3 for the four flaw locations. For Case 3, where the beam axis strikes the center of the flaw, the induced source magnitudes are largest, and Gaussian symmetry is closely maintained. The half-power incident beam width projected onto the flaw site is 0.05 in.

The scattered displacement fields on the upper surface of the plate are shown in Figs. 4 for the four flaw locations. The scattered shear waves become clearly visible, with details for Case 3 gleaned from Fig. 5. Note that the first peak occurs at the intersection of the direct scattered SV-beam and the surface, and that the shape about that peak has similarities with the shape of the induced source distribution. Subsequent peaks line up with the multiple reflected SV-beams.

The displacements on the surface generated by the input beam in the plate, by the scattering from the flaw, and by the combination of both have been plotted individually in Fig. 6 for Case 3. Clearly visible in a region which was initially quiescent is the disturbance attributable to the scatterer. The corresponding real displacement components are also shown. The observed periodicities turn out to be essentially those found in the guided mode whose upgoing reflected plane wave SV-spectra line up best with the dominant wave spectra of the incident P-beam; the amplitude modulation is due to interference between several of the modes in the surrounding cluster. The small oscillatory precursor in the total displacements, which appears to the left of the undisturbed region in the unflawed plate response, is explicitly tied to the flawed environment.

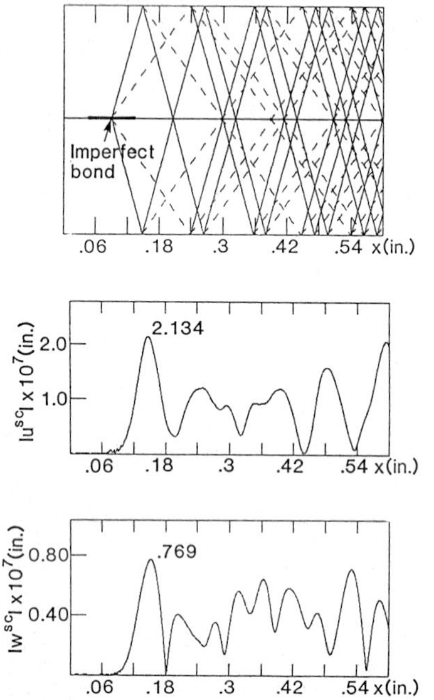

Fig. 5.
Enlarged scattered displacement fields for Case 3 of Fig. 4, and associated beam axis trajectories. Incident and reflected P-beams: dashed lines; coupled SV-beams: solid lines. Prominent features in the displacements are seen to occur near the beam axis intersections.

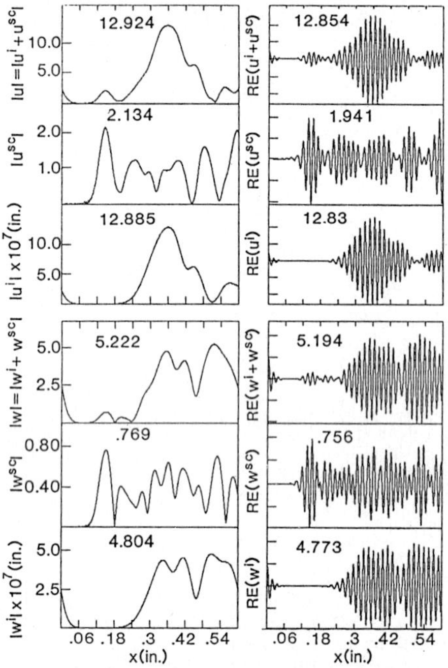

Fig. 6.
Displacements on plate surface for Case 3 of Fig. 4. $|u^i|$, $|w^i|$: generated by input beam. $|u^{sc}|$, $|w^{sc}|$: generated by scattering from flaw. $|u|=|u^i+u^{sc}|$, $|w|=|w^i+w^{sc}|$: combination of both. Also shown are the corresponding real displacement components; the imaginary components (not shown) exhibit similar characteristics. Phase plots (not shown) possess essentially the linear variation associated with the P-excited incident SV-plane wave observed at the boundary, except near the onset of the first disturbance where the field departs strongly from the plane wave prototype.

To link the measured displacements on the surface with the flaw that induces the perturbations, we are continuing the search for other observables, in addition to those described above. There are indications that the readily accessible ratio (u/w) contains information which can be related to the flaw location by simple ray construction, while inferences on flaw strength can be drawn from the data for u or w individually. Details of the inversion process, i.e., specifying flaw parameters from measured displacements at the surface, are still under investigation and will be reported separately.

5. SUMMARY

This study has been directed toward a "good parametrization" of observables when a Gaussian P-beam input in a perfectly bonded layered two-dimensional elastic plate encounters a transversely thin region with weak debonding of shearing stress. The observables are the horizontal and vertical displacements on the surface of the plate; the parametrization seeks to connect these observables with presence, location, and physical constitution of the flawed region. Emphasis here has been on how the plate environment, first without and then with the flaw, affects the incident beam; i.e., the parametrization is beam oriented. In this endeavor, the choice of input conditions that combines multiple beam resolution with strong shear excitation at the site of the flaw plays a critical role. The numerical results have shown that a good parametrization can be achieved. Some indications have been given on how to make the inverse connection between surface displacements and flaw identification but work on this aspect is still in progress.

Despite the beam oriented parametrization, the results have been computed by mode summation because the complex-source-point adapted mode algorithm has been available from a prior phase in this investigation [2]. Evidently, the numerical algorithm should be restructured in hybrid beam-mode form [2,5,6] to highlight the parametrized observables directly. Consideration of this aspect is postponed until a thorough understanding has been gained of the variety of phenomena that accompany the flaw identification problem in a multilayer bonded plate environment.

ACKNOWLEDGMENTS

This work has been supported by the U.S. Air Force Office of Scientific Research under Contract No. AFOSR-86-0318.

REFERENCES

[1] Felsen, L.B., "Geometrical Theory of Diffraction, Evanescent Waves, Complex Rays and Gaussian Beams," Geophys. J. Roy. Astron. Soc., 79, 77-88 (1984).

[2] Lu, I.T., Felsen, L.B., and Klosner, J.M., "Beam-To-Mode Conversion of a High Frequency Gaussian P-Wave Input in an Aluminum Plate in Vacuum," to be printed in ASME *Symposium on New Directions in the Ultrasonic NDE of Advanced Materials.*

[3] Pao, Y-H. and Varatharajulu, "Huygens' Principle, Radiation Conditions, and Integral Formulas for the Scattering of Elastic Waves," J. Acoust. Soc. Am., 59 (6), 1361-1371 (1976).

[4] Morse, P.M. and Feshbach, H., *Methods of Mathematical Physics* (McGraw Hill, N.Y. 1953).

[5] Lu, I.T. and Felsen, L.B., "Matrix Green's Functions for Array-Type Sources and Receivers in Multiwave Layered Media," Geophys. J. Roy. Astron., Soc., 84, 31-48 (1986).

[6] Lu, I.T. and Felsen, L.B., "Ray, Mode, and Hybrid Options for Source Excited Propagation in an Elastic Plate," J. Acoust. Soc. Am. 78 (2), 701-714 (1985).

ELASTIC WAVE PROPAGATION
M.F. McCarthy, M.A. Hayes, (Editors)

WAVE PROPAGATION ON THIN-WALLED ELASTIC CYLINDRICAL SHELLS

Allan D. PIERCE

School of Mechanical Engineering
Georgia Institute of Technology
Atlanta, Georgia 30332, USA

Equations governing the dynamics of thin-walled cylindrical shells are written in a form that is analogous to those governing wave propagation in a two-dimensional homogeneous anisotropic medium. With the understanding that excitation forces are periodic in one coordinate, the displacement field can be represented as a superposition of waves propagating out from localized sources into an unbounded medium. The types of waves that are excited are discussed in regards to their dependence on frequency and propagation direction.

1. INTRODUCTION

Vibrations of shells are customarily analyzed in terms of natural modes, and such an approach is ideal for most purposes. A wave viewpoint may, however, be helpful when one is seeking to understand how energy flows through the shell and how small perturbations affect this flow. Relevant prior literature on shells with the wave viewpoint explicitly or implicitly emphasized includes papers by Ross [1, 2], Germogenova [3], Steele [4, 5, 6], Feit [7], and Harari [8]. The present paper takes a fresh look at what is perhaps the simplest non-trivial model of a shell, that being a thin-walled circular cylinder. The basic equations, when written in the manner given below, formally resemble equations governing a structured wave disturbance in a two-dimensional unbounded homogeneous anisotropic medium. This allows one to freely adopt ideas concerning waves in anisotropic media which were introduced long ago by Hamilton [9], and which have been further developed in the present century by Synge [10], Lighthill [11, 12], Musgrave [13], Weinberg [14], and others.

2. THE DONNELL-YU EQUATIONS

The analysis of the present paper is concerned with a circular thin-walled cylinder with nominal radius R and unspecified length. (Reflection of waves from the cylinder ends can be incorporated into the theory, but a discussion of such is omitted here.) The wall thickness h is regarded as much less than R. The natural coordinate system appropriate to the geometry is that of cylindrical coordinates, with r denoting the radial distance from the axis and with the angle ϕ reckoned counterclockwise about the axis. The distance parallel to the axis can be denoted by X. However, for notational convenience and also to exhibit a formal similarity with a two-dimensional anisotropic medium, it is convenient to introduce dimensionless coordinates x and y, which represent distances along the nominal surface of the cylinder in units of R, such that $x = X/R$ and $y = \phi$. The displacement field of points on the center cylinder (half-way between the outer and inner surfaces) is characterized by variables u, v, and w (all with units of length), which correspond to displacements in the axial (X or x), tangential (ϕ or y) and outward radial (n) directions.

Also, rather than specifying quantities as functions of time t, it is convenient to use a dimensionless coordinate $\tau = \omega_r t$, where

$$\omega_r = R^{-1}(E/\rho)^{1/2}(1-\nu^2)^{-1/2} \tag{1}$$

is the "ring frequency" of the cylinder in radians per second. (The quantity ω_r is the natural frequency of the cylinder when the cylinder is undergoing purely radial oscillations.)

Considerations here are limited to the moderate frequency case when the frequencies of interest are substantially lower than the reciprocal of the transit time of an elastic wave across the thickness of the cylinder wall. Such implies that the standard Kirchhoff-Love assumptions of shell theory are applicable, and that the strain energy per unit surface area is analogous to that of a plate without the refinements associated with Timoshenko, Mindlin, and others.

The partial differential equations adopted here for the shell dynamics are essentially the same as those given previously by Yu [15], and also by Junger and Feit [16]. The static version of these equations apparently originated with Donnell [17]. With the allowance for a force excitation and with the notation described above, the dynamics equations take the form

$$u_{,\tau\tau} - \left[u_{,xx} + \tfrac{1-\nu}{2}u_{,yy} + \tfrac{1+\nu}{2}v_{,xy} + \nu w_{,x}\right] = 0 \tag{2a}$$

$$v_{,\tau\tau} - \left[v_{,yy} + \tfrac{1-\nu}{2}v_{,xx} + \tfrac{1+\nu}{2}u_{,xy} + w_{,y}\right] = 0 \tag{2b}$$

$$w_{,\tau\tau} + \left[\nu u_{,x} + v_{,y} + w\right] + \epsilon\nabla^4 w = F(x,y,\tau) \tag{2c}$$

Here $F(x,y,\tau)$ is the normal force per unit plate area divided by the product of the mass per unit area and ω_r^2. The quantity $\epsilon = h^2/12R^2$ is regarded as small.

One possible option for specifying the solution of the above equations is to require that the three field variables each be periodic in y with period 2π, with the force excitation's definition extended such that it is also periodic in y. Equivalently, one can regard the "medium" as unbounded in the y coordinate and the excitation field as a superimposition of responses to forces which are localized in various 2π intervals of y. This latter viewpoint allows one to write, for example, for the axial component of the displacement,

$$u(x,y,\tau) = \sum_{n=-\infty}^{\infty} u_{\mathrm{ub}}(x, y+2n\pi, \tau) \tag{3}$$

where $u_{\mathrm{ub}}(x,y,\tau)$ is the response of an "unbounded" (ub) medium, extending from $y = -\infty$ to $y = \infty$, to a force excitation that is nonzero only for y between $-\pi$ and π.

The analysis here follows the viewpoint expressed above, with emphasis on the basic field variables u_{ub}, v_{ub}, w_{ub} that correspond to an unbounded medium. Note that these fields are not periodic in y, but instead should satisfy radiation conditions at large y. The overall periodicity is insured by the fact that the overall field is composed of a sum of the form of Eq. (3).

Since any force field can be represented as a continuous smear of point forces and since the governing equations are linear, it is appropriate at the outset to limit consideration to a field generated by a point force, whose application point, without loss of generality, can be taken as the origin ($x = 0$, $y = 0$). For analogous reasons, one limits initial consideration to a force of constant frequency. Thus one can set

$$F = (2\pi)^2 G\delta(x)\delta(y)e^{-i\Omega\tau} \tag{4}$$

where G is a constant and $\delta(x)$ is the Dirac delta function. (Here it is understood that one should take the real part.) The "dimensionless frequency" Ω represents circular frequency in units of ω_r.

The unbounded medium response to the excitation represented by Eq. (4) can be written as a double Fourier transform, such that, with the $\exp(-i\Omega\tau)$ factor suppressed,

$$u_{\mathrm{ub}} = \lim_{\delta\to 0}\int_{-\infty}^{\infty}\int_{-\infty}^{\infty} \hat{u}(\alpha,\beta,\Omega^2+i\Omega\delta)e^{i\alpha x}e^{i\beta y}\,d\alpha\,d\beta \tag{5}$$

Here the use of the small positive parameter δ guarantees that there will be no poles on the real axis and that the limit will correspond to causal outgoing or exponentially decaying waves.

The quantities $\hat{u}(\alpha,\beta,\Omega^2)$, $\hat{v}(\alpha,\beta,\Omega^2)$, and $\hat{w}(\alpha,\beta,\Omega^2)$ may be regarded as the complex amplitudes of particle displacement components associated with a plane wave (or line wave) propagating in a two-dimensional anisotropic medium. The quantities α and β are the x and y components of the wave number vector $\vec{k}$. (With the nondimensionalization adopted in the present paper, α and β represent actual wavenumber components in units of $1/R$).

The algebraic equations governing $\hat{u}$, $\hat{v}$, and $\hat{w}$, with the overcarets suppressed, take the form

$$\begin{pmatrix} \alpha^2+\frac{1-\nu}{2}\beta^2-\Omega^2 & \frac{1+\nu}{2}\alpha\beta & \nu\alpha \\ \frac{1+\nu}{2}\alpha\beta & \frac{1-\nu}{2}\alpha^2+\beta^2-\Omega^2 & \beta \\ \nu\alpha & \beta & 1-\Omega^2+\epsilon(\alpha^2+\beta^2)^2 \end{pmatrix}\begin{pmatrix} u \\ v \\ -iw \end{pmatrix} = \begin{pmatrix} 0 \\ 0 \\ -iG \end{pmatrix} \tag{6}$$

and have the solutions

$$u(\alpha,\beta,\Omega^2) = \frac{U(\alpha,\beta,\Omega^2)}{D(\alpha^2,\beta^2,\Omega^2)}G \tag{7}$$

with analogous expressions (U being replaced by V or W) for $v(\alpha,\beta,\Omega^2)$ and $w(\alpha,\beta,\Omega^2)$. Here

$$D = \left[k^4\epsilon-\Omega^2\right]\left[\Omega^2-k^2\right]\left[\Omega^2-\tfrac{1-\nu}{2}k^2\right]+\tfrac{1-\nu}{2}\left[(1-\nu^2)\alpha^4-(2\nu+3)\alpha^2\Omega^2-\beta^2\Omega^2\right]+\Omega^4 \tag{8}$$

$$U = i(\alpha/k^2)\Phi - i(\beta/k^2)\Psi; \qquad V = i(\beta/k^2)\Phi + i(\alpha/k^2)\Psi \tag{9}$$

$$W = \left[k^2-\Omega^2\right]\left[\tfrac{1-\nu}{2}k^2-\Omega^2\right] \tag{10}$$

with

$$\Phi = (\nu\alpha^2+\beta^2)\left[\tfrac{1-\nu}{2}k^2-\Omega^2\right]; \qquad \Psi = (1-\nu)\alpha\beta(k^2-\Omega^2) \tag{11}$$

3. PLANE WAVES

Naturally occuring plane wave solutions of the shell dynamic equations are governed by the dispersion relation $D = 0$. A brief analysis suggests that, for sufficiently small h/R, or equivalently, for sufficiently small ϵ, the function D approximately factors to $D_{\mathrm{Flex}}D_{\mathrm{Mem}}$ where

$$D_{\mathrm{Flex}} = \epsilon k^4-\Omega^2+(1-\nu^2)\cos^4\theta_k; \qquad D_{\mathrm{Mem}} = \tfrac{1-\nu}{2}\left[I_4k^4+I_2k^2+I_0\right]/I_4 \tag{12}$$

with $I_0 = \Omega^4(1-\Omega^2)$, and

$$I_4 = \tfrac{1-\nu}{2}\left[(1-\nu^2)\cos^4\theta_k-\Omega^2\right]; \qquad I_2 = \Omega^2\left\{\tfrac{3-\nu}{2}\Omega^2-\tfrac{1-\nu}{2}\left[1+2(1+\nu)\cos^2\theta_k\right\}\right. \tag{13}$$

Here $\cos\theta_k$ and $\sin\theta_k$ correspond to α/k and β/k, respectively. (The angle θ_k is the phase velocity direction.) The two factors $D_{\rm Flex}$ and $D_{\rm Mem}$, when separately equated to zero, correspond to the dispersion relations for flexural waves and membrane waves, respectively. In the limit of high Ω, these factors are approximately

$$D_{\rm Flex} \to \epsilon k^4 - \Omega^2; \qquad D_{\rm Mem} = D_{\rm Long} D_{\rm Shear} \tag{14}$$

where

$$D_{\rm Long} \approx \Omega^2 - k^2 - [\sin^2\theta_k + \nu\cos^2\theta_k]^2 \to \Omega^2 - k^2 \tag{15a}$$

$$D_{\rm Shear} \approx \Omega^2 - \tfrac{1-\nu}{2}k^2 - 2(1-\nu)\sin^2\theta_k\cos^2\theta_k \to \Omega^2 - \tfrac{1-\nu}{2}k^2 \tag{15b}$$

The ultimate limiting cases, with the appropriate D-factor set to zero, correspond to flexural, longitudinal, and shear waves respectively, on flat plates. However, the extrapolation of these waves down to frequencies of the order of the ring frequency or lower may yield features markedly different than exhibited by their counterparts on flat plates.

When $\theta_k = 0$, such that the propagation is in the axial direction, an exact factorization of $D_{\rm Mem}$ yields

$$D_{\rm Long} = \frac{\Omega^2(\Omega^2-1)}{\Omega^2 - (1-\nu^2)} - k^2; \qquad D_{\rm Shear} = \Omega^2 - \tfrac{1-\nu}{2}k^2 \tag{16}$$

The longitudinal wave mode consequently has a stop band between $\Omega = (1-\nu^2)^{1/2}$ and $\Omega = 1$. The phase velocity $\omega_r R\Omega/k$ goes to zero at $\Omega = (1-\nu^2)^{1/2}$ and is infinite at $\Omega = 1$. Disturbances at frequencies within the stop band correspond to evanescent waves. When the finite value of ϵ is taken into account, one finds that the continuation of the longitudinal mode curve from somewhat below $\Omega = (1-\nu^2)^{1/2}$ to above that frequency is actually the flexural wave curve. This is because the exact mathematics prohibits dispersion curves from crossing or being tangential to each other, even though such may be a good practical approximation.

For frequencies somewhat lower than the ring frequency, and for arbitrary θ_k, the factor $D_{\rm Mem}$ is more appropriately approximated by

$$D_{\rm Mem} \approx \frac{\left[\Omega^2 - \tfrac{1-\nu}{2}a_1 k^2\right]\left[\Omega^2 - \tfrac{1-\nu}{2}a_2 k^2\right]}{(1-\nu^2)\cos^4\theta_k} \tag{17}$$

where a_1 and a_2 are the roots of

$$a^2 - \left[1 + 2(1+\nu)\cos^2\theta_k\right]a + 2(1+\nu)\cos^4\theta_k = 0 \tag{18}$$

When $\theta_k = 0$, one of these roots is $a_1 = 1$ and corresponds to a shear wave, while the other root is $a_2 = 2(1+\nu)$ and corresponds to a longitudinal wave. The phase velocity of the latter in this low frequency limit is different from what holds in the high frequency limit.

Equatiion (12) shows that plane waves of the flexural type do not propagate without attenuation if $\Omega^2 < (1-\nu^2)\cos^4\theta_k$. Similarly, two plane waves of the longitudinal type do not propagate if I_0/I_4 is negative. Such occurs when Ω^2 is between $(1-\nu^2)\cos^4\theta_k$ and 1.

The *polarization* of the plane waves of the various types is specified by the relative values and phases of u, v, and w. These are the same as those of U, V, and W, only with α and β replaced by the appropriate values at which the denominator factor D is zero. For *flexural waves*, one has approximately

$$\frac{u}{w} \approx -i\frac{\epsilon^{1/4}}{\Omega^{1/2}}\cos\theta_k\left[1-(1+\nu)\cos^2\theta_k\right]; \qquad \frac{v}{w} \approx i\frac{\epsilon^{1/4}}{\Omega^{1/2}}\sin\theta_k\left[1+(1+\nu)\cos^2\theta_k\right] \quad (19)$$

For *longitudinal waves* at frequencies somewhat higher than the ring frequency, the corresponding relations are

$$\frac{w}{u\cos\theta_k + v\sin\theta_k} = i\frac{\sin^2\theta_k + \nu\cos^2\theta_k}{\Omega} \quad (20a)$$

$$\frac{v\cos\theta_k - u\sin\theta_k}{u\cos\theta_k + v\sin\theta_k} = \left(\frac{1-\nu}{1+\nu}\right)\frac{2}{\Omega^2}\cos\theta_k\sin\theta_k(\sin^2\theta_k + \nu\cos^2\theta_k) \quad (20b)$$

while the analogous relations for *shear waves* are

$$\frac{w}{v\cos\theta_k - u\sin\theta_k} = i\left(\tfrac{1-\nu}{2}\right)^{1/2}(2/\Omega)\sin\theta_k\cos\theta_k \quad (21a)$$

$$\frac{u\cos\theta_k + v\sin\theta_k}{v\cos\theta_k - u\sin\theta_k} = -\left(\frac{1-\nu}{1+\nu}\right)\frac{2}{\Omega^2}\cos\theta_k\sin\theta_k(\sin^2\theta_k + \nu\cos^2\theta_k) \quad (21b)$$

The projections of the particle orbits on any plane will in general appear as ellipses when the corresponding complex amplitudes do not have the same phase. The analysis here suggests that, insofar as projections of orbits onto the tangent plane are concerned, the orbits will be straight lines, but not in the same direction as that of the wavefront normal. However, since w is 90° out of phase with the displacements within the tangent plane, orbits will appear as ellipses in planes that are perpendicular to the tangent plane.

4. WAVEFRONTS RESULTING FROM POINT EXCITATION

Techniques for evaluating integrals such as (5) in the limit of large distances (denoted by r in what follows) in the x-y "plane" have been described by Lighthill [11, 12]. Along a line making an angle θ with the x-axis, one has

$$u = \sum_{n=1}^{3} GA_{U\,n}r^{-1/2}e^{i\alpha_n x}e^{i\beta_n y} \quad (22)$$

where the three (or fewer) terms correspond to the outgoing waves of different wavelength excited by the force. The three parameters $A_{U\,n}$, α_n, and β_n may be regarded as functions of the angle θ as well as the frequency Ω. Here α_n and β_n are determined by the equations

$$D(\alpha_n, \beta_n, \Omega^2) = 0; \qquad \sin\theta\,(\partial D/\partial\alpha)_n - \cos\theta\,(\partial D/\partial\beta)_n = 0 \quad (23)$$

[Lighthill's theory also results in relatively simple expressions for the amplitude factors, $A_{U\,n}$, $A_{V\,n}$, and $A_{W\,n}$, but these are omitted in the present paper because of space limitations.]

For $\Omega > 1$ there are solutions of Eqs. (23) that correspond to each of the three fundamental wavetypes (such that each evolves continuously into one of the three wave-types as $\Omega \to \infty$).

The "ripples" excited by the shaking force correspond to lines of constant phase, such that

$$r = \frac{[\text{phase}]}{\alpha_n(\theta)\cos\theta + \beta_n(\theta)\sin\theta} \quad (24)$$

describes a given line in polar form. The wavefronts appear to move in a direction normal to themselves, this direction being that of $\vec{k} = \alpha\vec{e_x} + \beta\vec{e_y}$. The velocity (phase velocity) in this direction is $R\omega_r\Omega/k$ and the wavelength between successive ripples is $2\pi R/k$.

ACKNOWLEDGEMENTS

The research reported here was supported by the Office of Naval Research. The author thanks Jerry Ginsberg, Albert Tucker, and Gary Schwaiger for helpful discussions in regard to this work. Mr. Hyun-Gwon Kil assisted in the preparation of this manuscript.

REFERENCES

[1] Edward W. Ross, Jr. Transition solutions for axisymmetric shell vibrations," *Journal of Mathematics and Physics* **45**, 335–355 (1966).

[2] E. W. Ross, Jr., Asymptotic analysis of the axisymmetric vibrations of shells, *Journal of Applied Mechanics (Trans. ASME)* **33**, 85–92 (March 1966).

[3] O. A. Germogenova, Geometrical theory for flexure waves in shells, *J. Acoust. Soc. Am.* **53**(2), 535–540 (February 1973).

[4] C. R. Steele, A geometric optics solution for the thin shell equation, *Int. J. Engng. Sci.* **9**, 681–704 (1971).

[5] C. R. Steele, An asymptotic fundamental solution of the reduced wave equation on a surface, *Quarterly of Applied Mathematics* **29**, 509–524 (January 1972).

[6] C. B. Steele, Bending waves in shells, *Quarterly of Applied Mathematics* **34**, 385–392 (January 1977).

[7] David Feit, High-frequency response of a point-excited cylindrical shell, *J. Acoust. Soc. Am.* **49**(5 pt. 2), 1499–1504 (May 1971).

[8] A. Harari, Wave propagation in cylindrical shells with finite regions of structural discontinuity, *J. Acoust. Soc. Am.* **62**(5), 1196–1205 (November 1977).

[9] W. R. Hamilton, Third supplement to an essay on the theory of systems of waves, *Trans. Roy. Irish Soc.* **17**(pt. 1), 1–144 (1837). [Reprinted in *Mathematical Papers of* ..., Vol. I.]

[10] J. L. Synge, Elastic waves in anisotropic media, *Journ. of Math. and Phys. (MIT)* **35**, 323–334 (1956).

[11] M. J. Lighthill, Studies on magneto-hydrodynamic waves and other anisotropic wave motions, *Phil. Trans. Roy. Soc. London* **252A**, 397–410 (31 March 1960).

[12] M. J. Lighthill, Group velocity, *J. Inst. Maths. Applics.* **1**, 1–28 (1965).

[13] M. J. P. Musgrave, Elastic waves in anisotropic media, in *Progress in Solid Mechanics*, Vol. II, edited by I. N. Sneddon and R. Hill (North-Holland, Amsterdam, 1961), pp. 61–85.

[14] S. Weinberg, Eikonal method in magnetohydrodynamics, *Phys. Rev.* **126**, 1899–1909 (1962).

[15] Y.-Y. Yu, Free vibrations of thin cylindrical shells having finite lengths with freely supported and clamped edges, *J. Appl. Mech. (ASME)* **22**, 547–552 (December 1955).

[16] M. C. Junger and D. Feit, *Sound, Structures, and their Interaction*, 2nd Edition, (MIT Press, Cambridge, 1986), p. 217.

[17] L. H. Donnell, Stability of thin-walled tubes under torsion, NACA Technical Report No. 479, National Advisory Committee on Aeronautics, Washington, 1933.

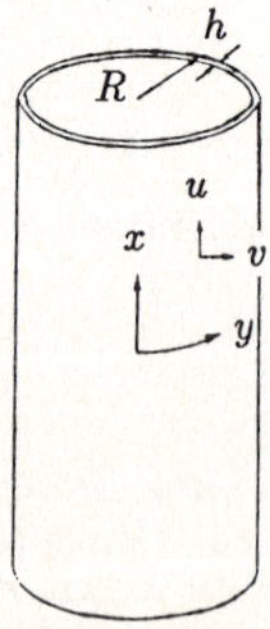

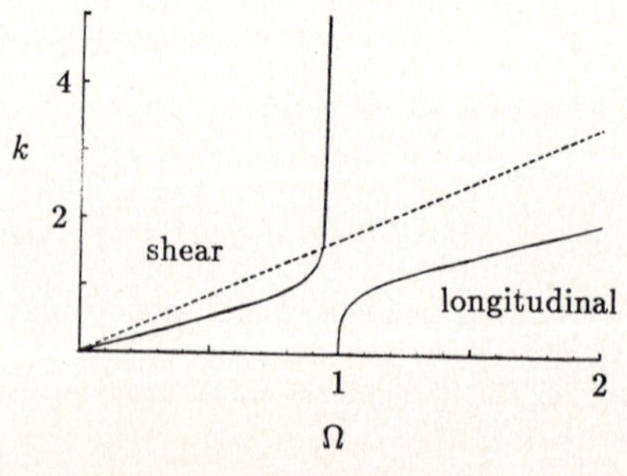

ELASTIC WAVE PROPAGATION
M.F. McCarthy, M.A. Hayes, (Editors)
Elsevier Science Publishers B.V. (North-Holland), 1989

NONSPECULAR REFLECTION OF BOUNDED ACOUSTIC BEAMS FROM LAYERED VISCO-ELASTIC HALFSPACES

Finn B. Jensen and Henrik Schmidt*

SACLANT Undersea Research Centre
Viale San Bartolomeo 400
19026 La Spezia, Italy

A recently developed numerical model of wave propagation in stratified fluid/solid layers is applied to the problem of non-specular beam reflection from a layered solid halfspace consisting of a thin chromium layer on steel. In this case nonspecular reflection is caused by the excitation of a leaky surface wave, which, however, will become doubly leaky (energy leaking back into the fluid as well as into the steel halfspace) when its phase speed exceeds the shear speed of the halfspace. It is demonstrated that the surface layer thickness can be readily determined from the reflectivity pattern for layer thicknesses up to approximately one shear wavelength.

1. INTRODUCTION

Aspects of bounded beam physics have been studied in considerable detail over the past 35 years within the fields of both optics and ultrasonics. A fairly complete list of references to the acoustics literature can be found in Ref. [1]. The phenomenon of primary interest is schematically illustrated in Fig. 1. When a beam of sound is incident upon a fluid/solid interface near the Rayleigh angle, a leaky surface wave (pseudo Rayleigh wave) is excited causing a complex reflectivity pattern with beam splitting and beam displacement. This nonspecular reflection phenomenon is now well understood and can be explained as an interference between the specularly reflected beam and the leaky wave field in the fluid.

Recently the beam reflection problem has been investigated in detail for a series of material combinations including situations with a thin surface layer on an elastic halfspace [1,2]. Two distinctly different cases have been considered: 1) a "soft" surface layer with a shear speed lower than that of the underlying halfspace, and 2) a "stiff" surface layer with a shear speed higher than that of the halfspace. While the case of a soft layer presents few problems and has been solved in detail by Nayfeh and Chimenti [2], the stiff surface layer generally leads to a condition where the surface wave with increasing surface-layer thickness becomes doubly leaky, i.e. the surface wave leaks energy into the substrate as well as into the water. The situation of double leakiness was not considered in Ref. [2], where it is stated that the surface mode ceases to propagate when its phase speed equals the shear speed of the substrate, and this despite the experimental evidence reported by the same authors [3] showing the existence of a leaky wave field in the fluid for surface layer thicknesses well beyond the one corresponding to "cutoff" of the surface mode.

We shall here demonstrate that a leaky surface wave can be identified for all layer thicknesses, as also shown by Bogy and Gracewski [1] in their analysis of

---

*Present address: Massachusetts Institute of Technology, Dept. of Ocean Engineering, Cambridge, MA 02139, USA.

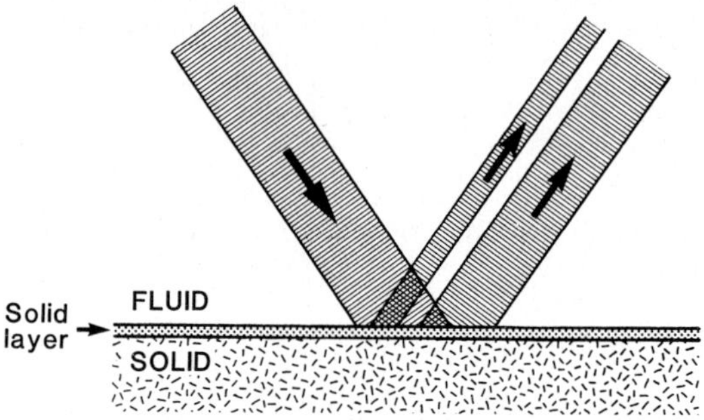

FIGURE 1
Schematic of nonspecular beam reflection from layered solid halfspace.

the plane-wave reflection coefficient at layered solid structures. Our analysis is carried out with a numerical model that provides complete wave-theory solutions for propagation in stratified viscoelastic media. It is also shown that the surface layer thickness can be determined from the nonspecularly reflected beam pattern for layer thicknesses up to approximately one shear wavelength.

2. THE NUMERICAL MODEL

A numerical implementation of a full-spectrum solution of the wave equation in horizontally stratified viscoelastic media is the modelling tool used in this study. The seismo-acoustic model SAFARI [4,5] is based on a Fourier integral representation of the field with respect to horizontal wavenumber. Particular attention was paid to numerical efficiency and stability in evaluating the depth-dependent Green's function. The standard propagator matrix formulation due to Thomson and Haskell [4] was thus substituted by a global matrix formulation, which assured numerical stability as well as an order-of-magnitude reduction of calculation time for problems of the type considered in this study. The SAFARI code has already successfully been applied to beam reflection studies at the interface between a fluid and a homogeneous solid [4] and at the interface between two fluid media [6]. The present application is merely an extension to the case of a fluid halfspace overlying a layered solid halfspace.

In the SAFARI code, a two-dimensional beam is generated by a vertical array of equidistantly spaced line sources. The beam direction is varied by phasing the source elements appropriately, and the intensity distribution across the beam is selected by applying an amplitude weighting across the array. By varying the distance of the array from the interface together with the number of source elements (at half-wavelength spacing), a beam of arbitrary width can be generated. Moreover beams can be focused or defocused by phasing the source elements appropriately. Finally it should be noted that the model fully accounts for diffraction effects as well as absorption losses in all layers.

3. RESULTS AND DISCUSSION

The wave speeds and densities used in the computations are the following: $c = 1480$ m/s, $\rho = 1.0$ g/cm^3 (water); $c_p = 6600$ m/s, $c_s = 4000$ m/s, $\rho = 7.2$ g/cm^3 (chromium); $c_p = 5690$ m/s, $c_s = 3100$ m/s, $\rho = 7.9$ g/cm^3 (steel). For consistency with the numerical results presented in Ref. [2], all materials are

ELASTIC WAVE PROPAGATION
M.F. McCarthy, M.A. Hayes, (Editors)

BEAM AND MODE ANALYSIS OF WEAK BONDING FLAWS IN A LAYERED ALUMINUM PLATE

by

I.T. Lu,* L.B. Felsen,* J.M. Klosner** and C. Gabay**
*Department of Electrical Engineering and Computer Science
Weber Research Institute

and

**Mechanical and Aerospace Engineering Department
Polytechnic University, Farmingdale, NY 11735 USA

Location and identification of faults in multilayer elastic materials by ultrasound is aided by a physically based parametrization of the input, scattered and detected fields. When the transducer input is beam-shaped, the beam-to-mode conversion in the unflawed layered environment suggests a "good" parametrization in terms of a self-consistent hybrid beam-mode format. The scattered field produced by interaction of this beam-mode field with a fault zone should then be parametrized in a similar manner. This strategy guides the present investigation of a weak bonding flaw in a multilayer aluminum plate. The horizontal and vertical displacements excited by a high frequency two-dimensional dilatational (P) Gaussian input beam have previously been tracked through successive cross sections in the perfectly bonded material. The resulting displacement profiles reveal clearly the beam-like character near the source, the deterioration of the successively reflected beam due to P-SV coupling at the boundaries, and the eventual evolution of oscillatory mode-like patterns. This input is now allowed to interact with an elongated weak bond zone. The induced equivalent forcing terms are modeled in the Born approximation, and the scattered field is evaluated accordingly. Depending on the flaw size, its location relative to the input and output transducers, and other variables, the detected response at the plate surface may contain beam-like or mode-like features. The beam-like phenomena are explored here with a view toward finding conditions through which the physical observables that should facilitate flaw location and identification are enhanced. Although, for convenience, the numerical data have been generated by normal mode summation, the results reveal clearly that the hybrid beam-mode format, to be developed next, furnishes the proper parametrization.

1. INTRODUCTION

Analytical modeling of flaw identification in a layered composite through response to an ultrasonic beam input poses a problem of substantial complexity. A good identification scheme must rely on a good parametrization of the propagation and scattering process in terms of physical observables. Translating such observables into analytic form is usually done best by selecting coordinates that are well matched to the phenomenology. This is easily accomplished when considering the input beam, the flaw induced scattering, and the layered environment in isolation but it is far less transparent when all of these constituents interact: the beam coordinates are generally not matched to those of the flaw nor to those of the layers. Thus, analytical basis ele-

here considered to be lossless. From the above material parameters we compute the Rayleigh wave speeds to be $c_{R,cr}$ = 3652 m/s for a water-loaded chromium halfspace, and $c_{R,st}$ = 2871 m/s for a water-loaded steel halfspace. The associated Rayleigh angles measured with respect to the fluid/solid interface are $\theta_{R,cr} = 66.1^{o}$ and $\theta_{R,st} = 59.0^{o}$.

The first step in this investigation is to determine the phase speed of the surface wave as a function of the chromium layer thickness. This is most conveniently done by computing the plane-wave reflection coefficient, noting that a rapid phase change is to be expected at the Rayleigh angle [2,4]. Figure 2 shows plots of the second derivative of the phase angle ϕ with respect to the grazing angle θ for different chromium layer thicknesses. Two distinct features are present in the upper curve for $(d/\lambda_s)_{cr} = 0$. There is a rapid phase change with $\partial^2\phi/\partial\theta^2 = 0$ at 59.0^{o} (Rayleigh angle for steel halfspace) and at 61.5^{o} (shear critical angle for steel). With increasing chromium layer thickness the Rayleigh angle is seen to move to the right and asymptotically approach the Rayleigh angle for a chromium halfspace (66.1^{o}). The point of particular interest is when the Rayleigh angle coincides with the shear critical angle of steel ($d/\lambda_s \simeq 0.2$), because this is where the surface wave becomes doubly leaky. Notice, however, that the Rayleigh angle can be identified for all layer thicknesses whether the surface wave is singly or doubly leaky. Figure 3 shows the detailed functional relationship between the Rayleigh angle and the chromium layer thickness.

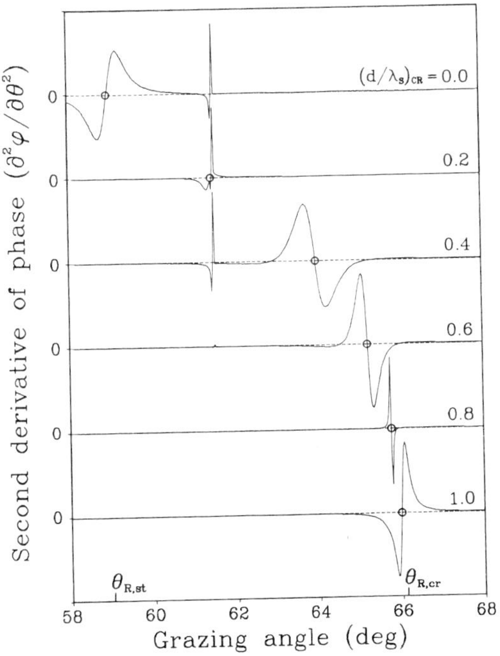

FIGURE 2
Plot of phase variation with grazing angle for different chromium layer thicknesses d/λ_s, where λ_s is the shear wavelength in chromium. A rapid phase change occurs in the plane-wave reflection coefficient at the Rayleigh pole, and the criterion $\partial^2\phi/\partial\theta^2 = 0$ has been used to determine the Rayleigh angle.

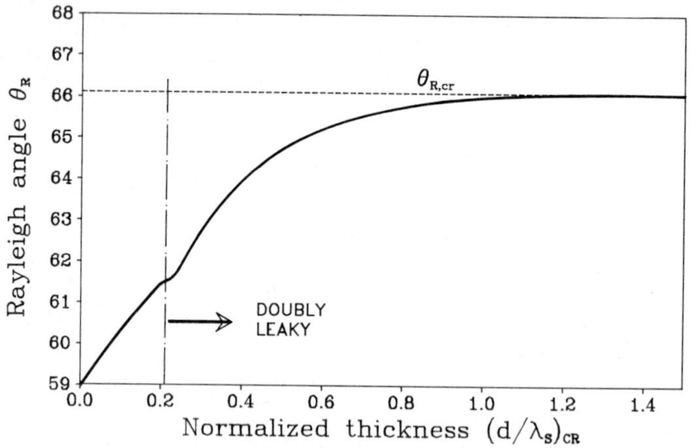

FIGURE 3
Plot of Rayleigh angle versus chromium layer thickness. The surface wave becomes doubly leaky when its phase speed exceeds the shear speed of the steel substrate, i.e. for $d/\lambda_s > 0.22$. Note that the Rayleigh angle varies smoothly between the value for a steel halfspace ($\theta_{R,st} = 59.0^0$) and the value for a chromium halfspace ($\theta_{R,cr} = 66.1^0$).

We next proceed to perform calculations of the reflected field associated with an incident 3.5-MHz Gaussian beam. The beamwidth is approximately 25λ measured across the beam between the 3-dB down points. This beam is generated in the SAFARI code with a vertical source array containing 381 elements spaced λ/2 apart. The computed reflected fields for three different chromium layer thicknesses are given in Fig. 4. Displayed here are field intensities (on a logarithmic scale) versus horizontal distance for a series of incident beam angles. Note in all instances the characteristic split-beam reflectivity pattern with a pronounced interference null between the two beams for incidence near the Rayleigh angle. Figure 4(a) shows the classical result for a singly leaky surface wave, while Figs. 4(b) and 4(c) show results for a doubly leaky surface wave. Note that the only qualitative difference between the three situations is the low level of the reradiated field associated with the Rayleigh wave in Fig. 4(b). Here most of the energy is radiated into the lower steel halfspace, which clearly affects the practical use of the reflectivity pattern for determining the Rayleigh angle to experimental setups with good signal-to-noise ratio. However, for thicker chromium layers, Fig. 4(c), the situation approaches the singly leaky case again with most energy being radiated into the upper fluid halfspace.

Finally we show full field solutions for the three situations discussed above. Each of the plots shown in Fig. 5 is generated in just 3 min on a VAX 8600. We first notice that Fig. 5(a) corresponds to a singly leaky situation with no energy being transmitted into the lower halfspace. On the other hand, Figs. 5(b) and 5(c) clearly involve a doubly leaky Rayleigh wave. The surface-wave energy leaking into the upper fluid halfspace can be identified to the right on the plots as a wave with an exponential decay in the horizontal direction. It is also seen that the isointensity contours of the leaky wave field form an angle with the horizontal that is equal to the Rayleigh angle. As demonstrated elsewhere [4] the only generally valid approach for determining the Rayleigh angle experimentally is to measure the inclination relative to horizontal of an isointensity line in the leaky wave field. When the Rayleigh angle has been determined, the surface layer thickness is found from the graph in Fig. 3.

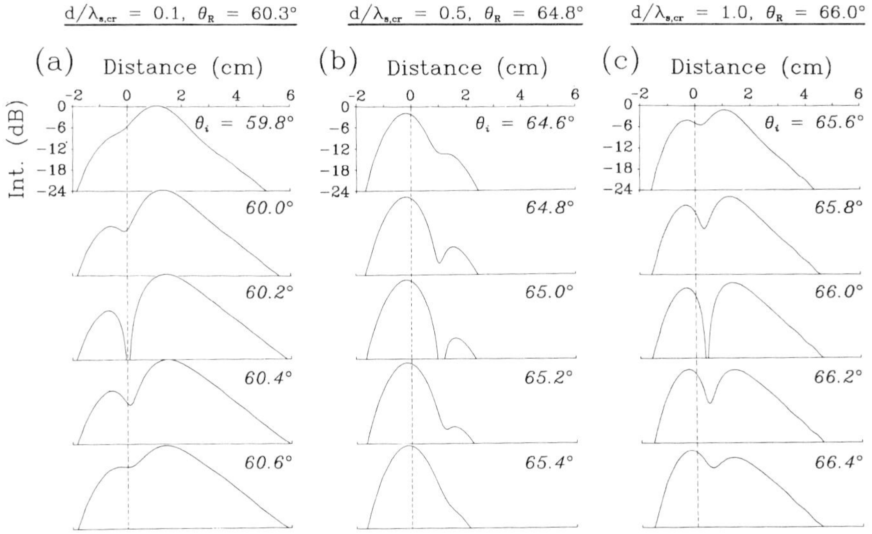

FIGURE 4

Computed reflectivity patterns for three different chromium layer thicknesses: a) $d/\lambda_s = 0.1$, b) $d/\lambda_s = 0.5$, c) $d/\lambda_s = 1.0$. The dashed vertical lines indicate the direction of specular reflection.

In summary, it has been demonstrated that the surface layer thickness can be determined from the reflectivity pattern independent of whether the excited surface wave is singly or doubly leaky. There are, however, experimental limitations on the utility of this technique. Firstly, an accurate determination of the surface layer thickness is possible only for layer thicknesses up to one shear wavelength. Thicker layers acoustically act as a halfspace. Secondly, there could be problems with the signal-to-noise ratio near the point of double leakiness, where most of the surface-wave energy is radiated into the lower halfspace.

REFERENCES

[1] Bogy, D.B. and Gracewski, S.M., On the plane-wave reflection coefficient and nonspecular reflection of bounded beams for layered half-spaces underwater, J. Acoust. Soc. Amer. **74**, 591-599 (1983).

[2] Nayfeh, A.H. and Chimenti, D.E., Reflection of finite acoustic beams from loaded and stiffened halfspaces, J. Acoust. Soc. Amer. **75**, 1360-68 (1984).

[3] Chimenti, D.E. and Nayfeh, A.H., Leaky Rayleigh waves at a layered halfspace-fluid interface, in Proceedings of Ultrasonic Symposium (IEEE, New York, 1981), pp. 291-294.

[4] Schmidt, H. and Jensen, F.B., A full wave solution for propagation in multilayered viscoelastic media with application to Gaussian beam reflection at fluid-solid interfaces, J. Acoust. Soc. Amer. **77**, 813-825 (1985).

[5] Schmidt, H., SAFARI: Seismo-acoustic fast field algorithm for range independent environments, Rep. SR-113, SACLANT Undersea Research Centre, La Spezia, Italy (1988).

[6] Jensen, F.B. and Schmidt, H., Subcritical penetration of narrow Gaussian beams into sediments, J. Acoust. Soc. Amer. **82**, 574-579 (1987).

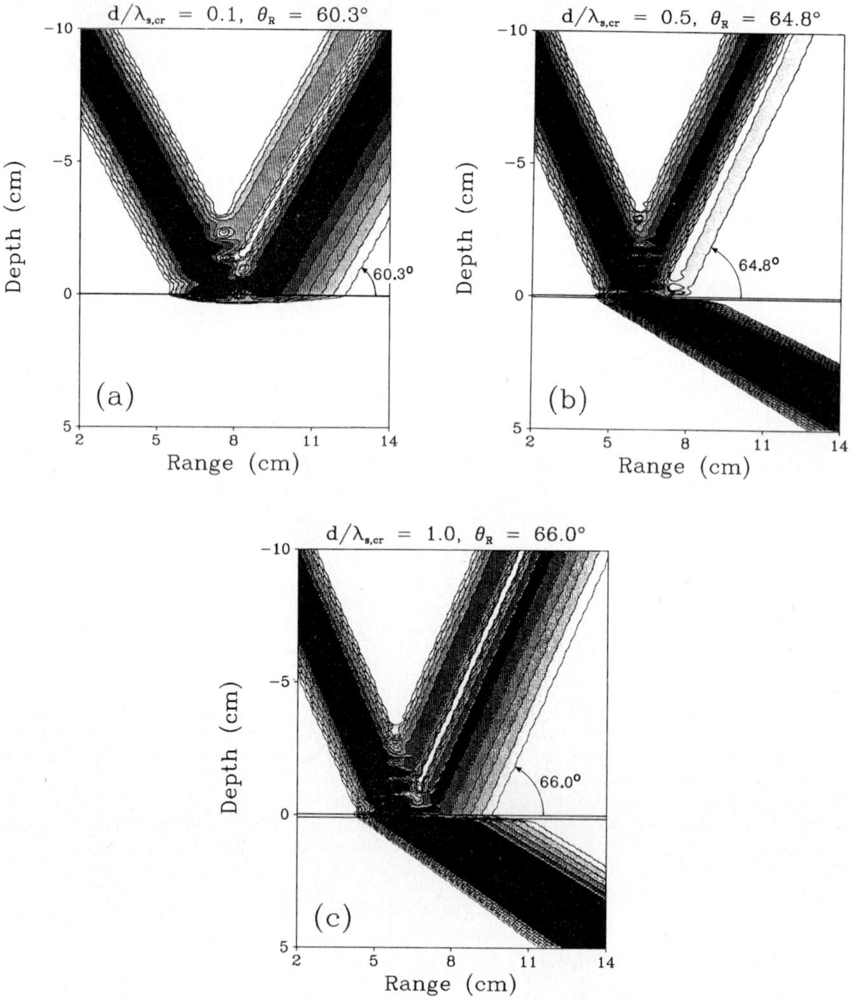

FIGURE 5
Full field solutions (normal stress) for acoustic beams incident at the Rayleigh angle onto a layered solid halfspace of chromium on steel: a) $d/\lambda_s = 0.1$, b) $d/\lambda_s = 0.5$, c) $d/\lambda_s = 1.0$. Note that the surface wave is singly leaky in case (a) where no energy is being transmitted into the lower halfspace. The contour shading is in 3 dB intervals with black indicating high intensity.

ELASTIC WAVE PROPAGATION
M.F. McCarthy, M.A. Hayes, (Editors)

SCATTERING OF POINT-SOURCE GENERATED TRANSIENT ELASTODYNAMIC WAVES BY A SEMI-INFINITE VOID CRACK IN A HALF-SPACE.

Bert Jan Kooij

Laboratory of Electromagnetic Research, Faculty of Electrical Engineering, Delft University of Technology, P.O. Box 5031, 2600 GA Delft, The Netherlands

The scattering of transient elastodynamic waves by a semi-infinite void crack in an isotropic, perfectly elastic half-space is investigated theoretically. Closed-form analytic expressions for the wave quantities are obtained through the application of the Wiener-Hopf technique and the Cagniard-de Hoop technique; they hold in a finite time window. The scheme presented is well suited for a first-arrival analysis of the wave field.

1. INTRODUCTION

The diffraction of elastodynamic waves by a crack in a half-space is of interest to the nondestructive evaluation of material structures. In this field of application there is a need for exact analytical space-time solutions for canonical problems; they can serve as a check of the accuracy of approximative or numerical procedures. In the geometry that we consider the crack is parallel to the boundary of the half-space. Diffraction problems of this kind require the solution of a matrix Wiener-Hopf equation with as the basic step the factorization of the kernel matrix. Considerable work has been done on the factorization of kernel matrices, but matrices of the type that occur in our problem have not been factorized yet. In our analysis an iterative analytical scheme is presented in which the factorization of the kernel matrix is circumvented. The procedure is based on the repeated application of the Wiener-Hopf technique to elastodynamic field constituents that are successively reflected at the boundary of the half-space. In each step of the iterative procedure the solution of the Wiener-Hopf equation involves a kernel matrix factorization that can be carried out analytically. Each iterate can be transformed back to the space-time domain with the aid of a modified version of the Cagniard-de Hoop method (see de Hoop [1], du Cloux [2], Kooij [3]). For a finite number of steps the iterative analytical scheme has the property that an exact analytic expression for the time-domain solution is obtained within a particular finite time window.

2. FORMULATION OF THE PROBLEM

In an isotropic, homogeneous, perfectly elastic medium occupying the half-space $D=\{\mathbf{x} \in R^3|x_3>0\}$ with a traction-free boundary a semi-infinite void crack is present parallel to the boundary. The domain D is divided into a part $\bar{D}$ that occupies the space between the crack and the traction-free boundary and a complementary part $\bar{\bar{D}}$. The elastodynamic properties are characterized by the tensorial stiffness c_{ijpq} and the scalar positive volume density of mass ρ. The crack S is denoted by

$$S = \{\mathbf{x} \in R^3|\ x_1 \in (0,\infty),\ x_2 \in R,\ x_3=d\}. \tag{1}$$

At the traction-free boundary of the half-space a pulsed force-type point source is present at $\mathbf{x} = \{x_{1,s}, x_{2,s}, 0\}$. The elastodynamic field in the sub-domains $\bar{D}$ and $\bar{\bar{D}}$ satisfies the elastodynamic equations

$$\partial_j\{\bar{\tau}_{ij},\bar{\bar{\tau}}_{ij}\} - \rho\partial_t\{\bar{v}_i,\bar{\bar{v}}_i\} = 0 \text{ in } \{\bar{D},\bar{\bar{D}}\}, \tag{2}$$

$$\partial_t\{\bar{\tau}_{ij},\bar{\bar{\tau}}_{ij}\} - c_{ijpq}\partial_q\{\bar{v}_p,\bar{\bar{v}}_p\} = 0 \text{ in } \{\bar{D},\bar{\bar{D}}\}. \tag{3}$$

At the traction-free boundary we have the boundary condition

$$\lim_{x_3\downarrow 0}\bar{\tau}_{i3}(\mathbf{x},t) = -f(t)a_i\delta(x_1-x_{1,s},x_2-x_{2,s}). \tag{4}$$

At the interface of the sub-domains $\bar{D}$ and $\bar{\bar{D}}$ the following boundary conditions hold:

$$\lim_{x_3\uparrow d}\{\bar{\tau}_{ij},\bar{v}_i\} = \lim_{x_3\downarrow d}\{\bar{\bar{\tau}}_{ij},\bar{\bar{v}}_i\} \text{ when } -\infty<x_1<0,\ x_2\in R, \tag{5}$$

$$\lim_{x_3\uparrow d}\bar{\tau}_{i3} = \lim_{x_3\downarrow d}\bar{\bar{\tau}}_{i3} = 0 \text{ when } 0<x_1<\infty,\ x_2 \in R. \tag{6}$$

The source starts to act at the instant t=0; the causality of the wave motion is ensured by taking the solid initially at rest.

3. FORMULATION OF THE SPECTRAL-DOMAIN ELASTODYNAMIC FIELD PROBLEM

To carry out our analysis, we cast the field representations in a particular form that is characteristic of the modified Cagniard technique. We subject the field quantities to a one-sided Laplace transformation with respect to time and a spatial Fourier transformation with respect to x_1 and x_2. This yields the following mapping for an arbitrary quantity $z(x_1,x_2,x_3,t)$ to the transform domain, while the sign conventions are chosen such that $\partial_t \rightarrow s$ and $\partial_{1,2} \rightarrow s\alpha_{1,2}$,

$$z(x_1,x_2,x_3,t) \rightarrow \tilde{z}(s\alpha_1,s\alpha_2,x_3,s), \text{ with } \alpha_1,\alpha_2 \in I \text{ and } s \in R,\ s>0. \tag{7}$$

Application of (7) to (2) and (3) yields

$$\partial_3\{\tilde{\bar{\tau}}_{i3},\tilde{\bar{\bar{\tau}}}_{i3}\} = s\rho\{\tilde{\bar{v}}_i,\tilde{\bar{\bar{v}}}_i\} + s\alpha_\zeta\{\tilde{\bar{\tau}}_{i\zeta},\tilde{\bar{\bar{\tau}}}_{i\zeta}\} \text{ when } \{0<x_3<d, d<x_3<\infty\}, \tag{8}$$

$$c_{ijp3}\partial_3\{\tilde{\bar{v}}_p,\tilde{\bar{\bar{v}}}_p\} = sc_{ijp\zeta}\alpha_\zeta\{\tilde{\bar{v}}_p,\tilde{\bar{\bar{v}}}_p\} + s\{\tilde{\bar{\tau}}_{ij},\tilde{\bar{\bar{\tau}}}_{ij}\} \text{ when } \{0<x_3<d, d<x_3<\infty\}, \tag{9}$$

in which the Greek subscripts are assigned the values 1 and 2. The equations (8) and (9) represent a system of ordinary first-order differential equations. By introducing an appropriate matrix formalism we solve this system in a standard manner. The motion-stress vector is chosen as the 6 by 1 matrix [4]

$$\{\bar{b}_J,\bar{\bar{b}}_J\} = (\{\tilde{\bar{v}}_i,\tilde{\bar{\bar{v}}}_i\},-\{\tilde{\bar{\tau}}_{i3},\tilde{\bar{\bar{\tau}}}_{i3}\})^T, \tag{10}$$

where T means transpose and in which the capital subscript is assigned the values 1 to 6. By eliminating the quantities $\{\tilde{\bar{\tau}}_{\zeta\nu},\tilde{\bar{\bar{\tau}}}_{\zeta\nu}\}$ from (8)-(9) we can write (8)-(9) in the matrix form

$$\partial_3\{\bar{b}_I,\bar{\bar{b}}_I\} = -sA_{IJ}\{\bar{b}_J,\bar{\bar{b}}_J\}, \tag{11}$$

in which A_{IJ} is the system's matrix. Eq. (11) is solved by transforming the system matrix A_{IJ} to a diagonal matrix Λ_{IJ} in which the diagonal elements are the corresponding eigenvalues of A_{IJ}:

$$\{\bar{b}_J,\bar{\bar{b}}_J\} = D_{JN}\{\bar{\psi}_N,\bar{\bar{\psi}}_N\}, \tag{12}$$

$$\partial_3\{\bar{\psi}_M,\bar{\bar{\psi}}_M\} = -sD^{-1}_{MI}A_{IJ}D_{JN}\{\bar{\psi}_N,\bar{\bar{\psi}}_N\} = -s\Lambda_{MN}\{\bar{\psi}_N,\bar{\bar{\psi}}_N\}. \tag{13}$$

The columns of the transformation matrix D_{JN} consist of the corresponding eigenvectors. The general solution of (13) is given by

$$\{\bar{\psi}_N,\bar{\bar{\psi}}_N\} = \delta_{NJ}\exp(-s\gamma^{(J)}x_3)\{\bar{m}_J,\bar{\bar{m}}_J\}, \tag{14}$$

with the eigenvalues $\gamma^{(J)}=\{\gamma^P,\gamma^S,\gamma^S,-\gamma^P,-\gamma^S,-\gamma^S\}$ in which

$$\gamma^{P,S} = (c^{-2}_{P,S}-\alpha_1^2-\alpha_2^2)^{1/2} \text{ with } \mathrm{Re}(\gamma^{P,S})\geq 0. \tag{15}$$

Decomposition of the general solution into upward propagating (J={1,2,3}) and downward propagating (J={4,5,6}) waves yields

$$\{\bar{\psi}^\pm_n,\bar{\bar{\psi}}^\pm_n\} = \delta_{nj}\exp(\pm s\gamma^{(j)}x_3)\{\bar{m}^\pm_j,\bar{\bar{m}}^\pm_j\} = Q^\pm_{nj}(x_3)\{\bar{m}^\pm_j,\bar{\bar{m}}^\pm_j\}, \tag{16}$$

in which the positive superscripts correspond to upward (i.e., decreasing x_3) propagating waves and the negative superscript to downward (i.e., increasing x_3) propagating waves. The transformation matrix D_{JN} is written as the block matrix [4, Eq.(3.46)]

$$D_{JN} = \begin{bmatrix} V^-_{kp} & V^+_{kr} \\ T^-_{ip} & T^+_{ir} \end{bmatrix} . \tag{17}$$

Substitution of (16)-(17) in (12) and use of the definition (10) yield for the particle velocity and the traction in the regions $0<x_3<d$ and $d<x_3<\infty$

$$\{\tilde{\bar{v}}_k,\tilde{\bar{\bar{v}}}_k\} = V^-_{kp}Q^-_{pj}(x_3)\{\bar{m}^-_j,\bar{\bar{m}}^-_j\} + V^+_{kr}Q^+_{rj}(x_3)\{\bar{m}^+_j,\bar{\bar{m}}^+_j\} \text{ when } \{0<x_3<d, d<x_3<\infty\}, \tag{18}$$

$$\{\tilde{\bar{\tau}}_{i3},\tilde{\bar{\bar{\tau}}}_{i3}\}=-T^{-}_{ip}Q^{-}_{pj}(x_3)\{\bar{m}^{--}_j,\bar{\bar{m}}^{-}_j\} - T^{+}_{ir}Q^{+}_{rj}(x_3)\{\bar{m}^{-+}_j,\bar{\bar{m}}^{+}_j\} \text{ when } \{0<x_3<d, d<x_3<\infty\}. \quad (19)$$

in which $\{\bar{m}^{\pm}_j,\bar{\bar{m}}^{\pm}_j\}$ represent the unknown spectral amplitudes of the P/SV/SH waves, respectively. Further, $\bar{\bar{m}}^{-}_j = 0$, since no sources are present in the region $d<x_3<\infty$. The amplitudes $\bar{m}^{--}_j$ and $\bar{m}^{-+}_j$ are expressed in terms of $\bar{\bar{m}}^{+}_j$ by applying the transform-domain counterparts of the boundary conditions (4)-(6) that do not depend on x_1. For those boundary conditions that only hold in the regions $-\infty<x_1<0$ or $0<x_1<\infty$ we obtain a system of mutually coupled dual integral equations of the Wiener-Hopf type. The Wiener-Hopf equation is given as

$$p_i - K_{ij}g^R_j = h^L_i \quad \text{with } \alpha_1 \in I, \quad (20)$$

$$p_i = K^{\infty}_{ij}T^{+}_{jr}Q^{-}_{ri}(d)[I-\Pi]^{-1}_{lk}(T^{+})^{-1}_{kn}u_n, \quad (21)$$

$$K_{ij} = K^{\infty}_{il}T^{+}_{lr}Q^{-}_{rk}(d)[I-\Pi]^{-1}_{kp}(Q^{-}(d))^{-1}_{ps}(T^{+})^{-1}_{sj}, \quad (22)$$

$$K^{\infty}_{ij} = V^{+}_{ik}(T^{+})^{-1}_{kj} - V^{-}_{in}(T^{-})^{-1}_{nj}, \quad (23)$$

$$\Pi_{ij} = (T^{+})^{-1}_{ik}T^{-}_{kn}Q^{-}_{nv}(d)(T^{-})^{-1}_{vd}T^{+}_{dl}Q^{-}_{lj}(d) \text{ with } ||\Pi_{ij}||<1, \quad (24)$$

$$u_i = F(s)a_i\exp[s(\alpha_1x_{1,s}+\alpha_2x_{2,s})]. \quad (25)$$

In (21)-(22), I_{ij} denotes the 3 by 3 unit matrix. The function $g^R_j(\alpha_1,\alpha_2,s)$ is an analytic function of α_1 in $D^R=\{\alpha_1 \in C; Re(\alpha_1)>0\}$ with $g^R_j=o(1)$ as $|\alpha_1|\to\infty$ in D^R and $h^L_i(\alpha_1,\alpha_2,s)$ is an analytic function of α_1 in $D^L=\{\alpha_1 \in C; Re(\alpha_1)<0\}$ with $h^L_i=o(1)$ as $|\alpha_1|\to\infty$ in D^L.

4. SOLUTION OF THE WIENER-HOPF PROBLEM

We solve the Wiener-Hopf equation (20) in an iterative way. To this end we write

$$g^R_j = \sum_{n=0}^{\infty} g^{R,n}_j, \quad (26)$$

$$h^L_i = \sum_{n=0}^{\infty} h^{L,n}_i, \quad (27)$$

in which the terms $g^{R,n}_j$ and $h^{L,n}_i$ will be shown to be of order Π^n as a function of s. We start by solving the equation

$$p^{(0)}_i - K^{\infty}_{ij}g^{R,0}_j = h^{L,0}_i \quad \text{with } \alpha_1 \in I, \quad (28)$$

$$p^{(0)}_i = K^{\infty}_{ik}T^{+}_{km}Q^{-}_{ms}(d)(T^{+})^{-1}_{sj}u_j, \quad (29)$$

which is found by expanding $[I-\Pi]^{-1}=\sum_{k=0}^{\infty}\Pi^k$ and cancelling all terms that are of order Π or higher as a function of s. Next, we relate $h^{L,n}_i$ to the function $p^{(n)}_i$ via

$$p^{(n)}_i - K^{\infty}_{ij}g^{R,n}_j = h^{L,n}_i \quad \text{with } \alpha_1 \in I \text{ and } n \in \{0,1,2,...\}, \quad (30)$$

$$p_i^{(n)} = K^{\infty}_{ik} T^{+}_{kl} Q^{-}_{lj}(d) \Pi_{jm} (Q^{-}(d))^{-1}_{ms} (T^{+})^{-1}_{sr} (K^{\infty})^{-1}_{rp} h_p^{L,n-1} \quad \text{with } n \geq 1,\ \alpha_1 \in I \cup D^L. \tag{31}$$

The function $p_i^{(n)} = p_i^{(n)}(\alpha_1, \alpha_2, s)$ can be viewed upon as the **excitation** function in the n-th Wiener-Hopf equation. The Wiener-Hopf equation (20) is reduced to an infinite set of Wiener-Hopf equations of the "standard" type with identical kernel functions. The "standard" type Wiener-Hopf equation can be related to an infinite medium in which a source-plane is present at the plane $x_3=0$ that excites downward propagating waves into D such that the boundary conditions at $x_3=0$ are satisfied and the field in the region $0<x_3<d$ is represented by

$$\tilde{\bar{v}}_j^{(n)} = \tilde{\bar{v}}_j^{(n),i} + \tilde{\bar{v}}_j^{(n),s} \quad \text{with } n \in \{0,1,2,\dots\}, \tag{32}$$

$$\tilde{\bar{\tau}}_{j3}^{(n)} = \tilde{\bar{\tau}}_{j3}^{(n),i} + \tilde{\bar{\tau}}_{j3}^{(n),s} \quad \text{with } n \in \{0,1,2,\dots\}, \tag{33}$$

$$\tilde{\bar{v}}_j^{(n),i} = V^{+}_{jn} Q^{+}_{nk}(x_3)(Q^{-}(d))^{-1}_{km}(T^{+})^{-1}_{mc}(K^{\infty})^{-1}_{cr} p_r^{(n)}, \tag{34}$$

$$\tilde{\bar{v}}_j^{(n),s} = V^{-}_{jg} Q^{-}_{gk}(x_3-d)\bar{\bar{m}}_k^{(n)}, \tag{35}$$

$$\tilde{\bar{\tau}}_{j3}^{(n),i} = -T^{+}_{jg} Q^{+}_{gk}(x_3)(Q^{-}(d))^{-1}_{km}(T^{+})^{-1}_{mc}(K^{\infty})^{-1}_{cr} p_r^{(n)}, \tag{36}$$

$$\tilde{\bar{\tau}}_{j3}^{(n),s} = -T^{-}_{jm} Q^{-}_{mk}(x_3-d)\bar{\bar{m}}_k^{(n)}, \tag{37}$$

$$p_j^{(n)} = -K^{\infty}_{jm} T^{+}_{mg} Q^{-}_{gk}(d)\Pi_{kl}(Q^{-}(d))^{-1}_{ld}(T^{+})^{-1}_{dv} T^{-}_{vc}\bar{\bar{m}}_c^{(n-1)} \quad \text{with } n \in \{1,2,3\dots\}, \tag{38}$$

in which the quantities with the superscript "i" represent the **incident** field and the quantities with the superscript "s" represent the **scattered** field. The superscript "n" refers to the n-th wave constituent, i.e., the different terms in the representation that are of order Π^n as a function of s. In the Eqs. (35),(37) and (38), $\bar{\bar{m}}_k^{(n)}$ and $\bar{\bar{m}}_k^{(n-1)}$ represent the spectral amplitudes of the upward propagating wave field constituent after the n-th and (n-1)-st iterate, respectively,in our recursive scheme. The problem that has to be analyzed now concerns the elastodynamic response of an unbounded solid containing a semi-infinite void crack whose surfaces are subjected to surface tractions equal and opposite to the incident traction. This leads to a natural decomposition of the problem into one of **symmetric** motion and one of **antisymmetric** motion with respect to the plane $x_3=d$ as noted by de Hoop [1], Miklowitz [5] and du Cloux [2, pp. 115-123]. This leads to two independent Wiener-Hopf problems, that have been solved by du Cloux [2, pp. 153-175]. Addition of the solutions for the unknowm amplitudes of the symmetric and antisymmetric motion yields the spectral amplitude $\bar{\bar{m}}_i^{(n)}$ of the n-th wave constituent

$$\bar{\bar{m}}_i^{(n)} = W_{i\zeta} z_{\zeta} + b r_i \quad \text{with } \alpha_1 \in I \cup D^L, \tag{39}$$

$$W_{i\zeta} = \begin{bmatrix} 1/(1+\chi/\alpha^2) & 0 \\ \chi/[\alpha^2\gamma^S(1+\chi/\alpha^2)] & 0 \\ 0 & 1 \end{bmatrix} \begin{bmatrix} -\alpha_1 & \alpha_2 \\ -\alpha_2 & -\alpha_1 \end{bmatrix}^{-1} \begin{bmatrix} -\alpha_1 & \alpha_2 \\ -\alpha_2 & -\alpha_1 \end{bmatrix}^{-1}, \tag{40}$$

$$z_1 = -2\Phi^L((\alpha_2\tilde{\tau}_{23}^{(n),i}(d)+\alpha_1\tilde{\tau}_{13}^{(n),i}(d))/[k^R(\Omega_R+\alpha_1)])\ \gamma^{S,L}/[\mu(1-c_S^2/c_P^2)(\Omega_R-\alpha_1)k^L], \tag{41}$$

$$z_2 = \Phi^L((\alpha_2\tilde{\tau}_{13}^{(n),i}(d)-\alpha_1\tilde{\tau}_{23}^{(n),i}(d))/\gamma^{S,R})\ /(\mu\gamma^{S,L}), \tag{42}$$

$$b = 2\gamma^{P,L}\Phi^L(\gamma^{P,R}k^R\ \tilde{\tau}_{33}^{(n),i}(d)/(\Omega_R+\alpha_1))\ /[\mu(c_S^2/c_P^2-1)(\Omega_R-\alpha_1)(1+\alpha^2/\chi)k^L\gamma^P], \tag{43}$$

$$r_i = (1,-\gamma^P/\chi,0)^T, \tag{44}$$

$$\alpha^2 = \alpha_1^2+\alpha_2^2;\ \chi = 1/(2c_S^2)-\alpha^2;\ k = 2(\chi^2+\alpha^2\gamma^S\gamma^P)/[(c_P^{-2}-c_S^{-2})(\Omega_R^2-\alpha_1^2)], \tag{45}$$

in which $\Omega_R=(c_R^{-2}-\alpha_2^2)^{1/2}$, and c_R represents the wave speed of the Raleigh wave along the surface of the void crack. The functions $\gamma^{P,R}, \gamma^{S,R}, k^R$ and $\gamma^{P,L}, \gamma^{S,L}, k^L$ are analytic in D^R and D^L, respectively. The functions k^R and k^L have been derived by du Cloux [2, pp. 170-175], while the other functions are obtained by inspection. The operator Φ^L represents Plemelj's integral decomposition operator, which maps the argument into an analytic function in D^L. The resulting expressions can, with the aid of a modified version of the Cagniard-de Hoop technique, be transformed to the time-domain (see du Cloux [2] and Kooij [3]). Since the matrix Π only contains exponential functions of the type $\exp(-2s\gamma^{P,S}d)$ or $\exp[-s(\gamma^{P,S}+\gamma^{S,P})d]$ as a function of s, it is anticipated that wave constituents with superscript "n", that are of order Π^n, only contain exponential functions of order "n". Application of the Cagniard-de Hoop technique shows that higher order wave constituents arrive later than the previous ones. This means that an exact solution is obtained within a certain time-window, the upper limit of which is determined by the minimum arrival time of the next wave constituent.

REFERENCES

[1] De Hoop, A.T., "Representation Theorems for the Displacement in an Elastic Solid and Their Application to Elastodynamic Diffraction Theory", Ph.D. Thesis, Delft University of Technology, Delft, The Netherlands (1958)

[2] Du Cloux, R.,"Symmetry properties of elastodynamic wave fields and their application to space-time scattering theory", Ph.D. Thesis, Delft University of Technology, Delft, The Netherlands (1986).

[3] Kooij, B.J., "Three-dimensional scattering of impulsive acoustic waves by a semi-infinite crack in the plane interface of a half-space and a layer", To be published (1988).

[4] Van der Hijden, J.H.M.T., "Propagation of Transient Elastic Waves in Stratified Anisotropic Media", North-Holland Vol.32, 48-56 (1987).

[5] Miklowitz, J., "The theory of elastic waves and wavequides", North-Holland Vol.22 , 496-517 (1978).

ELASTIC WAVE PROPAGATION
M.F. McCarthy, M.A. Hayes, (Editors)

ANALYSIS OF ELASTIC BAR METHOD FOR MATERIALS CHARACTERIZATION IN HIGH VELOCITY TENSION

Kozo Kawata and Masaaki Itabashi

Science University of Tokyo, Noda, Chiba-ken 278, Japan

Analysis of elastic bar methods for materials characterization in high velocity tension is reported, laying emphasis on the damping and dispersion of stress waves propagated in an elastic bar experimentally determined.

1. INTRODUCTION

Uniaxial dynamic tensile properties of solids up to breaking should be known as the most fundamental mechanical properties of solids from the standpoints of science and of advanced technology. Hitherto, such data have not been necessarily enough clarified. The reason is mainly attributed to the fact that some limits or difficulties existing in proposed experimental methods have limited their applicability in being utilized as the characterization method of dynamic mechanical properties in tension up to breaking applicable even for materials with large elongation. In the present paper, analysis of elastic bar methods for materials characterization in high velocity tension is reported, laying emphasis on the damping and the dispersion of stress waves propagated in an elastic bar, causing the deformation of real stress-strain curves and on the effective time duration for the methods.

2. COMPARISON OF VARIOUS ELASTIC BAR METHODS

Various elastic bar methods proposed for the experimental determination of pressure-time relation or stress-strain curve at high rates of strain are shown in FIG I. The original Hopkinson bar method (1914) [1] is combined with time piece of different lengths to measure the momentum trapped in them, giving the areas of the pressure-time curves for different intervals. The precise shape of the pressure-time curve cannot be deduced because the beginning points of the different intervals are not determined. In Davies' bar (1948) [2], a continuous record of the longitudinal displacement produced by the pressure pulse at the free end of the bar, u-t curve is obtained. This is converted to the longitudinal stress σ_x-t curve. A modified form of the Davies bar is the one used by Kolsky (1949) [3] to measure the stress-strain behaviour of disk-shaped specimens. This is the well known split Hopkinson pressure bar (or Kolsky apparatus), widely used experimental configurations for high strain-rate material measurements. The strain, stress and strain rate in the specimen are given by

$$\varepsilon = \frac{-2c}{l} \int_0^t \varepsilon_r \, dt$$

$$\sigma = E_b \frac{S_0}{S} \varepsilon_t$$

$$\dot{\varepsilon} = \frac{-2c}{l} \varepsilon_r$$

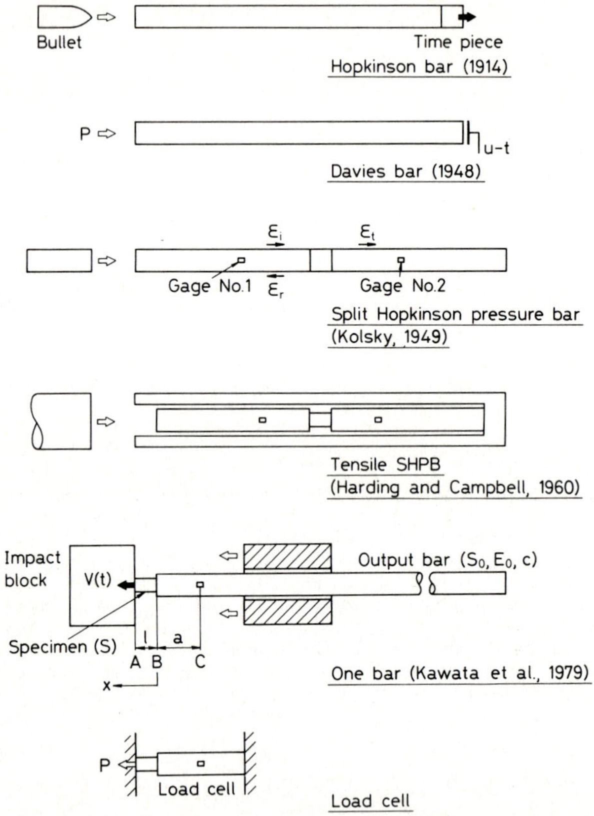

FIGURE 1
Various elastic bar methods proposed for the determination of pressure-time relation or stress-strain curve at high rates of strain

where l, S, S_0, c and E_b are the same with shown later, and ε_i, ε_r and ε_t are strains of incident, reflected and transmitted waves, respectively. Harding et al. (1960) [4] is the one of the first to demonstrate a tension version of SHPB technique, producing tension in an inner bar system by the reflection of a compression pulse in a tube surrounding it. The time duration for effective measurement may be affected by the dimension of bars, inevitably rather short. Kawata et al. (1979) [5,6] proposed a new testing method designated "one bar method," that can give precise tensile stress-strain curve up to breaking adopting enough length of output bar (ordinarily, 3000mm, effective time duration ≅1160μs) by using two measured values: $\varepsilon_g(t)$ and V(t), based upon the dynamic response analysis on 1) block to block type, 2') load-cell, 3) block to bar type and 4) bar to bar type. The stress, strain and strain rate of the specimen are given by

$$\sigma(t) = \frac{S_0}{S} E_b \, \varepsilon_g(t+a/c)$$

$$\varepsilon(t) = \frac{1}{l} \int_0^t [V(\tau) - c\ \varepsilon_g(\tau+a/c)]\ d\tau$$

$$\dot{\varepsilon}(t) = \frac{1}{l} [V(t) - c\ \varepsilon_g(t+a/c)]$$

where l and S are the length and cross-sectional area of the specimen, and S_0, E_b and c are the cross-sectional area, Young's modulus and elastic wave velocity of the output bar. V(t) and $\varepsilon_g(t)$ are the velocity of impact block and strain of the output bar at the distance a from the end obtained from the gage.

3. MEASUREMENT OF DAMPING AND DISPERSION OF STRESS WAVES PROPAGATED IN AN ELASTIC BAR

Waves propagated in solids are accompanied with damping and dispersion correspondence with the distance and path travelled. The former causes the decreasing of wave amplitude and the latter the variation of wave form, namely the variations in both ordinate and abscissa for amplitude-time relation. Measurement of damping and dispersion of stress waves propagated in an elastic bar is made, using Kawata et al.'s "one bar method" (FIGURE 2), where an output bar made of 18-8 stainless steel is used. To produce an original wave form, high velocity tension of an S45C structural carbon steel of typical carbon content of 0.45% specimen is used. To obtain $\varepsilon_g(t)$, three gage points are set at the distances of 50, 800 and 1600mm from the specimen-side end of the output bar. The records of stress-strain curve obtained at three points (FIGURE 3) clarify the varying of the form of stress-strain curve with increasing distance propagated in the output bar. FIGURES 4 and 5 show the variations of $\Delta\sigma_{UY}/\sigma_{UY}$ and $\Delta\varepsilon_T/\varepsilon_T$ with increasing x, where σ_{UY} and ε_T are upper yield stress and total elongation at breaking, respectively. The tendencies of $\Delta\sigma_{UY}/\sigma_{UY}$ and $\Delta\varepsilon_T/\varepsilon_T$ are formulated as follows:

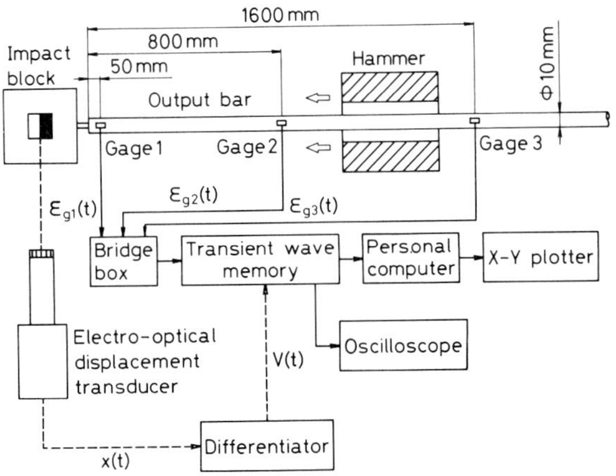

FIGURE 2
Measuring device of damping and dispersion of stress waves propagated in an elastic bar

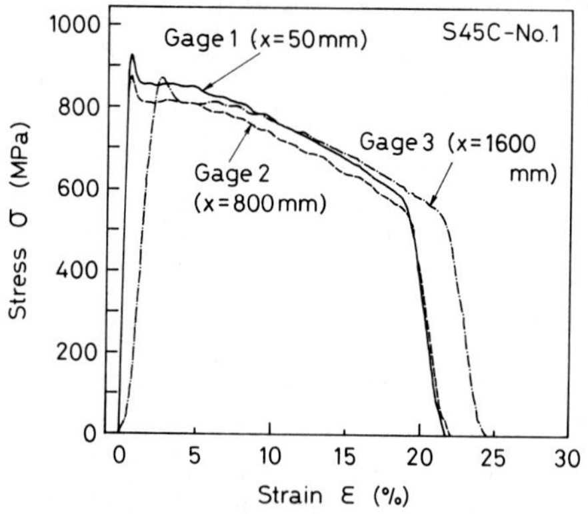

FIGURE 3
The records of stress-strain curve of S45C carbon steel obtained at three points with different distances from the specimen-side end in the output bar ($\dot{\varepsilon}$=1X10^3 s^{-1})

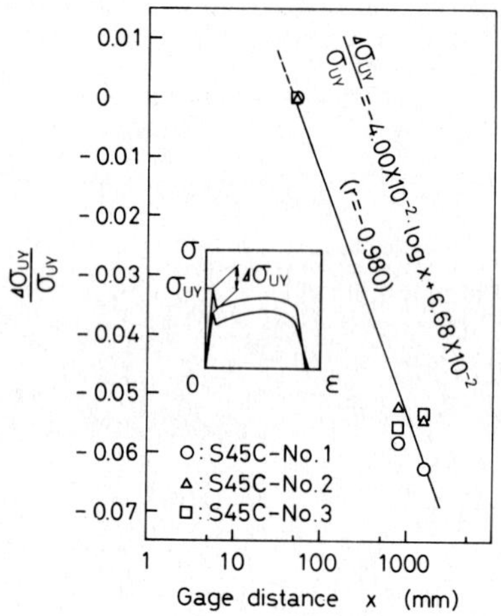

FIGURE 4
The damping of upper yield stress of S45C carbon steel with increasing travelling distance in elastic bar

$$\frac{\Delta\sigma_{UY}}{\sigma_{UY}} = -4.00\text{X}10^{-2} \log x + 6.68\text{X}10^{-2}$$

$$\frac{\Delta\varepsilon_T}{\varepsilon_T} = 3.00\text{X}10^{-11}\ x^3 - 2.41\text{X}10^{-3}.$$

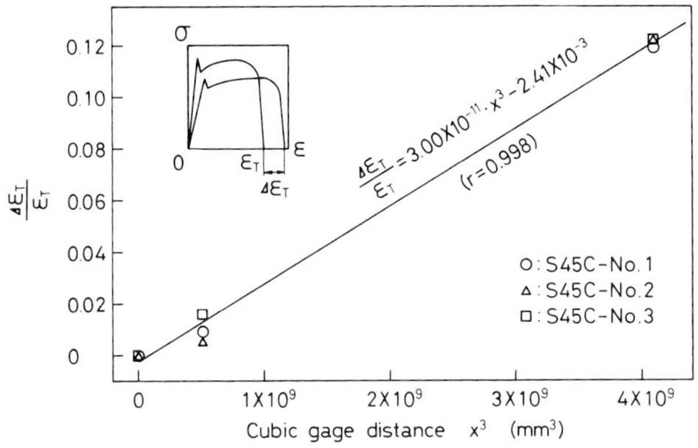

FIGURE 5
The dispersion of stress-strain curve of S45C carbon steel shown by the variation of total elongation with increasing travelling distance in elastic bar

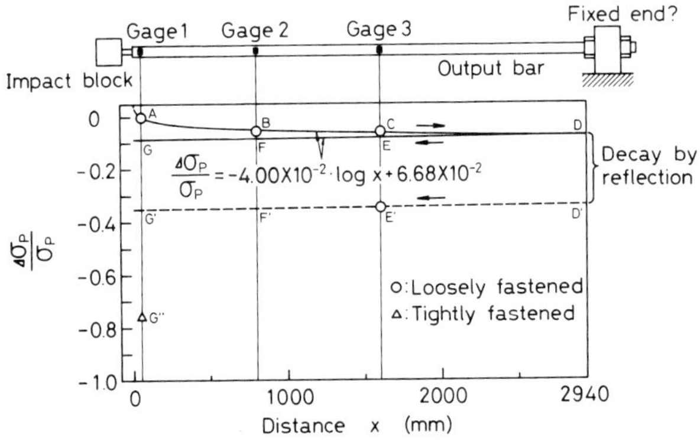

FIGURE 6
Experimental results on decay of maximum stress in stress-strain curve by reflection

In deriving the above expressions, it is assumed that the exact stress-strain curve is obtained using the signal measured at x=50mm (gage 1). This assumption is considered reasonable, because 50mm is approximately the shortest distance from the specimen-side end of the bar satisfying the condition of thin bar and at the same time St. Venant principle. The effects of reflection at the output bar end are also studied, using the same experimental arrangement, where both cases of loose and tight fastening of the output bar end, are adopted respectively (FIGURE 6). The decay of $\Delta\sigma_P/\sigma_P$ (where σ_P is maximum stress) by reflection is known as DD' for loosely fastened end. For the tightly fastened end, GG" is remarkably larger than GG'. Anyway the decay by reflection is serious for these cases. It seems important to take into account the effect of reflection in discussing the accuracy of wave forms obtained.

4. THE DEMANDS FOR ELASTIC BAR METHODS FOR THE DETERMINATION OF STRESS-STRAIN CURVE IN TENSION AT HIGH RATES OF STRAIN

The following demands are deduced considering the experimental results, for elastic bar methods for the determination of stress-strain curve in tension at high rates of strain.

I) It should be avoided to let the stress wave to be measured propagate too long before reaching the gage point in elastic bar.
2) It would be better to avoid the intervention of reflected waves in elastic bar
3) Effective time duration to be obtained by the method should be long enough to secure full stress-strain curve acquisition.

5. CONCLUSIONS

The following conclusions are obtained:

I) The damping and dispersion of stress waves propagated in a long elastic bar in correspondence with the distance and path travelled are measured and formulated.
2) The demands for elastic bar methods for the determination of the stress-strain curve in tension at high rates of strain, are discussed considering the damping and dispersion of stress waves propagated in an elastic bar.

ACKNOWLEDGEMENTS

The fact that this work is supported by Grant-in-Aid for Scientific Research of the Ministry of Education,Science and Culture (I986,87 and 88 FY) is stated with hearty thanks. The thanks are also due to the students who assisted eagerly with this work.

REFERENCES

I Hopkinson, B., Phil Trans. A 2I3 (I9I4) 437.
2 Davies, R.M., Phil. Trans. A 240 (I948) 375.
3 Kolsky, H.m Proc. Phys. Soc. B 62 (I949) 676.
4 Harding, J., Wood, E.D. and Campbell, J.D., J.Mech.Eng.Sci. 2(I960) 88.
5 Kawata, K., Hashimoto, S., Kurokawa, K. and Kanayama, N., A New Testing Method for the Characterization of materials in High-Veloctiy Tension, in: Harding, J., (ed), Mechanical Properties at High Rates of Strain I979, Conf. Ser. No.47 (Inst. of Phys., Bristol and London, I979) pp. 7I-80.
6 Kawata, K., Hashimoto, S., Sekino, S. and Takeda, N., Macro- and Micro-Mechanics of High-Velocity Brittleness and High-Velocity Ductility of Solids, in : Kawata, K. and Shioiri, J.m (eds), Macro- and Micro-Mechanics of High Velocity Deformation and Fracture (Springer-Verlag, Berlin and Heidelberg, I987) pp. I-25.

ELASTIC WAVE PROPAGATION
M.F. McCarthy, M.A. Hayes, (Editors)

PULSE EVOLUTION IN A MULTIMODE, ONE-DIMENSIONAL, HIGHLY DISCONTINUOUS MEDIUM

Robert BURRIDGE and Hung-Wen CHANG

Schlumberger Doll Research
Old Quarry Road
Ridgefield, CT 06877

A pulse propagates obliquely through a one-dimensional medium consisting of a large number N of homogeneous elastic layers. As it propagates, the directly transmitted principal arrival is reduced by transmission loss at each interface, but close to this arrival is a broad pulse, made up of multiply scattered energy, which ultimately appears to diffuse about a moving center. This may be regarded as an extension to multimode propagation of the phenomenon first studied by O'Doherty and Anstey in 1971 and further corroborated and elucidated by many authors since then. The broad pulse evolves according to an integrodifferential equation analogous to the Kolmogorov-Feller forward equation in probability theory, the kernel of which depends upon the auto and cross correlations of the material properties of the medium.

1. INTRODUCTION

It is the aim of this paper to show how a pulse evolves as it travels through a highly discontinuous medium consisting of a large number of uniform, isotropic, elastic layers whose density ρ, and P and S wave speeds α and β differ but slightly from the constant values $\bar{\rho}$, $\bar{\alpha}$, and $\bar{\beta}$. Such a system is a model of seismic pulse propagation through a geological formation.

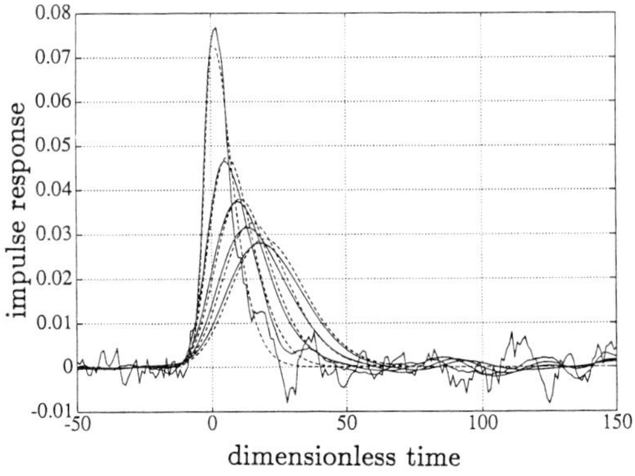

FIGURE 1.
Here we show an S-wave pulse, incident at 25.1°, after traveling through 2,000, 4,000, 6,000, 8,000, and 10,000 layers. The simulations are solid curves, the limiting solution is shown dashed.

To fix our ideas let us consider what happens when an impulsive plane wave traverses a very finely layered elastic medium. The continuous curves in Figure 1 show a plane S-wave pulse, incident at 25.1°, as it traverses a stack of 10,000 plane homogeneous layers. On entering the stack the pulse is impulsive and has the form of a Dirac delta function of time. We see the pulse shape as a function of time after it has traversed 2,000, 4,000, 6,000, 8,000, and finally 10,000 layers. Notice particularly the progressive broadening of the pulse and its increasing delay. After traversing 2,000 layers there are still considerable high-frequency 'fluctuations', which decrease as the pulse travels. We shall be concerned with understanding the evolution of the smooth pulse. The magnitude of the fluctuations are a measure of how well the smooth approximation fits the fluctuating wave.

Our theory predicts that the evolution of the smooth component of the pulse, as shown in the broken curves of Figure 1, depends upon the auto- and cross-correlations of the density and wave speeds of the medium, as described in the following sections.

In this work we follow a line of research initiated by the now classical paper of O'Doherty & Anstey [1] who noticed that a pulse is broadened and slightly delayed as it travels through a finely layered medium. In an appendix to their paper they also gave a quantitative explanation of the phenomenon in terms of the statistics of the reflection coefficients for a Goupillaud layered medium, i.e. one in which the transit time across every layer is the same. Indeed, up to now studies of this effect have been restricted to normally incident plane waves traveling through a Goupillaud layered medium. In the present work, however, we allow our layers to have arbitrary (but small) thicknesses, and we consider obliquely incident waves, with the restriction that we do not admit critical-angle phenomena – no evanescent waves are present in any layer.

Following the work of O'Doherty & Anstey many authors have elaborated on this phenomenon with numerical simulations and analysis; see Burridge, White, & Papanicolaou [2] and its references for further information; for a fuller account of this work see Burridge & Chang [3].

2. THE SCATTERING PROCESS

We shall consider only the one-dimensional problem in which the disturbance is made up of plane P and S waves in each layer. Let z and t be the space coordinate, representing depth into the ground, and the time. We shall assume that all waves derive from a single plane wave of P or S type incident on the stack of layers from the uniform half space $z < 0$ above the stack and thus all the plane wave components in every layer have the same transverse slowness (ray parameter) p. We shall investigate the plane transmitted wave which emerges into the uniform half space $z > L$ below the stack in the same mode.

To make the ensuing calculations less cumbrous we shall establish some notation:

$$\begin{array}{lll} 1 & \text{stands for down-going } P & (\grave{P}), \\ 2 & \text{stands for down-going } S & (\grave{S}), \\ 3 & \text{stands for up-going } P & (\acute{P}), \\ 4 & \text{stands for up-going } S & (\acute{S}). \end{array} \tag{1}$$

We shall denote the z-component of slowness in mode i by γ_i:

$$\gamma_1 = -\gamma_3 = q_\alpha \equiv \sqrt{\tfrac{1}{\alpha^2(z)} - p^2}, \quad \gamma_2 = -\gamma_4 = q_\beta \equiv \sqrt{\tfrac{1}{\beta^2(z)} - p^2}. \tag{2}$$

We shall let ρ_1, ρ_2, ρ_3 stand for the material properties ρ, α, β, and let S^k stand for the scattering matrix at interface z_k, where entry S^k_{ij} represents the reflection or transmission coefficient from incident mode i into scattered mode j and the waves are normalized by the square root of energy flux in the z-direction. When the contrasts in material properties across the interfaces are small and $j \neq i$, the S^k_{ij} are small and depend linearly upon the contrasts $\Delta\rho^k_\alpha$ in the ρ_α at interface k. When $j = i$ we have

$$S_{ii}^k \cong 1 - \tfrac{1}{2} \sum_{\{j|j\neq i\}} \left(S_{ij}^k\right)^2, \tag{3}$$

which is near unity. See Aki & Richards [4], section 5.2.

As it travels the pulse is modified by transmission loss and by double scattering events of the following type: an incident wave in mode i, say $u(t-\gamma_i z)$, impinges on interface z_ℓ where it produces a scattered wave in mode $j \neq i$,

$$S_{ij}^\ell u\Big(t - \gamma_j(z - z_\ell) - \gamma_i z_\ell\Big), \tag{4}$$

which produces at interface z_k a scattered wave in the original mode i:

$$S_{ji}^k S_{ij}^\ell u\Big(t - \gamma_i(z - z_k) - \gamma_j(z_k - z_\ell) - \gamma_i z_\ell\Big) = -\, S_{ij}^k S_{ij}^\ell u\Big(t - \gamma_i z - (\gamma_i - \gamma_j)(z_\ell - z_k)\Big). \tag{5}$$

Here, the equality follows from reciprocity and the fact that reversing the directions of the waves at an interface changes the sign of the $\Delta\rho_\alpha^k$ and hence changes the sign of the corresponding scattering coefficients in this approximation of small contrasts at the interfaces.

Now let h be small, but large enough that the interval $(z, z+h)$ contains many interfaces, and consider the incremental change to a pulse in mode i which enters the interval $(z,\ z+h)$ at z as $u(z,\ t-\gamma_i z)$ and leaves at $z+h$ as $u\,(z+h,\ t-\gamma_i(z+h))$. Writing τ for $t-\gamma_i z$, the increment is approximately

$$\begin{aligned} u(z+h,\ \tau) - u(z,\ \tau) = &-\tfrac{1}{2} \sum_{\{j|j\neq i\}} \sum_{\{\ell|z\leq z_\ell<z+h\}} \left(S_{ij}^\ell\right)^2 u(z,\ \tau) \\ &- \sum_{\{j|j\neq i\}} \sum_{\{k,\ell|z\leq z_\ell<z+h,\gamma_j(z_k-z_\ell)>0\}} S_{ij}^k S_{ij}^\ell u\left(z, \tau - (\gamma_i - \gamma_j)(z_\ell - z_k)\right). \end{aligned} \tag{6}$$

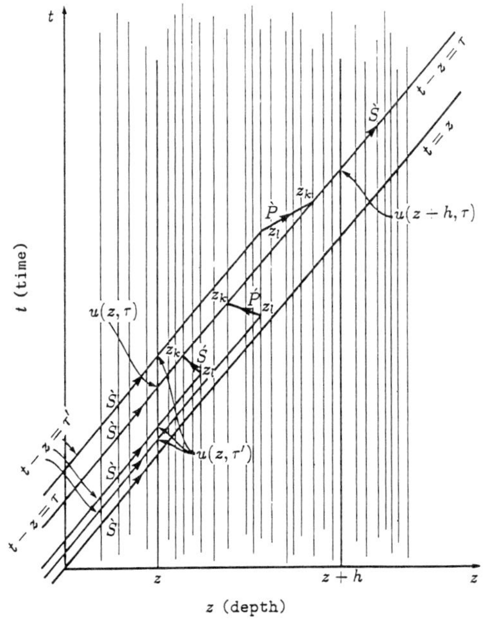

FIGURE 2.
This is a space-time diagram showing the kind of double scattering which scatter from $u(z,\tau')$ to $u(z+h,\tau)$. The transmitted pulse is made up of contributions from all such events.

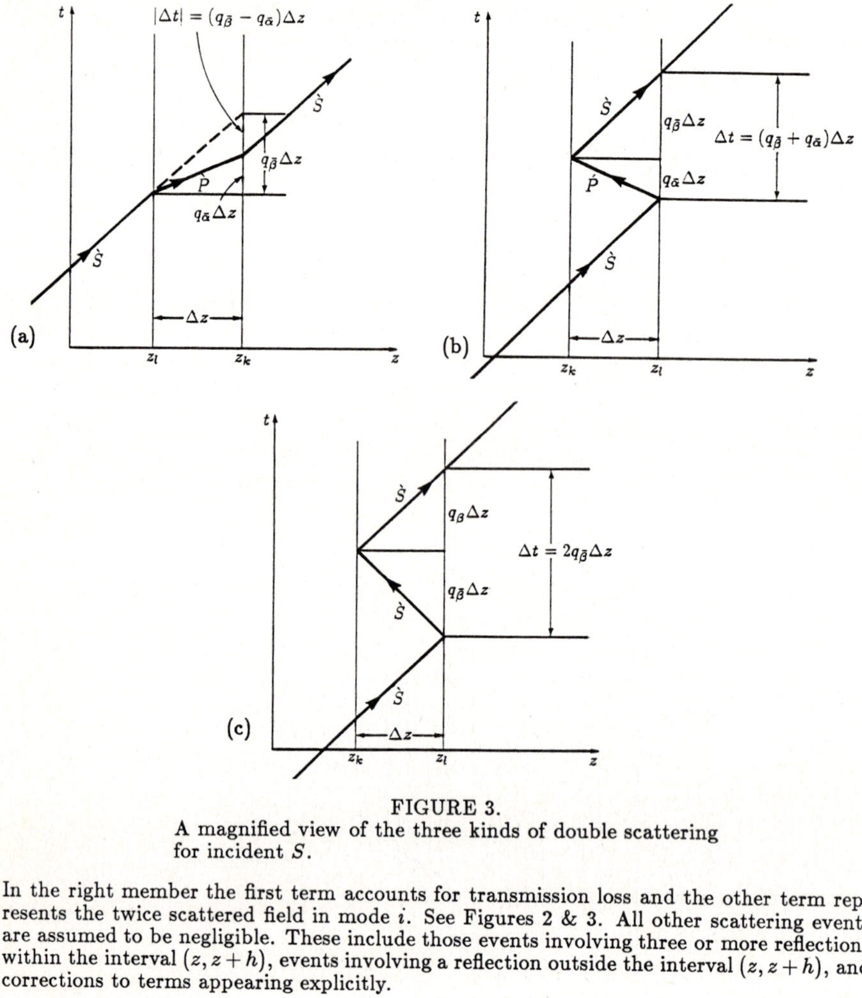

FIGURE 3.
A magnified view of the three kinds of double scattering for incident S.

In the right member the first term accounts for transmission loss and the other term represents the twice scattered field in mode i. See Figures 2 & 3. All other scattering events are assumed to be negligible. These include those events involving three or more reflections within the interval $(z, z+h)$, events involving a reflection outside the interval $(z, z+h)$, and corrections to terms appearing explicitly.

3. AVERAGING

We shall make the following statistical assumption concerning the S^k_{ij}. Let D be a domain in the $(z',\ z'')$ plane and ϕ a smooth function with support in D. We shall suppose that

$$\sum_{\{(z_k,\, z_\ell)\in D\}} S^k_{ij} S^\ell_{ij} \phi(z_k, z_\ell) = \int\int_D \sum_k S^k_{ij}\delta(z' - z_k)\sum_\ell S^\ell_{ij}\delta(z'' - z_\ell)\phi(z',\ z'')dz'dz''. \tag{7}$$

is well approximated by

$$\int\int_D \left[a^{ij}_0\,\delta(z' - z'') + a^{ij}(z' - z'')\right]\phi(z',\ z'')\,dz'dz'', \tag{8}$$

where a_0^{ij} is a scalar constant and a^{ij} is an even integrable function of one variable $(a^{ij}(-\nu) = a^{ij}(\nu))$. These are smoothed averages of the discrete terms in (7). Among other things, assumption (8) implies the wide-sense stationarity of the ij reflectivity function,

$$\sum_k S_{ij}^k \delta(z - z_k). \tag{9}$$

The quantity in square brackets in (8) is the autocorrelation of the reflectivity function, and we may write

$$a_0^{ij}\delta(z' - z'') + a^{ij}(z' - z'') = \langle \sum_k S_{ij}^k \delta(z' - z_k) \sum_\ell S_{ij}^\ell \delta(z'' - z_\ell) \rangle, \tag{10}$$

the angular brackets standing for average. Although it would be mathematically convenient if this were an ensemble average, it is clear from the analysis in [9] that $\langle \cdot \rangle$ represents a less manageable spatial average.

4. THE INTEGRODIFFERENTIAL EQUATION

Let us now use (7) and (8) in (6) to rewrite the right hand side:

$$\begin{aligned} u(z+h,\, \tau) - u(z,\, \tau) = &-\tfrac{1}{2} \sum_{\{j|j\neq i\}} \int_z^{z+h} dz' a_0^{ij} u(z,\, \tau) \\ &- \sum_{\{j|j\neq i\}} \int_z^{z+h} dz'' \int_{\{z''|\gamma_j(z'-z'')>0\}} dz'\, a^{ij}(z' - z'') u\left(z,\, \tau - (\gamma_i - \gamma_j)(z'' - z')\right). \end{aligned} \tag{11}$$

Changing variable of integration in (11) to $\tau' = (\gamma_i - \gamma_j)(z'' - z')$ and dividing by h we get

$$u_z(z,\, \tau) = -\tfrac{1}{2} \sum_{\{j|j\neq i\}} a_0^{ij} u(z,\, \tau) - \sum_{\{j|j\neq i\}} \frac{1}{|\gamma_i - \gamma_j|} \int_{I_{ij}} a^{ij}(\tfrac{\tau'}{\gamma_i - \gamma_j}) u(z,\, \tau - \tau') d\tau'. \tag{12}$$

Here the intervals of integration I_{ij} are either $(-\infty,\, 0)$ or $(0,\, \infty)$ depending on the signs of γ_j and of $\gamma_i - \gamma_j$. Set

$$a_0^i = \sum_{\{j|j\neq i\}} a_0^{ij}, \quad a^i(\tau) = \sum_{\{j|j\neq i\}} \frac{\chi_{I_{ij}}(\tau)}{|\gamma_i - \gamma_j|} a^{ij}\left(\tfrac{\tau}{\gamma_i - \gamma_j}\right), \tag{13}$$

where $\chi_{I_{ij}}(\tau)$ is the characteristic function of I_{ij}: $\chi_{I_{ij}}(\tau) = 1$ if $\tau \in I_{ij}$, and 0 otherwise. Then we may rewrite (12) as

$$u_z(z,\, \tau) = -\tfrac{1}{2} a_0^i u(z,\, \tau) - \int_{-\infty}^{\infty} a^i(\tau') u(z,\, \tau - \tau') d\tau'. \tag{14}$$

This integrodifferential equation, which is the main result of this section, governs the evolution of the pulse shape $u(z,\, \tau)$ as the disturbance travels through the stack of layers.

It is interesting to notice that equation (14) has the same structure as the Kolmogorov-Feller forward equation which arises in the theory of purely discontinuous stochastic processes, and its derivation is basically the same. See Gnedenko [5] chapter X.

5. COMPARISON OF SIMULATIONS WITH THE LIMITING SOLUTION

For the purpose of testing the usefulness of the integrodifferential equation we used a pseudo-random number generator to generate the layer parameters and a layer matrix code to simulate the pulse propagation. For comparison we also calculated the 'limiting' wave form by solving the integrodifferential equation. The layer thicknesses were exponentially distributed, and the parameters in each layer were uncorrelated with those in any other layer. See Burridge & Chang [3] for further details of the statistical model, of the layer code used in the simulations, and of the methods used to solve the integrodifferential equation. The comparison of the numerical simulations with the solution of the integrodifferential equation is not only of interest in its own right but, in the absence of a rigorous proof that the actual solution is well approximated by the limiting solution, it provides a corroboration that the integral equation does in fact govern the evolution of the pulse.

In Figure 4 we show the impulse response corresponding to a plane P wave incident at $\theta_P = 15°$ as it travels through a stack of 10,000 layers. The horizontal axis is time t measured in units of P-wave travel time across an average layer from the time of arrival of the directly transmitted impulse. The vertical scale is appropriate to the impulse response in that the pulses have unit area (approximately). The five curves represent the developed pulse after it has traversed 2,000, 4,000, 6,000, 8,000, and 10,000 layers. The further through the stack the pulse travels the lower and broader the pulse becomes.

For comparison we have plotted the limiting solution on the same graph as dashed curves. It is clear that the limiting solution captures the main features of the pulse and that the

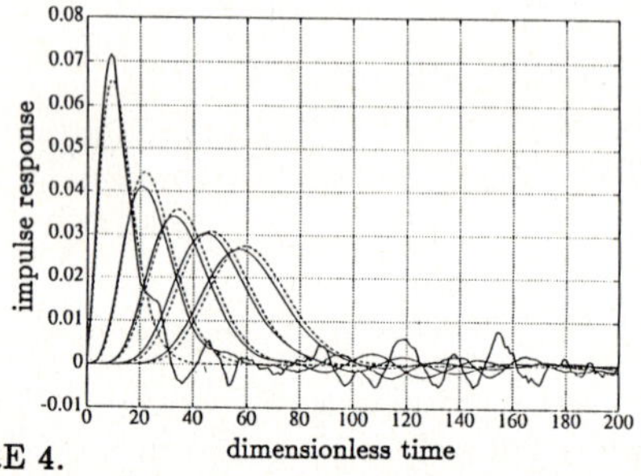

FIGURE 4.
This is the P impulse incident at $\theta_P = 45°$ after traveling through 2,000, 4,000, 6,000, 8,000, and 10,000 layers. The layer matrix code simulation is shown in full line and the limiting solution dashed.

agreement between the two is very good. The two solutions differ in that the simulation contains some high frequency fluctuations, which are most noticeable at the 2,000th interface. Actually, if it were possible to calculate the exact impulse response it would be found to consist of a discrete distribution of impulses (Dirac delta functions) to which the limiting solution approximates only in the weak sense, i.e. after integration. In Figures 1 and Figure 4 $\theta_S = 25.10°$ and $\theta_P = 45°$ correspond to the same horizontal slowness p.

Note that in these cases, as in others shown in [3], the simulated pulse is very closely approximated by the limiting solution.

REFERENCES

[1] R.F. O'Doherty and N.A. Anstey, "Reflections on amplitudes," *Geophysical Propsecting 19*, 430-458 (1971).

[2] R. Burridge, B. White, and G. Papanicolaou, "One-dimensional wave propagation in a highly discontinuous medium", *Wave Motion 10*, 19-44 (1988).

[3] R. Burridge and H.-W. Chang "Multimode, one-dimensional, wave propagation in a highly discontinuous medium", submitted *Wave Motion* (1988).

[4] K. Aki and P.G. Richards, Quantitative Seismology: Vol. I Theory and Methods, W. H. Freeman, San Francisco, (1980).

[5] B. V. Gnedenko, The Theory of Probability, (translated from the Russian by B. D. Seckler) Chelsea Publishing Company, New York, (1963).

ELASTIC WAVE PROPAGATION
M.F. McCarthy, M.A. Hayes, (Editors)

WAVE PROPAGATION FROM A POINT-SOURCE IN A WEDGE-SHAPED LAYER ON TOP OF A HALF-SPACE

Yih-Hsing PAO
Department of Theoretical and Applied Mechanics
Cornell University
Ithaca, N.Y., U.S.A.

Franz ZIEGLER
Civil Engineering Department
Technical University of Vienna
Vienna, Austria

The generalized ray integrals for the propagation of spherical acoustic waves from a point source to a (fixed) receiving point, following various paths of reflection and refraction are derived. The phase functions in the Weyl-Sommerfeld representation are constructively developed in terms of the apparent local slowness in two coordinate systems, one for each surface. Inversion of the Laplace-integrals requires the simultaneous transformation of time into two planes of complex slowness to render a pair of finite integrals for numerical evaluation.

1. THE SOURCE RAY

In order to apply the generalized ray theory, for parallel layers see e.g. the reviewing article [1], to a sloping fluid layer on top of a half-space, the scalar wave solution in Cartesian coordinates for a point source at $(x_0=0, y_0=0, z_0)$ in an infinite fluid space is needed. The Laplace transformed velocity potential $\bar{\phi}_0(\underline{x},s)$ with the relation to the hydrodynamic pressure $\bar{p} = -\rho s \bar{\phi}_0$ is expressed in terms of the two dimensional Fourier-Integrals by

$$\bar{\phi}_0(x,y,z,s) = \frac{a^2 s \bar{f}(s)}{8\pi^2} \int_{-\infty}^{\infty} d\eta \int_{-\infty}^{\infty} S(\xi,\eta) e^{s\{i\eta y + g(x,z;\xi,\eta)\}} d\xi \qquad (1)$$

The variables ξ and η are apparent wave slowness components along the x,y-directions, respectively, and $\zeta = (a^2+\xi^2+\eta^2)^{1/2}$ denotes the apparent slowness along the z-direction. Instead of the sound speed c, the reciprocal $a = c^{-1}$ is used. The emmittance function, $S(\xi,\eta) = 1/\zeta$, in case of a point of explosion, and f(t) denotes the time signature of the source function. In primed coordinates (x',y',z') which are rotated from the unprimed system about the y-axis through the (dipping) angle α, Fig. 1, the same velocity potential becomes

$$\bar{\phi}_0'(x',y,z',s) = \frac{a^2 s \bar{f}(s)}{8\pi^2} \int_{-\infty}^{\infty} d\eta \int_{-\infty}^{\infty} S'(\xi',\eta) e^{s\{i\eta y + g'(x',z';\xi',\eta)\}} d\xi' \qquad (2)$$

Considering one and the same infinitesimal plane wave in Eqs. (1) and (2) requires, see [2,3,4] for details with respect to cylindrical waves,

$$S(\xi,\eta)d\xi = S'(\xi',\eta)d\xi' = d\xi'/\zeta', \tag{3}$$

and the invariance of phase under coordinate rotation [5],

$$g(x,z;\xi,\eta)=i\xi x-\zeta|z-z_0|=g'(x'z';\xi',\eta)=i\xi'(x'-x_0')-\zeta'|z'-z_0'|, \tag{4}$$

respectively. Noting the unchanged η-slowness, the transformation of the two-dimensional (projected) source ray-slowness vector $\underline{\xi}^T=(i\xi,\zeta)$ is derived

$$\underline{\xi}'^T = (i\xi',\zeta') = \underline{\xi}^T\cdot\underline{D}, \qquad \underline{D} = \begin{Bmatrix} \cos\alpha & \sin\alpha \\ -\sin\alpha & \cos\alpha \end{Bmatrix} . \tag{5}$$

Inversion of the Laplace transform (1) or (2) gives the well-known spherical wave solution, see [6]

$$\phi_0(\underline{x},t) = a^2 f(t-aR)/4\pi R,. \quad R^2= x^2+ y^2+(z-z_0)^2, \tag{6}$$

which is valid also for waves inside the layer prior to the arrival of the waves reflected from either the horizontal or the inclined plane, Fig. 1.

2. INTEGRAL REPRESENTATION OF MULTI-REFLECTED RAYS

A key point in the derivation of the generalized ray integrals of spherical waves undergoing multiple reflections before arrival at a receiving station (x,y,z) within the fluid (source) layer is that the projection of the three dimensional ray path onto the plane y=0 is the same as the ray path from a line source, Fig. 1. Thus, following [1] or Kennett [7] in principle, the velocity potential of a partial wave, numbered ±k according to the number of successive reflections, the plus sign indicates a ray which is initiated downward (+ z-direction), the minus applies to a ray initiated upward from the source point undergoing (k+1) successive reflections, is expressed according to Eq. (1) by

$$\bar{\phi}_{\pm k}(\underline{x},s) = \frac{a^2 s\,\bar{f}(s)}{8\pi^2}\int_{-\infty}^{\infty} d\eta \int_{-\infty}^{\infty} S(\xi,\eta)\Pi_{\pm k}(\xi,\eta)e^{s\{i\eta y+g_{\pm k}(x,z;\xi,\eta)\}}\,d\xi \tag{7}$$

The change of amplitude is determined from reflection coefficients $R_{(j)}$, for plane waves at the free surface (j=1), thus, approximating the "air-water" interface as a pressure release surface, or at the interface (j=2). The product of k or (k+1) reflection coefficients, all to be expressed in local slowness form the factor $\Pi_{\pm k}$. The change of the wave normal is defined by the phase functions, essentially $g_{\pm k}$.

Considering a single reflection at the interface (2) first: Snell's law requires the angle of incidence of the source ray pointing downward to be equal to the angle of the emittance of the reflected ray when measured in the common plane including the normal to the interface which is parallel z'. Projection onto the plane y = 0 renders an affine geometry of the ray projections, i.e., the projected angle of incidence remains equal to the projected angle of emittance, Fig. 1. A second successive reflection at the free surface renders a pair of rays in a plane which contains the normal parallel to the z-axis, and where Snell's law is applicable. Again, projection on the plane y = 0 preserves the projected

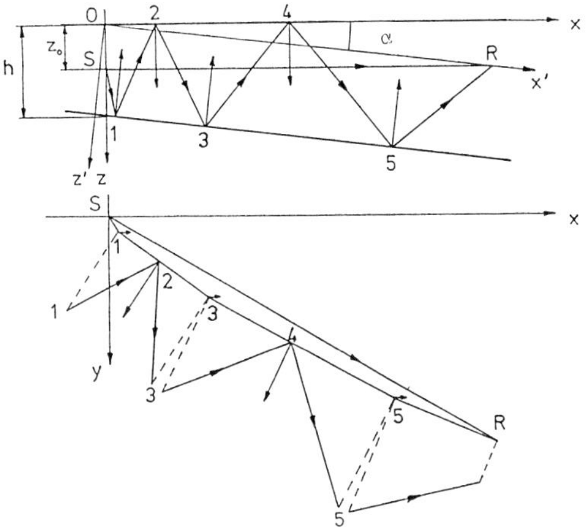

FIGURE 1
Direct ray path from S to R projected onto the planes y=0 and z=0, respectively: Source ray, straight. Curved ray with 5 reflections.

angle of incidence which is changed against the prior one by the dipping angle α according to the rotated normal vectors. The projected angles when measured in the other two coordinate planes, however, are unequal. From [2] the transformation of primed local slowness vectors is given by

$$\underline{\xi}'^T_k = \underline{\xi}'^T_{k-1} = \underline{\xi}^T \underline{D}^k \; \ldots \; k \text{ (odd)} \qquad \underline{D}^k = \begin{Bmatrix} \cos k\alpha & \sin k\alpha \\ -\sin k\alpha & \cos k\alpha \end{Bmatrix} \tag{8}$$
$$\underline{\xi}^T_k = \underline{\xi}^T_{k-1} = \underline{\xi}^T \underline{D}^k \; \ldots \; k \text{ (even)}$$

and, thus, the essential part of the phase functions are expressed in the source-ray slowness components ξ, ζ through

$$g_{\pm k}(x,z;\xi,\eta) = \underline{\xi}^T(-\underline{x}_{\pm 0} + \underline{D}^n \underline{x}), \qquad \underline{x}^T_{\pm 0} = (x_0, \mp z_0), \tag{9}$$
$$\underline{x}^T = (x+x_0, -\varepsilon z), \quad \varepsilon = \pm 1 \text{ is the direction factor.}$$

Above, x_0 is the epicentral tip-distance and n=k (even) or n=(k+1) for odd k. The product of reflection coefficients becomes a function of source ray slowness by substitution of Eq. (8) into,

$$\Pi_{\pm k} = (-1)^m \; R_{(2)}(\xi'_1, \eta'_1) R_{(2)}(\xi'_3, \eta'_3) \; \ldots \; R_{(2)}(\xi'_\kappa, \eta'_\kappa) \quad , \tag{10}$$

obviously, $\Pi_{\pm 0} = \pm 1$, and for k = odd, $\kappa = k$, for +k, m = (k-1)/2, for -k, m = (k+1)/2. For k = even, $\kappa = k-1$, for +k, m = k/2, for -k, m = (k+2)/2.

Final integrations are somewhat simplified when performed in (ξ',η) for +k rays and in (ξ,η) for -k, respectively. For convenience of subsequent derivations the $\pm$ sign and the primes are dropped and

$$E_k(\xi,\eta) = S\Pi_k \tag{11}$$

denotes the amplitude function which is even in both ξ and η variables.

3. INVERSE LAPLACE TRANSFORM

3.1. Transformation of the Wave Slowness

Denoting the x-component of $(-\underline{x}_0 + \underline{D}^n\underline{x})$ by X_k (in source ray slowness image plane, y=0), the pair of coordinates (X_k, y) is changed to (r_k, θ_k) through

$$X_k = r_k \cos\theta_k \quad , \; y = r_k \sin\theta_k \; . \tag{12}$$

Simultaneously, the slowness (ξ, η) is changed to (p,q) by

$$(\xi, \eta) = (p,q)\,\underline{D}(\theta_k), \tag{13}$$

and the ray integral (7) is changed to

$$\bar{\phi}_k = \frac{a^2 s\,\bar{f}(s)}{2\pi^2}\,\mathrm{Re}\int_0^\infty dq \int_0^\infty E_k(q,p)\,e^{s[ipr_k - \zeta Z_k]}dp, \tag{14}$$

Z_k is the z-component of $(-\underline{x}_0 + D^n\underline{x})$.

3.2. The Cagniard-de Hoop Method

Assuming time t real and positive in the mapping

$$-t = ipr_k - \zeta Z_k \tag{15}$$

renders the integration along one branch of hyperbola, Γ, a complex contour in the p-plane [6],

$$p(q,t) = \{itr_k + [t^2 - \frac{t_M^2}{a^2}(a^2+q^2)]^{1/2} Z_k\}\, a^2/t_M^2 \;, \; t_k < t < \infty,$$

$$t_k = t|_{p=0} = Z_k(a^2 + q^2)^{1/2} \tag{16}$$

The arrival time t_M is determined from the stationary value of t(q,p), where

$$\frac{\partial t}{\partial q} = 0, \; \frac{\partial t}{\partial p} = 0, \text{ with roots } q_M = 0, \; p_M = iar_k/(r^2 + Z_k^2)^{1/2}, \tag{17}$$

thus,

$$t_M = a(X_k^2 + y^2 + Z_k^2)^{1/2} \; . \tag{18}$$

In terms of the new variable t, Eq. (14) gives the inverse Laplace transform by "inspection" after interchanging integration, see Fig. 2, for $\bar{f}(s) = 1/s$. The general response is the time-convolution integral

$$\phi_k(t) = \dot{f}(t) * \phi_k^H(t), \tag{19}$$

where, upon elimination of p to render $\zeta(q,t)$ and upon substitution of $(dp/dt)|_q$ the final expression is

$$\phi_k^H(t) = \frac{a^2}{2\pi^2} H(t - aZ_k)\,\mathrm{Re}\int_0^{q_k(t)} \frac{E_k(q,t)\,\zeta(q,t)}{[t^2 - t_M^2(q^2+a^2)/a^2]^{1/2}}\,dq, \tag{20}$$

$$q_k(t)=(t^2/Z_k^2-a^2)^{1/2}>0,\quad \zeta(q,t) = \{tZ_k+i[t^2-(a^2+q^2)\frac{t_M^2}{a^2}]^{1/2} r_k\}\, a^2/t_M^2 \quad (21)$$

Further discussions and numerical results are included in a forthcoming paper [8], Fig. 3 shows the influence of the dipping angle α.

4. CONCLUSIONS AND OUTLOOK

The resulting hydrodynamic pressure contains all the contributions from body waves (direct rays), surface- and, in case of a fast bottom, head waves (refracted rays). The waves propagating in the penetrable wedge are to be compared to the solution assuming both boundaries perfectly reflecting, see Buckingham [9]. Extension of the solution to the solid elastic source layer with inclusion of wave mode conversion during reflection is visible, since the projection of the propagation onto the y-plane can be analyzed according to the two-dimensional case [10], [11].

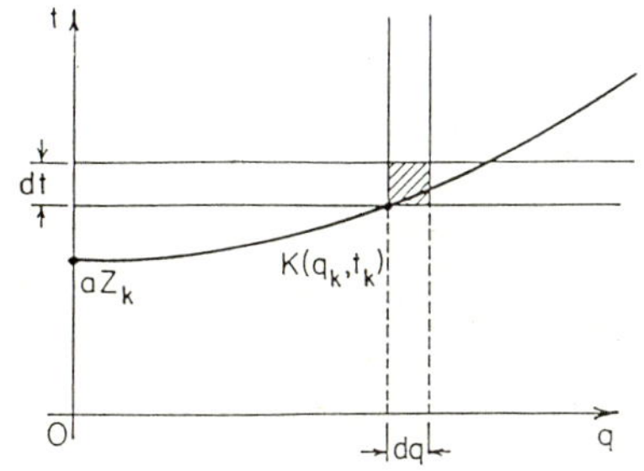

FIGURE 2

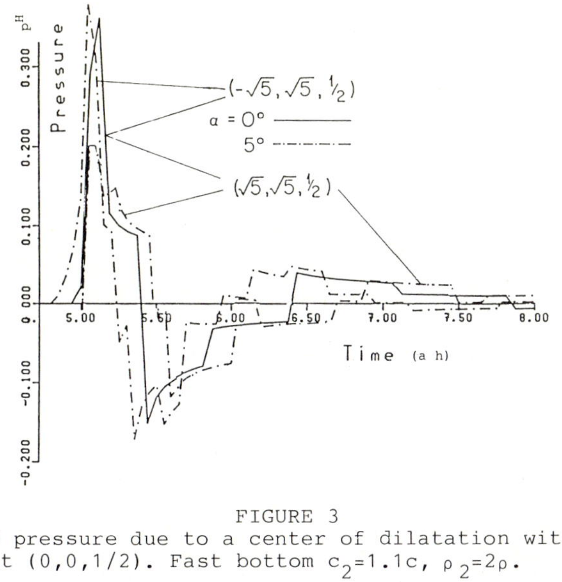

FIGURE 3

Normalized pressure due to a center of dilatation with a step time function at (0,0,1/2). Fast bottom $c_2=1.1c$, $\rho_2=2\rho$.

ACKNOWLEDGEMENT

Support at the early stages of this research by a grant from the National Science Foundation (CEE 8206758) and the support of F.Z. by the Austrian FWF, central project S30-01, are gratefully acknowledged.

REFERENCES

[1] Pao, Y.-H. and Gajewski, R., The Generalized Ray Theory and Transient Responses of Layered Elastic Solids, in Mason, P. and Thurston, R.N., (eds.), Physical Acoustics 13 (Academic Press, New York, 1977) 184-266.

[2] Pao, Y.-H. and Ziegler, F., Geophys. J.R. astr. Soc. 71 (1982) 57-77.

[3] Ziegler, F. and Pao, Y.-H., Wave Motion 7 (1985) 1-24.

[4] Ziegler, F., Pao, Y.-H. and Wang, Y.-S., Acta Mechanica 56 (1985) 1-15.

[5] Hong, T.L. and Helmberger, D.V., Bull. Seismolog. Soc. America 67 (1977) 995-1008.

[6] de Hoop, A.T., Applied Scientific Research (B) 8 (1960) 349-356.

[7] Kennett, B.L.N., Seismic Wave Propagation in Stratified Media (Cambridge University Press, Cambridge 1983).

[8] Pao, Y.-H., Ziegler, F. and Wang Y.-S., J. Acoustical Soc. America (forthcoming).

[9] Buckingham, M.J., Acoustic Propagation in a Wedge-Shaped Ocean with Perfectly Reflecting Boundaries, in: Felsen, L.B. (ed.), Hybrid Formulation of Wave Propagation and Scattering (Martinus Nijhoff, Dordrecht, 1984) 77-105.

[10] Ziegler, F. and Pao, Y.-H., Acta Mechanica 52 (1984) 133-163.

[11] Borejko, P. and Ziegler, F., this volume.

ELASTIC WAVE PROPAGATION
M.F. McCarthy, M.A. Hayes, (Editors)

ELASTIC GUIDED WAVES IN HETEROGENEOUS MEDIA - MATHEMATICAL ANALYSIS

A. Bamberger (1), Y. Dermenjian (2), P. Joly (3)

(1) 2,4 Avenue de Bois Préau, 92500 RUEIL-MALMAISON
(2) Université Paris-Nord - CSP - VILLETANEUSE
(3) INRIA - Domaine de Voluceau, 78153 LE CHESNAY Cédex

1. INTRODUCTION

In this paper we present some results concerning the theory of existence and behavior of guided waves in an elastic isotropic heterogeneous medium occupying the whole space $\mathbb{R}^3$ in which the coefficients are supposed to depend only on two space coordinates. These results are obtained by using the abstract spectral theory of self adjoint operators in Hilbert spaces ([6],[8]) and more specifically the well known Max Min principle that we shall recall in section. This approach has been already successfully applied to various problems of guided waves for instance in acoustic stratified media [9] or in optical fibers ([2],[3]) In a recent paper, we dealt with the question of elastic surface waves guided by the free boundary of a cylindrical cavity of arbitrary cross section [4] .
Our article is presented as follows : in section 2, we briefly present the equations of the problem and the main physical assumptions. We also set up the mathematical framework and show that the existence of guided waves is equivalent to an eigenvalue problem for a self adjoint operator. Sections 3 and 4 are devoted to the spectral theory of this operator. In section 3 we study the essential spectrum and obtain preliminary results for the study of the discrete spectrum described in section 4. Our main theorems are stated in this section.

2. PRESENTATION OF THE PROBLEM - MATHEMATICAL FORMULATION

The context is the classical context of elastodynamics equations in isotropic media ([1],[7]). We denote by $\rho(\underline{x})$ the density of the material and by $\lambda(\underline{x})$ and $\mu(\underline{x})$, its Lame's coefficients ($\underline{x}=(x_1,x_2,x_3) \in \mathbb{R}^3$). Our main assumption is that the functions ρ,λ,μ do no depend on x_3 :

(1.1) $$\rho(\underline{x}) = \rho(x_1,x_2) \quad \lambda(\underline{x}) = \lambda(x_1,x_2) \quad \mu(\underline{x}) = \mu(x_1,x_2) \quad , x = (x_1,x_2)$$

and that in each cross section x_3 = cste, they are constant at the exterior of a disc :

(1.2) $$\rho(x_1,x_2) = \rho_\infty \qquad \lambda(x_1,x_2) = \lambda_\infty \qquad \mu(x_1,x_2) = \mu_\infty \quad \text{for } |x| > R$$

More over we suppose that :

(1.3) $$\begin{cases} \rho_- = \inf \rho(x) > 0 & \rho_+ = \sup \rho(x) < +\infty \\ \mu_- = \inf \mu(x) > 0 & \mu_+ = \sup \mu(x) < +\infty \\ \lambda_- = \inf \lambda(x) > 0 & \lambda_+ = \sup \lambda(x) < +\infty \end{cases}$$

We are looking for particular solutions of the elastodynamics equations in the form of harmonic waves propagating in the direction x_3 :

(1.4) $$u(\underline{x},t) = u(x_1,x_2) \ \exp\ i(\omega t-\beta x_3)$$

where $u = (u_1,u_2,u_3)$ denotes the displacement field. By definition, such a solution is a guided wave if and only if :

(1.5) $$\int_{\mathbb{R}^2} |u(x)|^2 \, dx < +\infty$$

which means in practice that its energy is concentrated in the cylinder $|x| < R$. By definition :

. ω is the pulsation of the wave
. β is the wave number
. $c = \omega/\beta$ is the phase velocity

Plugging (1.4) in the equations of the problem leads to the equation :

(1.6) $$\begin{cases} A(\beta)u = \omega^2 u \\ u \in L^2(\mathbb{R}^2)^3 \end{cases}$$

where $A(\beta)$ is the differential operator defined by :

(1.7) $$A(\beta)u = -1/\rho \operatorname{div}(\sigma^\beta(u))$$

the stress tensor $\sigma^\beta(u)$ being given by :

(1.8) $$\begin{cases} \sigma^\beta_{ij}(u) = \lambda \operatorname{Tr}(\varepsilon^\beta(u))\delta_{ij} + 2\mu\, \varepsilon^\beta_{ij}(u) & \\ \varepsilon^\beta_{ij}(u) = 1/2((\partial u_i/\partial x_j) + (\partial u_j/\partial x_i)) & 1 \le i,j \le 2 \\ \varepsilon^\beta_{i3}(u) = 1/2((\partial u_3/\partial x_j) + i\beta u_j)) & 1 \le j \le 2 \\ \varepsilon^\beta_{33}(u) = i\beta u_3 & \end{cases}$$

One easily shows that $A(\beta)$ is a self adjoint operator in the Hilbert space $H = L^2(\mathbb{R}^2)^3$ with the inner product :

(1.9) $$(u,v) = \int \rho\, u.v\, dx$$

with domain :

$$D(A(\beta)) = \{u \in V = H^1(\mathbb{R}^2)^3 / A(\beta)u \in H\}$$

Therefore, equation (1.6) shows that (1.4) is a guided wave if and only if ω^2 is an eigenvalue of the operator $A(\beta)$ (β then appears as a parameter), the displacement field u being the corresponding eigenfunction. Note that, as the propagation domain is unbounded, the resolvent of $A(\beta)$ is not compact so that the question of the existence of eigenvalues is not obvious.

3. SOME PRELIMINARY RESULTS ABOUT THE SPECTRUM OF A(β)

Let us recall that the spectrum of $A(\beta)$ is defined by :

$$\sigma(A(\beta)) = \{\lambda \in \mathbb{R} \,/\, A(\beta)-\lambda I \text{ has no bounded inverse}\}$$

The point spectrum is the set of the eigenvalues of $A(\beta)$, namely :

$$\sigma_p(A(\beta)) = \{\lambda \in \mathbb{R} \,/\, \exists u \in H, u \neq 0 \,/\, Au = \lambda u\}$$

The discrete spectrum is the subset of finite multiplicity eigenvalues which are isolated in $\sigma(A(\beta))$.

$$\sigma_d(A(\beta)) = \{\lambda \in \sigma_p(A(\beta)) \,/\, \lambda \text{ has finite multipliciy an is isolated in } \sigma(A(\beta))\}$$

Finally, the essential spectrum of $A(\beta)$ is given by :

$$\sigma_{ess}(A(\beta)) = \sigma(A(\beta)) \setminus \sigma_d(A(\beta))$$

As we shall see in section 4, the essential spectrum plays a very important role in the analysis, because of the Max-Min principle.
The main result of the section is the following :

Theorem 1 :
(i) $\sigma(A(\beta)) \subset [\, \beta^2\mu_-/\rho_+, +\infty[$
(ii) $\sigma_{ess}(A(\beta)) = [\, \beta^2\mu_\infty/\rho_\infty, +\infty[$

Point (i) of the theorem is due to a generalized Korn's inequality. Point (ii) is obtained via a compact perturbation technique, the reference being given by the case of constant coefficients for which the theory is well known.
From theorem 1, we can deduce the a priori structure of $\sigma(A(\beta))$ as illustrated on the picture below :

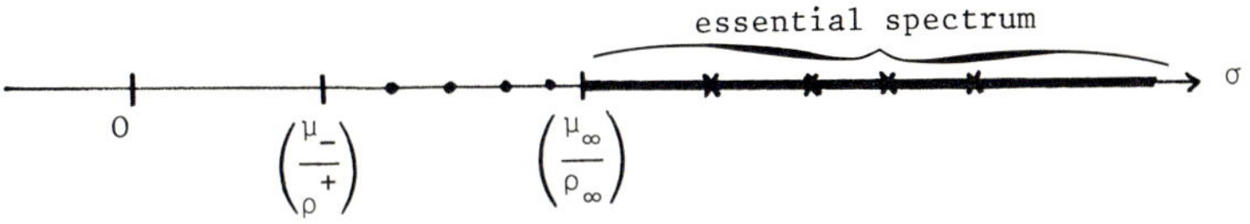

x : isolated eigenvlaues (discrete spectrum)
0 : eigenvalues embedded in the essential spectrum.

The discrete spectrum can be characterized by the Max-Min principle (see section 4). One can think that there is no eigenvalue (in any case) in the essential spectrum : this would mean physically that there can not exist a guided wave whose phase velocity is greater than the shear wave velocity in the exterior medium. Unfortunately this property remains a conjecture. Nevertheless we have some results in that direction that we can summarize in the :

<u>Theorem 2</u> :

(i) Assume that $A(\beta)u = \lambda u$, $u \neq 0$, $\lambda > \beta^2 \mu_\infty/\rho_\infty$ then λ has finite multiplicity. If $\lambda > \beta^2(\lambda_\infty+2\mu_\infty)/\rho_\infty$, u has compact support ; if $\beta^2 \mu_\infty/\rho_\infty < \lambda < \beta^2(\lambda_\infty+2\mu_\infty)/\rho_\infty$, $\partial u_3/\partial x_1 - i\beta u_1$ and $\partial u_3/\partial x_2 - i\beta u_2$ have compact support.

(ii) The only (possible) accumulation points of eigenvalues of $A(\beta)$ are $\beta^2 (\lambda_\infty+2\mu_\infty)/\rho_\infty$, and $+\infty$

(iii) When $\rho(x)$, $\lambda(x)$ and $\mu(x)$ are piecewise constant functions, $A(\beta)$ has no eigenvalue in the interval $]\beta^2(\lambda_\infty+2\mu_\infty)/\rho_\infty, +\infty[$.

4. STUDY OF THE DISCRET SPECTRUM OF A(β)

4.1. The Max-Min principle and its application

The principle characterizes the eigenvalues of $A(\beta)$ which are strictly smaller than the lower bound of the essential spectrum, i.e. in our case than $\beta^2\mu_\infty/\rho_\infty$. From a physical point of view, it will characterize the guided waves whose phase velocity is strictly smaller than the shear wave velocity at infinity.
Let us introduce the bilinear form $a(\beta)$ associated with the opertor $A(\beta)$.

(4.1) $$a(\beta;u,v) = (A(\beta)u,v) \quad (u,v) \in D(A(\beta))\times V$$

By Green's formula, we have :

(4.2) $$a(\beta;u,v) = \int_{\mathbb{R}^2} \sigma^\beta_{ij}(u)\, \overline{\varepsilon^\beta_{ij}(v)}\, dx$$

which proves that $a(\beta)$ extends to VxV. By definition the Rayleigh quotient of any test function u in V ($u \neq 0$) is given by :

(4.3) $$R(\beta;u) = a(\beta;u,u)/|u|^2$$

Now let us consider the sequence $S_m(\beta)$, $m \in \mathbb{N}^*$, of real numbers giuven by one of these two equivalent formulas.

(4.4) $$\begin{aligned} S_m(\beta) &= \max_{V_{m-1} \in \mathcal{V}_{m-1}} \ \min_{u \in V^\perp_{m-1}} R(\beta;u) \\ &= \min_{V_m \in \mathcal{V}_m} \ \max_{u \in V_m} R(\beta;u) \end{aligned}$$

where $\mathcal{V}_m$ denotes the set of m-dimensional subspaces of V and $V^\perp_m = (v \in H/(u,v)=0, \forall v \in V_m)$. A general fact is that $S_m(\beta)$ is non decreasing and tends to $\beta^2 \mu_\infty/\rho_\infty$:

(4.5) $S_m(\beta) \nearrow \beta^2 \mu_\infty/\rho_\infty$

Moreover, for each m≥1, one has the following alternative :

(i) $S_m(\beta) < \beta^2 \mu_\infty/\rho_\infty$

In that case, A(β) has at least m eigenvalues (including multiplicity) smaller than $\beta^2 \mu_\infty/\mu_\infty$ which are $S_1(\beta) \leq S_2(\beta) \leq ... \leq S_m(\beta)$

(ii) $S_m(\beta) = \beta^2 \mu_\infty/\rho_\infty$

In that case, A(β) has at most m-1 eigenvalues smaller than $\beta^2 \mu_\infty/\rho_\infty$.

The method consists then to study the sequence $S_m(\beta)$ and to study what case of the previous alternative holds. For example, to prove the existence of mth eigenvalue, it suffices to construct an m-dimensional subspace V_m of test functions such than :

$$\forall u \in V_m \quad a(\beta;u,u) - \beta^2 |u|^2 \mu_\infty/\rho_\infty < 0$$

Then the Min-Max characterization of $S_m(\beta)$ shows that $S_m(\beta) < \beta^2 \rho_\infty/\mu_\infty$.

We shall give in the next sections the results obtained by applying this method. For the complete proofs we refer the reader to [5] . Of course as β appears as a parameter in this analysis, the existence result will depend on β.

4.2. The two main existence theorems

We are now interested in the discrete spectrum of A(β). Let us note that this discrete spectrum can be empty. Indeed, from theorem we see that :

$$\left.\begin{array}{l} \mu(x) \geq \mu_- \text{ everywhere} \\ \rho(x) \leq \rho_- \text{ everywhere} \end{array}\right\} \Rightarrow \sigma_d(A(\beta)) \text{ is empty}$$

To state our first result, we introduce :

(4.5) $(\mu/\rho)_- = \inf (\mu(x)/\rho(x))$

and make the assumption :

(4.6) $(\mu/\rho)_- < (\mu_\infty/\rho_\infty)$

which means that the shear wave velocity presents a strict global minimum. Then we have :

<u>Theorem 3</u> :
When (4.6) holds, for each m≥1, there exists $\beta^*_m \geq 0$ such that for $\beta \geq \beta^*_m$, there exists at least m guided waves associated with the wave number β whose pulsations are $\omega^2 = S_j(\beta)$, $1 \leq j \leq m$.

<u>Remarks</u>
. The curves $\beta \rightarrow \omega = S_j(\beta)^{1/2}$ are the dispersion curves of the modes. Theorem 3 shows that there exists a countable set of such curves.
. It is interesting to note that the function λ(x) does not appear in the condition (4.6).

It is natural to wonder whether the condition (4.6) is necessary for the existence of eigenmodes or not. Our second existence result will proof that it is selves to the following particular case :

$$(4.7) \quad \left\{\begin{array}{ll} \lambda(x) = \lambda_0 \;\; \mu(x) = \mu_0 \;\; \rho(x) = \rho_0 & x \in \mathcal{O} \quad (\mathcal{O} \subset \mathbb{R}^2, \text{bounded}) \\ \lambda(x) = \lambda_\infty \;\; \mu(x) = \mu_\infty \;\; \rho(x) = \rho_\infty & x \in \mathbb{R}^2 \setminus \mathcal{O} \end{array}\right.$$

We shall denote by E_S the set of couples $((\rho_1,\lambda_1,\mu_1), (\rho_2,\lambda_2,\mu_2))$ for which there exists a Stoneley wave for a medium made up of two homogeneous half spaces of respective coefficients (ρ_1,λ_1,μ_1) and (ρ_2,λ_2,μ_2). We make the assumption :

(4.8) $\quad ((\lambda_0,\rho_0,\mu_0),(\lambda_\infty,\rho_\infty,\mu_\infty)) \in E_S$

We can state the :

Theorem 4 :
For the medium (4.7), if the condition (4.8) holds, for each $m \geq 1$, there exists $\beta^*_m \geq 0$ such that, for $\beta \geq \beta^*_m$, there exists at least m guided waves associated with the wave number β whose pulsations are $\omega^2 = S_j(\beta)$, $1 \leq j \leq m$.

Remarks :
. Such waves can be considered as generalized Stoneley waves. They are guided by the interface between the two media.
. A generalisation of theorem 4 is given in [5]
. As the set E_S has obviously the property :

$$((\rho_1,\lambda_1,\mu_1), (\rho_2,\lambda_2,\mu_2)) \in E_S \Rightarrow ((\rho_2,\lambda_2,\mu_2),(\rho_1,\lambda_1,\mu_1)) \in E_S$$

it is easy to construct a medium (4.7) for which (4.8) is true although (4.6) is not. Conversely, there exist media for which (4.6) is true and (4.8) is not. This proves that none of these two assumptions is necessary for the existence of guided waves.

4.3. The thresholds

In each of our two main theorems, the existence of an mth eigenmode is ensured if β exceeds a certain threshold or cut off wave number β^*_m. This leads us to define for any $m \in \mathbb{N}^*$.

$$(4.9) \quad \begin{cases} \beta^*_m = \inf \{\beta \in \mathbb{R} \,/\, S_m(\beta) < \beta^2 \mu_\infty/\mu_\infty\} \\ \beta^0_m = \sup \{\beta \in \mathbb{R} \,/\, S_m(\beta) = \beta^2 \mu_\infty/\mu_\infty\} \end{cases}$$

By definition β^*_m is the upper threshold :

- for $\beta \geq \beta^*_m$ the mth eigenmode always exists.

Conversely β^0_m is the upper threshold :

- for $\beta \leq \beta^0_m$ the mth eigenmode does not exist.

For each m, one has $\beta^0_m \leq \beta^*_m$ and, a priori, β^0_m and β^*_m could be distinct as illustrated below :

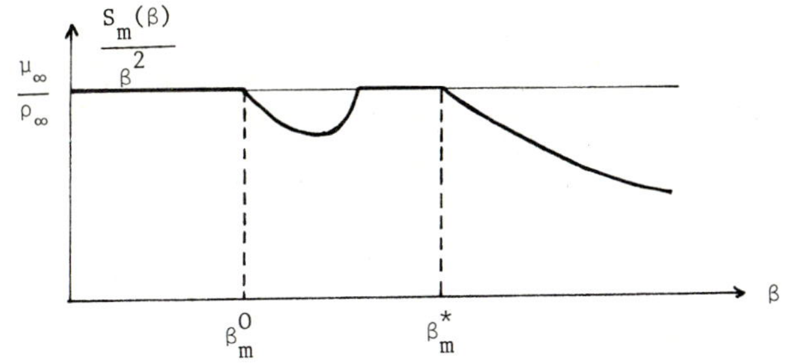

The max min principle allows us to obtain some results about the behaviour of these thresholds.

Theorem 5 :
(i) The sequence β^0_m tends to $+\infty$. More precisely there exists a positive constant C_1 such that :

$$\beta^0_m \geq C_1(m-3)^{1/2}$$

(ii) Under the assumption (4.6) ($(\mu/\rho)^- < (\mu_\infty/\rho_\infty)$), there exists a posititive constant C_2 such that :

$$\beta^*_m \leq C_2 \cdot m^{1/2}$$

(iii) Assuming that :

$$(4.10) \qquad \int_{\mathbb{R}^2} \rho\,(\mu/\rho - \mu_\infty/\rho_\infty)\, dx < 0$$

then $\beta^*_1 = \beta^*_2 = 0$

(iv) If we suppose that :

$$(4.11) \qquad (\mu/\rho)(x) < (\mu_\infty/\rho_\infty) \quad \text{everywhere}$$

then for each $m \geq 1$, $\beta^0_m = \beta^*_m$

Remarks and comments :

. An important consequence of (i) is that for each β the number $N(\beta)$ of corresponding guided waves is finite. It increases with β. In particular for $\beta \geq \beta^*_m$, $N(\beta) \geq m$.

. The result (iii) is interesting since it gives an example for which two guided waves exist for any value of the wave number β. Conversely one can find some conditions for which there exists a countable family of guided waves with $\beta^0_1 > 0$ (see [5]).

. (i) shows that in any case $\beta^0_4 > 0$. The question of β^0_3 remains open. Nevertheless one can show ([5]) the existence of media for which :

$$\beta^0_1 = \beta^0_2 = 0 \text{ and } \beta^0_3 > 0$$

We don't know if the strict positivity of β^0_3 is a general result.

. We illustrate in the picture below that assumptions (4.6), (4.10) and (4.11) are in order of increasing strength.

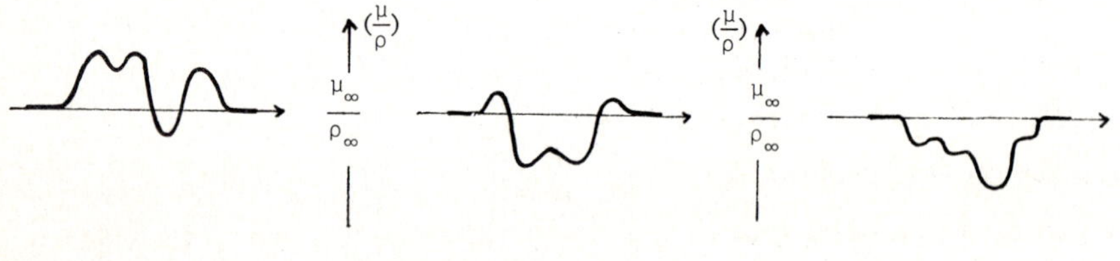

We still notice that the function $\lambda(x)$ does not appear in these conditions.

REFERENCES

[1] ACHENBACH J.D. : "Wave propagation in elastic solids", Noth Holland (1978).

[2] BAMBERGER A., BONNET A.S., DJELLOULI : "Calcul des modes guidés d'une fibre optique. Différentes formulations mathématiques du problème". Ecole Polytechnique. Report n° 142. (1986).

[3] BAMBERGER A., BONNET A.S. : "Calcul des modes guidés d'une fibre optique. Analyse Mathématique". Ecole Polytechnique. Report n° 143. (1986).

[4] BAMBERGER A., JOLY P., KERN M. : "Etude Mathématique des Modes Elastiques Guidés par l'extérieur d'une Cavité cylindrique de section arbitraire". INRIA Report n° 650. (1986).

[5] BAMBERGER A., DERMENJIAN Y., JOLY P. : "Ondes élastiques guidées en milieu hétérogène : analyse mathématique". INRIA Report. To appear.

[6] KATO T. : "Perturbation theory for linear operators". Springer. (1966).

[7] MIKLOWITZ J. : "The theory of elastic waves and waveguides". North-Holland. (1980).

[8] REED M., SIMON B. : "Methods of modern mathematical physics". Vol.I and IV, Academic Press. (1981).

[9] WILCOX C.H. : "Sound propagation in stratified media". Applied Mathematical Sciences, Vol. 50, Springer, (1984).

ELASTIC WAVE PROPAGATION
M.F. McCarthy, M.A. Hayes, (Editors)

SCATTERING BY FLUID-LOADED CO-PLANAR HALF-PLANE PLATES

P. R. Brazier-Smith

Topexpress Limited,
13-14 Round Church Street,
Cambridge, CB5 8AD.

1. INTRODUCTION

The scattering of fluid-coupled surface waves at the join of two co-planar half-plane plates has been studied by Brazier-Smith (1987). This is one of a class of mixed boundary value problems amenable to solution by methods based on Wiener-Hopf factorisation. Although the Wiener-Hopf method is not new, the number of solved mixed boundary problems involving acoustic fluid-loading of elastic structures is small, further examples being referenced in the reference above. We consider here the extension of the work referenced above to include insonification of the co-planar plate system by a plane wave. The solution of this problem may then be used to deduce the scattering of any incident sound field in the fluid incident on the plates.

2. SYSTEM RESPONSE TO FLUID COUPLED INCIDENT WAVES

The system is described with reference to a cartesian coordinate system (x,y,z). The half-plane, z=0; x<0 contains a thin plate of bending rigidity, B_-, and surface mass density, σ_-. The half-plane z=0; x>0 contains another thin plate of different bending rigidity, B_+, and surface mass density, σ_+. Subsequent reference to these plates will be as left hand and right hand plates respectively. The use of subscripts '-' and '+' is intended to indicate parameters associated with the left hand and right hand sides respectively. The semi-infinite space, z>0, is occupied by a fluid of density, ρ, and sound speed, c. The other side (z<0) is a vacuum.

All field variables are considered as time harmonic with convention, $e^{-i\omega t}$ and the convention adopted for the spatial Fourier transform pair in x is:

$$\tilde{\phi}(s) = \int_{-\infty}^{\infty} \phi(x)e^{isx}dx;\phi(x) = \frac{1}{2\pi}\int_{-\infty}^{\infty} \tilde{\phi}(s)e^{-isx}ds \qquad (2.1)$$

the tilde indicating quantities transformed into the wavenumber domain.

Consider now an incident acoustic wave of the form:

$$P = P_0(e^{ik_x x - k_z z}) \tag{2.2}$$

where k_x, k_z are positive and satisfy

$$k_\circ = \surd(k_x^2 + k_z^2); \; k_0 = \omega/c \text{ is the acoustic wavenumber.}$$

If this field were incident on an infinite full-plane plate of the same properties as the left hand plate, then the resulting net field would be

$$P_{in} = P_\circ\left(Re^{ik_z z} + e^{-ik_z z}\right)e^{ik_x x} \tag{2.3}$$

and the attendant surface normal displacement:

$$\xi_{in} = \frac{ik_z}{\rho\omega^2}P_\circ\left(Re^{ik_z z} - e^{-ik_z z}\right)e^{ik_x x} \tag{2.4}$$

where R is the reflection coefficient given by

$$R = \frac{ik_x(B_- k_x^4 - \sigma_- \omega^2) - \rho\omega^2}{ik_x(B_- k_x^4 - \sigma_- \omega^2) + \rho\omega^2}.$$

Returning to the half-plane plate problem, the total pressure and displacement fields (i.e. incident plus scattered fields) must satisfy the boundary conditions

$$\left(B_- \frac{\partial^4}{\partial x^4} - \sigma_- \omega^2\right)\xi(x) = -P(x,0); \; x<0 \tag{2.5}$$

$$\left(B_+ \frac{\partial^4}{\partial x^4} - \sigma_+ \omega^2\right)\xi(x) = -P(x,0); \; x>0 \tag{2.6}$$

for force, and
$$\rho\omega^2\xi(x) = \frac{\partial P}{\partial z}; \quad -\infty<x<\infty \tag{2.7}$$

for continuity. The point of constructing the solution for the net field over the left hand plate extended to full plane now becomes clear: it automatically satisfies the boundary conditions for $x<0$. Therefore (2.5) is also valid for the *scattered* fields, if we take this net field (eqns (2.3) and (2.4)) as the incident field, as indeed is (2.7) in the range $-\infty<x<0$. Thus the negative half-range Fourier transform (i.e. over the range $-\infty<x<0$)

of (2.3) and (2.7) will exist for the *scattered* field. Now, by making the fluid lossy by giving k_0 a small imaginary part, the positive half-range Fourier transform of (2.6) and (2.7) will also exist. These transforms are

$$(B_- S^4 - \sigma_- \omega^2)\tilde{\xi}_{sc-}(s) + \tilde{P}_{sc-}(s) = Q_-(s) \tag{2.8}$$

$$(B_+ s^4 - \sigma_+ \omega^2)\tilde{\xi}_{sc+}(s) + \tilde{P}_{sc+}(s) = Q_+(s) - i\left[\frac{-ik_z}{\rho\omega^2}\{R-1\}(B_+ k_x^4 - \sigma_+ \omega^2) + R\right]\frac{P_0}{k_x + s} \tag{2.9}$$

where $Q_- = B_-(-\xi_{sc}'''(0-) + is\xi_{sc}''(0-) + s^2\xi_{sc}'(0-) - is^3\xi_{sc}(0-))$

and $Q_+(s) = B_+(\xi_{sc}'''(0+) - is\xi_{sc}''(0+) - s^2\xi_{sc}'(0+) + is^3\xi_{sc}(0+))$

the edge displacement derivatives, denoted by primes, arising as a result of integration by parts at the x=0 limit of the respective half range transforms. Further the subscripts, + and -, attached to the transformed variables $\tilde{\xi}_{sc}$ and $\tilde{P}_{sc}$ refer to the additive Wiener-Hopf decomposition of the full range transformed variables. Note that these transforms include the incident field as a forcing term because the incident field does not satisfy the homogenous boundary condition as it did for the left hand plate. From (2.8), (2.9) and the transforms of (2.5)-(2.7), the analysis follows a similar line as in Brazier-Smith (1987), to give the solution for the scattered fields in transform space as

$$\tilde{\xi}_{sc}(s) = -\frac{1}{J_-(s)}\frac{\Phi}{[(B_+ s^4 - \sigma_+ \omega^2) - \rho\omega^2/\gamma]} P_0; \quad x>0 \tag{2.10}$$

$$\tilde{\xi}_{sc}(s) = -\frac{1}{J_+(s)}\frac{\Phi}{[(B_- s^4 - \sigma_- \omega^2) - \rho\omega^2/\gamma]} P_0; \quad x<0 \tag{2.11}$$

for normal displacement, and

$$\tilde{P}_{sc}(s,z) = \frac{\rho\omega^2\Phi e^{-\gamma z}}{J_-(s)[(B_+ s^4 - \sigma_+ \omega^2)\gamma - \rho\omega^2]} P_0; \quad x>0 \tag{2.12}$$

$$\tilde{P}_{sc}(s,z) = \frac{\rho\omega^2\Phi e^{-\gamma z}}{J_+(s)[(B_- s^4 - \sigma_- \omega^2)\gamma - \rho\omega^2]} P_0; \quad x>0 \tag{2.13}$$

for pressure, where $\gamma = \surd(s^2 - k_0^2)$. The cut from $-k_0$ is taken into the lower half plane and the cut from $+k_0$ into the upper half-plane. The functions, J_+, J_-, are the factorized split of a function, J, such that, in a strip in the s-plane containing the real axis:

$$J = \frac{(B_- s^4 - \sigma_- \omega^2) - \rho\omega^2/\gamma}{(B_+ s^4 - \sigma_+ \omega^2) - \rho\omega^2/\gamma} = \frac{J_-(s)}{J_+(s)} \tag{2.14}$$

with $J_-(s)$ being analytic in the lower half plane, and $J_+(s)$ being analytic in the upper half plane. These analyticity properties, combined with the requirement that (2.14) is satisfied and consideration of the algebraic behaviour of J as $s\to\infty$, uniquely define J_+ and J_- and are central to methods based on Wiener-Hopf factorization.

The function, Φ, is given by:

$$\begin{aligned}\Phi = {} & \frac{J_-(-\kappa)}{4}\left\{Q(-\kappa) + \frac{C}{(k_x-\kappa)}\right\}\left(\left(\frac{s}{\kappa}\right)^3 - \left(\frac{s}{\kappa}\right)^2 + \left(\frac{s}{\kappa}\right) - 1\right) \\ & + \frac{J_-(-i\kappa)}{4}\left\{Q(-i\kappa) + \frac{C}{(k_x-i\kappa)}\right\}\left(i\left(\frac{s}{\kappa}\right)^3 + \left(\frac{s}{\kappa}\right)^2 - i\left(\frac{s}{\kappa}\right) - 1\right) \\ & + \frac{J_+(\kappa)}{4}\left\{ Q(\kappa) + \frac{C}{(k_x+\kappa)}\right\}\left(- \left(\frac{s}{\kappa}\right)^3 - \left(\frac{s}{\kappa}\right)^2 - \left(\frac{s}{\kappa}\right) - 1\right) \\ & + \frac{J_+(i\kappa)}{4}\left\{Q(i\kappa) + \frac{C}{(k_x+i\kappa)}\right\}\left(-i\left(\frac{s}{\kappa}\right)^3 + \left(\frac{s}{\kappa}\right)^2 + i\left(\frac{s}{\kappa}\right) - 1\right) \\ & \frac{-J_-(-k_x)C(s^4-\kappa^4)}{(k_x^4-\kappa^4)(s+k_x)}\end{aligned} \tag{2.15}$$

where

$$\kappa = \left[(\sigma_+ - \sigma_-)\omega^2/(B_+ - B_-)\right]^{\frac{1}{4}}$$

$$C = -i\left[\frac{ik_z}{\rho\omega^2}(R-1)(B_+ k_x^4 - \sigma_+\omega^2) + R\right]P_0$$

and $Q(s) = Q_-(s) + Q_+(s)$.

As in Brazier-Smith (1987), the 8 unknown coefficients of Q, namely the values of that displacement field and its derivatives to third order at x = 0+ and 0- may be found from solution of 4 linear equations arising from the asymptotic behaviour of $\tilde{P}_{sc}(s,z)$ as $s \to \infty$ and a further 4 from the join conditions of the two plates.

Of course the inverse transform of (2.10)-(2.13) cannot be done analytically except in certain asymptotic limits. We will consider one, namely the far field pressure derived from inverse transforming (2.12), (2.13). Relevant features of the integrand in this case is a saddle point at $-k_0\cos\theta$ and a pole at $-k_x$ arising from the surface trace wavenumber of the insonifying wave. Interaction of this pole and saddle point can be worked through analytically the result of which models the reflected field from the RH plate *and* the the behaviour at the 'shadow' boundary. Subtracting off this field, the remnant field representing scattering from the join can be evaluated by method of steepest descent through the saddle point. A typical set of resulting directivity patterns for differing frequencies and a plate thickness ratio of 2 is shown below.

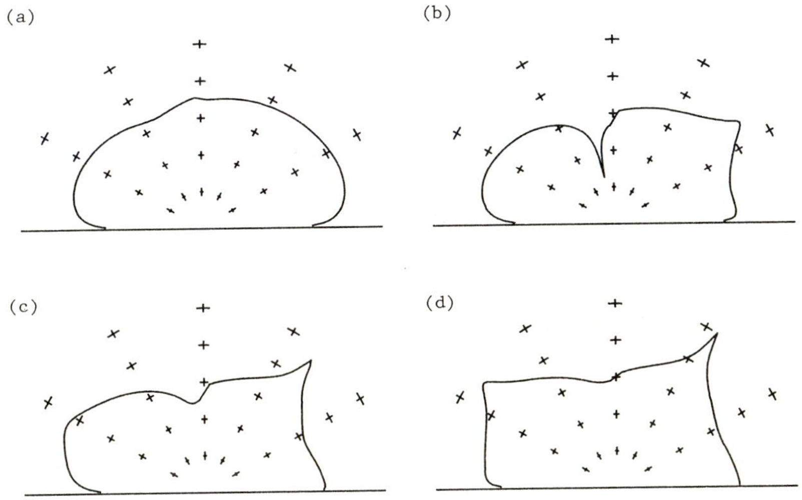

Figure 1. The directivity of the scattered field for a plane wave incident from the left at 30^0 to the normal of a two plate system, thickness ratio of 1:2. The four cases are for frequencies non-dimensionalised on the LH plate coincidence of (a) .4, (b) .8, (c) 1.2 and (d) 1.6.

Reference

Brazier-Smith, P. R., 1987. The acoustic properties of two co-planar half-plane plates. Proc. Roy. Soc., A409, 115-139.

ELASTIC WAVE PROPAGATION
M.F. McCarthy, M.A. Hayes, (Editors)

WAVE PROPAGATION IN THIN ELASTIC BODIES

Chiara Molina and Franco Pastrone

Dipartimento di Ingegneria Strutturale
Politecnico di Milano
Milano - Italy

Dipartimento di Matematica
Universita' di Torino
Torino - Italy

1. INTRODUCTION

The problem of wave propagation in thin Cosserat bodies, such as shells and rods, has been faced by a few authors, even if the Cosserat theory has received a resurgence of interest in recent years. For a general approach refer to the pioneering work of Ericksen & Truesdell [1] and, more recently, to Antman [2] for rods and to Naghdi [3] for shells. Ericksen [4,5] proposed an abstract setting for treating the non linear theory of shells and rods, according to the Cosserat model of oriented materials, which "rephrases the basic equations so that they become almost identical in form with corresponding non linear elasticity equations". Using this analogy relevant results can be derived mainly on problems concerning stability and wave propagation. Fundamental contributions to the general theory and, particularly, to the wave propagation in shells and rods have been made by Cohen [6,10,12,14].
The aim of this paper is to resume, coordinate and discuss some results obtained in [7,8,9] by Pastrone, et al., concerning to the study of the propagation of shock and acceleration waves in thin hyperelastic bodies, with particular attention to the problem of the linearization of the equations of motion and to the dependence of the wave speeds and modes on the geometrical properties of the body. An approximate (but general and in agreement with the model proposed) representation of this dependence is given, to be clarified by simple examples; the technique adopted leads to general results not obtained previously, but, even when they are analogous, the method is much simpler and heavy calculations are avoided.

2. NON LINEAR ELASTIC SHELLS

A Cosserat shell [3,5] is a smooth moving surface Σ with parametrical equations

$$OP = \mathbf{r}(y^\alpha, t), \quad \alpha = 1,2 \ , \tag{2.1}$$

where O is a fixed origin, $P \in \Sigma$, the $y^\alpha \in \mathcal{D} \subseteq \mathcal{R}^2$ are material coordinates on Σ and t is time. The coordinates y^α are defined on a smooth, bidimensional manifold, for example the reference configuration $\overset{\blacktriangle}{\Sigma}$ (as usual, Greek indices take the values 1,2; Latin indices take the values 1,2,3; repeated indices imply summation).
A vector field, called the field of *directors*, is also defined on the surface Σ

$$\mathbf{d} = \mathbf{d}(y^\alpha, t); \quad \mathbf{r}_{,1} \times \mathbf{r}_{,2} \cdot \mathbf{d} \neq 0 \quad (\mathbf{r}_{,\alpha} = \partial \mathbf{r}/\partial y^\alpha). \tag{2.2}$$

The motion of the shell is described by the evolution of the vectors $\mathbf{r}$ and $\mathbf{d}$ in time, while the deformation of the shell is described by the director $\mathbf{d}$ and the gradients $\mathbf{r}_{,\alpha}$, and $\mathbf{d}_{,\alpha}$. The shell is assumed hyperelastic, hence there exists a strain energy density

$$W = W(\mathbf{r}_{,\alpha}, \mathbf{d}, \mathbf{d}_{,\alpha}; y^\alpha). \tag{2.3}$$

The kinetic energy density K is of the form

$$2K = \rho_0 \dot{\mathbf{r}} \cdot \dot{\mathbf{r}} + 2\rho_1 \dot{\mathbf{r}} \cdot \dot{\mathbf{d}} + \rho_2 \dot{\mathbf{d}} \cdot \dot{\mathbf{d}}, \quad \rho_0\rho_2 - \rho_1^2 \geq 0\,, \tag{2.4}$$

where ρ_0, ρ_1 and ρ_2 are non negative smooth functions of the material coordinates (dots indicate partial derivatives with respect to time). The total energy is given by

$$E = \int_{\hat{\Sigma}} (K + W) dy^1 dy^2 + \mathcal{B} \tag{2.5}$$

and $\mathcal{B}$ is an integral over the boundary of $\hat{\Sigma}$ whose explicit form is not relevant in this study. For simplicity, assume that mass force density is zero, internal constraints are excluded and external loads are applied on the boundary only. The equation of motion is obtained as Lagrange equations of the functional (2.5):

$$\left(\frac{\partial W}{\partial \mathbf{r}_{,\alpha}}\right)_{,\alpha} = \frac{d}{dt}\frac{\partial K}{\partial \dot{\mathbf{r}}}; \qquad \left(\frac{\partial W}{\partial \mathbf{d}_{,\alpha}}\right)_{,\alpha} - \frac{\partial W}{\partial \mathbf{d}} = \frac{d}{dt}\frac{\partial K}{\partial \dot{\mathbf{d}}}. \tag{2.6}$$

The equations (2.6) can be reduced to the more compact form (2.7) if the 6-dimensional vector space $\mathcal{E}_6$ is introduced, generated by the vectors $\mathbf{P} = (\mathbf{r}, \mathbf{d})$

$$\left(\frac{\partial W}{\partial \mathbf{P}_{,\alpha}}\right)_{,\alpha} - \frac{\partial W}{\partial \mathbf{P}} = \mathcal{K}\ddot{\mathbf{P}}\,, \tag{2.7}$$

where $\mathcal{K}$ is a linear transformation induced by the kinetic energy [5]. The basic equations (2.7) are analogous in form to the corresponding non linear 3-dimensional equations and, using this similarity, it is possible to analize [5] the problem of static stability and wave propagation avoiding, or postponing, the introduction of the surface differential geometry and the heavy calculations typical of the literature in this field.
A few fundamental results in wave propagation will be recalled here.
An *acceleration wave* is defined as a moving curve C on Σ represented by the equation $\varphi(y^\alpha, t) = 0$, such that $\mathbf{P}, \mathbf{P}_{,\alpha}$ and $\dot{\mathbf{P}}$ are continuous on it, but $\mathbf{P}_{,\alpha\beta}$ and $\ddot{\mathbf{P}}$ have finite discontinuities. If some of the first derivatives have a discontinuity on C, C is called a *shock wave*. The kinematic conditions of compatibility for acceleration waves yield:

$$[\![\mathbf{P}_{,\alpha\beta}]\!] = \mathbf{A}\nu_\alpha\nu_\beta, \quad [\![\ddot{\mathbf{P}}]\!] = \mathbf{A}v^2\,, \tag{2.8}$$

where $[\![\bullet]\!]$ denotes the jump across C, ν_α are the components of the unit normal to C in the plane tangent to Σ in P, $\mathbf{A}$ is the amplitude 6-dimensional vector and v the wave speed. By imposing (2.8) on the equation of motion (2.7), the propagation conditions

$$\mathbf{QA} = \mathcal{K}\mathbf{A}v^2, \quad \mathbf{Q} = \frac{\partial^2 W}{\partial \mathbf{P}_{,\alpha}\,\partial \mathbf{P}_{,\beta}}\nu_\alpha\nu_\beta \tag{2.9}$$

are obtained; $\mathbf{Q}$ is the acoustic simmetric tensor. The eigenvalues v_H^2, $H = 1,..6$, of (2.9) are exactly 6 positive values if the tensor $\mathbf{Q}$ is positive semi-definite

$$\langle \mathbf{A}, \mathbf{QA}\rangle \geq 0, \quad \forall \mathbf{A} \in \mathcal{E}_6\,. \tag{2.10}$$

Shock wave propagation conditions are determined by applying the first order compatibility conditions to the balance equations of linear momentum [5]:

$$\frac{d}{dt}\int_{\Sigma_0} \mathcal{K}\dot{\mathbf{P}}dy^1dy^2 = \oint_{\partial\Sigma_0} \frac{\partial W}{\partial \mathbf{P}}ds_0 - \int_{\Sigma_0} \frac{\partial W}{\partial \mathbf{P}}dy^1dy^2 \tag{2.11}$$

where $ds_0 = \pm\epsilon_{\alpha\beta}dy^\beta, \epsilon_{\alpha\beta}$ is the Levi-Civita alternating tensor and Σ_0 is any portion of $\hat{\Sigma}$ containing the curve C. The jump conditions read

$$\left[\!\left[\frac{\partial W}{\partial \mathbf{P}_{,\alpha}}\right]\!\right]\nu_\alpha = \mathcal{K}\mathbf{A}v^2, \tag{2.12}$$

formally similar to the conditions for shock waves in 3-dimensional non linear elasticity.

3. NON LINEAR ELASTIC RODS

Wave propagation in Cosserat elastic rods has been extensively studied [8,11,12,13]. The approach is the same as in section for shells. Now only the topics of specific interest will be pointed out. A Cosserat rod [2,4] is a material curve $\mathbf{c}$ given by parametric equations

$$OP = \mathbf{r}(s,t) \quad s \in [0,l] \subseteq \mathcal{R}^+, \;\; t \in \mathcal{R}^+ \tag{3.1}$$

equipped with a pair of vector-valued functions, called the *directors*,

$$\mathbf{d}_\alpha = \mathbf{d}_\alpha(s,t), \;\; \mathbf{d}_1 \times \mathbf{d}_2 \cdot \mathbf{r}' \neq 0, \qquad \alpha = 1,2 \;, \tag{3.2}$$

where 0 is a fixed point, $P \in \mathbf{c}$ (the prime denotes differentiation with respect to s). The vectors $\mathbf{r}$ and $\mathbf{d}_\alpha$, as functions of t, describe the evolution of the rod. Equations (2.3), (2.4) and (2.7) in the case of rods become

$$W = W(\mathbf{r}', \mathbf{d}_\alpha, \mathbf{d}'_\alpha; s), \;\; 2K = \rho^{\alpha\beta}\dot{\mathbf{d}}_\alpha \cdot \dot{\mathbf{d}}_\beta + 2\rho^{\alpha 3}\dot{\mathbf{d}}_\alpha \cdot \dot{\mathbf{r}} + \rho^{33}\dot{\mathbf{r}} \cdot \dot{\mathbf{r}} \;, \tag{3.3}$$

$$\left(\frac{\partial W}{\partial \mathbf{P}'}\right)' - \frac{\partial W}{\partial \mathbf{P}} = \mathcal{K}\ddot{\mathbf{P}} \;\;, \;\; \mathbf{P} \equiv (\mathbf{r}, \mathbf{d}_1, \mathbf{d}_2) \;, \tag{3.4}$$

where $\rho^{ij} = \rho^{ji}$ are non negative smooth functions of s, depending on the structure of the rod, and $\mathcal{K}$ is a suitable linear transformation [13]. A point of equation $\varphi(s,t) = 0$, moving along $\mathbf{c}$, is called an *acceleration wave* if $\mathbf{P}, \dot{\mathbf{P}}, \mathbf{P}'$ are continuous on it, but second derivatives have finite discontinuties. The compatibility conditions take the form

$$[\![\ddot{\mathbf{P}}]\!] = \mathbf{A}v^2, \qquad [\![\mathbf{P}'']\!] = \mathbf{A} \;\;, \tag{3.5}$$

where $\mathbf{A}$ is the amplitude 9-dimensional vector. By imposing (3.5) to (3.4) it is obtained

$$\mathbf{QA} = \mathcal{K}\mathbf{A}v^2, \;\; \mathbf{Q} = \frac{\partial^2 W}{\partial \mathbf{P}'\partial \mathbf{P}'} \tag{3.6}$$

and $\mathbf{Q}$ is the acoustic tensor. The eingenvalues v_H^2, of (3.6) determine the squares of the possible wave speeds. They are exactly 9 if the acoustic tensor $\mathbf{Q}$ is positive semi-definite. Shock wave jump conditions read

$$\left[\!\left[\frac{\partial W}{\partial \mathbf{P}'}\right]\!\right] = \mathcal{K}\mathbf{A}v^2. \tag{3.7}$$

4. LINEARLY ELASTIC SHELLS

In this section we begin by linearizing the field equations (2.7). A static equilibrium configuration $\hat{\Sigma}$ (it is always possible to take this configuration as the reference configuration $\hat{\Sigma}$) is considered ($\hat{\mathbf{P}}$ denotes the corresponding field vectors) and a small deformation $\mathbf{u}$ is assumed superimposed upon it, i.e.: $\mathbf{P} = \hat{\mathbf{P}} + \mathbf{u}$. One could deal with this formalism by expanding W and its first order partial derivatives in Taylor's series about $\hat{\mathbf{P}}$ retaining only the linear terms in $\mathbf{u}$ and $\mathbf{u}'$. As a result it can be proved that the speeds are the same both for acceleration and shock waves. The following relations are introduced

$$\mathbf{r} = \hat{\mathbf{r}} + v^\alpha \hat{\mathbf{r}},_\alpha + v^3 \hat{\mathbf{d}}, \qquad \mathbf{d} = \hat{\mathbf{d}} + u^\alpha \hat{\mathbf{r}},_\alpha + u^3 \hat{\mathbf{d}} \ , \tag{4.1}$$

where u^i, v^i are assumed to be smooth functions of y^α and t, and $u^i = v^i = 0$ if and only if $\Sigma = \hat{\Sigma}$. Note that, using (4.1), detailed results can be obtained more easily than in [3,14], the calculation difficulties are avoided, thus obtaining the linearized equations, the propagation speeds and the decay equation of shock waves. Since it is assumed that W satisfies material obiectivity, it follows

$$W = W(u^i, u^i,_\alpha, v^i, v^i,_\alpha; y^\alpha). \tag{4.2}$$

Note that (4.2) contains also the metric $a_{ij} = \hat{\mathbf{d}}_i \cdot \hat{\mathbf{d}}_j$ and connection coefficients $c^i_{k\alpha} = \hat{\mathbf{d}}_{k,\alpha} \cdot \hat{\mathbf{d}}^i$ at $\hat{\Sigma}$, where $\hat{\mathbf{d}}_\alpha \equiv \hat{\mathbf{r}},_\alpha, \hat{\mathbf{d}}_3 \equiv \hat{\mathbf{d}}, \{\hat{\mathbf{d}}_i\}$ is a material basis, $\{\hat{\mathbf{d}}^i\}$ the dual basis. The kinetic energy density is then:

$$2K = a_{ij}(\rho_0 \dot{v}^i \dot{v}^j + 2\rho_1 \dot{v}^i \dot{u}^j + \rho_2 \dot{u}^i \dot{u}^j) \tag{4.3}$$

A linearization of the equations of motion is attained by assigning W as quadratic in its arguments, i.e. by expanding W in a neighbourhood of $\hat{\Sigma}$ and retaining the significant terms up to second degree. Assumptions of material symmetries will reduce the quadratic part of W to simpler forms and, as in linear theories, its quadratic part is retained as strain energy density. If the material shell is supposed to be transversally isotropic, W must be invariant with respect to rotations in the plane orthogonal to the unit vector $\mathbf{N}$ normal to $\hat{\Sigma}$. The isotropy is called holoedral [3] if there is a center of symmetry; the invariance under the reflection $\mathbf{d} \Rightarrow -\mathbf{d}$ is required. It should be remembered that $\hat{\mathbf{d}} = \mathbf{N}$ can always be chosen. To simplify the consequent expression of W, the covariant derivatives are introduced

$$u^i_{|\alpha} = u^i,_\alpha + c^i_{k\alpha} u^k, \quad v^i_{|\alpha} = v^i,_\alpha + c^i_{k\alpha} v^k \tag{4.4}$$

Thereafter the simplest expression for W can be written as

$$\begin{aligned} 2W = {} & \lambda_1 u^\alpha_{|\alpha} u^\beta_{|\beta} + \lambda_2 v^\alpha_{|\alpha} v^\beta_{|\beta} + \lambda_3 a^{\alpha\beta} u^3_{|\alpha} u^3_{|\beta} + \lambda_4 a^{\alpha\beta} v^3_{|\alpha} v^3_{|\beta} \\ & + \lambda_5 a^{\alpha\beta} a_{\rho\sigma} u^\rho_{|\alpha} u^\sigma_{|\beta} + \lambda_6 a^{\alpha\beta} a_{\rho\sigma} v^\rho_{|\alpha} v^\sigma_{|\beta} + \lambda_7 a_{\alpha\beta} u^\alpha u^\beta + \lambda_8 u^3 u^3 \end{aligned} \tag{4.5}$$

where λ_Γ are generalized elastic moduli given by second derivatives of W at $\hat{\Sigma}$. As a consequence, these moduli depend not only on the material properties of the reference configuration, but also on its geometrical structure:

$$\lambda_\Gamma = \lambda_\Gamma(a_{ij}, c^i_{k\alpha}), \qquad \Gamma = 1, ...8. \tag{4.6}$$

Wave speeds can be easily evaluated for *longitudinal, transverse, squeeze gradient, bending, twisting* and *kink* waves, respectively:

$$v^2_{\mathcal{L}} = \frac{\lambda_2 + \lambda_6}{\rho_0}, \quad v^2_{\mathcal{T}} = \frac{\lambda_6}{\rho_0}, \quad v^2_{\mathcal{S}} = \frac{\lambda_4}{\rho_0}, \quad v^2_{\mathcal{B}} = \frac{\lambda_1 + \lambda_5}{\mu\rho_0}, \quad v^2_{\mathcal{W}} = \frac{\lambda_5}{\mu\rho_0}, \quad v^2_{\mathcal{K}} = \frac{\lambda_3}{\mu\rho_0} \tag{4.7}$$

In [7] an attempt was made to obtain approximate, but explicit expressions for the moduli λ_Γ in terms of the metric and connection components of $\hat{\Sigma}$; the approximation is in agreement with the linear model and the hypothesis of material symmetries. So

$$\lambda_\Gamma = \mu_\Gamma + \mu_\Gamma^{\alpha\beta\rho\sigma} c^3_{\alpha\beta} c^3_{\rho\sigma} + \mu_\Gamma{}^{\alpha\beta\rho\sigma}{}_{\lambda\mu} c^\lambda_{\alpha\beta} c^\mu_{\rho\sigma}\,, \tag{4.8}$$

where

$$\mu_\Gamma^{\alpha\beta\rho\sigma} = \overset{1}{\mu}_\Gamma a^{\alpha\beta} a^{\rho\sigma} + \overset{2}{\mu}_\Gamma (a^{\alpha\rho} a^{\beta\sigma} + a^{\alpha\sigma} a^{\rho\beta}), \tag{4.9}$$

$$\mu_\Gamma{}^{\alpha\beta\rho\sigma}{}_{\lambda\mu} = \overset{3}{\mu}_\Gamma a_{\lambda\mu} a^{\alpha\beta} a^{\rho\sigma} + \overset{4}{\mu}_\Gamma a_{\lambda\mu} (a^{\alpha\rho} a^{\beta\sigma} + a^{\alpha\sigma} a^{\beta\rho}). \tag{4.10}$$

The terms μ_Γ and $\overset{n}{\mu}_\Gamma$ are constant coefficients depending only on the material properties of the shell in $\hat{\Sigma}$. If $\hat{\Sigma}$ is a plane, the elastic moduli are constant and given by $\lambda_\Gamma = \mu_\Gamma$; in this case the results obtained are in agreement with those in [15] for plates; for this reason the names associated to wave speeds in (4.7) are borrowed from [15].
If the reference configuration is a circular cylinder of radius R or a sphere of radius R

$$\lambda_\Gamma = \mu_\Gamma + (\overset{1}{\mu}_\Gamma + 2\overset{2}{\mu}_\Gamma) 1/R^2, \qquad \lambda_\Gamma = \mu_\Gamma + [1 + (\overset{3}{\mu}_\Gamma + 4\overset{4}{\mu}_\Gamma) cotg\theta] 1/R^2, \tag{4.11}$$

respectively (θ is the colatitude in spherical coordinates). In both cases the asymptotic behaviour for $R \Rightarrow \infty$ shows that, when the cylinder or the sphere tend to a plane, the limit speeds are those obtained above for a plane. For the sphere, if the initial wave curve is a circle of latitude, the speeds of propagation are constant along this curve hence the wave curves are a parallel system of circles of latitude; all speeds of propagation are constant, as shown in [14], at all points of the sphere, if and only if $\overset{3}{\mu}_\Gamma = \overset{4}{\mu}_\Gamma = 0$, condition that implies a further simplification in the model. The constants μ_Γ and $\overset{n}{\mu}_\Gamma$ can be related with classical elastic moduli, such as Young's modulus and Poisson's ratio, by comparing the strain energy density with the corresponding elastic potential used in [3], where different elastic coefficients are introduced and expressed in terms of the classical moduli. The difficulties arising from the different choice of deformation variables are particularly laborious for the shells, while one analogous problem for rod will be discussed in the next section with a different approach.

5. LINEARLY ELASTIC RODS

The technique used for linearizing the wave propagation problem for a Cosserat rod has many similarities to the shell linearized theory of the previous section. Only the essentials will be briefly outlined. Equations (4.1), (4.2), (4.3) and (4.5) in the case of rods become respectively

$$\mathbf{r} = \hat{\mathbf{r}} + v^i \hat{\mathbf{d}}_i, \qquad \mathbf{d}_\alpha = \hat{\mathbf{d}}_\alpha + v^i_\alpha \hat{\mathbf{d}}_i, \qquad \hat{\mathbf{d}}_3 \equiv \hat{\mathbf{r}}', \tag{5.1}$$

$$W = W(v^i, v^{i\prime}, v^i_\alpha, v^{i\prime}_\alpha, s) \;, \tag{5.2}$$

$$2K = \delta_{ij} (\rho^{\alpha\beta} \dot{v}^i_\alpha \dot{v}^i_\beta + \rho^{33} \dot{v}^i \dot{v}^j + 2\rho^{\alpha 3} \dot{v}^i_\alpha \dot{v}^j) \;, \tag{5.3}$$

$$\begin{aligned} 2W = {} & \lambda_1 \delta_{\alpha\beta} v^\alpha_{|s} v^\beta_{|s} + \lambda_2 v^3_{|s} v^3_{|s} + \lambda_3 \delta^{\alpha\beta} \delta_{\rho\sigma} v^\rho_{\alpha|s} v^\sigma_{\beta|s} + \\ & \lambda_4 \delta^{\alpha\beta} v^3_{\alpha|s} v^3_{\beta|s} + \lambda_5 \delta^{\alpha\beta} \delta_{\rho\sigma} v^\rho_\alpha v^\sigma_\beta + \lambda_6 \delta^{\alpha\beta} v^3_\alpha v^3_\beta \;, \end{aligned} \tag{5.4}$$

where, since the rod is homogeneus and transversally isotropic, the function W does not depend on s and it is invariant under rotations in the plane specified by $\hat{\mathbf{d}}_1$ and $\hat{\mathbf{d}}_2$ and under the transformation $s \Rightarrow -s$. The λ_Γ are the generalized elastic moduli. As shown in [8], the following explicit values for acceleration wave speeds are found, respectively for *shear, extensional, transverse shear, flexional* waves (along $\hat{\mathbf{d}}_1$ and $\hat{\mathbf{d}}_2$):

$$v^2_{\mathcal{S}} = \frac{\lambda_1}{\rho_0}, \quad v^2_{\mathcal{E}} = \frac{\lambda_2}{\rho_0}, \quad v^2_{\mathcal{T}_1} = \frac{\lambda_3}{\rho_0 \alpha_1}, \quad v^2_{\mathcal{T}_2} = \frac{\lambda_3}{\rho_0 \alpha_2}, \quad v^2_{\mathcal{F}_1} = \frac{\lambda_4}{\rho_0 \alpha_1}, \quad v^2_{\mathcal{F}_2} = \frac{\lambda_4}{\rho_0 \alpha_2}, \tag{5.5}$$

where α_1 and α_2 are the constant inertia moments of the rod section. Of course, the coefficients $\lambda_1, \lambda_2, \lambda_3$ and λ_4 must be positive. The elastic moduli λ_Γ depend on the material properties and the geometry of the rod at its reference configuration: in particular they depend on the metric tensor δ_{ij} and the connection coefficients c^i_j. If the reference configuration is straight and untwisted, all speeds are constant. Otherwise, following the procedure used for shells, it is possible to find approximate expressions for the moduli λ_Γ such as

$$\lambda_1 = \mu_1 + \mu_7 \delta^{\alpha\beta} c^3_\alpha c^3_\beta + 1/2 \mu_8 \delta^{\alpha\beta} \delta_{\rho\sigma} c^\rho_\alpha c^\sigma_\beta \ , \tag{5.6}$$

(analogously for $\Gamma = 2...9$) where the constant elastic coefficients $\mu_\Delta (\Delta = 1...18)$ appear. Approximate values for the wave speeds can be now obtained in which the effects of the geometry of the rod are disjointed from the effects of the material properties. The elastic coefficients μ_Δ can be related to the classical material constants of three dimensional theory. In practice, simple exact solutions of problems in three-dimensional isotropic elasticity for rods can be considered. A comparison between the displacements obtained there to the corresponding displacements arising in our model provides the required relations. For instance

$$\mu_1 = \frac{EAG^2}{2(1+\nu)}, \quad \mu_2 = EA, \quad \mu_3 = EI \quad , etc.$$

where E is Young's modulus and ν Poisson's ratio, A the area and I the moment of inertia (with respect to the $\hat{\mathbf{d}}_1$ axis) of the cross section, G^2 a constant depending on the form of the section.

REFERENCES

[1] Ericksen L., Truesdell C., ARMA, v.1, p.295, (1958).
[2] Antman S.S., Handbuch der Phys. Bd 6 a/2, Spring. Verlag, (1982).
[3] Naghdi P.M., Handbuch der Phys. Bd 6 a/2, Spring. Verlag, (1972).
[4] Ericksen J.L., Int. J. Solids Structures, v.6, p.371, (1970).
[5] Ericksen J.L., ARMA, v.43 (5), p.167, (1971).
[6] Cohen H., Whitman A.B., J. Sound and Vibration, v.51, p.283, (1977).
[7] Molina C., Pastrone F., Meccanica, v.22, p.92, (1987).
[8] Pastrone F., Tonon M., J. Mech. Th. Appl., v.5, p.615, (1986).
[9] Tonon M. L., Rend Sem. Math. Univ. Torino, (to appear).
[10] Cohen H., Suh S.L., J. of Math-Mech, v.19, p.1117, (1970).
[11] Cohen H., Iranian J. Sci. Techn., v.7, p.83, (1978).
[12] Cohen H., ARMA, v.67, p.151, (1978).
[13] Pastrone F., Atti Acc. Sci. TO, v.114, p.133, (1980).
[14] Cohen H., J. of Applied Mech. v.43, p.281, (1976).
[15] Cohen H., J. of Elasticity, v.6, p.245, (1976).

ELASTIC WAVE PROPAGATION
M.F. McCarthy, M.A. Hayes, (Editors)
© Elsevier Science Publishers B.V. (North-Holland), 1989

GROUP THEORETIC FORMULATION FOR ELASTIC PROPERTIES OF MEDIA COMPOSED OF ANISOTROPIC LAYERS

Michael SCHOENBERG* and Francis MUIR+

* Schlumberger-Doll Research, Ridgefield CT 06877-4108

+ Department of Geophysics, Stanford University, Stanford CA 94305-2215

A matrix formalism allows for the simple evaluation of the anisotropic homogeneous medium equivalent, in the long wavelength limit, to a distribution of fine layers, each layer itself being an elastic anisotropic medium. A given cumulative thickness of fine layers of an anisotropic constituent can be shown to be transformable to an element of a commutative group. Combining group elements $\mathbf{G}_A$ and $\mathbf{G}_B$ (corresponding to cumulative thickness H_A of thin layers of constituent A and cumulative thickness H_B of of constituent B, respectively) gives the group element corresponding to total thickness H_A+H_B of the homogeneous medium equivalent to the interleaved layers of constituents A and B. The inverse element gives us the notion of 'uncombining' layers by addition of the inverse; then, if the remaining layer is a stable anisotropic medium, a valid decomposition of the original anisotropic medium is revealed. Systems of parallel fractures and aligned microcracks (of assumed cumulative thickness zero) also are transformable to group elements which can be manipulated as easily as an element corresponding to any other set of layers. Such systems may be azimuthally anisotropic relative to their normal axis, the anisotropy being due either to fracture or crack geometry or to the anisotropy of the infilling material. Multiple fracture or crack systems are dealt with by successive coordinate system rotations and additions of group elements.

1. ELASTIC MODULI OF LAYERED MEDIA

Consider a layered medium composed of layers of arbitrary anisotropy in welded contact. Each layer is assumed to be very thin relative to a wavelength, but the cumulative thickness of a given constituent need not be small. For the properties of the calculated homogeneous medium to be equivalent to the layered medium over many wavelengths, a second assumption, stationarity, is invoked. Simply, the relative amount of each constituent is evenly spread over the layered medium down to some scale thickness that is assumed to be but a fraction of the shortest wavelength. The x_3-axis is taken perpendicular to the layering. Each constituent has a relative thickness h_i, $i=1,...,n$ (with $h_1+...+h_n=1$), a density ρ_i, and an elastic modulus tensor, c_{pqrs_i}, which relates stress, σ_{pq_i}, with strain, ϵ_{rs_i} according to the generalized Hooke's law, so that

$\sigma_{pq_i} = c_{pqrs_i}\epsilon_{rs_i}$. In condensed notation, where $c_{pqrs_i} \rightarrow c_{jk_i}$ according to $11\rightarrow1$, $22\rightarrow2$, $33\rightarrow3$, $23\rightarrow4$, $31\rightarrow5$ and $12\rightarrow6$, and the strain convention $[\epsilon_1, \epsilon_2, \epsilon_3, \epsilon_4, \epsilon_5, \epsilon_6] \equiv [\epsilon_{11}, \epsilon_{22}, \epsilon_{33}, 2\epsilon_{23}, 2\epsilon_{31}, 2\epsilon_{12}]$, the stress-strain relation in a layer of the ith constituent may be written

$$\sigma_{j_i} = \sum_{k=1}^{6} c_{jk_i}\epsilon_{k_i}, \qquad j=1,...,6 . \tag{1}$$

The stiffness matrix c_{jk_i} is symmetric and positive definite, with at most 21 independent elastic constants. The elastic moduli for the long wavelength equivalent anisotropic medium can be expressed in terms of thickness-weighted averages of functions of the moduli of the constituents. Following Backus [1] who treated a layered medium composed of several isotropic or transversely isotropic constituents, we assume all stress components acting on planes parallel to the layering are the same in all layers, i.e., $\sigma_{13_i} \equiv \sigma_{5_i} = \sigma_5$, $\sigma_{23_i} \equiv \sigma_{4_i} = \sigma_4$ and $\sigma_{33_i} \equiv \sigma_{3_i} = \sigma_3$. In addition all strain components in the plane of the layering are the same in all layers, i.e., $\epsilon_{11_i} \equiv \epsilon_{1_i} = \epsilon_1$, $\epsilon_{22_i} \equiv \epsilon_{2_i} = \epsilon_2$ and $2\epsilon_{12_i} \equiv \epsilon_{6_i} = \epsilon_6$. The other stress and strain components, $\sigma_{11_i} \equiv \sigma_{1_i}$, $\sigma_{22_i} \equiv \sigma_{2_i}$, $\sigma_{12_i} \equiv \sigma_{6_i}$, $\epsilon_{33_i} \equiv \epsilon_{3_i}$, $2\epsilon_{23_i} \equiv \epsilon_{4_i}$ and $2\epsilon_{13_i} \equiv \epsilon_{5_i}$, may differ from layer to layer but are taken to be constant within a layer *or* more precisely, are represented by average values within a layer.

Now define the following stress and strain vectors $\underline{S}_{1_i} = [\sigma_{1_i}\ \sigma_{2_i}\ \sigma_{6_i}]^T$, $\underline{S}_2 = [\sigma_3\ \sigma_4\ \sigma_5]^T$, $\underline{E}_1 = [\epsilon_1\ \epsilon_2\ \epsilon_6]^T$, and $\underline{E}_{2_i} = [\epsilon_{3_i}\ \epsilon_{4_i}\ \epsilon_{5_i}]^T$, where superscript T denotes the transpose. This allows the stress-strain relations (1) in the ith layer to be written as

$$\underline{S}_{1_i} = \mathbf{M}_i\,\underline{E}_1 + \mathbf{P}_i\,\underline{E}_{2_i}, \tag{2a}$$

$$\underline{S}_2 = \mathbf{P}_i^T\,\underline{E}_1 + \mathbf{N}_i\,\underline{E}_{2_i}, \tag{2b}$$

where

$$\mathbf{M}_i = \begin{bmatrix} c_{11_i} & c_{12_i} & c_{16_i} \\ c_{12_i} & c_{22_i} & c_{26_i} \\ c_{16_i} & c_{26_i} & c_{66_i} \end{bmatrix}, \quad \mathbf{N}_i = \begin{bmatrix} c_{33_i} & c_{34_i} & c_{35_i} \\ c_{34_i} & c_{44_i} & c_{45_i} \\ c_{35_i} & c_{45_i} & c_{55_i} \end{bmatrix}, \quad \mathbf{P}_i = \begin{bmatrix} c_{13_i} & c_{14_i} & c_{15_i} \\ c_{23_i} & c_{24_i} & c_{25_i} \\ c_{36_i} & c_{46_i} & c_{56_i} \end{bmatrix}, \tag{2c}$$

are 3×3 stiffness matrices. The moduli of the long wavelength equivalent homogeneous medium medium are found by averaging, but, because averages of products of any of the 3×3 stiffness matrices (which differ from layer to layer) with either $\underline{S}_{1_i}$ and $\underline{E}_{2_i}$ (which also differ from layer to layer) are indeterminate, it is necessary to solve Eqs.(2) for $\underline{S}_{1_i}$ and $\underline{E}_{2_i}$ before averaging. To do this, premultiply (2b) by $\mathbf{N}_i^{-1}$, solve for $\underline{E}_{2_i}$, substitute the result into (2a) and collect terms. The resulting expressions are

$$\underline{S}_{1_i} = \left[\mathbf{M}_i - \mathbf{P}_i\mathbf{N}_i^{-1}\mathbf{P}_i^T\right]\underline{E}_1 + \mathbf{P}_i\mathbf{N}_i^{-1}\,\underline{S}_2, \tag{3a}$$

$$\underline{E}_{2_i} = -\mathbf{N}_i^{-1}\mathbf{P}_i^T\,\underline{E}_1 + \mathbf{N}_i^{-1}\,\underline{S}_2, \tag{3b}$$

ready for averaging. Eq.(3a) gives the in-plane stress components in a layer. The average in-plane stress across many layers is the total in-plane force divided by the total thickness, i.e., the thickness weighted average of Eq.(3a). Similarly, the average strain across many layers is the total displacement of the top surface with respect to the bottom surface divided by the total thickness, i.e., the thickness weighted average of Eq.(3b). Taking the thickness weighted average of Eqs.(3), denoted by $<\cdot>$ gives,

$$<\underline{S}_1> = \left[<\mathbf{M}> - <\mathbf{PN}^{-1}\mathbf{P}^T>\right]\underline{E}_1 + <\mathbf{PN}^{-1}>\underline{S}_2\,, \tag{4a}$$

$$<\underline{E}_2> = -<\mathbf{N}^{-1}\mathbf{P}^T>\underline{E}_1 + <\mathbf{N}^{-1}>\underline{S}_2\,. \tag{4b}$$

Now premultiply (4b) by $<\mathbf{N}^{-1}>^{-1}$, solve for $\underline{S}_2$, substitute the result into (4a) and collect terms. Then comparison with Eqs.(2) enables us to write

$$<\underline{S}_1> = \mathbf{M}\ \underline{E}_1 + \mathbf{P}\ <\underline{E}_2>\,, \tag{5a}$$

$$\underline{S}_2 = \mathbf{P}^T\ \underline{E}_1 + \mathbf{N}\ <\underline{E}_2>\,, \tag{5b}$$

where

$$\mathbf{N} = <\mathbf{N}^{-1}>^{-1}\,, \qquad \mathbf{P} = <\mathbf{PN}^{-1}>\mathbf{N}\,, \tag{5c}$$

$$\mathbf{M} = <\mathbf{M}> - <\mathbf{PN}^{-1}\mathbf{P}^T> + <\mathbf{PN}^{-1}>\mathbf{N}<\mathbf{N}^{-1}\mathbf{P}^T>\,.$$

Eqs.(5) are the correctly averaged stress-strain relations a medium equivalent to the n interleaved constituent anisotropic media. The density of the equivalent medium is given by $<\rho>$, see Ref. [2].

2. GROUP THEORETIC APPROACH TO COMBINING ANISOTROPIC LAYERS

A given anisotropic constituent, say the ith, is distributed in fine layers throughout the total thickness H of a given region. Let the cumulative thickness of all layers of the ith constituent in the region be $H_i = h_i H$. The quantities H_i, ρ_i and the three 3×3 modulus matrices $\mathbf{N}_i$, $\mathbf{P}_i$ and $\mathbf{M}_i$ (with $\mathbf{N}$ and $\mathbf{M}$ symmetric and invertible) of the ith constituent medium are physical model parameters which map into a new set of two scalars and three 3×3 matrices, $G_i = [g_i(1), g_i(2), g_i(3), g_i(4), g_i(5)]$. The mapping is

$$\begin{bmatrix} H_i \\ \rho_i \\ \mathbf{N}_i \\ \mathbf{P}_i \\ \mathbf{M}_i \end{bmatrix} \rightarrow \begin{bmatrix} H_i \\ H_i\rho_i \\ H_i\mathbf{N}_i^{-1} \\ H_i\mathbf{P}_i\mathbf{N}_i^{-1} \\ H_i[\mathbf{M}_i - \mathbf{P}_i\mathbf{N}_i^{-1}\mathbf{P}_i^T] \end{bmatrix} \equiv \begin{bmatrix} g_i(1) \\ g_i(2) \\ \mathbf{g}_i(3) \\ \mathbf{g}_i(4) \\ \mathbf{g}_i(5) \end{bmatrix}, \tag{6}$$

Note that $g_i(1)$, the cumulative thickness of the ith constituent, has dimension of length, $g_i(2)$, the mass of the ith constituent in a column of unit area, has dimension of length times density, $\mathbf{g}_i(3)$ has dimension of length per stress, $\mathbf{g}_i(4)$ has dimension of length, and $\mathbf{g}_i(5)$ has dimension of length times stress. $\mathbf{g}_i(j), j=3,4,5$ are H_i times coefficients that occur in Eqs.(3a) and (3b). The set of all possible

$\mathbf{G} = [\mathbf{g}(1), \mathbf{g}(2), \mathbf{g}(3), \mathbf{g}(4), \mathbf{g}(5)]$ of two scalars and three 3×3 matrices with $\mathbf{g}(3)$ and $\mathbf{g}(5)$ symmetric, forms an Abelian group, called G, under addition. Eq.(6) maps a physical model to a group element and hereafter will be called the 'group mapping'.

For any $\mathbf{G}_i$ such that $g_i(1) \neq 0$ and $\mathbf{g}_i(3)$ is invertible, we can return to the set of physical model parameters by the 'inverse group mapping' from group element $\mathbf{G}_i$ to physical model. This inverse mapping is given by

$$\begin{bmatrix} g_i(1) \\ g_i(2) \\ \mathbf{g}_i(3) \\ \mathbf{g}_i(4) \\ \mathbf{g}_i(5) \end{bmatrix} \rightarrow \begin{bmatrix} g_i(1) \\ g_i(2)/g_i(1) \\ g_i(1)\mathbf{g}_i(3)^{-1} \\ \mathbf{g}_i(4)\mathbf{g}_i(3)^{-1} \\ [\mathbf{g}_i(5)+\mathbf{g}_i(4)\mathbf{g}_i(3)^{-1}\mathbf{g}_i(4)^T]/g_i(1) \end{bmatrix} \equiv \begin{bmatrix} H_i \\ \rho_i \\ \mathbf{N}_i \\ \mathbf{P}_i \\ \mathbf{M}_i \end{bmatrix} \qquad (7)$$

The combination operation in the group, addition of the respective scalars and matrices, corresponds to calculating the medium equivalent to interleaved fine layers in the space of the physical model parameters. The associative and commutative properties enable us to combine group elements, equivalent to interleaving constituent layers, knowing that the ordering of combination of the various constituents can have no effect on the final result. Adding group elements that correspond to stable elastic layers results in a new stable elastic layer. An elastic layer is stable if its 6×6 modulus matrix is positive definite. The existence in the group of an inverse element for every element enables us to attempt the decomposition of an anisotropic medium into a set of possible interleaved constituents. Consider an equivalent medium e with corresponding group element $\mathbf{G}_e$. Let $\mathbf{G}_c$ correspond to a set of layers, denoted by c. 'Subtracting' c means adding the inverse element $-\mathbf{G}_c$ to $\mathbf{G}_e$ yielding $\mathbf{G}_b = \mathbf{G} - \mathbf{G}_c$, but $\mathbf{G}_b$ need not correspond to a set of physical layers. That is, even if $g_b(1) \neq 0$ and $\mathbf{g}_b(3)$ is invertible so the inverse group mapping (7) can be performed, the result could have non-positive thickness, i.e., $H_b \leq 0$; non-positive density, i.e., $\rho_b \leq 0$; and/or be unstable, i.e. have a set of matrices $[\mathbf{N}_b, \mathbf{P}_b, \mathbf{M}_b]$ such that the overall 6×6 modulus matrix is not positive definite. However, if after subtraction, the inverse group mapping of $\mathbf{G}_b$ gives a stable medium b, then b and c are a valid layer decomposition of the original medium e.

3. SUBGROUPS OF G CORRESPONDING TO VARIOUS SYMMETRY SYSTEMS

Different constituent anisotropic layers may belong to different symmetry systems. Suppose constituents of a certain symmetry system give rise to a subset of G of a certain form. If any two constituents of that symmetry system combine to give an equivalent medium also of that symmetry system (closure), then the subset of G of that form is a subgroup of G. There are eight types of subgroups corresponding to various symmetry systems, two monoclinic subgroups ($MC_\perp$ with a two-fold axis perpendicular to the layering and $MC_\parallel$ with a two-fold axis parallel to the layering), an orthorhombic subgroup (OR), two trigonal subgroups (TR and TR', the second of these having a two-fold axis parallel to the layering in addition to the three-fold axis perpendicular to the layering), two tetragonal subgroups (TT and TT', the second of these having two-fold axes parallel to the layering in addition to the four-fold axis perpendicular to the layering), and a transversely isotropic subgroup (TI with a six-fold axis perpendicular to the layering). The relationships among the eight symmetry system subgroups of G are depicted in the Venn diagram, Figure 1.

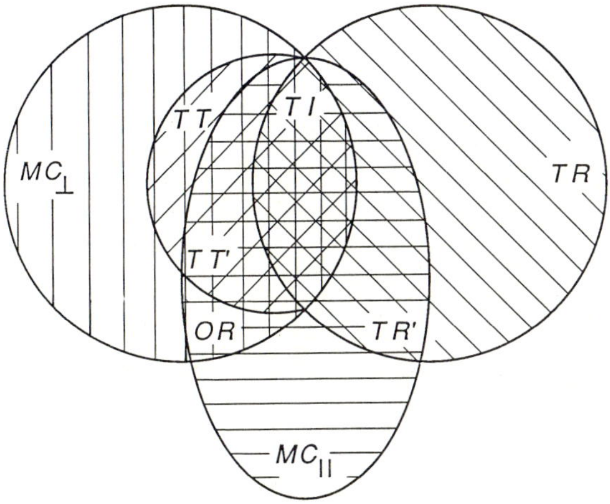

FIGURE 1
A Venn diagram of the eight subgroups of the complete group G corresponding to the various symmetry systems. The three primary subgroups that aren't subgroups of other subgroups are $MC\perp$, $MC_\parallel$ and TR. Note that $OR \equiv MC_\parallel \cap MC_\perp$ and $TR' \equiv TR \cap MC_\parallel$. In addition, $TI \equiv TR \cap MC_\perp$ which is contained in $MC_\parallel$, so TI is the intersection of all three primary subgroups. $TT \subset MC_\perp$ with $TT' \equiv TT \cap OR$.

4. GROUP ELEMENTS FROM SETS OF LONG PARALLEL FRACTURES

In a layered region of total thickness H, let a constituent, denoted with subscript f, with cumulative thickness H_f and hence relative thickness $h_f \equiv H_f/H$ have density $\rho_f = h_f \tilde{\rho}$ and 3×3 elastic stiffness matrices $\mathbf{N}_f = h_f \tilde{\mathbf{N}}$, $\mathbf{P}_f = h_f \tilde{\mathbf{P}}$ and $\mathbf{M}_f = h_f \tilde{\mathbf{M}}$. As $h_f \to 0$, this constituent approaches a soft medium with negligible inertia occupying a region of thickness H_f that becomes infinitesimally small. Hence the layers of this constituent become a set of infinitely long fractures, each of which behaves as a linear slip interface [3]. Across each slip interface, traction components σ_{3j} are continuous, as they are across any single layer in the long wavelength limit; but the components of displacement u_j need not be continuous, indicating that the strains in the fractures become infinite as the moduli approach zero. The physical properties of this fracture set approximating constituent are mapped by (6) into group element $\mathbf{G}_f$ given by

$$\begin{bmatrix} g_f(1) \\ g_f(2) \\ \mathbf{g}_f(3) \\ \mathbf{g}_f(4) \\ \mathbf{g}_f(5) \end{bmatrix} = \lim_{h_f \to 0} \begin{bmatrix} h_f H \\ h_f^2 H \tilde{\rho} \\ H\tilde{\mathbf{N}}^{-1} \\ h_f H \tilde{\mathbf{P}} \tilde{\mathbf{N}}^{-1} \\ h_f^2 H [\tilde{\mathbf{M}} - \tilde{\mathbf{P}}\tilde{\mathbf{N}}^{-1}\tilde{\mathbf{P}}^T] \end{bmatrix} = \begin{bmatrix} 0 \\ 0 \\ H\tilde{\mathbf{N}}^{-1} \\ \mathbf{O} \\ \mathbf{O} \end{bmatrix} \tag{8}$$

The fracture set is characterized by the symmetric 3×3 matrix $\tilde{\mathbf{N}}^{-1} \equiv \mathbf{Z}$, called the fracture set's compliance matrix. $\mathbf{Z}$ depends on, at most, six independent constants and specifies the fracture behavior, in general azimuthally anisotropic, discussed in detail by Schoenberg and Douma [4]. $\mathbf{G}_f$ can belong to symmetry system subgroups $MC_{||}$, *OR* or *TI*. Note that *TI* contains all group elements corresponding to sets of azimuthally isotropic fractures.

An unfractured background medium, denoted by subscript b, of thickness H_b is fractured by a set of parallel fractures with compliance matrix $\mathbf{Z}$. The group element of the fractured medium is $\mathbf{G}_b + \mathbf{G}_f$. The only component of $\mathbf{G}_b$ changed due to fracturing is $\mathbf{g}_b(3)$ which becomes $H_b(\mathbf{N}_b^{-1} + \mathbf{Z})$. The inverse group mapping gives the set of parameters of the fractured medium, in agreement with Ref. [4]. Defining $\mathbf{X}$ by $(\mathbf{N}_b^{-1} + \mathbf{Z})^{-1} = \mathbf{N}_b(\mathbf{I} + \mathbf{Z}\mathbf{N}_b)^{-1} \equiv \mathbf{N}_b(\mathbf{I} - \mathbf{X})$, the changes in the moduli due to the fractures may be written

$$\Delta\mathbf{N} = -\mathbf{N}_b\mathbf{X}\,, \qquad \Delta\mathbf{P} = -\mathbf{P}_b\mathbf{X}\,, \qquad \Delta\mathbf{M} = -\mathbf{P}_b\mathbf{X}\mathbf{N}_b^{-1}\mathbf{P}_b^T\,. \tag{9}$$

5. CONCLUSION

We have shown that constituent layers *of any anisotropy system* can be mapped to an element of an Abelian group. Expressing constituent properties as elements of a group enables us to use elementary arithmetic to form the group element from a sum of such elements, corresponding to the physical properties of the equivalent medium. Since systems of parallel fractures, in general anisotropic, are also representable by group elements, the group element of a fractured medium is the group element of the background medium plus the group element of the set of fractures. This formalism extends to modeling a medium with more than one set of fractures. To model two sets of fractures, we rotate (back in model space) the medium with a single set of fractures to a coordinate system appropriate to the second fracture system, and then "add" the second set. As an example, two or more oblique sets of parallel vertical fractures, with or without horizontal bedding planes give rise to a monoclinic medium with a vertical two-fold symmetry axis.

The inversion property of the group also allows for decomposition into constituents. The group element corresponding to a given set of layers subtracted from the group element corresponding to a homogeneous medium may provide a decomposition of the homogeneous medium into constituents.

REFERENCES

[1] Backus, G. E., "Long-Wave anisotropy produced by horizontal layering," J. Geophys. Res. 66 (1962) 4427.
[2] Schoenberg, M., "Reflection of elastic waves from periodically stratified media with interfacial slip," Geophys. Prospecting 31 (1983) 265.
[3] Schoenberg, M., "Elastic wave behavior across linear slip interfaces," J. Acoust. Soc. Am. 68 (1980) 1516.
[4] Schoenberg, M. and Douma J., " Elastic wave propagation in media with parallel fractures and aligned cracks," Geophys. Prospecting (1988) in print.

ELASTIC WAVE PROPAGATION
M.F. McCarthy, M.A. Hayes, (Editors)

STEADY PROPAGATION OF BUCKLE IN ELASTIC PIPELINES

Nobumasa SUGIMOTO

Department of Mechanical Engineering
Faculty of Engineering Science
Osaka University
Toyonaka, Osaka 560
Japan

This paper deals with steady propagation of buckle in elastic pipelines subjected to bending, axial tension and hydrostatic side pressure. Analysis is made on the basis of the approximate equations derived previously for the beam-flexural mode coupled with the ring-flexural mode. With a velocity smaller than that of the transverse wave due to 'effective tension', a kink can be propagated driven forward by both actions of axial tension and side pressure. Its propagation pressure and propagation velocity are given in terms of a maximum deflection, axial tension and side pressure.

1. INTRODUCTION

Buckle propagation is a very interesting phenomenon which occurs in a long and thin circular pipeline subjected to a combined action of bending, axial tension and hydrostatic side pressure [1,2]. Once a buckle, i.e., local kink, is formed even well below the buckling pressure, it can be propagated steadily along the axial direction, entailing a collapse of the pipeline in a catastrophic fashion. Up to the present, however, there seem to be no theoretical treatments for its mechanism of steady propagation.

From a collapsed pipe, one immediately observes traces of which severe plastic deformation has been propagated. But when the buckle is actually traveling, is the plasticity of primary importance in propagation? Since it occurs over a global scale of the pipeline, what mechanism will drive the buckle forward to form steady propagation? This question leads the author to conjecture that some elastic mechanism must be involved globally, whereas an ensuing collapse must be brought about plastically but locally. This 'globally elastic and locally plastic' assumption enables us to simplify the actual complicated phenomenon in two idealized problems.

Focusing on its global aspect and applying the theory of elasticity, the approximate equations have been derived for the beam-flexural mode coupled with the ring-flexural mode, where geometrically finite deformation has been taken into account. Based on these equations, it was presented at the IUTAM symposium in 1985 [3] that for pure bending, Brazier effect [4,5] is successfully explained and that the steady-progressive wave solutions have two types of propagation mode, one being the flexural wave mode and the other buckle propagation mode. This paper continues to discuss the steady buckle propagation and examines its propagation pressure and propagation velocity in detail.

With regards to Hooke's law assumed in deriving the approximate equations, it is important to remark here that its validity may still hold even when a kink appears. One might think that Hooke's law becomes invalid for the kink because of yield. But this is not always the case depending on a size of the kink, i.e, a radius of curvature. Suppose, for example, bending of a flexible steel measure in reel (see the inset in Figure 7). As it is bent, the concave cross-section tends to flatten out so that a 'kink' appears beyond a critical curvature. But a close observation shows that although its local radius of curvature becomes considerably small in comparison with a length of the measure, it is still much larger than the thickness and that it yet remains in the elastic range. It is this type of kink to be considered in this paper. Hence it should be remarked that even if the kink appears, it does not immediately mean a collapse.

2. GOVERNING EQUATIONS

At first, we present the approximate governing equations previously derived for an elastic buckle propagation [3]. Their derivation rests upon the following assumptions; (i) a thickness of the pipe h is sufficiently thin compared with a radius of the pipe a (at the middle surface), i.e., $h/a = \gamma \ll 1$, (ii) a characteristic deflection A_0 is comparable with the radius, i.e., $A_0/a \sim O(1)$, (iii) a characteristic axial length ℓ of a buckle is sufficiently long compared with the radius, i.e., $a/\ell = \varepsilon \ll 1$, and (iv) γ is comparable with ε^2, (e.g., $\gamma \sim O(10^{-2})$ and $\varepsilon \sim O(10^{-1})$). A cross-sectional deformation is shown in Figure 1 where A represents a deflection in the x direction due to finite bending, while B represents its resulting ovalized deformation. Note that $|B|$ must be less than a because no touching of the circumference is assumed. In the followings, B/a is defined as an ovalization B', whose maximum absolute value is designated by δ (<1).

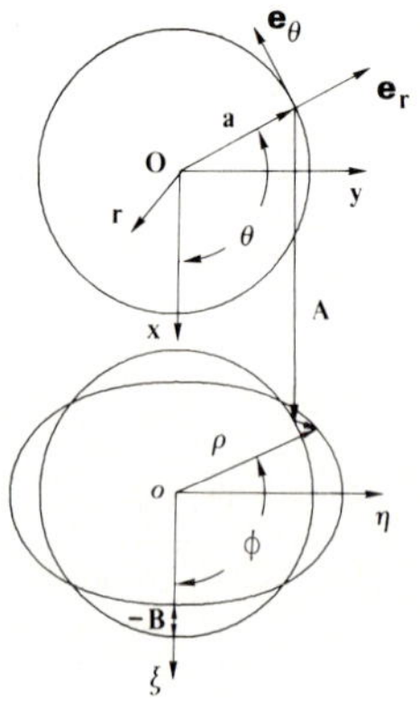

FIGURE 1
Cross-sectional configuration: the ovalized contour is given by
$\xi = \rho\cos\phi = a\cos\theta + B\cos^3\theta$,
$\eta = \rho\sin\phi = a\sin\theta - B\sin^3\theta$,
where ξ and η denote the current coordinates while $x(=r\cos\theta)$ and $y\,(=r\sin\theta)$ the original ones [3].

On the basis of the Lagrangian formulation, the governing equations are given by the coupled nonlinear wave equations for the beam-flexural mode A and the ring-flexural mode B' (=B/a):

$$\rho_0 \frac{\partial^2 A}{\partial t^2} + \frac{Ea^2}{2}\frac{\partial^2}{\partial z^2}\left[\left(1+\frac{3}{2}B'+\frac{1}{16}B'^2\right)\frac{\partial^2 A}{\partial z^2}\right] = T_e \frac{\partial^2 A}{\partial z^2}, \tag{1}$$

$$\rho_0 \frac{\partial^2 B'}{\partial t^2} + \frac{3Eh^{*2}}{5a^4}\left(1-\frac{p}{p_c}\right)B' - \bar{T}_e\frac{\partial^2 B'}{\partial z^2} + \frac{Ea^2}{20}\frac{\partial^4 B'}{\partial z^4} = -\frac{3E}{5}\left(1+\frac{5B'}{6}\right)\left(\frac{\partial^2 A}{\partial z^2}\right)^2, \tag{2}$$

where A and B' depend on the axial coordinate z and the time t; ρ_0 and E denote, respectively, the density in the undeformed state and Young's modulus, while p and p_c denote, respectively, the hydrostatic side pressure and the buckling pressure given by

$$p_c = E(h/a)^3/[4(1-\sigma^2)], \tag{3}$$

σ being Poisson's ratio. Use of this p_c and an 'effective thickness' h^* defined by $h/(1-\sigma^2)^{\frac{1}{2}}$ eliminates an explicit appearance of σ in (1) and (2), where T_e [=T+pa/(2h)] and $\overline{T}_e$ [=T+pa/(5h)] denote the 'effective tension' for the beam-flexural mode and that for the ring-flexural mode, respectively, T being the uniform axial tension applied externally. In the previous paper, T [$\lesssim O(E\varepsilon^2)$] was assumed larger than pa/h [$\sim O(E\gamma^2)$]. Particularly for the ring-flexural mode, since the tension term itself was assumed smaller than other terms, the correction due to the pressure was ignored. But when T becomes comparable with pa/h or even smaller, this inclusion of pa/(5h) in $\overline{T}_e$ is necessary, although this correction has no influence on the following analysis.

3. STEADY BUCKLE PROPAGATION

Let us reconsider the steady-progressive wave solutions to (1) and (2) discussed in [3]. Assuming that A and B' depend on z and t only through the combination ζ =z-vt where v(>0) is a constant velocity, it follows from (1) that

$$qC + \frac{a^2}{2}\frac{d^2}{d\zeta^2}[(1+\frac{3}{2}B' + \frac{1}{16}B'^2)C] = 0. \tag{4}$$

Here C stands for a dimensionless curvature $ad^2A/d\zeta^2$ [$\sim O(\varepsilon^2)$] and $q = (v/v_0)^2 - (v_t/v_0)^2$ where v_0 [$=(E/\rho_0)^{\frac{1}{2}}$] and v_t [$=(T_e/\rho_0)^{\frac{1}{2}}$] denote, respectively, the velocities of the longitudinal wave and the transverse wave due to the effective tension. Note that v_t is far smaller than v_0. Even for a high tension T/E [$\sim O(\varepsilon^2)$], $(v_t/v_0)^2$ still remains of $O(\varepsilon^2)$, while it becomes considerably small of $O(\gamma^2)$ in the absence of the axial tension. Thus $(v/v_0)^2$ and therefore q are also assumed to be small and comparable with $(v_t/v_0)^2$.

With such a small q, nevertheless, qC in (4) should not be ignored because it is the very effective tension and inertia that balance with the second term due to bending to drive the buckle steadily. Equation (4) suggests that this balance is realized over a longer characteristic axial length $|q|^{-\frac{1}{2}}\ell$ so that the derivative $ad/d\zeta$ is regarded as $O(|q|^{\frac{1}{2}}\varepsilon)$.

Expressing (2) in terms of ζ, on the other hand, it also follows that

$$\frac{a^4}{12}\frac{d^4B'}{d\zeta^4} + \frac{5}{3}\overline{q}a^2\frac{d^2B'}{d\zeta^2} + \frac{B'}{\alpha^*} = -(1+\frac{5}{6}B')C^2, \tag{5}$$

where $\alpha^* = (1-p/p_c)^{-1}(a/h^*)^2$ and $\overline{q} = (v/v_0)^2-(\overline{v}_t/v_0)^2$ with $\overline{v}_t$ [$=(\overline{T}_e/\rho_0)^{\frac{1}{2}}$]. Note that $\overline{q}$ is comparable with q. For the longer characteristic length $|q|^{-\frac{1}{2}}\ell$, the first two derivatives in (5) are of $O(q^2\varepsilon^4)$ and $O(|q\overline{q}|\varepsilon^2)$, respectively, whereas the third term is of $O(\gamma^2)$. Hence B/α^* balances mainly with the right-hand side to give the ovalization B' as

$$B' = -\alpha^*(1+5B'/6)C^2. \tag{6}$$

Note that when p is extremely close to p_c, the approximation (6) becomes invalid so that the original equation (5) should be used.

Substituting (6) into (4), we have a single equation for C. To simplify the analysis further, we assume a moderate ovalization δ (e.g., $\delta \lesssim 1/2$) so that the quadratic terms in B' may be neglected. Noting that 5B'/6 gives (4) a correction comparable with B'^2, we derive on neglecting the quadratic terms:

$$qC + \frac{a^2}{2}\frac{d^2}{d\zeta^2}[(1-\frac{3}{2}\alpha^*C^2)C] = 0. \tag{7}$$

In view of the fact that the critical ovalization occurs at $B'=-2/9$, as will be seen below, we define f by $9\alpha^* C^2/2$. On expressing (7) in terms of f and introducing a dimensionless variable $J=|q|^{\frac{1}{2}}\zeta/a$, it reduces to

$$(1-f)\frac{d^2f}{dJ^2} - \frac{1+f}{2f}\left(\frac{df}{dJ}\right)^2 + 4(\mathrm{sgn}\,q)f = 0. \qquad (8)$$

In terms of f, the deflection and the ovalization are given, respectively, from (4) and (6) as

$$A = \pm\frac{h^*}{q}\left(1-\frac{p}{p_c}\right)^{\frac{1}{2}}\tilde{A}, \text{ and } B' = -\frac{2}{9}f, \qquad (9)$$

where $\tilde{A}=2^{\frac{1}{2}}(1-f/3)f^{\frac{1}{2}}/6$. Note that the maximum deflection is attained at f=1. Since (8) can be integrated once to give $(df/dJ)^2=4\mathrm{sgn}q[(f-1)^2-\beta^2]f/(f-1)^2$ where $\beta(>0)$ is an integration constant, f is obtained by evaluating the following elliptic integral:

$$\frac{1}{2}\int_{f_0}^{f}\frac{f'-1}{\{\mathrm{sgn}\,q\,[(f'-1)^2-\beta^2]f'\}^{\frac{1}{2}}}df' = \pm(J-J_0), \qquad (10)$$

where f_0 and J_0 are such constants as $f_0=f(J_0)$. It is shown in [3] that depending on the sign of q, the types of solutions are classified into the flexural wave mode for q>0 and the buckle propagation mode for q<0. The flexural wave mode is nothing but a generalization of linear flexural waves to include the nonlinear cross-sectional ovalization due to bending. Since this has been fully discussed [3], we shall here concentrate on the buckle propagation mode only. For q<0, two types of bounded solutions are possible depending upon β. For $0<\beta<1$, we have a solution f satisfying $1-\beta\le f\le 1+\beta$, while for $\beta\ge 1$, we have f satisfying $0\le f\le 1+\beta$.

In the integral (10), however, we should note that the numerator f'-1 changes sign across f'=1 so that f takes multi-values over some region in J. To avoid this, the absolute value of f'-1 was taken in [3]. But was it really necessary? We should also remark that as f approaches unity, (6) becomes invalid. Since df/dJ diverges as $f\to 1$, the derivatives of $B'(=-2f/9)$ in (5) remain no longer to be small corrections to (6). Rather they now come into a primary role near the critical ovalization at f=1. Except its neighborhood, however, (6) still holds for the subcritical branch (f<1) and even for the supercritical branch (f>1), being apart from a question whether the latter exists stably or not. Hence the solutions to (8) cannot be valid across f=1 but they hold piecewise for f>1 or f<1. Thus for $0<\beta<1$, we have two separate branches in $1-\beta\le f<1$ and $1<f\le 1+\beta$, while for $\beta\ge 1$, $0\le f<1$ and $1<f\le 1+\beta$. Having such piecewise valid solutions in hand (whose analytical expressions are given in [3]), we can now construct various solutions.

In particular, when imposing the boundary conditions $f\to 0$ at infinite ends $|J|\to\infty$, we have two types of solutions for $\beta=1$. One type consists of both the sub- and the supercritical branches [3]:

$$(2-f)^{\frac{1}{2}} - 2^{-\frac{1}{2}}\tanh^{-1}(1-f/2)^{\frac{1}{2}} = \begin{cases} \pm J & (f>1), \\ 2J_c \pm J & (0<f<1), \end{cases} \qquad (11)$$

where $J_c=1-2^{-\frac{1}{2}}\tanh^{-1}(2^{-\frac{1}{2}})$, while the other consists of the subcritical branch only:

$$(2-f)^{\frac{1}{2}} - 2^{-\frac{1}{2}}\tanh^{-1}(1-f/2)^{\frac{1}{2}} = J_c \pm J. \tag{12}$$

Figures 2 and 3 show the respective profiles of f given by (11) and (12), and their associated deflections $\tilde{A}$. It is seen that $\tilde{A}$ has a pair of kinks or a single kink, respectively. This is the very kink mentioned in INTRODUCTION. By analogy of a kink in the measure, it is anticipated here that its fine structure, which now has a characteristic axial length ℓ, would be clarified by solving (4) and (5) without the approximation (6).

Although the solutions (11) and (12) are candidates satisfying the boundary conditions, it is an open question whether or not they can exist stably. To answer it, we have to wait until the full solutions to (4) and (5) are obtained and furthermore their stability is examined. This will be made clear in a future work.

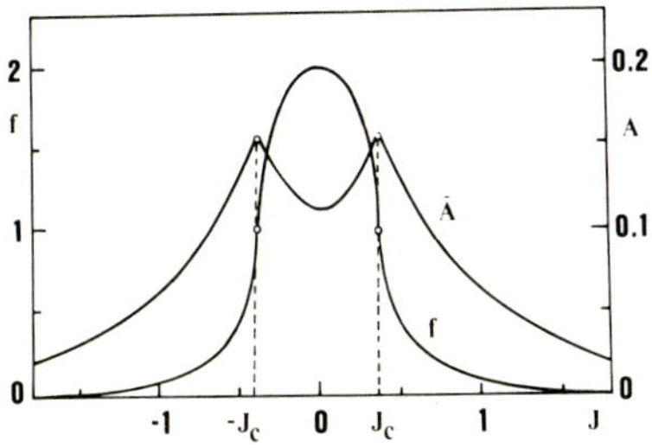

FIGURE 2
Profiles of f and $\tilde{A}$ consisting of the sub- and supercritical branches (11).

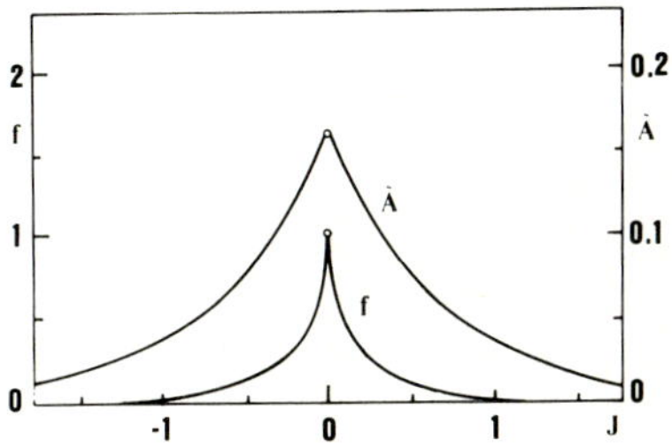

FIGURE 3
Profiles of f and $\tilde{A}$ consisting of the subcritical branch only (12).

4. PROPAGATION PRESSURE AND PROPAGATION VELOCITY

We now evaluate a propagation pressure and a propagation velocity derived from the steady-progressive wave solutions. From (9), the propagation velocity is given in terms of the maximum deflection A_m:

$$\left(\frac{v}{v_0}\right)^2 = \left(\frac{v_t}{v_0}\right)^2 - \frac{2^{\frac{1}{2}}h^*}{9A_m}\left(1-\frac{p}{p_c}\right)^{\frac{1}{2}}, \tag{13}$$

where $(v_t/v_0)^2 = [T+pa/(2h)]/E$. It is important to remark that the velocity changes depending on the magnitude of the maximum deflection.

To measure a relative magnitude of A_m, we introduce a dimensionless maximum deflection defined by

$$S = (A_m/D)(h^*/D), \tag{14}$$

where D (=2a) is the diameter of the pipe. In addition, a relative magnitude of the axial tension to the side pressure is measured by a dimensionless parameter N defined by

$$N = 2hT/(ap_c) = 2\pi ahT/(\pi a^2 p_c), \qquad (15)$$

where N implies a ratio of the axial force applied externally to an imaginary force as if the buckling pressure would be applied over the whole cross-section. In terms of S and N, (13) is expressed as

$$\left(\frac{v}{v_0}\right)^2 = \left(\frac{h^*}{D}\right)^2\left[\frac{1}{2}\left(N + \frac{p}{p_c}\right) - \frac{2^{\frac{1}{2}}}{9S}\left(1 - \frac{p}{p_c}\right)^{\frac{1}{2}}\right]. \qquad (16)$$

4.1. Propagation Pressure

Propagation pressure p_p is defined as a pressure such that a stationary buckle begins to propagate with an 'infinitely slow' velocity v=0 [6,7]. This pressure is obtained from (16) as

$$\frac{1}{2}\left(N + \frac{p_p}{p_c}\right) = \frac{2^{\frac{1}{2}}}{9S_p}\left(1 - \frac{p_p}{p_c}\right)^{\frac{1}{2}}, \qquad (17)$$

where S_p denotes S at $p=p_p$. For given values of S_p and N, p_p/p_c is obtained as follows:

$$\frac{p_p}{p_c} = -N - \frac{4}{81S_p^2} + \frac{4}{81S_p^2}\left[1 + \frac{81}{2}(N+1)S_p^2\right]^{\frac{1}{2}}. \qquad (18)$$

In particular, for a small value of S_p, (18) is approximated as

$$p_p/p_c = 1 - 81(N+1)^2 S_p^2/8. \qquad (19)$$

The semi-logarithmic graph of p_p/p_c vs. S_p is shown in Figure 4 for the several values of N. It is seen that as N increases, a smaller S_p can give rise to a buckle propagation. For N=0, the curve approaches asymptotically the abscissa $p_p/p_c=0$. This means that a buckle can never be propagated for whatever large deflection when both actions of axial tension and side pressure are absent. For N≠0, the curve crosses the abscissa at $S_p = 2^{3/2}/(9N)$ which gives the least S necessary for a buckle to be propagated in the absence of the side pressure. If a negative pressure is allowed, the curve approaches asymptotically $p_p/p_c = -N$. This negative pressure counteracts the tension to hinder the propagation.

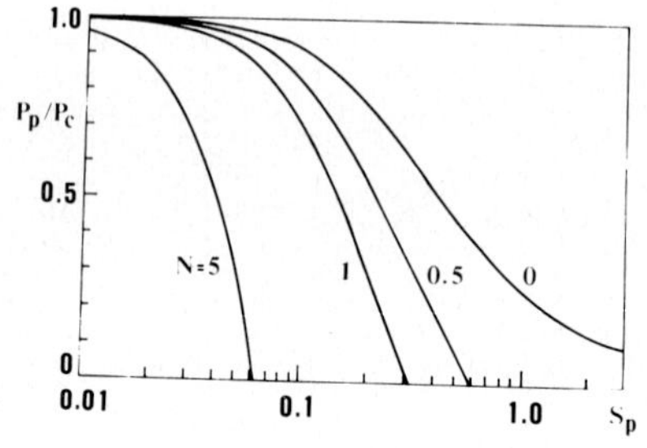

FIGURE 4
Semi-logarithmic graph of p_p/p_c vs. S_p with the tension N fixed.

Finally we remark that although S_p has been so far regarded as a maximum deflection of a given pipe, i.e., with h^*/D fixed, Figure 4 also indicates a decrease of p_p/p_c as h^*/D increases with A_m/D held constant.

4.2. Propagation Velocity

When the parameter N is eliminated from (16) by using (17), the propagation velocity can also be expressed as

$$(\frac{v}{v_0})^2 = (\frac{h^*}{D})^2 \{\frac{2^{\frac{1}{2}}}{9}[\frac{(1 - p_p/p_c)^{\frac{1}{2}}}{S_p} - \frac{(1 - p/p_c)^{\frac{1}{2}}}{S}] + \frac{1}{2}(\frac{p}{p_c} - \frac{p_p}{p_c})\} . \qquad (20)$$

Here S must be greater than S_p for the right-hand side to be positive. With N fixed (N=0.5), Figure 5 shows the propagation velocity vs. the pressure for the several values of S. The ordinate V $[=(v/v_0)/(h^*/D)]$ measures a velocity renormalized by h^*/D. Each curve of S crosses the abscissa V=0 at the respective propagation pressure with S_p=S. As p/p_c tends to unity, on the other hand, all curves are convergent toward $V=V_c$ with $V_c=[(N+1)/2]^{\frac{1}{2}}$, which corresponds to the velocity of transverse wave when $p=p_c$. Remark here that although V approaches V_c, the approximation (6) breaks down physically there.

For a fixed pressure, there always exists a threshold value of S, i.e., S_p. If the pressure is increased with S held constant, the velocity increases along the corresponding curve. With the pressure fixed, on the contrary, if the deflection is increased from its corresponding threshold value S_p, the velocity increases by traversing upward a family of curves having different values of S.

Figure 5 is drawn on the basis of p_c. But it is interesting to re-scale the abscissa by p_p to compare with a graph of experimental data [7]. Figure 6 shows that the velocity V vs. the pressure normalized by p_p for the same values of S as in Figure 5. All curves increase from $p/p_p=1$ toward $V/V_0=(h^*/D)V_c$ at the respective values of p_c/p_p. Note that the ordinate in [7] measures v/v_y with v_y $[=(\sigma_y/\rho_0)^{\frac{1}{2}}]$, σ_y being the yield stress. If σ_y is taken to be 0.005E, v_y is about $0.07v_0$. Since the factor 0.07 is comparable with h^*/D, Figure 6 seems consistent, at least qualitatively, with the experiment.

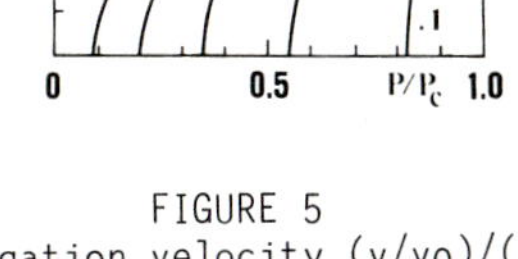

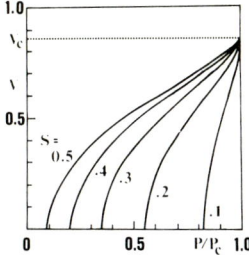

FIGURE 5
Propagation velocity $(v/v_0)/(h^*/D)$ vs. pressure p/p_c for N=0.5.

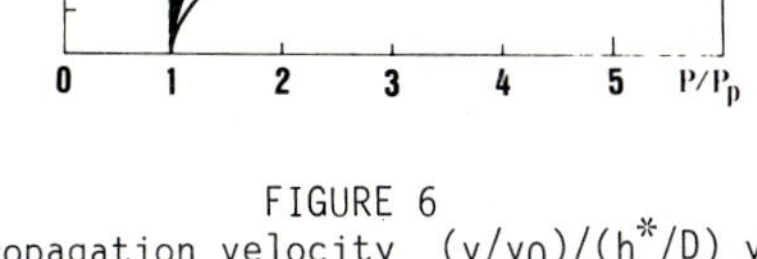

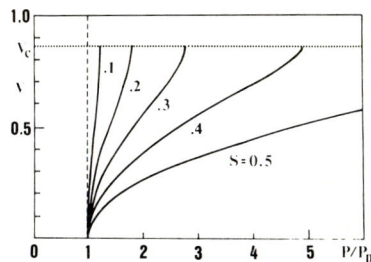

FIGURE 6
Propagation velocity $(v/v_0)/(h^*/D)$ vs. pressure p/p_p reproduced from Figure 5.

5. CONCLUDING REMARKS

In concluding this paper, we mention a simple experiment observing the propagation of a kink. Let the measure referred to in INTRODUCTION be hung, in a vertial plane, subjected to bending and axial tension due to its dead load (Figure 7). When a local kink is produced at an upper portion, it is easily seen that the kink can be propagated downwards (Figure 8). It seems that this phenomenon might be identified in nature as the one discussed for the circular pipe. It is expected that its behavior could be well modeled by the simplified equation (14) in [3] (but only for $\partial^2A/\partial z^2>0$, say because the measure lacks the symmetry in bending). Although it is within a rough qalitative analogy, this experiment facilitates a visual understanding of the pipeline-problem concerned here.

Within the elasticity, the present analysis assumes that the pipe resumes the original circular cross-section after the buckle has been propagated. In reality, however, it is conjectured that once a kink driven elastically is propagated along the pipe, a local plastic collapse would immediately follow the kink, which might be seen like a toppling of dominos.

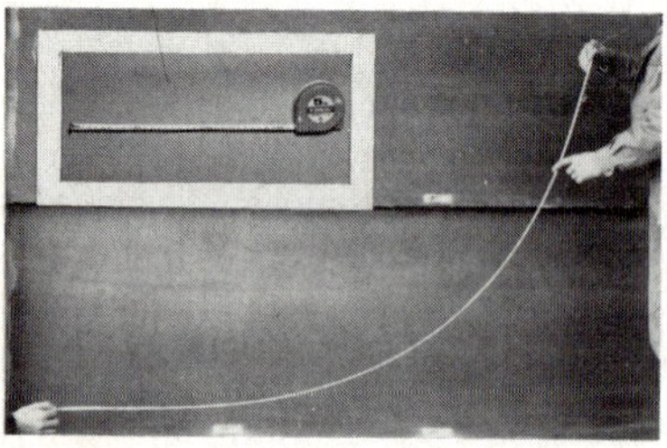

FIGURE 7
Experiment observing the propagation of a kink in the steel flexible measure in reel (shown inset) subjected to bending and tension. An initial kink is produced by flexing the measure locally with fingures.

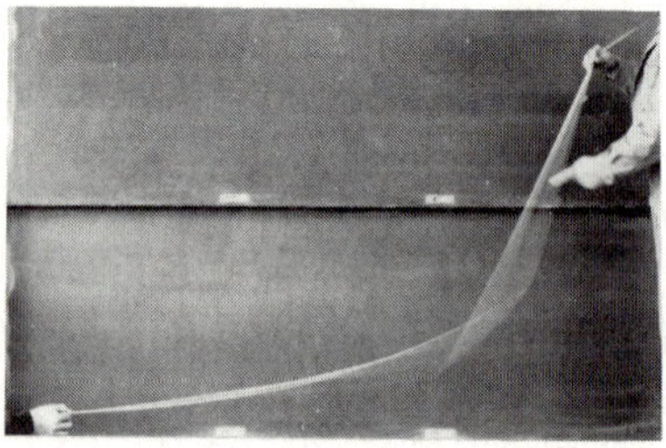

FIGURE 8
Propagation of a kink: Photographed is a locus of the kink during a short exposure (the measure itself is not twisted). Its propagation velocity is estimated to be a few meters per second.

ACKNOWLEDGEMENTS

The author would like to thank Professor T. Kakutani for his valuable comments on reading the manuscript. Assistance of Messrs. Tatsuoka and Umetani is also acknowledged.

REFERENCES

[1] Palmer, A. C. and Martin, J. H., Nature 254 (1975) 46.
[2] Mesloh, R., Johns, T. G. and Sorenson, J. E., The Propagating Buckle, in: Proc. BOSS'76 (The Norwegian Institute of Technology, Trondheim, 1976) pp. 787-797.
[3] Sugimoto, N., Buckle Propagation in Elastic Pipelines, in: Kawata, K. (ed.) Proc. IUTAM symposium on Macro- and Micro-mechanics of High Velocity Deformation and Fracture (Springer, Berlin, in print).
[4] Brazier, L. G., Proc. Roy. Soc. London, A-106 (1927) 104.
[5] Calladine, C. R., Theory of Shell Structures (Cambridge Univ. Press, Cambridge, 1983).
[6] Kyriakides, S. and Babcock, C. D., TRANS. ASME J. Pressure Vessel Tech. 103 (1981) 328.
[7] Kyriakides, S. and Babcock, C. D., Buckle Propagation Phenomena in Pipelines in: Thompson, J. T. and Hunt, G. W. (eds.) Proc. IUTAM symposium on Collapse: Buckling of Structures in Theory and Practice (Cambridge Univ. Press, Cambridge, 1983) pp.75-91.

ELASTIC WAVE PROPAGATION
M.F. McCarthy, M.A. Hayes, (Editors)

COUPLED WAVE MOTION OF A LAYERED PERIODIC BEAM

Huey-Ju Yang

Associate Professor
Institute of Applied Mechanics, National Taiwan University
Taipei, Taiwan, 10764, R. O. C.

The problem of analyzing coupled flexural, longitudinal and shear wave propagation in a layered periodic beam has been studied. The coupling effects due to damping, inertia and shear deformation are considered in the analysis. Principle of virtual work is used to derive the equation of motion and Floquet theory is employed to obtain the dispersion Floquet wave numbers and dispersion relations. Plots for frequency versus complex wave number are presented. Results for degenerated piece-wise homogeneous periodic beam and uniform two- and three- layered beams can be recovered. Discussions and suggestions for further studies are included.

Introduction

Wave motions in a periodic structure having identical material elements and coupling effects in a layered uniform beam have been studied in [1] and [2] respectively. However, as the periodically arranged composite materials have been used more and more extensively in engineering structures, we will study a layered beam with material rather than structural periodicity. A periodic beam consisting of two parts(A and B) while in each part it allows up to three layers with variable thicknesses is considered in this work. The geometry, displacement and coordinate system of a longitudinal section of the beam is shown in figure 1. The transverse cross-section is rectangular of width b and the longitudinal axis coincides with the center of gravity (CG) of the beam. Any two adjacent parts comprise a unit cell, and this unit cell is extended to infinity along the positive and negative x-axis. Perfect bonding is assumed between adjacent parts and layers respectively. The material properties of each layer in each part has density ρ ,Young's modulus E and shear modulus G with its identification subscript , such as ρ_{A1} or G_{B3}. To simplify the problem , the CG axis should maintain a straight line through the structure, and the thickness h of each layer may be adjusted accordingly. Either elastic ,viscoelastic or ceramic material can be used for each layer of the structure with relevant material properties for specific purposes. For instance, a linear viscoelastic layer may be considered to construct damped beams and gives rise to the damping coupling effects. Its moduli may be complex numbers while the constitutive law is the same as for linear elastic material. For some other structures, one ceramic layer can be assumed to act as an extra non-structural mass with negligible Young's and shear moduli which vibrates with the beam, and adds up inertia coupling effect significantly.

It is assumed that plane sections of the beam remain plane, but not necessarily perpendicular to the longitudinal axis, the shear deformations are therefore included, but lateral strains through the thickness of all layers are neglected. Inertia couplings between both flexural-longitudinal and flexural-rotational motion are considered.

Problem Formulation

A time-harmonic wave propagating along a infinite beam will be analyzed in this paper. Following a similar procedure as in [1] and [2], the governing equations for either part A or B can be obtained by using the virtual work principle as :

$$\begin{aligned}
&Pu_{,xx} - mu_{,tt} + S\psi_{,xx} = 0, \\
&Su_{,xx} + Q\psi_{,xx} - I\psi_{tt} - R\psi - Rw_{,x} = 0, \qquad (1)\\
&-R\psi_{,x} - \overset{*}{R}w_{,xx} + mw_{,tt} = 0,
\end{aligned}$$

where

$P \equiv \sum_{i=1}^{3} bh_i E_i$ = normal stiffness,
$Q \equiv \sum_{i=1}^{3} E_i I_i$ = flexural stiffness about the CG axis,
$R \equiv \sum_{i=1}^{3} bh_i G_i$ = shear stiffness,
$S \equiv \sum_{i=1}^{3} bh_i E_i H_i$ = first moment of direct stiffness about CG axis,
$m \equiv \sum_{i=1}^{3} bh_i \rho_i$ = mass per unit length of beam,
$I \equiv \sum_{i=1}^{3} I_i \rho_i$ = moment of inertia about the CG axis.

From the fact that the coefficients of the equations of motion have periodic variation from cell to cell (i.e. with period $\tau = l_A + l_B$), it is known from the theory of Floquet [3] that solutions of the differential equations with periodic coefficients have the quasi-periodic recurrence relation

$$q(x+\tau, t) = q(x,t)\lambda,$$

where $\lambda = e^{ik_f\tau}$, k_f is said to be the wave number of the Floquet wave, and $q = u$, ψ or w.
Considering the harmonic wave motion, we substitute

$$\begin{Bmatrix} u \\ \psi \\ w \end{Bmatrix} = \begin{Bmatrix} U \\ \Psi \\ W \end{Bmatrix} e^{-i(kx-\omega t)} \tag{2}$$

into eqn.(1), and obtain

$$\begin{bmatrix} -Pk^2 + m\omega^2 & -Sk^2 & 0 \\ -Sk^2 & -Qk^2 - R + i\omega^2 & iRk \\ 0 & iRk & Rk^2 - m\omega^2 \end{bmatrix} \begin{Bmatrix} U \\ \Psi \\ W \end{Bmatrix} = 0. \tag{3}$$

For nontrivial solutions, the determinant in eqn.(3) must be zero and the dispersion relation is obtained as :

$$R(PQ - S^2)k^6 - \omega^2[PRI - m(P + R - S^2)]k^4$$

$$+m\omega^2[\omega^2(RI + mQ) - P(R - I\omega^2)]k^2 + m^2\omega^4(R - I\omega^2) = 0. \tag{4}$$

This is a bi-cubic equation in k, so that both $\pm k$ are solutions. For any given real frequency ω, wave numbers k_n of complex values will be solutions for eqn.(4), which are indicative of wave attenuation.
Because of the existence of the connecting faces, traveling waves in the positive and negative x-direction must both be considered. Thus we have

$$\begin{Bmatrix} u \\ \psi \\ w \end{Bmatrix}_{AorB} = \sum_{n=1}^{3} \begin{Bmatrix} U_n e^{-ik_n x} + U_{-n} e^{ik_n x} \\ \Psi_n e^{-ik_n x} + \Psi_{-n} e^{ik_n x} \\ W_n e^{-ik_n x} + W_{-n} e^{ik_n x} \end{Bmatrix}_{AorB} e^{i\omega t} \tag{5}$$

Substituting $U_{\pm n}$, $\Psi_{\pm n}$ and $W_{\pm n}$ into eqn.(3), we obtain the following relations :

$$\Psi_{\pm} = \beta_{\pm n} W_{\pm n}, \qquad \text{where} \qquad \beta_{\pm n} \equiv \pm \frac{(Rk_n^2 + m\omega^2)}{Rk_n} \tag{6}$$

$$U_{\pm n} = \gamma_{\pm n} W_{\pm n}, \qquad \text{where} \qquad \gamma_{\pm n} \equiv \pm \frac{Sk_n(Rk_n^2 + m\omega^2)}{R(Pk_n^2 + m\omega^2)}. \tag{7}$$

At any connecting face, three displacement(angle) and three force(moment) continuity conditions must be satisfied, i.e.

$$u|_- = u|_+ \quad , \quad (P\tfrac{\partial u}{\partial x} + S\tfrac{\partial \psi}{\partial x})|_- = (P\tfrac{\partial u}{\partial x} + S\tfrac{\partial \psi}{\partial x})|_+ \quad ,$$

$$\psi|_- = \psi|_+ \quad , \quad (S\tfrac{\partial u}{\partial x} + Q\tfrac{\partial \psi}{\partial x})|_- = (S\tfrac{\partial u}{\partial x} + Q\tfrac{\partial \psi}{\partial x})|_+ \quad , \tag{8}$$

$$w|_- = w|_+ \quad , \quad R(\psi + \tfrac{\partial w}{\partial x})|_- = R(\psi + \tfrac{\partial w}{\partial x})_+ \quad ,$$

where $_-$ and $_+$ denote the beam cross sections just to the left and right of the connecting face.

Applying continuity conditions at faces f_1 and f_2 to eqn.(5) with local coordinate x' instead of global coordinate x as in Fig.1.(a) and using eqns. (6) and (7), matrix equations for part A1, B1 and A2 are obtained :

$$[B]\{W\}_{B1} = [A][D]_A\{W\}_{A1}, \tag{9}$$

$$[A]\{W\}_{A2} = [B][D]_B\{W\}_{B2}, \tag{10}$$

where

$$[D]_j \equiv diag(\ e^{-i(k_1l)_j} \quad e^{-i(k_2l)_j} \quad e^{-i(k_3l)_j} \quad e^{i(k_1l)_j} \quad e^{i(k_2l)_j} \quad e^{i(k_3l)_j}\ ), \tag{11}$$

$$j = A \text{ or } B,$$

$$\{W\} \equiv (\ W_1 \quad W_2 \quad W_3 \quad W_{-1} \quad W_{-2} \quad W_{-3}\ )^T, \tag{12}$$

$$[A]\text{or}[B] = \begin{bmatrix} 1 & 1 & 1 & 1 & 1 & 1 \\ \beta_1 & \beta_2 & \beta_3 & \beta_{-1} & \beta_{-2} & \beta_{-3} \\ \gamma_1 & \gamma_2 & \gamma_3 & \gamma_{-1} & \gamma_{-2} & \gamma_{-3} \\ \varsigma_1 & \varsigma_2 & \varsigma_3 & \varsigma_1 & \varsigma_2 & \varsigma_3 \\ \eta_1 & \eta_2 & \eta_3 & \eta_1 & \eta_2 & \eta_3 \\ \xi_1 & \xi_2 & \xi_3 & \xi_{-1} & \xi_{-2} & \xi_{-3} \end{bmatrix} \text{(evaluated for part A or B)} \tag{13}$$

where $\beta_{\pm n}, \gamma_{\pm n}$ are defined in eqns.(6) and (7),

$$\begin{aligned} \varsigma_n &\equiv -k_n(Q\beta_n + S\gamma_n) \\ \eta_n &\equiv -k_n(S\beta_n + P\gamma_n), \quad \text{for} \quad n = 1, 2, 3. \\ \xi_{\pm n} &\equiv \pm m\omega^2/k_n \end{aligned} \tag{14}$$

Combining equations (9) and (10) and using Floquet theory, we have the eigenvalue problem :

$$[T]\{W\}_{A1} = \{W\}_{A2} = \lambda\{W\}_{A1}, \tag{15}$$

where

$$[T] = [A]^{-1}[B][D]_B[B]^{-1}[A][D]_A. \tag{16}$$

After solving eqn.(15),we can obtain the eigenvalue λ for any given frequency ω. In a similar way, applying continuity conditions at connecting face f_3,

$$[B]\{W\}_{B2} = [A][D]_A\{W\}_{A2}. \tag{17}$$

Applying Floquet theory again, eqns. (10) and (17) yield the second eigenvalue problem :

$$[T']\{W\}_{B1} = \{W\}_{B2} = \lambda\{W\}_{B1}, \tag{18}$$

where

$$[T'] = [B]^{-1}[A][D]_A[A]^{-1}[B][D]_B = [L]^{-1}[T][L] \quad , \quad [L] \equiv [D]_A^{-1}[A]^{-1}[B]. \tag{19}$$

Note that $[T']$ and $[T]$ are similar matrices, thus the eigenvalue λ is the characteristic parameter from cell to cell. The relation between eigenvectors $\{X\}_A$ and $\{X\}_B$ of parts A and B respectively is

$$\{X\}_B = [B]^{-1}[A][D]_A\{X\}_A = [L]^{-1}\{X\}_A. \tag{20}$$

The wave amplitudes can then be expressed as linear combinations of eigenvectors $\{X_n\}$:

$$\begin{aligned}\{W\}_{A1} &= [\{X\}]_{A1}\{C\}_{A1} \\ \{W\}_{B1} &= [\{X\}]_{B1}\{C\}_{B1}\end{aligned} \tag{21}$$

where $\{W\}$ is defined in eqn.(12) and

$$[\{X\}] \equiv [\ \{X_1\}\ \ \{X_2\}\ \ \{X_3\}\ \ \{X_{-1}\}\ \ \{X_{-2}\}\ \ \{X_{-3}\}\ ].$$

$\{C\}$ is an arbitrary constant vector, that should be determinated from given boundary conditions (we are concerned with waves propagating in infinite beams only, thus the boundary conditions can not specified here). After substituting eqn.(20) and eqn.(21) into eqn.(9), we obtain

$$[A][D]_A[\{X\}]_{A1}\{C\}_{A1} = [B][B]^{-1}[A][D]_A[\{X\}]_{A1}\{C\}_{B1} \tag{22}$$

which gives $\{C\}_{A1} = \{C\}_{B1}$, i.e. the wave amplitudes of part A and part B in a unit cell can be expressed as linear combinations of their eigenvectors respectively and these linear combinations have the same coefficient vector $\{C\} = \{C\}_{A1} = \{C\}_{B1}$.

When materials of each layer in parts A and B are the same, i.e. $[A] = [B]$, the uniform three-layered beam is recovered. The transfer matrices for eigenvalue problems $[T]$ and $[T']$ in eqns.(16) and (19) reduce to $[T] = [D]_B[D]_A = [D]_A[D]_B = [T']$. The eigenvalues thus become $e^{-ik_1\tau}$, $e^{-ik_2\tau}$, $e^{-ik_3\tau}$, $e^{ik_1\tau}$, $e^{ik_2\tau}$ and $e^{ik_3\tau}$. This result identifies the wave propagation characteristics for non-periodic structures, and the Floquet wave numbers coincide with regular wave numbers as expected.

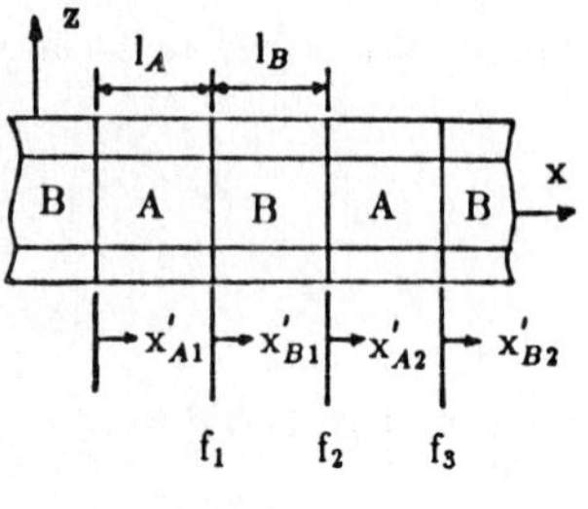

Fig. 1 (a) Illustrating diagram for a layered periodic infinite beam.

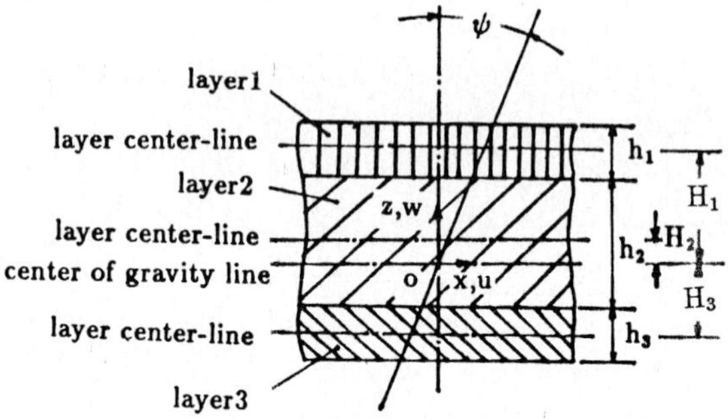

Fig. 1 (b) Longitudinal cross-section diagram for part A or part B.

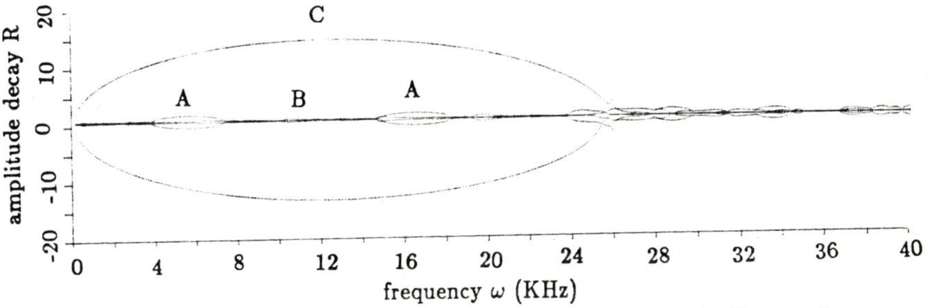

Fig. 2 (a) Amplitude decay of waves versus frequency for periodic materials.

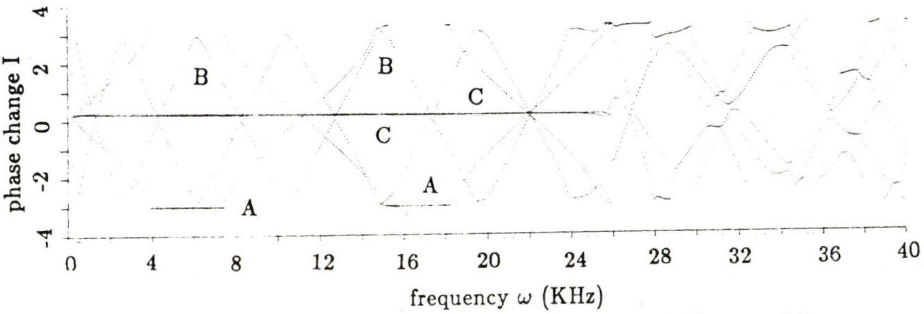

Fig. 2 (b) Phase change of waves versus frequency for periodic materials.

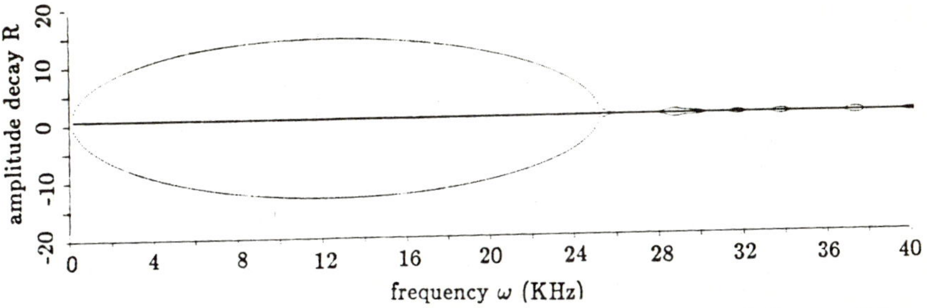

Fig. 2(c) Amplitude decay of waves versus frequency for non-periodic materials.

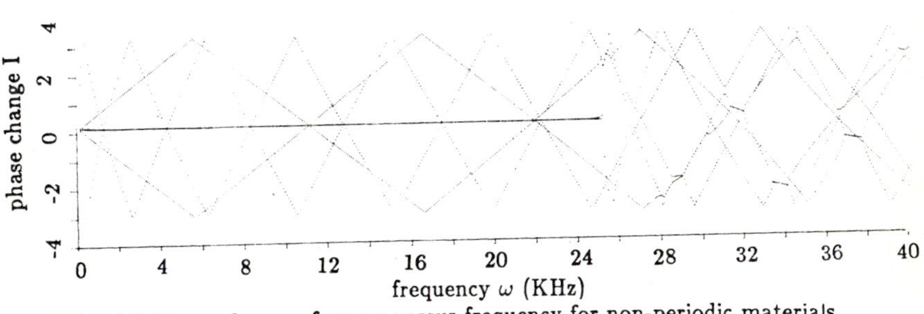

Fig 2(d) Phase change of waves versus frequency for non-periodic materials.

Results and Discussions

There are six dispersion curves corresponding to six wave numbers. Dispersion relations of Floquet wave numbers versus frequency are shown in Fig. 2. We write λ as e^{R+iI} where e^R and I give amplitude attenuation ratio and phase change of the three propagating waves for any two adjacent cells. R and I versus frequency ω are plotted in Figs. 2(a) to 2(d) while in 2(c) and 2(d) the two periodic materials are assumed to be equal in order to make comparisons with 2(a) and 2(b). In these figures, curves A, B and C represent the longitudinal, shear and flexural waves respectively.

In the low frequency range these three waves propagate independently, yet under high frequency vibrations, they are highly coupled as is a well understood phenomenon. In Figs. 2(a) and 2(b) curve A has two small attenuation zones for the low frequency region since particle motion of the longitudinal wave is parallel to the wave normal and produces attenuation upon impacting on any material discontinuity at the interface. Note also that the shear and flexural waves are almost identical for periodic and non-periodic materials and this implies that periodicity of the structure does not affect the transverse waves seriously.

Once the eigenvalues are found they can also be applied to periodic structures with periodic supports and the propagation constants can be obtained for such structures. Reductions to homogeneous material with periodic supports should be recovered to agree with previous results[1]. Similarly, results for degenerated piece-wise homogeneous periodic beam can be derived by having only one material in each part and should be compared with previous work for such cases. Further studies for forced vibrations of finite beams as well as for periodically supported beams and plates are both planned for the future.

References

[1] D. J. Mead, "A new method of analyzing wave propagation in periodic structures: applications to periodic Timoshenko beams and stiffened plates ", *J. of Sound And Vibration*, vol. 104(1), pp. 9-27, 1986.

[2] D. J. Mead and S. Markus, "Coupled flexural, longitudinal and shear wave motion in two- and three-layered damped beams", *J. of Sound and Vibration*, vol. 99(4), pp. 501-519, 1985.

[3] E. I. Ince, "Ordinary differential equations", Dover Publications, New York, pp. 381-384, 1956.

[4] M. G. Faulkner and D. P. Hong, "Free vibrations of a mono-coupled periodic system", *J. of Sound and Vibration*, vol. 99(1), pp. 29-42, 1985.

[5] D. J. Mead and S. Markus, "Coupled flexural-longitudinal wave motion in a periodic beam", *J. of Sound and Vibration*, vol. 90(1), pp. 1-24, 1983.

[6] T. J. Delph , G. Herrmann, and R. K. Kaul, "Harmonic wave propagation in a periodically layered, infinite elastic body: antiplane strain ", *J. of Applied Mechanics*, vol. 45, pp. 343-349, 1978.

[7] D. J. Mead, "Vibration response and wave propagation in periodic structures", *J. of Engineering for Industry*, pp. 783-792, August, 1971.

[8] T. C. Huang, "The effect of rotatory inertia and of shear deformation on the frequency and normal mode equations of uniform beams with simple end conditions", *J. of Applied Mechanics*, pp. 579-584, December, 1961.

[9] K. F. Graff, "Wave motion in elastic solid", *Ohio State University Press*, pp. 431-479, 1975.

[10] C. T. Chung, "A study of wave propagation in periodic plates", *master thesis, National Taiwan University*, 1987.

ELASTIC WAVE PROPAGATION
M.F. McCarthy, M.A. Hayes, (Editors)

The Normal Impact of a Rod-Mass System on a Viscoelastic Layered Half Space

Harold A. Downey, Jr.

IBM Corp.
1001 W. T. Harris Blvd.
Charlotte, NC 28257

and

David B. Bogy

Department of Mechanical Engineering
University of California, Berkeley, CA 94720

A rod with a lumped mass attached to its trailing end travels axially with a uniform velocity and strikes an elastic half space that is covered with an adhering viscoelastic layer. The problem is reduced to integral equations for the average contact stress and the displacement of the rod tip into the contact surface. The kernels of these integral equations are composed of temporal Green's functions for the rod and the layered half space, which represent the response of each to an impulsive uniform normal traction. The Green's function for the rod is obtained in closed form, while that for the layered half space is obtained through a numerical Laplace transform inversion. The integral equations are solved numerically with a second-order stable scheme. Solutions are computed for a wide variety of materials and configurations, providing the stress and displacement history, as well as the stress-displacement response. The results show the effects of changes in rod material and length, lumped mass, layer material, substrate material, and viscoelastic material parameters.

To appear in *ASME Journal of Applied Mechanics*, 1988.

ELASTIC WAVE PROPAGATION
M.F. McCarthy, M.A. Hayes, (Editors)

LOVE WAVES IN A FLUID-SATURATED POROUS ANISOTROPIC LAYER RESTING ON A HETEROGENEOUS ELASTIC HALF-SPACE

Z. Kończak

Technical University of Poznań, Institute of Applied Mechanics, Piotrowo 3, P.B. 5, 60-965 Poznań, Poland

The paper deals with the propagation of Love waves in a transverse-isotropic fluid-saturated porous layer resting on an elastic non-homogeneous half-space. Using the Biot's theory for the porous layer and the theory of elasticity for the lower medium, the dispersion equation has been derived. This complex transcendental equation that relates frequency, phase velocity, anisotropy factor of the layer and the inhomogeneity character of the half-space has been solved by the use of approximation method.

1. INTRODUCTION

The propagation of Love waves in layered media, due to its important applications in geophysics and seismology, has been the subject of investigation by a number of researchers. However, most of the work done in this field does not concern porous layer saturated with fluid. There are few papers only in which the propagation of Love waves in such media is studied. Deresiewicz [1], basing on the Biot's theory [2] for wave propagation in an isotropic fluid-saturated porous medium, has studied the propagation of Love waves in a porous layer resting on an elastic, homogeneous and isotropic half-space. Chakraborty and Dey [3], have examined the same problem, but under the assumption that the water-saturated porous layer possesses, so called, Weiskopf-type anisotropy and rests on a non-homogeneous, elastic semi-infinite space. Chattopadhyay and De [4] have investigated the propagation of Love waves in an isotropic, fluid-saturated porous layer with irregular interface between the layer and the lower homogeneous and isotropic half space.

It must be pointed out, however, that in [3,4] the dissipation caused by the relative fluid flow has been omitted.

In the present paper a homogeneous transverse-isotropic fluid-saturated porous layer has been considered over the elastic heterogeneous semi-infinite space, the inhomogeneity of which varies exponentially with depth. No simplifying assumptions concerning the dissipation terms have been made. The dispersion equation has been derived. This complex transcendental equation has been solved by the use of successive approximation method. It has been found that the phase velocityof Love waves is influenced by the non-homogeneity characteristic of the medium as well as by the fluid-saturated porous medium.

2. GOVERNING EQUATIONS

Consider a two-layer medium. The layer $-h \leqslant x_3 \leqslant 0$ is a fluid-saturated porous transverse-isotropic medium resting on an elastic non-homogeneous half-space $x_3 \geq 0$. The x_3-axis is taken vertically downward in the lower medium and x_1-axis is chosen parallel to the layer in the direction of propagation of the disturbance. It is assumed that the propagating wave changes harmonically in time.

The inhomogeneity in the half-space has been taken as

(1) $$\mu(x_3) = \mu_0 \exp(q x_3), \quad \varrho^*(x_3) = \varrho_0^* \exp(q x_3)$$

where μ_0 and ϱ_0^* are the constant values of shear modulus μ and of mass density ϱ^* at the interface ($x_3 = 0$), q is a constant.

The equations of motion for Love wave propagation have been derived in paper [5] and are quoted here without derivation. We have:

- for the porous layer ($u_1 = 0$, $u_3 = 0$, $u_2 = u(x_1, x_3, t)$, $U_1 = 0$, $U_3 = 0$, $U_2 = U(x_1, x_3, t)$) [5,6]

(2) $$N\frac{\partial^2 u}{\partial x_1^2} + G\frac{\partial^2 u}{\partial x_3^2} = \varrho_{11}\frac{\partial^2 u}{\partial t^2} + \varrho_{12}\frac{\partial^2 U}{\partial t^2} - b_{11}\frac{\partial}{\partial t}(U-u)$$
$$0 = \varrho_{12}\frac{\partial^2 u}{\partial t^2} + \varrho_{22}\frac{\partial^2 U}{\partial t^2} + b_{11}\frac{\partial}{\partial t}(U-u)$$

- for the lower heterogeneous semi-infinite space ($v_1 = 0$, $v_3 = 0$, $v_2 = v(x_1, x_3, t)$) [5]

(3) $$\nabla^2 v + \frac{1}{\mu}\frac{d\mu}{dx_3}\frac{\partial u}{\partial x_3} = \frac{1}{\beta}\frac{\partial^2 v}{\partial t^2}$$

Here u, U, v are the displacement of the solid skeleton, the fluid and the elastic base, respectively, N, G are the shear moduli of the porous skeleton, b_{11} is the component of flow resistance tensor for the fluid, $\beta^2 \equiv \beta_0^2 = \mu_0/\varrho_0^*$ is the shear wave velocity in lower medium. The dynamics coefficients ϱ_{11}, ϱ_{22}, ϱ_{12} take into account the inertia effects of the moving fluid and are related to the mass densities of the solid ϱ_s and fluid ϱ_f by the equations [2]

$$\varrho_{11} + \varrho_{12} = (1-f)\varrho_s, \quad \varrho_{12} + \varrho_{22} = f\varrho_f, \quad \varrho = \varrho_{11} + 2\varrho_{12} + \varrho_{22} = \varrho_s + f(\varrho_f - \varrho_s)$$

The details concerning these parameters can be found in [2,5].

The equations of motion (2) and (3) have to be solved subjected to continuity conditions on stresses and displacements at the interface between the porous layer and the subjacent half-space. Also the surface $x_3 = -h$ of the layer is required to be free of traction. Thus at $x_3 = -h$ we should have

(4) $$G\frac{\partial u}{\partial x_3} = 0$$

and at $x_3 = 0$

(5) $$u = v, \quad G\frac{\partial u}{\partial x_3} = \mu_0\frac{\partial v}{\partial x_3}$$

Equations (1) to (5) are the governing equations of the problem.

3. DISPERSION EQUATION

The solution of the equations (2) and (3), for the wave

changing harmonically with time, have been given for $-h \leqslant x_3 \leqslant 0$ by

$$\text{(6)} \qquad \begin{aligned} u(x_1,x_3,t) &= (A_1 \cos \varkappa_1 x_3 + A_2 \sin \varkappa_1 x_3)\exp(i\phi), \quad i^2=-1, \quad \phi = kx_1 - \omega t \\ U(x_1,x_3,t) &= (\bar{A}_1 \cos \varkappa_1 x_3 + \bar{A}_2 \sin \varkappa_1 x_3)\exp(i\phi) \end{aligned}$$

and for $x_3 \geqslant 0$, where the disturbance is required to vanish with depth, by

$$\text{(7)} \qquad v(x_1,x_3,t) = A_3 \exp(-\eta x_3)\exp(i\phi)$$

where A_1, A_2, A_3 are arbitrary constants and

$$\text{(8)} \qquad \varkappa_1^2 = \xi_1^2 - \frac{N}{G}k^2, \quad \eta = \frac{1}{2}\left(q + \sqrt{q^2 + 4\varkappa_2^2}\right), \quad \varkappa_2^2 = k^2 - \frac{\omega^2}{\beta_0^2}$$

$$\text{(9)} \qquad \begin{aligned} &\xi_1^2 = \alpha_1 + i\alpha_2, \quad k^2 = a_1 + ia_2, \quad a_1 > 0, \quad a_2 \geqslant 0, \quad \alpha_1 = \mathcal{F}\frac{\omega^2}{c_G^2}, \quad \alpha_2 = \mathcal{R}\frac{\omega^2}{c_G^2} \\ &\mathcal{F} \equiv \mathcal{F}(\Omega) = \frac{1+\Omega^2\gamma_{22}C_1}{1+(\Omega\gamma_{22})^2}\cdot\frac{\gamma_{22}}{C_1}, \quad \mathcal{R} = \mathcal{R}(\Omega) = \frac{(\gamma_{22}-C_1)\Omega}{1+(\Omega\gamma_{22})^2}\cdot\frac{\gamma_{22}}{C_1} \end{aligned}$$

$$c_G^2 = \frac{G}{\varrho_z}, \quad \Omega = \frac{\varrho\omega}{b_{11}}, \quad \gamma_{ij} = \frac{\varrho_{ij}}{\varrho}, \quad (j=1,2), \quad C_1 = \gamma_{11}\gamma_{22} - \gamma_{12}^2 > 0, \quad \varrho_z = \varrho_{11} - \frac{\varrho_{12}^2}{\varrho_{22}}$$

Here k is a complex wave number, ω is the angular frequency, Ω is the dimensionless frequency and c_G is the velocity of shear wave in the porous layer.
The relations between A_j and $\bar{A}_j$ $(j=1,2)$ result from fulfillment the equation (2).

Substitution of the solutions (6) and (7) into the conditions (4) and (5) leads to a system of equations which has non-zero solution if and only if

$$\text{(10)} \qquad \tan(k_1 + ik_2)h = A_r + iA_i$$

where

$$\text{(11)} \qquad A_r = \frac{\mu_0}{G}\frac{\zeta_1^* k_1 + \zeta_2 k_2}{k_1^2 + k_2^2}, \quad A_i = \frac{\mu_0}{G}\frac{k_1\zeta_2 - k_2\zeta_1^*}{k_1^2 + k_2^2},$$

$$\text{(12)} \qquad k_{1,2} = \left\{\frac{1}{2}\left[\left(\left(\alpha_1 - \frac{N}{G}a_1\right)^2 + \left(\alpha_2 - \frac{N}{G}a_2\right)^2\right)^{1/2} \pm \left(\alpha_1 - \frac{N}{G}a_1\right)\right]\right\}^{1/2}$$

$$\text{(13)} \qquad \zeta_{1,2} = \left\{\frac{1}{2}\left[(B^2 + a_2^2)^{1/2} \pm B\right]\right\}^{1/2}, \quad B = \frac{q^2}{4} + a_2 - \frac{\omega^2}{\beta_0^2}, \quad \zeta_1^* = \frac{q}{2} + \zeta_1$$

The complex transcendental equation (10) is referred to as the dispersion equation.

4. APPROXIMATE SOLUTION

In order to investigate the dispersion equation (10), an approximation method, namely, a small parameter expansion has been applied. Keeping in mind that, for most porous materials found in nature, the ratio α_2/α_1 (see (9)) is small as compared with one, we assume that $\alpha_2 \ll \alpha_1$ and also $a_2 \ll a_1$. Thus, making use of these assumptions the expressions (10)-(13) can be written in an approximate form. Similarly, the left-hand side of (10) can be expressed in an approximate form. Thus, making use of the obtained expressions, upon separation of real and imaginary parts of (10) and after applying some calculations, we come to the first approximation of the dispersion equation (10)

$$\tan \lambda_{r(1)} h \left(\mathcal{F}\frac{c^2}{c_G^2} - \frac{N}{G}\right)^{1/2} = \frac{\mu_0}{G} \frac{\left[\frac{q^2}{4} + \lambda_{r(1)}^2\left(1 - \frac{c^2}{\beta_0^2}\right)\right]^{1/2} + q}{\lambda_{r(1)}\left(\mathcal{F}\frac{c^2}{c_G^2} - \frac{N}{G}\right)^{1/2}} \tag{14}$$

Here $\lambda_{r(1)}$ is the first approximation of real part of wave number k, c is the phase velocity and $\mathcal{F}$ is given by the formula (9). Now, for given values ω, q, N/G and $\lambda_{r(1)}h$, we can find the values of the phase velocity c and, hence, of $a_{1(1)}$. The first approximation to a_2, $a_{2(1)}$ is found from equation

$$a_2 = \mathcal{L}(\omega, a_1)\cdot\alpha_2 \ , \quad \mathcal{L}(\omega, a_1) = \left\{\frac{N}{G} + \frac{1}{\left[1 + \frac{r}{s}\left(1 + s^2 + \frac{\mu_0 h}{2Gr^2} q\right)\right]\left(\frac{sG}{\mu_0}\right)^2}\right\}^{-1}$$

where

$$r = h\left(\alpha_1 - a_1\frac{N}{G}\right)^{1/2} \ , \quad s = \frac{\mu_0}{G}\left(\frac{\frac{q^2}{4} + a_1 - \frac{\omega^2}{\beta_0^2}}{\alpha_1 - a_1\frac{N}{G}}\right)^{1/2}$$

The attenuation coefficient in this order of approximation is given by

$$\lambda_{i(1)} = \frac{a_{2(1)}}{2\sqrt{a_{1(1)}}}$$

Making use of $a_{2(1)}$ (already found) we can determine the dispersion equation in the second approximation. It has the following form

$$\tan \lambda_{r(1)} h \left[\mathcal{F}\frac{c^2}{c_G^2} - (1-\chi)\frac{N}{G}\right]^{1/2} =$$

$$= \frac{\mu_0}{G\lambda_{r(1)}}\left(\frac{\frac{q^2}{4} + (1-\chi)\lambda_{r(1)}\left(1 - \frac{c^2}{\beta_0^2}\right)}{\mathcal{F}\frac{c^2}{c_G^2} - (1-\chi)\frac{N}{G}}\right)^{1/2}(1+\psi_1^*) + \psi_2^* \tag{15}$$

where χ, ψ_1^* and ψ_2^* depend on α_1, α_2, β_0, q, $a_{1(1)}$, $a_{2(1)}$, μ_0/G and N/G.
A more detailed account can be found in [5].

5. NUMERICAL RESULTS AND CONCLUSIONS

In order to ilustrate the approximate solution obtained in the previous section the numerical calculations have been made. These calculations have been confined to the solution given by the formula (14), that means to the first approximation only. The investigation of this equation leads to the conclusion that it must additionally satisfy the inequality

$$c_G\sqrt{\frac{N}{G\,\mathcal{F}(\Omega)}} < c < \beta_0$$

The phase velocity c/c_G has been computed for different wave numbers $\lambda_{r(1)}h$, and for some particular values of N/G and q describing the influence of anisotropy and heterogeneity.

The following values

$$\frac{\mu_0}{G} = 4, \quad \beta_0 = 3600 \text{ ms}^{-1}, \quad \omega = 40000 \text{ s}^{-1}, \qquad h = 30 \text{ m},$$

have been considered for the purpose of graphical representation of c/c_G.

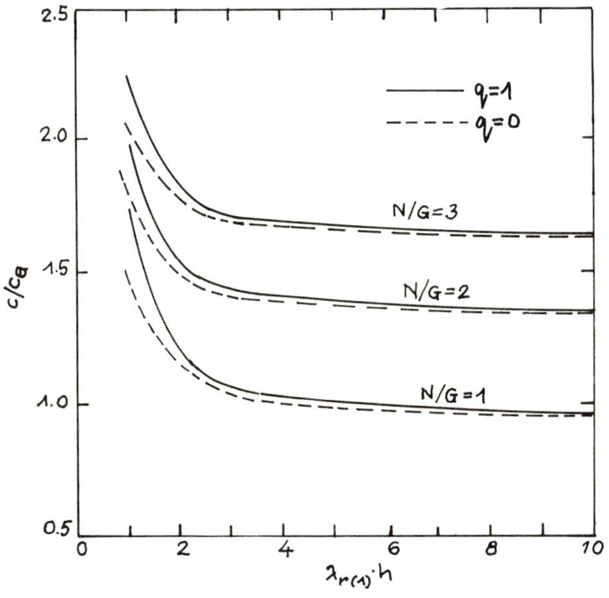

FIGURE 1
Variation of phase velocity with reduced wave number.

Fig. 1 shows that the magnitude of phase velocity c/c_G decreases as the wave number $\lambda_{r(1)}h$ increases. It is interesting to note that for a fixed values of $\lambda_{r(1)}h$ the magnitude of phase velocity is greater in the anisotropic case than that in the isotropic case. The magnitude of c/c_G is greater for the heterogeneous half-space than for the homogeneous one. The difference are quite distinct for the smallest wave numbers. The magnitude of phase velocity increases as the anisotropy factor N/G increases.

On the basis of the results of numerical calculations we can draw the following conclusions:

(i) The real roots of the dispersion equation can be obtained

only for a limited range of phase velocity which depends on the anisotropy factor N/G, shear wave velocity of the half-space β_0, inhomogeneity parameter q and the wave number.
(ii) The Love waves are attenuated in contrast to the Love waves in a non-porous elastic layer.

REFERENCES

[1] Deresiewicz, H., Bull. Seism. Soc. Am. 51 (1961) 51.
[2] Biot, M.A., J. Acoust. Soc. Am. 28 (1956) 168.
[3] Chakraborty, S.K. and Dey, S., Acta Mechanica 44 (1982) 169.
[4] Chattopadhyay, A. and De, R.K., Int. J. Engng Sci. 21 (1983) 1295.
[5] Kończak, Z., Acta Mechanica (to appear).
[6] Biot, M.A., J. Appl. Phys. 26 (1955) 182.

ELASTIC WAVE PROPAGATION
M.F. McCarthy, M.A. Hayes, (Editors)

COUPLED THERMOELASTIC WAVES IN PERIODICALLY LAMINATED PLATES

Sen Y. Lee and Liang C. Chin

Department of Mechanical Engineering
National Cheng Kung University
Tainan, Taiwan, R.O.C.

Huey J. Yang

Institute of Applied Mechanics
National Taiwan University
Taipei, Taiwan, R.O.C.

The theory of coupled thermoelastic plane strain wave propagation in an unbounded, periodically layered elastic plate is developed in terms of Floquet Waves. The dispersion spectrum is shown to be governed by the six roots of the dispersion relation which is presented in the form of a determinant of order twelve. The spectrum shows the typical band structure, consisting of stopping and passing bands, of wave propagation in a periodic medium. For the special case of wave propagation normal to the layering, the dispersion relation degenerates into the product of a fourth-order determinant and an eighth-order determinant. For the case of wave propagation at an arbitrary angle, it is shown that if there exists one coordinate system to impart symmetry to the structure, the dispersion relations along both ends of Brillouin zone can be factorized into the product of two determinants of order six. The significance of this uncoupling is examined.

1. INTRODUCTION

Wave propagation in periodic laminated composite has been the focus of considerable studies in recent yeras, mainly because of its significance in the analysis and design of composite materials. It is known that the theory of thermoelasticity is well established and the temperature field is, in general, coupled with the elastic strain. Deresiewicz [1] reduces the dynamic equations of coupled thermoelasticity to three uncoupled differential equations by applying the Helmholtz theorem. Sve [2] adapts Deresiewicz's formulation and studies the problem of coupled thermoelastic waves in a periodically laminated composite in a state of plane strain. The exact dispersion relation is presented in the form of a determinant of order twelve.

In this paper, we use a different formulation to study and extend Sve's work. Special attention will be paid to the structure for which there exist one coordinate system to impart symmetry to the structure. The dispersion relations along both ends of the Brillouin zone are examined in great detail.

2. ANALYSIS

Consider the problem of coupled thermoelastic waves in a periodically laminated plate in a state of plain strain. The plate is the union of denumerable infinite number of cells with identical properties and has real minimal period d. In each primitive cell, there consist two layers. Each of the two layers in the unit cell is assumed to be homogeneous, isotropic, and perfectly bonded to the adjoing layers. The lamellae of a primitive unit cell have Lame's constants $\{(\lambda_1, \mu_1); (\lambda_2, \mu_2)\}$, thickness $\{2h_1; 2h_2\}$, and mass densities $\{\rho_1; \rho_2\}$, respectively.

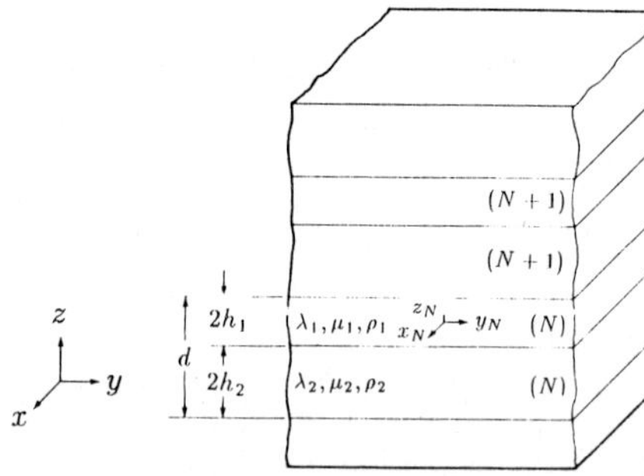

Fig. 1 Geometry of periodically laminated plate

Let u, v and w be the three displacement components in the x, y and z directions, respectively. For plane strain motion in the layer with nonsubscript constants in the N-th unit cell, we take

$$u = u(x, z_N; t), \quad w = (x, z_N; t), \quad v = 0, \qquad (1)$$
$$-h < z_N < h, \quad |x| < \infty,$$

where z_N is a local coordinate with its origin at the midplane of the layer (Fig. 1). The three governing equations of coupled thermoelasticity are

$$\rho \frac{\partial^2 u}{\partial t^2} = (2\mu + \lambda)\frac{\partial^2 u}{\partial x^2} + \mu \frac{\partial^2 u}{\partial z^2} + (\mu + \lambda)\frac{\partial^2 w}{\partial x \partial z} + \beta \frac{\partial \theta}{\partial x},$$
$$\rho \frac{\partial^2 w}{\partial t^2} = (\mu + \lambda)\frac{\partial^2 u}{\partial x \partial z} + \mu \frac{\partial^2 w}{\partial x^2} + (2\mu + \lambda)\frac{\partial^2 w}{\partial z^2} + \beta \frac{\partial \theta}{\partial z}, \qquad (2)$$
$$\frac{\partial^2 \theta}{\partial x^2} + \frac{\partial^2 \theta}{\partial z^2} - \frac{1}{a^2}\frac{\partial \theta}{\partial t} + \beta \frac{T_o}{K}\left(\frac{\partial^2 u}{\partial x \partial t} + \frac{\partial^2 w}{\partial z \partial t}\right) = 0,$$

where $a^2 = \frac{K}{\rho C_v}$, $\beta = \frac{-E\alpha_t}{1-2\nu}$, θ is the temperature deviation from equilibrium temperature T_o, a^2 is the thermal diffusivity, K is the thermal conductivity, C_v is the specific heat at constant volumn, β is the thermal moduli, E is the Young's moduli, α_t is the coefficient of thermal expansion and ν is the Poisson's ratio of the materials.

For harmonic waves with angular frequency ω, propagating in the $x - z$ plane, the displacement components u, w and the temperature deviation θ can be written as

$$u = Ue^{i(k_x + k_z z - \omega t)},$$
$$w = We^{i(k_x + k_z z - \omega t)}, \qquad i^2 = -1 \tag{3}$$
$$\theta = \Theta e^{i(k_x + k_z z - \omega t)},$$

where k_x and k_z are the wave numbers in x and z directions respectively. Upon substituting (3) into (2), it leads to a set of three homogeneous linear equations in the coefficients U, W and Θ. The requirement that the determinant of the matrix of coefficients must vanish for a nontrivial solution then leads to the dispersion equation of thermoelastic waves in homogeneous, isotropic medium

$$A \cdot \Big\{X^4 - \big[C_1 + C_2(1+\varepsilon)\big]X^2 + C_1C_2\Big\} = 0, \tag{4}$$

where X, C_1, C_2, ε and A are defined as

$$X = (k_x^2 + k_z^2)^{\frac{1}{2}}, \quad C_1 = \frac{\rho\omega^2}{\lambda + 2\mu}, \quad C_2 = \frac{i\rho C_v\omega}{K},$$
$$\varepsilon = \frac{\beta^2 T_o}{\rho C_v(2\mu + \lambda)}, \quad A = \rho\omega^2 - \mu X^2, \tag{5}$$

where ε is the coupling parameter. For $A = 0$, it governs the transverse motion, which is not coupled with the thermal field. When $A \neq 0$, then

$$X_{1,2}^2 = \frac{\rho\omega^2}{2(2\mu + \lambda)}\Big\{1 - \frac{1+\varepsilon}{\Lambda} \pm \sqrt{1 + \frac{2(1-\varepsilon)}{\Lambda} + (\frac{1+\epsilon}{\Lambda})^2}\Big\} \tag{6}$$

where $\Lambda = \frac{ia^2\omega\rho}{\lambda+2\mu}$. This equation is exactly the same as the dispersion equation (6) given in Sve's paper. In addition, when $\varepsilon = 0$, the first solution of (6), X_1^2, yields to the dispersion equation for longitudinal motion in Delph, Herrmann and Kaul's paper [3], in which the thermal effect is not considered.

From the three homogeneous linear equations in the coefficients U, W, Θ and the dispersion relation, we can establish the displacement and temperature field of two different layers $\{u_1, w_1, \theta_1); (u_2, w_2, \theta_2)\}$ in N-th cell

$$\begin{aligned}
u_j =& (-\xi A_{1j}C_{\alpha j} - \xi A_{2j}S_{\alpha j} + \beta_j A_{3j}C_{\beta j} + \beta_j A_{4j}S_{\beta j} - \xi A_{5j}C_{\gamma j} - \xi A_{6j}S_{\gamma j})e^{i(\frac{\xi x - c\Omega t}{2h_2})},\\
w_j =& (-\alpha_j A_{1j}S_{\alpha j} + \alpha_j A_{2j}C_{\alpha j} - \xi A_{3j}S_{\beta j} + \xi A_{4j}C_{\beta j} - \gamma_j A_{5j}S_{\gamma j} + \gamma_j A_{6j}C_{\gamma j})e^{i(\frac{\xi x - c\Omega t}{2h_2})},\\
\theta_j =& \frac{1-\xi_j}{\alpha_{tj}(1+\gamma_j)}(H_{1j}^2 A_{1j}C_{\alpha j} + H_{1j}^2 A_{2j}S_{2j} + H_{2j}^2 A_{5j}C_{\gamma j} + H_{2j}^2 A_{6j}S_{\gamma j})e^{i(\frac{\xi x - c\Omega t}{2h_2})},\\
& j = 1, 2
\end{aligned} \tag{7}$$

where

$$C_{\alpha j} = \cos(\frac{\alpha_j z_{Nj}}{2h_2}), \quad S_{\alpha j} = \sin(\frac{\alpha_j z_{Nj}}{2h_2}), \quad C_{\beta j} = \cos(\frac{\beta_j z_{Nj}}{2h_2}),$$

$$S_{\beta j}=\sin(\frac{\beta_j z_{Nj}}{2h_2}),\quad C_{\gamma j}=\cos(\frac{\gamma_j z_{Nj}}{2h_2}),\quad S_{\gamma j}=\sin(\frac{\gamma_j z_{Nj}}{2h_2}),$$
$$\xi=2h_2k_x,\quad \chi_{1j}=2h_2X_{1j},\quad \chi_{2j}=2h_2X_{2j},\ \sigma=K_2/K_1,$$
$$\alpha_j=(\chi_{1j}^2-\xi^2)^{1/2},\quad \gamma_j=(\chi_{2j}^2-\xi^2)^{1/2},\quad \Omega=\frac{2h_2\omega}{(\mu_1/\rho_1)^{1/2}},$$
$$G=\rho_2/\rho_1,\quad R=\mu_2/\mu_1,\quad \beta_1=(\Omega^2-\xi^2)^{1/2},\quad \beta_2=(\frac{G\Omega^2}{R}-\xi^2)^{1/2},$$
$$H_{i1}^2=\frac{\Omega^2}{D_1}-\chi_{i1}^2,\quad H_{i2}^2=\frac{G\Omega^2}{RD_2}-\chi_{i2}^2,\quad D_j=\frac{2(1-\nu_j)}{(1-2\nu_j)},$$
$$i=1,2;\ j=1,2.$$

Here, F_j stands for the physical quality F of jth layer in one unit cell, and X_{ij} represents the ith solution in (6) of jth layer.

3. DISPERSION RELATIONS

Being a perfect bonded period laminated plate with period d, we require the displacement, stresses, temperature and heat flux to be continuous at all the common interfaces and at least quasi-periodic from cell to cell. Therefore

$$\begin{aligned} u_1(x,h_1)&=\tau u_2(x,-h_2), & w_1(x,h_1)&=\tau w_2(x,-h_2),\\ \sigma_{xz1}(x,h_1)&=\tau\sigma_{xz2}(x,-h_2), & \sigma_{zz1}(x,h_1)&=\tau\sigma_{zz2}(x,-h_2),\\ \theta_1(x,h_1)&=\tau\theta_2(x,-h_2), & K_1\tfrac{\partial\theta_1}{\partial z}(x,h_1)&=\tau K_2\tfrac{\partial\theta_2}{\partial z}(x,-h_2). \end{aligned}\tag{8}$$

If we take the transformation

$$\tau=e^{ik_zd}=e^{i\eta(1+\frac{1}{\xi})},\quad \eta\equiv 2k_zh_2\tag{9}$$

then k_z is the Floquet wave number. Using (7), the stress-strain and strain-displacement relationships, six continuity conditions at $z=-h_1$ and six quasi-periodic conditions, we have a set of twelve linear homigeneous equations in the twelve constants $A_{ij}, i=1,2,\cdots,6; j=1,2$. For nontrivial solutions, the determinant of the matrix coefficients must vanish. This leads to the dispersion equation in the form of a determinant of order twelve. The dynamic nature of the system thus depends on the six roots of τ of this dispersion equation.

When $|\tau|<1$, then from (9), we observe that k_z is real. Therefore the wave will propagate and the frequency intervals those correspond to $|\tau|<1$ are known as passing bands. Otherwise, the frequency intervals those correspond to $|\tau|>1$ are stopping bands. When $\tau=1$ or $\tau=-1$, the waves are standing and the solution is periodic with period d and $2d$ respectively. A lucid description of the properties of the Floquet wave can be found in Lee's paper [4].

4. DECOMPOSITION OF THE DISPERSION EQUATION

For normal incident, $\xi = 0$, the dispersion equation can be degenerated into the product of a determinant of order four

$$\begin{vmatrix} \beta_1 C_{\beta_1} & -\beta_1 S_{\beta_1} & -\beta_2 C_{\beta_2} & \beta_2 S_{\beta_2} \\ -\beta_1^2 S_{\beta_1} & -\beta_1^2 C_{\beta_1} & -R\beta_2^2 S_{\beta_2} & R\beta_2^2 C_{\beta_2} \\ \beta_1 C_{\beta_1} & \beta_1 S_{\beta_1} & -\beta_2^2 C_{\beta_2}\tau & \beta_2 S_{\beta_2}\tau \\ \beta_1^2 S_{\beta_1} & -\beta_1^2 C_{\beta_1} & R\beta_2^2 S_{\beta_2}\tau & R\beta_2^2 C_{\beta_2}\tau \end{vmatrix} = 0, \tag{10}$$

and a determinant of order eight

$$\begin{vmatrix} -\alpha_1 S_{\alpha 1} & -\alpha_1 C_{\alpha 1} & -\gamma_1 S_{\gamma 1} & -\gamma_1 C_{\gamma 1} & -\alpha_2 S_{\alpha 2} & \alpha_2 C_{\alpha 2} & -\gamma_2 S_{\gamma 2} & -\gamma_2 C_{\gamma 2} \\ -\beta_1^2 C_{\alpha 1} & -\beta_1^2 S_{\alpha 1} & -\beta_1^2 C_{\gamma 1} & \beta_1^2 S_{\gamma 1} & R\beta_2^2 C_{\alpha 2} & R\beta_2^2 S_{\alpha 2} & R\beta_2^2 C_{\gamma 2} & R\beta_2^2 S_{\gamma 2} \\ H_{11}^2 C_{\alpha 1} & -H_{11}^2 S_{\alpha 1} & H_{21}^2 C_{\gamma 1} & -H_{21}^3 S_{\gamma 1} & -\Delta H_{12}^2 C_{\alpha 2} & -\Delta H_{22}^2 C_{\gamma 2} & -\Delta H_{22}^2 S_{\gamma 2} & -\Delta H_{22}^2 S_{\gamma 2} \\ H_{11}^2 \alpha_1 S_{\alpha 1} & H_{11}^2 \alpha_1 C_{\alpha 1} & H_{21}^2 \gamma_1 S_{\gamma 1} & H_{21}^2 \gamma_1 C_{\gamma 1} & \Delta\sigma H_{12}^2 \alpha_2 S_{\alpha 2} & -\Delta\sigma H_{12}^2 \alpha_2 C_{\alpha 2} & \Delta a H_{22}^2 \gamma_2 S_{\gamma 2} & -\Delta a H_{22}^2 \gamma_2 C_{\gamma 2} \\ \alpha_1 S_{\alpha 1} & -\alpha_1 C_{\alpha 1} & \gamma_1 S_{\gamma 1} & -\gamma_1 C_{\gamma 1} & \alpha_2 S_{\alpha 2}\tau & \alpha_2 C_{\alpha 2}\tau & \gamma_2 S_{\gamma 2}\tau & \gamma_2 C_{\gamma 2}\tau \\ -\beta_1^2 C_{\alpha 1} & -\beta_1^2 S_{\alpha 1} & -\beta_1^2 C_{\gamma 1} & -\beta_1^2 S_{\gamma 1} & R\beta_2^2 C_{\alpha 2}\tau & -R\beta_2^2 S_{\alpha 2}\tau & R\beta_2^2 C_{\gamma 2}\tau & R\beta_2^2 S_{\gamma 2}\tau \\ H_{11}^2 C_{\gamma 1} & H_{11}^2 S_{\alpha 1} & H_{21}^2 C_{\gamma 1} & H_{21}^2 S_{\gamma 1} & -\Delta H_{12}^2 C_{\alpha 2}\tau & \Delta H_{12}^2 S_{\alpha 2}\tau & -\Delta H_{22}^2 C_{\alpha 2}\tau & \Delta H_{22}^2 S_{\gamma 2}\tau \\ -H_{11}^2 \alpha_1 S_{\alpha 1} & H_{11}^2 \alpha_1 C_{\alpha 1} & -H_{21}^2 \gamma_1 S_{\gamma 1} & H_{21}^2 \gamma_1 C_{\gamma 1} & -\Delta\sigma H_{12}^2 \alpha_2 S_{\alpha_2}\tau & -\Delta\sigma H_{12}^2 \alpha_2 C_{\alpha_2}\tau & -\Delta\sigma H_{22}^2 \gamma_2 S_{\gamma_2}\tau & -\Delta\sigma H_{22}^2 \gamma_2 C_{\gamma_2}\tau \end{vmatrix}$$

$$= 0, \tag{11}$$

where $\Delta = \frac{\alpha_{t1}}{\alpha_{t2}}\left(\frac{1+\nu_1}{1-\nu_1}\right)\left(\frac{1+\nu_2}{1-\nu}\right)$. Equation (10) governs the transverse motion in a periodic plate, which is not coupled with the thermal field. Equation (11) governs the coupled longitudinal motion and thermal variations.

We now investigate the form of the dispersion equation on the end points of the Brillouin zones, i.e., $\tau = \pm 1$. If we set $\tau = 1$ in (8), then the dispersion equation in determinant of order twelve will factor into the product of following two (6×6) determinants

$$\begin{vmatrix} \xi C_{\alpha 2} & -\xi C_{\alpha 1} & -\beta_2 C_{\beta_2} & \beta_1 C_{\beta_1} & \xi C_{\gamma_2} & -\xi C_{\gamma_1} \\ \alpha_2 S_{\alpha_2} & \alpha_1 S_{\alpha_1} & \xi S_{\beta_2} & \xi S_{\beta_1} & \gamma_2 S_{\gamma_2} & \gamma_1 S_{\gamma_1} \\ -R(\xi^2 - \beta_2^2)C_{\alpha 2} & (\xi^2 - \beta_1^2)C_{\alpha 1} & 2R\xi - \beta_2 C_{\beta 2} & -2\xi\beta_1 C_{\beta 1} & -R(\xi^2 - \beta_2^2)C_{\gamma 2} & (\xi^2 - \beta_1^2)C_{\gamma 1} \\ -2R\xi_{\alpha 2} S_{\alpha 2} & -2\xi_{\alpha 1} S_{\alpha 1} & -R(\xi^2 - \beta_2^2)S_{\beta_2} & -(\xi^2 - \beta_1^2)S_{\beta 1} & -2R\xi\gamma_2 S_{\gamma_2} & -2\xi\gamma_1 S_{\gamma 1} \\ \Delta H_{12}^2 C_{\alpha 2} & H_{11}^2 C_{\alpha 1} & 0 & 0 & -\Delta\sigma H_{22}^2 \gamma_2 S_{\gamma 2} & -H_{21}^2 \gamma_1 S_{\gamma 1} \end{vmatrix}$$

$$= 0, \tag{12}$$

$$\begin{vmatrix} \xi S_{\alpha 2} & \xi S_{\alpha 1} & -\beta_2 S_{\beta 2} & -\beta_1 S_{\beta 1} & \xi S_{\gamma 2} & \xi S_{\gamma 1} \\ \alpha_2 C_{\alpha 2} & -\alpha_1 C_{\alpha 1} & \xi C_{\beta 2} & -\xi C_{\beta 1} & \gamma_2 C_{\gamma 2} & -\gamma_1 C_{\gamma 1} \\ -R(\xi^2 - \beta_2^2)S_{\alpha 2} & -(\xi^2 - \beta_1^2)S_{\alpha 1} & 2R\xi\beta_2 S_{\beta 2} & 2\xi\beta_1 S_{\beta 1} & -R(\xi^2 - \beta_2^2)S_{\gamma 2} & -(\xi^2 - \beta_1^2)S\gamma_1 \\ -2R\xi - \alpha_2 C_{\alpha 2} & 2\xi - \alpha 1 C_{\alpha 1} & -R(\xi^2 - \beta_2^2)C_{\beta 2} & (\xi^2 - \beta_1^2)C_{\beta 1} & -2R\xi\gamma_2 C_{\gamma 2} & 2\xi\gamma_1 C_{\gamma 1} \\ -\Delta H_{12}^2 S_{\alpha 2} & -H_{11}^2 S_{\alpha 1} & 0 & 0 & -\Delta H_{22}^2 S_{\gamma 2} & -H_{21}^2 S_{\gamma 1} \\ -\Delta\sigma H_{12}^2 \alpha_2 C_{\alpha 2} & H_{11}^2 \alpha_1 C_{\alpha 1} & 0 & 0 & -\Delta\sigma H_{22}^2 \gamma_2 C_{\gamma 2} & H_{21}^2 \gamma_1 C_{\gamma 1} \end{vmatrix}$$

$$= 0. \tag{13}$$

Similarly, if we set $\tau = -1$, then the dispersion equation factors into the product of two (6×6) determinants, which take the form

$$\begin{vmatrix} -2\xi C_{\alpha 1} & 2\xi S_{\alpha 2} & 2\beta_1 C_{\beta 1} & -2\beta_2 S_{\beta 2} & -2\xi C_{\gamma 1} & 2\xi S_{\gamma 2} \\ 2\alpha_1 S_{\alpha 1} & -2\alpha_2 C_{\alpha 2} & 2\xi S_{\beta 1} & -2\xi C_{\beta 2} & 2\gamma_1 S_{\gamma 1} & -2\gamma_2 C_{\gamma 2} \\ 2(\xi^2 - \beta_1^2)C_{\alpha 1} & -2R(\xi^2 - \beta_2^2)S_{\alpha 2} & -4\xi\beta_1 C_{\beta 1} & 4R\xi\beta_2 S_{\beta 2} & 2(\xi^2 - \beta_1^2)C_{\gamma 1} & -2R(\xi^2 - \beta_2^2)S_{\gamma 2} \\ -4\alpha_1 S_{\alpha 1} & 4R\xi\alpha_2 C_{\alpha 2} & -2(\xi^2 - \beta_1^2)S_{\beta 1} & 2R(\xi^2 - \beta_2^2)C_{\beta 2} & -4\xi\gamma_1 S_{\gamma 1} & 4R\xi\gamma_2 C_{\gamma 2} \\ 2H_{11}^2 C_{\alpha 1} & -2\Delta H_{12}^2 S_{\alpha 2} & 0 & 0 & 2H_{21}^2 C_{\gamma 1} & -2\Delta H_{22}^2 S_{\gamma 2} \\ -2H_{11}^2\alpha_1 S_{\alpha 1} & 2\Delta H_{12}^2\alpha_2 C_{\alpha 2} & 0 & 0 & -2H_{21}^2\gamma_1 S_{\gamma 1} & 2\Delta\sigma H_{22}^2\gamma_2 C_{\gamma 2} \end{vmatrix}$$
$$= 0, \tag{14}$$

$$\begin{vmatrix} \xi C_{\alpha 2} & \xi S_{\alpha 1} & -\beta_2 C_{\beta 2} & -\beta_1 S_{\beta 1} & \xi C_{\gamma 2} & \xi S_{\gamma 1} \\ -\alpha_2 S_{\alpha 2} & -\alpha_1 C_{\alpha 1} & -\xi S_{\beta 2} & -\xi C_{\beta 1} & -\gamma_2 S_{\gamma 2} & -\gamma_1 C_{\gamma 1} \\ -R(\xi^2 - \beta_2^2)C_{\alpha 2} & -(\xi^2 - \beta_1^2)S_{\alpha 1} & 2R\xi\beta_2 C_{\beta 2} & 2\xi\beta_1 S_{\beta 1} & -R(\xi^2 - \beta_2^2)C_{\gamma 2} & -(\xi^2 - \beta_1^2)S_{\gamma 2} \\ 2R\xi - \alpha 2 S_{\alpha 2} & 2\xi_{\alpha 1} C_{\alpha 1} & R(\xi^2 - \beta_2^2)S_{\beta 2} & (\xi^2 - \beta_1^2)C_{\beta 1} & 2R\xi\gamma_2 S_{\gamma 2} & 2\xi\gamma_1 C_{\gamma 1} \\ -\Delta H_{12}^2 C_{\alpha 2} & -H_{11}^2 S_{\alpha 1} & 0 & 0 & -\Delta H_{22}^2 C_{\gamma 2} & -H_{21}^2 S_{\gamma 1} \\ -\Delta\sigma H_{12}^2\alpha_2 S_{\alpha 2} & H_{11}^2\alpha_1 C_{\alpha 1} & 0 & 0 & \Delta\sigma H_{22}^2\gamma_2 S_{\gamma 2} & H_{21}^2\gamma_1 C_{\gamma 1} \end{vmatrix}$$
$$= 0. \tag{15}$$

The significance of these uncouplings can be easily explained. Since the laminated structure we considered is symmetric with respect to the coordinate system we choose, if we set coefficients $A_{2j} = A_{4j} = A_{6j} = 0$, $j = 1,2$ in (7), then, the displacements u_j and temperatrure field θ_j, $j = 1,2$ are symmetric with respect to the layer midplanes, and the displacements w_j, $j = 1,2$ are antisymmetric. We may call this a symmetric-symmetric motion emphasizing only the character of u_j displacements. The dispersion equation for this symmetric-symmetric motion thus leads to a determinant of order six which gives us equation (12). Equation (12) is thus the dispersion equation for symmetric-symmetric motion with period d. Similarly, (13) is found to be the dispersion equation for antisymmetric-antisymmetric motion ($A_{1j} = A_{3j} = A_{5j} = 0$, $j = 1,2$) with period d, (14) is the dispersion equation for symmetric-antisymmetric motion ($A_{21} = A_{41} = A_{61} = A_{12} = A_{32} = A_{52} = 0$) with period $2d$ and (15) governs the antisymmetric-symmetric motion ($A_{11} = A_{31} = A_{51} = A_{22} = A_{42} = A_{62} = 0$) with period $2d$.

REFERENCES

[1] Deresiewicz, H., Solution of the Equations of Thermoelasticity, *Proc. 3rd U.S. Natl. Congr. Appl. Mech.*, (1958) pp.287-291.

[2] Sve, C., Thermoelastic Waves in a Periodically laminated Medium, *Int. J. Solids Structures*, 7, (1971) pp.1363-1373.

[3] Delph, T.J., Herrmann, G. and Kaul, R.K. Harmonic Wave propagation in a Periodically Layered, Infinite Elastic Body : Plane Strain, Analytical Results, *J. Appl. Mech.*, (1979) pp.113-119.

[4] Lee, E.H., Wave Propagation in Composites With Periodic Structure, *Proc. 5th Canadian Congr. Appl. Mech.*, (1975) pp.45-59.

ELASTIC WAVE PROPAGATION
M.F. McCarthy, M.A. Hayes, (Editors)

DIRECT INVERSION FOR A LAYERED ELASTIC MEDIUM

J.A.H. Alkemade & J.W. de Maag

c/o Koninklijke/Shell Exploratie & Produktie Laboratorium
P.O. Box 60
2280 AB RIJSWIJK
The Netherlands

Prestack direct inversion for a 3-D plane stratified elastic earth is a well-established method. Recent publications for a variety of full bandwidth sources and data have proved its potential. Unfortunately, direct application to real seismic data still presents severe problems. Real seismic data are corrupted with noise and lack low and high frequencies.

This problem is investigated for a smoothly varying elastic earth. We show which parameters can be reliably estimated from the data by studying their sensitivity to noise. The basic principles of elastic inversion are discussed and it is demonstrated that these can be retained for band-limited data by incorporating a few simple assumptions.

1. INTRODUCTION

Direct layer-stripping inversion methods for the recovery of the elastic parameters of a stack of plane homogeneous isotropic elastic layers from its seismic reflection response have been studied extensively [1]. Recent work [1,2] revealed inversion algorithms for which the processing and storage requirements are of the order of forward modelling algorithms. However, they failed to account for such practical aspects as noise and band limitation.

We have investigated these aspects in view of the clear computational advantages of direct inversion compared to iterative inversion. A practical approach to a solution will be discussed.

2. PLANE WAVES AT OBLIQUE INCIDENCE

The equations governing the propagation of plane waves incident at an interface between plane homogeneous isotropic layers yield well-known reflection and transmission coefficients. Despite their simple derivation, these coefficients depend in a complicated way on the compressional velocity α, the shear velocity β, the density ρ and the horizontal ray parameter p. Fortunately, in seismic applications the velocities and the densities differ only slightly from layer to layer. For these minor differences, denoted by $\Delta\alpha$, $\Delta\beta$ and $\Delta\rho$, it is permissible to linearise the expressions for the coefficients [3]. For example, the linearised P-P and P-SV reflection coefficients for potentials (normalised with respect to the energy flux normal to the interface [4]) are:

$$r_{p-p} = \frac{1}{2(1-p^2\alpha^2)}\frac{\Delta\alpha}{\alpha} - 4p^2\beta^2\frac{\Delta\beta}{\beta} + \left(\frac{1}{2} - 2p^2\beta^2\right)\frac{\Delta\rho}{\rho} \qquad (1)$$

$$r_{p-s} = \frac{1}{2}\,p(\alpha_p\beta_p)^{1/2}\left\{\left(4p^2\beta^2 - \frac{4\,\beta^2}{\alpha_p\beta_p}\right)\frac{\Delta\beta}{\beta} - \left(1-2p^2\beta^2 + \frac{2\,\beta^2}{\alpha_p\beta_p}\right)\frac{\Delta\rho}{\rho}\right\} \qquad (2)$$

where $\alpha_p = \alpha(1-p^2\alpha^2)^{-1/2}$ and $\beta_p = \beta(1-p^2\beta^2)^{-1/2}$.

The elastic reflection coefficients, supplemented with layer-delay matrices, are the basis of Kennett's recursion algorithm for the computation of seismic reflection data [5]. In every step of this recursion a subsequent layer is added on top of all the preceding ones, until the stack of layers is complete.

3. THE DIRECT INVERSION ALGORITHM

Seismic reflection data, emanating from a point-source generating P-waves, are collected at the surface. These data are decomposed by the Radon transform into a set of plane wave data, characterised by the horizontal ray parameter p. Proceeding along similar lines to published methods (e.g. Ref. [2]), we apply a layer-stripping direct inversion method to reconstruct elastic parameters (or rather parameter contrasts) from these data. The method consists of two major steps per layer:

a) Wave update.
Determination of the fields at depth $z+\Delta z$ from the fields and the elastic parameters at depth z. This is achieved by solving the wave equation with an explicit scheme [2].

b) Parameter update.
Imaging and determination of the parameter contrasts at $z+\Delta z$. The p-dependent reflection coefficients r_{p-p} and r_{p-s} follow from an imaging principle [2]. The parameter contrasts are determined by making a least squares fit of the expressions (1) and (2) to the imaged reflection coefficients.

By alternate application of steps a) and b) a recursive layer-stripping inversion algorithm results. This algorithm is similar to the forward algorithm of Section 2 but reversed in direction.

4. PRACTICAL PROBLEMS

Direct application of the described algorithm to real seismic data is hampered by practical problems. Two major problems that we investigate are band limitation and noise.

4.1. Band limitation

If the data lack low frequencies, as is the case in seismic applications, the trend in the velocities will not be recovered by the algorithm. As these trend or background velocities are indispensable for the algorithm to succeed, we assume them to be known. They can be obtained by conventional move-out analysis. The remaining unknowns for the wave update step a) are the parameter contrasts, which are reconstructed in step b).

For the interpretation of the results it is important to observe that data with temporal bandwidth $[\omega_l,\omega_h]$ contain information about the parameter contrasts in a vertical-wavenumber band that depends on the ray parameter p. Therefore, a least squares determination of the parameter contrasts from these traces is reliable only in the intersection of these wavenumber bands. An estimate of this intersection is given by:

$$[1+(\alpha/\beta)_{max}]\omega_l \leq \omega \leq 2[1-(p\alpha)^2_{max}]^{1/2}\omega_h. \tag{3}$$

Our results adhere to this estimate.

4.2. Noise

To recognise whether the extra processing effort is effective and whether the reconstructed parameters of step b) are sufficiently accurate for step a) of the algorithm, we investigated the sensitivity of the reconstructed parameters to noise.

Let us assume that, for a compressional source, both the compressional and the converted shear wave are measured at the surface. For a typical seismic experiment incidence angles up to approximately 30°, corresponding to $p\alpha = 1/2$, occur. Expansion of Eqs. (1) and (2) near $p\alpha = 0$ gives:

$$r_{p-p} = A + B(p\alpha)^2 + O((p\alpha)^4), \tag{4}$$

$$r_{p-s} = D(p\alpha) + O((p\alpha)^3), \tag{5}$$

where

$$A = \frac{1}{2}\left(\frac{\Delta\alpha}{\alpha} + \frac{\Delta\rho}{\rho}\right), \tag{6}$$

$$B = \frac{1}{2}\frac{\Delta\alpha}{\alpha} - 2\left(\frac{\beta}{\alpha}\right)^2\left(\frac{2\Delta\beta}{\beta} + \frac{\Delta\rho}{\rho}\right), \tag{7}$$

$$D = -\frac{1}{2}\left(\frac{\beta}{\alpha}\right)^{1/2}\left(\frac{4\beta}{\alpha}\frac{\Delta\beta}{\beta} + \left(1 + 2\frac{\beta}{\alpha}\right)\frac{\Delta\rho}{\rho}\right), \tag{8}$$

For small $p\alpha$ the parameter A constitutes the main body of r_{p-p} (see Eq. (4)). It is intuitively clear that the determination of B from measured reflection coefficients is then more sensitive to noise than A. A similar argument holds for D. This intuitive picture has been rigorously extended to the concept of optimal parameters, which is based on asymptotic analysis for $p\alpha \to 0$ (see the appendix). It rates the information about the elastic parameters that is contained in the (imaged) reflection amplitudes. It can be shown that A,D and $C=\Delta\alpha/(2\alpha)$ are optimal for pressure and converted shear data. Here A is the least and C the most sensitive to noise. In contrast, $\Delta\alpha/\alpha$, $\Delta\beta/\beta$ and $\Delta\rho/\rho$ all have noise-sensitivities comparable to C.

5. INVERSION RESULTS

The inversion algorithm was tested for a continuous model in order to avoid linearisation errors as much as possible. The model is depicted in Fig. 1. Synthetic plane wave data were calculated by a forward modelling program based on the recursion formula by Kennett mentioned above [5].

The 0-2 Hz filtered model velocities were used as background velocities.

For data band-limited to 3-30 Hz the inversion results are plotted in Fig. 2. The model and reconstructed A,D and C values have been filtered to the band given by Eq. (3), and these filtered versions coincide quite well. Hence, we are able to reconstruct band-limited versions of the model from band-limited data, provided background velocities are used.

To illustrate the effect of noise, the data were contaminated with white Gaussian noise with standard deviation $\sigma = 0.001$. The inversion algorithm was applied twice, with two different maximum incidence angles. The results are plotted in Fig. 3 for $\theta_{max} = 20^{\circ}$ and in Fig. 4 for $\theta_{max} = 18^{\circ}$. Note the difference in vertical scale. The fact that A,D and C are optimal is nicely illustrated, since A is indeed much better determined than C. For the smaller maximum incidence angle the inversion breaks down. It can be concluded that the success of the inversion of P-P and P-SV data depends greatly on the amount of noise and the maximum incidence angle. A precise relation between the noise level and the maximum incidence angle to ensure successful inversion is not known to us.

6. CONCLUSIONS

We discussed a simple practical approach to direct inversion of band-limited data and data contaminated by noise. By incorporating experimentally accessible background velocities a computationally efficient inversion algorithm is obtained that can cope with bandlimited pre-stack data. We have shown that the success of this algorithm depends greatly on the amount of noise and the maximum incidence angle. The transparent sensitivity analysis in terms of optimal parameters sets limits for the application of this algorithm to real seismic data.

REFERENCES

[1] Yagle, A.E., Layer stripping solutions of inverse seismic problems, Ph.D. Thesis, MIT, Cambridge 1985.
[2] Yagle, A.E. and Levy, B.C., Geophysics 50 (1985) 425.
[3] Richards, P.G. and Frasier, C.W., Geophysics 41 (1976) 441.
[4] Frasier, C.W., Geophysics 35 (1970) 197.
[5] Kennett, B.L.N., Advances in applied mechanics 21 (1981) 79.

APPENDIX

Assume $r(p) \in \mathbf{R}^L$ $(p \in \mathbf{R})$ is related to the parameters $x \in \mathbf{R}^K$ by $r(p) = M_p x$, where M_p is an L×K matrix. Let

$$M_p = \sum_{k=0}^{\infty} p^k m_k ,$$

with m_k L×K matrices. From measurements $\tilde{r}(p)$ of $r(p)$ for n different p-values $0 \le p_1 < \dots < p_n$, an estimate of x is obtained by solving

$$M x \equiv \begin{pmatrix} M_{p_1} \\ \vdots \\ M_{p_n} \end{pmatrix} x = \begin{pmatrix} \tilde{r}(p_1) \\ \vdots \\ \tilde{r}(p_n) \end{pmatrix}$$

in a least squares sense. The concept of optimal parameters is based on an asymptotic analysis for $p_n \to 0$. Let the K eigenvalues of $M^T M$ satisfy

$$\lambda_i = a_i(n)\, p_n^{b_i} (1 + o(1)) \qquad (p_n \to 0),$$

ordered such that $0 \le b_1 \le \dots \le b_k$ and let N be a K×K non-singular matrix not depending on n and the p_i's. The parameters $y = N x$ are called optimal if the variances behave as

$$\sigma_{y_i}^2 = c_i(n)\, p_n^{-b_i} (1 + o(1)) \qquad (p_n \to 0).$$

Note that optimal parameters are not unique.

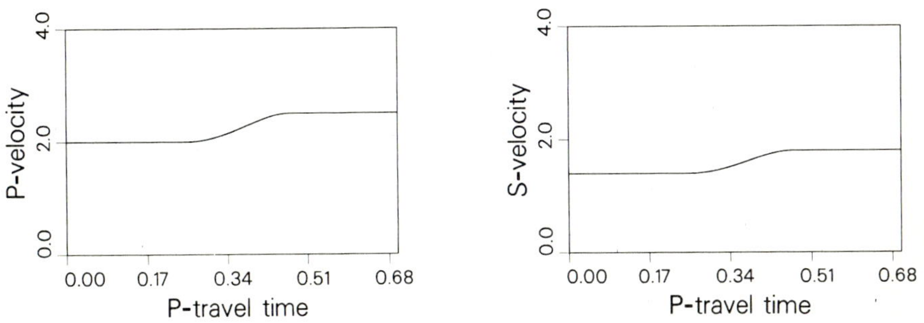

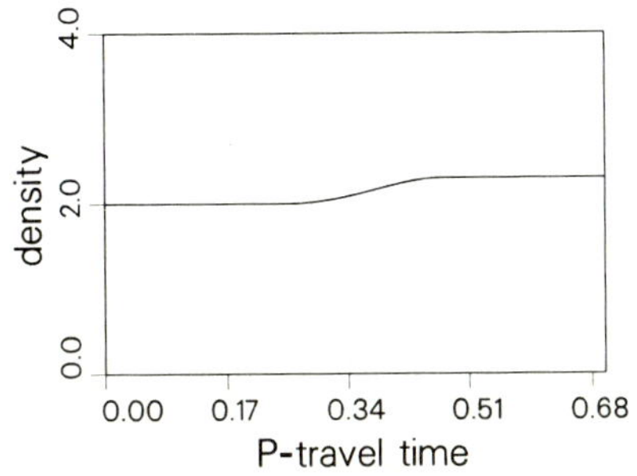

FIGURE 1

The P - and S- velocity and the density of the model.

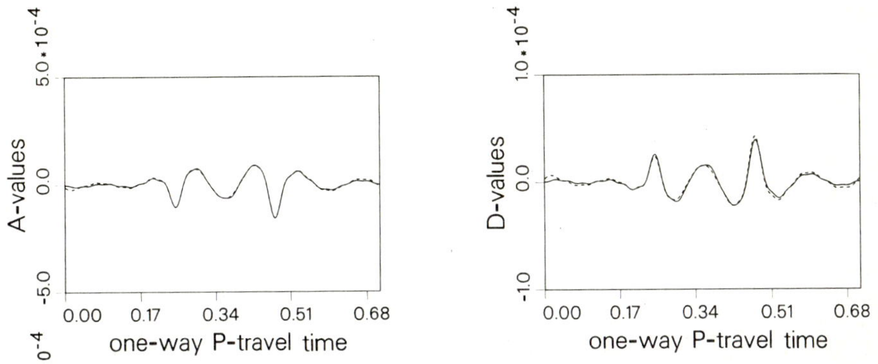

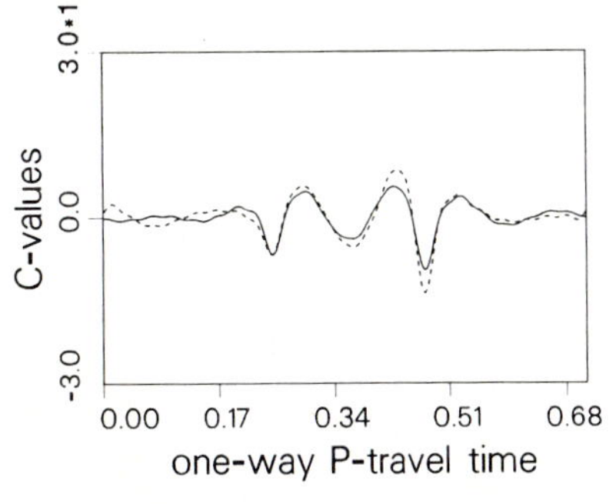

FIGURE 2

Inversion result for band-limited data (3-30 Hz). The band-limited model (solid line) and recovered (dashed line) A, D and C values.

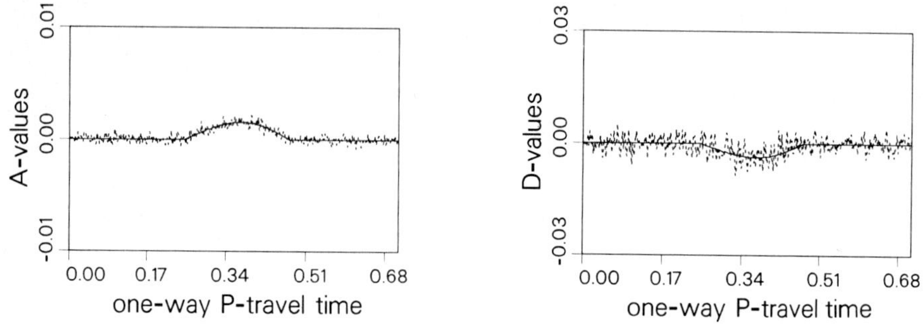

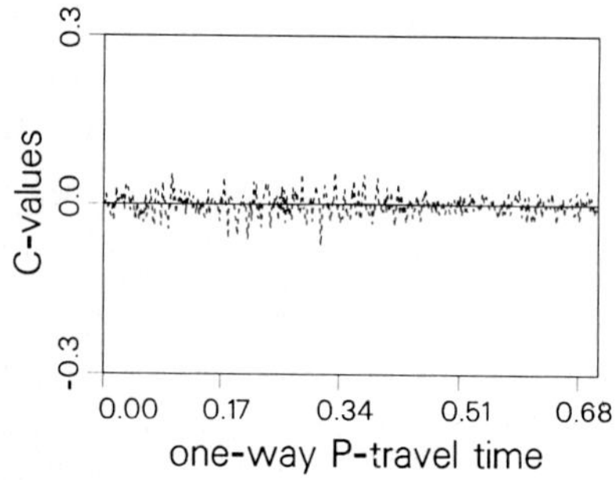

FIGURE 3

Inversion for noisy data with $\theta_{max} = 20°$
The model (solid line) and recovered (dashed line) A, D and C values.
Note the difference in scale.

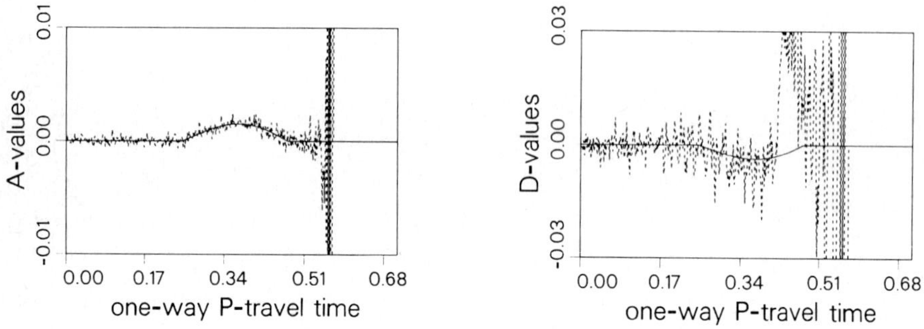

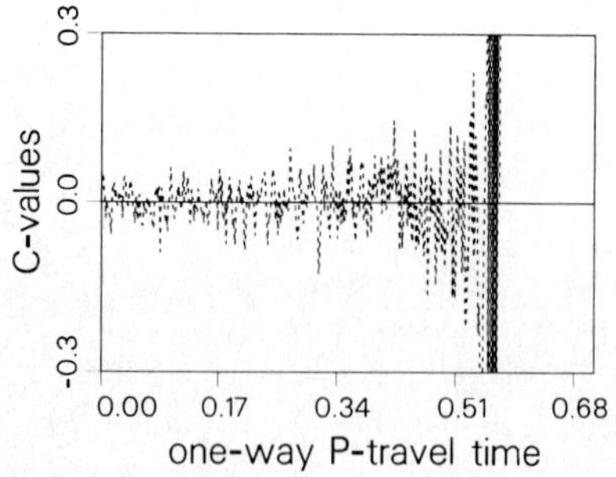

FIGURE 4

Inversion for noisy data with $\theta_{max} = 18°$
The model (solid line) and recovered (dashed line) A, D and C values.
Same scaling as in Fig. 3.

D
FLUID/SOLID WAVE INTERACTION

ELASTIC WAVE PROPAGATION
M.F. McCarthy, M.A. Hayes, (Editors)

WAVE PROPAGATION IN SUBMERGED LAYERED COMPOSITES

Arthur M. B. BRAGA and George HERRMANN

Division of Applied Mechanics
Stanford Universiy
Stanford, California 94305-4040

The propagator matrix method is applied to study the propagation of plane strain waves in periodically isotropic layered composites. A closed form algebraic solution for the dispersion equation of such waves is derived. The general motion of the laminated body is described by a linear combination of partial waves (eigenwaves). Quantitative results are presented for a periodically layered half-space interacting with an acoustic fluid.

1. INTRODUCTION

The propagation of elastic waves in periodically layered composites has been investigated by many researchers. Most of the contributions have focused on the strongly dispersive character of such waves and, with a few exceptions, these studies have been restricted to unbounded regions. In the present paper, we describe a methodology based on a six-dimensional theory, which has been developed to investigate problems involving the dynamic interaction of a periodically layered solid with an acoustic fluid. In this sextic formalism the equations of linear elastodynamics are written in matrix form, where the dependent variables, grouped in a six-dimensional array, are the displacements and tractions acting across planes parallel to the layers interfaces.

For harmonic motions, the governing equation is reduced to a sixth order matrix ordinary differential equation, and the idea of a propagator matrix appears naturally. When the layers are made of isotropic materials, the problem uncouples into a second order and a fourth order systems of equations describing, respectively, antiplane and plane-strain motions. Only the latter is treated in this paper. The dispersion equation for plane-strain waves is shown to be the characteristic equation for the propagator matrix of a unit cell. As a consequence of properties related to the isotropy of the layers, this fourth degree polynomial admits reciprocal roots, and we obtain a closed form algebraic solution for the dispersion equation. This solution is applied to show that a set of four eigenwaves suffices to describe the general wave motion of the periodically layered solid. Finally, the problem of fluid/solid interaction is solved by combining these four waves in such a way that the continuity conditions at the interface are satisfied.

2. GENERAL EQUATIONS

2.1. Homogeneous elastic medium

In the present investigation we are concerned with plane harmonic motions. Therefore, the following assumption is made

$$\boldsymbol{u}(x,z,t) = \bar{\boldsymbol{u}}(z)e^{i(\kappa_x x - \omega t)} \quad \text{and} \quad \boldsymbol{t}(x,z,t) = \bar{\boldsymbol{t}}(z)e^{i(\kappa_x x - \omega t)}, \tag{1}$$

where $\boldsymbol{u}$ is the displacement field and $\boldsymbol{t}$ is the traction vector acting across planes perpendicular to the Cartesian coordinate axis z. It can be shown that, for such motions, the field equations governing the elastodynamic behavior of a homogeneous, elastic medium may be recast in the form of a matrix differential equation

$$\frac{d}{dz}\bar{\boldsymbol{\xi}} = i\kappa_x \boldsymbol{N}\bar{\boldsymbol{\xi}}\ , \quad \text{where} \quad \bar{\boldsymbol{\xi}}(z) = \left\{ \begin{array}{c} \bar{\boldsymbol{u}}(z) \\ \frac{i}{\kappa_x}\bar{\boldsymbol{t}}(z) \end{array} \right\} . \tag{2}$$

The linear operator $\boldsymbol{N}$ is the *fundamental elasticity tensor* [1], which possesses the symmetry

$$\boldsymbol{NK} = (\boldsymbol{NK})^T\ , \quad \text{where} \quad \boldsymbol{K} = \begin{bmatrix} 0 & \boldsymbol{I} \\ \boldsymbol{I} & 0 \end{bmatrix} . \tag{3}$$

The solution of equation (2) is given by

$$\bar{\boldsymbol{\xi}}(z) = \boldsymbol{M}(z)\bar{\boldsymbol{\xi}}(0)\ , \quad \text{where} \quad \boldsymbol{M}(z) = e^{i\kappa_x \boldsymbol{N} z} \tag{4}$$

is the so-called *propagator matrix* [2].

For a homogeneous isotropic solid, with elastic constants λ and μ, the system (2) decouples into two equations, describing antiplane and plane-strain motions. In this paper we are concerned only with plane-strain waves. For such waves, equation (2) corresponds to

$$\frac{d}{dz}\left\{ \begin{array}{c} \bar{u}_x \\ \bar{u}_z \\ i\frac{\bar{\sigma}_{xz}}{\kappa_x} \\ i\frac{\bar{\sigma}_{zz}}{\kappa_x} \end{array} \right\} = i\kappa_x \begin{bmatrix} 0 & -1 & -1/\mu & 0 \\ -\lambda/(\lambda+2\mu) & 0 & 0 & -1/(\lambda+2\mu) \\ 4\mu\frac{\lambda+\mu}{\lambda+2\mu} - \rho(\frac{\omega}{\kappa_x})^2 & 0 & 0 & -\lambda/(\lambda+2\mu) \\ 0 & -\rho(\frac{\omega}{\kappa_x})^2 & -1 & 0 \end{bmatrix} \left\{ \begin{array}{c} \bar{u}_x \\ \bar{u}_z \\ i\frac{\bar{\sigma}_{xz}}{\kappa_x} \\ i\frac{\bar{\sigma}_{zz}}{\kappa_x} \end{array} \right\} , \tag{5}$$

and the sextic formalism is reduced to a fourth order theory. The propagator matrix has the properties

$$\boldsymbol{M}(z)\boldsymbol{K}\boldsymbol{M}^{\dagger}(z) = \boldsymbol{K}\ , \quad \text{and} \quad \det \boldsymbol{M}(z) = 1\ . \tag{6}$$

Equation $(6)_1$ is obtained from (3) and definition (4), while property $(6)_2$ is a consequence of the trace of $\boldsymbol{N}$ being zero for isotropic media.

2.2. Periodically layered media

We now consider a laminated body made of an arbitrary number of homogeneous, isotropic elastic layers — say J layers — stacked in a unit cell, of thickness d, that repeats periodically. The thickness of each layer is h_m ($m = 1, 2, \ldots, J$) and the spatial period is given by $d = h_1 + h_2 + \cdots + h_J$. To each layer is associated a matrix $\boldsymbol{N}_m$, which depends on the material properties as well as on the pair (ω, κ_x). We also define the matrix $\boldsymbol{M}_m(z)$ as

$$\boldsymbol{M}_m(z) = e^{i\kappa_x \boldsymbol{N}_m z}\ . \tag{7}$$

The harmonic elastic motions of such a body are still described by equation (2), but now the linear operator $\boldsymbol{N} = \boldsymbol{N}(z)$ is a periodic (piece-wise constant) function of the coordinate z. The value of $\boldsymbol{N}(z)$ along the mth layer of any unit cell is $\boldsymbol{N}_m$.

The solution of equation (2), with $\boldsymbol{N} = \boldsymbol{N}(z)$ being a periodic function, can still be written in terms of a propagator matrix $\boldsymbol{P}(z)$, *i.e.*

$$\bar{\boldsymbol{\xi}}(z) = \boldsymbol{P}(z)\bar{\boldsymbol{\xi}}(0)\ , \quad \text{but now} \quad \boldsymbol{P}(z) = \boldsymbol{F}(z)\, e^{i\boldsymbol{A}z}\ , \tag{8}$$

where $\boldsymbol{F}(z)$ exhibits the same periodicity as the layering, *i.e.* $\boldsymbol{F}(z+d) = \boldsymbol{F}(z)$, and $\boldsymbol{A}$ is a constant matrix [3]. This is an extension of Floquet's theory to higher dimensional spaces. The eigenvalues of $\boldsymbol{A}$ are the Floquet wave-numbers.

If we define the matrix $\boldsymbol{D}$ as the product

$$\boldsymbol{D} = \boldsymbol{M}_J(h_J)\boldsymbol{M}_{J-1}(h_{J-1})\cdots \boldsymbol{M}_2(h_2)\boldsymbol{M}_1(h_1) \,, \tag{9}$$

it can be shown that the following identity must hold

$$e^{i\boldsymbol{A}d} = \boldsymbol{D} \,. \tag{10}$$

Let τ_j $(j = 1, 2, \ldots, 4)$ be the eigenvalues of $\boldsymbol{D}$. From equation (10), the Floquet wave-numbers κ_z are related to the eigenvalues of $\boldsymbol{D}$ through $\tau = \exp\{i\kappa_z d\}$. Hence, the dispersion equation for the periodically layered medium corresponds to the characteristic equation of the matrix $\boldsymbol{D}$. This is a polynomial equation in τ of degree 4 and roots τ_j. It can be verified that the matrix $\boldsymbol{D}$ possesses the properties

$$\boldsymbol{D}\boldsymbol{K}\boldsymbol{D}^{\dagger} = \boldsymbol{K} \quad \text{and} \quad \det \boldsymbol{D} = 1 \,. \tag{11}$$

It follows, from (11), that if τ is an eigenvalue of $\boldsymbol{D}$ so is $1/\tau$. In other words, the characteristic equation of $\boldsymbol{D}$ or, as pointed out above, the dispersion equation for the periodically layered media, admits reciprocal roots. This property expresses mathematically the fact that the wave speed is the same in the z-positive or in the z-negative directions. Hence, the dispersion equation for plane-strain motions has the form

$$(\tau^2 + \frac{1}{\tau^2}) - I_1\,(\tau + \frac{1}{\tau}) + I_2 = 0 \,, \tag{12}$$

where I_1 and I_2 are, respectively, the first and the second invariants of the matrix $\boldsymbol{D}$. In general, equation (12) produces a set of four Floquet wave-numbers for a given pair ω, κ_x. Each κ_z is associated with a partial wave propagating freely in the unbounded medium. These eigenwaves are proper combinations of longitudinal and transverse modes propagating in each homogeneous layer. They can propagate independently in the infinite medium but, when boundaries are present, they must be coupled in order to satisfy the boundary conditions.

We now introduce a local coordinate z_n defined as $z_n = z - (n-1)d$, where the subscript n represents an integer labeling the unit cell such that $0 \le z_n \le d$. The state vector at $z = 0$ can be written as a linear combination of the eigenvectors of $\boldsymbol{D}$, *i.e.*

$$\bar{\boldsymbol{\xi}}(0) = \boldsymbol{\Xi}\boldsymbol{c} \,, \quad \text{where} \quad \boldsymbol{c}^T = (\,c_1 \quad c_2 \quad c_3 \quad c_4\,) \,, \tag{13}$$

and $\boldsymbol{\Xi}$ is the matrix whose columns are the eigenvectors of $\boldsymbol{D}$. The general solution may then be given by

$$\bar{\boldsymbol{\xi}}(z) = \boldsymbol{G}(z_n)\boldsymbol{\Xi}\,\{\mathrm{diag}\,(\tau_1^{n-1}, \tau_2^{n-1}, \tau_3^{n-1}, \tau_4^{n-1})\}\,\boldsymbol{c} \,, \tag{14}$$

rather than in terms of the propagator matrix $\boldsymbol{P}(z)$. The constant vector $\boldsymbol{c}$ is obtained from the boundary conditions, while

$$\boldsymbol{G}(z_n) = \begin{cases} \boldsymbol{M}_1(z_n) \,, & \text{if } 0 \le z_n \le h_1 \;; \\ \boldsymbol{M}_2(z_n - h_1)\boldsymbol{M}_1(h_1) \,, & \text{if } h_1 \le z_n \le h_1 + h_2 \;; \\ \vdots & \vdots \\ \boldsymbol{M}_J(z_n - d + h_J)\cdots \boldsymbol{M}_2(h_2)\boldsymbol{M}_1(h_1) \,, & \text{if } d - h_J \le z_n \le d \,. \end{cases} \tag{15}$$

3. DISPERSION EQUATION

In order to solve the dispersion equation (12), we introduce a new variable $\chi = \tau + 1/\tau$. We are then left with a system of two quadratic equations,

$$\chi^2 - I_1\,\chi + (I_2 - 2) = 0 \quad \text{and} \quad \tau^2 - \chi\tau + 1 = 0 \,. \tag{16}$$

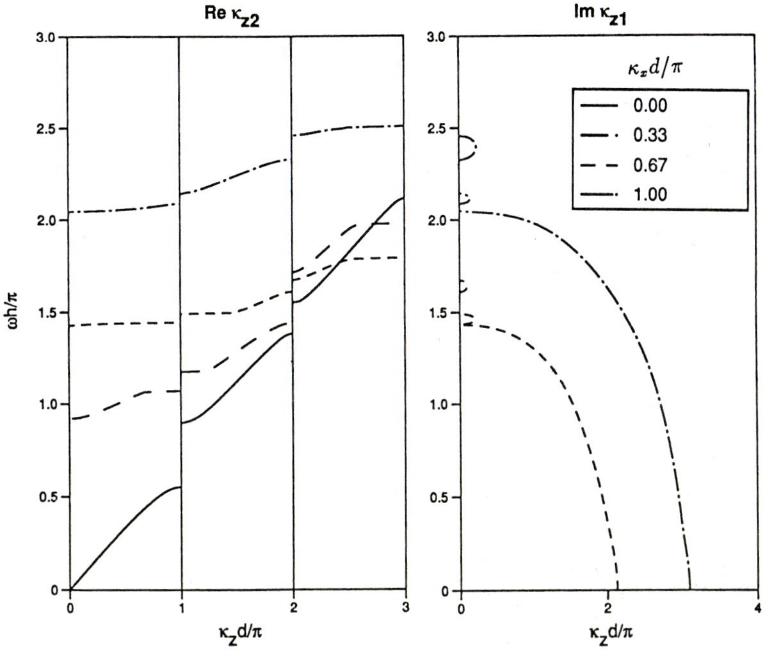

Figure 1. Dispersion curves at planes of constant κ_x.

real, or with positive imaginary part when complex. These conditions may be expressed in terms of the eigenvalues of $\boldsymbol{D}$, respectively, as $|\tau| = 1$ and $\text{Im}(\tau)/\text{Re}(\tau) \geq 0$, or $|\tau| < 1$. Assuming that τ_1 and τ_2 are the roots of the dispersion equation satisfying these conditions, the components c_3 and c_4 of the constant vector $\boldsymbol{c}$ vanish, and only the eigenvectors of $\boldsymbol{D}$ associated with τ_1 and τ_2 are needed to describe the plane-strain motions of a periodically layered half-space.

We now consider the problem of a semi-infinite, periodically layered body interacting with an acoustic fluid. Let the fluid, with density ρ_f and sound speed c_f, occupy the region $z < 0$, with $z = 0$ being the plane separating the semi-infinite layered solid from the fluid half-space. The interface conditions to be satisfied at $z = 0$ are then

$$\partial u_z/\partial t = v_z \ , \quad \sigma_{xz} = 0 \quad \text{and} \quad \sigma_{zz} = -p \ , \tag{20}$$

where v_z is the z-component of the fluid velocity, and p is the pressure field.

When a plane harmonic wave, propagating through the fluid with frequency ω and amplitude $\rho_f c_f^2$, impinges upon the layered solid surface, we have

$$\begin{aligned} p &= \rho_f c_f^2 \left(e^{i\kappa_0 z} + Re^{-i\kappa_0 z}\right) e^{i(\kappa_x x - \omega t)} \\ v_z &= \frac{\kappa_0 c_f^2}{\omega} \left(e^{i\kappa_0 z} - Re^{-i\kappa_0 z}\right) e^{i(\kappa_x x - \omega t)} \ , \end{aligned} \tag{21}$$

where $\kappa_0 = (\omega^2/c_f^2 - \kappa_x^2)^{1/2}$, and R is the reflection coefficient. The boundary conditions (20) can be written in terms of the transformed variables $\bar{u}_z$, $\bar{\sigma}_{xz}$, and $\bar{\sigma}_{zz}$ as

$$-i\omega\bar{u}_z(0) = \frac{\kappa_0 c_f^2}{\omega}(1 - R) \ , \quad \bar{\sigma}_{xz}(0) = 0 \ , \quad \text{and} \quad \bar{\sigma}_{zz}(0) = -\rho_f c_f^2 (1 + R) \ . \tag{22}$$

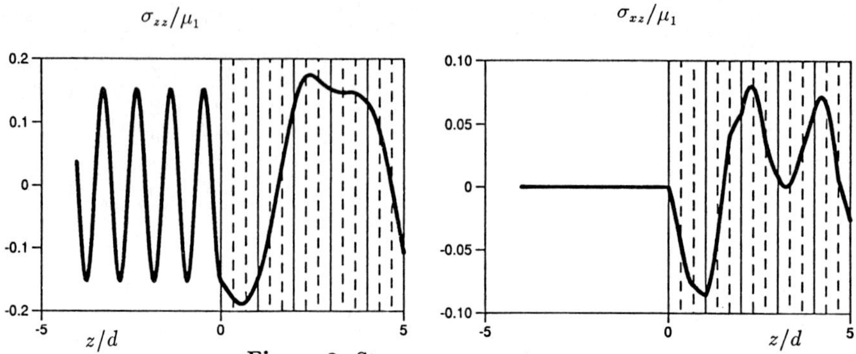

Figure 2. Stresses at a passing band.

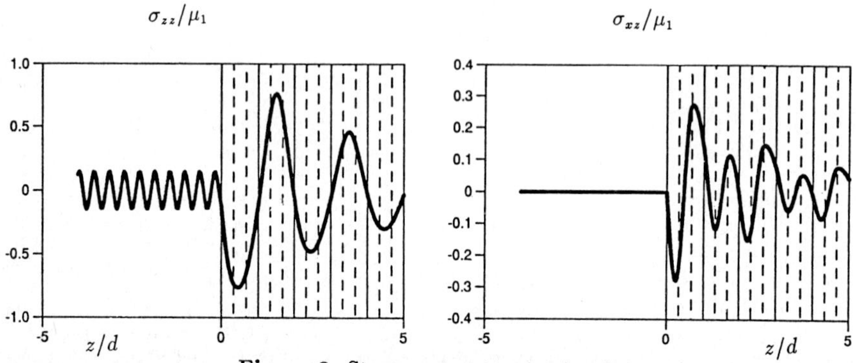

Figure 3. Stresses at a stopping band.

Equations (22) allow us to determine the reflection coefficient together with the components c_1 and c_2 of the constant vector $\mathbf{c}$. Figs. 2 and 3 show quantitative results for a three-phase periodically layered material (dispersion curves shown in Fig. 1) in contact with water. Fig. 2 presents the stresses for a frequency within a passing band, while results plotted in Fig. 3 are for a frequency in a stopping band.

ACKNOWLEDGEMENTS

During the course of this work A. M. B. Braga was partially supported by the Brazilian Government through a grant CAPES Proc. 6579-84/5 , and by the Catholic University of Rio de Janeiro. Support through ONR contract N00014-85-K-0471 to Stanford University is also gratefully acknowledged.

REFERENCES

[1] Chadwick, P., and Smith, G. D., " Foundations of The Theory of Surface Waves in Anisotropic Elastic Materials," *Adv. Appl. Mech.*, Vol. 17, pp 303–376 (1977).

[2] Gilbert, F., and Backus, G. E., " Propagator Matrices in Elastic Wave and Vibrations Problems," *Geophysics*, Vol. 31, No. 2, pp 326–332 (1966).

[3] Pease, M. C., *Methods of Matrix Algebra*, Academic Press, New York (1965).

ELASTIC WAVE PROPAGATION
M.F. McCarthy, M.A. Hayes, (Editors)

Elastic Wave Scattering from an Interface Crack in a Layered Half Space Submerged in Water

D. B. Bogy

Department of Mechanical Engineering
University of California, Berkeley CA 94720

and

S. M. Gracewski

Department of Mechanical Engineering
University of Rochester, Rochester, NY 14627

The analytical plane strain solution of time harmonic elastic wave scattering by an interface crack in a layered half space submerged in water is presented. The solution of the problem leads to a set of coupled singular integral equations for the jump in displacements across the crack. The kernels of these integrals are represented in terms of the Green's functions for the structure without the crack. Analysis of the integral equations yields the form of the singularities of the unknown functions at the crack tip. These singularities are taken into account to arrive at an algebraic approximation for the integral equations that can then be solved numerically.

Numerical results are presented for plane waves incident from the liquid onto the solid structure and for an incident Gaussian beam. The study is related to the use of ultrasonic techniques for detecting interface cracks in layered elastic media.

Published as "Elastic Wave Scattering from an Interface Crack in a Layered Half Space Submerged in Water: Part II: Incident Plane Waves and Bounded Beams," *ASME Journal of Applied Mechanics*, Vol. **53**, 1986, pp. 333-338.

ELASTIC WAVE PROPAGATION
M.F. McCarthy, M.A. Hayes, (Editors)

VIBRATIONS OF A POINT-EXCITED INFINITE ELASTIC PLATE INTERACTING WITH A FLUID.

Marie–Josèphe GHALEB and Monique PIAU

Institut de Mécanique de Grenoble (USTMG–INPG–CNRS)
Domaine Universitaire
BP 68
38402 SAINT–MARTIN–D'HÈRES CEDEX
(France)

In this paper, we consider the vibrations of an infinite plate interacting with a fluid and subjected to a normal point–force.

We show that the plate's displacement and deformation can be found by an analytic method using integral transforms. An inversion of the calculus should permit to characterize the force, knowing the measured deformation, particularly in the case of impacts due to the collapse of cavitation bubbles.

The method is tested in the case of an infinite plate in vacuum. The deformations are compared to those of a square plate under a ball impact, which are determined numerically and experimentally.

Many authors have studied the problem of the harmonic response, but most of them have concentrated on the acoustic radiation [1-7]; the subject of this paper is to calculate the structural vibrations of the plate and to develope a method giving the deformation under a general impact, especially for a steel plate interacting with water.

Free time–harmonic waves are studied [8] in connection with the dispersion relation and radiation conditions. Wave numbers to be considered for the problem are determined for a steel plate interacting with water. Using contour integrals in the complex k–plane (where k is the conjugate variable for the Hankel transform in space), the displacement due to a time–harmonic point–force is then obtained as the sum of a branch integral plus residue contributions due to the poles connected to free waves included in the contour path. The displacement obtained depends on the frequency domain considered.

The response of the half–submerged plate to any impact can then be obtained, using the above method with a Fourier transform in time.

I. EQUATIONS OF MOTION OF THE COUPLED SYSTEM.

Consider an infinite plate in the (r, θ) plane, coupled to an infinite fluid contained in the half–space $(z < 0)$.

The equation of motion governing the flexural vibrations of the plate excited by an external normal pressure $p_e(r, 0, t)$ is :

$$D(\Delta\Delta + \frac{m}{D}\frac{\partial}{\partial t^2})w(r,t) = -p(r,0,t) + p_e(r,0,t) \tag{1.1}$$

where

- $w(r,t)$ is the plate flexural displacement
- $p(r,0,t)$ is the acoustic pressure in the plane $(z=0)$
- m is the mass per unit area, and D is the bending stiffness of the plate.

The relation between the pressure and the potential in the fluid of density ρ_0 is :

$$p(r,z,t) = -\rho_0 \frac{\partial \Phi}{\partial t}(r,z,t) \ , \tag{1.2}$$

and the normal velocity of the plate statisfies :

$$\frac{\partial \Phi}{\partial z}(r,0,t) = \frac{\partial w}{\partial t}(r,t) \ . \tag{1.3}$$

For a given frequency ω , the potential must satisfy the Helmholtz equation :

$$(\Delta + k_0^2)\Phi(r,z,t) = 0 \tag{1.4}$$

where $k_0 = \dfrac{\omega}{c_0}$, and c_0 is the speed of sound in the fluid

II. RESPONSE OF THE PLATE TO A NORMAL POINT–FORCE. GENERAL METHOD.

The equations relating w, p, Φ are then (1.1 to 1.4), with

$$p_e(r,0,t) = F(t)\frac{\delta(r)}{2\pi r} \ .$$

These variables may be calculated by an analytic method consisting in three steps :

1) For a harmonic–force : $F(t) = F_0 e^{-i\omega t}$, using a Hankel transform in space $(r \to k)$;

2) For a triangular symmetrical force : $F(t) = F_0 \varphi_0(t) e^{-i\omega t}$

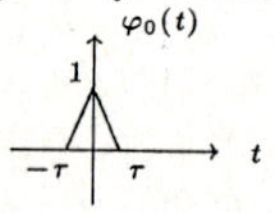

using a Fourier transform in time and the harmonic response.

3) For a force $F(t)$, using a linear decomposition of $F(t)$ on a basis of triangular symmetrical functions $\varphi_i(t)$.

III. VALIDATION OF THE ANALYTIC METHOD.

The analytic method was tested for a plate without any fluid, by comparing its results with those obtained numerically for a square plate using an improved method.

The equation of motion of an infinite plate in vacuum is :

$$D(\Delta\Delta + \frac{m}{D}\frac{\partial^2}{\partial t^2})w(r,t) = F(t)\frac{\delta(r)}{2\pi r} \ . \tag{3.1}$$

For a harmonic force :

$$F(t) = F_0 \bar{e}^{i\omega t}\frac{\delta(r)}{2\pi r} \ , \quad \text{with} \quad \omega > 0 \ ,$$

the displacement will be harmonic too :

$$w(r,t) = W(r)e^{-i\omega t} \ .$$

Using a Hankel transform in space :

$$W(k) = \int_0^\infty W(r)J_0(kr)rdr \;; \quad W(r) = \int_0^\infty W(k)J_0(kr)kdk$$

the Hankel transformed version of equation (3.1) can be written :

$$D(k^4 - k_f^4)W(k) = \frac{F_0}{2\pi} \quad \text{with} \quad k_f = \left(\frac{m\omega^2}{D}\right)^{1/4}$$

then the spatial component of the displacement is :

$$W(r) = \frac{F_0}{2\pi D}\int_0^\infty \frac{J_0(kr)}{k^4 - k_f^4} kdk \;.$$

This integral is defined in the complex k plane with two physical conditions : the waves must propagate away from the source, and the amplitude of the displacement must be finite for large values of r .

A damping term is then added to the plate operator, which becomes :

$$D\left(\Delta\Delta + \frac{m\alpha}{D}\frac{\partial}{\partial t} + \frac{m}{D}\frac{\partial^2}{\partial t^2}\right) \;, \alpha > 0$$

so that :

$$W(r) = \frac{F_0}{8Dk_f^2}\left\{-\frac{2}{\pi}K_0(k_f r) + iH_0^1(k_f r)\right\} .$$

Now for a triangular force $F_0\ \varphi_0(t)$, with a Fourier transform $F_0\ \varphi_0(\omega)$, the displacement can be written :

$$w(r,t) = \frac{F_0}{2\pi}\int_0^\infty \varphi_0(\omega)e^{i\omega t}d\omega \int_0^\infty \frac{J_0(kr)}{2\pi D(k^4 - k_f^4)} kdk$$

with

$$\int_0^\infty \frac{J_0(kr)}{(k^4 - k_f^4)} kdk = \frac{1}{8Dk_f^2}\left\{-(Y_0(k_f r) + \frac{2}{\pi}K_0(k_f r)) - i\ \mathrm{sign}(\omega)J_0(k_f r)\right\}$$

$$\varphi_0(\omega) = \tau\left(\frac{\sin\omega\tau/2}{\omega\tau/2}\right)^2 .$$

Any general force $F(t)$ may be decomposed on a basis of triangular functions :

$$F(t) = \sum_{n=0}^{N} F_n\varphi_n(t) \quad \text{where} \quad F_n = F(n\tau) \;.$$

The response to this force is, accounting for the linearity of the equations :

$$w(r,t) = \sum_{n=0}^{N} F_n R_n(r,t)$$

where $R_n(r,t)$ is the displacement under a force $1 \cdot \varphi_n(t)$.

The response of a square steel plate to a steel ball impact is calculated using a modal decomposition and a Hertz–type contact law, and verified by experimental measurement of the deformations of the plate.

By comparing the results obtained for a square plate and an infinite plate of the same thickness, under the same force, using the analytic method, a very good agreement is found, especially in the first moments, when the square plate may be supposed to be infinite.

Figure (1) shows the deformations at a distance of 5cm from the impact point, for the square plate and the infinite plate.

The impact considered is due to a steel ball falling on the centre of a steel plate of dimensions $50 \times 50 \times 0.2 cm^3$. The mass of the ball is 11g and its velocity at the beginning of the shock is 10m/s .

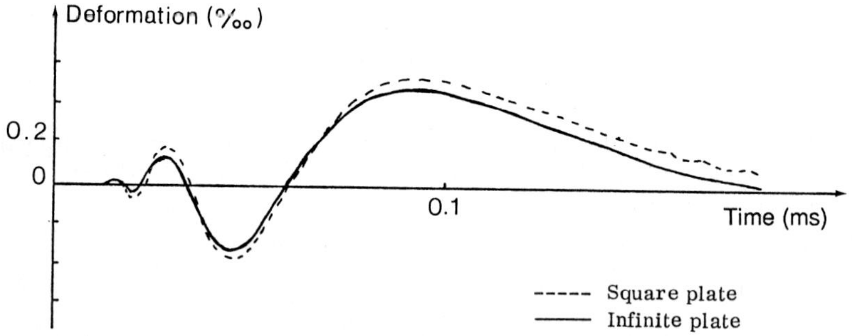

Figure (1)

IV. THE HALF-SUBMERGED PLATE.

Using a Hankel transform in space for the coupled equations (1.1 to 1.4) with $p_e(r,0,t) = F_0 \bar{e}^{i\omega t} \frac{\delta(r)}{2\pi r}$, the plate's displacement and the acoustic pressure in the fluid are given by :

$$w(r,t) = W(r)e^{-i\omega t}$$
$$p(r,z,t) = P(r,z)e^{-i\omega t}$$

while the integral representations of their spatial components are :

$$W(r) = \frac{iF_0}{2\pi\omega} \int_0^\infty \frac{1}{Z_F(k) + Z_P(k)} J_0(kr) k dk$$
$$P(r,z) = \frac{F_0}{2\pi} \int_0^\infty \frac{Z_F(k)}{Z_F(k) + Z_P(k)} e^{-\sqrt{k^2-k_0^2}z} J_0(kr) k dk$$

with the conditions : $\sqrt{k^2 - k_0^2} > 0$ and $\mathrm{Im}\sqrt{k^2 - k_0^2} < 0$. Z_F and Z_P are respectively the impedances of the fluid and the plate, defined by :

$$Z_F(k) = \frac{P(k,0)}{\dot{W}(k)} = -\frac{i\omega\rho_0}{\sqrt{k^2 - k_0^2}}$$
$$Z_P(k) = \frac{P_e(k,0)}{\dot{W}(k)} = \frac{iD(k^4 - k_f^4)}{\omega}$$

for the thin plate, where $k_f = (\frac{m\omega^2}{D})^{1/4}$ is the flexural wave number of the plate in the vacuum.

The free wavenumbers satisfy the dispersion equation :

$$Z_F(k) + Z_P(k) = 0$$
$$(k^4 - k_f^4)\sqrt{k^2 - k_0^2} - \frac{\rho_0\omega^2}{D} = 0 .$$

Solving this system for τ, two pairs of solutions are obtained

$$\tau_{1,3} = \frac{1}{4}\left(I_1 + \Delta_0^{1/2} \pm \Delta_1^{1/2}\right) \quad \text{and} \quad \tau_{2,4} = \frac{1}{4}\left(I_1 - \Delta_0^{1/2} \pm \Delta_2^{1/2}\right) , \tag{17}$$

where

$$\Delta_0 = I_1^2 - 4(I_2 - 2) ,$$
$$\Delta_1 = \left(I_1 + \Delta_0^{1/2}\right)^2 - 16 , \quad \text{and} \quad \Delta_2 = \left(I_1 - \Delta_0^{1/2}\right)^2 - 16 . \tag{18}$$

It can be verified that $\tau_1\tau_3 = \tau_2\tau_4 = 1$, where τ_j $(j = 1, \ldots, 4)$ may assume real or complex values according to the signs of Δ_0, Δ_1, and Δ_2. From (10) follows

$$\kappa_{z_j} = \frac{1}{d} \arctan\left(\frac{\mathrm{Im}(\tau_j)}{\mathrm{Re}(\tau_j)}\right) - i\frac{\ln|\tau_j|}{d} , \quad j = 1, \ldots, 4 . \tag{19}$$

Hence, the wave-number κ_z is real if $|\tau| = 1$ or complex otherwise. Real wave-numbers correspond to propagating waves, while complex values of κ_z are associated with waves whose amplitudes either grow or decay along the coordinate z.

The dispersion spectrum can be studied by examining the signs of the discriminants Δ_0, Δ_1, and Δ_2. Degeneracies occur when these discriminants vanish, causing the dispersion equation to have multiple roots. All possible cases are summarized below:

(a) $\Delta_0 > 0$, $\Delta_1 > 0$, $\Delta_2 > 0$, (τ_1 and τ_2 real) ; two pairs of complex κ_x.

(b) $\Delta_0 > 0$, $\Delta_1 < 0$, $\Delta_2 > 0$, (τ_1 complex , $|\tau_1| = 1$, and τ_2 real) ; One pair of real and one pair of complex wave-numbers κ_z.

(c) $\Delta_0 > 0$, $\Delta_1 > 0$, $\Delta_2 < 0$, (τ_1 real and τ_2 complex , $|\tau_2| = 1$) ; same as above.

(d) $\Delta_0 > 0$, $\Delta_1 < 0$, $\Delta_2 < 0$, (τ_1 and τ_2 complex , $|\tau_1| = |\tau_2| = 1$) ; all four κ_x are real.

(e) $\Delta_0 < 0$, $\Delta_1 = \Delta_2^*$, (τ_1 and τ_2 complex, $\tau_1 = \tau_2^*$, and $|\tau_1| \neq 1$) ; four complex Floquet wave-numbers.

(f) $\Delta_0 = 0$, $\Delta_1 = \Delta_2 > 0$,($\tau_1 = \tau_2$ real, $\tau_1 \neq \pm 1$) ; two pairs of complex Floquet wave-numbers coalesce.

(g) $\Delta_0 = 0$, $\Delta_1 = \Delta_2 < 0$, ($\tau_1 = \tau_2$ complex , $|\tau_1| = 1$) ; two pairs of real Floquet wave-numbers coalesce.

(h) $\Delta_0 \geq 0$, $\Delta_1 = 0$ and/or $\Delta_2 = 0$, ($\tau_1 = \pm 1$ and/or $\tau_2 = \pm 1$) ; $\kappa_{zj} = k\pi/d$ where $k = 0, \pm 1, \pm 2, \ldots$.

The roots of the dispersion equation generate a set of two surfaces in the frequency-wave-number space. Fig. 1 shows dispersion curves that correspond to cuts in one of these surfaces along planes of constant κ_x. The curve originating at the origin corresponds to $\kappa_x = 0$, propagation normal to the layering, when the Floquet waves decouple into pure longitudinal and transverse shear modes. These results are obtained for a three-phase periodically layered body ($J = 3$). A more complete discussion of the dispersion spectrum for plane-strain waves in periodically layered media will be presented in a forthcoming paper by the authors.

4. HALF-SPACE WITH AN OVERLYING ACOUSTIC FLUID

Not all Floquet wave-numbers are physically acceptable in the case of a half-space without an internal source. Only waves propagating away from the surface or, in the case of complex wave-numbers, those with decaying amplitude, are allowed in the semi-infinite media. Thus, it is required that the admissible wave-numbers be either positive, when

Multiplying by the conjug-ate expression, a fifth degree polynomial in k^2 with real coefficients is obtained. The nature of the roots depends on the frequency : there are one real root k_+^2 and four complex roots : $k_\bullet^2$, $\overline{k}_\bullet^2$, $k_\times^2$, $\overline{k}_\times^2$ or three real roots k_+^2 , $k_{\times_1}^2$, $k_{\times_2}^2$ and two complex roots $k_\bullet^2$, $\overline{k}_\bullet^2$. The values of k in the first quadrant are plotted in figure (2) for a steel plate of $2mm$ thickness interacting with water.

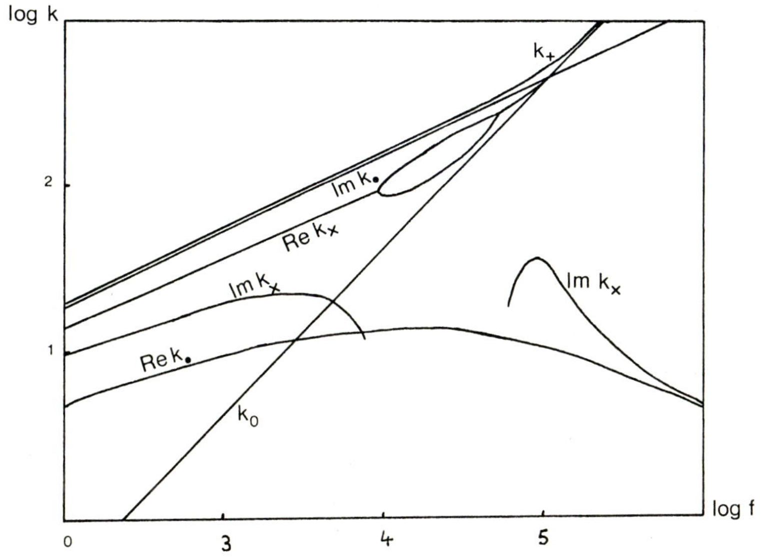

Figure (2)

The problem of the free waves has been studied by many authors [1-7]. This paper concentrates on the interaction between a steel plate and water.

Using contour integrals in the k plane, the spatial component of the displacement $W(r)$ in farfield ($k_0 r \gg 1$) is obtained as the sum of a branch integral and residue contributions due to the poles of the integrand included in the contour path. The expression of the displacement depends on the frequency domain considered [8].

CONCLUSION.

In this paper, an analytic method using integral transforms is developed for the response of an infinite half-submerged plate subjected to a normal impact. The method is tested with an infinite plate in vacuum, by comparison with the results of an improved calculus for a square plate.

The harmonic response, which is necessary for applying the method, is found for the half-submerged plate and for larges distances from the excitation point, especially for a steel plate interacting with water : this response is the sum of a contour integral and different terms of residues, and depends on the frequency domain considered.

The method above could then be used to find the response of the half-submerged plate to any normal impact.

AKNOWLEDGMENT.

M.-J. GHALEB is grateful to ALSTHOM-ACB, 38800 PONT-DE-CLAIX (France) for their support during this study.

REFERENCES.

[1] S. LOWENTHAL. — *Etude théorique et expérimentale de l'interaction entre une plaque élastique plane et son rayonnement acoustique*, Annales de Radioélectricité, T.14, n°77, 1964.

[2] M.C. JUNGER–D. FEIT. — *Sound, structures and their interactions*, MIT Press, 1972.

[3] D. FEIT. — *Pressure radiated by a point-excited elastic plate*, J.A.S.A., **40** (6) (1966), 1489–1494.

[4] D.G. CRIGHTON. — *The free and forced waves on a fluid-loaded elastic plate*, J.S.V., **91** (2) (1979), 225–235.

[5] D.G. CRIGTHON–D. INNES. — *Low frequency acoustic radiation and vibration response of locally excited fluid-loaded structures*, J.S.V., **91** (2) (1983), 293–314.

[6] F. SHROTER–F.J. FAHY. — *Point-force excited vibrations of a thin, Infinite panel separating a fluid-layer from a fluid half-space*, J.S.V., **74** (4) (1981), 465–476.

[7] A.D. STUART. — *Acoustic Radiation from submerged plates*, J.A.S.A., **59** (5) (1976), 1160–1174.

[8] M.J. GHALEB. — *Vibrations d'une plaque élastique infinie couplée avec un fluide sous l'effet d'un impact ponctuel*, Thèse de troisième cycle de Mécanique, Grenoble, 1987.

ELASTIC WAVE PROPAGATION
M.F. McCarthy, M.A. Hayes, (Editors)

ELASTODYNAMIC RADIATION FROM AN IMPULSIVE POINT SOURCE IN A POROELASTIC FLUID/SOLID MEDIUM

Sijtze M. de Vries

Laboratory of Electromagnetic Research, Faculty of Electrical Engineering, Delft University of Technology, P.O. Box 5031, 2600 GA Delft, The Netherlands

Elastodynamic radiation from an impulsive point source in an unbounded, homogeneous, isotropic, poroelastic fluid/solid medium is investigated with the aid of a temporal Laplace and spatial Fourier transforms. To describe the macroscopic propagation of linear acoustic waves through a fluid-saturated porous solid we use four coupled partial differential equations that follow upon applying a volume averaging procedure to the two constituents of the porous medium, viz. a perfectly elastic solid and an ideal fluid.

1. INTRODUCTION

The macroscopic theory of the acoustic wave motion in a porous material is taken up in the context of averaging [1]. For pore sizes much less than the acoustic wavelengths, it can be expected that some volume average of the microscopic geometry shows up in (some of) the acoustic wave phenomena on the macroscopic scale. By using a volume averaging procedure, macroscopic acoustic equations follow from the equations that apply to the two constituents, while coupling between the two phases takes place via interaction terms at the microscopic boundary. The presence of sources is accounted for by the introduction of volume source densities of body force and of strain rate, for each of the constituents. The fluid and solid volume averaged particle velocities of the wave motion generated by a point source (of body force) are obtained with the aid of a temporal Laplace and spatial Fourier transforms.

2. AVERAGING CONSIDERATIONS

The basic assumption underlying our analysis is that the macroscopic properties of the porous medium under consideration follow from a procedure of spatial averaging applied to the local solid and fluid quantities over a so-called representative elementary domain D_ε, whose maximum diameter is such that the obtained average is insensitive to small changes in this diameter. If a small variation in the diameter of D_ε would result into a varying averaged quantity, the averaging procedure is not meaningful. The latter happens when the diameter of D_ε is either so small that it approaches the value of the diameter of the pores, or so large that macroscopic inhomogeneities affect the averaging procedure. The domain D_ε is taken to be shift- and time-invariant; its position is specified by the position vector $\underline{x}$ of its "center". The macroscopic quantities are then assigned to the position of this "center". In the two-component medium, $D_\varepsilon = D_\varepsilon(\underline{x})$ is the union of the subdomain $D^f_\varepsilon(\underline{x})$ where fluid is present and the subdomain $D^s_\varepsilon(\underline{x})$ where solid is present. The volumes of $D_\varepsilon(\underline{x})$, $D^f_\varepsilon(\underline{x})$, $D^s_\varepsilon(\underline{x})$ are denoted by V_ε, $V^f_\varepsilon(\underline{x})$, and $V^s_\varepsilon(\underline{x})$, respectively. It is clear that the volume fractions $\phi^{f,s}_\varepsilon(\underline{x}) = V^{f,s}_\varepsilon(\underline{x})/V_\varepsilon$ occupied by the fluid or solid phase in general vary with position. Since $V_\varepsilon = V^f_\varepsilon(\underline{x}) + V^s_\varepsilon(\underline{x})$ it follows that $\phi^f(\underline{x}) + \phi^s(\underline{x}) = 1$ for all $\underline{x}$. For any quantity Ψ^f (Ψ^s) associated with the fluid (solid) phase in D_ε we define the corresponding "fluid (solid) average" as

$$\langle\Psi^{f,s}\rangle(\underline{x},t) = V_\varepsilon^{-1} \int_{\underline{x}'\in D^{f,s}_\varepsilon(\underline{x})} \Psi^{f,s}(\underline{x}',t)dV. \tag{1}$$

In deriving the expressions for the spatial derivatives of the volume averages, we also encounter surface integrals over the interface between fluid and solid as far as this is located in the interior of the representative elementary domain. This interface will be denoted by $\Sigma_\varepsilon = \Sigma_\varepsilon(\underline{x})$. The relevant relations are

$$\langle\partial'_i\Psi^{f,s}\rangle(\underline{x},t) = \partial_i\langle\Psi^{f,s}\rangle(\underline{x},t) + V_\varepsilon^{-1} \int_{\underline{x}'\in\Sigma_\varepsilon(\underline{x})} \nu_i^{f,s}\, \Psi^{f,s}(\underline{x}',t)dA, \tag{2}$$

$$\langle\partial_t\Psi^{f,s}\rangle(\underline{x},t) = \partial_t\langle\Psi^{f,s}\rangle(\underline{x},t), \tag{3}$$

where ν_i^f (ν_i^s) is the unit vector pointing away from the subdomain D^f_ε (D^s_ε).

3. VOLUME AVERAGING OF THE BASIC EQUATIONS OF ACOUSTIC WAVE THEORY

To derive the basic equations that govern the acoustical behaviour of a porous medium, we apply the method of local volume averaging to the basic equations of acoustic wave theory for an ideal fluid and the basic equations of acoustic wave theory for a perfectly elastic solid. We start with the linearized equations of motion and the linearized deformation rate equations (together with their respective linearized constitutive equations) for a fluid and a solid. For the fluid phase we write

$$\delta_{km}\partial_m\sigma - \rho^f_{kr}\partial_t v^f_r = -f^f_k, \tag{4}$$

$$\delta_{mr}\partial_m v^f_r - \kappa^f\partial_t\sigma = h^f, \tag{5}$$

and for the solid phase

$$\Delta_{kmpq}\partial_m\tau_{pq} - \rho^s_{kr}\partial_t v^s_r = -f^s_k, \tag{6}$$

$$\Delta_{ijmr}\partial_m v^s_r - s_{ijpq}\partial_t\tau_{pq} = h^s_{ij}, \tag{7}$$

where δ_{km} is the symmetrical unit tensor of rank two (Kronecker tensor), ∂_m is the partial derivative with respect to x_m (m^{-1}), σ is the fluid traction (Pa),

ρ^f_{kr} (ρ^s_{kr}) is the tensorial fluid (solid) volume density of mass (kg/m^3), ∂_t is the partial derivative with respect to t (s^{-1}), v^f_r (v^s_r) is the fluid (solid) particle velocity (m/s), f^f_k (f^s_k) is the volume source density of fluid (solid) body force (N/m^3), κ^f is the fluid compressibility (Pa^{-1}), h^f is the volume source density of fluid injection rate (s^{-1}), $\Delta_{kmpq} = (\delta_{kp}\delta_{mq} + \delta_{kq}\delta_{mp})/2$ is a symmetrical unit tensor of rank four, τ_{pq} is the solid stress (Pa), s_{ijpq} is the solid compliance (Pa^{-1}), and h^s_{ij} is the volume source density of solid strain rate (s^{-1}). The most general kind of anisotropy has been included, but we restrict ourselves to an instantaneous reaction, i.e., we neglect losses (also the ones of the viscous type). Further we consider media that are time-invariant, i.e., the constitutive coefficients are time-independent. In general, the constitutive coefficients depend on position (inhomogeneous medium).

The boundary conditions at the interfaces Σ_ε between the fluid and the solid phases, require the equality of the normal component of the traction, the vanishing of the tangential components of the traction of the solid, and the continuity of the normal component of the particle velocity. Note that the tangential components of the particle velocities in the solid and the fluid may differ at either side of the interface Σ_ε (slip can occur parallel to the interface since viscosity has been neglected).

We start the averaging procedure by applying the volume averaging operator (cf. Eq. (1)) to Eqs. (4) - (7). In view of the properties (2) - (3) we obtain

$$\delta_{km}\partial_m\langle\sigma\rangle - \rho^f_{kr}\partial_t\langle v^f_r\rangle + V_\varepsilon^{-1}\int_{\underline{x}'\in\Sigma_\varepsilon(\underline{x})} \delta_{km}\nu^f_m\sigma dA = -\langle f^f_k\rangle, \tag{8}$$

$$\delta_{mr}\partial_m\langle v^f_r\rangle - \kappa^f\partial_t\langle\sigma\rangle + V_\varepsilon^{-1}\int_{\underline{x}'\in\Sigma_\varepsilon(\underline{x})} \delta_{mr}\nu^f_m v^f_r dA = \langle h^f\rangle, \tag{9}$$

and

$$\Delta_{kmpq}\partial_m\langle\tau_{pq}\rangle - \rho^s_{kr}\partial_t\langle v^s_r\rangle + V_\varepsilon^{-1}\int_{\underline{x}'\in\Sigma_\varepsilon(\underline{x})} \Delta_{kmpq}\nu^s_m\tau_{pq}dA = -\langle f^s_k\rangle, \tag{10}$$

$$\Delta_{ijmr}\partial_m\langle v^s_r\rangle - s_{ijpq}\partial_t\langle\tau_{pq}\rangle + V_\varepsilon^{-1}\int_{\underline{x}'\in\Sigma_\varepsilon(\underline{x})} \Delta_{ijmr}\nu^s_m v^s_r dA = \langle h^s_{ij}\rangle. \tag{11}$$

The surface integrals describe the acoustic interaction between the fluid and solid phases. In Eqs. (8) and (9) this interaction is related to stresses and traction and can, therefore, be expected to represent the inertia properties of the coupled system. In Eqs. (10) and (11) the interaction is related to the deformation rates and can, therefore, be expected to represent the compliance properties of the coupled system. Interaction terms in general consist of a self induced part and a mutually induced part. In this respect we observe that in many branches of physics linear coupling between subsystems occurs; we mention the magnetic coupling between circuits in electromagnetics, and coupled springs in mechanics. In all these cases experience learns that linear coupling terms can adequately explain the physical behaviour of such coupled systems. Therefore, such an approach seems promising in the analysis of acoustic wave motion in the coupled fluid/solid system as well. The values of the coupling coefficients could, in principle, be deduced from a detailed statistical analysis of the geometry of the fluid/solid interface at the pore-size level, and the mechanical properties of the fluid and the solid. This is beyond the scope of our analysis. Within the realm of a linear interaction theory we shall introduce the most general phenomenological coupling coefficients that are compatible with such a theory, and leave aside how their values are related to the microscopic structure of the composite. We write

$$V_\varepsilon^{-1}\int_{\underline{x}'\in\Sigma_\varepsilon(\underline{x})}\delta_{km}\nu_m^f\sigma\,dA = -V_\varepsilon^{-1}\int_{\underline{x}'\in\Sigma_\varepsilon(\underline{x})}\Delta_{kmpq}\nu_m^s\tau_{pq}\,dA = m_{kr}^{fs}\partial_t\langle v_r^s\rangle - m_{kr}^{sf}\partial_t\langle v_r^f\rangle, \quad (12)$$

$$V_\varepsilon^{-1}\int_{\underline{x}'\in\Sigma_\varepsilon(\underline{x})}\delta_{mr}\nu_m^f v_r^f\,dA = \kappa^{fs}\partial_t\langle\tau_{kk}\rangle/3 - \kappa^{sf}\partial_t\langle\sigma\rangle, \quad (13)$$

$$V_\varepsilon^{-1}\int_{\underline{x}'\in\Sigma_\varepsilon(\underline{x})}\Delta_{ijmr}\nu_m^s v_r^s\,dA = \kappa^{sf}\partial_t\langle\sigma\rangle\delta_{ij}/3 - K_{ijpq}\partial_t\langle\tau_{pq}\rangle, \quad (14)$$

where $m_{kr}^{fs}=m_{kr}^{sf}$ is the mutually-induced tensorial volume density of mass (kg/m^3), $\kappa^{fs}=\kappa^{sf}$ is the mutually-induced scalar compressibility (Pa^{-1}), and K_{ijpq} is the mutually-induced tensorial compliance (Pa^{-1}) (with $K_{iipq} = \kappa^{fs}\delta_{pq}/3$). Upon substituting Eqs. (12) - (14) in Eqs. (8) - (11) we end up with

$$\delta_{km}\partial_m\langle\sigma\rangle - m_{kr}^{ff}\partial_t\langle v_r^f\rangle + m_{kr}^{fs}\partial_t\langle v_r^s\rangle = -\langle f_k^f\rangle, \quad (15)$$

$$\delta_{mr}\partial_m\langle v_r^f\rangle - \kappa^{ff}\partial_t\langle\sigma\rangle + \kappa^{fs}\partial_t\langle\tau_{pp}\rangle/3 = \langle h^f\rangle, \quad (16)$$

and

$$\Delta_{kmpq}\partial_m\langle\tau_{pq}\rangle - m_{kr}^{ss}\partial_t\langle v_r^s\rangle + m_{kr}^{sf}\partial_t\langle v_r^f\rangle = -\langle f_k^s\rangle, \quad (17)$$

$$\Delta_{ijmr}\partial_m\langle v_r^s\rangle - \kappa_{ijpq}^{ss}\partial_t\langle\tau_{pq}\rangle + \kappa^{sf}\partial_t\langle\sigma\rangle\,\delta_{ij}/3 = \langle h_{ij}^s\rangle, \quad (18)$$

where $m_{kr}^{ff}=\rho_{kr}^f+m_{kr}^{sf}$ is the self-induced tensorial fluid volume density of mass (kg/m^3), $\kappa^{ff}=\kappa^f+\kappa^{sf}$ is the self-induced scalar fluid compressibility (Pa^{-1}), $m_{kr}^{ss}=\rho_{kr}^s+m_{kr}^{fs}$ is the self-induced tensorial solid volume density of mass (kg/m^3), and $\kappa_{ijpq}^{ss}=s_{ijpq}+K_{ijpq}$ is the self-induced tensorial solid compliance (Pa^{-1}). Eqs. (15) - (18) constitute the basic partial differential equations that describe the propagation of acoustic waves in the anisotropic, inhomogeneous, fluid/solid composite (note that the volume fractions do not explicitly occur). The macroscopic picture is that the fluid and solid phases are fully mixed and simultaneously present, while their interaction (on the basis of a linear theory) is incorporated in the mutually induced coefficients $\{m_{kr}^{fs}, m_{kr}^{sf}, \kappa^{fs}, \kappa^{sf}\}$.

4. RADIATION FROM AN IMPULSIVE POINT SOURCE

For an isotropic, homogeneous, and unbounded porous medium Eqs. (15) - (18) can be solved with the aid of a temporal Laplace and spatial Fourier transformation method (where the sign conventions are chosen such that $\partial_t \to p$ and $\partial_m \to -jk_m$). For such a medium the mass tensors are writen as $m^{\cdot\cdot}_{kr} = m^{\cdot\cdot}\delta_{kr}$ where $m^{\cdot\cdot}$ is a self- or mutually-induced scalar volume density of mass, while the self-induced stiffness tensor is written as $(\kappa^{ss}_{ijpq})^{-1}=\lambda^{ss}\delta_{ij}\delta_{pq}+2\mu^{ss}\Delta_{ijpq}$ where λ^{ss} and μ^{ss} denote the self-induced Lamé coefficients (the self-induced scalar solid compressibility κ^{ss}_{iipp} is denoted by κ^{ss}). Because most transducers respond to the particle velocity of the surrounding media, $\langle\sigma\rangle$ and $\langle\tau_{pq}\rangle$ are eliminated, so $\langle v^f_k\rangle$ and $\langle v^s_k\rangle$ can be written as a function of macroscopic sources. We assume there is no action due to an expansion source, so $\langle h^f\rangle=\langle h^s_{ij}\rangle=0$; further the action of a macroscopic force source f_k inside the porous medium is represented by the equations $\langle f^f_k\rangle=\phi^f f_k$ and $\langle f^s_k\rangle=\phi^s f_k$. For this force source acting at the origin we have $f_k=a_k g(t)\delta(\underline{x})$, where a_k is the unit vector in the direction of the point force, $g(t)$ is its normalized pulse shape, and $\delta(\underline{x})$ is the three-dimensional Dirac delta function. The inverse transforms are relatively easily performed if in the $\underline{k}$,p domain the method of partial-fraction decomposition is used such that the result can be written in terms of elementary Green's functions, i.e., $(k_i k_i - p^2 s_{\bullet}^2)^{-1}$, where $s_{\bullet}$ denotes the slowness of either the fast compressional wave s_+, or the slow compressional wave s_-, or the shear wave s_S. We finally obtain

$$(\langle v^f_k\rangle+\langle v^s_k\rangle)(\underline{x},t) = (\langle v^f_k\rangle+\langle v^s_k\rangle)^D+(\langle v^f_k\rangle+\langle v^s_k\rangle)^{NF}+(\langle v^f_k\rangle+\langle v^s_k\rangle)^{IF}+(\langle v^f_k\rangle+\langle v^s_k\rangle)^{FF}, \quad (19)$$

where

$$(\langle v^f_k\rangle+\langle v^s_k\rangle)^D(\underline{x},t) = a_r\,(m^{ff})^{-1}\phi^f\delta(\underline{x})\int_0^t g(t')dt' \quad (20)$$

is the direct source contribution,

$$(\langle v^f_k\rangle+\langle v^s_k\rangle)^{NF}(\underline{x},t) = (3\Xi_k\Xi_r-\delta_{kr})a_r\int_0^t[\alpha_+ g(t'-Rs_+)H(t'-Rs_+)+\alpha_- g(t'-Rs_-)H(t'-Rs_-)$$
$$+\alpha_S g(t'-Rs_S)H(t'-Rs_S)]dt'/4\pi R^3 \quad (21)$$

is the near field contribution (proportional to R^{-3}),

$$(\langle v^f_k\rangle+\langle v^s_k\rangle)^{IF}(\underline{x},t) = (3\Xi_k\Xi_r-\delta_{kr})a_r\,[\alpha_+ g(t-Rs_+)H(t-Rs_+)s_+ + \alpha_- g(t-Rs_-)H(t-Rs_-)s_-$$
$$+\alpha_S g(t-Rs_S)H(t-Rs_S)s_S]/4\pi R^2 \quad (22)$$

is the intermediate field contribution (proportional to R^{-2}), and

$$(\langle v^f_k\rangle+\langle v^s_k\rangle)^{FF}(\underline{x},t) = \Xi_k\Xi_r a_r\,\partial_t[\alpha_+ g(t-Rs_+)H(t-Rs_+)s_+^2 + \alpha_- g(t-Rs_-)H(t-Rs_-)s_-^2]/4\pi R$$
$$+(\Xi_k\Xi_r-\delta_{kr})a_r\,\partial_t\alpha_S g(t-Rs_S)H(t-Rs_S)s_S^2]/4\pi R \quad (23)$$

is the far field contribution (proportional to R^{-1}), where R is the distance between source and receiver and H(t) is the unit step function. All time-domain expressions have a travel time delay $R\dot{s}$ in common. Further, each acoustic wave constituent exhibits a particular directional pattern (see Figure 1).

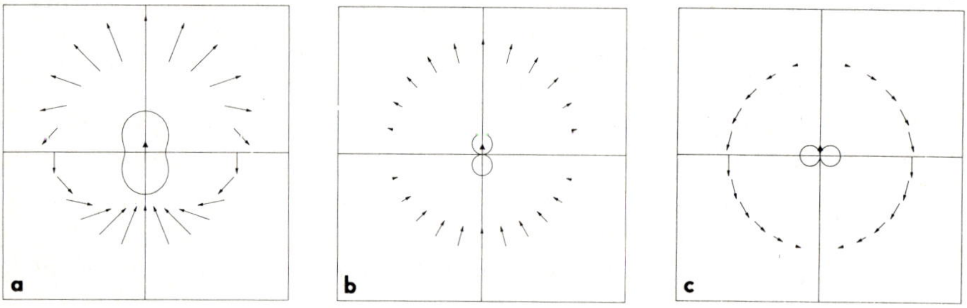

Fig. 1. Radiation patterns $3a_r\Xi_r\Xi_k - a_k$ in the near-field and intermediate-field regions (a), $a_r\Xi_r\Xi_k$ in the far-field region for the compressional waves (b), and $a_r\Xi_r\Xi_k - a_k$ in the far-field region for the shear wave (c).

In these expressions, $\Xi_k = x_k/R$ is the unit vector in the direction of $\underline{x}$. Further, by using the abbreviations

$$m^{ss*} = m^{ss}(1 - m^{sf}m^{fs}/m^{ff}m^{ss}), \tag{24}$$

$$\kappa^{ff*} = \kappa^{ff}(1 - \kappa^{sf}\kappa^{fs}/\kappa^{ff}\kappa^{ss}), \tag{25}$$

$$\zeta_1 = m^{ff}\kappa^{ff*} + (m^{sf}/m^{ff} + \kappa^{sf}/\kappa^{ss})^2\, m^{ff}/(\lambda^{ss} + \mu^{ss}), \tag{26}$$

$$\zeta_2 = m^{ss*}/(\lambda^{ss} + 2\mu^{ss}) + (m^{sf}/m^{ff} + \kappa^{sf}/\kappa^{ss})(m^{ff} + 2m^{sf})/(\lambda^{ss} + 2\mu^{ss}), \tag{27}$$

the inverse mass coefficients $\{\alpha_+, \alpha_-, \alpha_S\}$ can be written as

$$\alpha_{\pm} = [m^{sf}/m^{ff} + \kappa^{sf}/\kappa^{ss} + (1 + m^{sf}/m^{ff})(1 + \lambda^{ss}/\mu^{ss})(\zeta_1 - s_{\pm}^2)/(s_S^2 - s_{\pm}^2)]\phi^s/(\lambda^{ss} + 2\mu^{ss})(s_{\mp}^2 - s_{\pm}^2)$$
$$+ [\zeta_2 - s_{\pm}^2 + (1 + m^{sf}/m^{ff})m^{sf}(1 + \lambda^{ss}/\mu^{ss})(\zeta_1 - s_{\pm}^2)/(s_S^2 - s_{\pm}^2)(\lambda^{ss} + 2\mu^{ss})]\phi^f/m^{ff}(s_{\mp}^2 - s_{\pm}^2), \tag{28}$$

$$\alpha_S = -(\phi^s + \phi^f m^{sf}/m^{ff})(m^{ff} + m^{sf})/m^{ff}m^{ss*}. \tag{29}$$

The slowness of the shear wave is given by

$$s_S = (m^{ss*}/\mu^{ss})^{1/2}, \tag{30}$$

while the slownesses of the compressional waves follow from

$$s_+^2 + s_-^2 = m^{ff}\kappa^{ff*} + m^{ss*}/(\lambda^{ss} + 2\mu^{ss}) + (m^{sf}/m^{ff} + \kappa^{fs}/\kappa^{ss})^2 m^{ff}/(\lambda^{ss} + 2\mu^{ss}), \tag{31}$$

$$s_+^2 s_-^2 = \kappa^{ff*}m^{ff}m^{ss*}/(\lambda^{ss} + 2\mu^{ss}). \tag{32}$$

REFERENCES

[1] De Vries, S.M. and De Hoop, A.T., "Theory of linear acoustic waves in a fluid-saturated porous solid", submitted to J. Acoust. Soc. Am.

ELASTIC WAVE PROPAGATION
M.F. McCarthy, M.A. Hayes, (Editors)

ON VARIATIONAL FORMULATIONS OF BOUNDARY VALUE PROBLEMS FOR FLUID-SOLID INTERACTIONS

G.C. HSIAO and R.E. KLEINMAN

Department of Mathematical Sciences
University of Delaware
Newark, Delaware 19716 USA

L.S. SCHUETZ

Acoustics Division, Code 5130
Naval Research Laboratory
Washington, D.C. 20375 USA

I. INTRODUCTION

The problem of determining the manner in which an incoming acoustic wave is scattered by an elastic body immersed in a fluid is one of central importance in detecting and identifying submerged objects. The problem is more complicated but of greater practical interest when the scattering body is a thin elastic shell containing complex substructures.

In this paper, we present weak formulations of the coupled problem using the field equations in the solid and boundary integral equations for the fluid paying particular attention to the appropriate function spaces in which the solutions should be sought. The guiding idea behind the weak formulation consists of deriving a bilinear form defined on an appropriate function space on a bounded domain which generates a positive definite quadratic form. This strong coerciveness property will guarantee existence of a unique solution. This is a most desirable case but one that is not always realizable. However, even if strong coercivity is not available, if the quadratic form can be shown to satisfy a Garding inequality then the Fredholm alternative may be applied which means that uniqueness implies existence.

We consider first the case when the surrounding fluid is viscous. We present the weak formulation in terms of a sesquilinear form for which strong coercivity is not established but for which Garding's inequality holds. When the fluid is inviscid two weak formulations are given and in both cases a sesquilinear form is obtained for which a Garding's inequality holds, exactly as in the viscous case, but on a different space. Then existence can be established provided the classical fluid-solid problem has at most solution. This is not always the case, however when it is true the theoretical justification for the numerical procedure based on this weak formulation is therefore established.

2. COMPRESSIBLE VISCOUS FLUID

We consider a closed bounded simply connected elastic body $\Omega \subset \mathbf{R}^3$ with smooth boundary Γ and exterior Ω^c. Denote position vectors by $\mathbf{x}$ and $\mathbf{y}$ in $\mathbf{R}^3$ and the exterior unit normal to Γ at $\mathbf{y}$ by $\mathbf{n}_y$. The problem under consideration consists of determining the pressure and velocity in Ω^c and displacement in Ω due to a given excitation in the fluid. The usual linearized time dependent equations are

$$\hat{\rho}\frac{\partial \mathbf{V}}{\partial t} = -\nabla P + \mu_o[\nabla^2\mathbf{V}+\frac{1}{3}\nabla(\nabla\cdot\mathbf{V})] \qquad \text{(momentum)} \qquad (1)$$

$$\frac{\partial \rho}{\partial t} + \hat{\rho}\nabla\cdot\mathbf{V} = 0 \qquad \text{(continuity)} \qquad (2)$$

$$P = \hat{c}^2\rho \qquad \text{(state)} \qquad (3)$$

where P, ρ, $\mathbf{V}$ are pressure, density and velocity respectively, $\hat{\rho}$ is the time independent equilibrium density, μ_o is the viscosity and $\hat{c}$ is the speed of sound in the fluid. These equations may be combined in a straightforward fashion to obtain uncoupled equations for velocity and pressure. In order to obtain equations in a form suitable for coupling with the displacement of the elastic body in the time harmonic case we define

$$\mathbf{V}(\mathbf{x}, t) = -i\omega\,\hat{\mathbf{u}}(\mathbf{x})e^{-i\omega t}\ ,\ P(\mathbf{x}, t) = p(\mathbf{x})e^{-i\omega t}\ ,\ \rho(\mathbf{x}, t) = \rho(\mathbf{x})e^{-i\omega t} \tag{4}$$

$$\hat{\mu} = -i\omega\,\mu_o\ ,\ \hat{\lambda} = \hat{\rho}\hat{c}^2 + \frac{2}{3}\,i\omega\mu_o\ , \tag{5}$$

$$\hat{k}_p^2 = \frac{\omega^2}{\hat{c}^2}\,\frac{1}{1+\frac{4}{3}\frac{\hat{\mu}}{\hat{\rho}\hat{c}^2}} \qquad \hat{k}_s^2 = \frac{\hat{\rho}\omega^2}{\hat{\mu}}\ . \tag{6}$$

Then equations (1)-(3) may be combined to yield the following equations in Ω^c

$$p = -\hat{\rho}\hat{c}^2\,\nabla\cdot\hat{\mathbf{u}} \tag{7}$$

$$(\nabla^2 + \hat{k}_p^2)p = 0 \tag{8}$$

$$\hat{\mu}\,\nabla^2\,\hat{\mathbf{u}} + (\hat{\lambda} + \hat{\mu})\,\nabla(\nabla\cdot\hat{\mathbf{u}}) + \omega^2\,\hat{\rho}\,\hat{\mathbf{u}} = \mathbf{0}\ . \tag{9}$$

In the elastic body the displacement is governed by the linearized elasticity equation

$$\mu\,\nabla^2\,\mathbf{u} + (\lambda+\mu)\,\nabla(\nabla\cdot\mathbf{u}) + \omega^2\,\rho\,\mathbf{u} = 0 \tag{10}$$

where μ and λ are the Lame constants and ρ is the density. The traction operator on the boundary is defined by

$$\mathbf{T}(\mathbf{u}) = 2\,\mu\,\frac{\partial\mathbf{u}}{\partial n} + \lambda\,\hat{\mathbf{n}}\,\nabla\cdot\mathbf{u} + \mu\,\hat{\mathbf{n}}\times\nabla\times\mathbf{u}$$

and $\hat{\mathbf{T}}$ is similarly defined with $\hat{\mu}$, $\hat{\lambda}$ replacing μ, λ.

Now the classical fluid-solid interaction problem is formulated as follows. For a specified incident velocity field $\mathbf{u}^{inc}$ which satisfies equ.(9) almost everywhere in $\mathbf{R}^3$ find $\hat{\mathbf{u}}$ in Ω^c and $\mathbf{u}$ in Ω satisfying equs.(9) and (10) respectively such that on Γ

$$\mathbf{u}^{inc} + \hat{\mathbf{u}} = \mathbf{u} \tag{11}$$

$$\hat{\mathbf{T}}(\mathbf{u}^{inc}) + \hat{\mathbf{T}}(\hat{\mathbf{u}}) = \mathbf{T}(\mathbf{u})\ , \tag{12}$$

and as $r = |x| \to \infty$,

$$\left(\frac{\partial\hat{\mathbf{u}}_s}{\partial r} - i\hat{k}_s\right)\hat{\mathbf{u}}_s = o\left(\frac{1}{r}\right)\ ,\ \left(\frac{\partial\hat{\mathbf{u}}_p}{\partial r} - i\hat{k}_p\ \hat{\mathbf{u}}_p\right) = o\left(\frac{1}{r}\right) \tag{13}$$

$$\text{where}\quad \hat{\mathbf{u}}_p := -\frac{1}{\hat{k}_p^2}\,\nabla(\nabla\cdot\hat{\mathbf{u}})\ ,\ \hat{\mathbf{u}}_s := \hat{\mathbf{u}} - \hat{\mathbf{u}}_p\ . \tag{14}$$

That this problem has at most one solution was established in [8], [9] in a classical setting under some positively contraints on the Lame constants however the result remains valid in the present case. Following [1], [9], [10] we introduce wave numbers for the elastic body

$$k_s^2 = \frac{\rho\omega^2}{\mu} \qquad k_p^2 = \frac{\rho\omega^2}{\lambda+2\mu} \tag{15}$$

and the fundamental solution of the Helmholtz equation as

$$\gamma_k = \gamma_k(\mathbf{x}, \mathbf{y}) = -\frac{e^{ik|\mathbf{x}-\mathbf{y}|}}{4\pi|\mathbf{x}, \mathbf{y}|}\ , \tag{16}$$

then define the fundamental displacement tensor

$$\mathbf{G} := -\frac{1}{\mu}\,\gamma_{k_s}\,\mathbf{I} + \frac{1}{\rho\omega^2}\,\nabla\nabla(\gamma_{k_p} - \gamma_{k_s}) \tag{17}$$

with $\hat{\mathbf{G}}$ similarly defined with $\hat{\mu}$, $\hat{\rho}$, $\hat{k}_s$, $\hat{k}_p$ replacing μ, ρ, k_s, k_p. Then for non-zero μ and $\hat{\mu}$ we have the following representations

$$\int_\Gamma \{\hat{\mathbf{u}}(\mathbf{y})\cdot\hat{\mathbf{T}}_y(\hat{\mathbf{G}}(\mathbf{x},\mathbf{y})) - \hat{\mathbf{G}}(\mathbf{x},\mathbf{y})\cdot\hat{\mathbf{T}}_y(\hat{\mathbf{u}}(\mathbf{y}))\}ds_y = \begin{cases} \hat{\mathbf{u}}(\mathbf{x}) , & \mathbf{x}\in\Omega^c \\ \frac{1}{2}\hat{\mathbf{u}}(\mathbf{x}) , & \mathbf{x}\in\Gamma , \\ 0 , & \mathbf{x}\in\Omega \end{cases} \tag{18}$$

$$\int_\Gamma \{\mathbf{G}(\mathbf{x},\mathbf{y})\cdot\mathbf{T}_y(\mathbf{u}(\mathbf{y})) - \mathbf{u}(\mathbf{y})\cdot\mathbf{T}_y(\mathbf{G}(\mathbf{x},\mathbf{y}))\}ds_y = \begin{cases} 0 , & \mathbf{x}\in\Omega^c \\ \frac{1}{2}\mathbf{u}(\mathbf{x}) , & \mathbf{x}\in\Gamma \\ \mathbf{u}(\mathbf{x}) , & \mathbf{x}\in\Omega \end{cases} \tag{19}$$

where the subscript on T indicates the differentiation variables. Operating with the traction operators leads to the additional boundary integral equations

$$\hat{\mathbf{T}}_x \int_\Gamma \hat{\mathbf{u}}(\mathbf{y})\cdot\hat{\mathbf{T}}_y(\hat{\mathbf{G}}(\mathbf{x},\mathbf{y}))ds_y - \int_\Gamma \hat{\mathbf{T}}_x(\hat{\mathbf{G}}(\mathbf{x},\mathbf{y}))\cdot\hat{\mathbf{T}}_y(\hat{\mathbf{u}}(\mathbf{y}))ds_y = \frac{1}{2}\hat{\mathbf{T}}_x\ (\hat{\mathbf{u}}(\mathbf{x})) , \tag{20}$$

$$-\mathbf{T}_x \int_\Gamma \mathbf{u}(\mathbf{y})\cdot\mathbf{T}_y(\mathbf{G}(\mathbf{x},\mathbf{y}))ds_y + \int_\Gamma \mathbf{T}_x(\mathbf{G}(\mathbf{x},\mathbf{y}))\cdot\mathbf{T}_y(\mathbf{u}(\mathbf{y}))ds_y = \frac{1}{2}\mathbf{T}_x\ (\mathbf{u}(\mathbf{x})) . \tag{21}$$

The boundary integral equations given in (18)-(21) may be written symbolically as

$$\frac{1}{2}\ \hat{\mathbf{u}} = \hat{K}_y\ \hat{\mathbf{u}} - \hat{S}\ \hat{\mathbf{t}} \quad ; \quad \frac{1}{2}\ \mathbf{u} = -K_y\ \mathbf{u} + S\mathbf{t} , \tag{22}$$

$$\frac{1}{2}\hat{t} = -\hat{N}\ \hat{\mathbf{u}} - \hat{K}_x\ \hat{\mathbf{t}} \quad ; \quad \frac{1}{2}t = N\ \mathbf{u} + K_x\mathbf{t} , \tag{23}$$

where $\mathbf{t} = \mathbf{T}(\mathbf{u})$, $\hat{\mathbf{t}} = \hat{\mathbf{T}}(\hat{\mathbf{u}})$ and the integral operators are defined implicitly in equations (18)-(21). We remark that these equations together with the transmission conditions (11) and (12) may be combined to yield coupled boundary integral equations for a pair of unknown functions, e.g. $\hat{\mathbf{u}}$ and $\mathbf{t}$.

Here however we consider the weak formulation of the problem following [2] and [4]. The first Betti formula is employed for this purpose yielding

$$0 = \int_\Omega \overline{\mathbf{v}}\cdot(\mu\nabla^2\mathbf{u}+(\lambda+\mu)\nabla(\nabla\cdot\mathbf{u})+\rho\omega^2\mathbf{u})d\mathbf{x} = -a(\mathbf{u},\mathbf{v}) + \int_\Gamma \overline{\mathbf{v}}\cdot\mathbf{T}(\mathbf{u})ds \tag{24}$$
$$+ \rho\omega^2 \int_\Omega \overline{\mathbf{v}}\cdot\mathbf{u}\ d\mathbf{x}$$

where

$$a(\mathbf{u},\mathbf{v}) = \lambda\int_\Omega \nabla\cdot\overline{\mathbf{v}}\ \nabla\cdot\mathbf{u}\ d\mathbf{x} + \frac{\mu}{2}\int_\Omega (\nabla\overline{\mathbf{v}}+(\nabla\overline{\mathbf{v}})^T) : (\nabla\mathbf{u}+(\nabla\mathbf{u})^T)d\mathbf{x} . \tag{25}$$

Introducing the usual complex inner product in $L_2(\Omega) = H^o(\Omega)$ which we denote by $(\cdot,\cdot)$ and denoting by $<\cdot,\cdot>$ the dual pairing of functions defined on Γ, equ.(24) becomes

$$a(\mathbf{u},\mathbf{v}) - \rho\omega^2(\mathbf{u},\mathbf{v}) = <\mathbf{t},\overline{\mathbf{v}}> = <\hat{\mathbf{t}},\overline{\mathbf{v}}> + <\mathbf{t}^{inc},\overline{\mathbf{v}}> \tag{26}$$

where the transmission condition (12) has been used and $\mathbf{t}^{inc} = \hat{\mathbf{T}}(\mathbf{u}^{inc})$.

Now employ (23) with $\hat{\mathbf{u}}$ replaced by $\mathbf{u} - \mathbf{u}^{inc}$ to obtain

$$a(\mathbf{u},\mathbf{v}) - \rho\omega^2(\mathbf{u},\mathbf{v}) - \frac{1}{2}<\hat{\mathbf{t}},\overline{\mathbf{v}}> + <\hat{N}\mathbf{u},\overline{\mathbf{v}}> + <\hat{K}_x\ \hat{\mathbf{t}},\overline{\mathbf{v}}> = <\hat{N}\mathbf{u}^{inc} + \mathbf{t}^{inc},\overline{\mathbf{v}}> . \tag{27}$$

The weak form of (22) is simply

$$\frac{1}{2} < \hat{\mathbf{u}},\overline{\boldsymbol{\tau}}> - <\hat{K}_y\hat{\mathbf{u}},\overline{\boldsymbol{\tau}}> + <\hat{S}\hat{\mathbf{t}},\overline{\boldsymbol{\tau}}> = 0 \tag{28}$$

and again replacing $\hat{\mathbf{u}}$ by $\mathbf{u} - \mathbf{u}^{inc}$

$$\frac{1}{2} <\mathbf{u},\overline{\boldsymbol{\tau}}> - <\hat{K}_y\mathbf{u},\overline{\boldsymbol{\tau}}> + <\hat{S}\hat{\mathbf{t}},\overline{\boldsymbol{\tau}}> = <(\frac{1}{2}I - \hat{K}_y)\mathbf{u}^{inc},\overline{\boldsymbol{\tau}}> \tag{29}$$

Now we combine (27) and (29) to obtain the following weak formulation of the fluid-elastic problem: $(\mathbf{u}, \hat{\mathbf{t}}) \in (H^1(\Omega))^3 \times (H^{-1/2}(\Gamma))^3$ is a weak solution of the fluid-solid interaction problem if

$$A((\mathbf{u}, \hat{\mathbf{t}}), (\mathbf{v}, \boldsymbol{\tau})) = F((\mathbf{v}, \boldsymbol{\tau})) \quad \forall\ (\mathbf{v}, \boldsymbol{\tau}) \in (H^1(\Omega))^3 \times (H^{-1/2}(\Gamma))^3 \tag{30}$$

where

$$A((\mathbf{u}, \hat{\mathbf{t}}), (\mathbf{v}, \boldsymbol{\tau})) := a(\mathbf{u}, \mathbf{v}) - \rho\omega^2(\mathbf{u}, \mathbf{v}) - \frac{1}{2} <\hat{\mathbf{t}}, \overline{\mathbf{v}}> + <\hat{N}\mathbf{u}, \overline{\mathbf{v}}> \tag{31}$$

$$+ <\hat{K}_x\ \hat{\mathbf{t}}, \overline{\mathbf{v}}> + \frac{1}{2} <\mathbf{u}, \overline{\boldsymbol{\tau}}> - <\hat{K}_y\ \mathbf{u}, \overline{\boldsymbol{\tau}}> + <\hat{S}\hat{\mathbf{t}}, \overline{\boldsymbol{\tau}}>,$$

$$F((\mathbf{v}, \boldsymbol{\tau})) := <\hat{N}\ \mathbf{u}^{inc} + \mathbf{t}^{inc}, \overline{\mathbf{v}}> + <(\frac{1}{2} I - \hat{K}_y)\mathbf{u}^{inc}, \overline{\boldsymbol{\tau}}>. \tag{32}$$

Using standard arguments [5] it may be shown that the sesquilinear form A satisfies Garding's inequality in the sense that

$$\operatorname{Re} A((\mathbf{u},\hat{\mathbf{t}}),(\mathbf{u},\hat{\mathbf{t}})) \geq \gamma\{\ \|\mathbf{u}\|^2_{H^1(\Omega)} + \|\hat{\mathbf{t}}\|^2_{H^{-1/2}(\Gamma)}\} - |K((\mathbf{u},\hat{\mathbf{t}}),(\mathbf{u},\hat{\mathbf{t}}))| \tag{33}$$

where K is a compact quadratic form on $H^1(\Omega))^3 \times (H^{-1/2}(\Gamma))^3$. This means that the Fredholm alternative is applicable hence existence of a solution follows from uniqueness. Uniqueness for the weak problem follows from Kupradze's result in the classical case using regularity theory which implies that the weak solution of the homogeneous problem is also a classical solution. Thus the classical uniqueness result suffices to establish both existence and uniqueness of the weak solution of the fluid-solid interaction problem.

3. ZERO VISCOSITY ($\mu_o = 0$)

The situation changes markedly in the inviscid case. The original problem, equs.(9)-(13), is no longer well-posed if we set $\hat{\mu} = 0$ e.g. [8]. The field equation becomes

$$\hat{\lambda}\ \nabla(\nabla\cdot\hat{\mathbf{u}}) + \omega^2\hat{\rho}\ \mathbf{u} = \mathbf{0} \quad \text{in} \quad \Omega^c \tag{34}$$

which, with (7), implies that

$$\hat{\mathbf{u}} = \frac{1}{\omega^2\hat{\rho}} \nabla p \tag{35}$$

that is, the velocity is completely determined by the pressure. Assuming a similar form for $\mathbf{u}^{inc}$, it follows that the transmission conditions (11), (12) become

$$-(p + p^{inc})\hat{\mathbf{n}} = \mathbf{t} \tag{36}$$

$$\nabla(p + p^{inc}) = \hat{\rho}\ \omega^2\ \mathbf{u}\ . \tag{37}$$

Both equations are thus seen to imply conditions on the tangential derivatives of $p + p^{inc}$ which over specifies the problem. What is generally done is to keep the first condition but only the equality of the normal components of the second. The integral representation for $\hat{\mathbf{u}}$, (18) and the boundary integral equation (20) are no longer valid when $\hat{\mu} = 0$. In this case, with (35) it is more convenient to express fluid displacement in terms of pressure for which the following standard integral equations hold:

$$\frac{1}{2}\frac{\partial p}{\partial n_x} = -\frac{\partial}{\partial n_x}\int_\Gamma \frac{\partial}{\partial n_y}\gamma_{\hat{k}_p} p(y)ds_y + \int_\Gamma \frac{\partial p}{\partial n_y}\frac{\partial\gamma_{\hat{k}_p}}{\partial n_x} ds_y := -\dot{N}p - \dot{K}_x\frac{\partial p}{\partial n} \tag{38}$$

$$\frac{1}{2}p = -\int_\Gamma \frac{\partial}{\partial n_y}\gamma_{\hat{k}_p} p(y)ds_y + \int_\Gamma \frac{\partial p}{\partial n}\gamma_{\hat{k}_p} ds_y := \dot{K}_y\ p - \dot{S}\frac{\partial p}{\partial n} \tag{39}$$

where the operators $\dot{N}$, $\dot{K}_x$, $\dot{K}_y$, $\dot{S}$ are defined implicitly by (38) and (39). We could attempt to set $\hat{\mu} = 0$ in the weak formulation of the previous section however there is considerable difficulty in ascribing meaningful limits to the terms $\hat{N}\mathbf{u}$, $\hat{K}_x\hat{\mathbf{t}}$, $\hat{K}_y\ \mathbf{u}$ and $\hat{S}\hat{\mathbf{t}}$. Alternatively we could follow [7] to obtain a weak formulation from (24) using the transmission condition (36) to obtain

$$a(\mathbf{u}, \mathbf{v}) - \rho\omega^2(\mathbf{u}, \mathbf{v}) + <p, \hat{\mathbf{n}}\cdot\overline{\mathbf{v}}> = -<p^{inc}, \hat{\mathbf{n}}\cdot\overline{\mathbf{v}}> \tag{40}$$

and the weak form of (38) with the transmission condition (37) i.e.

$$\frac{1}{2} <\hat{\mathbf{n}}\cdot\mathbf{u},\bar{q}> + <\dot{K}_x\hat{\mathbf{n}}\cdot\mathbf{u},\bar{q}> + \frac{1}{\hat{\rho}\omega^2}<\dot{N}p,\bar{q}> = \frac{1}{\hat{\rho}\omega^2} < (\frac{1}{2}I+K_x)\frac{\partial p^{inc}}{\partial n},\bar{q}> . \quad (41)$$

As in the viscous case we combine (40) and (41) to obtain the weak formulation of the inviscid fluid-elastic problem: $(\mathbf{u},p) \in (H^{(1)}(\Omega))^3 \times H^{1/2}(\Gamma)$ is a weak solution of the fluid solid interaction problem if

$$\dot{A}((\mathbf{u},p)\ ,\ (\mathbf{v},q)) = \dot{F}((\mathbf{v},q)) \quad \forall (\mathbf{v},q) \in (H^1(\Omega))^3 \times H^{1/2}(\Gamma) \quad (42)$$

where

$$\dot{A}((\mathbf{u},p)\ ,\ (\mathbf{v},q)) := a(\mathbf{u},\mathbf{v}) - \rho\omega^2(\mathbf{u},\mathbf{v}) + < p ,\hat{\mathbf{n}}\cdot\bar{\mathbf{v}} > + < \hat{\mathbf{n}}\cdot\mathbf{u}\ ,\ \bar{q} >$$

$$+ \frac{2}{\hat{\rho}\omega^2} < \dot{N}p ,\bar{q} > + 2< \dot{K}_x\hat{\mathbf{n}}\cdot\mathbf{u},\bar{q} > \quad (43)$$

and

$$\dot{F}((\mathbf{v},q)) := - < p^{inc}\ ,\ \hat{\mathbf{n}}\cdot\bar{\mathbf{v}} > + \frac{1}{\hat{\rho}\omega^2}< (I+2\dot{K}_x)\frac{\partial p^{inc}}{\partial n}\ ,\ \bar{q} > . \quad (44)$$

We are able to establish the Garding inequality

$$\mathrm{Re}\ \dot{A}((\mathbf{u},p)\ ,\ (\mathbf{v},p)) \geq \gamma\{\|\mathbf{u}\|^2_{H^1(\Omega)} + \|p\|^2_{H^{1/2}(\Gamma)}\} - |\dot{\mathbf{K}}((\mathbf{u},p)\ ,\ (\mathbf{u},p))| \quad (45)$$

where $\dot{\mathbf{K}}$ is a compact quadratic form on $(H^{(1)}(\Omega))^3 \times H^{1/2}(\Gamma)$ (see [6] for details).

Another way to derive a weak formulation parallels the treatment in the case of $\hat{\mu} \neq 0$ which, when $\hat{\mu} = 0$, was employed in [3]. This leads to the system (41) and

$$a(\mathbf{u},\mathbf{v}) - \rho\omega^2(\mathbf{u},\mathbf{v}) + \frac{1}{2}<p,\bar{\mathbf{v}}\cdot\hat{\mathbf{n}}> + <\dot{K}_y p ,\hat{\mathbf{n}}\cdot\bar{\mathbf{v}}> - \hat{\rho}\omega^2<\dot{S}\,\hat{\mathbf{n}}\cdot\mathbf{u},\hat{\mathbf{n}}\cdot\bar{\mathbf{v}}> = -<p^{inc} + \dot{S}(\frac{\partial p^{inc}}{\partial n}),\hat{\mathbf{n}}\cdot\bar{\mathbf{v}}> \quad (46)$$

Here again we may combine these to obtain a positive definite operator with a compact perturbation on the same space as before. Note that K_y is compact on $H^{1/2}(\Gamma)$ in contract to the case when $\hat{\mu} \neq 0$ where $\hat{K}_y$ is not compact on $(H^{1/2}(\Gamma))^3$.

In both cases the Fredholm alternative is applicable and existence of a weak solution follows from uniqueness. Uniqueness for the weak formulation follows from uniqueness of the original interaction problem. However it is known that the original problem is not always uniquely solvable, e.g. [8], [9]. The existence of traction free rotational oscillations for certain values of $\rho\omega^2$ and certain geometries is a fundamental difficulty of the formulation of the interaction problem.

Thus we have provided an alternative to the usual practice of basing numerical procedures on equations (40) and (39). There what is done is to employ a two stage procedure which is useful when the quantity of interest is the fluid pressure. Using equation (40) we may define a sesquilinear form on $(H^1(\Omega))^3 \times (H^1(\Omega))^3$

$$B(\mathbf{u},\mathbf{v}) := a(\mathbf{u},\ \mathbf{v}) - \rho\omega^2(\mathbf{u},\mathbf{v}) \quad (47)$$

for which the following Garding inequality may be established: $\mathrm{Re}\ B(\mathbf{u},\mathbf{u}) \geq \gamma\|\ \mathbf{u}\ \|^2_{H^1(\Omega)} - \delta\|\ \mathbf{u}\ \|^2_{H^0(\Omega)}$ for some constants $\gamma > 0$, $\delta \geq 0$. This means that if the interior homogeneous problem has at most one solution then the standard finite element scheme can be employed to determine $\hat{\mathbf{n}}\cdot\mathbf{u}$ in terms of p on the boundary. This then may be used with the boundary integral equation (39) ((38) could also be used) for p which may then be solved numerically by any suitable method. Numerical experiments based on this approach have shown good results.

Numerical results based on the new weak formulations are not yet available. It is hoped that such results together error estimates and comparison with the two stage procedure will appear shortly.

ACKNOWLEDGEMENT:
This work was supported by the Naval Research Laboratory under contract N00014-87-C-2121. The work of GCH was also supported in part by the Center for Advanced Study of the University of Delaware.

REFERENCES

[1] Ahner, J.F., Hsiao, G.C., A Neumann Series Representation for Solutions to Boundary-Value Problems in Dynamic Elasticity, Quart. J. Appl. Math. (1975), pp. 78-80.

[2] Costabel, M., Stephan, E.P., Coupling of Finite Elements and Boundary Elements for Transmission Problems of Elastic Waves, in: Symposium on Advanced Boundary Element Methods, San Antonio, Cruse et al. (eds.), (San Antonio, 1987), in print.

[3] Hamdi, M.A., Jean, P., A Mixed Functional for the Numerical Resolution of Fluid-Structure Interaction Problems, in: Aero- and Hydro-Acoustics IUTAM Symposium, Comte-Bellot, G. and Williams, J.E. (eds.), (Springer-Verlag, 1985) pp. 269-276.

[4] Han, H., A New Class of Variational Formulations for the Coupling of Finite and Boundary Element Methods, Tsinghua University, pre-print, (1987).

[5] Hsiao, G.C., Wendland, W.L., On a Boundary Integral Equation Method for Some Exterior Problems in Elasticity, Proceedings of Tbilisi University, Tbilis: University Press, Tbilisi, (1985), pp. 31-60.

[6] Hsiao, G.C. and Kleinman, R.E., Weak Solutions of Fluid-Solid Interaction Problems, in preparation.

[7] Johnson, C., Nedelec, J.C., On the Coupling of Boundary Integral and Finite Element Methods., Math. Comp. 35 (1980) 1063-1079.

[8] Jones, D.S., Low-Frequency Scattering by a Body in Lubricated Contact, Quart. J. Mech. Appl. Math. 36 (1983), pp. 111-137.

[9] Kupradze, V.D., Potential Methods in the Theory of Elasticity (Israel Program for Scientific Translations, Jerusalem, 1965).

[10] Pao, Y.H., Varatharajulu, V., Huygen's Principle, Radiation Conditions, and Integral Formulas for the Scattering of Elastic Waves, J. Acoust. Soc. Am. 59 (1976), 1361-1371.

ELASTIC WAVE PROPAGATION
M.F. McCarthy, M.A. Hayes, (Editors)
Elsevier Science Publishers B.V. (North-Holland), 1989

A SIMPLE SELF-CONSISTENT ANALYSIS OF DISPERSION AND ATTENUATION OF ELASTIC WAVES IN A POROUS MEDIUM

Federico J. SABINA

Instituto de Investigaciones en Matemáticas Aplicadas y en Sistemas
Universidad Nacional Autónoma de México
Apartado Postal 20-726
01000 México, D.F., México

John R. WILLIS

School of Mathematical Sciences
University of Bath
Bath BA2 7AY, United Kingdom

A porous material is considered whose pores are homogeneously and isotropically distributed at random in an elastic matrix. When the spatial correlations of the pores are unknown, a simple self-consistent embedding scheme is developed for the approximate analysis of elastic waves in that medium. An approximate solution for the scattering of a single void is used, yielding simple explicit equations which are easily solved by iteration. The dispersion and attenuation of longitudinal and transverse waves are calculated as functions of the concentration of the pores for various materials.

1. INTRODUCTION

Wave propagation in a matrix containing either discrete particles or voids has been investigated recently by several authors [1–5]. Extensive computations have usually been necessary to solve the component problem of scattering by a single inclusion, and thus results for S-waves have only recently been presented [5]. In contrast, Sabina and Willis [6] compromised on the solution of the single scattering problem to obtain an approximate solution simple enough to permit the development of a self-consistent scheme, which directly generalized to dynamics the static scheme of Hill [7]. This scheme, which permits the study of S-waves along with P-waves and requires only a small iterative computation,is summarized below. New results are then presented by specializing the inclusions to be void.

2. FORMULATION

A two-phase composite is considered comprising a matrix with tensor of elastic moduli L_2 and mass density ρ_2, in which are embedded inclusions of the same size and shape having tensor of elastic moduli L_1 and mass density ρ_1. A typical inclusion occupies the domain $x' + \Omega_1$, where Ω_1 defines the shape and size of the inclusion and its 'centre' x' is distributed randomly. The probability density for finding the inclusion centred at x' is taken as n_1, independent of x', so that n_1 is the number density of the inclusions. The probability that a point lies within the inclusion is then the volume fraction c_1 given by n_1 times the volume of

Ω_1. The volume fraction of the matrix material is $c_2 = 1 - c_1$. The mean field, or ensemble average, $< u >$ of the displacement u is sought here.

With no body forces, the response of the composite is governed by the equation of motion

$$\text{div}\, \sigma = \dot{p} \tag{2.1}$$

understood in a weak sense. The stress σ and momentum density p are related to strain e and velocity $\dot{u}$ by the constitutive relations

$$\sigma = Le, \quad p = \rho \dot{u} \tag{2.2}$$

written in symbolic notation. The tensor of elastic moduli L and mass density ρ are functions of x, which are given in the forms

$$L(x) = L_1 f_1(x) + L_2 f_2(x), \quad \rho(x) = \rho_1 f_1(x) + \rho_2 f_2(x), \tag{2.3}$$

where $f_i(x)$ is the indicator function equal to 1 if $x \in$ material i and 0 otherwise for $i = 1, 2$.

If the exact configuration of the inclusions were known, it would be possible to find u for any particular loading. However we proceed as follows. The mean response of the composite is obtained by averaging (2.1) as

$$\text{div} \, < \sigma > = < \dot{p} > . \tag{2.4}$$

The mean displacement $< u >$ could be found if effective constitutive relations, similar to (2.2), could be established to relate $< \sigma >$ and $< p >$ to $< e >$ and $< \dot{u} >$.

Let e_i be defined as the ensemble mean of e, conditional on $x \in$ material i, for the point x at which e_i is evaluated. A similar definition for $\dot{u}_i$ is given. Thus it follows that

$$< e > = c_1 e_1 + c_2 e_2, \quad < \dot{u} > = c_1 \dot{u}_1 + c_2 \dot{u}_2, \tag{2.5}$$

$$< \sigma > = c_1 L_1 e_1 + c_2 L_2 e_2, \quad < p > = c_1 \rho_1 \dot{u}_1 + c_2 \rho_2 \dot{u}_2. \tag{2.6}$$

Following Hill (1965) in his analysis of the static case, $c_2 e_2$ is eliminated between (2.5a) and (2.6a), and $c_2 \dot{u}_2$ between (2.5b) and (2.6b). This yields the *exact* relations

$$< \sigma > = L_2 < e > + c_1 (L_1 - L_2) e_1, \quad < p > = \rho_2 < \dot{u} > + c_1 (\rho_1 - \rho_2) \dot{u}_1. \tag{2.7}$$

Thus, the sought effective constitutive relations would follow if e_1 and $\dot{u}_1$ could be found in terms of $< e >$ and $< \dot{u} >$. A simple procedure for estimating e_1 and $\dot{u}_1$ approximately is to consider a single inclusion embedded in homogeneous material whose properties are chosen to match the effective properties of the composite. This defines a scattering problem: the mean field $< u >$ is taken as the far field from the inclusion. The problem having been solved for an inclusion centred at x', e_1 and $\dot{u}_1$ are estimated, at position x and time t, as

$$e_1(x,t) = \frac{1}{|\Omega_1|} \int_{U_1(x)} e(x,t;x') dx', \quad \dot{u}_1(x,t) = \frac{1}{|\Omega_1|} \int_{U_1(x)} \dot{u}(x,t;x') dx', \tag{2.8}$$

where $U_1(x) = \{x' : x - x' \in \Omega_1\}$ and $|\Omega_1|$ denotes the volume of Ω_1. The estimates e_1 and $\dot{u}_1$ depend on $< e >$, $< \dot{u} >$ and on the properties assumed for the homogeneous material in which the inclusion is considered to be embedded. Then equations for the properties follow by requiring that the effective constitutive relations are consistent with (2.7).

3. THE SINGLE SCATTERING PROBLEM

The scattering problem mentioned above will be solved for the inclusion of properties L_1, ρ_1 in an elastic (or viscoelastic) matrix whose constitutive response at radian frequency ω is

$$\sigma = L_0 e, \quad p = -i\omega\rho_0 u, \tag{3.1}$$

where L_0 is a fourth-order tensor and ρ_0 is a scalar; both L_0 and ρ_0 could be complex.

The stress and momentum density within the inclusion can be expressed in terms of 'polarizations' τ, π as [8]

$$\sigma = L_0 e + \tau, \quad p = -i\omega\rho_0 u + \pi, \tag{3.2}$$

where τ and π are non-zero only over the inclusion. Substituting (3.2) into the (time-reduced) equation of motion (2.1), then shows that u could be generated in a homogeneous material with moduli L_0 and density ρ_0, if it were driven by a body force div $\tau + i\omega\pi$. Thus, in terms of Green's function G for the homogeneous material

$$u(x) = u_0 + G * (\text{div}\, \tau + i\omega\pi) = u_0 - S * \tau - M * \pi \tag{3.3}$$

where the symbol $*$ represents the operation of convolution with respect to x and, in component form,

$$S_{ikl}(x) = -G_{i(k,l)}(x), \quad M_{ij} = -i\omega G_{ij}(x), \tag{3.4}$$

the brackets on the suffixes denote symmetrization and the comma notation for differentiation is used. The field u_0 is the one incident on the inclusion, which will be equated to the mean harmonic plane wave

$$< u >= m \exp\left[i(k \cdot x - \omega t)\right]; \tag{3.5}$$

here, k is the mean wave-vector. From differentiating (3.3) a representation for e follows. Then, by eliminating e and u in favour of τ and π, the integral equations

$$(L_1 - L_0)^{-1}\tau + S_x * \tau + M_x * \pi = e_0, \quad (\rho_1 - \rho_0)^{-1}\pi + S_t * \tau + M_t * \pi = -i\omega u_0 \tag{3.6}$$

are obtained. In (3.6), e_0 is the strain associated with u_0 and S_x, S_t, M_x, M_t are operators given explicitly in [6]. Equations (3.6) are solved approximately, by a Galerkin method, taking τ, π as constant over the inclusion. In the particular case when the inclusion has a centre of symmetry, it follows [6], since M_x, S_t are odd functions of x, that the equations uncouple to yield

$$\tau = [(L_1 - L_0)^{-1} + \overline{S}_x]^{-1}\overline{e}_0, \quad \pi = -i\omega[(\rho_1 - \rho_0)^{-1} + \overline{M}_t]^{-1}\overline{u}_0. \tag{3.7}$$

In (3.7), $\overline{S}_x$, $\overline{M}_t$ are mean values, obtained by integrating over the inclusion the functions $S_x\tau$, $M_t\pi$,when τ, π are constant over the inclusion. Here u_0 is taken as a plane wave of the form (3.5), where k and m satisfy

$$[(L_0)_{ijkl}k_j k_l - \rho_0\omega^2\delta_{ik}]m_k = 0. \tag{3.8}$$

If the inclusion is centred at x', it follows that

$$\overline{e}_0 = e_0(x')h_1(k), \quad \overline{u}_0 = u_0(x')h_1(k), \tag{3.9}$$

where

$$h_1(k) = \frac{1}{|\Omega_1|}\int_{\Omega_1} e^{ik\cdot x}dx. \tag{3.10}$$

This solves the inclusion problem. The field within the inclusion can be estimated from the relations

$$e = (L_1 - L_0)^{-1}\tau, \quad -i\omega u = (\rho_1 - \rho_0)^{-1}\pi. \tag{3.11}$$

Evaluation of e_1 and $-i\omega u_1$, now requires the calculation of the corresponding mean values τ_1 and π_1 from an equation like (2.8). Explicitly for τ_1, from (3.11a) and (3.13a),

$$\begin{aligned}\tau_1(x) &= h_1(k)[(L_1 - L_0)^{-1} + \overline{S}_x]^{-1}\frac{1}{|\Omega_1|}\int_{U_1(x)} e_0(x')dx' \\ &= h_1(k)h_1(-k)[(L_1 - L_0)^{-1} + \overline{S}_x]^{-1}e_0(x),\end{aligned} \tag{3.12}$$

since $e_0(x') = e_0(x)\exp[-ik\cdot(x - x')]$. Hence, from (3.9)

$$e_1(x) = h_1(k)h_1(-k)[I + \overline{S}_x(L_1 - L_0)]^{-1}e_0(x). \tag{3.13}$$

Similarly,

$$-i\omega u_1(x) = -i\omega h_1(k)h_1(-k)[I + \overline{M}_t(\rho_1 - \rho_0)]^{-1}u_0(x). \tag{3.14}$$

Finally, using (2.7) the self-consistent equations are

$$L_0 = L_2 + c_1h_1(k)h_1(-k)(L_1 - L_2)[I + \overline{S}_x(L_1 - L_0)]^{-1}, \tag{3.15}$$

$$\rho_0 = \rho_2 + c_1h_1(k)h_1(-k)(\rho_1 - \rho_2)[I + \overline{M}_t(\rho_1 - \rho_0)]^{-1}. \tag{3.16}$$

It should be noted that equations (3.15) and (3.16) are compatible with the requirement of self-consistency as outlined earlier but represent an extension through the 'cancellation' of the post-multiplicative factors $< e >$ and $< u >$. They provide, however, a convenient prescription for finding L_0, ρ_0 uniquely for a chosen plane wave.

4. EXAMPLE: A MATRIX CONTAINING SPHERICAL PORES

The general formulae of the previous section are now illustrated for a porous material consisting of a matrix with properties L_2, ρ_2 in which is embedded a random array of spherical pores, each of radius a, with properties $L_1 = 0$, $\rho_1 = 0$. The matrix material is taken isotropic with L_2 characterized by bulk modulus κ_2 and shear modulus μ_2. With L_0 taken isotropic with bulk modulus κ_0 and shear modulus μ_0, $\overline{S}_x$ and $\overline{M}_t$ become [6]

$$\overline{S}_x = (\epsilon_\alpha/(3\kappa_0 + 4\mu_0), (1/5)(2\epsilon_\alpha/(3\kappa_0 + 4\mu_0) + \epsilon_\beta/\mu_0)), \tag{4.1}$$

$$\overline{M}_t = I(3 - \epsilon_\alpha - 2\epsilon_\beta)/3\rho_0, \tag{4.2}$$

where

$$\epsilon_\gamma = (3 - ik_\gamma a)(\sin k_\gamma a - k_\gamma a\cos k_\gamma a)e^{ik_\gamma a}/(k_\gamma a)^3, \tag{4.3}$$

I is the identity and $\gamma = \alpha$ or β given by

$$\alpha = [(3\kappa_0 + 4\mu_0)/3\rho_0]^{1/2}, \quad \beta = (\mu_0/\rho_0)^{1/2} \tag{4.4}$$

The moduli κ_0, μ_0 and density ρ_0 may be complex but are assumed to be such that the square roots in (4.4) can both be chosen with positive real parts and imaginary parts so that, when ω is real and positive the waves correspond to outgoing waves that decay as $|x|$ increases.

Hence, substituting into equations (3.15) and (3.16), it is found that

$$\kappa_0/\kappa_2 = 1 - c_1 h_1(k)h_1(-k)[1 - 3\kappa_0\epsilon_\alpha/(3\kappa_0 + 4\mu_0)]^{-1}, \tag{4.5}$$

$$\mu_0/\mu_2 = 1 - c_1 h_1(k)h_1(-k)\{1 - 2[2\mu_0\epsilon_\alpha + (3\kappa_0 + 4\mu_0)\epsilon_\beta/[5(3\kappa_0 + 4\mu_0)]\}^{-1}, \tag{4.6}$$

$$\rho_0/\rho_2 = 1 - 3c_1 h_1(k)h_1(-k)/(\epsilon_\alpha + 2\epsilon_\beta). \tag{4.7}$$

The function $h_1(k)$ for the spherical void is

$$h_1(k) = 3(\sin ka - ka \cos ka)/(ka)^3. \tag{4.8}$$

Equations (4.5–7) can be solved by iteration. It should be noted that the wave number k, in view of isotropy, treated here as a scalar, strictly equal to the magnitude of the wave vector, is calculated from L_0 and ρ_0. Thus for an incident P- or S-wave

$$k = k_\alpha = \omega[(3\kappa_0 + 4\mu_0)/3\rho_0]^{-1/2}, \quad k = k_\beta = \omega(\mu_0/\rho_0)^{-1/2}, \tag{4.9}$$

respectively. The alternative choices (4.9) give different equations (4.5–7) for P- or S-waves, and thus generate estimates for κ_0, μ_0, ρ_0 which are the "most appropriate" for the wave under consideration. In either case κ_0 and μ_0 depend on one matrix parameter only, for instance, α_2/β_2.

The scheme embodied in equations (4.5–7) has the virtue of being simple and explicit. It deals equally well with P- or S-waves, and since κ_0, μ_0 and ρ_0 are in any case complex, there is no limitation to taking κ_2, μ_2 complex, corresponding to viscoelastic behaviour.

5. RESULTS

Equations (4.5–7) have been solved, by iteration, for several cases. The iteration was started by taking κ_0, μ_0 equal to the matrix values κ_2, μ_2, respectively.

The figures discussed below show plots of the fractional change $k_2/Re\{k_\gamma\} - 1$, in the phase velocity divided by the concentration of pores c_1 and attenuation divided by c_1 against normalized frequency for P- and S-waves. The phase velocity is taken as $\omega/Re\{k_\gamma\}$, where $\gamma = \alpha$ for P-waves or β for S-waves and k_γ is defined by (4.9a) or (4.9b). Phase velocities are normalized to the phase velocity of the some type of wave in the matrix material. The measure of attenuation that is used is $Im\{k_\gamma a\}$. Normalized frequency is $k_2 a$, where $k_2 = Re\{k_\alpha\}$, evaluated for $\kappa_0 = \kappa_2$, $\mu_0 = \mu_2$, $\rho_0 = \rho_2$.

Figures 1 to 4 give results for four porous media containing spherical voids. The porosity value $c_1 = 0.1$ is taken and values of $\alpha_2/\beta_2 = 1.45$, 1.84, 2.09 and 2.57 are considered which correspond to beryllium, steel, aluminum and lead respectively. Figure 1 plots $[k_2/Re\{k_\alpha\} - 1]/c_1$ against $k_2 a$. As the ratio α_2/β_2 increases, the values at zero frequency and the minimum values decrease. It is interesting to note that for lead the phase velocity is faster than that of the matrix material for about $k_2 a \geq 1.4$. The results can be considered reliable, however, only for frequencies for which the approximation of τ_1 and π_1 as constants is plausible: an approximate restriction is $k_2 a < \pi/2$, corresponding to waves in the matrix having wavelength at least $4a$. Figure 2 shows attenuation for P-waves divided by c_1. Here the attenuation maximum increases as α_2/β_2 increases. Figure 3 shows a plot similar to Figure 1 but for S-waves. As α_2/β_2 increases, the values at zero frequency and minimum values increase. The attenuation for S-waves divided by c_1 is shown in Figure 4. The maximum values decrease as α_2/β_2 increases. In Figures 3 and 4 the curves shown are very similar. Note that for P- and S-waves the dependences with α_2/β_2 show opposite trends except for the frequency shift at minimum values in the scaled phase velocities and at maximum values in the scaled attenuation plots, which show the same behaviour.

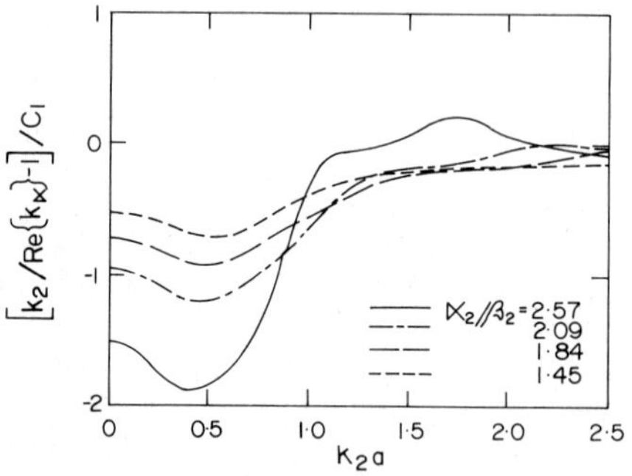

FIGURE 1

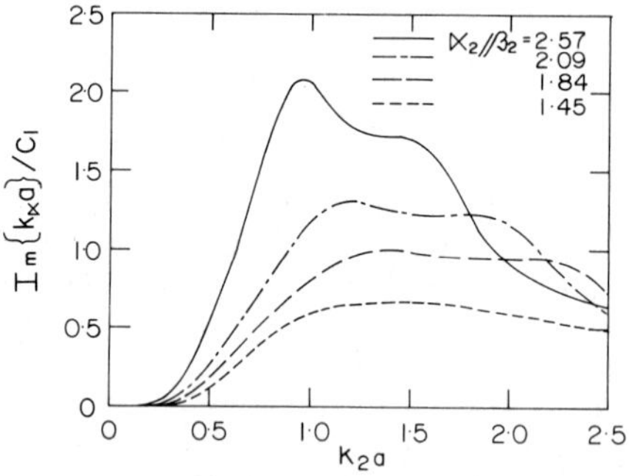

FIGURE 2

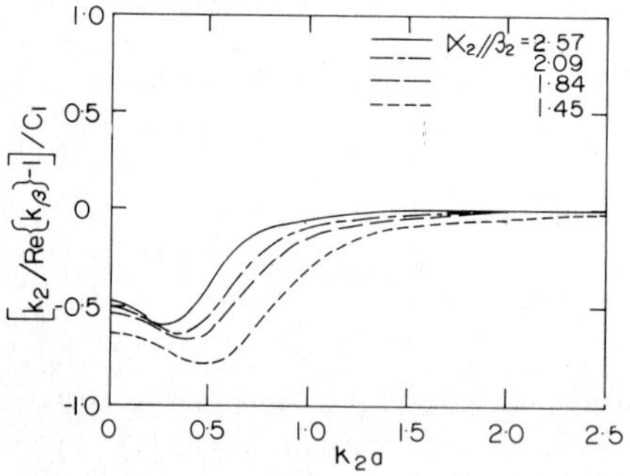

FIGURE 3

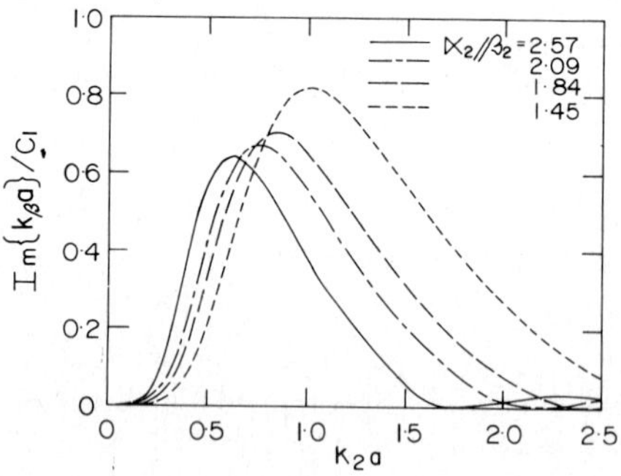

FIGURE 4

REFERENCES

[1] Gubernatis, J.E. and Domany, E., Wave Motion 6 (1984) 579.

[2] Gubernatis, J.E. and Domany, E., Effects of Microstructure on the Speed and Attenuation of Elastic Waves: Formal Theory and Simple Approximations, in: Thompson, D.O. and Chimenti, D.E., (eds.), Review of Progress in Quantitative Nondestructive Evaluation (Plenum Publishing Corporation, New York, 1983) pp. 833–850.

[3] Sayers, C.M., J. Phys. D: Appl. Phys. 14 (1981) 413.

[4] Sayers, C.M. and Smith, R.L., Ultrasonics 20 (1982) 201.

[5] Datta, S.K.,Ledbetter, H.M., Shindo, Y. and Shah, A.H., Wave Motion 10 (1988) 171.

[6] Sabina, F.J. and Willis, J.R., Wave Motion 10 (1988) 127.

[7] Hill, R., J. Mech. Phys. Solids 13 (1965) 89.

[8] Willis, J.R., J. Mech. Phys. Solids 28 (1980) 287.

E
SCATTERING OF ELASTIC WAVES

ELASTIC WAVE PROPAGATION
M.F. McCarthy, M.A. Hayes, (Editors)
Elsevier Science Publishers B.V. (North-Holland), 1989

RESONANCE ACOUSTIC SCATTERING FROM UNDERWATER ELASTIC BODIES

Guillermo C. Gaunaurd

Naval Surface Warfare Center
White Oak Laboratory (R-43)
Research and Technology Department
Silver Spring, Maryland 20903-5000 U.S.A.

SUMMARY: We review some progress made in our analysis of the interaction of waves emitted by active acoustic and elastic wave-projectors, with material media and target obstacles. Our work addresses the case of penetrable scatterers, particularly in the resonance region of their scattering cross-sections, thus, the name of the resonance scattering theory becomes appropriate.

INTRODUCTION

We review some progress made in our analysis of the interaction of waves emitted by active acoustic and elastic wave-projectors, with material media and target obstacles. Our work has always included the most difficult, middle, or resonance region of the backscattering cross-sections (BSCS) of various scatterers. The pertinent targets have various shapes, material compositions, are embedded in fluid or solid surroundings, and in most relevant cases, have to be modeled as penetrable (i.e., admitting interior fields) bodies. Scatterers modeled as impenetrable behave quite unrealistically, particularly in underwater acoustics applications. Under the direct scattering mode of operation of our approach, we have predicted the echoes returned by targets, using various levels of complexity in the analytic description of their shape and various degrees of penetrability. Under the inverse scattering mode of operation of the method, we have extracted some size, shape, and composition information (i.e., aspect ratio, wall-thickness, diameter, density, elastic constants, etc...) from selected "resonance features" that we have learned to isolate and pull-out of the echoes that these targets return. Accounting for the elasticity of a target, considerably broadens the resonance region of its BSCS. Elastic bodies exhibit resonance features in all bands of their spectra for which: $\lambda \lesssim a$. Our method has repeatedly shown that it is the interaction of resonances with "backgrounds" that causes or that generates the BSCS, or the differential scattering cross section (DSCS), of each target. Since the resonance region of an elastic target becomes so broadened, it follows that resonance scattering seems to be the dominant mechanism for scattering above the initial Rayleigh spectral region. Associated with each resonance is a complex eigenfrequency or pole in the corresponding scattering amplitude, which can be identified and located in the complex-frequency plane x ($\equiv$ka). From these pole-position diagrams we can extract the dispersion plots for the phase (or group) velocities and attenuation "constants," of the surface waves that the resulting scattering process launches around any smooth target. We also display many of these dispersion plots in selected cases. Either transient or c.w. incident waves excite the same resonances in targets, but the resonance features communicated to the returned echoes differ in each case. We conclude by briefly describing novel experimental techniques that have validated many of the theoretical predictions, for both modes of operation of our approach.

I. EARLY CLASSICAL THEORY: SCATTERING BY ELASTIC TARGETS

The expression for the acoustic BSCS of an impenetrable (i.e., soft/rigid) sphere in a fluid was first derived by Rayleigh [Ref. 1] in the long-wavelength region. It was first computed over a wide spectral band by Stenzel [Ref. 2] who invented the later called [Ref. 3] Stenzel functions for this purpose. The result of this calculation is shown in Fig. 1a, which displays the modulus of the form-function $f_\infty(\pi)$, whose square is the BSCS, versus ka, in the band: $0 \leqslant ka \leqslant 50$. This spectral plot shows the oscillating resonance-region in between the long-wavelength (Rayleigh) region, and the high-frequency (geometrical acoustics) regime. The classical expression for the BSCS of an <u>elastic</u> sphere was first given by Faran [Ref. 4], and later corrected by Hickling [Ref. 5]. Since then it has been re-computed by many researchers [Ref. 6-8] for various materials and in many frequency bands, such as that in Fig. 1b, which here is computed [Ref. 8] for a tungsten carbide (WC) ball-bearing in water at 20°C. Except for their Rayleigh regions, there is no similarity between Figs. 1a and 1b. The resonance region now contains, not only oscillations, but additional deep dips and spiky peaks spread all over the spectrum, with no apparent upper bound. Such broadening of the resonance region for <u>elastic</u> scatterers is an indication of the dominance of resonance scattering as a scattering mechanism over almost the entire spectrum. Experimental observations [Ref. 9] have confirmed Fig. 1b quite accurately, and totally discredited Fig. 1a. Hence, in underwater acoustics, the impenetrability assumption is quite unrealistic and unadmissible, since it does not describe the observed physical reality. The situation in Fig. 1 could be reversed by considering a fluid-filled spherical cavity in an elastic (metal) matrix. Fluid and solid now reverse their locations and roles. If a compressional (p) elastic wave, travelling through the aluminum matrix falls upon the water-filled cavity, and we consider the case of $p \rightarrow p$ scattering (i.e., $f^{pp}(\pi)$), the resulting form-function is not too different from that shown in Fig. 1b. This situation was first studied by Knopoff [Ref. 10], and later computed by Ying and Truell [Ref. 11]. The case of an incident shear (s) wave was considered later [Ref. 12]. This early work is mentioned merely to state the problem, and to introduce the modern approaches to be described next. Most of this early work ended with the determination (and later the calculation) of form-functions, such as $f^{pp}(\pi)$. Modern approaches have started from this point on.

II. MODERN APPROACHES TO SCATTERING: THE SEARCH FOR PHYSICAL INTERPRETATIONS

The form-function of the water-filled cavity in aluminum mentioned above, can be decomposed [Ref. 13, 14] into its partial-waves or normal mode contributions. The first four (i.e., n=0,1,2,3) such partial-waves $f_n^{pp}(\pi)$ are plotted [Ref. 14]. vs. $k_{d_1}a \equiv x$ in the band: $0 \leqslant x \leqslant 10$, in the left column of Fig. 2. Observation of these plots shows that the ever present dips/peaks seem to be riding on top of smooth "background"

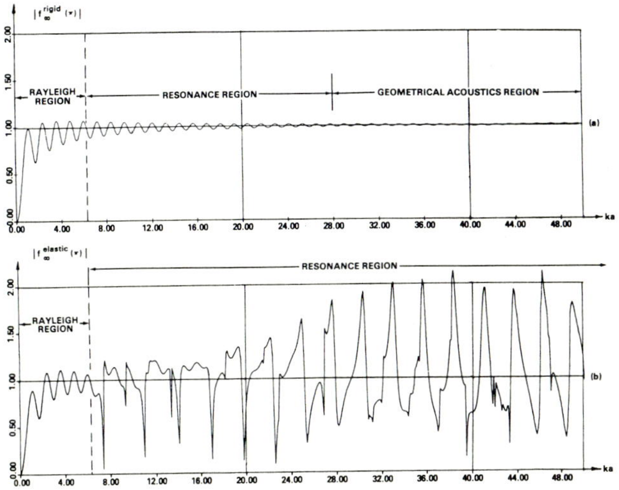

Fig. 1: (a) Form-function of a rigid sphere in water. (b) Form function of a tungsten carbide sphere in water. Deep dips and sharp peaks due to the sphere's elasticity develop and show no trend to disappear. (From Ref. 30).

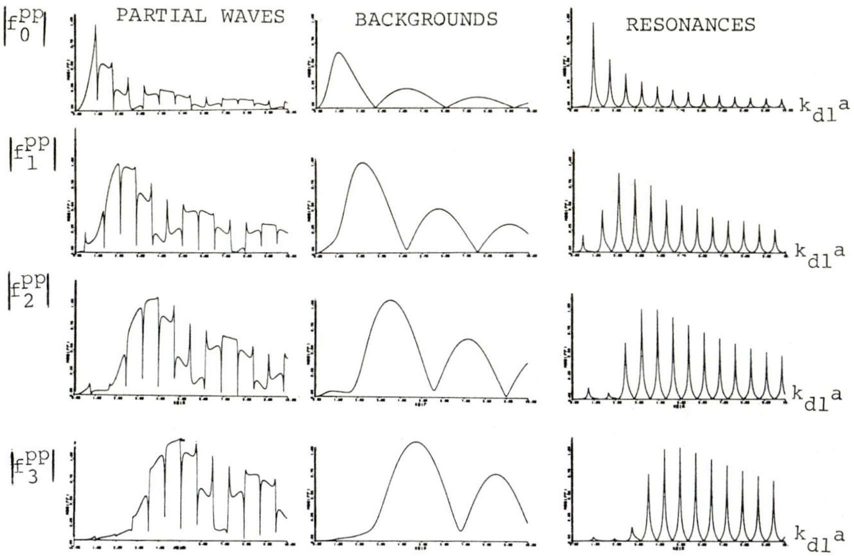

Fig. 2. First four partial-waves (left) contained within the form-function $|f^{pp}|$ of a water-filled spherical cavity within an aluminum matrix. Each partial-wave subdivides into a smooth background (center) and a set of resonances (right). (Ref.14).

curves. These are the partial-waves of the totally evacuated or empty cavity, shown in the central column of Fig. 2. The modulus of the difference between the complex quantities shown in absolute value in the first two columns, yields the set of spiky resonances displayed in the right-most column. This decomposition of the BSCS into partial waves, and the further split of each one of these into backgrounds and resonances, is one of the basic principles of the resonance scattering theory (RST) [Ref. 14-16]. This idea was patterned after a similar situation in nuclear physics, in the study of resonance reactions [Ref. 17]. To actually carry out this decomposition, which can be done either analytically, computationally, or experimentally, is very informative, simplifies calculations, and always allows clear physical interpretation of the scattering process. Resonances can be displayed vs. n at fixed frequencies [Ref. 14-16]. They can also be displayed vs. both, frequency x <u>and</u> mode-order n, in 3-dimensional plots such as that in Fig. 3. This plot is for mode-converted ($s \to p$) scattering, from a water-filled spherical cavity in an aluminum matrix [Ref. 18]. We have generated these plots for many target shapes and compositions. Figure 4 shows that plot for an air-filled steel shell, covered with a layer of viscoelastic material, submerged in water. Some of the resonances group in ridges, some are due to the shell, some to the filler, and some to the covering layer. The analysis of these "response surfaces" as we have called [Ref. 18] these plots, can be quite involved. The resonances so extracted and displayed can be used to obtain all the necessary information to uniquely characterize a target's shape [Ref. 19] and composition [Ref. 20]. An alternative and quite useful representation of the resonances, is by means of the <u>poles</u> in the scattering amplitude of their echoes, and we will study these next. Our previous emphasis was on the prediction, interpretation, and disentangling of the "wiggles" in the scattering cross-sections of targets, which is another route to the same goal.

III. POLES OF SCATTERING AMPLITUDE: TARGET-IDENTIFICATION

Any wave, either pulsed or c.w., incident on an elastic cylinder submerged in a fluid will excite the same resonances. This follows because resonances are a property of each target, just as its elastic constants. If the target composition is <u>close</u> to the <u>soft</u> limit (i.e., if it is tenuous compared to its surroundings) it will have poles close to those shown in Fig. 5, which are [Ref. 21] the roots of $H_n^{(1)}(z) = 0$. Near the <u>rigid</u> limit (i.e., a metal cylinder in water), the poles will be very close to those shown in Fig. 6, which are those of a perfectly rigid cylinder [Ref. 21], given by the zeros of $H_n^{(1)\prime}(z)$. These poles characterize the cylindrical shape of the body in its two impenetrable extremes. They all lie in the lower-half of the z-plane. For a cylinder, they are asymmetric with respect to the Imz-axis. From the pole-positions we can compute the dispersion plots for the phase velocities c_ℓ, phase attenuation "constants" θ_ℓ, group velocities $c_\ell^{(g)}$, and group attenuation "constants" $\theta_\ell^{(g)}$, of the circumferential waves revolving around the cylinder, by the relations [Ref. 14, 21]:

$$\frac{c^l(x)}{c_1} = \frac{x}{\operatorname{Re} \hat{n}_l}, \quad \frac{c^l_{(g)}(x)}{c_1} = \left[\operatorname{Re}\left(\frac{d\hat{n}_l}{dx}\right)\right]^{-1}, \tag{1}$$

$$\theta_l(x) = \frac{1}{\operatorname{Im} \hat{n}_l}, \quad \theta_l^{(g)}(x) = \left[x \operatorname{Im}\left(\frac{d\hat{n}_l}{dx}\right)\right]^{-1}. \tag{2}$$

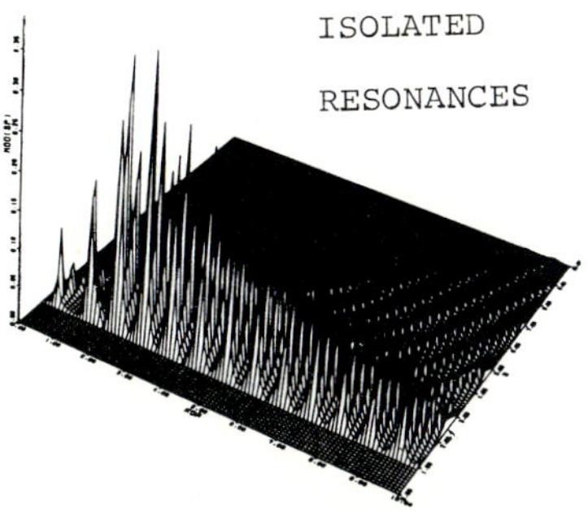

Fig. 3: Response surface of a water-filled cavity in an Al. matrix. (From Ref. 18).

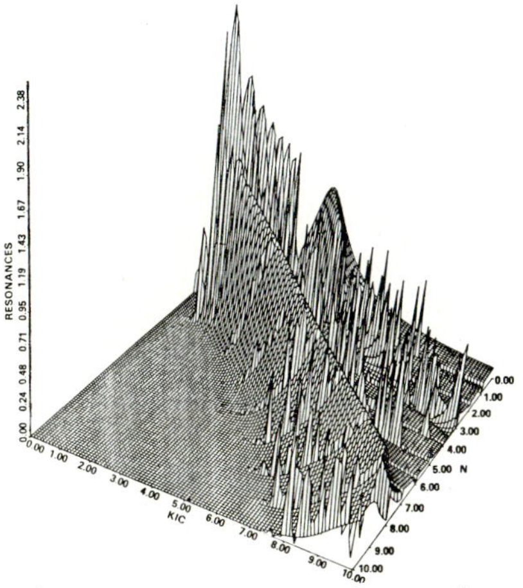

Fig. 4: Response surface of an air-filled steel shell covered with a rubber coating layer and inmersed in water.

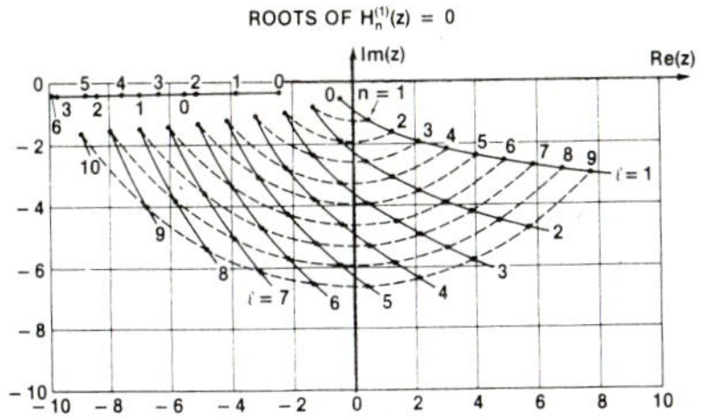

Fig. 5: The "shape" poles of a soft cylinder in water in the z=ka plane. Families of constant n (or ℓ) values are shown in dashed (solid) lines. (Ref.21).

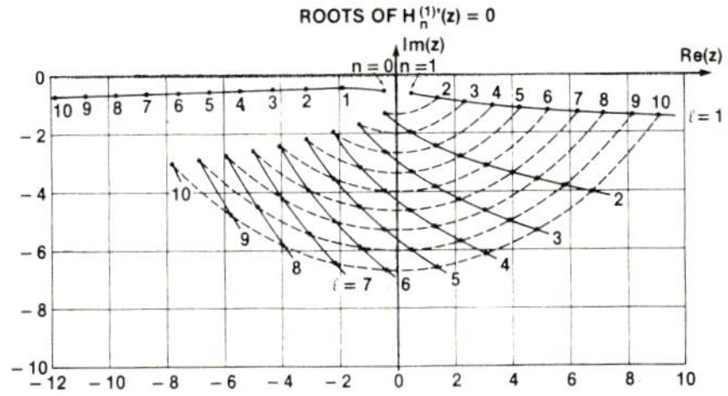

Fig.6: Shape poles of rigid cyl.

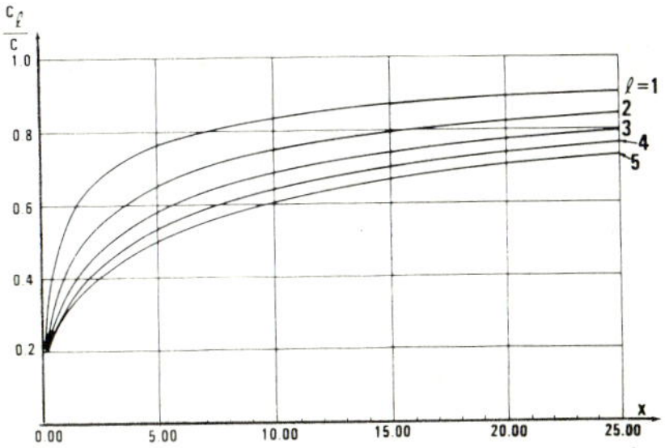

Fig.7: c_ℓ for soft cyl/sphere.

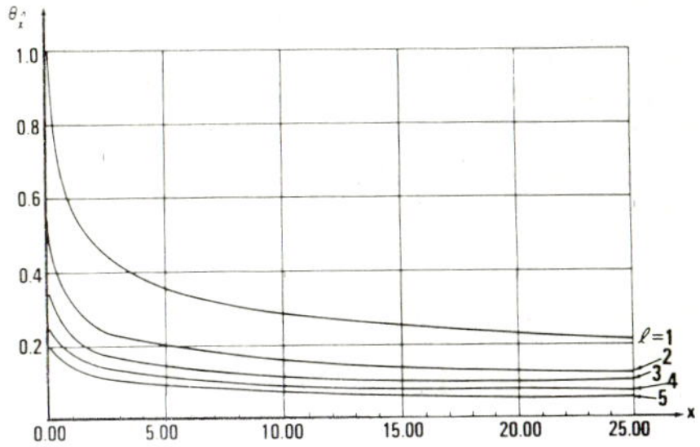

Fig.8: Phase attenuations for a soft cyl/sphere vs. x=ka. (From Ref. 24).

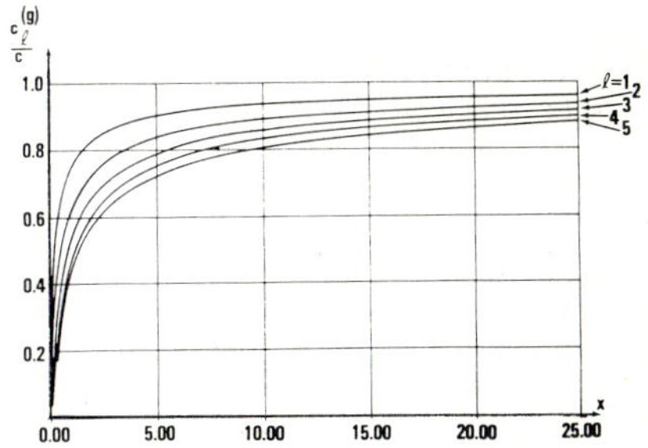

Fig.9: Group velocities of a soft cyl/sphere vs.x. (Ref.24)

The results are respectively shown in Figs. 7-10, for the first five (i.e., ℓ=1,2,3,4 & 5) external circumferential waves revolving around a soft cylinder. Using the rigid poles in Fig. 6 we can also compute these curves for the rigid cylinder limit, and the results are shown in Figs. 11-14, respectively. For an elastic (i.e., aluminum) cylinder in water, all its "shape" poles are very close to those for a rigid cylinder given in Fig. 6. Minute differences appear after the third or fourth decimal place of the coordinates of each pole. Target elasticity, thus, hardly alters the "shape" poles of Fig. 6. However, a new family of poles [Ref. 22] now appears, very close to the Rex (or Rez) axis. These are denoted the "composition" poles, and they are shown in Fig. 15. They align in families labeled by constant values of the index ℓ, and to evaluate them constitutes a formidable computational task. The dispersion plots for the (internal) surface waves around the cylinder, now inside the metal, are also found by means of Eqs. (1) and (2), and the poles of Fig. 15. The results for the phase and group velocities of the first seven surface waves are shown in Figs. 16 and 17, respectively.

Analogous arguments can be given for a spherical elastic target in water [Ref. 15,8]. The soft poles, roots of $h_n^{(1)}(z)=0$, are displayed in Fig. 18. The rigid poles, zeros of $h_n^{(1)\prime}(z)$, are shown in Fig. 19. For the sphere, they are symmetrically located [Ref. 23] with respect to the Imz-axis. The dispersion curves for c_ℓ, θ_ℓ, $c_\ell^{(g)}$ and $\theta_\ell^{(g)}$, are analogous to Eqs. (1) and (2), but now with $\hat{n}_\ell$ replaced by $\hat{n}_\ell + 1/2$. The dispersion plots for the case of the rigid sphere are exhibited [Ref. 8] in Figs. 20-23, respectively. For an elastic (viz., say, tungsten carbide) sphere in water, the "shape"-poles remain very close to the rigid set shown in Fig. 19. A new family of "composition" poles now appears inside a strip just below the Rez-axis, plotted [Ref. 23,8] in Fig. 24. This is the spherical counterpart of Fig. 15. We have given [Ref. 24,21] asymptotic formulas for these poles and for the associated dispersion curves in Figs. 7-14 and 20-23. We note that the dispersion plots in Figs. 7-10, for the soft cylinder, are the same as those for the soft sphere. We have given many explanations [Ref. 25] for these poles, which are the complex representations of the target resonances. We note that the phase attenuation "constants" θ_ℓ, are the reciprocal of the half-width of the resonances (viz., $\theta_\ell = 2/\Gamma_{n_\ell \ell}$), and the wavelength of the ℓth surface wave is,

$$\lambda_\ell = 2\pi a/(\hat{n}_\ell + 1/2) . \qquad (3)$$

The locations of the rigid/soft poles for the case of a cylinder or a sphere, although used repeatedly by us in the past [i.e., Refs. 21 and 31], have not appeared tabulated yet. Tables 1 and 2 list the locations of the rigid cylinder and rigid sphere poles, respectively. They are listed with five decimal places, as obtained by the root-finding computer routine we have developed to obtain them.

For elongated bodies, such as spheroids, the "shape" poles can be of two basic types. For broadside incidences, they are not too different from those of a cylinder. For end-on incidences, they tend to coincide with those of the sphere for small aspect-ratio, L/D. As L/D increases, they all move downward [Ref. 26], into the bottom-half of the z-plane. This indicates greater attenuations for the surface waves they generate, which then become even more damped than the

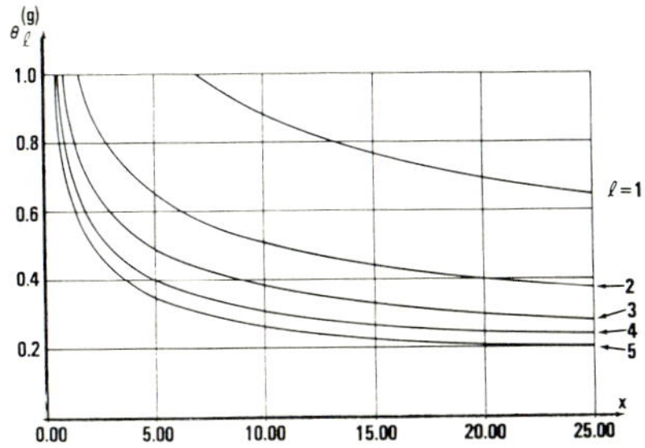

Fig.10: Group attenuations for a soft cyl/sphere. (Ref. 24).

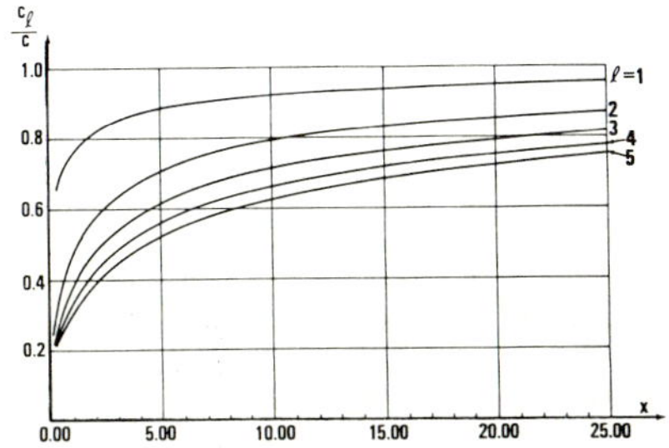

Fig.11: Phase velocities for a rigid cylinder. (Ref. 21).

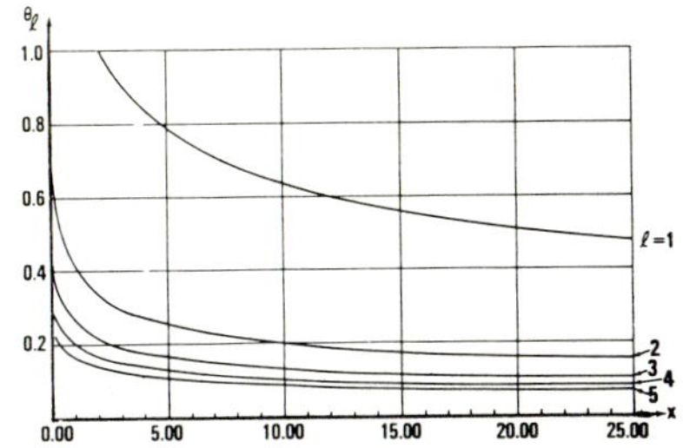

Fig.12: Phase attenuations for the creeping waves around a rigid cylinder. (Ref. 21).

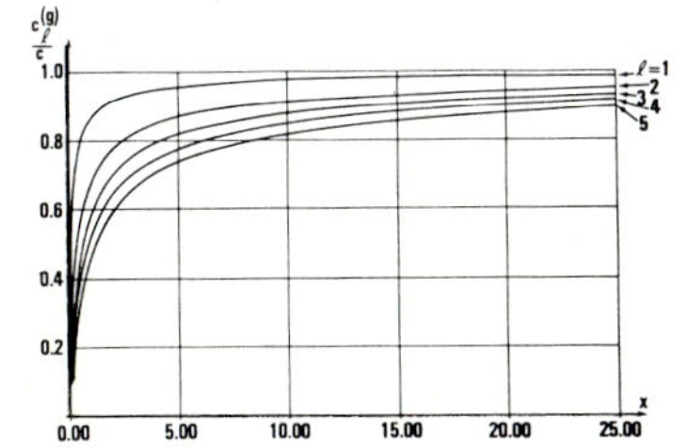

Fig.13: Group velocities for a rigid cylinder vs.x. (Ref. 21)

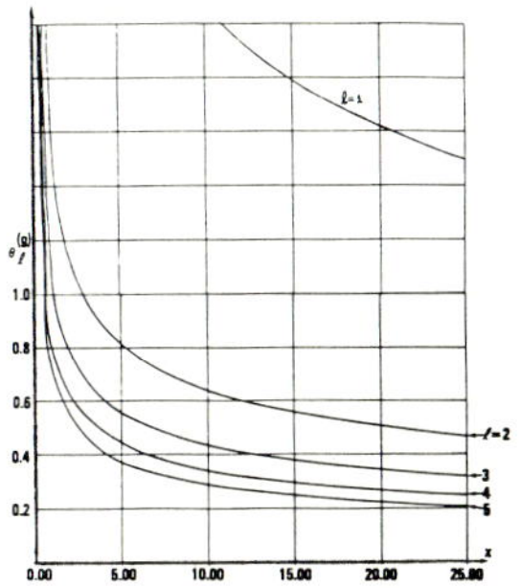

Fig.14: Group attenuations for a rigid cylinder in water vs. x. (From Ref. 21). The various creeping waves are labeled by ℓ.

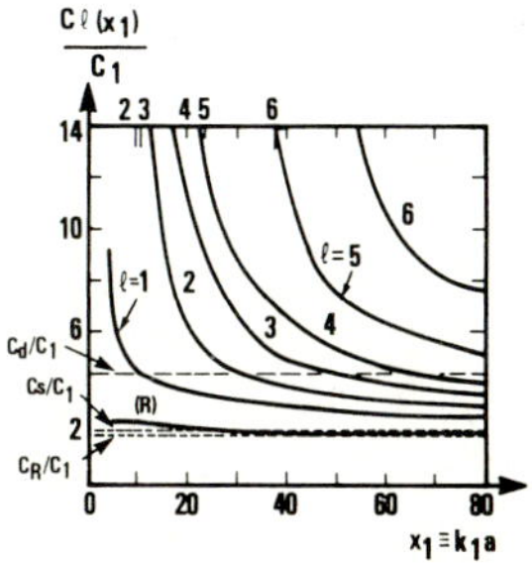

Fig. 16: Phase velocities of the first seven surface waves inside an elastic (aluminum) cylinder in water. (Here R→ Rayleigh wave).(Ref. 16).

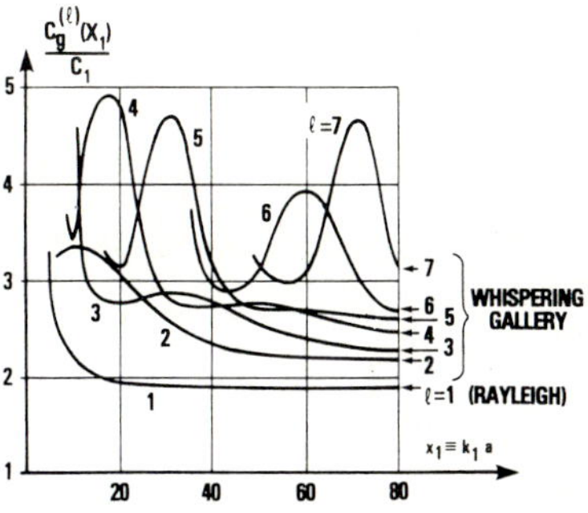

Fig. 17: Group velocities of the first 7 surface waves in the aluminum cylinder in water.(Rayleigh→ℓ=1) (Ref.23)

corresponding ones for the sphere and thus, become much harder to notice. Composition poles have not been studied for spheroids, however, there have been other inversion techniques [Ref. 20] that have succeeded in extracting composition information from the resonances in an echo.

IV. EXPERIMENTAL VERIFICATIONS OF THE THEORY: NSWC TESTS

A long acoustic pulse backscattered from an elastic cylinder in water takes on a very informative, or a very useless, shape depending on whether the pulse has excited, or not, any of the target resonances. The situation is illustrated [Ref. 27] in Fig. 25. The central graph shows the behavior at resonance. The initial transient (#1), is followed by the forced (or steady-state) vibration region (#2), and then by the final transient (#3), which in turn is followed by the exponential decay of the free-vibration (or "ringing") region. If the received echoes are sampled in their region #2, and the results are plotted vs. frequency (or angle), we obtain very complicated frequency (or angular) plots, which contain the obscuring and undesirable modal backgrounds of the RST. Figure 26 (upper curve) shows this complicated frequency dependence. If we sample the returned pulses in their free-vibration region (#3) as the "ringing" starts, the resulting frequency plot, (keeping $\theta=\pi$), contains no backgrounds, and we end-up with the "spectrogram" or set of isolated resonances. This is shown in Fig. 26 (lower measured curve), which verifies the theoretical prediction displayed in the central graph, (for n=1), of Fig. 27. Details have been given elsewhere [Refs. 28, 29]. If we plot the returned amplitude versus angle (viz., $0^\circ \leq \theta \leq 360^\circ$) around the cylinder, but keeping the frequency coincident with any of the target resonances (viz., say, the (4,2)-resonance occurring at $ka \cong 16.200$), we then observe the simple daisy or "rosetta" pattern shown in Fig. 28, which has 2n-petals (eight in this case). The theory, hence, has been experimentally confirmed and implemented, and can be used to characterize underwater elastic targets by simple in-situ measurements. Work continues in various fronts [Ref. 30].

ACKNOWLEDGEMENTS

The author thanks the computational assistance of J. Barlow, and the support received over the years by various Codes of the Office of Naval Research, and the IR Board of NSWC. Thanks are also due to a number of faculty members at various universities, and colleagues in several Naval installations who over the years have received support from this Task, and have provided part-time assistance with the various aspects of this work. A partial (alphabetical) list follows: V. Ayres (NSWC), D. Brill (U.S. Naval Acad.), J. Barlow (NSWC), E. Callen (American Univ.), J. Coughlin (Towsen State Univ.), G. Conde (NSWC), P.D. Jackins (NSWC), A. Kalnins (Lehigh Univ.), M. McCarthy (Nat. Univ. of Ireland), E. Tanglis (NSWC), W. Tobocman, (Case-Western), C.Y. Tsui (Univ. of MD), H. Ueberall (Catholic Univ.), M. F. Werby (NORDA), W. Wertman (NSWC).

LIST OF REFERENCES

1) J. Strutt, (Lord Rayleigh), Philos. Mag., 41, 107-120, (1871; ibid, 47, 375-384, (1899).
2) H. Stenzel, Elektr. Nachrichtentec., 15, 71-78, (1938).

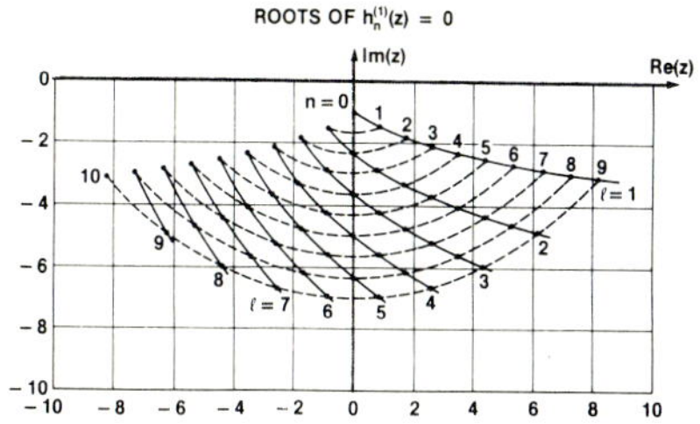

Fig. 18: "Shape" poles of a soft sphere. (From Ref.24).

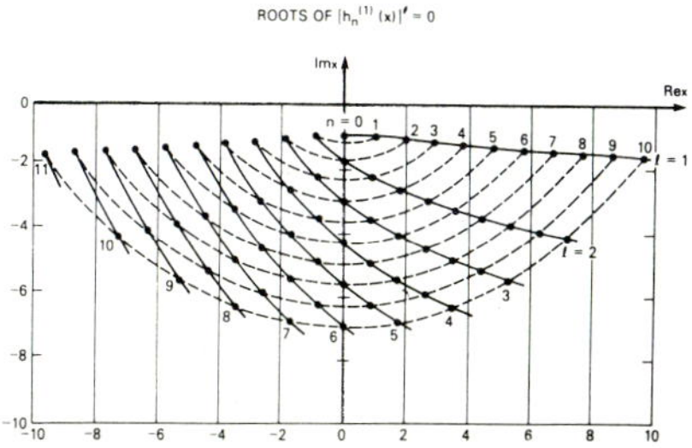

Fig. 19: "Shape" poles of a rigid sphere. (From Ref. 8).

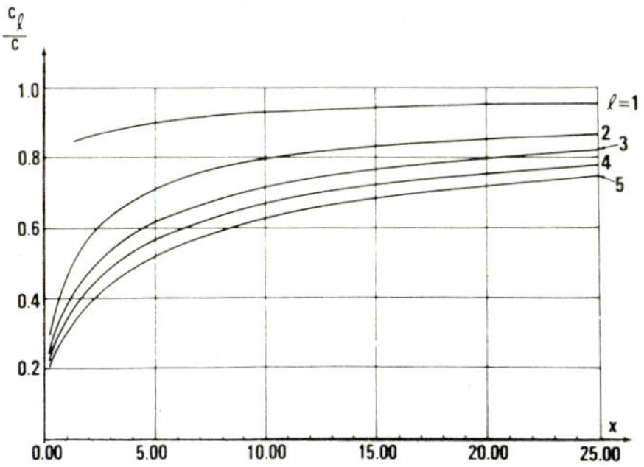

Fig. 20: Phase velocities of the waves around a rigid sphere. (23)

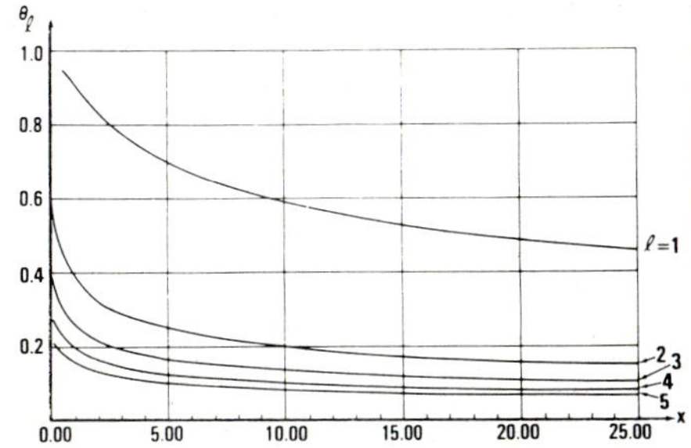

Fig. 21: Phase attenuations of the waves around a rigid sphere. (23).

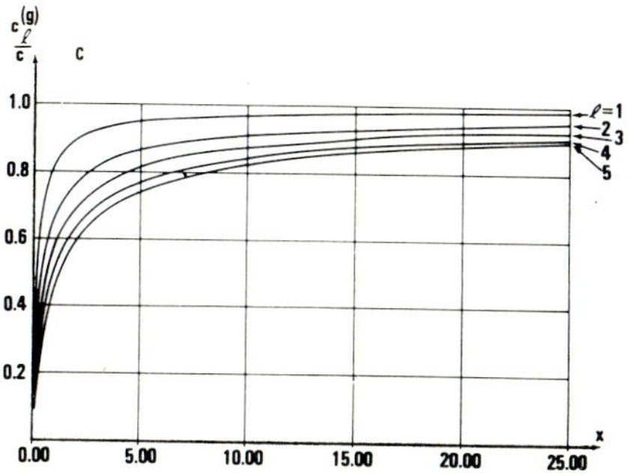

Fig.22: Group velocities of the first five external "creeping" waves circumnavigating a rigid sphere in water. (From Ref.23).

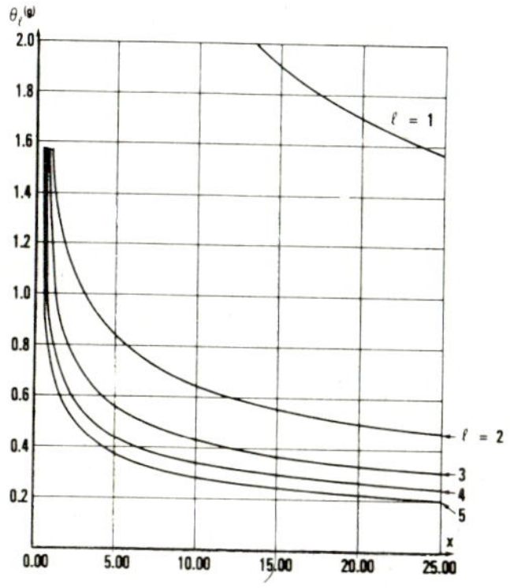

Fig. 23: Group attenuations of the first five creeping waves around a rigid sphere. The $\ell=1$ wave is the least attenuated. (From Ref. 23).

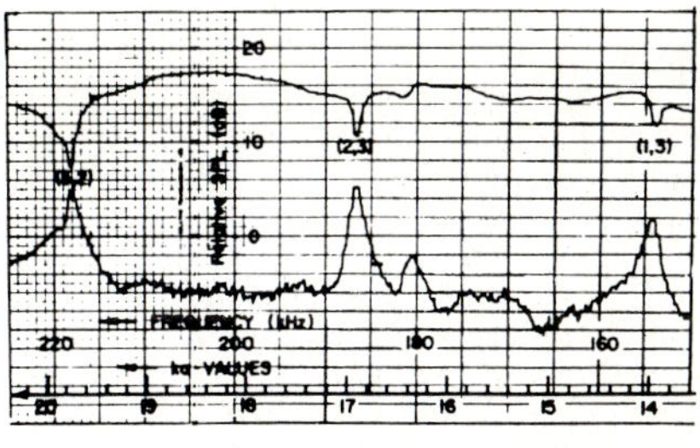

Fig.26: Measured BSCS of an Al-cylinder in water in :$13 \leq ka \leq 21$. (a) Upper curve: With backgrounds included. (b) Lower curve: Isolated resonances. (Ref. 28).

3) E. Skudrzyk, Foundations of Acoustics, Springer, New York, (1971).
4) J.J. Faran, J. Acoust. Soc. Amer., 23, 405-418, (1951).
5) R. Hickling, J. Acoust. Soc. Amer., 34, 1582-1592, (1962).
6) R. Chivers and L. Anson, J. Ultrasonics, 20, 25-34, (1982).
7) L. Flux, J. Acoust. Soc. Amer., 62, 1502-1503, (1977).
8) G.C. Gaunaurd and H. Ueberall, "RST analysis of monostatic and bistatic acoustic echoes from an elastic sphere",J. Acoust. Soc. America, Volume 73, 1-12, (1983).
9) W.G. Neubauer et al., J. Acoust. Soc. Amer., 55, 1123-1129, (1974).
10) L. Knopoff, Geophysics, 24, 30, Feb. 1959.
11) C. Ying and R. Truell, J. Appl. Phys., 27, 1086-1097, (1956).
12) N. Einspruch et al., J. Appl. Phys., 31, 806-818, (1960).
13) G.C. Gaunaurd, in Modern Problems in Elastic Wave Propagation, p. 550, J. Wiley, (1978). (Invited).
14) G.C. Gaunaurd and H. Ueberall, J. Acoust. Society Amer., Vol.63, 1699-1712, (1978); ibid., J. Appl. Phys., 50, 4642-4660, (1979); ibid., Science, Vol.206, 61-64, Oct. 5, 1979.
15) D. Brill and G.C. Gaunaurd, "Resonance Theory of Elastic Waves Ultrasonically Scattered from an Elastic Sphere", J. Acoustical Soc. America, Vol. 81, 1-21, (1987).
16) L. Flax, G.C. Gaunaurd and H. Ueberall, "The Theory of Resonance Scattering," in Physical Acoustics, 15, 191-294, Ch. 3, Academic Press, (1981).
17) G. Breit and E. Wigner, Phys. Rev., 49, 519-531, (1936).
18) D. Brill et. al., J. Acoust. Soc. Amer., 67, 414-424, (1980); also, ibid, J. Appl. Phys., 52, 3205-3214, (1981).
19) D. Brill et al., Acustica, 53, 11-18, (1983).
20) V. Ayres and G.C. Gaunaurd, J. Acoust. Soc. Amer., 82, 1291-1302, (1987); also, ibid, 81, 301-311, (1987).
21) G.C. Gaunaurd and D. Brill, J. Acoust. Soc. Amer., 75, 1680-1693, (1984).
22) D. Brill and G.C. Gaunaurd, J. Acoust. Soc. Amer., Vol. 73, 1448, (1983).
23) G.C. Gaunaurd, "Techniques for Sonar Target Identification", IEEE J. Ocean. Engr., Vol. OE-12, 419-423, (Special Issue on Scattering),(1987).
24) G.C. Gaunaurd et al., Il Nuovo Cimento, Vol. 76B, 153-175, (1983); also, Vol. 77B, 73-86, (1983).
25) G.C. Gaunaurd and M. F. Werby, "Lamb waves in submerged spherical shells resonantly excited by elastic waves", J. Acoust. Soc. Amer., 82, 2021-2033, (1978).
26) J.D'Archangelo et al., J. Acoust. Soc. Amer., 77, 6-10, (1985).
27) G.C. Gaunaurd and C. Y. Tsui, "Transient and steady-state target resonance excitation by sound scattering", Applied Acoustics, (to be published, 1988); also, ibid, Revista de Acústica, (to be published, 1988).
28) C. Y. Tsui, G. N. Reid and G. C. Gaunaurd, J. Acoust. Soc. Amer., 80, 382-390, (1986), and ibid., "Bistatic measurements of target scattering at resonance", J. Acoust. Soc. Amer., 83, (to be published, 1988).
29) G. Maze et al., Phys. Lett., Vol. 84A, 309-312, (1981).
30) D. Brill, V. Ayres and G. C. Gaunaurd, "The Influence of Natural Resonances on Scattering and Radiation Processes", J. Wash. Academy of Sciences, Vol. 77, 55-65, (1987).
31) G. C. Gaunaurd and H. Ueberall, J. Acoust. Soc. Amer., 78, 234-243, (1985); also, 72, 1014-1017, (1982); and also, IEEE Transactions, Vol. AP-32, 1071-1079, (1984).

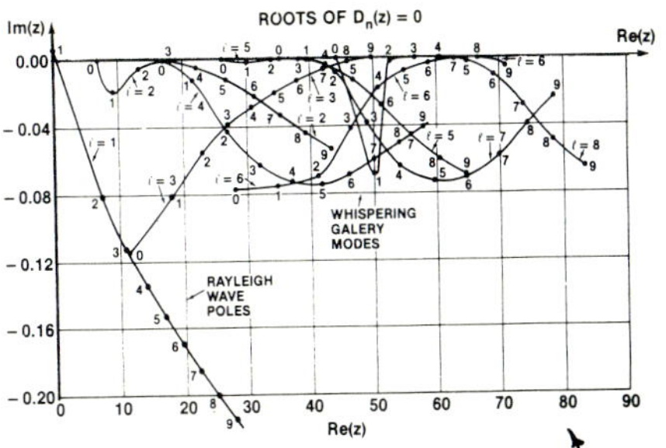

Fig. 24: "Composition" poles of a tungsten carbide sphere in water. Companion plot to Fig.19.(23, 8).

Fig. 15: "Composition" poles of the scattering a plitude of an Al cylinder in water. Companion plot to Fig. 6. (From Ref. 22).

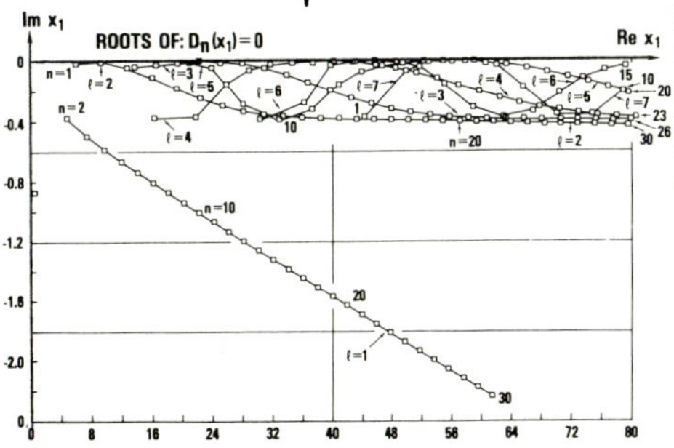

Fig. 28: Isolated resonances in the first 3 partial-waves in the BSCS of an Al. cylinder in water. The n=1, ℓ=2 resonance at $k_1a \cong$ 6.059 is the one used in Fig. 25. (From Ref. 27).

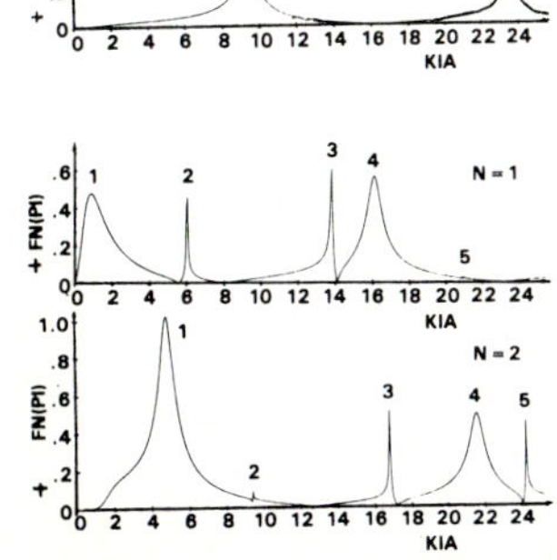

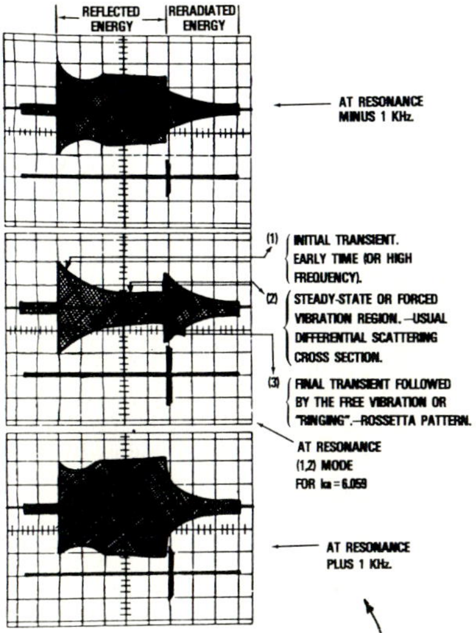

Fig. 25: Backscattered pulse from an Aluminium cylinder in water. Center: At resonance (i.e., the (1,2)-resonance). Top/Bottom: Sligtly away from resonance, (Ref.27).

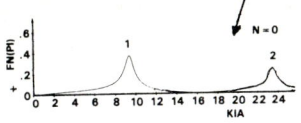

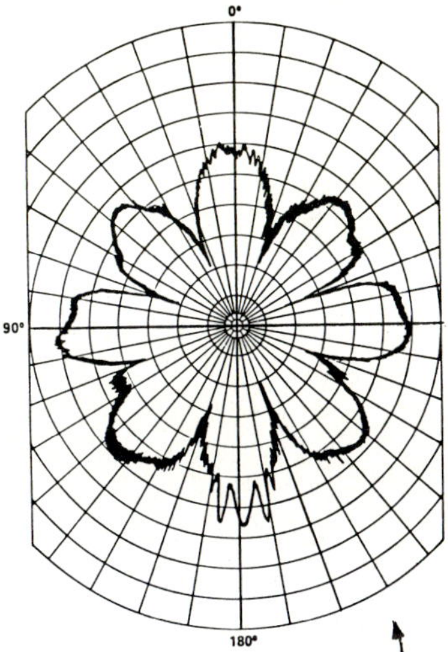

Fig. 27: Symmetric rosetta or daisy pattern measured at the (4,1)-resonance. Here we sample the returned pulses in their free-vibration region, just after #3 in Fig. 25. Eight lobes result. (From Ref. 28).

Table 1.
RIGID-CYLINDER ZEROS: $H_n^{(1)\prime}(z_{n\ell})$

n	ℓ	Re($z_{n\ell}$)	Im($z_{n\ell}$)
1	1	+0.497264	-0.526783
2	1	+1.412426	-0.746427
	2	-0.456975	-1.270221
3	1	+2.348607	-0.891683
	2	+0.447859	-1.953039
	3	-1.340713	-1.661922
4	1	+3.296183	-1.004301
	2	+1.323988	-2.416482
	3	-0.443861	-2.625097
	4	-2.240004	-1.951729
5	1	+4.251087	-1.097908
	2	+2.210818	-2.781794
	3	+0.441636	-3.293240
	4	-1.316818	-3.129564
	5	-3.150864	-2.186933
6	1	+5.211174	-1.178853
	2	+3.106971	-3.088045
	3	+1.312949	-3.824514
	4	-0.440212	-3.959513
	5	-2.197514	-3.543757
	6	+4.070500	-2.387293
7	1	+6.175160	-1.250666
	2	+4.010907	-3.354141
	3	+2.190107	-4.273371
	4	+0.439233	-4.624746
	5	-1.310559	-4.509748
	6	-3.085672	-3.899239
	7	-4.997028	-2.563184
8	1	+7.142205	-1.315528
	2	+4.921366	-3.590867
	3	+3.073324	-4.665533
	4	+1.308962	-5.189123
	5	-0.438511	-5.289341
	6	-2.185474	-4.984602
	7	-3.980485	-4.212901
	8	-5.929127	-2.720803

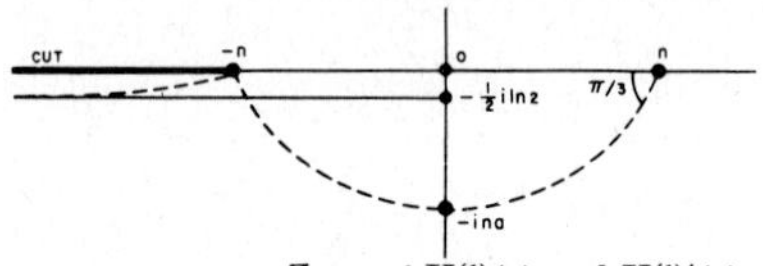

Zeros of $H_n^{(1)}(z)$ *and* $H_n^{(1)\prime}(z)$
$|\arg z| \leq \pi$.

Table 2
RIGID-SPHERE ZEROS: $h_n^{(1)\prime}(z_{n\ell})$

n	ℓ	Re($z_{n\ell}$)	Im($z_{n\ell}$)
1	1	-1.	-1.
	2	+1.	-1.
2	1	-1.954093	-1.108378
	2	+1.954093	-1.108378
	3	+0.	-1.783243
3	1	-0.890601	-2.298134
	2	+0.890601	-2.298134
	3	-2.903917	-1.201866
	4	+2.903917	-1.201866
4	1	+0.	-3.051595
	2	-1.779352	-2.691643
	3	+1.779352	-2.691643
	4	-3.857450	-1.282559
	5	+3.857450	-1.282559
5	1	-0.876830	-3.630292
	2	+0.876830	-3.630292
	3	-2.674458	-3.015904
	4	+2.674458	-3.015904
	5	-0.481517	-1.353803
	6	+4.815177	-1.353803
6	1	+0.	-4.355613
	2	-1.754967	-4.109728
	3	+1.754967	-4.109728
	4	-3.576502	-3.294578
	5	+3.576502	-3.294578
	6	-5.776583	-1.417888
	7	+5.776583	+1.417888
7	1	-0.872607	-4.959488
	2	+0.872607	-4.959488
	3	-2.637550	-4.523537
	4	+2.637550	-4.523537
	5	-4.484872	-3.540616
	6	+4.484872	-3.540616
	7	-6.741092	-1.476359
	8	+6.741092	-1.476359
8	1	+7.708211	-1.530299
	2	-7.708211	-1.530299
	3	+5.398801	-3.761963
	4	-5.398801	-3.761963
	5	+3.525179	-4.890099
	6	-3.525179	-4.890099
	7	+1.746581	-5.482734
	8	-1.746581	-5.482734
	9	+0.	-5.669810

For the case of the rigid (and soft) cylinder, there are also some zeros in a "solitary" branch which has a horizontal asymptote given by Im(z)= -(ln2)/2 = -0.34657. These zeros have been asymptotically studied (for large n) in detail by F. W. J. Olver, in Phil. Trans. Royal Soc., London, Vol. A-247, 328-368, (1954), and also in his book Asymptotics and Special Functions, Academic Press, New York, London, (1974). Prof. Olver is presently at the University of Maryland, U.S.A. The Handbook of Mathematical Functions edited by M. Abramowitz and I. Stegun has a section (pp. 371-374), briefly discusing the Olver results. Tables such as the above, plotted in Figs. 6 and 19 (and analogously for the soft cylinder and sphere) have never been computed for finite (and small) values of n. END.

ELASTIC WAVE PROPAGATION
M.F. McCarthy, M.A. Hayes, (Editors)

TRANSIENT DIRECT AND INVERSE SCATTERING

James Corones

Applied Mathematical Sciences
Ames Laboratory - USDOE
Iowa State University
Ames, Iowa 50011 U.S.A.

This paper reviews some recent work in transient direct and inverse scattering for elastic and viscoelastic media. Wave splitting and invariant imbedding concepts are employed to obtain scattering operators for stratified media.

1. INTRODUCTION

This paper gives a brief review of some recent work in direct and inverse scattering for elastic and viscoelastic media. All the analysis is carried out in the time domain. Attention is restricted to problems involving the plane wave illumination of stratified media. In the approach presented here attention is focused on the scattering operators (reflection and transmission operators) of the media rather than the incident and scattered fields. These scattering operators resemble Green's functions and, as with Green's functions, knowledge of the scattering operators allows the computation of scattered fields for any incident field.

Due to limitations of space the presentation will be somewhat telegraphic. However references are given that allow the interested reader to obtain a full and detailed discussion of all that is sketched here. The general approach presented here has been developed over several years. The list of references [1-16] gives some of the relevant papers, including applications to electrodynamics. These references contain many numerical examples. Again, due to limitations of space no numerical results are presented here.

2. VISCOELASTIC MEDIA

As a first example consider the viscoelastic wave equation appropriate for normally incident waves on a stratified medium [1]. Due to the normal incidence a scalar field is sufficient to describe the scattering process. The wave equation is

$$\partial_x w(x,t) = \rho(x) u_{tt}(x,t) \tag{1}$$

This work was supported by the Applied Mathematical Sciences subprogram of the Office of Energy Research, U.S. Department of Energy under contract No. W-7405-ENG-82.

where $u(x,t)$ is the displacement, $w(x,t)$ is the traction and $\rho(x)$ is the density. The constitutive relation between displacement and traction is given by

$$w(x,t) = \partial_t[G(x,t)*u_x(x,t)] \tag{2}$$

where

$$f(t)*g(t) = \int_0^t f(t-t')g(t')dt' \tag{3}$$

and $G(x,t)$ is the relaxation modulus. $G(x,t) = \lambda + 2\mu$ for longitudinal strain and $G(x,t) = \mu$ for shear waves. It is assumed that both $u(x,t)$ and $w(x,t)$ are everywhere continuous. It is further assumed that the slab upon which the waves are incident lies in the half-space $x > 0$ and that prior to the arrival of the incident wave front at $t=0$ the medium is quiescent so that $u(x,0) = u_t(x,0) = 0$ for $x > 0$. Further, the medium that occupies $x < 0$ is elastic (not viscoelastic) and has constant material parameters.

If the slab that occupies $x > 0$ is illuminated by a right-going plane wave, a left-going plane wave is reflected back into the region $x < 0$. It is useful to express the wave equation in variables that are most suitable to this situation. To do this let

$$u^{\pm}(x,t) = \frac{1}{2}\left[u(x,t) \pm \frac{1}{\rho(x)c(x)}\partial_t^{-1}w(x,t)\right]. \tag{4}$$

Hence

$$c(x) = \left[\frac{G(x,0)}{\rho(x)}\right]^{\frac{1}{2}}, \qquad G(x,0) \equiv G_0 \neq 0 \tag{5}$$

where $\partial_t^{-1}f(t) = \int_0^t f(t')dt'$. If this change of variables is used in (1) it follows that

$$\frac{\partial}{\partial x}\begin{pmatrix}u^+\\u^-\end{pmatrix} = \begin{pmatrix}\tilde\alpha & \tilde\beta\\ \tilde\gamma & \tilde\delta\end{pmatrix}\begin{pmatrix}u^+\\u^-\end{pmatrix} \tag{6}$$

with

$$\tilde\alpha\, u^+ = \frac{1}{2c}(A - 2 + h*)\partial_t u^+ \tag{7}$$

$$\tilde\beta\, u^- = -\frac{1}{2c}(A + h*)\partial_t u^- \tag{8}$$

$$\tilde{\gamma}\, u^+ = \frac{1}{2c}\,(-A + h*)\partial_t u^+ \qquad (9)$$

$$\tilde{\delta}\, u^- = \frac{1}{2c}\,(A + 2 - h*)\partial_t u^- \qquad (10)$$

where

$$A(x) = -c\,\partial_x \ln(\rho c) \qquad (11)$$

and $h(x,t)$ is the resolvent of $G_t(x,t)$ so that

$$(G_0 + G_t*)h = G_t \qquad (12)$$

Note that if the medium is elastic $G_t = 0$ so that $h = 0$, provided $G_0 \neq 0$ as assumed. Note further that if the medium is both elastic and has constant material parameters, as is the case here when $x < 0$, then

$$\partial_x u^{\pm}(x,t) = \mp \frac{1}{c}\,\partial_t u^{\pm} \qquad (13)$$

which justifies the notation $u^{\pm}(x,t)$ indicating right (left) going waves.

Although in the analysis that follows it is not necessary to restrict attention to the case when the viscoelastic slab is impedance matched to the host elastic medium, suppose that this is the case. That is, suppose that the viscoelastic slab extends from $x = 0$ to $x = L$ and that the elastic impedance $Z(x) = c(x)\rho(x)$ is continuous along with its first derivative at $x = 0$ and $x = L$. Under these circumstances it can be shown that, due to causality and the time translation invariance of the wave equation (1), that

$$\begin{aligned} u^-(0,t) &= \int_0^t R(0,L,t-s)u^+(0,s)ds \\ &\equiv [\tilde{R}(0,L,t)u^+(0,s)](t) \end{aligned} \qquad (14)$$

This means that the reflected (left-going) field just to the left of the slab is related to the incident (right-going) field by a convolution operator with kernel $R(0,L,t)$. Note the homogeneous elastic medium to the right of $x = L$ does not contribute to the reflected field due to the impedance match.

The operator $\tilde{R}(0,L,t)$, or in this case equivalently the kernel $R(0,L,t)$, fully characterizes the reflective properties of the slab extending from $0 \leq x \leq L$. It is possible to think about this operator as the boundary value of a family of operators each associated with a subslab of the original slab. Hence $\tilde{R}(x,y,t)$, where $x \leq y$ and $0 \leq x \leq L$, $0 \leq y \leq L$ but x and y otherwise arbitrary, is an element of the family. By doing this the problem of interest, finding $\tilde{R}(0,L,t)$, is imbedded in a larger problem, finding $\tilde{R}(x,y,t)$. The reason that this is useful is that it is possible to fully specify solutions to the larger problem by writing an integro-differential equation and initial/boundary values for the family.

Here this characterization of $R(x,y,t)$ will be obtained in two steps. First a generic result depending only on the form of the field equation and the reflection operator will be derived. The derivation is completed by using the particular form of the field equations and reflection operator appropriate for the given problem. In general field equations of the form (6) can be used in (14) to obtain the following operator equation for the reflection operator [3,4,7]:

$$\partial_x \tilde{R} = \tilde{\gamma} + \tilde{\delta}\,\tilde{R} - \tilde{R}\,\tilde{\alpha} - \tilde{R}\,\tilde{\beta}\,\tilde{R} \tag{15}$$

This result is obtained by differentiating (14) and using the field equations. If, as is the case here, $\tilde{R}(x,y,t)$ is a convolution operator it can be shown from (15) that the kernel $R(x,y,t)$ satisfies

$$2cR_x - 4R_t = AR*R + \partial_t[(1-R*)(1-R*)h] \tag{16}$$

$$R(x,L,0^+) = \frac{1}{4}[A(x) - h(x,0)] \tag{17}$$

$$R(L,L,t) = 0 \tag{18}$$

These equations can be used in two different ways. If $A(x)$ and $G(x,t)$ (or $h(x,t)$) are known the set (16-18) can be used to determine $R(0,L,t)$ for $0 < t < T$, where T is the time it takes for a wave front to transverse the slab and return to the front edge. [The formalism presented here can be used to find R for longer times but certain discontinuities in $R(0,L,t)$ must be taken into account; see [4,14]]. Alternatively knowledge of $R(0,L,t)$ together with the initial condition (17) and integro-differential equation (16) can be used to obtain information about $A(x)$ and $G(x,t)$ (or $h(x,t)$).

To see how this works, first note that $R(0,L,t)$ can be obtained by deconvolution from knowledge of the incident field $u^+(0,t)$ and the reflected field $u^-(0,t)$. Assuming that $R(0,L,t)$ is known consider two limiting cases.

First assume that the slab is purely elastic so that $G(x,t) = 0$ ($h(x,t) = 0$). The time derivative on the right hand side of (16) vanishes and the initial condition (17) becomes

$$R(x,L,0^+) = \frac{1}{4} A(x) \tag{19}$$

Using this when $x = 0$ yields

$$R(0,L,0^+) = \frac{1}{4} A(0) \tag{20}$$

That is, the leading edge of the reflection kernel yields information about the front edge of the slab. Surely not a suprising result. However, equation (16) makes it possible to step by Δx into the medium and compute

$$R(\Delta x,L,0^+) = \frac{1}{4} A(\Delta x) \tag{21}$$

The process can then be repeated allowing the computation of $A(x)$ for $0 \leq x \leq L$. The algorithm for doing this is given in [5]. Note that this approach allows the computation of a function of space $A(x)$ from a given time trace $R(0,L,t)$.

As a second limiting case assume that the region $x > 0$ is occupied by a spatially homogeneous viscoelastic medium so that $G(x,t) = G(t)$, $A(x) = 0$. Under these conditions $R_x = 0$ and (16) can be reduced to the Volterra equation of the second kind

$$4R(t) + (1-R^*)(1-R^*)h(t) = 0 \tag{22}$$

This equation can be solved for $h(t)$ given $R(t)$. In this problem a function of time $h(t)$ is recovered from a second function of time $R(t)$. As is shown in [1] the relaxation modulus for a finite homogeneous slab can also be recovered from $R(0,L,t)$. This reference also deals with the non-impedance matched case.

The finite homogeneous viscoelastic slab can be approached in another way [13]. If a wave $u^+(0,t)$ is incident on the slab, a transmitted field is produced as well as a reflected field. A transmission kernel and an attenuation factor characterize the transmission as $R(0,L,t)$ characterizes the reflection. Karlsson has shown how knowledge of the transmission kernel can be used to reconstruct the relaxation modulus of a finite viscoelastic slab, including the case of no impedance matching.

In each of the special cases discussed a single function of space or time is recovered from a single function of time. In the general case the viscoelastic medium is characterized by $\rho(x)$ and $G(x,t)$. Clearly more data are needed if the goal is to deduce all material properties from a scattering experiment. Several approaches are possible, among them illuminating a slab from one side but at various angles in order to obtain independent pieces of information.

A first step in this direction has been taken in [11], where the direct and inverse scattering problem associated with the oblique incidence of an SH mode on a viscoelastic slab is treated. In particular a plane wave propagating in the direction $\hat{n} = (\cos(\psi), \sin(\psi), 0)$ impinges on the slab at $(x,y) = (0,0)$ at time $T = 0$. The slab extends from $x = 0$ to $x = L$ and is inhomogeneous in the x direction but homogeneous in the y and z directions. The illuminating wave is assumed to be an SH mode. The displacement is

$$\vec{u} = \hat{z}\, u(\hat{n}\cdot\hat{r} - c_0 T) \tag{23}$$

where c_0 is the S wave speed in the host medium. The scalar wave equation governing the z component of the equation of motion is

$$w_x = \rho\, u_{TT} - \partial_y \partial_T (G * u_y)* \tag{24}$$

Since the target medium is stratified and the incident field is a plane wave this equation can be reduced to an equation in one spatial variable. This is done by employing the change of variables

$$t = T - \frac{\mu}{c_0} y \tag{25}$$

$$t' = T + \frac{\mu}{c_0} y, \qquad \mu = \sin(\psi) \tag{26}$$

The incident field does not depend upon t' and the resulting field equation is

$$w_x(x,t) = \frac{G(x,0)}{c^2(x)} u_{tt}\,(x,t) - \left(\frac{\mu}{c_0}\right)^2 \left[G_t + u_{tt}\right](x,t) \tag{27}$$

where the wave speed $c(x)$ is

$$c^{-2}(x) = \frac{1}{G(x,0)} \left[\rho(x) - G(x,0)\,\left(\frac{\mu}{c_0}\right)^2\right] \tag{28}$$

Using methods similar to those above it can be shown that the wave equation expressed in terms of right and left going fields is of the same form as (6) but now with the $\tilde{\alpha}$, $\tilde{\beta}$, $\tilde{\gamma}$, $\tilde{\delta}$ given by

$$\left[\tilde{\alpha}u^+\right](x,t) = \frac{1}{2}\left[A(x) - \frac{2}{c(x)} - Z(x)J_t* + \frac{1}{Z(x)}\left(\frac{\mu}{c_0}\right)^2 G_t*\right]u^+(x,t) \qquad (29)$$

$$\left[\tilde{\beta}u^-\right](x,t) = \frac{1}{2}\left[Z(x)J_t* + \frac{1}{Z(x)}\left(\frac{\mu}{c_0}\right)^2 G_t* - A(x)\right]u^-(x,t) \qquad (30)$$

$$\left[\tilde{\gamma}u^+\right](x,t) = -\frac{1}{2}\left[Z(x)J_t* + \frac{1}{Z(x)}\left(\frac{\mu}{c_0}\right)^2 G_t* + A(x)\right]u^+(x,t). \qquad (31)$$

$$\left[\tilde{\delta}u^-\right](x,t) = \frac{1}{2}\left[A(x) + \frac{2}{c(x)} + Z(x)J_t* - \frac{1}{Z(x)}\left(\frac{\mu}{c_0}\right)^2 G_t*\right]u^-(x,t). \qquad (32)$$

where $J(x,t)$ is the creep compliance satisfying

$$\partial_t[G(x,t)*J(x,t)] = 1 \qquad (33)$$

Assuming impedance matching, the reflection operator is of the same form as in the normal incidence case. It can be shown that the reflection kernel satisfies

$$\begin{aligned} 2c(x)R_x(x,t) - 4\,R_t(x,t) &= -\,c(x)Z(x)\partial_t\left[(1-R*)(1-R*)J_t\right](x,t) \\ &\quad - c(x)\,\frac{1}{Z(x)}\left(\frac{\mu}{c_0}\right)^2 \partial_t\left[(1+R*)(1+R*)G_t\right](x,t) \qquad (34) \\ &\quad + c(x)A(x)[R*R](x,t) \end{aligned}$$

$$\left.\begin{aligned} R(x,0) &= \frac{1}{4}\,c(x)\left(A(x) + Z(x)J_t(x,0)\left(1 - \left(c(x)\mu/c_0\right)^2\right)\right) \\ R(L,t) &= 0 \end{aligned}\right\}. \qquad (35)$$

These equations are valid up to the time necessary for a wave front to travel across the slab from x = 0 to x = L and return.

If $\rho(x)$ and G(x,t) are known the reflection kernel for the slab can be calculated. The additional flexibility of allowing the incidence angle, and therefore μ, to vary offers the possibility of using information at various angles to recover more information about the slab than is possible from a single normal incidence experiment. Two general considerations, considerations that extend beyond this particular case, must be kept in mind. First, care must be taken to utilize any inversion algorithm only in spatial regions in which the velocity in equation (28) is everywhere non-zero: the critical angle at any depth must be avoided. Second, the inverse problem will become ill-conditioned in an inversion for several functions if the data being used are nearly dependent. In this example this means that the incidence angles used should not be too close together.

A detailed treatment of the use of this approach to recover several functions that characterize a viscoelastic slab is beyond the scope of this paper [11]. It is sufficient to say that the data requirements necessary to recover fully $\rho(x)$ and G(x,t) from plane wave data are excessive, making plane wave illumination unrealistic for this case. However there are more restricted problems that can be treated. For example consider a material that is thermorheologically simple so that

$$G(x,t) = g(a(T(x))t)$$

where now T is temperature and a(T(x)) is a scaling function, the WLF scaling function for example. It can be shown, and tested numerically, that if a known temperature gradient is placed across the slab that both g(y) and a(T(x)) can be recovered from data obtained by illuminating the slab at two different angles. As might be expected the accuracy of the depend upon the angles of illumination.

3. ELASTIC MEDIA

The method, or perhaps better, approach to direct and inverse scattering discussed above can be utilized in the study of stratified elastic media illuminated at oblique incidence. Indeed this was done in [12]. Space restrictions prohibit even a brief description of the details of this work; however several broad statements can be made.

The interesting part of the problem, the P-SV part, is a problem involving a two component wave equation. In terms of the first order system analogous to (6) $\tilde{\alpha}$, $\tilde{\beta}$, $\tilde{\gamma}$ and $\tilde{\delta}$ are each 2x2 matrices since now there are right and left going waves of two distinct types, P and SV. The reflection operator is a 2 x 2 matrix of convolution operators including mode to same mode and mode coupling components. It turns out, not surprisingly, that the P to SV and SV to P reflection operators are related by a scaling. This means that in terms of reflection data only three pieces of information are available at a single angle of illumination, the two mode to same mode information and the single piece of mode conversion information. Fortunately this turns out to be enough information to recover $\lambda(x) + 2\mu(x)$, $\mu(x)$ and $\rho(x)$.

4. REFERENCES

[1] E. Ammicht, J.P. Corones and R.J. Krueger, "Direct and inverse scattering for viscoelastic media," J. Acoust. Soc. Am. 81, No. 4, pp. 827-834. 1987.

[2] R.S. Beezley, "Electromagnetic direct and inverse problems for absorbing media," Ph.D. thesis, University of Nebraska (1985).

[3] R.S. Beezley and R.J. Krueger, "An electromagnetic inverse problem for dispersive media," J. Math. Phys. 26, 317-325 (1985).

[4] J.P. Corones, M.E. Davison and R.J. Krueger, "Wave splittings, invariant imbedding and inverse scattering," in Inverse Optics, Proc. SPIE 413, edited by A.J. Devaney, SPIE, Bellingham, WA, p. 102-106 (1983)

[5] J.P. Corones, M.E. Davison and R.J. Krueger, "The effects of dissipation in one-dimensional inverse problems," in Inverse Optics, Proc. SPIE 413, edited by A.J. Devaney, SPIE, Bellingham, WA, p. 107-114 (1983).

[6] J.P. Corones and R.J. Krueger, "Obtaining scattering kernels using invariant imbedding," J. Math. Anal. Appl. 95, 393-415 (1983).

[7] J.P. Corones, "Wave splitting and invariant imbedding in direct and inverse scattering," in Wave Propagation in Homogeneous Media and Ultrasonic Nondestructive Evaluation, AMD-Vol. 62, edited by G.C. Johnson, ASME, New York, NY, p. 31-35 (1984)

[8] J.P. Corones, R.J. Krueger and V.H. Weston, "Some recent results in inverse scattering theory," in Inverse Problems of Acoustic and Elastic Waves, edited by Santosa, Pao, Sytmes and Holland, SIAM, Philadelphia, PA (1984), p. 65-81.

[9] J.P. Corones, M.E. Davison and R.J. Krueger, "Dissipative inverse problems in the time domain," in Inverse Methods in Electromagnetic Imaging, NATO ASI series, Series C, vol. 143, edited by W-M Boerner, Reidel, Dordrecht, Holland, p. 121-130, (1985).

[10] J.P. Corones, R.J. Krueger and C.R. Vogel, "The effects of noise and bandlimiting in a one dimensional time dependent inverse scattering technique," in Review of Progress in Quantitative Nondestructive Evaluation, Vol. 4, edited by D.O. Thompson and D.E. Chimenti, Plenum, NY (1985), p. 551-558.

[11] J.P. Corones and A. Karlsson,"Transient Direct and Inverse Scattering for Inhomogeneous Viscoelastic Media: Obliquely Incident SH mode" to appear in Inverse Problems.

[12] R. Dougherty, Ph.D. Thesis, Iowa State University (1986).

[13] A. Karlsson, "Inverse scattering for viscoelastic media using transmission data," to appear in Inverse Problems.

[14] G. Kristensson and R.J. Krueger, "Direct and inverse scattering in the time domain for a dissipative wave equation. Part 1: Scattering operators," J. Math. Phys., 27, no. 6, pp. 1667-1682 (1986).

[15] G. Kristensson and R.J. Krueger, "Direct and inverse scattering in the time domain for a dissipative wave equation. Part 2: Simultaneous reconstruction of dissipation and phase velocity profiles," J. Math. Phys., 27, no. 6, pp. 1683-1693 (1986).

[16] G. Kristensson and R.J. Krueger, "Direct and inverse scattering in the time domain for a dissipative wave equation. Part 3: Scattering operators in the presence of a phase velocity mismatch," J. Math. Phys., 28, no. 2, pp. 360-369 (1987).

ELASTIC WAVE PROPAGATION
M.F. McCarthy, M.A. Hayes, (Editors)

Propagation Modeling Based on Wave Field Factorization and Path Integration

JOHN J. MCCOY
THE CATHOLIC UNIVERSITY OF AMERICA
WASHINGTON, DC

INTRODUCTION

The introductions of the "parabolic equation" method and the "split-step Fourier" algorithm to ocean acoustic propagation modeling by Tappert and Hardin in 1973[1], represented a significant advance in our ability to simulate realistic, deep-ocean, acoustic experiments. These also provided impetus for a fair amount of research with the goals of (i) understanding the limits of applicability of the theory and of its solution algorithm, (ii) extending these limits to accomodate more severe experimental scenarios, and (iii) generalizing the theory and its solution algorithm to incorporate additional "physics" of interest to ocean acoustics. In this paper and presentation I shall review and summarize the results obtained in one research effort with these goals. The research is continuing, however, so the story is not complete.

The parabolic wave equation for narrow-band acoustic signals—a time-dependent Shrödinger equation with a distinguished direction, the range direction, in the role of time—is an example of a one-way equation which implicitly incorporates a radiation condition that precludes the possibility of any back-reflection, as this term applies to the range direction. It is intuitive and readily demonstrated by a solution representation in terms of modes, that, the solutions of the governing two-way equation for narrow-band acoustic waves—a Helmholtz equation—can in the absence of variability of the propagation medium with the range coordinate, be collected in two uncoupled sets according to the sense of the projection of the propagation direction on the range direction. The parabolic wave equation is intended to approximately describe the solutions collected in one of the two uncoupled sets. A point to be emphasized is, there is no approximation in the concept of "splitting" the governing two-way equation into a pair of uncoupled, exact one-way equations, in the absence of range variability in the propagation medium. For range-independent media, then, it is reasonable to argue the correct framework for addressing the limits of validity of the parabolic wave equation for reproducing solutions of the Helmholtz equation, to be an exact one-way wave equation that implicitly incorporates a radiation condition. Moreover, this exact one-way formulation also provides the framework for the systematic construction of approximate one-way wave theories in a variety of asymptotic limits, only one of which results in the parabolic equation method.

In this interpretation any range variability is to be reintroduced into the framework of a pair of exact one-way equations through coupling terms. This two-step approach for experiments in which range variability is a significant input, is motivated by experience in realistic ocean acoustic envirous. The experience is that range variability when encountered, is usually much less severe than is variability over the depth coordinate.

The exact one-wave wave theory equivalent to the elliptic Helmholtz equation, is usually motivated by a formal factorization process. This factorization follows from the algebraic structure of the Helmholtz equation, written using an operator notation. In this way the exact parabolic equation contains a certain square root (of a sum of an algebraic and a differential) operator that is identified as pseudodifferential. The algorithmic definition of a pseudodifferential operator is given prescription as an integral operator with a kernel function that requires the introduction of distributions. These distributions are determined in terms of ordinary functions in a Fourier space; the transformed function is referred to as the operator symbol. Fishman and McCoy[2] provide a composition equation that determines the Weyl operator symbol of the square root operator obtained on factoring the Helmholtz equation. McCoy and Frazer[3] provide a further discussion of pseudodifferential operators in the context of one-way equations.

An alternate derivation of one-way wave theory equivalents to the two-way Helmholtz equation, by McCoy and Wales[4], emphasizes that the equivalence, when it can be argued, is an identical integral expressions for the radiation field at a generic location in the propagation medium, taken over an initial plane across which acts the forcing of the problem. The equivalence is on the level of a complete boundary value problem specification. Thus, in the elliptic formulation we must specify not only the problem forcing, over the initial plane, and a radiation condition for the direction normal to the initial plane—the range direction—but also homogeneous auxiliary conditions for crossrange directions. These latter conditions are necessary to complete the specification of a well-posed elliptic equation formulation. In the exact corresponding parabolic wave theory, these crossrange conditions are incorporated in the square root operator obtained in the formal factorization.

The two derivations and interpretations of one-way wave theory equivalents to a governing two-way formulation are complementary in emphasizing different aspects of the equivalence. Brief summaries of the derivations are provided in the next two sections.

A principal advantage of propagation models described by one-way wave equations is that they admit of solution algorithms by marching. That the crossrange operators in the equations to be marched are pseudodifferential complicates the numerics somewhat, because of the singular nature of these operators expressed as integral operators. The derivation of a marching algorithm that is accomplished in a combined physical/Fourier space referred to a Fourier space is summarized in Section 4. The obtained marching algorithm is of particular interest for two reasons. In the special case of a Shrödinger equation as the governing one-way equation, the marching algorithm is identified with the split-step Fourier algorithm of Tappert and Hardin. Also, for any one-way wave equation, the cascading of the marching algorithm over N range steps of range increment Δx results in a representation of a fundamental propagator of the equation as a functional integral interpreted as an integral over paths in phase space. This identification has not been exploited in any of the work we have accomplished to date. However, there are some interesting speculations that the line of inquiry suggested by the identification will prove fruitful both in addressing questions of theoretical interest and as a basis for computational algorithms.

The framework for constructing exact and approximate one-way wave equations and for marching the radiation field governed by them, is hardly the end results for the goal of propagation models expressed as computer programs. This latter requires the construction of specific approximate theories and the implementations of the marching algorithm for these theories. Finally, the proof of the efficacy of any general purpose propagation model is in a test program of its predictions for a series of problems designed to mirror realistic experimental scenarios, and to explore those scenarios for which the predictions begin to degrade significantly. This last task is not a trivial one and is often not carried out very enthusiastically. Space requirements are such that the results of such a program, by Fishman, McCoy and Wales[5], are not reproduced here although they were part of the oral presentation.

The brief history of the research accomplished and summarized in the following pages suggest that a fair amount has been accomplished toward satisfying the first two of the three goals of the efforts. The ongoing work is mostly directed toward generalizing the theory to accomodate additional physics of interest to ocean acoustics. Three particular "additional physics" that have attracted most of our attention are (i) the accomodation of shear waves and the possibility of P and SV coupling in a one-way wave theory equivalent to the equations of linear elastodynamics[6–8] (ii) the accomodation of a back-scattering in the presence of range variability in the two-step approach referred to above[9–11], and (iii) a one-way wave theory for broad-band signal propagation. The first two of these are of particular interest to ocean acoustic experiments in which the ocean bottom plays a significant role.

Factorization of the Helmholtz Equation

The more usual methods of deriving parabolic formulations are based on the algebraic structure of the governing elliptic equation. Thus, consider the Helmholtz equation to be governing,

$$[\partial_x^2 + \nabla_q^2 + k_0^2 K^2(\vec{q})]\, \phi(\vec{q}, x) = 0. \tag{1}$$

The inhomogeneous wave speed field is shown to be independent of one direction, the range direction x. The algebraic structure is emphasized with the introduction of an operator notation. We rewrite Eq. (1) as

$$[\partial_x^2 + k_0^2 \mathbf{B}^2(\vec{q})]\, \phi(\vec{q}, x) = 0, \tag{2}$$

where the operator $\mathbf{B}^2(q)$ is given by,

$$\mathbf{B}^2(\vec{q}) = (1/k_0^2)\nabla_q^2 + K^2(\vec{q}). \tag{3}$$

The factoring of the Helmholtz equation is formally accomplished in writing,

$$[(i/k_0)\partial_x + \mathbf{B}]\,[(-i/k_0)\partial_x + \mathbf{B}]\, \phi(\vec{q}, x) = 0. \tag{4}$$

To make this formal equation algorithmic requires an algorithmic definition be given the *square root* operator, $\mathbf{B}$. At this point, it is common to introduce approximation, by formally expanding the square root. One suggested expansion which leads to the ordinary parabolic wave theory as a first approximation, would be in powers of $(K^2(\vec{q}) - K_0^2) + (1/k_0^2)\nabla_q^2$. Other expansions (in terms of Pade's, for example) have also been used. Fishman and McCoy[1] explicitly showed how $\mathbf{B}$ could be given an exact algorithmic definition as a pseudodifferential operator. We briefly present the following calculation, which is equivalent to Fishman and McCoy but which contains an intermediate step that might make the derivation more accessible to the reader not familiar with the language of pseudodifferential equations.

What is required to make the operator algorithmic is first a representation and second a condition expressed as an equation. It is readily demonstrated (See Fishman and McCoy[1].) that the square root of a differential operator is, in general, nonlocal. Thus, an integral representation is indicated,

$$\mathbf{B}f(\vec{q}) \Leftrightarrow \int B(\vec{q}, \vec{q}_1)\, f(\vec{q}_1)\, d\vec{q}_1. \tag{5}$$

The condition that results in an equation is that applying the square root operator twice results in an operator for which we have an algorithmic prescription. Thus the equation is

$$\mathbf{B} \times \mathbf{B} = \mathbf{B}^2. \tag{6}$$

Using Eq. (6) we write an equation on $B(\vec{q}, \vec{q}_1)$ by substituting the representation of Eq. (5) into the left hand side and by expressing the known $\mathbf{B}^2$ operator on the right hand side in this same representation. Since the $\mathbf{B}^2$ is local, the second part of this substitution requires the introduction of distributions. The equation on $B(q, q_1)$ is written,

$$\int B(\vec{q}, \vec{q}\,')B(\vec{q}\,', \vec{q}_1)\, d\vec{q}\,' = K^2(\vec{q})\delta(\vec{q} - \vec{q}_1) + (1/k_0^2)\delta^{(2)}(\vec{q} - \vec{q}_1), \tag{7}$$

where δ is the delta function distribution and $\delta^{(2)}$ is the second derivative of the delta function distribution. Equation (7) is, of course, not directly algorithmic until we have a procedure for solving integral equations forced by distributions. Fortunately, we do have a procedure for the representation of distributions by ordinary functions, i.e. by taking their Fourier transform. If we first introduce a coordinate transformation to average and difference coordinates and then Fourier transform with respect to the difference coordinate, we remove the distributions from Eq. (7) and obtain the following equation of the Fourier transform of $B(\vec{q}, \vec{q}_1)$

$$\begin{aligned}(\frac{1}{\pi^4}) \iiiint \Omega_B^{(W)}\, (\vec{p} + \vec{u}, \vec{q} + \vec{r})\Omega_B^{(W)}\, (\vec{p} + \vec{v}, \vec{p} + \vec{s})e^{[2ik_0(\vec{u}\cdot\vec{s} - \vec{r}\cdot\vec{v})]}\, d\vec{r}\, d\vec{s}\, d\vec{u}\, d\vec{v} \\ = K^2(\vec{q}) - p^2.\end{aligned} \tag{8}$$

The function $B(\vec{q}, \vec{q}_1)$ is obtained from the function $\Omega_B^{(W)}(\vec{p}, \vec{q})$ using the Fourier transform relation

$$B(\vec{q}, \vec{q}_1) = (\frac{k_0}{2\pi}) \int \Omega_B^{(W)}(\vec{p}, (\vec{q} + \vec{q}_1)/2)e^{[ik_0(\vec{p}\cdot(\vec{q} - \vec{q}_1))]}\, d\vec{p}. \tag{9}$$

The function $\Omega_B^{(W)}(\vec{p}, \vec{q})$ is termed the Weyl symbol of the operator $\mathbf{B}$. The name Weyl refers to the keeping of the averaged coordinate fixed in accomplishing the Fourier transform relation to obtain $B(\vec{q}, \vec{q}_1)$.

DERVIATION OF AN EQUIVALENT PARABOLIC FORMULATION OF AN ELLIPTIC BOUNDARY VALUE PROBLEM

Consider the following boundary value problem. The governing equation is the Helmholtz equation, Eq. (1); the domain is rectangular, the problem forcing is through an inhomogeneous boundary condition specified across one of the boundary planes. We will assume for definiteness that this inhomogeneous boundary condition is expressed as specified values for the radiation field at all points across the plane. Other type boundary conditions will be discussed after the derivation is completed. To complete the problem specification, homogeneous boundary conditions are required to be satisfied across the remaining boundary planes. On making use of Green's theorem it is possible to write a representation of the solution of the boundary value problem as an integral taken over the *source plane*, i.e.

$$\phi(\vec{x}) = \int G^{+}(\vec{x};\vec{q},0)\phi(\vec{q},0)d\vec{q}, \tag{10}$$

where $\vec{x}$ denotes a generic three dimensional position vector and $\vec{q}$ denotes a two dimensional position vector. The function $G^{+}(\vec{x};\vec{q},0)$ is a surface Green's function for the specified boundary value problem. It is the solution of the originally specified problem, modified to have a forcing across the source plane that is described by a distribution. Thus, $G^{+}(\vec{x};\vec{q},0)$ contains all of the possible multiple reflections that would occur at the boundary surfaces or as a result of an inhomogeneous propagation wave speed throughout the domain.

Consider that the propagation medium in the above described problem is characterized by a wave speed field that is translationally invariant with respect to position measured normal to the forcing plane; i.e., a range independent medium; and that the homogeneous boundary condition specified for the plane opposite to the source plane is a radiation condition; i.e., there is no reflection of energy at points across this plane. For these special conditions, the solution field possesses a *semigroup* property that allows another representation of the solution field,

$$\phi(\vec{x}) = \iint G^{+}(\vec{x};\vec{q}_1,x_1)G^{+}(\vec{q}_1,x_1;\vec{q},0)\phi(\vec{q},0)d\vec{q}_1\, d\vec{q}. \tag{11}$$

Using Eq. (11) and integrating first over the intermediate plane identified by $0<x_1<x$ reproduces the representation of Eq. (10). Integrating first over the source plane however, results in a two stage solution representation. As indicated by the representation, the same surface Green's function applies for the two integrations required by Eq.(11). A demonstration of the semigroup property is trivially obtained from a normal mode representation of the field.

The semigroup property expressed in Eq. (11) can next be used to construct a solution algorithm in the form of a marching over a sequence of intermediate range steps. It is clear that this algorithm has the property that the solution field across one of the arbitrarily chosen intermediate range planes is determined in terms of the solution field across the immediately preceding arbitrarily chosen range plane. One can write the relation

$$\phi(\vec{q},x+\Delta x) = \int G^{+}(\vec{q},x+\Delta x;\vec{q}_1,x)p(\vec{q}_1,x)d\vec{q}_1. \tag{12}$$

Again it is to be noted that the same boundary surface Green's function applies to all range planes. The marching property, when taken to the limit in which the separation distance between the incremented range plane vanishes, has an equivalent representation as a first order integro-differential equation, which can be obtained on taking the derivative of Eq. (12) with respect to $x + \Delta x$ and taking the limit of Δx vanishing. We write,

$$(i/k_0)\partial_x \phi(\vec{q}, x) = \int B(\vec{q}, \vec{q}_1)\phi(\vec{q}_1, x)dq_1, \tag{13}$$

where

$$B(\vec{q}, \vec{q}_1) = (i/k_0) \lim_{x \to 0} \partial_x G^+(\vec{q}, x; \vec{q}_1, 0). \tag{14}$$

In writing the expression for $B(\vec{q}, \vec{q}_1)$ use has been made of the fact G^+ does not depend on absolute position in range.

The two derivations are complementary in emphasizing different aspects of the equivalence of one- and two-way formulations. The factorization derivation enjoys a very significant advantage in representing a complete theory that provides an explicit equation to be solved to make the exact square root operator explicit. The advantage can be claimed principally for a number of mathematical tasks required to provide a rigorous basis for the entire process and for obtaining error estimates for approximate factorizations. On the other hand, since the factorization explicitly treats only the governing equation it obscures the fact that the resulting one-way wave equation actually incorporates the boundary conditions of a specific elliptic formulation. It is, of course, obvious that the one-way equation incorporates a radiation condition as this applies to the range coordinate. Less obvious is that the specific square root operator that effects the factorization also incorporates the nature of homogeneous, crossrange boundary conditions. Moreover, there is some freedom (ambrguity) in the choice of the specific field measure in terms of which the one-way formulation is to be expressed. All of these aspects are seen much more clearly in the construction procedure of the last section.

Although the conceptual importance of the role of boundary conditions in a specific one-way formulation is clear, the practical importance may be less so. The reason is that the practical importance may be limited so long as one considers only scalar radiation fields. Extending the results achieved to apply to the governing equations of linear elastodynamics has provided a challenge that to date, has not been successfully met. A key to a future successfuly approach could be the manner of incorporating crossrange boundary conditions and of the choice of a field measure to be marched. Preliminary thoughts on these issues are given in McCoy and Wales[5].

Derivation of a Marching Algorithm

The most straightforward approach to the deriviation of algorithms for marching a wave field that is governed by a one-way equation, is based on a forward difference approximation of the range derivative. The marching algorithm is written

$$\begin{aligned}\phi(\vec{q}, x + \Delta x) &= \phi(\vec{q}, x) + ik_0 \Delta x \mathbf{B}(\vec{q})\phi(\vec{q}, x) \\ &= \phi(\vec{q}, x) + ik_0 \Delta x \int B(\vec{q}, \vec{q}_1)\phi(\vec{q}_1, x)d\vec{q}_1 \ .\end{aligned} \tag{15}$$

Two obvious questions must be addressed in considering the implementation of this marching algorithm. What estimate can be given to the errors involved in marching the field? And, how does one numerically evaluate the required integration, given the singular nature of the kernel function, $B(\vec{q},\vec{q}_1)$? Accepting that the error in replacing a first-order derivative by a forward difference approximation to be of the order of $(\Delta x)^2$, the error in marching the wave field through one range step can be taken to be of this order. This, however, is not the error estimate that is important. The real question applies to the error in marching N range steps where N must become unbounded as Δx becomes infinitesimal, if the wave field is to be marched a finite distance. Thus the question of the manner of accumulation of error becomes as important as an estimate of the error for a single range step. Further with regard to the particular application that motivates our investigations, the one-way wave equation to be marched in any specific algorithm is itself only an approximation to an exact one-way equation that applies to the governing two-way (Helmholtz) theory. Thus, in specific algorithms the **B** operator that determines the one-way theory must be considered to be in error. An estimate of the error in marching the radiation field over one range step, which is a result of the error in the prescription of **B** vanishes as Δx to the first power. The task of developing rigorous global error estimates for approximate one-way wave theories and for a marching algorithm for the radiation field raises questions of some mathematical subtlety. To this point we have not attempted to address these issues directly, accepting a more pragmatic approach in which we test the numerical predictions of the end product of the effort—a computer program—in a number of experimental scenarios.

The motivation presented in an earlier section, for describing the **B** operator using an operator symbol instead of the kernel function $B(\vec{q},\vec{q}_1)$, was the singular nature of the kernel function. This same motivation now suggests a Fourier transformation be introduced in the equation that describes the marching. In doing this, we first introduce a delta function distribution so as to represent the first term on the right-hand side as an integral and further introduce the following representation for the delta function distribution:

$$\delta(\vec{q}-\vec{q}_1) = (\frac{k_0\pi}{2})^2 \int e^{ik_0\vec{p}\cdot(\vec{q}-\vec{q}_1)}d\vec{p}\,. \tag{16}$$

This and the expression for $B(\vec{q},\vec{q}_1)$ in terms of the Weyl symbol, for example, gives

$$\phi(\vec{q},x+\Delta x) = (\frac{k_0\pi}{2})^2 \iint [1+ik_0\Delta x\Omega_B^{(W)}(\vec{p},\frac{\vec{q}+\vec{q}_1}{2})]e^{ik_0\vec{p}\cdot(\vec{q},-\vec{q}_1)}\phi(\vec{q}_1,x)d\vec{q}\,d\vec{p}\,. \tag{17}$$

Equation (17) with the $\vec{p}$ integral accomplished first clearly reproduces Eq. (15), a marching algorithm requiring a single-fold integration for each cross-range coordinate. Equation (17) with the $\vec{q}_1$ integral accomplished first leads to a marching algorithm that is easily distinguished from Eq. (15)—it requires a two-fold integration for each cross-range coordinate.

Continuing, we next exponentiate the terms in Eq. (17). The result is written

$$\phi(\vec{q},x+\Delta x) = (\frac{k_0\pi}{2})^2 \iint e^{ik_0[\vec{p}\cdot(\vec{q}-\vec{q}_1)+\Delta x\Omega_B^{(W)}(\vec{p},\frac{\vec{q}+\vec{q}_1}{2})]}\phi(\vec{q}_1,x)d\vec{q}_1\,d\vec{p} \tag{18}$$

The "error" introduced by the exponentiation can be estimated by noting that an expansion and truncation of Eq. (18) reproduces Eq. (17), if one retains only terms of order Δx. The error in marching a single range step remains one that behaves as $(\Delta x)^2$.

This marching algorithm, Eq. (18) with the $\vec{q}_1$ integration accomplished first, is made somewhat inconvenient by the appearance of the $\vec{q}_1$ variable in the Weyl symbol. This inconvenience can be removed, however, on recognizing that the Weyl symbol is obtained from $B(\vec{q}, \vec{q}_1)$ as a Fourier transformation with respect to $\vec{q} - \vec{q}_1$ *while holding* $(\vec{q} + \vec{q}_1)/2$ *constant*. An alternate representation of the same **B** operator can be expressed by another symbol, the Standard symbol, obtained as a Fourier transform of $B(\vec{q}, \vec{q}_1)$ with respect to $\vec{q} - \vec{q}_1$ *while holding* $\vec{q}$ *a constant*. The functional dependence of the Standard symbol, $\Omega_B^{(S)}(\vec{p}, \vec{q})$, on the $\vec{p}$ and $\vec{q}$ coordinates, is of course different than is the dependence of $\Omega_B^{(W)}(\vec{p}, (\vec{q} + \vec{q}_1)/2)$ on the $\vec{p}$ and $(\vec{q}_1 + \vec{q}_1)/2$ coordinates. There is a precise relation connecting the two functional dependences, which follows from the two definitions in terms of the singular kernel function $B(\vec{q}, \vec{q}_1)$. McCoy and Frazer[3] discuss this aspect from this perspective in greater detail. The advantage in removing the $\vec{q}_1$ variable from the operator symbol in Eq. (18) is that the $\vec{q}_1$ integration can be accomplished; it is reduced to a Fourier transform applied to $\phi(\vec{q}, x)$. The marching solutions reported by Fishman, McCoy and Wales[5] and discussed in the oral version of this presentation, are based on using the Standard symbol in Eq. (18).

It is to be noted that there is no general way of similarly removing the $\vec{q}$ coordinate when accomplishing the integration over $\vec{p}$. The special case in which the one-way wave equation is a Shrödinger equation is an exception, in that this equation corresponds to an operator symbol that is a sum of a function of $\vec{p}$ plus a function of $\vec{q}$. This separable form enables removal of the $\vec{q}$ variable when accomplishing the $\vec{p}$ integration, which is reduced thereby to an inverse Fourier transform. This special algorithm is the split-step Fourier algorithm of Tappert and Hardin[1].

A final comment pertains to the derivation of a phase path integral representation of a full-range propagator of the one-way wave equation, based on Eq. (18). This derivation is provided by McCoy and Frazer[3].

REFERENCES

[1] Tappert, F.D., "The parabolic approximation method," *Wave Propagation in Underwater Acoustics*, Lecture notes is Physics No. 70, edited by J. B. Keller and J. S. Papadakis (Springer-Verlag, New York 1977). This article contains a brief historical survey of the earlier work on the parabolic approximation method.

[2] Fishman, L. and McCoy, J.J., "Derivation and application of extended parabolic wave theories. Part I. The factorized Helmholtz equation," *J. Math. Phys.* Vol. 25 (Z) pp. 285-296 (1984).

[3] McCoy, J.J. and Frazer, L.N., "Pseudodifferential operators, operator orderings, marching algorithms and path integrals for one-way wave equations," *Wave Motion*, Vol. 9, pp. 413-427 (1987).

[4] McCoy, J.J. and Wales, S.C., "Another interpretation and derivation of one-way wave theories," paper to be submitted.

[5] Fishman, L., McCoy, J.J. and Wales. S.C., "Factorization and path integration of the Helmholtz equation: Numerical algorithms," *J. Acoust. Soc. Am.* Vol. 81(5), pp. 1355-1376 (1987).

[6] McCoy, J.J., "A parabolic theory of stress wave propagation through inhomogeneous linearly elastic solids," *J. Appl. Mech.*, Vol. 44, pp. 462-468 (1977).

[7] Wales, S.C. and McCoy, J.J., "A comparison of parabolic wave theories for linearly elastic solids," *Wave Motion* Vol. 5, pp. 99-113 (1983).

[8] Wales, S.C., "Wide-angle propagation in inhomogeneous fluids and elastic solids," Doctoral dissertation, (May 1988).

[9] McCoy, J.J., Fishman, L. and Frazer, L.N., "Reflection and transmission at an interface separating transversely inhomogeneous acoustic half-spaces," *Geophys. J. Royal Astro. Soc.*, Vol. 85, pp. 543-562 (1986).

[10] McCoy, J.J. and Frazer, L.N., " Propagation modelling based on wave field factorization and invariant imbedding," *Geophys. J. Royal Astro. Soc.*, Vol. 86, pp. 703-717 (1986).

[11] McCoy, J.J. and Wales, S.C., " A range-dependent propagation model that incorporates back-reflection," to be submitted.

ELASTIC WAVE PROPAGATION
M.F. McCarthy, M.A. Hayes, (Editors)

ELASTODYNAMIC TIME-DOMAIN RECIPROCITY THEOREMS FOR SOLIDS WITH RELAXATION

Adrianus T. de Hoop and Hendrik J. Stam

Delft University of Technology, Faculty of Electrical Engineering,
Laboratory of Electromagnetic Research, P.O. Box 5031, 2600 GA Delft,
The Netherlands

Time-domain reciprocity theorems of the time-convolution and the time-correlation type for elastodynamic wave fields in linear, time-invariant, and locally reacting solids are discussed. Inhomogeneity, anisotropy, and arbitrary relaxation effects, both of the active (anti-causal) and passive (causal) kind, are included. The analysis is entirely carried out in space-time, without intermediate recourse to the frequency or the wavevector domains. The application to inverse source problems is briefly indicated.

1. INTRODUCTION

A wave field reciprocity theorem interrelates, in a specific manner, the quantities that characterize two admissible physical states that could occur in one and the same domain in space-time. The present investigation deals with time-convolution and time-correlation reciprocity theorems for elastodynamic wave fields in time-invariant configurations that are linear and locally reacting in their elastodynamic behavior. The space-time geometry in which the two admissible states occur, is the Cartesian product $D \times R$ of a time-invariant spatial domain $D \subset R^3$ and the real time axis R. Further, the constitutive parameters of the media present in the two states are time invariant and independent of the wave field values. No further restrictions are imposed. Inhomogeneity and arbitrary anisotropy are included, as well as arbitrary relaxation effects. Both the time-convolution and the time-correlation type of reciprocity theorem have an important field of application in inverse source and related problems. These applications will, from a general point of view, briefly be indicated.

The position of observation in R^3 is specified by the coordinates $\{x_1,x_2,x_3\}$ with respect to a fixed, orthogonal, Cartesian reference frame with origin $\underline{O}$ and the three mutually perpendicular base vectors $\{\mathbf{i}_1,\mathbf{i}_2,\mathbf{i}_3\}$ of unit length each. In the indicated order the base vectors form a right-handed system. The subscript notation for Cartesian vectors and tensors in R^3 is employed and the summation convention applies. The corresponding lower-case Latin subcripts are to be assigned the values $\{1,2,3\}$. Whenever appropriate, the position vector will be denoted by $\mathbf{x} = x_p\mathbf{i}_p$. The time coordinate is denoted by t. Partial differentiation is denoted by ∂; ∂_p denotes the differentiation with respect to x_p, ∂_t denotes the differentiation with respect to t.

The reciprocity theorems will be derived for bounded domains D. In the analysis also the boundary ∂D of D occurs. The unit vector along the normal to ∂D is denoted by ν_m; it points away from D.

2. SOME PROPERTIES OF THE TIME CONVOLUTION AND THE TIME CORRELATION OF SPACE-TIME FUNCTIONS

Let $f_1 = f_1(\mathbf{x},t)$ and $f_2 = f_2(\mathbf{x},t)$ be two transient space-time functions. By this we mean that the functions are absolutely integrable on the entire $t \in R$. Then, the time convolution of f_1 and f_2 is defined as

$$C(f_1,f_2;\mathbf{x},\tau) = \int_{t\in R} f_1(\mathbf{x},t)f_2(\mathbf{x},\tau - t)dt \tag{2.1}$$

and the time correlation of f_1 and f_2 as

$$R(f_1,f_2;\mathbf{x},\tau) = \int_{t\in R} f_1(\mathbf{x},t)f_2(\mathbf{x},t - \tau)dt. \tag{2.2}$$

Let $\bar{f}$ denote the time-reversed of f, i.e.,

$$\bar{f}(\mathbf{x},t) = f(\mathbf{x},-t), \tag{2.3}$$

then, it follows from (2.1) - (2.3) that

$$R(f_1,f_2;\mathbf{x},\tau) = C(f_1,\bar{f}_2;\mathbf{x},\tau). \tag{2.4}$$

Using (2.1), we obtain the property

$$C(\bar{f}_1,f_2;\mathbf{x},\tau) = \bar{C}(f_1,\bar{f}_2;\mathbf{x},\tau). \tag{2.5}$$

For the time derivative of the time convolution the rules

$$\partial_\tau C(f_1,f_2;\mathbf{x},\tau) = C(f_1,\partial_t f_2;\mathbf{x},\tau) = C(\partial_t f_1,f_2;\mathbf{x},\tau) \tag{2.6}$$

apply. In view of the property

$$\overline{\partial_t f} = - \partial_t \bar{f}, \tag{2.7}$$

the time derivatives of the time correlation are taken care of by using

$$\partial_\tau C(f_1,\bar{f}_2;\mathbf{x},\tau) = C(\partial_t f_1,\bar{f}_2;\mathbf{x},\tau) = - C(f_1,\overline{\partial_t f_2};\mathbf{x},\tau). \tag{2.8}$$

For the incorporation of relaxation effects in the reciprocity theorems we also need the time convolution of three space-time functions. For this, either of the definitions

$$C(f_1,f_2,f_3;\mathbf{x},\tau) = C(f_1,C(f_2,f_3);\mathbf{x},\tau) = C(C(f_1,f_2),f_3;\mathbf{x},\tau) \tag{2.9}$$

holds. In view of its simpler properties, the time convolution concept is used throughout the entire subsequent derivations, i.e., both for the time convolution and for the time correlation reciprocity theorems.

3. PROPERTIES OF THE ELASTODYNAMIC WAVE FIELD IN THE CONFIGURATION

In each subdomain of the configuration where the elastodynamic properties vary continuously with position, the elastodynamic wave field quantities are continuously differentiable and satisfy the equations

$$-\Delta_{k,m,p,q}\partial_m\tau_{p,q} + \dot{\Phi}_k = f_k, \tag{3.1}$$

$$\Delta_{i,j,m,r}\partial_m v_r - \dot{e}_{i,j} = h_{i,j}, \tag{3.2}$$

where $\tau_{p,q}$ = stress (Pa), v_r = particle velocity ($m\cdot s^{-1}$), $\dot{\Phi}_k$ = mass flow density rate ($kg\cdot m^{-2}\cdot s^{-2}$), $\dot{e}_{i,j}$ = deformation rate (s^{-1}), f_k = volume source density of force ($N\cdot m^{-3}$), $h_{i,j}$ = volume source density of strain rate (s^{-1}), $\Delta_{i,j,m,r} = (\delta_{i,m}\delta_{j,r} + \delta_{i,r}\delta_{j,m})/2$, and $\delta_{i,m}$ is the symmetrical unit tensor of rank two (Kronecker tensor). Equations (3.1) and (3.2) are supplemented by the constitutive relations. For a linear, time-invariant, locally reacting solid these are, using the notation of (2.1),

$$\dot{e}_{i,j}(\mathbf{x},t) = \partial_t C(\kappa_{i,j,p,q},\tau_{p,q};\mathbf{x},t), \tag{3.3}$$

$$\dot{\Phi}_k(\mathbf{x},t) = \partial_t C(\gamma_{k,r},v_r;\mathbf{x},t), \tag{3.4}$$

where $\kappa_{i,j,p,q}$ = compliance relaxation function ($Pa^{-1}\cdot s^{-1}$), and $\gamma_{k,r}$ = inertia relaxation function ($kg\cdot m^{-3}\cdot s^{-1}$). In (3.3) and (3.4), inhomogeneity, anisotropy and relaxation of the solid are included. If $\{\kappa_{i,j,p,q},\gamma_{k,r}\}(\mathbf{x},\tau) = 0$ when $\tau<0$, the solid at $\mathbf{x}$ is causal. If $\kappa_{i,j,p,q}(\mathbf{x},\tau) = s_{i,j,p,q}(\mathbf{x})\delta(\tau)$, $\gamma_{k,r}(\mathbf{x},\tau) = \rho_{k,r}(\mathbf{x})\delta(\tau)$, where $\delta(\tau)$ is the unit impulse (Dirac distribution), the solid is instantaneously reacting, and $s_{i,j,p,q}$ and $\rho_{k,r}$ are its compliance and its (tensorial) volume density of mass, respectively. If $\{\kappa_{i,j,p,q},\gamma_{k,r}\}(\mathbf{x},\tau) = 0$ when $\tau>0$, the solid is anticausal or effectual. For our reciprocity theorems no specific type of relaxation function is presupposed. It is assumed that $\kappa_{i,j,p,q}$ and $\gamma_{k,r}$ are piecewise continuous functions of position. At an interface between two different solids, at which we assume the solids to be in rigid contact, the constitutive parameters jump by finite amounts, but the traction (i.e., the normal component of the stress) and the particle velocity are continuous. If an elastically impenetrable object is present, either the traction (at a void) or the particle velocity (at an immovable rigid object) has zero value at its boundary.

The two states that occur in the reciprocity theorems will be denoted by the superscripts 'a' and 'b', respectively.

4. THE RECIPROCITY THEOREM OF THE TIME-CONVOLUTION TYPE

The reciprocity theorem of the time-convolution type follows upon considering the interaction quantity $\Delta_{m,r,p,q}[C(-\tau^a_{p,q},v^b_r;\mathbf{x},\tau) - C(-\tau^b_{p,q},v^a_r;\mathbf{x},\tau)]$. Using (3.1) - (3.4) for each of the two states we arrive at

$$\Delta_{m,r,p,q}\partial_m[C(-\tau^a_{p,q},v^b_r;\mathbf{x},\tau) - C(-\tau^b_{p,q},v^a_r;\mathbf{x},\tau)]$$
$$= \partial_\tau C(\gamma^b_{r,k} - \gamma^a_{k,r},v^a_r,v^b_k;\mathbf{x},\tau) - \partial_\tau C(\kappa^b_{p,q,i,j} - \kappa^a_{i,j,p,q},\tau^a_{p,q},\tau^b_{i,j};\mathbf{x},\tau)$$
$$+ C(f^a_r,v^b_r;\mathbf{x},\tau) + C(-\tau^a_{p,q},h^b_{p,q};\mathbf{x},\tau) - C(f^b_r,v^a_r;\mathbf{x},\tau) - C(-\tau^b_{p,q},h^a_{p,q};\mathbf{x},\tau). \tag{4.1}$$

Equation (4.1) is the local form of the time-convolution reciprocity theorem. The first two terms at the right-hand side are representative for the differences in the properties of the solids present in the two states. If $\gamma^b_{r,k}(\mathbf{x},\tau) = \gamma^a_{k,r}(\mathbf{x},\tau)$ and $\kappa^b_{p,q,i,j}(\mathbf{x},\tau) = \kappa^a_{i,j,p,q}(\mathbf{x},\tau)$ for all $\tau \in R$, these terms vanish and the two media are denoted as each other's adjoints. Note that the adjoint of a causal (effectual) medium is a causal (effectual) one. Integrating (4.1) over the subdomains of D where both sides are continuously differentiable, applying Gauss' divergence theorem to the resulting left-hand sides, and adding the results, we obtain

$$\int_{\mathbf{x}\in\partial D} \Delta_{m,r,p,q}\nu_m[C(-\tau^a_{p,q},v^b_r;\mathbf{x},\tau) - C(-\tau^b_{p,q},v^a_r;\mathbf{x},\tau)]dA$$
$$= \int_{\mathbf{x}\in D} [\partial_\tau C(\gamma^b_{r,k} - \gamma^a_{k,r},v^a_r,v^b_k;\mathbf{x},\tau)$$
$$- \partial_\tau C(\kappa^b_{p,q,i,j} - \kappa^a_{i,j,p,q},\tau^a_{p,q},\tau^b_{i,j};\mathbf{x},\tau)]dV$$
$$+ \int_{\mathbf{x}\in D} [C(f^a_r,v^b_r;\mathbf{x},\tau) + C(-\tau^a_{p,q},h^b_{p,q};\mathbf{x},\tau)$$
$$- C(f^b_r,v^a_r;\mathbf{x},\tau) - C(-\tau^b_{p,q},h^a_{p,q};\mathbf{x},\tau)]dV. \tag{4.2}$$

Equation (4.2) is the global form, for the domain D, of the time-convolution reciprocity theorem. Note that in the left-hand side the contributions from interfaces between different solids present in D have cancelled and that the contributions from the boundaries of elastically impenetrable objects present in D have vanished in view of the boundary conditions stated in Section 3.

5. THE RECIPROCITY THEOREM OF THE TIME-CORRELATION TYPE

The reciprocity theorem of the time-correlation type follows upon considering the interaction quantity $\Delta_{m,r,p,q}[R(-\tau^a_{p,q},v^b_r;\mathbf{x},\tau) + R(-\tau^b_{p,q},v^a_r;\mathbf{x},-\tau)]$ $= \Delta_{m,r,p,q}[C(-\tau^a_{p,q},\bar{v}^b_r;\mathbf{x},\tau) + C(-\bar{\tau}^b_{p,q},v^a_r;\mathbf{x},\tau)]$. Using (3.1) - (3.4) for each of the two states, we arrive at

$$\Delta_{m,r,p,q}\partial_m[C(-\tau^a_{p,q},\bar{v}^b_r;\mathbf{x},\tau) + C(-\bar{\tau}^b_{p,q},v^a_r;\mathbf{x},\tau)]$$
$$= \partial_\tau C(\bar{\gamma}^b_{r,k} - \gamma^a_{k,r},v^a_r,\bar{v}^b_k;\mathbf{x},\tau) + \partial_\tau C(\bar{\kappa}^b_{p,q,i,j} - \kappa^a_{i,j,p,q},\tau^a_{p,q},\bar{\tau}^b_{i,j};\mathbf{x},\tau)$$
$$+ C(f^a_r,\bar{v}^b_r;\mathbf{x},\tau) + C(-\tau^a_{p,q},\bar{h}^b_{p,q};\mathbf{x},\tau) + C(\bar{f}^b_r,v^a_r;\mathbf{x},\tau) + C(-\bar{\tau}^b_{p,q},h^a_{p,q};\mathbf{x},\tau). \tag{5.1}$$

Equation (5.1) is the local form of the time-correlation reciprocity theorem. The first two terms at the right-hand side are representative for the differences in the properties of the solids present in the two states. If $\bar{\gamma}^{b}_{r,k}(\mathbf{x},\tau) = \gamma^{a}_{k,r}(\mathbf{x},\tau)$ and $\bar{\kappa}^{b}_{p,q,i,j}(\mathbf{x},\tau) = \kappa^{a}_{i,j,p,q}(\mathbf{x},\tau)$ for all $\tau \in R$, these terms vanish and the two media are denoted as each other's time-reverse adjoints. Note that the time-reverse adjoint of a causal (effectual) medium is an effectual (causal) one. Integrating (5.1) over the subdomains of D where both sides are continuously differentiable, applying Gauss' divergence theorem to the resulting left-hand sides, and adding the results, we obtain

$$\int_{\mathbf{x}\varepsilon\partial D} \Delta_{m,r,p,q}\nu_m[C(-\tau^{a}_{p,q},\bar{v}^{b}_{r};\mathbf{x},\tau) + C(-\bar{\tau}^{b}_{p,q},v^{a}_{r};\mathbf{x},\tau)]dA$$
$$= \int_{\mathbf{x}\varepsilon D} [\partial_\tau C(\bar{\gamma}^{b}_{r,k} - \gamma^{a}_{k,r},v^{a}_{r},\bar{v}^{b}_{k};\mathbf{x},\tau)$$
$$+ \partial_\tau C(\bar{\kappa}^{b}_{p,q,i,j} - \kappa^{a}_{i,j,p,q},\tau^{a}_{p,q},\bar{\tau}^{b}_{i,j};\mathbf{x},\tau)]dV$$
$$+ \int_{\mathbf{x}\varepsilon D} [C(f^{a}_{r},\bar{v}^{b}_{r};\mathbf{x},\tau) + C(-\tau^{a}_{p,q},\bar{h}^{b}_{p,q};\mathbf{x},\tau)$$
$$+ C(\bar{f}^{b}_{r},v^{a}_{r};\mathbf{x},\tau) + C(-\bar{\tau}^{b}_{p,q},h^{a}_{p,q};\mathbf{x},\tau)]dV. \qquad (5.2)$$

Equation (5.2) is the global form, for the domain D, of the time-correlation reciprocity theorem. Note that in the left-hand side the contributions from interfaces between different media present in D have cancelled and that the contributions from the boundaries of elastically impenetrable objects present in D have vanished in view of the boundary conditions stated in Section 3.

6. APPLICATION TO INVERSE PROBLEMS

In this section we briefly indicate the application of (4.2) and (5.2) to the elastodynamic inverse source problem. In the inverse source problem the elastodynamic wave field in State 'a' is taken to be one that is radiated by the unknown source distributions $\{f^{T}_{r},h^{T}_{p,q}\}$. Let $D^{T} \subset R^{3}$ be their spatial support. The radiated wave field $\{-\tau^{T}_{p,q},v^{T}_{r}\}$ is measured in some, accessible, observational domain $D^{\Omega} \subset R^{3}$. The intersection of D^{T} and D^{Ω} is empty (Figure 1). State 'b' is taken to be a computational state, denoted as the 'observational' one. The corresponding wave field $\{-\tau^{\Omega}_{p,q},v^{\Omega}_{r}\}$ that would be radiated by known sources with distributions $\{f^{\Omega}_{r},h^{\Omega}_{p,q}\}$ is computed and its interaction with the measured elastodynamic wave field in D^{Ω} is evaluated. In general, one could say that the introduction of the observational state is representative for the processing of the measured data. Since only the interaction in D^{Ω} is considered, it makes no sense to take the support of $\{f^{\Omega}_{r},h^{\Omega}_{p,q}\}$ larger than D^{Ω}. Finally, the solid in the observational state is taken to be either the adjoint (for the application of (4.2)) or the time-reverse adjoint (for the application of (5.2)) of the one in which the unknown sources radiate. The reciprocity relations are now applied to the domain interior to the closed surface S^Ω that is taken such that D^T and D^Ω are located in its interior. Then, through the reciprocity relations, the known interactions in D^Ω are related to the source distributions to be reconstructed. Exploiting these relationships, several reconstruction algorithms can be developed. As to the role of S^Ω, we observe that in practice one is as a rule interested only in causal media. Then, it is advantageous to choose, in the application of (4.2), the wave fields causal as well. Given the fact that S^Ω surrounds all sources, the integral over S^Ω can be shown to be zero. In the application of (5.2), however, effectual (or anticausal) wave fields are involved in all cases and the integral over S^Ω differs from zero. This difference in the roles of the surface integrals in the two cases has been pointed out by Bojarski [1].

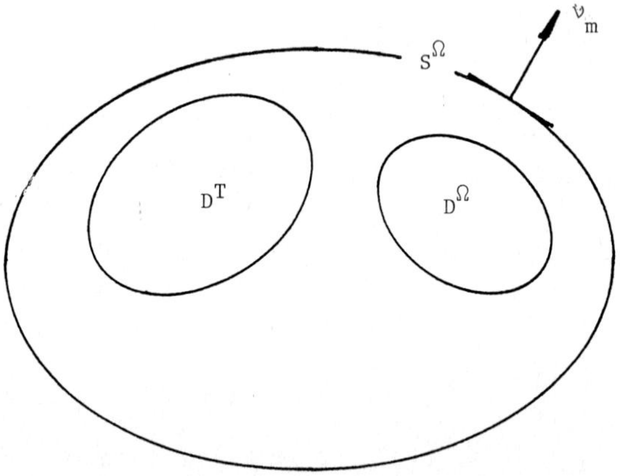

FIGURE 1

Configuration ilustrative for the inverse source problem: unknown acoustic sources radiate in D^T; the elastodynamic wave field is measured in D^Ω and on S^Ω.

REFERENCES

[1] Bojarski, N.N., "Generalized reaction principles and reciprocity theorems for the wave equations, and the relationship between the time-advanced and time-retarded fields", Journal of the Acoustical Society of America 74 (1), 281-285, 1983.

ELASTIC WAVE PROPAGATION
M.F. McCarthy, M.A. Hayes, (Editors)
Elsevier Science Publishers B.V. (North-Holland), 1989

ELASTODYNAMIC SCATTERING AND MATRIX FACTORISATION

Erhard MEISTER and Frank-Olme SPECK*

Fachbereich Mathematik
Technische Hochschule Darmstadt
Darmstadt, West-Germany

The scattering of a time-harmonic elastic wave by a half-plane-shaped crack Σ is modelled by a Dirichlet boundary value problem. This is shown to be equivalent to a system of Wiener-Hopf equations. The reduced symbol matrix function admits a right canonical factorisation that yields the correctness and explicit solution of the problem. Our factoring technique is based on the spirit of paired singular operators and decomposing commutative but non-rational matrix function algebras.

1. INTRODUCTION

Let $\Sigma := \{x \in \mathbb{R}^3 : x_1 > 0, x_3 = 0\}$, $\Omega = \mathbb{R}^3 - \overline{\Sigma}$, and g^+, g^- be given in the Sobolev trace space $H^{1/2}(\Sigma)^3$. We look for a weak solution $u \in H^1(\Omega)^3$ of

$$(\Delta + \frac{\lambda+\mu}{\mu} \text{ grad div} + \frac{\omega^2\rho}{\mu})\, u = 0 , \quad \text{in } \Omega \tag{1}$$

$$u_0^{\pm} = g^{\pm} \qquad \text{on } \Sigma^{\pm} , \tag{2}$$

where $u_0^{\pm}$ denote the Dirichlet data on the upper/lower bank of Σ in the sense of the trace theorem, and $\lambda, \mu, \omega, \rho$ are known constants, which satisfy $\mu, \rho > 0$, $\lambda + 2\mu > 0$, $\mathrm{Re}\,\omega$, $\mathrm{Im}\,\omega > 0$ [1].

The paper consists of two parts. Firstly we give a derivation of the Wiener-Hopf system, which proceeds similar to the case of elliptic boundary and transmission problems for the (scalar) Helmholtz equation and a half-plane screen [6,8,9]. Some of these results are known from [2,3], for instance, where variational principles have been used. In the second part the factorisation problem will be analysed and solved by an approach, which works very effectively in the elastodynamic theory where a distinctive type of non-rational matrix function appears.

2. THE WIENER-HOPF SYSTEM

Proposition 1 The general solution $u^+ \in H^1(\Omega^+)^3$ of Equation (1) in $\Omega^+ : x_3 > 0$ reads - by use of the notation $x' = (x_1, x_2)$, $\xi = (\xi_1, \xi_2)$ and

$$\hat{\varphi}_j(\xi) = F_{x' \mapsto \xi}\, \varphi_j(x') = \int_{\mathbb{R}^2} e^{ix'\xi} \varphi_j(x')\, dx' , \tag{3}$$

$$t_j = t_j(\xi) = (\xi^2 - k_j^2)^{1/2} , \quad k_1^2 = \frac{\omega^2\rho}{\lambda+2\mu} , \quad k_2^2 = \frac{\omega^2\rho}{\mu}, \tag{4}$$

* Sponsored by the Deutsche Forschungsgemeinschaft under grant number Me 261/4-2

(with $t_j \to +\infty$ as $\xi \to +\infty$ and vertical branch cuts connecting $\pm k_j$ over ∞)

$$u^+(x) = F^{-1}_{\xi \mapsto x'} \begin{pmatrix} \hat{\varphi}_1(\xi) e^{-t_2(\xi) x_3} + \frac{i\xi_1}{t_1(\xi)} \hat{\varphi}_3(\xi) e^{-t_1(\xi) x_3} \\ \hat{\varphi}_2(\xi) e^{-t_2(\xi) x_3} + \frac{i\xi_2}{t_1(\xi)} \hat{\varphi}_3(\xi) e^{-t_1(\xi) x_3} \\ -\{\frac{i\xi_1}{t_2(\xi)} \hat{\varphi}_1(\xi) + \frac{i\xi_2}{t_2(\xi)} \hat{\varphi}_2(\xi)\} e^{-t_2(\xi) x_3} + \hat{\varphi}_3(\xi) e^{-t_1(\xi) x_3} \end{pmatrix} 1_+(x_3) \quad (5)$$

We briefly write (dropping the dependence on x' and ξ)

$$u^+ = F^{-1} \Phi_1 \cdot \begin{pmatrix} \hat{\varphi}_1 e^{-t_2 x_3} \\ \hat{\varphi}_2 e^{-t_2 x_3} \\ \hat{\varphi}_3 e^{-t_1 x_3} \end{pmatrix} 1_+(x_3) \; , \; \Phi_1(\xi) = \begin{pmatrix} 1 & 0 & \frac{i\xi_1}{t_1} \\ 0 & 1 & \frac{i\xi_2}{t_1} \\ \frac{-i\xi_1}{t_2} & \frac{-i\xi_2}{t_2} & 1 \end{pmatrix} . \quad (6)$$

In addition the Dirichlet data u_0^+ on $x_3 = +0$ must satisfy

$$u_0^+ = \begin{pmatrix} u_{01}^+ \\ u_{02}^+ \\ u_{03}^+ \end{pmatrix} = B_1 \varphi^+ = F^{-1} \Phi_1 \cdot F \begin{pmatrix} \varphi_1 \\ \varphi_2 \\ \varphi_3 \end{pmatrix} \in H^{1/2}(\mathbb{R}^2)^3 . \quad (7)$$

<u>Proof</u> Taking the partially Fourier transformed Equations (1) for functions u in a dense subspace $S(\Omega^+)^3$, one obtains a system of ordinary differential equations. The general solution yields Formula (5), which can be extended to data $u_0^+ \in H^{1/2}(\mathbb{R}^2)^3$ according to the continuous dependence $u_0^+ \mapsto u$ #

Note that $\varphi^+ \in H^{1/2}(\mathbb{R}^2)^3$ does not hold in general. Although the (translation invariant) boundary operator B_1 , that relates the Dirichlet data on $\mathbb{R}^2$ to φ^+, is of order zero, since $\Phi_1 \in L^\infty(\mathbb{R}^2)^{3\times 3}$ is bounded, the inverse is not bounded as we learn from

$$\Phi_1^{-1}(\xi) = \frac{1}{t_1 t_2 - \xi^2} \begin{pmatrix} t_1 t_2 - \xi_2^2 & \xi_1 \xi_2 & -i\xi_1 t_2 \\ \xi_1 \xi_2 & t_1 t_2 - \xi_1^2 & -i\xi_2 t_2 \\ i\xi_1 t_1 & i\xi_2 t_1 & t_1 t_2 \end{pmatrix} \quad (8)$$

$$= O(\xi^2) \; , \; |\xi| \to \infty$$

in the meaning of an element-wise estimate. However, the ansatz functions φ^+ are important for a simple representation, but belong to a strange subspace of

$H^{-3/2}(\mathbb{R}^2)^3$, which causes some topological trouble (this phenomenon makes the situation different from the Helmholtz equation reasoning).

Let us also consider the following boundary data: $u_0^- \in H^{1/2}(\mathbb{R}^2)^3$ as the trace of a solution $u^- \in H^1(\Omega^-)^3$ of Eq. (1) in $\Omega^-: x_3 < 0$, the Dirichlet data jump $f_0 = u_0^+ - u_0^-$ and, if possible, the Neumann data jump $f_1 = u_1^+ - u_1^-$, $u_1^\pm = \frac{\partial}{\partial x_3} u \Big|_{x_3 = \pm 0}$. Formula (3) and its analogue for Ω^- yields the following results, cf. [6].

<u>Corollary 1</u> The general solution $u \in H^1(\Omega)^3$ of Eq. (1) is given by $u = u^\pm$ in $\Omega^\pm$,

$$u^+ = F^{-1}\Phi_1 \cdot \begin{pmatrix} e^{-t_2x_3} & 0 & 0 \\ 0 & e^{-t_2x_3} & 0 \\ 0 & 0 & e^{-t_1x_3} \end{pmatrix} \Phi_1^{-1}\, 1_+(x_3)\hat{u}_0^+ ,$$
$$u^- = F^{-1}\Phi_2 \cdot \begin{pmatrix} e^{t_2x_2} & 0 & 0 \\ 0 & e^{t_2x_3} & 0 \\ 0 & 0 & e^{t_1x_3} \end{pmatrix} \Phi_2^{-1}\, 1_-(x_3)\hat{u}_0^- , \tag{9}$$

$$\Phi_2 = \begin{pmatrix} 1 & 0 & \frac{-i\xi_1}{t_1} \\ 0 & 1 & \frac{-i\xi_2}{t_1} \\ \frac{i\xi_1}{t_2} & \frac{i\xi_2}{t_2} & 1 \end{pmatrix} = M\Phi_1 M \quad , \quad M = \begin{pmatrix} 1 & 0 & 0 \\ 0 & 1 & 0 \\ 0 & 0 & -1 \end{pmatrix} , \tag{10}$$

provided two conditions hold:

$$f_0 = 0 \quad , \quad f_1 = 0 \qquad \text{on } \Sigma': x_1 < 0 \, , \, x_3 = 0 \tag{11}$$

Note that the products of three matrices in (9) are bounded, each despite of the increase of Φ_j^{-1} . Similarly, Formulae (9) imply (for $S(\mathbb{R}^2)^3$ data in the first instant and subsequent continuous extension)

$$\hat{u}_1^+ = \Phi_3 \hat{\varphi}^+ \quad , \quad \hat{u}_1^- = \Phi_4 \hat{\varphi}^- , \tag{12}$$

$$\Phi_3 = \begin{pmatrix} -t_2 & 0 & -i\xi_1 \\ 0 & -t_2 & -i\xi_2 \\ i\xi_1 & i\xi_2 & -t_1 \end{pmatrix} = -\Phi_1 D \quad , \quad D = \begin{pmatrix} t_2 & 0 & 0 \\ 0 & t_2 & 0 \\ 0 & 0 & t_1 \end{pmatrix} ,$$

$$\Phi_4 = \begin{pmatrix} t_2 & 0 & -i\xi_1 \\ 0 & t_2 & -i\xi_2 \\ i\xi_1 & i\xi_2 & t_2 \end{pmatrix} = \Phi_2 D \quad , \quad M\Phi_1 M D ,$$

This yields (despite of the increase of Φ_j^{-1})

$$u_1^+ = -\Phi_1 D \Phi_1^{-1} u_0^+ \in H^{-1/2}(\mathbb{R}^2)^3 \quad ,$$

$$u_1^- = \Phi_2 D \Phi_2^{-1} u_0^- \in H^{-1/2}(\mathbb{R}^2)^3 \quad . \tag{13}$$

Thus (11) can be written as

$$f_0 \in \tilde{H}^{1/2}(\Sigma)^3 \quad , \quad f_1 \in \tilde{H}^{-1/2}(\Sigma)^3 \quad , \tag{14}$$

where $\tilde{H}^s(\Sigma)$ denotes the closed subspace of $H^s(\mathbb{R}^2)$ distributions supported on $\overline{\Sigma}$ (identified with a half-plane in $\mathbb{R}^2$). It is convenient to introduce further data arrangements

$$\varphi = \begin{pmatrix} \varphi^+ \\ \varphi^- \end{pmatrix}, \quad u_0 = \begin{pmatrix} u_0^+ \\ u_0^- \end{pmatrix}, \quad u_1 = \begin{pmatrix} u_1^+ \\ u_1^- \end{pmatrix},$$

$$f = \begin{pmatrix} f_0 \\ f_1 \end{pmatrix} = \begin{pmatrix} u_0^+ - u_0^- \\ u_1^+ - u_1^- \end{pmatrix}, \quad h = \begin{pmatrix} u_0^+ - u_0^- \\ u_0^+ + u_0^- \end{pmatrix}, \tag{15}$$

which are in one-to-one correspondence to each other via the above formulae; f is unknown but supported on Σ, h is known on Σ (but $u_0^+ + u_0^-$ not supported there) for the Dirichlet problem. Define the boundary operators

$$B_- = F^{-1}\Psi_{B_-} \cdot F \quad : \varphi \mapsto f$$
$$B_+ = F^{-1}\Psi_{B_+} \cdot F \quad : \varphi \mapsto h \tag{16}$$

given by the 6×6 block Fourier symbol matrix functions

$$\Psi_{B_-} = \begin{pmatrix} \Phi_1 & -\Phi_2 \\ \Phi_3 & -\Phi_4 \end{pmatrix} = \begin{pmatrix} \Phi_1 & -M\Phi_1 M \\ -\Phi_1 D & -M\Phi_1 M D \end{pmatrix},$$

$$\Psi_{B_+} = \begin{pmatrix} \Phi_1 & -\Phi_2 \\ \Phi_1 & \Phi_2 \end{pmatrix} = \begin{pmatrix} \Phi_1 & -M\Phi_1 M \\ \Phi_1 & M\Phi_1 M \end{pmatrix}. \tag{17}$$

<u>Theorem 1</u> 1. The Dirichlet problem (1),(2) is equivalent to the Wiener-Hopf system

$$Wf = \chi_\Sigma \cdot F^{-1}\Psi \cdot F f = h \,, \tag{18}$$

with $\Psi = \Psi_{B_+}\Psi_{B_-}^{-1}$, $h = (g^+ - g^-, g^+ + g^-)$ and $\chi_\Sigma(x') = 1_+(x_1)$.

2. A compatibility condition $g^+ - g^- \in \tilde{H}^{1/2}(\Sigma)$ is necessary for the existence of a solution.

3. The operator

$$W : \tilde{H}^{1/2}(\Sigma)^3 \times \tilde{H}^{-1/2}(\Sigma)^3 \to \tilde{H}^{1/2}(\Sigma)^3 \times H^{1/2}(\Sigma)^3 \tag{19}$$

is linear , continuous and of normal type [7,8,9] , i.e.

$$B = F^{-1}\Psi \cdot F : H^{1/2}(\mathbb{R}^2)^3 \times H^{-1/2}(\mathbb{R}^2)^3 \to H^{1/2}(\mathbb{R}^2)^6 \tag{20}$$

is bijective.

Proof The first statement is now obvious, the second one is a consequence of (14). The rest follows from a representation of the symbol matrix Ψ. We considered previously

$$\Psi^{-1} = \Psi_{B_-} \Psi_{B_+}^{-1} = \begin{pmatrix} \Phi_1 & -M\Phi_1 M \\ -\Phi_1 D - M\Phi_1 MD \end{pmatrix} \cdot \frac{1}{2} \begin{pmatrix} \Phi_1^{-1} & \Phi_1^{-1} \\ -M\Phi_1^{-1}M & M\Phi_1^{-1}M \end{pmatrix}$$

$$= \begin{pmatrix} I & 0 \\ -\Pi_1(\Phi_1 D \Phi_1^{-1}) & -\Pi_2(\Phi_1 D \Phi_1^{-1}) \end{pmatrix}, \tag{21}$$

using the notation for 3 x 3 matrices

$$\Phi = \Pi_1(\Phi) + \Pi_2(\Phi) = \frac{1}{2}(\Phi - M\Phi M) + \frac{1}{2}(\Phi + M\Phi M)$$

$$= \begin{pmatrix} 0 & 0 & x \\ 0 & 0 & x \\ x & x & 0 \end{pmatrix} + \begin{pmatrix} x & x & 0 \\ x & x & 0 \\ 0 & 0 & x \end{pmatrix} \tag{22}$$

which means an obvious decomposition by the replacement of matrix elements by zeroes. We abbreviate $\Pi_j(\Phi_1 D \Phi_1^{-1})$ by Π_j and obtain easily the inverse symbol matrix

$$\Psi = \begin{pmatrix} I & 0 \\ -\Pi_2^{-1}\Pi_1 & -\Pi_2^{-1} \end{pmatrix}, \tag{23}$$

where

$$-\Pi_2^{-1} = \frac{1}{k_2^2 t_1 t_2} \begin{pmatrix} t_2(t_1t_2 - \xi_1^2) - t_1\xi_2^2 & (t_1-t_2)\xi_1\xi_2 & 0 \\ (t_1-t_2)\xi_1\xi_2 & t_2(t_1t_2-\xi_2^2) - t_1\xi_1^2 & 0 \\ 0 & 0 & t_1(t_1t_2-\xi^2)\frac{k_2^2}{k_1^2} \end{pmatrix},$$

$$\Pi_1 = \frac{t_2-t_1}{t_1t_2-\xi^2} \begin{pmatrix} 0 & 0 & -i\xi_1 t_2 \\ 0 & 0 & -i\xi_2 t_2 \\ -i\xi_1 t_1 & -i\xi_2 t_1 & 0 \end{pmatrix} \tag{24}$$

hold with obvious mapping properties of A and W due to $\Pi_j = O(|\xi|)$ $\Pi_2^{-1} = O(|\xi|^{-1})$ as $|\xi| \to \infty$

#

According to (23) it is easier to treat the reduced WH system

$$W_1 f_1 = \chi_\Sigma \cdot F^{-1}(-\Pi_2^{-1}) \cdot F f_1 = g^+ + g^- + \Pi_2^{-1}\Pi_1(g^+ - g^-) . \tag{25}$$

But note that the method so far applies also to boundary and transmission problems where the system does not decompose, i.e. to other normal type conditions prescribed on the crack Σ, and to arbitrarily shaped plane Lipschitz domains $\Sigma \subset \mathbb{R}^2$.

3. MATRIX FACTORISATION BY THE USE OF PAIRED OPERATORS

The remainding work consists mainly in the factorisation with respect to ξ_1 of the 2 x 2 block of Π_2^{-1} into lower/upper holomorphic function matrices at least with algebraic growth at infinity in order to deal with the Wiener-Hopf technique. Since the scalars t_j are simply factored by $t_j(\xi) = (\xi_1 - i\sqrt{\xi_2^2-k_j^2})^{1/2} \cdot (\xi_1 + i\sqrt{\xi_2^2-k_j^2})^{1/2} = t_{j-}(\xi) \cdot t_{j+}(\xi)$, we consider the *lifted* [6] matrix $-t_1 k_2^2 \Pi_2^{-1}$ where the 2 x 2 block reads

$$G(\xi) = \frac{1}{t_2} \begin{pmatrix} t_2(t_1t_2-\xi_1^2)-t_1\xi_2^2 & (t_1-t_2)\xi_1\xi_2 \\ (t_1-t_2)\xi_1\xi_2 & t_2(t_1t_2-\xi_2^2)-t_1\xi_1^2 \end{pmatrix}. \tag{26}$$

This matrix function is obviously bounded and invertible in $C(\dot{\mathbb{R}})^{2\times 2}$ with respect to ξ_1 ($\dot{\mathbb{R}}$ denotes the one-point compactification, ξ_2 plays the role of a fixed parameter). The elements are even in the Wiener algebra $W = \mathbb{C} \dot{+} F L^1(\mathbb{R})$ (replace $\xi_1^2 = t_1^2 - \xi_2^2 + k_1^2$ e.g., and make use of $t_j^{-1} \in W$, the Wiener-Levy theorem, and $\frac{\xi_1}{\xi_1^2+1} = \frac{1}{2i}\int_{\mathbb{R}} e^{ix_1\xi_1} e^{-|x_1|} \operatorname{sgn} x_1 dx_1$. Now we rewrite G as

$$\begin{aligned} G(\xi) &= (t_1t_2-\xi^2)\cdot\frac{1}{\xi^2}\begin{pmatrix}\xi_1^2 & \xi_1\xi_2\\ \xi_1\xi_2 & \xi_2^2\end{pmatrix} - k_2^2\frac{t_1}{t_2}\cdot\frac{1}{\xi^2}\begin{pmatrix}\xi_2^2 & -\xi_1\xi_2\\ -\xi_1\xi_2 & \xi_1^2\end{pmatrix} \\ &= a(\xi)\cdot R_1(\xi) + b(\xi)\cdot R_2(\xi). \end{aligned} \tag{27}$$

With complementary projection matrices, i.e. $R_1^2 = R_1$, $R_1 + R_2 = I$ hold, consisting of rational functions, and scalar (non-rational) coefficients, which are regular Wiener algebra elements.

Theorem 2 For fixed $\xi_2 \in \mathbb{R}$, G admits a Wiener-Hopf factorisation $G = \tilde{G}_-\tilde{G}_+$ with simple poles at $\xi_1 = \pm i|\xi_2|$ in the commutative subalgebra $A(\mathbb{R}_1)$ of $C(\dot{\mathbb{R}})^{2\times 2}$ matrix functions of type (27) with fixed R_j i.e. $\tilde{G}_\pm$ and $\tilde{G}_\pm^{-1}$ are of the same type and holomorphically extendable into the upper/lower complex half plane $\mathbb{C}_\pm = \{\xi_1 \in \mathbb{C} : \operatorname{Im}\xi_1 \gtrless 0\}$, respectively, $\xi_1 = \pm i|\xi_2|$ excluded. Explicitly one may take

$$\tilde{G}_\pm = a_\pm R_1 + b_\pm R_2, \tag{28}$$

$$a_\pm(\xi) = \sqrt{a(\infty)}\exp\{F_{x_1\mapsto\xi_1} 1_\pm(x_1)\cdot F^{-1}_{\xi_1\mapsto x_1}\ln\frac{a(\xi)}{a(\infty)}\},$$

$$b_\pm(\xi) = ik_2\left(\frac{\xi_1 \pm i\sqrt{\xi_2^2-k_1^2}}{\xi_1 \pm i\sqrt{\xi_2^2-k_2^2}}\right)^{1/2}.$$

A *strong* Wiener-Hopf factorisation $G = G_-G_+$ (without poles) can be obtained in the case $\lambda a_+(i|\xi_2|, \xi_2) = b_+(i|\xi_2|, \xi_2)$ with $\lambda = \lambda(\xi_2) \neq 1$ by factoring a rational matrix function $R(\xi_1,\xi_2,\lambda) = R_1(\xi_1,\xi_2) + \lambda^2 R_2(\xi_1,\xi_2)$ with a standard technique [7] as

$$R = R_-R_+ = \frac{1}{1+\lambda^2}\begin{pmatrix}\frac{\xi_1 - i\lambda^2|\xi_2|}{\xi_1 - i|\xi_2|} & \pm i \\ \pm i\frac{\lambda^2\xi_1 - i|\xi_2|}{\xi_1 - i|\xi_2|} & 1\end{pmatrix}\begin{pmatrix}\frac{\xi_1 + i\lambda^2|\xi_2|}{\xi_1 + i|\xi_2|} & \mp i\frac{\lambda^2\xi_1 + i|\xi_2|}{\xi_1 + i|\xi_2|} \\ \mp i\lambda^2 & \lambda^2\end{pmatrix} \tag{29}$$

for $\xi_2 \gtrless 0$ and putting

$$G = G_- G_+ = (a_- R_1 + \frac{b_-}{\lambda} R_2)\; R_- R_+ (a_+ R_1 + \frac{b_+}{\lambda} R_2) \tag{30}$$

Proof The scalar factoring of a and b in W is possible (and unique up to constant factors) due to $a(\xi) \neq 0$ and $b(\xi) \neq 0$ for $\xi_1 \in \mathbb{R}$, $\xi_2 \in \mathbb{R}$, and vanishing winding numbers, since a and b are even functions in ξ_1, see [7]. Thus $\tilde{G}_\pm$ and $\tilde{G}_\pm^{-1} = a_\pm^{-1} R_1 + b_\pm^{-1} R_2$ have all the desired properties but, so far, $\tilde{G}_+, \tilde{G}_+^{-1}$ may have a simple pole in $\xi_1 = i|\xi_2|$ and $\tilde{G}_-, \tilde{G}_-^{-1}$ may have one in $\xi_1 = -i|\xi_2|$. The polynomial matrices in (27) satisfy

$$\begin{pmatrix} \xi_1^2 & \xi_1\xi_2 \\ \xi_1\xi_2 & \xi_2^2 \end{pmatrix} = - \begin{pmatrix} \xi_2^2 & -\xi_1\xi_2 \\ -\xi_1\xi_2 & \xi_1^2 \end{pmatrix} \quad \text{in } \xi_1 = \pm\, i|\xi_2|, \tag{31}$$

so that the pole of G_+ is cancelled, if $a_+(i|\xi_2|, \xi_2) = b_+(i|\xi_2|, \xi_2)$ holds. Simultaneously, the poles of $\tilde{G}_+^{-1}$ in $\xi_1 = i|\xi_2|$ and of $\tilde{G}_-^{\pm 1}$ in $\xi_1 = -i|\xi_2|$ cancel out, since $a_-(\xi_1, \xi_2) = a_+(-\xi_1, \xi_2)$, and equally for b_+, holds in the corresponding (overlapping) half-planes. In the second case, formula (31) follows by inspection.
#

Corollary 2 The reduced Wiener-Hopf operator $W_1 : \tilde{H}^{-1/2}(\Sigma)^3 \to H^{1/2}(\Sigma)^3$ is invertible by

$$W_1^{-1} = A_+^{-1}\, \chi_\Sigma \cdot A_-^{-1}\, \ell, \tag{32}$$

where $\ell : H^{1/2}(\Sigma)^3 \to H^{1/2}(\mathbb{R}^2)^3$ is any extension operator (e.g. even extension) and

$$A_\pm^{-1} = F^{-1} \left(\begin{array}{cc|c} k_2 t_{1\pm} G_\pm^{-1} & & 0 \\ & & 0 \\ \hline 0 & 0 & k_1 t_{2\pm}\, a_\pm^{-1} \end{array} \right) \cdot F, \tag{33}$$

in obvious block matrix notation. Due to the boundedness of $G_\pm^{-1}$, see (28)-(30) and $t_{j\pm}(\xi) = O(|\xi|^{1/2})$, we have order $A_\pm^{-1} = \frac{1}{2}$, i.e.

$$A_-^{-1} : H^{1/2}(\mathbb{R}^2)^3 \to L^2(\mathbb{R}^2)^3, \qquad A_+^{-1} : L^2(\mathbb{R}^2)^3 \to H^{-1/2}(\mathbb{R}^2)^3, \tag{34}$$

and $\chi_\Sigma \cdot$ is a projector in $L^2(\mathbb{R}^2)^3$.

The correct solution of the Dirichlet problem, provided $g^+ - g^- \in \tilde{H}^{1/2}(\Sigma)^3$ holds, can now be concluded from (32), (25), (23) and (9). Consequently [9], there holds $\nabla u = O(|x|^{-1/2})$ as $|x| \to 0$, if $g^\pm$ are smooth. Finally we would like to remark that the crucial idea of factoring G cannot be managed by means of arguments, which are common in system theory [4].

REFERENCES

[1] Achenbach, J.D. , Wave Propagation in Elastic Solids (North-Holland, Amsterdam, 1984)

[2] Bamberger, A., Approximation de la Diffraction d'Ondes Élastiques. Une Novelle Approche I, in: Brezis, H. and Lions, J.L., (eds.), Nonlinear Partial Differential Equations and their Applications (Pitman, London, 1984)pp.

[3] Bamberger, A., Approximation de la Diffraction d'Ondes Élastiques. Une Novelle Approche III, Rapport interne 98, Centre de Mathématiques

Appliquées (Ecole Politechnique, Palaiseau, 1983)
[4] Bart, H., Gohberg, I. and Kaashoek, M.A., Minimal Factorization of Matrix and Operator Functions (Birkhäuser, Basel, 1979)
[5] Costabel, M. and Stephan, E.P., An Improved Boundary Element Galerkin Method for Three-Dimensional Crack Problems, Int. Equs. Op. Th. 10 (1987) 467-504.
[6] Meister, E. and Speck, F.-O., Boundary Integral Equation Methods for Canonical Problems in Diffraction Theory, in: Brebbia, C.A., Wendland, W.L. and Kuhn, G., (eds.) Boundary Elements IX, vol. 1 (Springer, Berlin, 1987) pp. 59-77.
[7] Mikhlin, S.G. and Prössdorf, S., Singular Integral Operators (Springer, Berlin, 1986).
[8] Speck, F.-O., Mixed Boundary Value Problems of the Type of Sommerfeld's Half-plane Problem, Proc. Royal Soc. Edinburgh 104 A (1986) 261-277.
[9] Speck, F.-O., Sommerfeld Diffraction Problems with First and Second Kind Boundary Conditions, to appear.

ELASTIC WAVE PROPAGATION
M.F. McCarthy, M.A. Hayes, (Editors)

SCATTERING OF ELASTIC WAVES BY INCLUSIONS SURROUNDED BY THIN INTERFACE LAYERS

Peter OLSSON[a†], Subhendu K. DATTA[b*], and Anders BOSTRÖM[a†]

[a]Division of Mechanics
Chalmers University of Technology
S-412 96 GÖTEBORG
Sweden

[b]Department of Mechanical Engineering and CIRES
University of Colorado
Boulder, CO 80309-0427
USA

The problem of scattering of elastodynamic waves by elastic inclusions surrounded by interface layers is one which is of interest for ultrasonic nondestructive evaluation of interfaces in composites. In the present paper we study scattering by a single elastic inclusion with a thin interface layer in an elastic matrix. The thin interface layer is modelled as an elastic shell, the equation of motion of which enters the boundary condition on the surface of the elastic inclusion. The scattering problem is solved by means of the null field approach, and the results are compared with those that have recently been obtained [1] by the use of a hybrid finite element and wave funtion expansion technique. Numerical results are presented showing the dependence of the scattering cross section on frequency and the elastic properties of constituents. Results show significant dependence on parameters containing the thickness, density, and stiffnesses of the interface layer.

1. INTRODUCTION

Three-dimensional elastodynamic scattering problems form a subject which has been studied for a long time. [2] contains a comprehensive review of the history of the subject. Still it is only in recent years that solutions for non-spherical scatterers have been obtained. These are mainly approximate solutions valid at low frequencies and numerical solutions useful at arbitrary frequencies. Some examples are contained in [3-9].

An important application of the theory of elastodynamic wave scattering is the non-destructive characterization of composite materials. The present paper addresses a particular problem in this area. In metal-matrix composites reinforced by fibers or particles, the processing conditions often induce an interface layer surrounding the inclusions. This interface layer, having elastic properties different than the inclusion and the matrix, significantly influences the strength and fracture behaviour of the composite. The purpose of the present paper is to analyze the effect of an interface layer on the total scattering cross section of a spherioidal inclusion.
The method employed is the null field approach or T matrix method [8,9]. The interface layer is

†Sponsored by the National Swedish Board for Technical Deveopment (STU)

*Supported in part by grants from the National Science Foundation (INT-8521422 and INT-8610487) and a grant from the Office of Naval Research (#N00014-86-K0280; Programme Manager: Dr. Y. Rajapakse).

modelled as a thin shell. An approximate equation of motion for the shell is used to formulate boundary conditions on the surface of the scatterer. We give numerical results for incident plane P waves, and our results are for low frequencies found to agree well with the results of [1], a study of the same type of problem performed by means of a hybrid finite element method. The latter study used the full elastic equations for both the inclusion and the layer, and showed that the cross section is measurably influenced by the presence of the interface layer. In the present paper we also study the effect of changing the thickness, density, and stiffness of the layer.

An earlier example of the combination of shell theory and the null field approach to elastodynamic scattering in the case of two-dimensional scattering can be found in [10].

2. THE NULL FIELD APPROACH

Let us consider an inclusion with a thin interface layer of thickness h, residing within an elastic matrix. The inclusion is of density ρ_1 and has Lamé constants λ_1 and μ_1, while the corresponding parameters for the matrix and interface layer are ρ, λ, μ and ρ_o, λ_o, and μ_o, respectively.The pressure and shear wavenumbers in the matrix are denoted by k_p and k_s, respectively. The surface of the inclusion inside the layer we denote by S, and its outward pointing unit normal by $\hat{n}$. We assume that the layer is sufficiently thin to allow us to apply all boundary conditions on S.

Expanding the incident and scattered displacement fields as

$$\mathbf{u}^{in} = \sum_n a_n \, \mathrm{Re}\psi_n \tag{1}$$

$$\mathbf{u}^{s} = \sum_n f_n \, \psi_n \tag{2}$$

in terms of regular and outgoing spherical vector waves, we have the relations

$$a_n = - ik_s\mu^{-1}\int_S [\mathbf{u}_+ \cdot \mathbf{t}(\psi_n) - \mathbf{t}_+ \cdot \psi_n] \, dS \tag{3}$$

$$f_n = ik_s\mu^{-1}\int_S [\mathbf{u}_+ \cdot \mathbf{t}(\mathrm{Re}\psi_n) - \mathbf{t}_+ \cdot \mathrm{Re}\psi_n] \, dS \tag{4}$$

Here $\mathbf{u}_+$ is the limit of the total displacement field as S is approached from the outside, and $\mathbf{t}_+$ similarly the limit of the surface traction. $\mathbf{t}(\cdot)$ is the surface traction operator for the medium on the outside of S.

Denoting the limits from the inside of S by $\mathbf{u}_-$ and $\mathbf{t}_-$, it can be shown that if $\mathbf{u}_-$ is expanded in terms of the regular spherical waves $\mathrm{Re}\psi_n^1$ of the material of the inclusion as

$$\mathbf{u}_- = \sum_n \alpha_n \, \mathrm{Re}\psi_n^1 \tag{5}$$

then $\mathbf{t}_-$ has the expansion

$$\mathbf{t}_- = \sum_n \alpha_n \, \mathbf{t}^1(\mathrm{Re}\psi_n^1) \tag{6}$$

where $\mathbf{t}^1(\cdot)$ is the surface traction operator for the medium inside S.

To solve (3) and (4) for f_n in terms of a_n we need relations between $\mathbf{u}_+$, $\mathbf{t}_+$ and $\mathbf{u}_-$, $\mathbf{t}_-$, that allows us to insert the expansions (5) and (6) into (3) and (4). These can be furnished by the equation of motion of the interface layer and Hooke's law for the layer medium. In the present study we use a simplified equation of motion for the layer, namely the membrane shell type equation

$$\nabla_s \cdot \sigma^s(\mathbf{u}) + \rho_o\, \omega^2\, \mathbf{u} = -\, h^{-1}(\mathbf{t}_+ - \mathbf{t}_-) \tag{7}$$

where $\nabla_s \cdot$ is the surface divergence operator, and σ^s only contains tangential stresses. For a rotationally symmetric inclusion, with tangent vectors $\hat{\varphi}$ and $\hat{\tau} = \hat{\varphi} \times \hat{n}$, we use the components

$$\sigma^s_{\tau\tau}(\mathbf{u}) = \lambda_o \nabla_s \cdot \mathbf{u} + 2\mu_o\, \hat{\tau} \cdot (\hat{\tau} \cdot \nabla \mathbf{u}) \tag{8}$$

$$\sigma^s_{\varphi\varphi}(\mathbf{u}) = \lambda_o \nabla_s \cdot \mathbf{u} + 2\mu_o\, \hat{\varphi} \cdot (\hat{\varphi} \cdot \nabla \mathbf{u}) \tag{9}$$

$$\sigma^s_{\tau\varphi}(\mathbf{u}) = \mu_o[\hat{\varphi} \cdot (\hat{\tau} \cdot \nabla \mathbf{u}) + \hat{\tau} \cdot (\hat{\varphi} \cdot \nabla \mathbf{u})] \tag{10}$$

Regarding the relation between $\mathbf{u}_+$ and $\mathbf{u}_-$, we make the simple approximation

$$\mathbf{u}_+ = \mathbf{u}_- \tag{11}$$

Equations (7) and (11) now give us the sought relations. Collecting the above equations, we have the relation

$$f_n = \sum_{n'} T_{nn'}\, a_{n'} \tag{12}$$

where

$$T = -\, \mathrm{Re}Q\, Q^{-1} \tag{13}$$

and

$$Q_{nn'} = k_s \mu^{-1} \int_S [\mathbf{t}(\psi_n) \cdot \mathrm{Re}\psi^1_{n'} - \psi_{n} \cdot (\mathbf{t}^1(\mathrm{Re}\psi^1_{n'}) - h\, \nabla_s \cdot \sigma^s(\mathrm{Re}\psi^1_{n'}) - h\, \rho_o \omega^2\, \mathrm{Re}\psi^1_{n'})]\, dS \tag{14}$$

This solves the scattering problem for an inclusion with a thin interface layer, but as we have only used an approximate equation of motion , we now turn to numerical comparisons to see how well the solution performs.

3. NUMERICAL RESULTS

In the numerical examples presented in the present section we consider spherical and prolate 2:1 spheroidal inclusions. The matrix material we take to be Al, with parameters

$$\lambda + 2\mu = 110.5 \text{ GPa}$$
$$\mu = 26.7 \text{ GPa}$$
$$\rho = 2.706 \text{ kg m}^{-3}$$

and the inclusion we take to be an SiC particle with

$$\lambda_1 + 2\mu_1 = 474.2 \text{ GPa}$$
$$\mu_1 = 188.1 \text{ GPa}$$
$$\rho_1 = 3.181 \text{ kg m}^{-3}$$

The interface layer will be of one of the following types:

Type 1: $\lambda_o = (\lambda+\lambda_1)/2$, $\mu_o = (\mu+\mu_1)/2$, $\rho_o = (\rho+\rho_1)/2$
Type 2: $\lambda_o + 2\mu_o = 36.8$ GPa, $\mu_o = 8.9$ GPa, $\rho_o = \rho$
Type 3: $\lambda_o + 2\mu_o = 36.8$ GPa, $\mu_o = 8.9$ GPa, $\rho_o = 1.353$ kg m^3

The scattering problem for an SiC particle with a type 1 interface layer in an Al matrix has been solved in [1] by means of a hybrid FEM and wave function expansion technique.

As a first check on the present method, we compare it to the exact separation of variables solution for a spherical body of radius a, surrounded by a type 1 layer of thickness h. Figure 1 contains such a comparison of the total scattering cross section for an incident P wave for two values of h/a. We see that we get rather good agreement with the exact analytical soluton in the low frequency region. That this is not due simply to a good representation of the P-P coupling can be shown by considering the S wave contribution to the total scattering cross section separately. In fact roughly 80% of the total scattering cross section in Figure 1 is due to the scattered S wave.

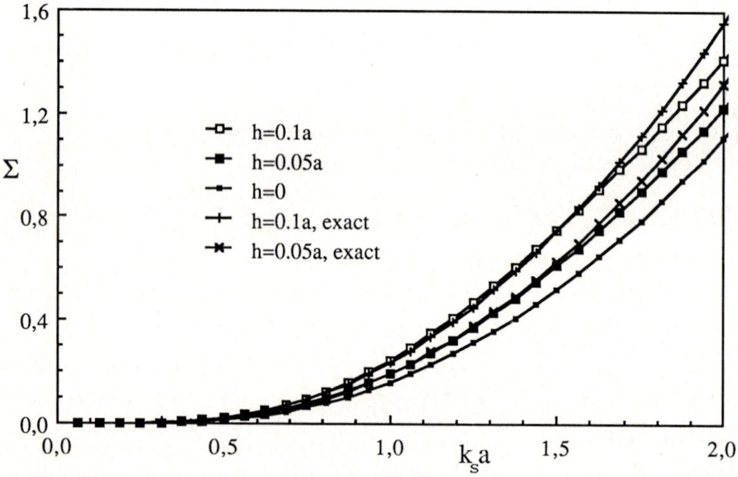

FIGURE 1
Total scattering cross section for P wave incident on a sphere with an interface layer of type 1. The same normalization as in [1].

An interesting feature of Figure 1 is that the difference between the results of the present approach and the exact solution starts to become noticeable at roughly the same frequency for both values of $h > 0$ considered. (For $h = 0$ the null field solution and the separation of variables solution coincide.) However, the main conclusion to be drawn from Figure 1 is of course that the thickness of the interface layer is a significant parameter influencing the total scattering cross section.

Turning now to non-spherical bodies, we can check our method against the FEM calculations of [1]. The angle of incidence of the plane P wave we take to be at 45° to the symmetry axis. In FIg. 2 we have plotted the total scattering cross section for a type 1 interface on a prolate 2:1 spheroid. a is the length of the shorter semiaxes, and b the length of the longer semiaxis, the thickness of the interface layer not included. b is thus the radius of the circumscribing sphere of the inclusion.

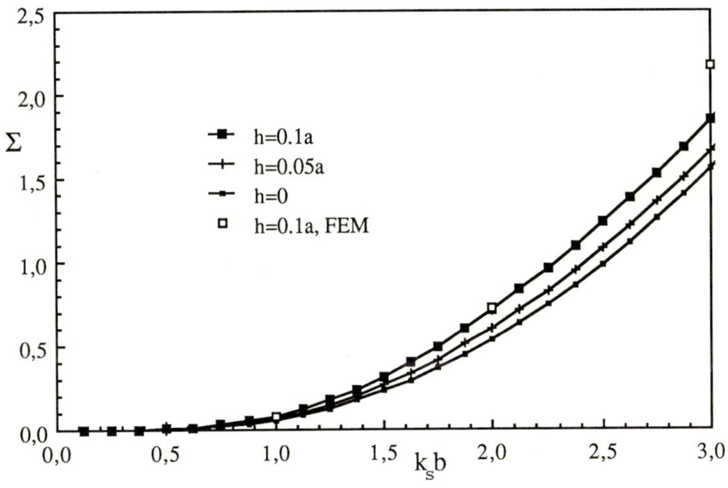

FIGURE 2

Total scattering cross section for P wave incident on a prolate 2:1 spheroid with an interface layer of type 1. Data points from [1] included.

The influence of the shell thickness is again very prominent in the figure. Three data points from the FEM calculation of [1] are included, indicating that the calculations presented here give good results up to at least $k_s b = 2$, but not up to $k_s b = 3$.

In Figure 3 we show the influence of varying the stiffness and density of the interface layer, for the case of P wave incidence on a prolate spheroid. The angle of incidence is again 45°. The change from type 1 to type 2 interface material is a change to a much softer interface. The effect of this is seen to be rather small, especially when compared to the result of also lowering the density of the interface layer, taking the interface to be of type 3. This suggests that the main contribution to $\mathbf{t}_+ - \mathbf{t}_-$ in (7) comes from the inertial term.

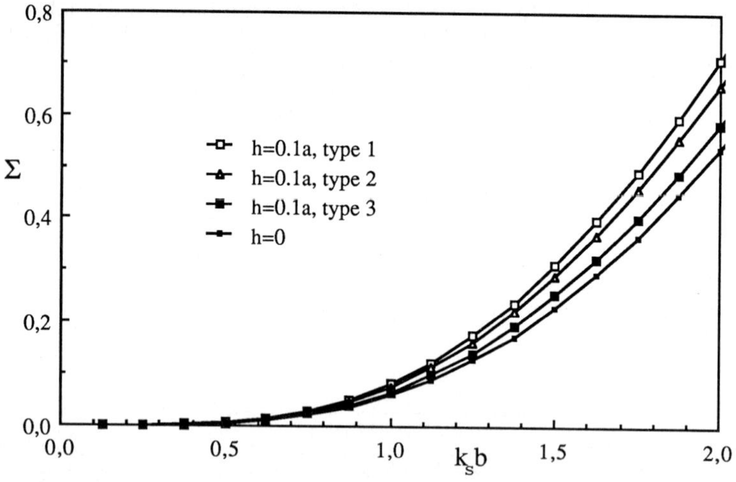

FIGURE 3
Total scattering cross section for P wave incident on a prolate spheroid with interface layers of three different types.

The present calculations are somewhat preliminary, since we have not yet included all terms to order h in the expressions for $\mathbf{t}_+ - \mathbf{t}_-$ and $\mathbf{u}_+ - \mathbf{u}_-$. This can be done, and will be the subject of a forthcoming paper.

ACKNOWLEDGEMENT

S.K.D. gratefully acknowledges the hospitality extended to him at Chalmers University of Technology during his visit there when this work was initiated.

REFERENCES

[1] Paskaramoorthy, R., Datta, S.K., and Shah, A.H., J. Appl. Mech (to appear).
[2] Pao, Y.-H. and Mow, G.G., Diffraction of Elastic Waves and Dynamic Stress Concentration (Crane and Russak, New York, 1973).
[3] Datta, S.K. and Sangster, J.D., SIAM J. Appl. Math. 26 (1974) 350.
[4] Datta, S.K., J. Acoust. Soc. Am. 61 (1977) 1432.
[5] Gubernatis, J.E., J. Appl. Phys. 50 (1979) 4046.
[6] Willis, J.R., J. Mech. Phys. Solids 28 (1980) 287.
[7] Opsal, J.L. and Visscher, W.M., J. Appl. Phys. 58 (1985) 1102.
[8] Waterman, P.C., J. Acoust. Soc. Am. 60 (1976) 567.
[9] Varatharajulu, V. and Pao, Y.-H., J. Acoust. Soc. Am. 60. (1976) 560.
[10] Simon, M.M. and Radlinsky, R.P., J. Acoust. Soc. Am. 71 (1982) 273.

ELASTIC WAVE PROPAGATION
M.F. McCarthy, M.A. Hayes, (Editors)

SINGLE INTEGRAL EQUATIONS FOR SCATTERING OF ELASTIC WAVES BY AN ELASTIC INCLUSION

P.A. MARTIN

Department of Mathematics
University of Manchester
Manchester M13 9PL, England

The 'basic contact problem' of elastodynamics, namely the scattering of time-harmonic elastic waves by an elastic inclusion with different material properties, is usually formulated as a pair of coupled boundary integral equations over the interface S between the inclusion and the exterior solid. In this paper, however, we consider methods for solving the basic contact problem using a single integral equation over S for a single unknown. We use a hybrid of the direct and indirect methods, and obtain several different equations. In particular, we obtain a quasi-Fredholm integral equation of the second kind. This equation appears to be new, and is attractive computationally.

1. INTRODUCTION

Consider the scattering of time-harmonic stress waves by a smooth bounded obstacle in an otherwise unbounded homogeneous isotropic elastic solid. If the obstacle is a cavity, the corresponding displacement vector solves an exterior boundary-value problem. There are two familiar methods for reducing this problem to a boundary integral equation (BIE), namely the direct method (Green's theorem) and the indirect method (layer ansatz); see, e.g. [2], [4], [6]. Both methods yield a single BIE for a single unknown function.

Suppose now that the obstacle is an elastic inclusion, whose material properties differ from those of the surrounding solid. Waves can propagate through the interface, S: the single boundary condition for the cavity is replaced by a pair of conditions, guaranteeing the continuity of the displacement and traction across S. This leads to a problem, called the *basic contact problem* (BCP) by Kupradze [4], that we study here.

The BCP can be reduced to a pair of coupled BIE for a pair of unknowns: for 2-D statics, a direct method was used by Rizzo and Shippy [7]; an elaborate indirect method was used by Kupradze [4, chpt. 12, §2], leading to a quasi-Fredholm system of the second kind.

In this paper, we develop methods for solving the BCP using a single BIE for a single unknown. Two such equations were obtained by Kennett [1]. We show, formally, how many different single BIE can be derived. This is an extension to elastodynamics of some previous work on the corresponding problem in acoustics (scattering of sound in a compressible fluid by a fluid inclusion) [3]. In [3], we obtained various single BIE, including some Fredholm integral equations of the second kind; standard theory was used to prove the solvability of these latter equations, and hence of all the other BIE. Here, the same method is not immediately available.

2. THE BASIC CONTACT PROBLEM

Let B_i denote a bounded domain, with a smooth closed boundary S and simply-connected unbounded exterior, B_e. We seek displacement vectors $\underline{u}_e(P)$ and $\underline{u}_i(P)$ so that

$$L_e\underline{u}_e(P) = \underline{0}, \quad P \in B_e; \quad L_i\underline{u}_i = \underline{0}, \quad P \in B_i; \tag{2.1}$$

$$\underline{u}(p) = \underline{u}_i(p) \text{ and } T_e\underline{u}(p) = T_i\underline{u}_i(p), \quad p \in S; \tag{2.2}$$

$$\underline{u}(P) = \underline{u}_e(P) + \underline{u}_{inc}(P), \quad P \in B_e; \tag{2.3}$$

and $\underline{u}_e$ satisfies a radiation condition at infinity. The given incident wave, $\underline{u}_{inc}$, satisfies $(2.1)_1$ everywhere, except possibly at isolated points in B_e. The operator L_β is defined by

$$L_\beta\underline{u} = k_\beta^{-2} \text{ grad div } \underline{u} - K_\beta^{-2} \text{ curl curl } \underline{u} + \underline{u}$$

where $\rho_\beta\omega^2 = (\lambda_\beta + 2\mu_\beta)k_\beta^2 = \mu_\beta K_\beta^2$ and $\beta = e$ or i. ω is the radian frequency, ρ_β is the density of the solid in B_β, whereas λ_β and μ_β are the Lamé moduli and ν_β is Poisson's ratio: $2\nu_\beta = \lambda_\beta/(\lambda_\beta + \mu_\beta)$. The traction operator T_β is defined on S by

$$(T_\beta\underline{u})_m(p) = \lambda_\beta n_m \operatorname{div}\underline{u} + \mu_\beta n_\ell(\partial u_m/\partial x_\ell + \partial u_\ell/\partial x_m) \tag{2.4}$$

where $\underline{n}(p)$ is the unit normal at $p \in S$, pointing into B_e. It is known that the BCP has at most one solution [4, p.137].

3. ELASTIC POTENTIALS

We introduce two fundamental tensors $\underline{\underline{G}}_\beta(P;Q)$, $\beta = e$ or i: for 2-D motion (plane strain), we have

$$(\underline{\underline{G}}_\beta(P;Q))_{ij} = \mu_\beta^{-1}\{\Psi_\beta\delta_{ij} + K_\beta^{-2}\frac{\partial^2}{\partial x_j \partial x_j}(\Psi_\beta - \Phi_\beta)\}$$

where $\Psi_\beta = -\tfrac{1}{2}iH_0^{(1)}(K_\beta R)$, $\Phi_\beta = -\tfrac{1}{2}iH_0^{(1)}(k_\beta R)$ and $R = |\underline{r}_p - \underline{r}_Q|$.

Next, we define elastic singe-layer and double-layer potentials by

$$(S_\beta\underline{\mu})(P) = \int_S \underline{\mu}(q).\underline{\underline{G}}_\beta(q;P)ds_q \tag{3.1}$$

and

$$(D_\beta \underline{\mu})(P) = \int_S \underline{\mu}(q) . T^q{}_\beta \underline{\underline{G}}_\beta(q;P) ds_q \qquad (3.2)$$

respectively, where $T^q{}_\beta \equiv T_\beta$ applied at $q \in S$. $(S_\beta \underline{\mu})(P)$ is continuous in P as P crosses S, whereas both D_β and $T^p{}_\beta S_\beta$ exhibit jumps, given by

$$D_\beta \underline{\mu} = (\mp I + \bar{K}^*_\beta)\underline{\mu} \quad \text{and} \quad T^p{}_\beta S_\beta \underline{\mu} = (\pm I + K_\beta)\underline{\mu}, \qquad (3.3)$$

where the upper (lower) sign corresponds to $P \longrightarrow p \in S$ from $B_e(B_i)$. Here, K_β and $\bar{K}^*_\beta$ are singular integral operators, defined by

$$K_\beta \underline{\mu} = \int_S \underline{\mu}(q) . T^p{}_\beta \underline{\underline{G}}_\beta(q;p) ds_q, \qquad \bar{K}^*_\beta \underline{\mu} = \int_S \underline{\mu}(q) . T^q{}_\beta \underline{\underline{G}}_\beta(q;p) ds_q$$

where $\bar{K}^*_\beta$ is the Hermitian adjoint of K_β: the asterisk denotes the adjoint with respect to the inner product in $L_2(S)$, and the overbar denotes the complex conjugate. We also require the tractions corresponding to $D_\beta \underline{u}$, defined by

$$N_\beta \underline{\mu} = \mu_\beta^{-1} T^P_\beta D_\beta \underline{\mu}. \qquad (3.4)$$

We note that $\bar{S}^*_\beta = S_\beta$ and $\bar{N}^*_\beta = N_\beta$. N_β is a hypersingular operator, but we can prove the regularisations

$$\mu_\beta N_\beta S_\beta = -I + K^2_\beta \quad \text{and} \quad \mu_\beta S_\beta N_\beta = -I + (\bar{K}^*_\beta)^2. \qquad (3.5)$$

4 SINGLE INTEGRAL EQUATIONS

The basic idea is to use a layer ansatz in one region (B_i, say) and Green's theorem in the other: different combinations lead to different single integral equations. Assume that $\underline{u}_i(P)$ can be represented in B_i as an elastic single-layer potential, with density $\underline{\mu}(q)$:

$$\underline{u}_i(P) = (S_i \underline{\mu})(P), \quad P \in B_i \qquad (4.1)$$

Letting $P \longrightarrow p \in S$, we find that

$$\underline{u}_i(p) = S_i \underline{\mu} \quad \text{and} \quad T_i \underline{u}_i = (-I + K_i)\underline{\mu} \qquad (4.2)$$

From Green's theorem in B_e, we have

$$2\underline{u}_e(P) = (S_e(T_e \underline{u}))(P) - (D_e \underline{u})(P), \quad P \in B_e. \qquad (4.3)$$

Letting $P \longrightarrow p \in S$, this and (2.3) give

$$(I + \bar{K}^*_e)\underline{u} - S_e T_e \underline{u} = 2\underline{u}_{inc}, \quad p \in S. \qquad (4.4)$$

We now substitute (4.2) into (4.4), using (2.2), to give

$$\{(I + \bar{K}^*_e)S_i + S_e(I - K_i)\}\underline{\mu} = 2\underline{u}_{inc} \qquad (4.5)$$

This is a BIE for $\underline{\mu}(q)$, $q \in S$; having found $\underline{\mu}$, $\underline{u}_e$ and $\underline{u}_i$ are to be constructed from

$$2\underline{u}_e(P) = (S_e((-I + K_i)\underline{\mu}))(P) - (D_e(S_i \underline{\mu}))(P) \qquad (4.6)$$

and (4.1), respectively. Kennett (1984; his eqn. (2.17)) would have obtained (4.5) if he had noted the jump relations (3.3).

Equation (4.5) is a Fredholm integral equation of the first kind, with a weakly-singular kernel: $\pm I + K_\beta$ and $\pm I + \bar{K}^*_\beta$ are all singular integral operators with index zero [5, §130], [6, p.215], but composition with S_α gives the stated result [5, §45].

To derive other BIE, we could use the tractions corresponding to (4.3), we could use an elastic double-layer potential in B_i, or we could use a layer potential in B_e and Green's theorem in B_i. However, unlike in the corresponding (scalar) acoustic problem [3], none of these BIE is of the second kind. One route to such an equation is to use (4.3) in B_e and

$$\underline{u}_i(P) = (\tilde{D}_i\underline{\mu})(P), \qquad P \in B_i, \tag{4.7}$$

where

$$(\tilde{D}_i\underline{\mu})(P) = \int_S \underline{\mu}(q).\tilde{T}^q{}_i\underline{\underline{G}}_i(q;P)ds_q \tag{4.8}$$

and $\tilde{T}_i$ is a *generalised traction operator* [4, chpt. 1, §13], [6, p.129], defined by (cf. (2.4))

$$(\tilde{T}_\beta\underline{u})_m(p) = (\lambda_\beta + \mu_\beta - \alpha)n_m \,\mathrm{div}\underline{u} + n_\ell\,(\alpha\,\partial u_m/\partial x_\ell + \mu_\beta\partial u_\ell/\partial x_m)$$

and α is a parameter at our disposal; $\tilde{T}_\beta = T_\beta$ when $\alpha = \mu_\beta$. With an obvious notation, we find that

$$\{(I + \bar{K}^*_e)(I + \bar{\tilde{K}}^*_i) - \mu_i S_e\tilde{N}_i\}\underline{\mu} = 2\underline{u}_{inc}. \tag{4.9}$$

Now,

$$S_e\tilde{N}_i = \gamma S_e N_e + S_e(\tilde{N}_i - \gamma N_e),$$

where γ is another disposable constant. By (3.5) and [5, p.419], the first term is singular with index zero. In general $\tilde{N}_i - \gamma N_e$ is hyper-singular, but we can choose α and γ so that it is merely singular, whence (4.9) will be a quasi-Fredholm integral equation of the second kind. The required values are

$$\alpha = \mu_i\{2(1-\nu_i) + (\nu_e-\nu_i)\}/\Delta \quad \text{and} \quad \gamma = 2(1-\nu_e)/\Delta,$$

where $\Delta = 2(1-\nu_i) + (3-4\nu_i)(\nu_e-\nu_i)$. In the special case of equal Poisson's ratios ($\nu_e = \nu_i$), we find that $\alpha = \mu_i$ and $\gamma = 1$, whence $\tilde{N}_i - \gamma N_e = N_i - N_e$ is not hypersingular; the corresponding problem is well known to be simpler [4, chpt. 12, §5], [6, §36].

5 CONCLUSION

We have sketched the derivation of several different single BIE. Formally, at least, any of these can be used to solve the BCP, i.e. the problem of elastic-wave scattering by an elastic inclusion. We also showed that, by introducing the generalised elastic double-layer potential (4.8), we can obtain a singular integral equation of the second kind, with index zero, i.e. a quasi-Fredholm equation. (Note that Kupradze used generalised elastic double-layer potentials in both B_e and B_i [4, chpt. 12, §2].) Standard theory is now available to treat the solvablity of this equation.

REFERENCES

[1] KENNETT, B.L.N.: Reflection operator methods for elastic waves I-irregular interfaces and regions. Wave Motion 6(1984) 407-418.

[2] Kitahara, M. Boundary integral equation methods in eigenvalue problems of elastodynamics and thin plates. Amsterdam: Elsevier 1985.

[3] Kleinman, R.E.; Martin, P.A.: On single integral equations for the transmission problem of acoustics. SIAM J. Appl. Math., to appear.

[4] Kupradze, V.D. (ed.) Three-dimensional problems of the mathematical theory of elasticity and thermoelasticity. Amsterdam: North-Holland 1979.

[5] Muskhelishvili, N.I. Singular integral equations. Groningen: Noordhoff 1953.

[6] Parton, V.Z.; Perlin, P.I. Integral equations in elasticity. Moscow: Mir 1982.

[7] Rizzo, F.J.; Shippy, D.J.: A formulation and solution procedure for the general non-homogeneous elastic inclusion problem. Int. J. Solids Struct. 4 (1968) 1161-1179.

ELASTIC WAVE PROPAGATION
M.F. McCarthy, M.A. Hayes, (Editors)
Elsevier Science Publishers B.V. (North-Holland), 1989

RESONANCES IN THE BACKSCATTERED ECHOES FROM ELASTIC SHELLS NEAR AN INTERFACE

Michael F. Werby
Naval Ocean R&D Activity
Numerical Modeling Division (221)
NSTL Station, Mississippi
39529-5000 U.S.A.

Guillermo C. Gaunaurd
Naval Surface Warfare Center
Research Dept. (R-43)
Silver Spring, Maryland
20903-5000 U.S.A.

SUMMARY: We study the scattering of c.w. waves and long sound pulses incident on an elastic shell of revolution located near an interface. The method we use is based on a novel version of the extended boundary condition (EBC) approach, combined with the method of images. The model describes, computes, and displays the echoes that are backscattered from the shell, and in some instances, from a solid elastic body of the same shape, as functions of frequency and time. One aim is to assess the influence of the nearby interface on the location of the resonance features, and the smooth "background" that they are superimposed on, as the distance to the interface increases. The present T-Matrix formulation allows the treatment of any arbitrary target-shape. The time-domain graphs, display the individual resonance effects as a succession of delayed wave-packets which follow the specular return.

INTRODUCTION

Acoustical scattering from submerged impenetrable objects, has been known to be only of purely academic value for four decades. Sound scattering for penetrable objects, such as elastic shells and solid bodies, has received much attention in the last thirty years. Much work has addressed the case of the "separable" target [1-6]. Non-separable scatterers have been the subject of more recent interest [7-13]. Both geometrical (i.e., shape) and resonance (i.e., composition) effects have been considered [6,13], with the later ones yielding a powerful means for target-identification [14-16]. Scatterers in bounded media are harder to treat. There have been some instances in which sound scattering by an elastic body in a half-space [17] has been tackled. The same is true for elastic objects in waveguides [18]. To extract numbers out of these theoretical analyses has proven to be quite formidable. The present study develops a theoretical method to compute and predict the scattering patterns of bodies near an impenetrable smooth interface, particularly for elastic bodies and shells of revolution. We also perform a parametric study using this method, to determine the effect of the interface on the "resonance features" and the "background" of the target's backscattering cross-section (BSCS). "Backgrounds" and "resonances" are terms of the resonance scattering theory (RST), reviewed elsewhere [19] in these Proceedings, and elsewhere [6, 13:pp. 413-430, & 20].

I. SCATTERING FROM ELASTIC BODIES NEAR INTERFACES (EBC APPROACH)

Figure 1 depicts a spheroidal body near a soft/rigid interface. A long c.w.-pulse is incident on the body from above. The body is either an elastic spheroid, or sphere, or a spherical shell. The EBC arising from Waterman's work [21,13] on the displacement field is

$$\vec{u}_r(\vec{r}) = \vec{u}_i(\vec{r}) + \iint_S [\vec{u}(\vec{r}')\vec{\nabla} g_h - g_h \vec{\nabla}\vec{u}(\vec{r}')] \cdot d\vec{S} \quad , \tag{1}$$

where $\vec{u}_r$ vanishes inside the object. This leads to three conditions, in the interior of the object, in the interior of its image, and at the exterior domain of both. The associated Green's function is,

$$g_h(\vec{r},\vec{r}') = g(\vec{r},\vec{r}') \pm g(\vec{r},\vec{r}'') \tag{2}$$

where, $\vec{r}' = (x', y', z'-d)$ and $\vec{r}'' = (x', y', -z'-d)$. The $\pm$ sign refers to a soft/rigid interface. The jump B.C. for the displacement and traction vectors at the target's surface are,

$$\vec{u}_+ \cdot \hat{n} = \vec{u}_- \cdot \hat{n} \quad , \quad \vec{t}_+ \cdot \hat{n} = \vec{t}_- \cdot \hat{n} \quad , \quad \vec{t} \times \hat{n} = 0 \tag{3}$$

and for an internally evacuated shell one also has: $\vec{t}_{in} = 0$. The outward normal is $\hat{n}$, and $\vec{t}$ is the traction vector on $S = S_1 + S_2$. There are several sets of null-field equations inside S which together with Eqs. (3) yield additional constraints. We write [21],

$$\vec{u}_s(\vec{r}) = \vec{u}(\vec{r}) - \vec{u}_i(\vec{r}) = \sum_i f_i \vec{\varphi}_i - \sum_i a_i \, \mathrm{Reg}\, \vec{\varphi}_i \, , \tag{4}$$

where φ_i is an outgoing (velocity potential) partial-wave, f_i are the partial-wave scattering coefficients, Reg φ_i is the incident partial-wave, and a_i are its known (partial-wave) coefficients. This leads to 3 equations:

$$\begin{cases} r(-\vec{d})\, a_j = i \sum_k Q_{jk}\, \alpha_{1k} + i \sum_k \sigma(-2\vec{d})\, \mathrm{Reg}\, Q_{jk}\, \alpha_{2k} \\ R(\vec{d})\, a_k = i \sum_j \sigma(2\vec{d})\, \mathrm{Reg}\, Q_{kj}\, \alpha_{1j} + i \sum_j Q_{kj}\, \alpha_{2j} \\ f_k = -i \sum_j R(\vec{d})\, \mathrm{Reg}\, Q_{jk}\, a_{1j} - i \sum_j R(-\vec{d})\, \mathrm{Reg}\, Q_{jk}\, \alpha_{2j} \, , \end{cases} \tag{5}$$

where the Q's are: $Q_{jk} = -k \iint_{S_1} \mathrm{Reg}\, \varphi_n(\vec{r}_1{}') \vec{\nabla} \varphi_n(\vec{r}_1{}') \cdot \hat{n}\, dS$.

We note that the Q's are in an nx3n space, and must be coupled with additional constraints to yield the appropriate T-Matrix for an elastic object. The translation operators $R(\vec{d})$ and $\sigma(\vec{d})$ are borrowed from Solid State Physics [22], in order to translate Bessel and Hankel functions to a common origin as required in the integrals. If $\vec{f}$ and $\vec{a}$ are the scattered and incident fields, then $\vec{f} = \mathbb{T}\vec{a}$, where T is the T-Matrix. In this case it turns out to be given by:

$$\mathbb{T} = R(\vec{d})\, T\, D_+ M_- + R(-\vec{d})\, T\, D_- M_+ \quad , \tag{6}$$

where

$$\begin{cases} D_\pm = [1 - \sigma(\mp 2\vec{d})\, T\, \sigma(\pm 2\vec{d})\, T\,]^{-1} \\ M_\mp = R(\mp\vec{d}) + \sigma(\mp 2\vec{d})\, T R(\pm\vec{d}) . \end{cases} \quad \text{and}$$

The incident field (from above) will be taken as the exact solution of Helmholtz equation for a scatterer near an infinite <u>rigid</u> half-space. In this case the displacement field is given by

$$u_i(r) = \exp(i\vec{k}' \cdot \vec{r}') + \exp(i\vec{k}'' \cdot \vec{r}'') , \tag{7}$$

for a linear combination of fields due to a point source and its image in the upper and lower half-planes, respectively.

II. NUMERICAL CALCULATIONS AND ANALYSIS OF OUR RESULTS.

Figure 2 displays $|f_\infty|$ for an aluminum spherical shell in water in: $3 \leqslant ka \leqslant 10$. Here, h/a=0.1%, thus, only two resonances appear in this band. The shell is in free-space, viz., $R=\infty$. Since the shell is so thin, the background [6,19] is soft in this case [10]. Figure 3 shows $|f_\infty|$ for a solid WC-sphere in a boundless water medium for: $3 \leqslant ka \leqslant 10$. Now, the background is rigid. This BSCS displays an initial rigid-like oscillatory background behavior, and two resonances [viz., (2,1) and (1,2)], are seen to be superimposed on it. These are the first Rayleigh and the first whispering gallery resonances, respectively, [22]. Figure 4 shows $|f_\infty|$ for an aluminum shell in water in: $3 \leqslant ka \leqslant 10$. Here R=2, i.e., the shell is two diameters above the interface, in the upper half-space. This is a crude model for an elastic target (solid/hollow) above a sea bottom. All plots for the shell (solid sphere) require use of the soft (rigid) back-ground. Figure 5 displays $|f_\infty|$ for the same aluminum shell, now for: R=4. Figure 6 exhibits $|f_\infty|$ for the solid WC sphere, at two diameters above the interface (viz., R=2). We can now study the effect of the interface on the target's BSCS.

For the WC sphere we compare Fig. 3 (sphere infinitely-far away from the interface), to Fig. 6, computed for R=2. For the shell we compare Fig. 2 ($R=\infty$), to Fig. 5 (R=4), to Fig. 4 (R=2; the closest to the interface). For both sphere and shell, the following observations hold. The resonance features remain at about the same spectral locations close or far away from the boundary. The resonances manifest themselves as prominent spiky peaks (rather than deep dips) as the target approaches the boundary. The prominent size of these peaks is more noticeable for shells. These peaks, particularly the first one, stick out considerably above the background. The width of these peaks tends to be broader (narrower) for solid targets (hollow targets) as the body approaches the interface. Finally, the background level increases as the target approaches the interface. For the sphere it increases from level unity (Fig. 3, $R=\infty$) to $\sim$ 2.5 (Fig. 6, R=2). For the shell it increases from unity (Fig. 2, $R=\infty$), to two (Fig. 5, R=4), and eventually to three (Fig. 4, R=2). For the sphere, the low-frequency oscillatory behavior of the $|f_\infty|$ in free-space is washed-out near the interface.

Time-domain calculations are shown in Figs. 7 and 8, for the sphere in free-space ($R=\infty$), or at four diameters from the interface (R=4). We use a non-dimensional time $\tau=ct/a$. Fig. 7, for example, is the inverse Fourier transform (generated numerically) of the spectrum in Fig. 2. We actually use this spectrum over a much broader band such as we show it in the bottom graph of Fig. 9. The successive resonances appear as a "tail" following the specular return in: $0 \leqslant \tau \leqslant 50$ (not shown). The specular return is over ten times larger than the first Rayleigh (2,1) resonance, and goes way beyond the scale of these plots. The similarity between Figs. 7 and 8 is an indication that in the time-domain, R=4 is already "distant enough" to give the same result as $R=\infty$ (free-space). Plots closer to the interface will be generated later to verify the nature of this apparent "skin effect." Figure 9 contrasts the $|f_\infty|$ of an elastic (lower graph) sphere to that of a rigid (top graph) sphere. The differences are striking. This confirms the inadmissibility of the impenetrability assumption, as we stated at the start of the INTRODUCTION. Further details will appear elsewhere.

CONCLUSIONS

Proximity to the interface tends to: (1) smooth out the low frequency oscillatory behavior of the BSCS, (2) increase the overall background level, (3) make the resonances more "spiky" and of more prominent amplitude than that of the corresponding background. (4) In the time-domain, the boundary influence seems to be confined to a "skin layer" bounded by $R \leqslant 4$.

REFERENCES

1. R. Goodman & R. Stern, J. Acoust.Soc.Am., 34, 338-344, (1962).
2. R. Hickling, J. Acoust. Soc. Amer., 36, 1124-1137, (1964).
3. C. Horton et. al., J. Acoust. Soc. Am., 34, 1929-1932, (1962).
4. M. Junger, J. Acoust. Soc. Amer., 69, 1568-1572, (1981).
5. N. Veksler, Information Analysis in Hydroelasticity, Acad. Sc. Estonian SSR, Tallin-Valgus, (In Russian), (1982).
6. G. Gaunaurd & A. Kalnins, Intern. J. Solids and Struct.,18, 1083-1093, (1982).
7. M. Werby & G. Gaunaurd, Acoust. Letters 9, 89-93, (1986).
8. M. Werby & G. Gaunaurd, J.Acoust.S.Am., 82, 1369-1377, (1987).
9. G. Gaunaurd & M. Werby, J.Acoust.S.Am., 82, 2021-2011, (1987).
10. G. Gaunaurd & M. Werby, Intern. J. Solids & Struct., 22, 1149, (1986).
11. G. Gaunaurd & M. Werby, J.Acoust.S.Am.,77, 2081-2093,(1985).
12. G. Gaunaurd & D. Brill, JASA 75, 1680-1693, (1984).
13. V.K. Varadan & V.V. Varadan, Editors, Acoustic, EM and Elastic Wave-Scattering, Pergamon Press, New York, (1979).
14. G. Gaunaurd, IEEE J. Ocean. Engr., OE-12, 419-423, (1987).
15. V. Ayres & G. Gaunaurd, JASA 82, 1291-1302, (1987).
16. D. Brill et. al., Acustica, Vol. 53, 11-18, (1983).
17. G. Gaunaurd & M. McCarthy, IEEE J. Ocean. Engr., OE-12, 395-404 (Special Issue on Scattering, 1987).
18. M. Werby & R. Evans, IEEE J.O.E., OE-12, 380-394, (1987).
19. G. Gaunaurd, "Acoustic scattering from underwater elastic bodies," This Conference, pp. , (1988).
20. G. Gaunaurd & H. Ueberall, JASA 73, 1-12, (1983).
21. P. Waterman, JASA 60, 567, (1976); ibid., 45, 1417, (1969).
22. M. Danos & L. Maximon, J. Math. Phys., 6, (#5), 766, (1964).

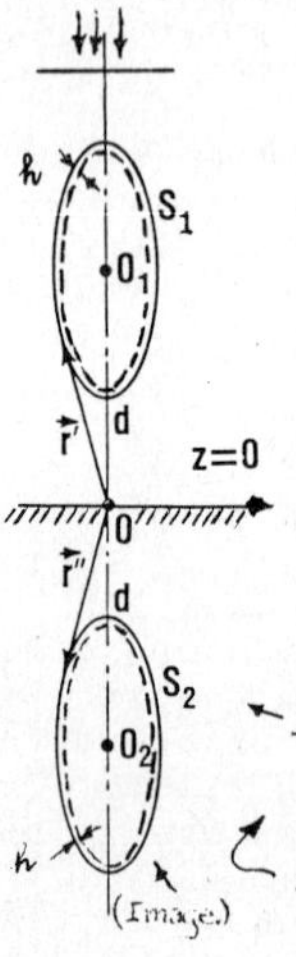

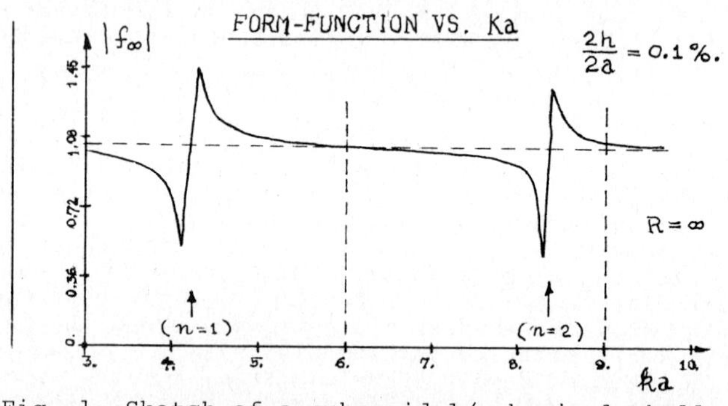

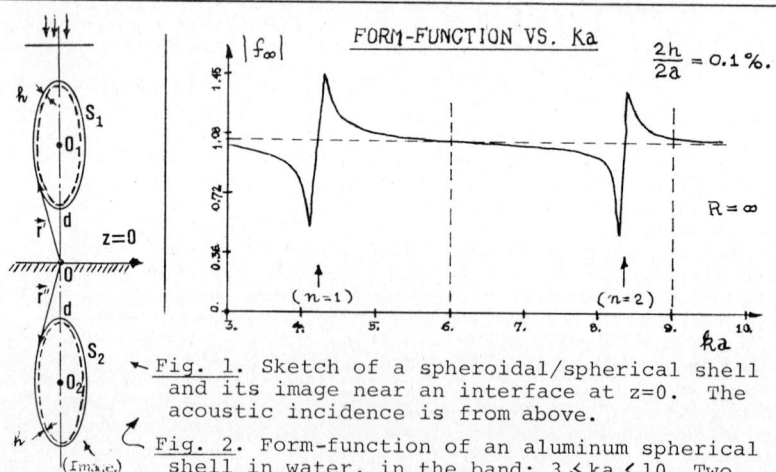

Fig. 1. Sketch of a spheroidal/spherical shell and its image near an interface at z=0. The acoustic incidence is from above.

Fig. 2. Form-function of an aluminum spherical shell in water, in the band: $3 \leqslant ka \leqslant 10$. Two resonance features are noticeable here.

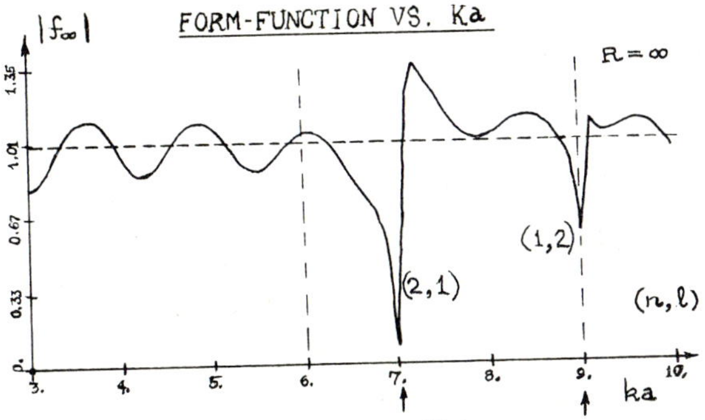

Fig. 3. Form-function of a solid WC (tungsten carbide) sphere in water in the same band: 3 ⩽ ka ⩽10. (In a boundless medium). The two observable resonance features are at different spectral locations and are superimposed on a different background.

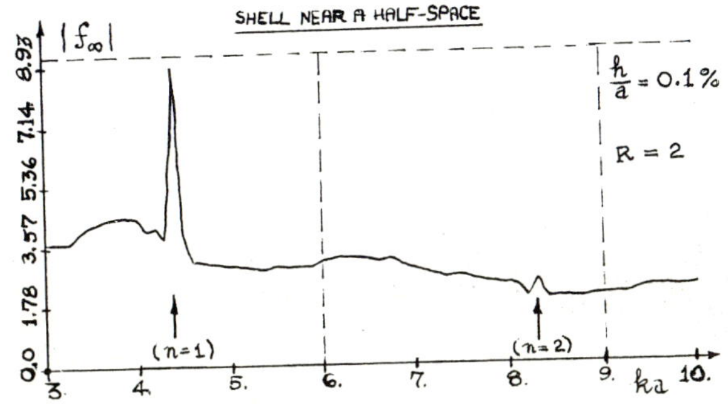

Fig. 4. Form-function of an aluminum spherical shell in water in the band: 3 ⩽ ka ⩽ 10. Here, the shell is two diameters away from the interface, i.e., R=2.

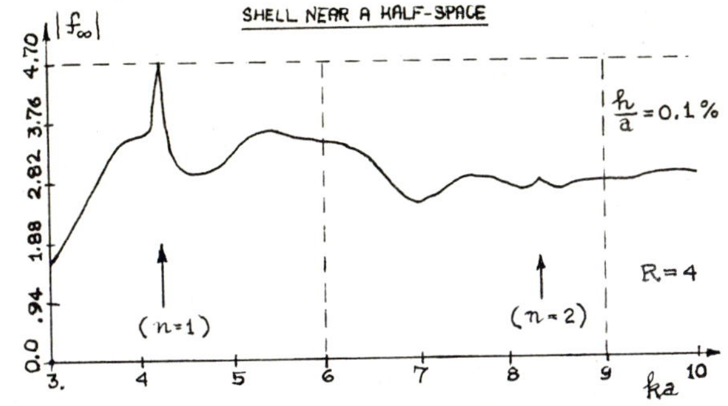

Fig. 5. Form-function for an aluminum spherical shell in water in the band: 3 ⩽ ka ⩽10. Here R=4 i.e., the shell is four diameters away from the interface.

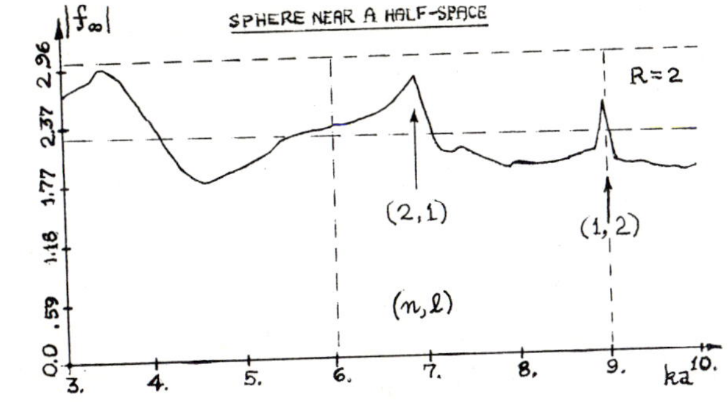

Fig. 6. Form-function for a solid, WC sphere in a water half-space in the band: 3 ⩽ ka ⩽ 10. Here R=2, i.e., the depth is 2 diameters from the interface.

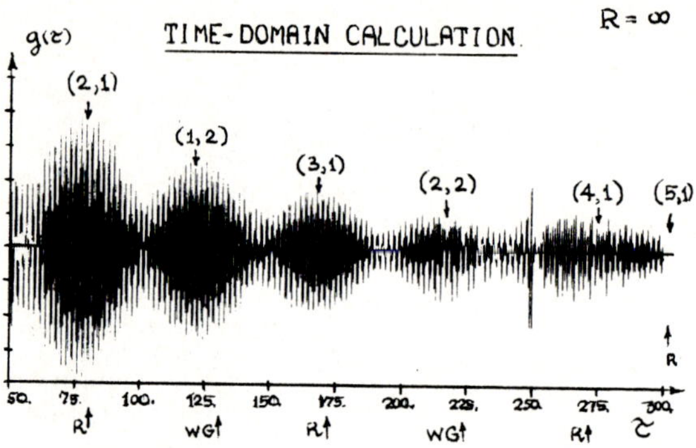

Fig. 7. Time-domain response of a solid elastic (WC) sphere in a boundless water medium. The time is non-dimensionalized to ct/a The early-time region (not shown), contains the specular return which is large and goes out of this scale. The spectrum of this waveform is given in Fig. 3.

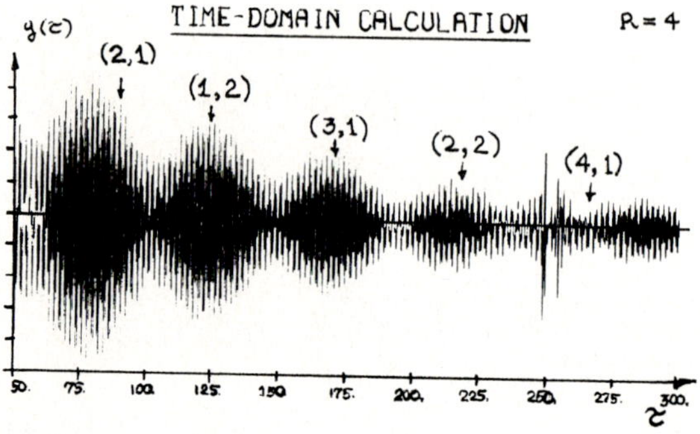

Fig. 8. Time-domain response of an elastic WC sphere inmersed in a bounded water medium. Here, R=4 (i.e., four diameters away from the interface.) The spectrum of this waveform is given in Fig. 5. This is the resonance "tail" following the specular return (not shown). The various resonances appear as succesive wavepackets.

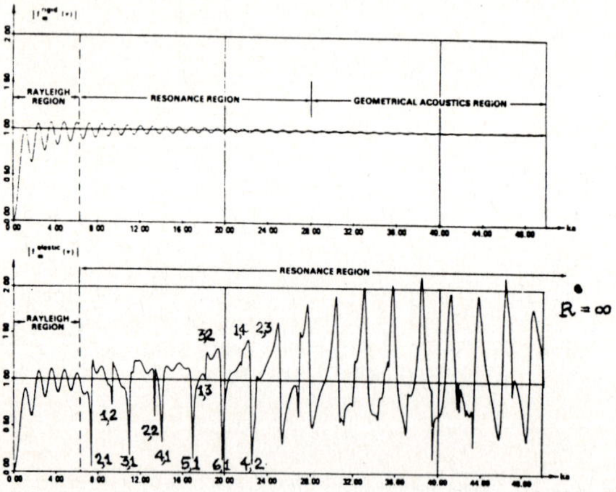

Fig. 9. Form-function of an elastic (WC) sphere in a boundless sea (bottom plot) contrasted with that of a rigid sphere (top). The Rayleigh and "resonance" regions are clearly distinguishable in both instances. The bottom graph is an enlargement of Fig. 3 over a much broader band. Rayleigh and "whispering gallery" waves are observable here.

ELASTIC WAVE PROPAGATION
M.F. McCarthy, M.A. Hayes, (Editors)

DYNAMIC STRESS INTENSITY FACTORS FOR 3D NON–PLANAR CRACKS

Peter OLSSON* and Anders BOSTRÖM*

Division of Mechanics
Chalmers University of Technology
S-412 96 Göteborg
Sweden

By means of a recent modification of the null field approach, dynamic stress intensity factors are computed for non-planar cracks. The cracks considered are of two types, open and fluid-filled, respectively. As incident fields plane P and S waves are chosen. While the cracks for simplicity are taken to be rotationally symmetric, the plane waves are allowed to have any direction of incidence. Two orthogonal polarizations of the shear waves are considered.

1. INTRODUCTION

The scattering of ultrasonic waves by cracks in elastic materials is a problem that has invoked much interest lately. It is mostly the crack-scattered far fields that are computed, although the dynamic stress intensity factors are also of importance. For the penny-shaped crack Martin and Wickham [1] and Krenk and Schimdt [2] define and show how to compute the stress intensity factors, but no numerical examples are given. The elliptic crack has been treated by Roy [3] for low frequencies but no numerical results are given.

In the present paper the dynamic stress intensity factors for non-planar open or fluid-filled cracks are computed by the null field approach. In Boström and Olsson [4] and Olsson [5] the corresponding scattering problems are considered and results for the far fields are given, and the presentation will therefore be very cursory here.

2. THE NULL FIELD APPROACH

Consider a non-planar crack with surface S_c in an otherwise homogeneous elastic matrix with shear modulus μ, Poisson ratio ν, and longitudinal and transverse wavenumbers k_p and k_s, respectively (the time factor $\exp(-i\omega t)$ is omitted throughout). A closed surface S (with outward-pointing normal $\hat{n}$) is formed by adjoining to S_c an, in principle arbitrary, fictitious surface S_w. For the null field approach to be numerically useful the surface S should not be to far from spherical.

*Sponsored by the National Swedish Board for Technical Development (STU).

In the null field approach all pertinent fields are expanded in spherical vector waves. The regular spherical vector waves are denoted $\mathrm{Re}\psi_n$, the outgoing ones are denoted ψ_n, and the vector spherical harmonics are denoted $\mathbf{A}_n$, where n is a quadruple index. The incoming displacement field is expanded in the regular waves

$$\mathbf{u}^{in} = \sum_n a_n \mathrm{Re}\psi_n \tag{1}$$

where the expansion coefficients a_n are assumed known. As the null field approach employs surface integral representations, expansions for the surface fields on S also play an important role. These are chosen as

$$\mathbf{u}_+ = \sum_n \alpha_n \mathbf{A}_n \tag{2}$$

$$\mathbf{v} = \sum_n \beta_n \mathbf{A}_n \tag{3}$$

where $\mathbf{u}_+$ is the displacement field on S in the limit from the outside and

$$\mathbf{v} = \begin{cases} Z^{-1}\cdot(\mathbf{u}_+ - \mathbf{u}_-) & \text{on } S_c \\ 1/\mu k_s \; Z^{-1}\cdot \mathbf{t} & \text{on } S_w \end{cases} \tag{4}$$

Here $(\mu k_s)^{-1}$ is introduced for dimensional purposes and the dyadic Z will later be chosen so that **v** becomes continuous, i. e. Z takes care of the singularity at the crack edge. **v** is thus essentially equal to the crack opening displacement on the cracked part S_c and equal to the traction **t** on the welded part. In (2) and (3) other expansion functions are also possible, the regular waves $\mathrm{Re}\psi_n$, for instance.

Starting from the surface integral representations it is now straightforward to derive the following relations between the expansion coefficients in (1)–(3) (in obvious vector and matrix notation):

$$a = -i(Q^0\alpha + Q^1\beta) \tag{5}$$

$$0 = \mathrm{Re}Q^0\alpha + (\mathrm{Re}Q^1 - \mathrm{Re}Q^2)\beta \tag{6}$$

For the open crack the matrices are

$$Q^0_{nn'} = k_s/\mu \int_S \mathbf{t}(\psi_n) \cdot \mathbf{A}_{n'}\, dS \tag{7}$$

$$Q^1_{nn'} = -k_s^2 \int_{S_w} \psi_n \cdot Z\cdot \mathbf{A}_{n'}\, dS \tag{8}$$

$$\mathrm{Re}Q^2_{nn'} = k_s/\mu \int_{S_c} \mathbf{t}(\mathrm{Re}\psi_n)\cdot Z \cdot \mathbf{A}_{n'}\, dS \tag{9}$$

while for the fluid-filled crack Q^0 is the same and

$$Q^1_{nn'} = -k_s^2 \int_{S_c} \hat{n} \cdot \psi_n \hat{n} \cdot Z \cdot A_{n'} dS - k_s^2 \int_{S_w} \psi_n \cdot Z \cdot A_{n'} dS \tag{10}$$

$$\mathrm{Re}Q^2_{nn'} = k_s/\mu \int_{S_c} (\mathbf{t}(\mathrm{Re}\psi_n))_{tan} \cdot Z \cdot (\mathbf{A}_{n'})_{tan} dS \tag{11}$$

$\mathrm{Re}Q^0$ and $\mathrm{Re}Q^1$ contain regular instead of outgoing wave functions. **t** is the traction operator that gives the partial traction corresponding to a partial wave. To compute the stress intensity factors the surface field **v** is needed and (5) and (6) are therefore solved for β:

$$\beta = (Q^1 - Q^0(\mathrm{Re}Q^0)^{-1}(\mathrm{Re}Q^1 - \mathrm{Re}Q^2))^{-1} a \tag{12}$$

To achieve a satisfactory numerical convergence it is essential to choose the dyadic Z so that the surface field is continuous at the crack edge. This point is discussed and exemplified in detail in Boström and Olsson [4] and for determining the stress intensity factors it is even more important. For a rotationally symmetric crack the dyadic Z is for the open crack chosen as

$$Z(\mathbf{r}) = \begin{cases} I\,(k_s|\mathbf{r}-\mathbf{r}_C|)^{1/2} & \text{on } S_c \\ [\hat{n}\hat{n} + \hat{\tau}\hat{\tau} + \hat{\phi}\hat{\phi}(1-\nu)] \,/\, [4(1-\nu)(k_s|\mathbf{r}-\mathbf{r}_C|)^{1/2}] & \text{on } S_w \end{cases} \tag{13}$$

and for the fluid-filled crack as

$$Z(\mathbf{r}) = \begin{cases} \hat{n}\hat{n} + (\hat{\tau}\hat{\tau} + \hat{\phi}\hat{\phi})(k_s|\mathbf{r}-\mathbf{r}_C|)^{1/2} & \text{on } S_c \\ \hat{n}\hat{n} + [\hat{\tau}\hat{\tau}(1-\nu)^{-1} + \hat{\phi}\hat{\phi}] \,/\, [4(k_s|\mathbf{r}-\mathbf{r}_C|)^{1/2}] & \text{on } S_w \end{cases} \tag{14}$$

where I is the unit dyadic, $\hat{\tau} = \hat{\phi} \times \hat{n}$, and $\mathbf{r}_C$ is that point on the crack edge C (which is a circle with tangent vector $\hat{\phi}$) that is closest to **r**. With this choice of Z the surface field **r** becomes both bounded and continuous, but the first derivatives of **v** can in general be expected to be discontinuous on C.

The dynamic stress intensity factors as defined by Krenk and Schmidt [1] have the disadvantages of not being dimensionless, of going to zero for low frequencies, and of being dependent on the amplitude of the incoming wave. To remedy these deficiences their expressions are divided by $u_o \mu k_s \sqrt{c}$, where u_o is the amplitude of the incoming wave (which will be taken as a plane wave in the numerical examples) and c is the radius of the crack. In terms of **v** the dynamic stress intensity factors are thus defined as

$$K_1(\phi) = \hat{n} \cdot \mathbf{v}\, [u_o(1-\nu)(8k_sc)^{1/2}]^{-1} \tag{15}$$

$$K_2(\phi) = \hat{\tau} \cdot \mathbf{v}\, [u_o(1-\nu)(8k_sc)^{1/2}]^{-1} \tag{16}$$

$$K_3(\phi) = \hat{\phi} \cdot \mathbf{v}\, [u_o\,(8k_sc)^{1/2}]^{-1} \tag{17}$$

where ϕ is the azimuthal angle along the circular crack edge and the value of **v** on the crack edge is taken.

3. NUMERICAL EXAMPLES

A few examples of stress intensity factors will now be given. The numerical procedure will not be discussed at all as some details can be found in Boström and Olsson [4] and further references given therein.

Only one shape of the non-planar crack is considered, namely a spherical cap with half-angle 60° and crack-edge radius c. The Poisson ratio is fixed to $\nu = 0.25$. The incoming plane waves are longitudinal (P), transverse, polarized in the plane of the incoming direction and the symmetry axis of the crack (SV), or transverse, polarized perpendicular to this plane (SH). The direction of propagation of the plane wave makes the angle α with the symmetry axis of the crack, and the plane wave hits the crack edge before the crack surface for $\alpha = 0^o$.

Figure 1 shows the absolute value of the stress intensity factor K_2 for open and fluid-filled cracks with incident P waves along the symmetry axis from both directions ($\alpha = 0^o$ and $\alpha = 180^o$) as a function of frequency $k_s c$. Note that K_2 is independent of ϕ for these cases. In the static limit the open and fluid-filled cracks have different limits for K_2 that are, however, independent of the direction of incidence. Increasing the frequency there are resonance peaks in K_2, but in all cases these peaks are less than twice the static limit.

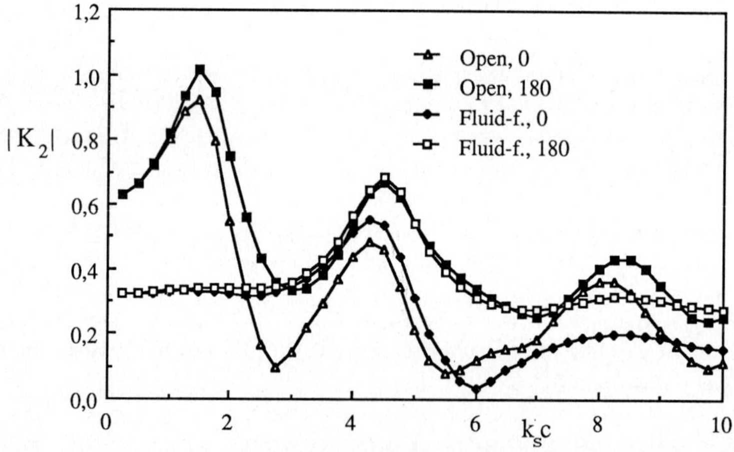

FIGURE 1
The stress intensity factor K_2 as a function of frequency for incident P waves along the symmetry axis on open and fluid-filled cracks in the shape of a spherical cap with half-angle 60° and radius c.

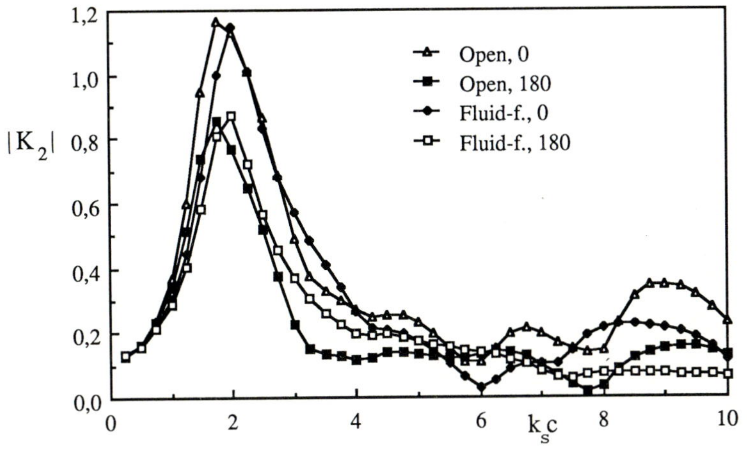

FIGURE 2
Same as Fig. 1 but for incident S waves.

Figure 2 shows the absolute value of the stress intensity factor K_2 for open and fluid-filled cracks with incident S (SV or SH) waves along the symmetry axis from both directions ($\alpha = 0^o$ and $\alpha = 180^o$) as a function of frequency $k_s c$. K_2 varies as cos ϕ for these cases and the value for

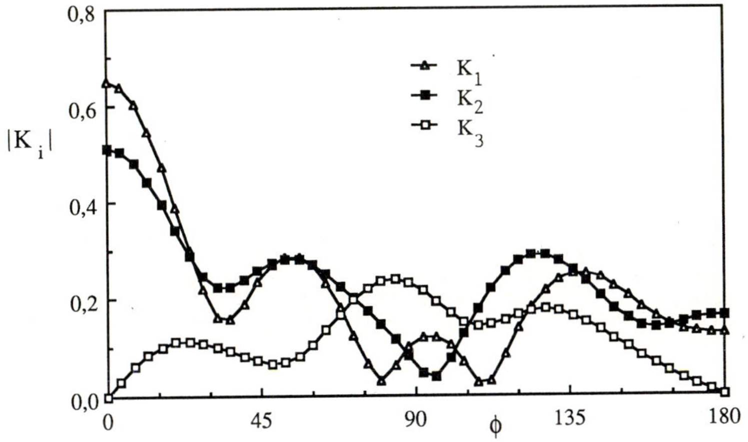

FIGURE 3
The stress intensity factors as a function of the angle ϕ along the crack edge for the frequency $k_s c$ = 5 and an SV wave incident perpendicular to the symmetry axis on an open crack in the shape of a spherical cap with half-angle 60° and radius C.

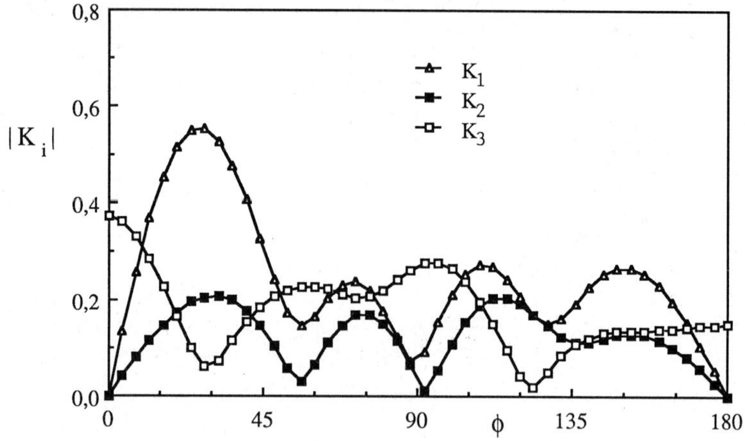

FIGURE 4
Same as Fig. 3 but for an incident SH wave.

$\phi = 0^o$ is shown. As opposed to the incident P waves the static limit is the same for open and fluid-filled cracks. There is also a high resonance peak around $k_sc = 2$ that for $\alpha = 0^o$ is nearly ten times the static limit and for $\alpha = 180^o$ is about seven times the static limit.

Figures 3 and 4 show how the absolute values of K_1, K_2, and K_3 for an open crack varies around the edge for frequency $k_sc = 5$ and incident SV and SH waves, respectively, with $\alpha = 90^o$. In both cases K_1 shows the largest peak value.

REFERENCES

[1] Martin, P.A. and Wickham, G.R., Proc. Roy. Soc. London Ser. A 390 (1983) 91.

[2] Krenk, S. and Schmidt, H., Phil. Trans. Roy. Soc. London Ser. A 308 (1982) 167.

[3] Roy, A., Int. J. Engng. Sci. 25 (1978) 155.

[4] Boström, A. and Olsson, P., Wave Motion 9 (1987) 61.

[5] Olsson, P., J. Nondestructive Eval. 5 (1986) 161.

ELASTIC WAVE PROPAGATION
M.F. McCarthy, M.A. Hayes, (Editors)

OPTIMAL GEOMETRICAL AND PHYSICAL BOUNDS FOR ELASTIC RAYLEIGH SCATTERING

GEORGE DASSIOS

Department of Mathematics, University of Patras, GR 261 10 Patras - Greece.

An incident longitudinal or transverse plane elastic wave is scattered by a rigid body whose characteristic dimension is much smaller than the wavelength of the incident wave. The leading low-frequency approximation, known as Rayleigh scattering, provides an approximation for the spherical scattering amplitudes in terms of a surface traction integral. Upper and lower geometrical bounds for the Rayleigh traction integral are obtained through the best fitting interior and exterior confocal triaxial ellipsoids. These bounds are optimal in the sense that they both become equalities when the scatterer degenerates to an ellipsoid. Another pair of bounds is also derived in terms of the capacity of the scatterer.

1. INTRODUCTION

Rayleigh scattering deals with the effect of a small (compared to the wavelength) discontinuity upon the propagation of a time harmonic wave. We discuss here the corresponding problem whenever the incident wave is an elastic longitudinal or transverse plane wave and the scatterer is an elasticaly rigid body. An extensive analysis of this and other scattering problems is given in [1] while in [2] and [3] the general technique has been applied to the case of a rigid ellipsoid and an ellipsoidal cavity, respectively.

The most important quantity in Scattering Theory is the scattering amplitude. It is a function of two directions $\hat{\underset{\sim}{k}}$ and $\hat{\underset{\sim}{r}}$ and describes the effect of the scattering process in the far-field, in the direction $\hat{\underset{\sim}{r}}$, which is due to a plane wave excitation in the direction $\hat{\underset{\sim}{k}}$. In [1], it is proved that the radial as well as the angular part of the scattering amplitude have leading low - frequency terms that are proportional to the integral of the traction field over the surface of the scatterer. In other words, it is this particular integral that determines all of the physically important quantities associated with the low - frequency scattering by a rigid body, on the boundary of which the displacement field vanishes. The present work aims toward a study of this integral as far as lower and upper bounds are concerned. The detailed analysis and proofs can be found in a joint paper with L.Payne [4]. The well known principles of Dirichlet and Thomson are extended to their counterparts in the theory of elasticity and they are used to produce upper and lower bounds for the basic surface traction integral.

2. STATEMENT OF THE PROBLEM

Let $V \subset \mathbb{R}^3$ be a regular exterior domain, in the sense that its complement V^c is a compact and connected set with PC^1- smooth boundary S. The set V^c describes the scatterer, while the exterior domain V is filled with a homogeneous and isotropic elastic material characterized by the Lamé constants μ and λ as well as the constant mass density ρ=1. We suppress the harmonic time dependence $e^{-i\omega t}$ and assume an incident plane wave $\underset{\sim}{\Phi}(\underset{\sim}{r})$, having the form

$$\underset{\sim}{\Phi}^p(\underset{\sim}{r}) = \hat{\underset{\sim}{k}}\, e^{ik_p \hat{\underset{\sim}{k}}\cdot\underset{\sim}{r}} \tag{1}$$

in the case of a longitudinal incident field, or

$$\underset{\sim}{\Phi}^s(\underset{\sim}{r}) = \hat{\underset{\sim}{b}}\, e^{ik_s \hat{\underset{\sim}{k}}\cdot\underset{\sim}{r}} \tag{2}$$

in the case of a transverse incident field. The unit vector $\hat{\underset{\sim}{k}}$ denotes the direction of propagation while the unit vector $\hat{\underset{\sim}{b}}$ is orthogonal to $\hat{\underset{\sim}{k}}$ and indicates the direction of polarization of the transverse wave $\underset{\sim}{\Phi}^s$. The constants k_p, k_s are the wave numbers of the longitudinal and the transverse wave, respectively. They are connected with the corresponding phase velocities c_p and c_s via $\omega = c_p k_p = c_s k_s$, $c_p = \sqrt{\lambda+2\mu}$, $c_s = \sqrt{\mu}$, where ω is the constant angular frequency of the incident wave.

The rigid scattering problem consists in finding the unique scattered field $\underset{\sim}{u}(\underset{\sim}{r})$ in V, such that

$$(\Delta^* + \omega^2)\underset{\sim}{u}(\underset{\sim}{r}) = \underset{\sim}{0}, \quad \underset{\sim}{r} \in V \tag{3}$$

$$\underset{\sim}{u}(\underset{\sim}{r}) = -\underset{\sim}{\Phi}(\underset{\sim}{r}), \quad \underset{\sim}{r} \in S \tag{4}$$

and $\underset{\sim}{u}$ satisfies the Kupradze's radiation conditions as $r \to +\infty$ (5). The operator

$$\Delta^* = c_s^2\Delta + (c_p^2 - c_s^2)\nabla\nabla. \tag{5}$$

is the basic operator of the linear theory of static elasticity, which is strongly elliptic for $\mu > 0$ and $\lambda + 2\mu > 0$.

In [1], it has been shown that the scattered field $\underset{\sim}{u}$ has the following spherical decomposition in the far-field region as $r \to +\infty$

$$\underset{\sim}{u}(\underset{\sim}{r}) = g_r(\hat{\underset{\sim}{r}},\hat{\underset{\sim}{k}})\,\hat{\underset{\sim}{r}}\,h(k_p r) + \left[g_\theta(\hat{\underset{\sim}{r}},\hat{\underset{\sim}{k}})\hat{\underset{\sim}{\theta}} + g_\varphi(\hat{\underset{\sim}{r}},\hat{\underset{\sim}{k}})\hat{\underset{\sim}{\varphi}}\right]h(k_s r) + 0\left(\frac{1}{r^2}\right), \tag{6}$$

where $h(x) = \frac{e^{ix}}{ix}$ is the zeroth order radiative spherical Hankel function, $\hat{\underset{\sim}{r}}$, $\hat{\underset{\sim}{\theta}}$, $\hat{\underset{\sim}{\varphi}}$ are the unit vectors of the spherical coordinate system, while the radial g_r and the tangential g_θ, g_φ normalized (dimensionless) spherical scattering amplitudes have the low-frequency expansions

$$g_r(\hat{\underset{\sim}{r}},\hat{\underset{\sim}{k}}) = -\frac{ik}{4\pi\mu}\,\tau^3\,\hat{\underset{\sim}{r}}\cdot\oint_S T\,\underset{\sim}{u}_0(\underset{\sim}{r})ds + 0(k^2) \tag{7}$$

$$g_\theta(\hat{\underset{\sim}{r}},\hat{\underset{\sim}{k}}) = -\frac{ik}{4\pi\mu}\,\hat{\underset{\sim}{\theta}}\cdot\oint_S T\,\underset{\sim}{u}_0(\underset{\sim}{r})ds + 0(k^2) \tag{8}$$

$$g_\varphi(\hat{\underset{\sim}{r}},\hat{\underset{\sim}{k}}) = -\frac{ik}{4\pi\mu}\,\hat{\underset{\sim}{\varphi}}\cdot\oint_S T\,\underset{\sim}{u}_0(\underset{\sim}{r})ds + 0(k^2) \tag{9}$$

as $k \to 0+$. The dimensionless parameter τ^2 is defined to be the ratio of the phase velocity of the transverse to the phase velocity of the longitudinal wave. The operator

$$T = 2\mu\,\hat{\underset{\sim}{n}}\cdot\nabla + \lambda\,\hat{\underset{\sim}{n}}\,\nabla\cdot + \mu\,\hat{\underset{\sim}{n}}\times\nabla\times \tag{10}$$

in (7)-(9) stands for the surface stress operator on S [5], and $\hat{\underset{\sim}{n}}$ is the outer unit normal. Furthermore, the field $\underset{\sim}{u}_0$ solves the static boundary value problem

$$\Delta^*\,\underset{\sim}{u}_0(\underset{\sim}{r}) = \underset{\sim}{0}, \qquad \underset{\sim}{r}\in V, \tag{11}$$

$$\underset{\sim}{u}_0(\underset{\sim}{r}) = -\hat{\underset{\sim}{a}}_0, \qquad \underset{\sim}{r}\in S, \tag{12}$$

$$\underset{\sim}{u}_0(\underset{\sim}{r}) = 0(\tfrac{1}{r}), \qquad r\to+\infty, \tag{13}$$

where

$$\hat{\underset{\sim}{a}}_0 = \begin{cases}\hat{\underset{\sim}{k}}, & \text{for } \underset{\sim}{\Phi} = \underset{\sim}{\Phi}^p\\ \hat{\underset{\sim}{b}}, & \text{for } \underset{\sim}{\Phi} = \underset{\sim}{\Phi}^s\end{cases}. \tag{14}$$

Relations (7) - (9) show that the leading term approximation of the scattering amplitudes g_r, g_θ and g_φ depend only on the value of the total traction over the surface of the scatterer that is created by the displacement $\underset{\sim}{u}_0$, given by (11) - (13). In fact, the scattering cross-section, which is defined to be the ratio of the total time-averaged energy that the scatterer receives from the incident field and reradiates in all directions divided by the time-averaged energy of the incident wave that crosses a unit area perpendicular to the direction of propagation, is given by [1], i.e.

$$\sigma^p = \frac{\tau(\tau^3+2)}{12\pi\mu^2}\left|\oint_S T\,\underset{\sim}{u}_0(\underset{\sim}{r})ds\right|^2 + 0(k^2) \tag{15}$$

for longitudinal incidence and by

$$\sigma^s = \frac{\tau^3+2}{12\pi\mu^2}\left|\oint_S T\,\underset{\sim}{u}_0(\underset{\sim}{r})ds\right|^2 + 0(k^2) \tag{16}$$

for transverse incidence, where $\tau^2 = \mu/(\lambda+2\mu)$ and $\tau^2 > 0$ by the assumed strong ellipticity of Δ^*.

3. DIRICHLET'S AND THOMSON'S PRINCIPLES IN ELASTICITY

In [4] one can find the proof of the following elastic analogue of Dirichlet's principle. If $\underset{\sim}{u}_0 \in \ker \Delta^*$, $\underset{\sim}{u}_0 = -\hat{\underset{\sim}{a}}_0$ on S and $\underset{\sim}{u}_0 = O(r^{-1})$ as $r \to +\infty$, then

$$\hat{\underset{\sim}{a}}_0 \cdot \oint_S T\, \underset{\sim}{u}_0(\underset{\sim}{r})ds(\underset{\sim}{r}) \le 2 \int_V W(\underset{\sim}{v},\underset{\sim}{v})dv, \tag{17}$$

where T is the surface traction operator, $\hat{\underset{\sim}{a}}_0$ the polarization vector given in (14),

$$W(\underset{\sim}{u},\underset{\sim}{v}) = \frac{\mu}{4}(\nabla\underset{\sim}{u} + \nabla\underset{\sim}{u}^T) : (\nabla\underset{\sim}{v} + \nabla\underset{\sim}{v}^T) + \frac{\lambda}{2}(\nabla\cdot\underset{\sim}{u})(\nabla\cdot\underset{\sim}{v}) \tag{18}$$

is the strain energy density product and $\underset{\sim}{v} \in [C^1(V) \cap C(\bar{V})]^3$, with $\underset{\sim}{v} = -\hat{\underset{\sim}{a}}_0$ on S.

Similarly, the proof of the following elastic analogue of Thomson's principle is included in [4]. For $\underset{\sim}{u}_0$ as above

$$\frac{\left[\hat{\underset{\sim}{a}}_0 \cdot \oint_S \left[\mu(\tilde{V} + \tilde{V}^T) + \lambda\tilde{V} : \mathbb{I} \otimes \mathbb{I}\right] \cdot \hat{\underset{\sim}{n}}ds\right]^2}{\int_V \left[\mu\tilde{V} : (\tilde{V} + \tilde{V}^T) + \lambda(\tilde{V} : \mathbb{I})^2\right]dv} \le \hat{\underset{\sim}{a}}_0 \cdot \oint_S T\,\underset{\sim}{u}_0(\underset{\sim}{r})ds \tag{19}$$

where $\tilde{V}$ is any continously differentiable dyadic field such that $\tilde{V}(\underset{\sim}{r}) = O(r^{-2})$ as $r \to +\infty$.

4. GEOMETRICAL ESTIMATES

Using a monotony principle and the generalized Dirichlet's principle (17) it is possible to show [4] that

$$\hat{\underset{\sim}{a}}_0 \cdot \oint_S T\underset{\sim}{u}_0(\underset{\sim}{r})ds(\underset{\sim}{r}) \le 8\pi\mu\, \hat{\underset{\sim}{a}}_0 \cdot \tilde{A} \cdot \hat{\underset{\sim}{a}}_0, \tag{20}$$

where

$$\tilde{A} = -\frac{\hat{\underset{\sim}{x}}_1 \otimes \hat{\underset{\sim}{x}}_1}{L_0^1(a_1)} - \frac{\hat{\underset{\sim}{x}}_2 \otimes \hat{\underset{\sim}{x}}_2}{L_0^2(a_1)} - \frac{\hat{\underset{\sim}{x}}_3 \otimes \hat{\underset{\sim}{x}}_3}{L_0^3(a_1)} \tag{21}$$

$$L_0^n(a_1) = (\tau^2-1)a_n^2 I_1^n(a_1) - (\tau^2+1) I_0^1(a_1), \quad n = 1,2,3 \tag{22}$$

$$I_0^1(a_1) = \frac{1}{2}\int_0^{+\infty} \frac{dx}{\sqrt{x+a_1^2}\sqrt{x+a_2^2}\sqrt{x+a_3^2}} \tag{23}$$

$$I_1^n(a_1) = \frac{1}{2}\int_0^{+\infty} \frac{dx}{(x+a_n^2)\sqrt{x+a_1^2}\sqrt{x+a_2^2}\sqrt{x+a_3^2}}, \quad n = 1,2,3 \tag{24}$$

and a_1, a_2, a_3 are the principal semiaxes of the best circumscribed ellipsoid [2].

On the other hand, an appropriate use of the generalized Thomson's principle [4] provides the lower bound

$$8\pi\mu\, \hat{\underset{\sim}{a}}_0 \cdot \tilde{B} \cdot \hat{\underset{\sim}{a}}_0 \le \hat{a}_0 \cdot \oint_S T\underset{\sim}{u}_0(\underset{\sim}{r})ds(\underset{\sim}{r}) \tag{25}$$

where the dyadic $\tilde{B}$ is the same with the dyadic $\tilde{A}$ given by (21) where the a_1, a_2, a_3 semiaxes are replaced with the semiaxes b_1, b_2, b_3 of the confocal inscribed ellipsoid.

The principal elements of the dyadics $\tilde{A}$ and $\tilde{B}$, appearing in the estimates (20) and (25), can be expressed [4] via the principal polarizations, or the principal virtual masses [6,7] of the ellipsoid.

In terms of our scattering problem the above geometrical estimates can be stated as

$$2\, \hat{\underset{\sim}{a}}_0 \cdot \tilde{B} \cdot \hat{\underset{\sim}{a}}_0 \le \hat{\underset{\sim}{a}}_0 \cdot \underset{\sim}{G}^1(\hat{\underset{\sim}{r}},\hat{\underset{\sim}{k}}) \le 2\, \hat{\underset{\sim}{a}}_0 \cdot \tilde{A} \cdot \hat{\underset{\sim}{a}}_0, \tag{26}$$

where

$$\underset{\sim}{G}^1(\hat{\underset{\sim}{r}},\hat{\underset{\sim}{k}}) = \lim_{k\to 0+} \frac{i}{k}\left[\tau^{-3}\, \hat{\underset{\sim}{r}}\, g_r(\hat{\underset{\sim}{r}},\hat{\underset{\sim}{k}}) + \hat{\underset{\sim}{\theta}}\, g_\theta(\hat{\underset{\sim}{r}},\hat{\underset{\sim}{k}}) + \hat{\underset{\sim}{\varphi}}\, g_\varphi(\hat{\underset{\sim}{r}},\hat{\underset{\sim}{k}})\right]. \tag{27}$$

In the case of best fitting spheres (26) is simplified to

$$\frac{3}{\tau^2+2}\, b \le \hat{\underset{\sim}{a}}_0 \cdot \underset{\sim}{G}^1(\hat{\underset{\sim}{r}},\hat{\underset{\sim}{k}}) \le \frac{3}{\tau^2+2}\, a\,. \tag{28}$$

5. CAPACITY ESTIMATES

The generalized Dirichlet's principle (17) can also be used to obtain an upper estimate for the scattering amplitudes in terms of the capacity of the scatterer. In fact, it can be shown [4] that

$$\hat{\underset{\sim}{a}}_0 \cdot \oint_S T\underset{\sim}{u}_0(\underset{\sim}{r})ds(\underset{\sim}{r}) \le \frac{4\pi\mu}{\tau^2}\, C \tag{29}$$

where C stands for the capacity of S.

A more elaborate set of calculations shows [4] that

$$4\pi\mu C \leq \hat{\underset{\sim}{a}}_0 \cdot \oint_S T\underset{\sim}{u}_0(\underset{\sim}{r})ds(\underset{\sim}{r}). \tag{30}$$

Consequently,

$$C \leq \hat{\underset{\sim}{a}}_0 \cdot \underset{\sim}{G}^1(\underset{\sim}{r},\underset{\sim}{k}) \leq \frac{1}{\tau^2} C. \tag{31}$$

Although the above estimate depends upon the exact shape of the scatterer, via the value of its capacity, we do not have equalities for any shape of the scatterer. Note that the upper bound depends on the elastic properties of the medium of propagation, i.e., the ratio of the phase velocities of the transverse and the longitudinal wave. The smaller the phase velocity of the transverse wave the larger the number τ^{-2} and in the limit of an incompressible medium (fluid) the upper bound tends to infinity.

Therefore, the choice of the appropriate estimate depends not only upon the geometrical characteristics of the scatterer, as in the case of acoustic scattering, but also upon the elastic properties of the medium in which the rigid scatterer is embedded.

REFERENCES

[1] Dassios, G. and Kiriaki, K., Quarterly of Applied Mathematics, 42 (1984) 225.

[2] Dassios, G. and Kiriaki, K., Quarterly of Applied Mathematics, 43 (1986) 435.

[3] Dassios, G. and Kiriaki, K., Quarterly of Applied Mathematics, 44 (1987) 709.

[4] Dassios, G. and Payne, L.E., Estimates for Low-Frequency Elastic Scattering by a Rigid Body. Journal of Elasticity (in press).

[5] Kupradze, V.D., Three-Dimensional Problems of the Mathematical Theory of Elasticity and Thermoelasticity (North-Holland, Amsterdam, 1979).

[6] Schiffer, M and Szegö, G., Transactions of A.M.S., 67(1949) 130.

[7] Payne, L.E., SIAM Review, 9(1967) 453.

ELASTIC WAVE PROPAGATION
M.F. McCarthy, M.A. Hayes, (Editors)
Elsevier Science Publishers B.V. (North-Holland), 1989

BEYOND THE BORN APPROXIMATION

R.T. Coates and C.H.Chapman

Bullard Laboratories, Madingley Road, Cambridge CB3 OEZ, U.K.

The description of scattered waves in an inhomogeneous elastic medium is frequently treated using the Born formalism. In most previous studies this involved ray tracing through a smooth approximation to the medium under consideration, and the use of the deviation of the true medium from this smooth approximation as the 'potential' in the Born scattering integral.

A formalism is developed here for describing body waves scattered from parameter gradients in the medium, which obviates the necessity for an approximate smooth medium. The substitution of an asymptotic ray theory (ART) Green's function into the elastodynamic wave equation produces terms due to the error in the ART Green's function. These terms are functions of the velocity and the density gradients of the medium. They are treated as 'source' terms and the solution for the scattered waves obtained as an integral over these terms. The expression is similar in form to the Born formalism but is superior in that both the ART Green's function and the scattering integral are calculated using ray tracing through the true medium, not some smooth approximation to it.

1.INTRODUCTION

The Born approximation has long been established as a method for modelling the scattering of energy by inhomogeneities in elastic or acoustic media, both for forward modelling and as a basis for inversion, e.g. [1]. The Born formalism is based on the division of the medium into a smooth reference model, in which the medium parameters vary over length scales much greater than the wavelength of the disturbance being modelled, and a rapidly varying perturbation to the reference model, forming the difference between the true medium and the smooth reference model. By assumption no energy is scattered when the disturbance propagates through the reference model. All scattering in the true medium is assumed to be due to the short wavelength perturbation.

For brevity and clarity the theory will be developed for a constant density, acoustic medium. The extension to variable density and elastic media is straightforward though algebraically complicated.

In an acoustic medium with a constant density we can describe the medium by its slowness u at every point. In the Born approach we divide the true medium $u(\mathbf{x})$ into a smooth reference slowness field $u_0(\mathbf{x})$ and a perturbation $\delta u(\mathbf{x})$:

$$u(\mathbf{x}) = u_0(\mathbf{x}) + \delta u(\mathbf{x}). \tag{1}$$

The Green's function of the true medium $g(\mathbf{x},\mathbf{x_s}:\omega)$ is a solution of the acoustic wave equation for an ideal point source:

$$\frac{1}{\rho_0}(\nabla^2+\omega^2u^2(\mathbf{x}))\,g(\mathbf{x},\mathbf{x_s}:\omega)=-\delta(\mathbf{x}-\mathbf{x_s}). \quad (2)$$

We divide g into two parts:

$$g(\mathbf{x},\mathbf{x_s}:\omega)=g_0(\mathbf{x},\mathbf{x_s}:\omega)+g_s(\mathbf{x},\mathbf{x_s}:\omega) \quad (3)$$

where g_0 is the solution to (2) in the *reference medium*, that is where u_0 replaces u in (2), and g_s represents the scattered energy and is defined by (3).

Combining the equations of motion for the true medium and for the reference medium, using Green's Theorem we obtain an expression for the scattered energy g_s. To first order in the perturbation this is [2]:

$$g_s(\mathbf{x},\mathbf{x_s}:\omega)\approx 2\omega^2\iiint d^3x\,u_0(\mathbf{x})\,\delta u(\mathbf{x})g_0(\mathbf{x},\mathbf{x_s}:\omega)g_0(\mathbf{x},\mathbf{x_r}:\omega). \quad (4)$$

We now approximate the Green's function g_0 by the ART solution:

$$g_0(\mathbf{x},\mathbf{x_s}:\omega)=A_0(\mathbf{x},\mathbf{x_s})\exp\{\,i\omega\,\tau_0(\mathbf{x},\mathbf{x_s})\}. \quad (5)$$

The travel-time phase τ_0 satisfies the eikonal equation *in the reference medium:*

$$\nabla\tau_0(\mathbf{x},\mathbf{x_s}).\nabla\tau_0(\mathbf{x},\mathbf{x_s})=u_0^2(\mathbf{x}). \quad (6)$$

The amplitude A_0 satisfies the transport equation *in the reference medium:*

$$2\nabla\tau_0(\mathbf{x},\mathbf{x_s}).\nabla A_0(\mathbf{x})+A_0(\mathbf{x})\nabla^2\tau_0(\mathbf{x},\mathbf{x_s})=0. \quad (7)$$

Using (5) we may write (4) as:

$$g_s(\mathbf{x},\mathbf{x_s}:\omega)\approx 2\omega^2\iiint d^3x\,u_0(\mathbf{x})\,\delta u(\mathbf{x})A_0(\mathbf{x},\mathbf{x_s})A_0(\mathbf{x},\mathbf{x_r}) \times\exp\{\,i\omega(\tau_0(\mathbf{x},\mathbf{x_s})+\tau_0(\mathbf{x},\mathbf{x_r}))\}. \quad (8)$$

This is one of the most commonly used forms of the Born approximation.

The validity of the Born method has been investigated by many authors, e.g. [3,4]. Simplifying their conclusions we find four conditions which must be satisfied for the method to be valid:

1. $[\delta u(\mathbf{x})/u_0(\mathbf{x})]\ll 1$ everywhere.
2. $[g_s(\mathbf{x},\mathbf{x_s}:\omega)/g_0(\mathbf{x},\mathbf{x_r}:\omega)]\ll 1$ everywhere.
3. The linear dimensions of the volume in which δu is finite are small compared to the wavelength of the disturbance.
4. Single scattering is the dominant process.

These conditions are restrictive, and are not satisfied in a large range of geophysical situations [3].

A further disadvantage of the method should be noted. Both the amplitudes A_0 and more importantly the travel-time phases τ_0 in (8) are calculated in the *reference medium* not the true medium. This means the method may miscalculate the travel-times and the

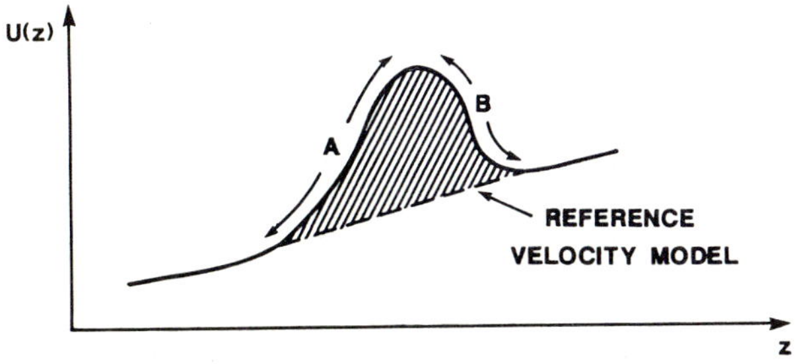

Figure 1: The 1D case. The shaded region shows the Born perturbation. The slowness gradients A and B are the major contributors (with opposite phase) to the gradient scattering integrals.

amplitudes of scattered arrivals. The error in arrival times is particularly important in reflection seismology and tomography.

We demonstrate below that a partial, first-order correction of these effects is implicit in the integral, so the problem is not as acute as it appears. We emphasize this implicit correction is only partial.

Furthermore we know from physical considerations that scattering in the true medium is due to gradients in the medium's parameters. This has been quantitatively described for 1D media, e.g. [5]. Similarly in elastic media coupling between P and S waves occur in regions of high parameter gradients [6]. These studies describe the scattered waves in terms of integrals over the gradients of the medium's parameters. Such terms are not explicitly present in the Born formalism. Fig. 1 shows the different volumes contributing to the two approximations. In section 3 we develop a formalism which models gradient dependent scattering in 3D inhomogeneous media.

2. RAY PERTURBATION AND THE BORN APPROXIMATION

It is widely appreciated that a high frequency disturbance may be regarded as propagating through a small volume surrounding the ART ray [7]. This volume is in many ways a generalisation of the Fresnel zone of classical optics. The boundary of this volume is defined by the fact that a disturbance initiated at $\mathbf{x}_s$, scattered from a point on the boundary of the volume and then propagating to $\mathbf{x}_r$ has a total phase lag at the receiver of π with respect to the direct ray from $\mathbf{x}_s$ to $\mathbf{x}_r$.

In view of this we express (3) in ray centred coordinates [8], see Fig.2. The volume integral in (3) is now an integral along the ray, and over the coordinates q_1, q_2. In order to investigate the behaviour of (3) to highest order in ω we evaluate (3) using the stationary phase method. Fermat's principle implies that the stationary phase point lies on the ray joining $\mathbf{x}_s$ to $\mathbf{x}_r$. Consequently we expand the Green's function about the central ray, $\mathbf{x}_\sigma$, using the paraxial approximation. This expresses the Green's function at a small distance away from the ray in terms of quantities calculated on it [7]:

$$g_0(\mathbf{x}, \mathbf{x}_s : \omega) \approx A_0(\mathbf{x}_\sigma, \mathbf{x}_s) \exp\{ i\omega(\tau_0(\mathbf{x}_\sigma, \mathbf{x}_r) + \frac{1}{2}\mathbf{q}^T \mathbf{M}\mathbf{q})\} \tag{9}$$

where $\mathbf{M}$ is the matrix of second-order derivatives of τ_0 with respect to $\mathbf{q}$. We note here the amplitude is independent of $\mathbf{q}$ in this approximation [7]. Similarly we expand $\delta u(\mathbf{x})$ in

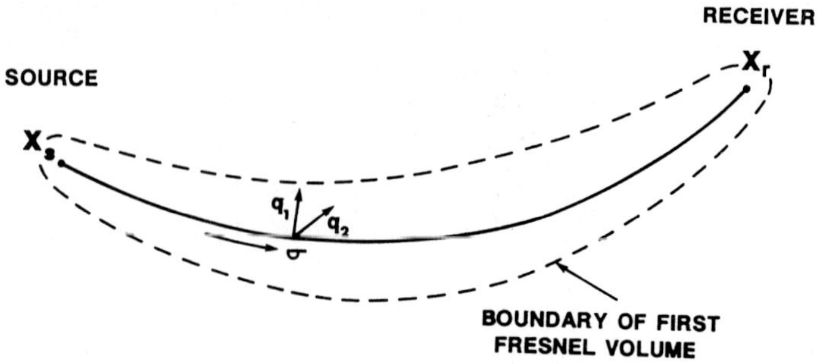

Figure 2: Ray centred coordinates. σ measures distance along the ray, q_1 and q_2 are mutually orthogonal and lie in the plane of the wavefront. The first Fresnel zone is the volume bounded by a surface of constant total travel-time.

a Taylor series to second order:

$$\delta u(\mathbf{x}) \approx \delta u(\mathbf{x}_\sigma) + \delta \mathbf{u}'.\mathbf{q} + \frac{1}{2}\mathbf{q}^{\mathrm{T}}\delta\mathbf{u}''\mathbf{q} \tag{10}$$

where $\delta\mathbf{u}'$ is the vector of first derivatives of δu, and $\delta\mathbf{u}''$ is the matrix of second derivatives of δu with respect to $\mathbf{q}$. We neglect variations in u_0 as these are by assumption negligible compared to those of δu. Substituting (9) and (10) into (11) and performing stationary phase evaluations over q_1 and q_2 we obtain after adding in the solution in the reference medium g_0:

$$g(\mathbf{x_r}, \mathbf{x_s} : \omega) \approx A_0(\mathbf{x_r}, \mathbf{x_s}) \exp\{ i\omega\tau_0(\mathbf{x_r}, \mathbf{x_s})\}[1 + \Phi - \frac{F}{2}] \tag{11}$$

where Φ and F are given by (14).

In ray perturbation theory the small changes in the position and slowness vectors $\delta\mathbf{q}$ and $\delta\mathbf{p}$ relative to the reference ray, are related at different points $\mathbf{x}_\sigma$, $\mathbf{x_s}$ on the ray by

$$\begin{pmatrix} \delta\mathbf{q}(\mathbf{x}_\sigma) \\ \delta\mathbf{p}(\mathbf{x}_\sigma) \end{pmatrix} = \begin{pmatrix} \mathbf{Q}_1(\mathbf{x}_\sigma, \mathbf{x_s}) & \mathbf{Q}_2(\mathbf{x}_\sigma, \mathbf{x_s}) \\ \mathbf{P}_1(\mathbf{x}_\sigma, \mathbf{x_s}) & \mathbf{P}_2(\mathbf{x}_\sigma, \mathbf{x_s}) \end{pmatrix} \begin{pmatrix} \delta\mathbf{q}(\mathbf{x_s}) \\ \delta\mathbf{p}(\mathbf{x_s}) \end{pmatrix}. \tag{12}$$

This equation is used to define the matrices $\mathbf{Q}_1$ etc. [9]. The ray perturbation result for the arrival in the true medium is [9]:

$$g(\mathbf{x_r}, \mathbf{x_s} : \omega) = \frac{A_0(\mathbf{x_r}, \mathbf{x_s})}{(1 + F + G)^{1/2}} \exp\{ i\omega(\tau_0(\mathbf{x_r}, \mathbf{x_s}) + \Phi)\} \tag{13}$$

where

$$\begin{aligned} \Phi &= \int_{ray} \delta u(\mathbf{x}_\sigma) d\sigma \\ F &= \int_{ray} tr(\mathbf{Q}_2^{-1}(\mathbf{x_r}, \mathbf{x_s})\mathbf{Q}_2(\mathbf{x_r}, \mathbf{x}_\sigma)\delta\mathbf{u}''\mathbf{Q}_2(\mathbf{x}_\sigma, \mathbf{x_s}))\, d\sigma \\ G &= -\int_{ray} tr(\mathbf{Q}_2^{-1}(\mathbf{x_r}, \mathbf{x_s})\mathbf{Q}_1(\mathbf{x_r}, \mathbf{x}_\sigma)(\frac{\delta u\mathbf{I}}{u_0^2})\mathbf{P}_2(\mathbf{x}_\sigma, \mathbf{x_s}))\, d\sigma. \end{aligned} \tag{14}$$

We see that for scatterers lying close to the direct ray the Born and ray perturbation theories agree to first order in ω and δu, excepting the term G which is frequently much smaller than F. The integral F depends on the transverse curvature of the velocity perturbation, whilst G depends on the perturbation along the direct ray path.

This important result implies the arrival times of direct rays predicted by Born theory are more accurate than was previously thought. Sneider [10] proved the analogous result for surface waves, in a laterally homogeneous reference medium. This result holds for an arbitary 3D inhomogeneous reference medium with continuous second-order derivatives of the slowness. However the correction is only partial and does not apply for energy scattered far from the direct rays joining source to receiver. In the next section, we outline a more accurate new scattering theory.

3. GRADIENT SCATTERING

The acoustic wave equation (2) is an exact equation for the unknown Green's function g. We know that for a high frequency disturbance an approximate expression for g is the ART Green's function $\tilde{g}$:

$$\tilde{g}(\mathbf{x}, \mathbf{x_s} : \omega) = A(\mathbf{x}, \mathbf{x_s}) \exp\{ i\omega\tau(\mathbf{x}, \mathbf{x_s})\} \tag{15}$$

where τ is a solution to the eikonal and A a solution to the transport equation in the *true medium* under consideration. Direct substitution of (15) into (2) reveals (15) is an exact solution to:

$$\frac{1}{\rho_0}(\nabla^2 + \omega^2 u^2(\mathbf{x}))\tilde{g}(\mathbf{x}, \mathbf{x_s}; \omega) = -\delta(\mathbf{x} - \mathbf{x_s}) + \frac{\nabla^2 A(\mathbf{x}, \mathbf{x_s})}{\rho_0 A(\mathbf{x}, \mathbf{x_s})}\tilde{g}(\mathbf{x}, \mathbf{x_s} : \omega). \tag{16}$$

Combining (16) and (2) using Green's Theorem gives:

$$g(\mathbf{x_r}, \mathbf{x_s} : \omega) = \tilde{g}(\mathbf{x_r}, \mathbf{x_s} : \omega) + \iiint d^3x\, \tilde{g}(\mathbf{x_r}, \mathbf{x} : \omega)\frac{\nabla^2 A(\mathbf{x}, \mathbf{x_s})}{\rho_0 A(\mathbf{x}, \mathbf{x_s})}\tilde{g}(\mathbf{x}, \mathbf{x_s} : \omega). \tag{17}$$

Here we have neglected a surface integral which tends to zero as the volume enclosed by the surface tends to infinity. We have also approximated the true Green's function g by the ART approximation $\tilde{g}$ inside the integral. This is physically equivalent to considering single scattering only.

The solution to the transport equation is [11]:

$$A(\mathbf{x_1}, \mathbf{x_s}) = A(\mathbf{x_2}, \mathbf{x_s})\sqrt{\frac{\alpha(\mathbf{x_2})}{\alpha(\mathbf{x_1})J(\mathbf{x_1}, \mathbf{x_2})}}. \tag{18}$$

Where α is the acoustic velocity of the medium and $J(\mathbf{x_1}, \mathbf{x_2})$ is the geometrical spreading function of a elementary ray tube propagating from $\mathbf{x_2}$ to $\mathbf{x_1}$.

Substituting (18) in (17) and using the divergence theorem, the integral term in (17) becomes

$$\begin{aligned} \tilde{g}_s(\mathbf{x_r}, \mathbf{x_s} : \omega) = \iiint d^3x \frac{A(\mathbf{x}, \mathbf{x_s})A(\mathbf{x}, \mathbf{x_r})}{\rho_0} & \exp\{ i\omega(\tau(\mathbf{x}, \mathbf{x_s}) + \tau(\mathbf{x}, \mathbf{x_r}))\} \\ \times[(\frac{\nabla\alpha(\mathbf{x})}{2\alpha(\mathbf{x})} + \frac{\nabla J(\mathbf{x}, \mathbf{x_s})}{2J(\mathbf{x}, \mathbf{x_s})}).(\frac{\nabla\alpha(\mathbf{x})}{2\alpha(\mathbf{x})} & + \frac{\nabla J(\mathbf{x}, \mathbf{x_r})}{2J(\mathbf{x}, \mathbf{x_r})}) \\ -i\omega\nabla(\tau(\mathbf{x}, \mathbf{x_s}) + \tau(\mathbf{x}, \mathbf{x_r})).(\frac{\nabla\alpha(\mathbf{x})}{2\alpha(\mathbf{x})} & + \frac{\nabla J(\mathbf{x}, \mathbf{x_s})}{2J(\mathbf{x}, \mathbf{x_s})})]. \end{aligned} \tag{19}$$

The relative magnitudes of the terms dependent upon ω and those independent of it are of the order of the ratio of velocity variations in the medium, to the wavelength of the disturbance. In almost all cases the maximum value of the second term is significantly larger than the first term, giving rise to a strong reflection when the angle between the velocity gradient and the incoming ray equals the angle between the gradient and the scattered ray, see Fig. 3.

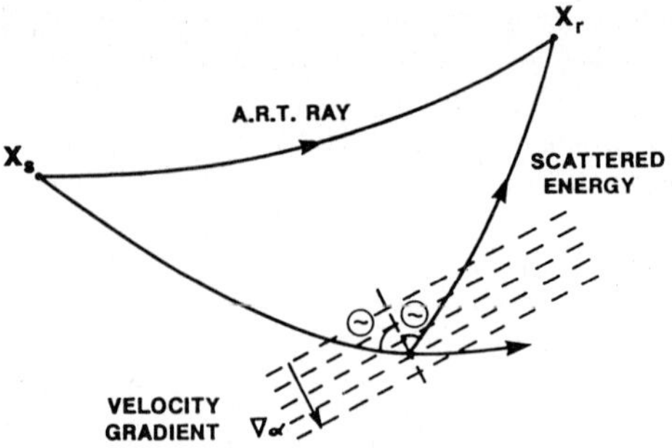

Figure 3: The scattered energy modelled by the new theory. Large amplitude scattering occurs when the angle of incidence and of reflection are equal.

However far from this angle, in regions of high velocity gradients, the terms may be of comparable magnitude.

Equation (19) appears to be both non-symmetrical and non-zero in a homogeneous medium due to the terms in J. However performing a stationary phase evaluation of these terms in (19) shows that the highest order terms in ω are $O(1/\omega^2)$ which we may safely neglect. This allows us to re-write (19) as

$$\tilde{g}_s(\mathbf{x_r},\mathbf{x_s}:\omega) \approx \iiint d^3x A(\mathbf{x},\mathbf{x_s})A(\mathbf{x},\mathbf{x_r})\exp\{i\omega(\tau(\mathbf{x},\mathbf{x_s})+\tau(\mathbf{x},\mathbf{x_r}))\} \times[\frac{\nabla\alpha(\mathbf{x})}{2\alpha(\mathbf{x})}.\frac{\nabla\alpha(\mathbf{x})}{2\alpha(\mathbf{x})} - i\omega\frac{\nabla\alpha(\mathbf{x})}{2\alpha(\mathbf{x})}.\nabla(\tau(\mathbf{x},\mathbf{x_s})+\tau(\mathbf{x},\mathbf{x_r}))] \quad (20)$$

This expression now models energy scattered from velocity gradients. The phase and amplitude functions are calculated in the *true medium* not a reference medium. Thus arrival time in particular will be correctly predicted. In addition the potential has a more physically suitable form. Thus although the integral maintains the same *form* as the Born integral the authors believe it forms a significant improvement in modelling the scattering of energy by inhomogeneities.

References

[1] Cohen J.K., Hagin F.G., Bleistein N., Geophys. (1986), **51**, 1552–1558.
[2] Aki K., Richards P., Quantitative Seismology (Freeman : San Fransisco, 1980).
[3] Hudson J.A., Heritage J.R., Geophys. J. R. astr. Soc. (1981), **56**, 221–240.
[4] Kennett B.L.N., Geophys. J. R. astr. Soc. (1972), **27**, 301–336.
[5] Chapman C.H., Orcutt J.A., Rev. Geophys. (1985), **23**, 105–163.
[6] Ben-Menahem A., Beydoun W.B., Geophys. J. R. astr. Soc. (1985), **82**, 207–234.
[7] Červený V., Pšenčík I., J. Geophys. (1983), **53**, 1–15.
[8] Babich V.M., Buldyrev V.S., Asymptotic Methods In Short Wave Diffraction Problems. (Nauka, Moscow, 1972).
[9] Farra V., Madariaga R., J. Geophys. Res. (1987), **92**, 2697–2712.
[10] Sneider R., On The Connection Between Ray Theory And Scattering Theory For Surface Waves, in; Modern Approaches In Geophysics (D.Reidel, Dortrecht, 1988).
[11] Červený V., Ravindra R., Theory Of Seismic Head Waves (Univ. of Toronto Press, Toronto, 1971).

ELASTIC WAVE PROPAGATION
M.F. McCarthy, M.A. Hayes, (Editors)

THE EFFECT OF NEAR-TIP RESIDUAL COHESIVE TRACTIONS

M. K. Kuo†, J. D. Achenbach‡, and H. J. Yang†

†Institute of Applied Mechanics
National Taiwan University
Taipei, Taiwan 10764, R. O. C.

‡Department of Civil Engineering
Northwestern University
Evanston, Illinois 60201, U. S. A.

The diffraction of elastic waves by a crack tip is investigated for the case that cohesive tractions on the crack faces in the immediate vicinity of the crack tip have not been completely released. In this paper it is assumed that such residual cohesive tractions are proportional to the local crack-opening displacement. This assumption is equivalent to a spring-constrained near-tip zone. The crack is subjected to normal incidence of a longitudinal wave.

For the time interval before signals from the other crack tip arrive, approximate expressions have been obtained for the spring tractions and the elastodynamic scattering field. The results are valid for the case of small range of constraints and small spring constant. In this approximation, the spring tractions are reduced to a separable form of a time dependence multiplied by a space dependence. The space dependence is found first. The time dependence is then expressed in a form of perturbation series for small $\hat{k}$ in terms of powers of $\hat{k}$, where $\hat{k}$ is the nondimensional spring constant. The leading order term result is then obtained.

Introduction

A perfect mathematical crack is an infinitesimally thin crack with smooth faces. The effect of crack-face interaction are ignored in considerations based on the perfect mathematical crack. The failure processes that result in a crack generally produce rough crack faces. Once a crack has been opened and the crack faces have undergone some relative sliding displacement, it must be expected that the crack will never completely close again due to incompatibility of the rough crack faces. Under subsequent loading conditions the faces of the crack generally may not be free of surface tractions, as is assumed for a perfect mathematical crack. In addition residual cohesive tractions might also present due to not-yet-complete fracture process. It can be stated that the perfect-mathematical-crack results are valid for a real crack if the ratio of the crack-opening displacement to the amplitude of the incident wave is larger than unity, and if the wavelength is much larger than a characteristic spacing length of the crack-face roughness.

Achenbach and Norris [1]-[2] addressed the question of interaction between the crack faces by considering the nonlinear interaction effect. The results of [1]-[2] show that the specular reflection can be reduced considerably due to crack-face interactions. In an alternate approach, Thompson and Fiedler [3] computed the stress and deformation fields in the body by taking a linear relation between the tractions and the crack-opening displacements across a perfectly flat surface. This surface may be considered as the median plane of the actual flaw surface. In this approximation the interaction between the crack faces is effectively described by a distribution of springs. It was shown by Angel and Achenbach [4]-[5], that with regard to reflection and transmission coefficients, a flaw plane consisting of an array of equally spaced microcracks can be represented with excellent accuracy by a spring layer, at least at low frequency.

The previous studies on both nonlinear [1]-[2], and linear [3] crack face interactions are only applicable to specular reflections. In this paper, we consider the corresponding effects on crack tip diffractions. These diffractions play an essential role in the crack-sizing method that has so far proven to be the most successful [6]. For simplicity, we present only the results for the case of normal incidence of a longitudinal wave. The cases of oblique wave incidence will appear elsewhere.

1 Statement of Problem

A stationary, semi-infinite crack is struck by a normal incident longitudinal stress wave at time $t = 0$. In the immediate vicinity of the crack tip, the cohesive tractions have not been completely released. The length of the cohesive zone, which is denoted by a, could be very small. It is then assumed that such residual cohesive tractions are proportional to the local crack-opening displacement. This assumption is equivalent to a spring-constrained near-tip zone. The effects of cohesive tractions on the tip-diffracted field are investigated by analyzing the elastodynamic field of a problem with spring-constrained crack.

The incident longitudinal wave is taken to be of the form

$$\sigma_y^{in} = \sigma_0\, f(t - s_L y)\, H(t - s_L y), \qquad \epsilon_x^{in} \equiv 0, \tag{1}$$

where $f(t)$ is a nondimensional loading time function, s_L and s_T are slownesses of longitudinal waves and transverse waves, respectively. The incident wavefront, the coordinate system (x, y), and the crack geometry are shown in Fig.1.

The original diffraction problem can be considered as a superposition of the problem of a uncracked plane subjected to the incident wave and a superposition problem. The superposition problem concerns an initially quiescent solid which contains a semi-infinite crack with spring constraints near the crack tip. At time $t = 0$ the crack faces are subjected to crack-face tractions which are equal in magnitude but opposite in sign to the incident stresses. Boundary conditions of the superposition problem are along the crack faces, $x < 0, y = 0$,

$$\sigma_y(x,0,t) = -\sigma_0\, f(t)\, H(t)\, H(-x) + k[u_y]\, H(x+a)\, H(-x), \tag{2}$$

$$\sigma_{xy}(x,0,t) = 0, \tag{3}$$

where

$$[u_y] \equiv u_y(x,0^+,t) - u_y(x,0^-,t)\ ,$$

k is the spring constant and $H(\cdot)$ is the Heaviside step function, respectively.

2 Method of Solution

Owing to the existence of the characteristic length a, the method of solution to the problem defined by eqns. (2)-(3) is more difficult than one without the characteristic length, since the direct application of integral transforms will not lead to a standard Wiener-Hopf equation as defined in Noble's monograph [7]. Hence some other method might be needed.

Consider two related semi-infinite crack problems without any spring constraints. The first problem is subjected to a prescribed crack-face tractions which are equal in magnitude but opposite in sign to the incident stress σ_y^{in}. The crack-face boundary condition of this problem is exactly the same as that defined by eqns. (2)-(3), except that the second term in the right hand side of eq. (2) will not appear. The solution of this problem can easily be constructed by convolving the crack-face loading time function $\sigma_0 f(t)$ with the solution to a fundamental problem. The corresponding crack-face displacement can be expressed as

$$u_{y1}(x,t) = \sigma_0 \int_0^t f(t-\eta)\, u_y^*(x,\eta)\, d\eta, \tag{4}$$

where $u_y^*(x,t)$ is the crack-face displacement of the fundamental problem which is subjected by an impulsive normal loading on the crack faces. The boundary conditions of this fundamental problem are, along $y = 0, x < 0$,

$$\sigma_y^*(x,0,t) = -\delta(t)\, H(-x)\ , \qquad \sigma_{xy}^*(x,0,t) = 0 \quad , \tag{5}$$

where $\delta(\cdot)$ is the Dirac delta function.

Due to the absence of the characteristic length a, the solution of the fundamental problem can be obtained by combining the method of integral transform, the Wiener-Hopf technique and the Cagniard- de Hoop method. The details can be found in [8]. The mode I stress intensity factor and the near-tip crack-face displacement, which is dominated by the term with square root behavior, of the first problem can the be expressed as

$$K_{I1}(t) = (\frac{2}{\pi})^{1/2} \frac{\sigma_0 (s_L)^{1/2}}{s_R \Gamma_+(0)} \int_0^t f(t-\eta)\eta^{-1/2} d\eta, \quad (6)$$

and

$$u_{y1}(x,t) = (2\pi)^{-1/2} \frac{s_T^2}{\mu(s_T^2 - s_L^2)} K_{I1}(t) (-x)^{1/2} + O(x), \quad (7)$$

where s_R and μ are the slowness of Rayleigh wave and the shear modulus, respectively. Expressions $\Gamma_\pm(\xi)$ are defined as

$$\Gamma_\pm(\xi) = Exp\left\{-\frac{1}{\pi}\int_{s_L}^{s_T} tan^{-1}\left[\frac{4z^2(z^2-s_L^2)^{1/2}(s_T^2-z^2)^{1/2}}{(2z^2-s_T^2)^2}\right]\frac{dz}{z \pm \xi}\right\}. \quad (8)$$

The second problem is the same as the first one except that it is subjected to an unknown crack-face traction, $\sigma_y^{sp}(x,t)$, over $-a < x < 0$, which can be regarded as the reaction stress of the spring constraints on the crack faces. The corresponding crack-face displacement $u_{y2}(x,t)$ can be related to $\sigma_y^{sp}(x,t)$ through an appropriate fundamental solution as

$$u_{y2}(x,t) = \int_0^t d\eta \int_0^a \sigma_y^{sp}(-\xi,\eta) \frac{\partial}{\partial t}[G_v(x,t-\eta;\xi,0)] d\xi, \quad (9)$$

where $G_v(x,t;\xi,0)$ is the crack-face displacement of a fundamental problem subjected to a unit step point force acting on the crack faces at position $x = -\xi$ starting from time $t = 0$. The mode I stress intensity factor of this fundamental problem can be found in [9] as

$$K_{I_g}(t;\xi) = -\frac{1}{\pi}(\frac{2}{\pi\xi})^{1/2} H(t - s_L\xi) \int_{s_L}^{t/\xi} \frac{(\eta - s_L)^{1/2}}{(t/\xi - \eta)^{1/2}(\eta - s_R)} Re\left[\frac{1}{\Gamma_-(\eta)}\right] d\eta. \quad (10)$$

As stated in [9], for the case of $t/\xi > s_T$, equation (10) can be evaluated in the complex η-plane and yielded to a closed form expression

$$K_{I_g}(t;\xi) = -(\frac{2}{\pi\xi})^{1/2}\left[1 - \frac{\xi^{1/2}(s_R - s_L)^{1/2}}{(s_R\xi - t)^{1/2}\Gamma_-(s_R)} H(s_R\xi - t)\right]. \quad (11)$$

Notice that there were some misprints in this expression of [9]. The near-tip crack-face displacement due to the step point force on the crack faces can then be approximated by

$$G_v(x,t;\xi,0) = (2\pi)^{-1/2} \frac{s_T^2}{\mu(s_T^2 - s_L^2)} K_{I_g}(t;\xi)(-x)^{1/2} + O(x). \quad (12)$$

Superposition of these first and second problems leads to the problem with spring constraints which is defined by eqns.(2)-(3) as long as the compatibility condition of the spring is satisfied as

$$\sigma_y^{sp}(x,t) = 2k(u_{y1} + u_{y2}) \quad , -a < x < 0. \quad (13)$$

Here the properties of symmetry of the above two related problems, namely $[u_{yi}] = 2u_{yi}, i = 1,2$, have been used. The combination of eqns. (9) and (13) yields a two-dimensional integral equation for the unknown crack face traction $\sigma_y^{sp}(x,t)$. For the case that the spring-constrained range, a, is very small, the displacement fields $u_{y1}(x,t)$ and $G_v(x,t;\xi,0)$ can be approximated by the first term of the near-tip asymptotic expansion for the crack-face displacements as eqns. (7) and (12). This two-dimensional integral equation can then be written in a nondimensional form as

$$(2\pi)^{1/2}(s_T^2 - s_L^2)(s_T^2)^{-1} \hat{\sigma}_y^{sp} = 2\hat{k}\hat{K}_{I1}(t)(-x/a)^{1/2} +$$

$$2\hat{k}(-x/a)^{1/2}\int_0^t d\eta \int_0^a \hat{\sigma}_y^{sp}(-\xi,\eta)\frac{\partial}{\partial t}[\hat{K}_{I_g}(t-\eta;\xi)]d\xi \quad , -a < x < 0, \quad (14)$$

where

$$\hat{\sigma}_y^{sp}(x,t) = \sigma_y^{sp}(x,t)/\sigma_0, \qquad \hat{k} = ka/\mu, \tag{15}$$

$$\hat{K}_{I1}(t) = K_{I1}(t)/(\sigma_0 a^{1/2}), \qquad \hat{K}_{Ig}(t) = K_{Ig}(t)/a^{1/2}. \tag{16}$$

Equation (14) holds at every point within the region $-a < x < 0, y = 0$, it is then certainly to expect that $\hat{\sigma}_y^{sp}$ has the square root behavior in x of the form

$$\hat{\sigma}_y^{sp}(x,t) = \tau(t)\,(-x/a)^{1/2}. \tag{17}$$

The first derivative of τ is needed for the calculation of tip-diffracted field. Taking the differentiation with respect to t to eq.(14), making use of eq.(17) and rearranging terms lead to a Votlerra integral equation of the second kind for τ'

$$(2\pi)^{1/2}(s_T^2 - s_L^2)(s_T^2)^{-1}\,\tau'(t) = 2\hat{k}\,\frac{d\hat{K}_{I1}(t)}{dt} + $$

$$2\hat{k}\,a^{-1/2}\int_0^t \tau'(\eta)\left[\frac{\partial}{\partial t}\int_0^a \xi^{1/2}\hat{K}_{Ig}(t-\eta;\xi)d\xi\right]d\eta. \tag{18}$$

Equation (18) can be solved by a usual numerical method. We note that the kernel itself of this integral equation is actually in a double integral form, since $\hat{K}_{Ig}$ is in the integral form as in (10). Change the order of integration and evaluate the inner integral analytically leads to

$$\frac{\partial}{\partial t}\int_0^a \xi^{1/2}\hat{K}_{Ig}(t;\xi)d\xi = -(\frac{1}{2\pi a})^{1/2}\int_{s_L}^{\infty}\frac{(\xi - s_L)^{1/2}}{\xi^{3/2}(\xi - s_R)}Re\left[\frac{1}{\Gamma_-(\xi)}\right]d\xi + \frac{1}{\pi}(\frac{2}{\pi a})^{1/2}.$$

$$\int_{s_L}^{t/a}\frac{(\xi - s_L)^{1/2}}{\xi\,(\xi - s_R)}Re\left[\frac{1}{\Gamma_-(\xi)}\right]\cdot\left[a^{1/2}(t - a\xi)^{-1/2} + \xi^{-1/2}(\frac{\pi}{2} - tan^{-1}\sqrt{\frac{a\xi}{t - a\xi}})\right]d\xi. \tag{19}$$

For the case of $t/a > s_R$, we find directly from eq. (11) that

$$\frac{\partial}{\partial t}\int_0^a \xi^{1/2}\hat{K}_{Ig}(t;\xi)d\xi = 0. \tag{20}$$

By using of (19) and (20), the integral equation (18) is then readily to be solved by numerical methods directly. One of the simplest method of solving eq. (18) is by constructing the Neumann series of $\tau'(t)$ (method of successive approximation). Alternatively, this is equivalent to solving for $\tau'(t)$ in an asymptotic sense. Construct a perturbation series of $\tau'(t)$ for small $\hat{k}$ as

$$\tau'(t) = \hat{k}\,\tau_1'(t) + \hat{k}^2\,\tau_2'(t) + \cdots \tag{21}$$

On substituting (21) into (18) and equating the same powers of $\hat{k}$, we have

$$\tau_1'(t) = 2(2\pi)^{-1/2}(s_T^2 - s_L^2)^{-1}s_T^2\,\hat{K}_{I1}'(t), \tag{22}$$

and

$$\tau_j'(t) = 2(2\pi a)^{-1/2}(s_T^2 - s_L^2)^{-1}\,s_T^2\int_0^t \tau_{j-1}'(\eta)\left[\frac{\partial}{\partial t}\int_0^a \xi^{1/2}\hat{K}_{Ig}(t-\eta;\xi)d\xi\right]d\eta$$

$$,j = 2,3,\ldots \tag{23}$$

3 Radiation Field

The diffraction field of the incident wave by the spring-constrained crack can be expressed as the sum of the diffraction field of the same incident wave by a unconstrained crack and the radiation field of a correction problem. The correction problem concerns a unconstrained crack subjected to the prescribed crack-face loading $\sigma_0\,\tau(t)(-x/a)^{1/2}$ over $-a < x < 0$. For the incident wave of the form defined in (1), the diffraction field by the unconstrained crack can be expressed, for $-\pi/2 < \theta < \pi/2$, as

$$\sigma_{ij}(x,y,t) = -\sigma_0 \left\{ H(t - s_L r) \int_{s_L r}^{t} f(t-\eta) Im[D_1(\xi_L) g_{ij}^L(\xi_L)] d\eta + \right.$$

$$\left. H(t - s_T r) \int_{s_T r}^{t} f(t-\eta) Im[D_1(\xi_T) g_{ij}^T(\xi_T)] d\eta \right\} \quad ,i,j = x,y, \tag{24}$$

where

$$D_1(\xi) = \frac{(s_L)^{1/2}(s_L - \xi)^{1/2}}{2\pi s_R \xi (s_T^2 - s_L^2)\Gamma_-(0)\,(\xi - s_R)\,\Gamma_-(\xi)} \frac{d\xi}{d\eta}. \tag{25}$$

Also

$$g_{xx}^L(\xi) = (s_L^2 - \xi^2)^{-1/2}(s_T^2 - 2\xi^2)(s_T^2 - 2s_L^2 + 2\xi^2), \quad g_{xx}^T(\xi) = -4\xi^2\,(s_T^2 - \xi^2)^{1/2} \tag{26}$$

$$g_{yy}^L(\xi) = (s_L^2 - \xi^2)^{-1/2}(s_T^2 - 2\xi^2)^2, \quad g_{yy}^T(\xi) = 4\xi^2\,(s_T^2 - \xi^2)^{1/2} \tag{27}$$

$$g_{xy}^L(\xi) = -2\xi(s_T^2 - 2\xi^2), \quad g_{xy}^T(\xi) = 2\xi\,(s_T^2 - 2\xi^2) \tag{28}$$

and

$$\xi_\gamma = -\frac{\eta}{r}\cos\theta + i[(\frac{\eta}{r})^2 - s_\gamma^2]^{1/2}\sin\theta \quad ,\gamma = L,T. \tag{29}$$

The branch cuts have been taken along the $Re(\xi_\gamma)$ axis from $\xi_\gamma \to -\infty$ to $\xi_\gamma = -s_\gamma$ and from $\xi_\gamma = s_\gamma$ to $\xi_\gamma \to \infty$, $\gamma = L,T$.

Owing to the existence of the characteristic length a, again the direct application of integral transform to the correction problem will not lead to standard Wiener-Hopf equation. On the other hand, the solution can easily be constructed by the method of superposition, as by Freund [9], Brock [10] and Brock & Rossmanith [11]. The elastodynamic radiation field can be expressed as

$$\sigma_{ij}(x,y,t) = -\sigma_0 \int_0^t \tau'(t-\eta) \left[\int_{-a}^{0} (-\ell/a)^{1/2}\,(\Lambda_{ij}^L + \Lambda_{ij}^T) d\ell \right] d\eta, \tag{30}$$

where

$$\Lambda_{ij}^\gamma(x,y,\eta;\ell) = -\frac{1}{\pi} H(\eta - s_\gamma r_1) Im \left\{ \frac{(s_L^2 - \xi_{\gamma 1}^2)^{1/2}}{R(\xi_{\gamma 1})} g_{ij}^\gamma(\xi_{\gamma 1}) \frac{d\xi_{\gamma 1}}{d\eta} \right\} - \frac{1}{\pi^2} H(\eta + s_L \ell - s_\gamma r) \cdot$$

$$\int_{-s_L}^{s^*} Im \left[\frac{(s_L + z)^{1/2}}{(z + s_R)\,\Gamma_+(z)} \right] Im \left[E(\xi_{\gamma 2})\, g_{ij}^\gamma(\xi_{\gamma 2}) \frac{1}{z - \xi_{\gamma 2}} \frac{d\xi_{\gamma 2}}{d\eta} \right] dz \quad ,\gamma = L,T. \tag{31}$$

Here $s^* = (\eta - s_\gamma r)/\ell$, g_{ij}^L and g_{ij}^T are defined in eqns. (26)-(28). The Rayleigh function R and the expression E are defined as

$$R(\xi) = (2\xi^2 - s_T^2)^2 + 4\xi^2\,(s_L^2 - \xi^2)^{1/2}\,(s_T^2 - \xi^2)^{1/2}, \tag{32}$$

$$E(\xi) = -\frac{(s_L - \xi)^{1/2}}{2\,(s_T^2 - s_L^2)\,(\xi - s_R)\,\Gamma_-(\xi)}. \tag{33}$$

In equation (31) $\xi_{\gamma 1}$ and $\xi_{\gamma 2}$ are defined as

$$\xi_{\gamma 1} = -\frac{\eta}{r_1} cos\theta_1 + i\left[(\frac{\eta}{r_1})^2 - s_\gamma^2\right]^{1/2} sin\theta_1, \tag{34}$$

$$r_1^2 = (x-\ell)^2 + y^2 \ , \qquad \theta_1 = tan^{-1}\frac{y}{x-\ell}, \tag{35}$$

and

$$\xi_{\gamma 2} = -\frac{\eta_1}{r} cos\theta + i\left[(\frac{\eta_1}{r})^2 - s_\gamma^2\right]^{1/2} sin\theta \ , \qquad \eta_1 = \eta - z\ell. \tag{36}$$

4 Numerical Results

For numerical examples the nondimensional time function is taken to be of the form

$$f(t) = \frac{1}{2}\left[1 - cos\frac{2\pi t}{T}\right] H(t)\, H(T-t) \tag{37}$$

The Lamé constants of the elastic medium are chosen to be $\lambda = \mu$ (i.e. Poisson's ratio =0.25) and the nondimensional cohesive zone to be $s_L a/T = 0.1$. The effects of nondimensional spring constant $\hat{k}$ on the derivative of spring tractions and is investigated by parametrical studies on $\hat{k}$. The derivative of nondimensional spring traction are needed in the tip-diffracted field calculation, they can be computed by solving the Votlerra integral equation, directly and the results are shown in Fig.2 for $0 < t/T < 2$. For $t/T > 2$, the derivative of $\tau(t)$ is of the order of 10^{-3}.Other results for the case of oblique incidence will appear elsewhere.

Acknowledgment

The work of M. K. K. was carried out in the course of research sponsored by the National Science Council of R.O.C. under Grant NSC77-0401-E002-08 to National Taiwan University

References

[1] Achenbach, J. D. and Norris, A. N., J. Nondestructive Evaluation, Vol. 3, 1982, 229-239.

[2] Achenbach, J. D. and Norris, A. N., "Specular reflection by contacting crack faces", in *Review of Progress in Quantitative Nondestructive Evaluation*, Vol. 3A, 1984, 163-174, Thompson, D. O. and Chimenti, D. E. (ed.), Plenum Press, New York.

[3] Thompson, R. B. and Fiedler, C. J., "The effects of crack closure on ultrasonic scattering measurements", in *Review of Progress in Quantitative Nondestructive Evaluation*, Vol. 3A, 1984, 207-215, Thompson, D. O. and Chimenti, D. E. (ed.), Plenum Press, New York.

[4] Angel, Y. C. and Achenbach, J. D., J. Applied Mechanics, Vol. 52, 1985, 33-41.

[5] Angel, Y. C. and Achenbach, J. D., Wave Motion, Vol. 7, 1985, 375-397.

[6] Norris, A. N., J. Acoust. Soc. Am., Vol. 73, 1983, 421.

[7] Noble, B., *Methods based on the Wiener-Hopf Technique*, Pergamon Press, New York, 1958.

[8] Achenbach, J. D., *Wave Propagation in Elastic Solids*, North-Holland Pub. Co., New York, 1973, 380-388.

[9] Freund, L. B., Int. J. Engng. Sci., Vol. 12, 1974, 179-189.

[10] Brock, L. M., Int. J. Solid Structures, Vol. 18, 1982, 467-477.

[11] Brock, L. M. and Rossmanith, H. P., J. Applied Mechanics, Vol. 52, 1985, 57-61.

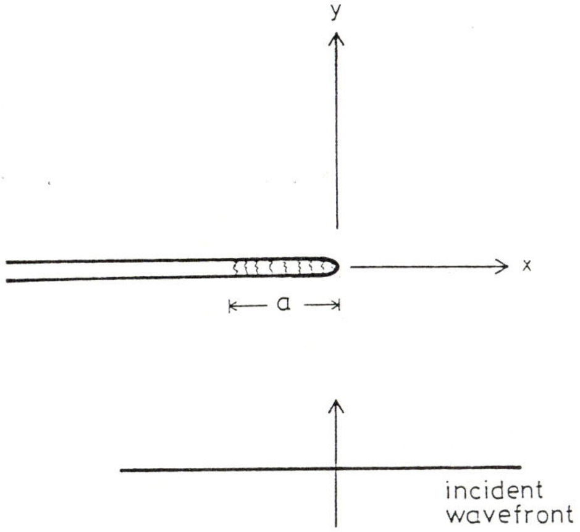

Figure 1: Geometry of the tip-constrained semi-infinite crack, where the length of the tip-constrained zone is denoted by a. The wavefront of the incident longitudinal wave is parallel to the crack faces.

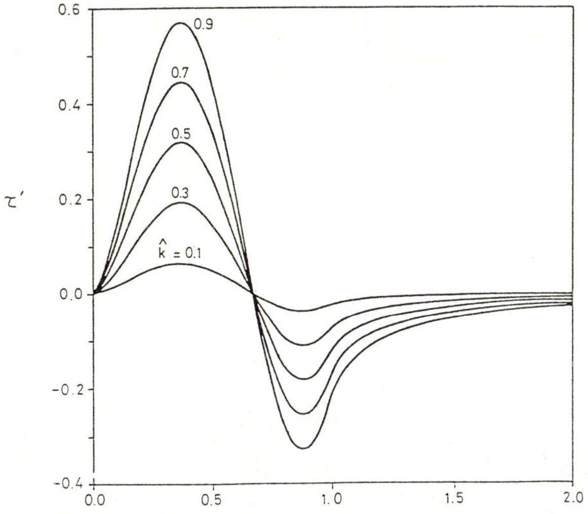

Figure 2: The derivative of nondimensional spring traction, τ', versus nondimensional time, t/T, where T is the duration of the incident wave, for various values of nondimensional spring constant $\hat{k} = 0.1, 0.3, 0.5, 0.7,$ and 0.9.

ELASTIC WAVE PROPAGATION
M.F. McCarthy, M.A. Hayes, (Editors)
© Elsevier Science Publishers B.V. (North-Holland), 1989

THE DYNAMIC RESPONSE OF RANDOM VISCOELASTIC COMPOSITES

A.I. Beltzer*

Dept. of Mechanical Engineering
University of Alberta
Edmonton, Alberta, Canada T6G 2G8

Wave propagation in random discrete media is a classical but rather complicated subject. The continuing interest in this problem is motivated by applications to the effective behavior of composites, ultrasonics, and remote sensing.

A recently developed causal approach appeals from the outset to linearity, causality, and passivity of the effective medium. The objective of this paper is to extend this approach to viscoelastic composites, which is of particular interest for biomedical and geophysical applications, and to review the pertinent works. The relevant concept of generalized extinction cross section is first discussed, which paves a way to analysis of the effects of interplay between scattering and viscoelastic losses. Harmonic and transient waves in random media are then considered.

1. INTRODUCTION

Random elastic composites and polycrystals appear as attenuating materials when subject to wave propagation. Despite their conservative nature, they show spatial decay of progressive disturbances, which depends on the ratio of the wavelength to a typical microstructure dimension. The wave speed is also sensitive to this ratio. Scattering by randomly dispersed inclusions (or randomly oriented grains) induces these effects, which render the static theories of effective behavior inapplicable to a dynamic case. On the other hand, the well-known theory of configurational averaging has been shown to be unsatisfactory for moderate and high frequencies (see, for example [1]).

The situation becomes even more complicated if one of the components, say, the matrix is viscoelastic, as in many biomedical and seismological applications. Indeed, the interplay between the two kinds of losses may become substantial. It suffices to consider, as an example, a perfectly elastic inclusion in a viscoelastic homogeneous matrix. When a wave strikes the inclusion, the energy loss at the expense of scattering may be compensated to a certain degree by the absence of dissipation in the region occupied by inclusion. Furthermore, normalization of the losses by the incident energy flux, becomes ambiguous in the viscoelastic case, since this time it varies in space. Clearly, analysis of such systems may require reconsideration of some of the concepts of wave scattering.

In what follows we first consider the energy perturbations taking place when a viscoelastic wave strikes the inclusion. Then a fairly general approach to an approximate analysis of the effective dynamic behavior of viscoelastic composites is given.

* On leave from Holon Tech. Inst., Holon 58102, ISRAEL

2. GENERALIZED EXTINCTION CROSS SECTION

We consider the wave interaction with a single obstacle, bearing in mind that this investigation may provide a basis for a more general analysis. The above problem is one of the classical subjects of the wave theory. Main efforts however have been directed towards conservative systems consisting of a lossless (elastic or dielectric) inclusion in a lossless background medium and evaluation of scattering cross sections. These investigations yield either the angular distribution of the scattered energy (the differential cross section) or its total value (the scattering cross section) relatively to the incident energy. In some cases the analysis dealt with a lossy (viscous or conducting) obstacle in a lossless medium and subsequent evaluation of the energy absorbed by the scatterer in addition to scattering. The total energy withdrawn from the incident beam is then estimated with the help of the extinction cross section.

To extend the analysis to the case of a lossy surrounding medium consider, following [2], a plane harmonic wave propagating in the direction β_i and striking an inclusion with the boundary Ω. The presence of an inclusion gives rise to the scattered field with the average power flux $< I^{sc} >_T$, given by

$$< I^{sc} >_T = \frac{1}{T}\int_o^T \int_\Omega P_i^{sc} n_i \, d\Omega dt \quad , \tag{1}$$

where n_i denotes an outward unit vector normal to Ω and P_i^{sc} the Poynting vector

$$P_i^{sc} = -\sigma_{ij}^{sc} \dot{u}_j^{sc} \quad , \tag{2}$$

with σ_{ij}^{sc} and u_j^{sc} being stresses and displacements of the scattered field. Similarly, for the absorption within the obstacle we get

$$< I^{ab} >_T = -\frac{1}{T}\int_o^T \int_\Omega P_i^{tot} n_i \, d\Omega dt \quad , \tag{3}$$

where P_i^{tot} is the Poynting vector of the total field. Consequently, the energy loss averaged over the period is

$$< I >_T = < I^{sc} >_T + < I^{ab} >_T \quad , \tag{4}$$

Now we need to normalize (4) with respect to the incident energy, since it is the relative loss which is a physically meaningful quantity. It is usually done by dividing (4) by the average incident intensity

$$< e^{in} >_T = \frac{1}{T}\int_0^T P_i^{in} \beta_i \, dt \tag{5}$$

However, for a lossy matrix this entity depends on a spatial coordinate, which makes the analysis ambiguous. Taking into account that the entire treatment deals with averaged (integral) values, we may remedy this by introducing its spatial average as a normalizing factor. The simplest possibility is given by

$$< e^{in} >_{T\Omega} = \frac{1}{T\Omega} \int_{\Omega} \int_{0}^{T} P_i^{in} \beta_i \, dt d\Omega \quad . \tag{6}$$

This leads to the following definition of the generalized extinction cross section

$$\gamma^{ex} = \frac{< I^{sc} >_T + < I^{ab} >_T}{< e^{in} >_{T\Omega}} \quad . \tag{7}$$

Eq. (7) can, of course, be split into two parts, the scattering cross section and the absorption cross section

$$\gamma^{ex} = \gamma^{sc} + \gamma^{ab} \quad . \tag{8}$$

While the above generalization appears straightforward it enables one to approximately investigate the interplay between the two kinds of losses and show a possibility of the "negative" extinction. Indeed it can be shown that for the above case of elastic inclusion in a viscoelastic medium we get

$$\gamma^{ex} = \frac{\int_o^T \int_{\Omega} (P_i^{sc} - P_i^{in}) \, n_i \, d\Omega \, dt}{\frac{1}{\Omega} \int_o^T \int_{\Omega} P_i^{in} \beta_i \, d\Omega \, dt} \tag{9}$$

Indeed, using the definition of the Poynting vector,

$$P_i = - \sigma_{ij} \dot{u}_j \tag{10}$$

we get for (3),

$$< I^{ab} >_T = \frac{1}{T} \int_o^T \int_{\Omega} \sigma_{ij}^{in} \dot{u}_j^{in} n_i \, d\Omega dt$$

$$+ \frac{1}{T} \int_o^T \int_{\Omega} (\sigma_{ij}^{in} \dot{u}_j^{sc} + \sigma_{ij}^{sc} \dot{u}_j^{in} + \sigma_{ij}^{sc} \dot{u}_j^{sc}) n_i \, d\Omega dt \, . \tag{11}$$

The second integral term vanishes for a purely elastic obstacle because of the optical theorem [3], which leads to (9).

Since both terms in the numerator are positive, γ^{ex} may be negative in extreme cases and the term "extinction" should not therefore be taken literally. Indeed, scattering losses may be compensated to a certain extent by the absence of viscoelastic losses, which would occur at the site if the lossless obstacle were not present. Computations made for the case of SH-wave diffraction by an elastic cylinder embedded in a viscoelastic medium confirmed this conclusion [4]. Eq. (7) may thus serve as a measure of the total energy perturbation when both of the above losses take place.

While these quantities reduce to the conventional ones provided the matrix is elastic, in general, they show a different behavior. In particular, γ^{sc} may not obey the Rayleigh law for low frequencies [4]. Furthermore, the spatial integrals involved must be taken precisely over Ω, unlike the elastic case, to avoid the interference of scattering and elastic losses.

3. HARMONIC AND TRANSIENT WAVES IN COMPOSITES

Now we may turn to a viscoelastic random composite. Taking into account that the energy effectively withdrawn from the incident viscoelastic beam by a single inclusion is evaluated by the above generalized extinction cross section, γ^{ex}, we get $N\gamma^{ex}$ for the case of N independent scatterers per unit volume. This is appropriate only for a sufficiently dilute mixture, when multiple interactions are negligible. Thus, the amplitude attenuation, α, is

$$\alpha = \alpha_V + N\gamma^{ex}/2 = \alpha_V + \alpha_{ex} \quad . \tag{12}$$

Here the first term in the right-hand side accounts for the viscoelastic losses and the second for the losses induced by the wave-obstacle interactions. Note that (12) incorporates the coupling between the two kinds of energy losses for γ^{ex} depends on α_V. It thus differs from the widely used linear rule

$$\alpha = \alpha_V + N\gamma^{sc}/2 \ , \tag{13}$$

which holds for small losses only.

Having defined α, one applies the Kramers-Kronig relations to find the associated wave velocity, c, and completes thus the identification of the effective dispersive wave number k

$$k(\omega) = \omega/c(\omega) + i\alpha(\omega) . \tag{14}$$

This simple theory of independent scatterers (with $\alpha_V = 0$, [5]) has allowed for a discovery of a pattern exhibited by the attenuation and dispersion curves, despite the randomness of the microstructure.

To account for effects of multiple scattering we invoke the concept of differential effective media, well-known in the statics, and extend it to a dynamic case by making use of the above approach of independent scatterers as a building block. Consider a two-phase solid mixture with a homogeneous background medium and randomly dispersed identical lossless inclusions at the volume concentration ϕ. We begin the construction process with a matrix of phase 1 and imbedded inclusions of phase 2 in dilute concentration, which is chosen so as to ensure slight rescattering. Next, a homogenization is carried out by employing the above theory of independent scatterers, and then new inclusions are imbedded into the homogenized matrix. This approach puts no frequency limitation. Denoting the current entities by tilde we get the evolution equation for the total attenuation, $\tilde{\alpha}$,

$$\tilde{\alpha}(\tilde{\phi} + \Delta\tilde{\phi}) = \alpha_V + \tilde{\alpha}_{ex}(\tilde{\phi}) + \Delta\tilde{\phi}\tilde{\gamma}^{ex}/(2V) , \tag{15}$$

where $\Delta\tilde{\phi}$ is the increment in the current volume concentration and V the inclusion volume. The associated $\tilde{c}(\omega)$ follows from the Kramers-Kronig relation.

Having specified $k(\omega)$ for the entire frequency interval, $0 \leq \omega < \infty$, one may proceed with evaluation of a transient wave propagation by a classical Fourier transform technique, which is a remarkable feature of the approach. Computations made with the help of a standard FFT subroutine show good convergence, and indicate how both of the above two attenuation mechanisms affect the pulse propagation [6]. Figure 1 shows the pulse evolution during propagation in a random viscoelastic medium [6]. The scattering losses have

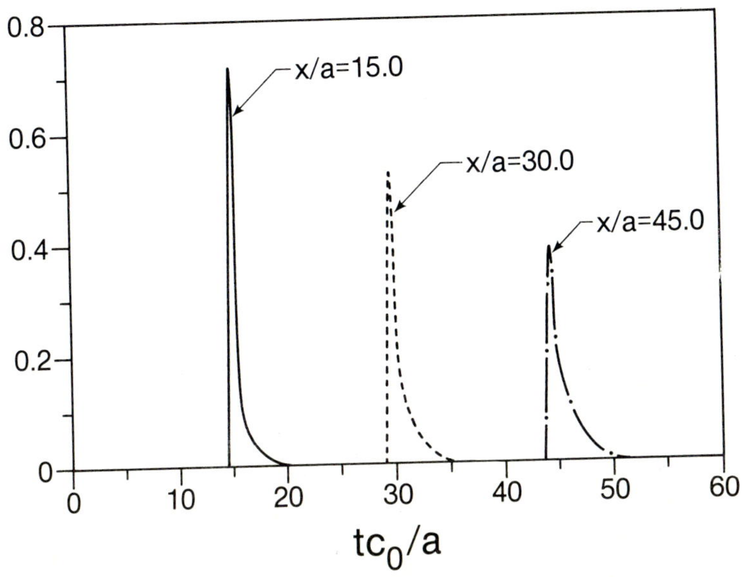

FIGURE 1
Pulse evolution in a scattering viscoelastic medium, x is the spatial coordinate and a the microstructure dimension

been described by Wu's model, while the viscoelastic losses by a modified Azimi's law, both of which find applications in geophysics.

Recently, a good deal of work has been carried out in geophysics on pulse propagation in a randomly discontinuous medium (see, for example [7]). Numerical simulatons showed the pulse broadening, reinforcing and delaying effects of multiple scattering. The causal approach predicts similar effects [2,6]. Furthermore, the pulse reinforcing and delaying as compared to a single scattering approach appear to be interrelated. Indeed, the static limit, c_0, and the geometric limit, c_∞, of the phase speed are related by [1,5]

$$\frac{1}{c} = \frac{1}{c_\infty} + \frac{2}{\pi} \int_0^\infty \frac{\alpha(\omega)}{\omega^2} \, d\omega \, . \qquad (16)$$

Since $\alpha(\omega) > 0$ by definition and c_∞ is the pulse speed it follows that any overall decrease in attenuation leads to the pulse delay.

4. CONCLUSIONS

As other random systems, composite media are not amenable to the exact analysis. There always exists uncertainty concerning, say, the phase real configuration and distribution. Further, a resort to formal statistical operations, such as the ensemble averaging, may lead to ambiguous predictions. This is why incorporation of *a priori* given features of the effective medium, such as its causality, appears necessary while formulating approximate theoretical models. It has been shown that the effective dynamic response may be investigated by appeal to the above basic features of the effective medium and energy considerations.

REFERENCES

[1] A.I. Beltzer, Dispersion of Seismic Waves by a Causal Approach, Pure and Appl. Geophys. (accepted).

[2] A.I. Beltzer and N. Brauner, The Dynamic Response of Random Composites by a Causal Differential Method, Mechanics of Materials (accepted).

[3] J.E. Cubernatis, E. Domany and J.A. Krumhansl, Formal Aspects of the Theory of the Scattering of Ultrasound by Elastic Materials, J. Appl. Phys, 48, 2804-2811, 1977.

[4] N. Brauner and A.I. Beltzer, Wave-obstacle Interaction: Energy Perturbations and Negative Extinction (to appear).

[5] A.I. Beltzer and N. Brauner, SH-waves of an Arbitrary Frequency in Random Fibrous Composites via the Kramers-Kronig Relations, J. Mech. Phys. Solids, 33, 5, 471-487, 1985.

[6] A.I. Beltzer, J. Wegner, B.R. Tittmann and J.B. Haddow, Pulse Propagation in Random Media by a Causal Approach (to appear).

[7] P.G. Richards and W. Menke, The Apparent Attenuation of a Scattering Medium, Bull. Seism. Soc. Am. 73, No. 4, 1005-1021, 1983.

ELASTIC WAVE PROPAGATION
M.F. McCarthy, M.A. Hayes, (Editors)

ELASTIC WAVE PROPAGATION IN ANISOTROPIC MICROCRACKED ROCKS

Colin M. SAYERS

Koninklijke/Shell Exploratie & Produktie Laboratorium
Volmerlaan 6
2288 GD Rijswijk ZH
The Netherlands

The elastic wave velocities in rock are reduced significantly by the presence of microcracks. As a result of the stress history of the rock these microcracks frequently exhibit some degree of alignment. The rock therefore displays an elastic anisotropy determined by the shape and content of the cracks and by the crack orientation distribution function. The coefficients $W_{\ell mn}$ of a series expansion of this function in generalised spherical harmonics can be obtained to order $\ell = 4$ from the angular variation of the elastic wave velocities. This allows construction of microfracture pole figures, which may be compared with those obtained by petrofabric examination. Since the anisotropy in strength properties originates from the preferred orientation of microstructural defects, the prediction of the microfracture orientation distribution will have an important application in the field of rock fracture.

1. INTRODUCTION

Laboratory measurements of ultrasonic wave velocities in rock samples as a function of confining pressure have demonstrated that the presence of microcracks significantly decreases the elastic wave velocities in the rock [1,2]. As the pressure is increased the measured velocities increase markedly. This behaviour is attributed to the closure of cracks with increasing pressure until they have no effect on the elastic properties of the rock. In general, these microcracks are not randomly orientated and the rock displays an elastic anisotropy determined by the shape and content of the cracks and by the crack orientation distribution function (CODF). This function gives the probability of a crack having a given orientation with respect to a reference frame fixed in the rock and can be used to derive expressions for the anisotropy of elastic wave velocities in the long wavelength limit. Recently it has been shown that the coefficients $W_{\ell mn}$ of a series expansion of the CODF in generalised spherical harmonics can be obtained to order $\ell = 4$ from the angular variation of the elastic wave velocities [3]. In this paper these developments are reviewed and the theory is applied to ultrasonic measurements reported by Thill et al. [4] and Nur and Simmons [5].

2. THEORY

For an ellipsoidal crack with principal axes 2a, 2b, 2c ($a \geq b \geq c$) it is convenient to introduce a set of axes $OX_1X_2X_3$ with origin at the centre of the ellipsoid and OX_1, OX_2 and OX_3 along the a, b, and c axes, respectively. If the axes $OX_1X_2X_3$ for all cracks are aligned, the rock will exhibit orthorhombic (orthotropic) symmetry with three orthogonal planes of mirror symmetry having normals along OX_1, OX_2 and OX_3. It is convenient to write the fourth-order elastic stiffness tensor of the cracked material $C_{ijk\ell}$ in the form

$$C_{ijk\ell} = C^0_{ijk\ell} + \gamma_{ijk\ell}. \tag{1}$$

$\gamma_{ijk\ell}$ is the difference between $C_{ijk\ell}$ and the elastic stiffness tensor $C^0_{ijk\ell}$ of the uncracked material which is assumed to be isotropic. For aligned ellipsoidal cracks let the $\gamma_{ijk\ell}$ be denoted by $\Gamma_{ijk\ell}$. In the crack reference frame $OX_1X_2X_3$ the non-zero $\Gamma_{ijk\ell}$ are Γ_{11}, Γ_{22}, Γ_{33}, $\Gamma_{12}=\Gamma_{21}$, $\Gamma_{23}=\Gamma_{32}$, $\Gamma_{31}=\Gamma_{13}$, Γ_{44}, Γ_{55} and Γ_{66} in the Voigt (two-index) notation.

In general it cannot be assumed that the cracks will be perfectly aligned and a quantitative description of the elastic constants and elastic wave velocities in the rock requires a knowledge of the orientation distribution of cracks. The orientation of an ellipsoidal crack with principal axes $OX_1X_2X_3$ with respect to a set of axes $Ox_1x_2x_3$ fixed in the rock may be specified by three Euler angles ψ, θ and ϕ [6] shown in Figure 1. The orientation distribution of cracks is then given by the crack orientation distribution function $W(\xi,\psi,\phi)$ where $\xi = \cos\theta$. $W(\xi,\psi,\phi)d\xi d\psi d\phi$ gives the fraction of cracks between ξ and $\xi + d\xi$, ϕ and $\phi + d\phi$ and ψ and $\psi + d\psi$. Clearly,

$$\int_0^{2\pi}\int_0^{2\pi}\int_{-1}^{1} W(\xi,\psi,\phi)d\xi d\psi d\phi = 1 \tag{2}$$

It will be assumed that the orientation distribution of cracks is orthotropic with symmetry axes coincident with the reference axes Ox_1, Ox_2 and Ox_3. Expressions for the elastic constants $C_{ijk\ell}$ of the cracked material can be derived from the $\Gamma_{ijk\ell}$ and the crack orientation distribution as follows [7]. If $T_{ijk\ell mnpq}$ is given by the transformation rule for tensors of rank four, i.e.

$$T_{ijk\ell mnpq} = (\partial x_i/\partial X_m)(\partial x_j/\partial X_n)(\partial x_k/\partial X_p)(\partial x_\ell/\partial X_q) \tag{3}$$

then, taking into account the orientation distribution of cracks, the first-order correction $\gamma_{ijk\ell}$ in equation (1) is given by

$$\gamma_{ijk\ell} = \bar{T}_{ijk\ell mnpq}\,\Gamma_{mnpq} \tag{4}$$

where

$$\bar{T}_{ijk\ell mnpq} = \int_0^{2\pi}\int_0^{2\pi}\int_{-1}^{1} T_{ijk\ell mnpq}(\xi,\psi,\phi)\, W(\xi,\psi,\phi)d\xi d\psi d\phi \tag{5}$$

These integrals can be evaluated by expanding $W(\xi,\psi,\phi)$ as a series of generalised spherical harmonics [6,7] and using the orthogonality relations between these functions. Since the elastic stiffness tensor is of fourth rank, the expressions for the $\gamma_{ijk\ell}$ involve the $\Gamma_{ijk\ell}$ and the coefficients $W_{\ell mn}$ of the expansion of $W(\xi,\psi,\phi)$ for $\ell \le 4$. For orthorhombic material symmetry and ellipsoidal cracks, the non-zero $W_{\ell mn}$ are all real and are restricted to even values of ℓ, m and n. In the following circular cracks are assumed with a=b>>c for which $W_{\ell mn} = 0$ unless $n = 0$. The γ_{ijkl} are therefore determined in this case by W_{200}, W_{220}, W_{400}, W_{420}, W_{440} and three anisotropy factors a_1, a_2 and a_3. In terms of the $\Gamma_{ijk\ell}$ defined above the a_i are given by

$$\begin{aligned} a_1 &= \Gamma_{11} + \Gamma_{33} - 2\,\Gamma_{13} - 4\,\Gamma_{44} \\ a_2 &= \Gamma_{11} - 3\,\Gamma_{12} + 2\,\Gamma_{13} - 2\,\Gamma_{44} \\ a_3 &= 4\,\Gamma_{11} - 3\,\Gamma_{33} - \Gamma_{13} - 2\,\Gamma_{44} \end{aligned} \tag{6}$$

The elastic wave velocities in the rock may be obtained from equation (1) allowing the $W_{\ell mn}$ to be determined from velocity measurements provided that the a_i given by equation (6) are known and the crack-free velocities can be estimated. The $W_{\ell mn}$ can be used to calculate the microcrack normal orientation distribution $q(\zeta,\eta)$ defined by

$$q(\zeta,\eta) = n(\zeta,\eta) \,/\, \int_0^{2\pi}\int_{-1}^{1} n(\zeta,\eta)\,d\zeta\,d\eta \tag{7}$$

where $n(\zeta,\eta)d\zeta d\eta$ is the number of cracks with normal between ζ and $\zeta + d\zeta$ and η and $\eta + d\eta$. Here $\zeta = \cos\chi$, χ and η being the polar and azimuthal angles in the reference frame $Ox_1x_2x_3$. $q(\zeta,\eta)$ may be expanded as a series of spherical harmonics

$$q(\zeta,\eta) = \sum_{\ell=0}^{\infty} \sum_{m=-\ell}^{\ell} Q_{\ell m}\, P_\ell^m(\zeta)\, \exp(-im\eta) \tag{8}$$

where the $P_\ell^m(\zeta)$ are the normalised associated Legendre functions [6]. For circular cracks the $Q_{\ell m}$ are given in terms of the $W_{\ell mn}$ by $Q_{\ell m} = 2\pi W_{\ell m0}$. Only the coefficients $W_{\ell mn}$ for $\ell \leq 4$ can be obtained from the angular variation of the elastic wave velocities. Expanding equation (8) to order $\ell = 4$ gives

$$4\pi q(\zeta,\eta) = 1 + 4\pi[\, Q_{20}P_2^0(\zeta) + 2Q_{22}P_2^2(\zeta)\cos 2\eta \tag{9}$$
$$+ Q_{40}P_4^0(\zeta) + 2Q_{42}P_4^2(\zeta)\cos 2\eta + 2Q_{44}P_4^4(\zeta)\cos 4\eta\,]$$

3. INVERSION OF ELASTIC WAVE VELOCITIES

Thill et al. [4] have measured longitudinal ultrasonic wave velocities in a large number of directions on a spherical sample of Salisbury granite. Figure 2 shows a contoured equal-area projection of the measured longitudinal velocities. The velocity distribution exhibits orthotropic symmetry, the coordinate planes being approximately parallel to the three mirror planes. The orientation distribution of microcrack normals was also measured using a petrofabric microscope and correlates closely with the longitudinal velocity symmetry pattern. Two sets of microfractures were found to be present. The principal set consisted of well developed microcracks frequently extending through many of the grains. For this set the crack normals showed a marked alignment along the direction Ox_1 in Figure 2. The second set had crack normals aligned in the direction Ox_3 in Figure 2 and consisted of relatively few, discontinuous, weakly defined microcracks. Thill et al. [4] concluded that the effect of this second set on the measured velocity is negligible, the principal set controlling the velocity anisotropy of the rock.

The anisotropy factors a_i occurring in equations (6) can be calculated from the theory of Hudson [8] who obtained the Γ_{ij} by considering the interaction of an elastic wave with a random spatial distribution of cracks in the long-wavelength limit. For the case of dry cracks appropriate to the measurements of Thill et al. [4] the results are plotted in Figure 3. Because of the small value of a_1, the coefficients W_{4mn} only make a small contribution to the measured longitudinal wave velocity [3] and cannot therefore be obtained accurately from the data of Thill et al. [4] for which an error in the measured velocities of 1.4% is quoted. Therefore, only W_{200} and W_{220} were evaluated and the microfracture pole figure, plotted in Figure 4, was calculated using the values $W_{200} = -0.0105$ and $W_{220} = 0.00646$ determined from the measurements. The ultrasonic pole figure shows a preferred orientation of crack normals along Ox_1 in agreement with the results of Thill et al. [4].

Nur and Simmons [5] subjected a cylinder of Barre granite to a uniaxial compressive stress normal to the axis of the cylinder. Compressional and shear wave velocities were measured for propagation normal to the cylinder axis at various angles to the stress axis. It is convenient in this case to choose the reference axes $Ox_1x_2x_3$ with Ox_3 along the cylinder axis and Ox_1 in the direction of applied stress [9]. Upon application of the stress, an initially isotropic distribution of open microcracks will develop axial symmetry with symmetry axis along Ox_1. Figure 5 shows the orientation distribution of open microfractures calculated from equation (9) using the values of $Q_{\ell m}$ determined from the measured velocities for an applied stress of 30 MPa [9]. The two pole figures obtained independently from the longitudinal and shear wave measurements are in rather good agreement and display an axial symmetry about the x_1 axis with cracks with plane normals parallel to the applied stress being closed preferentially. Because of the axial symmetry it is sufficient to plot a cross section of the pole figure in the x_3x_1 plane. The actual density of open microcracks with a specific orientation is given by $n(\zeta,\eta)$ (equation(7)). This is plotted at intervals of 100 MPa in Figure 6. Cracks with normals aligned along the stress direction are closed preferentially as expected. However, for crack normals perpendicular to the applied stress there is some evidence of crack opening.

4. CONCLUSION

The ultrasonic pole figures presented in this paper are approximations to those obtained by petrofabric analysis since only the first few coefficients $W_{\ell mn}$ in an expansion of the crack orientation distribution in generalised spherical harmonics are included. $W_{\ell mn}$ is given by

$$W_{\ell mn} = (4\pi^2)^{-1} \int_0^{2\pi} \int_0^{2\pi} \int_{-1}^{1} W(\xi,\psi,\phi) Z_{\ell mn}(\xi) \exp(im\psi) \exp(in\phi) d\xi d\psi d\phi$$

where the $Z_{\ell mn}(\xi)$ are the generalised Legendre functions defined by Roe [6]. $W_{\ell mn}$ therefore represents the value of a polynomial of trignometrical functions of θ, ψ and ϕ averaged over all crack orientations. The evaluation of a limited number of $W_{\ell mn}$ therefore corresponds to the specification of the distribution function by its first few moments [6] and represents the maximum information that can be obtained using elastic waves with wavelengths greater than the crack size. For the pole figures shown only W_{200} and W_{220} were used; more detail could be obtained by including the terms W_{400}, W_{420} and W_{440}. However, this would require more accurate measurements of the wave velocities.

ACKNOWLEDGEMENT

This paper is published by permission of Shell Research B.V.

REFERENCES

[1] Birch, F., J. Geophys. Res. 65 (1960) 1083.
[2] Birch, F., J. Geophys. Res. 66 (1961) 2199.
[3] Sayers, C.M., Ultrasonics 26 (1988) 73.
[4] Thill, R.E., Willard, R.J. and Bur, T.R., J. Geophys. Res. 74 (1969) 4897.
[5] Nur, A. and Simmons, G., J. Geophys. Res. 74 (1969) 6667.
[6] Roe, R.J., J. Appl. Phys. 36 (1965) 2024.
[7] Morris, P.R., J. Appl. Phys. 40 (1969) 447.
[8] Hudson, J.A., Geophys. J. Roy. astr. Soc. 64 (1981) 133.
[9] Sayers, C.M., Submitted to Ultrasonics.

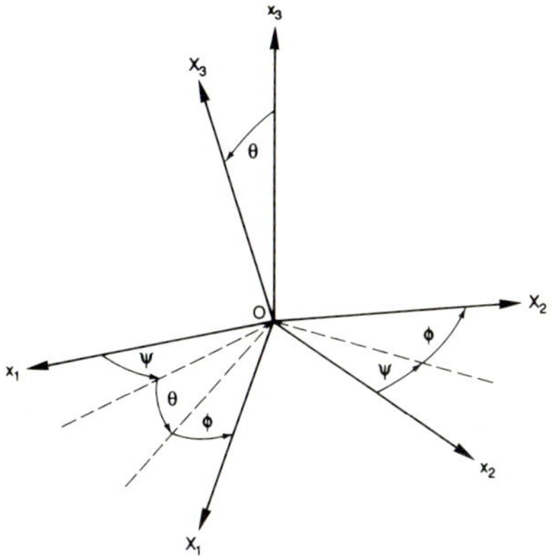

Figure 1

The orientation of coordinate system $OX_1X_2\ X_3$ with respect to the material coordinate system $Ox_1\ x_2\ x_3$ specified by the Euler angles ψ, θ and ϕ [6].

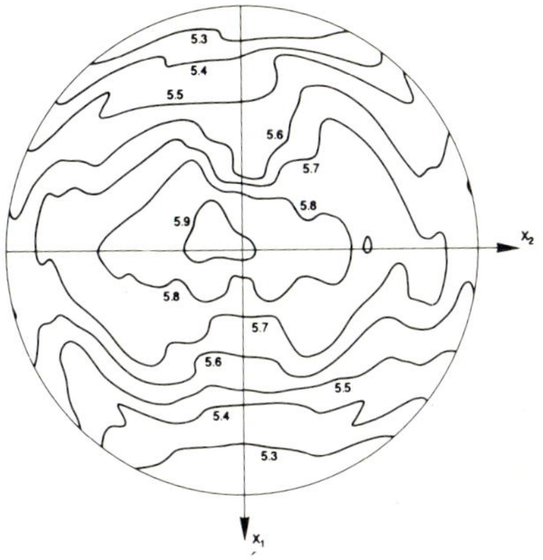

Figure 2

Contoured equal-area projection of the longitudinal velocities (kms^{-1}) measured by Thill et al. [4] on Salisbury granite.

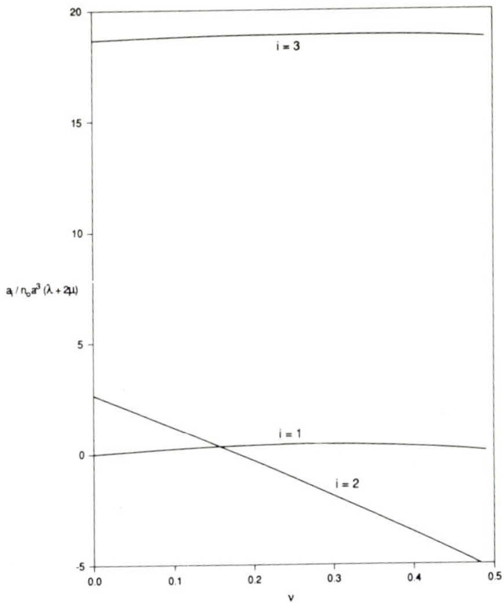

Figure 3

Variation of the anisotropy factors a_i defined by equations (6) with Poisson's ratio υ. λ and μ are the second order (Lame) elastic constants of the uncracked rock, n_0 is the density of cracks and a their radius.

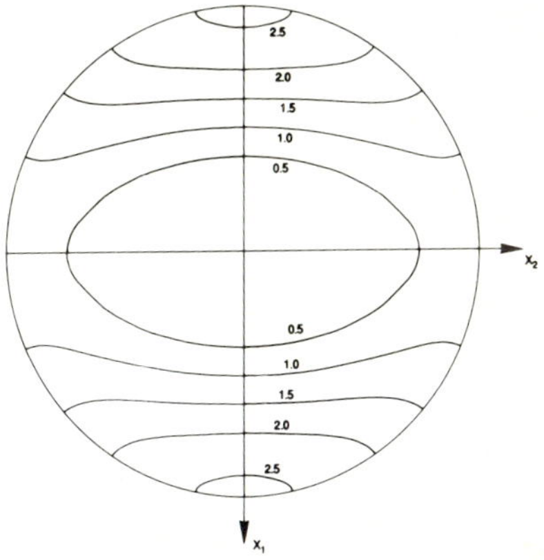

Figure 4

Contoured equal-area projection of the orientation distribution of microcrack normals calculated from equation (9) using the values W_{200} = -0.0105, W_{220} = 0.00646 obtained from the ultrasonic measurements of Thill et al. [4] on Salisbury granite.

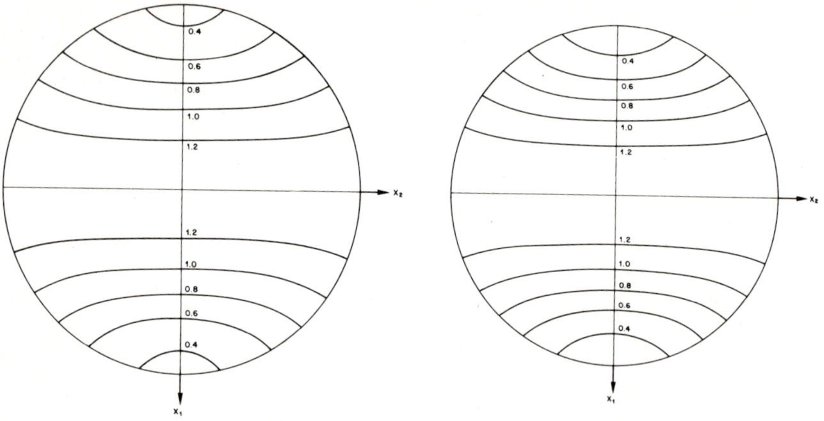

Figure 5
Contoured equal-area projection of the orientation distribution of microcrack normals $4\pi q(\zeta,\eta)$ calculated from equation (9) using the values of W_{200} and W_{220} obtained from (a) the longitudinal measurements and (b) the shear wave measurements of Nur and Simmons [5] for Barre granite at an applied stress of 30MPa.

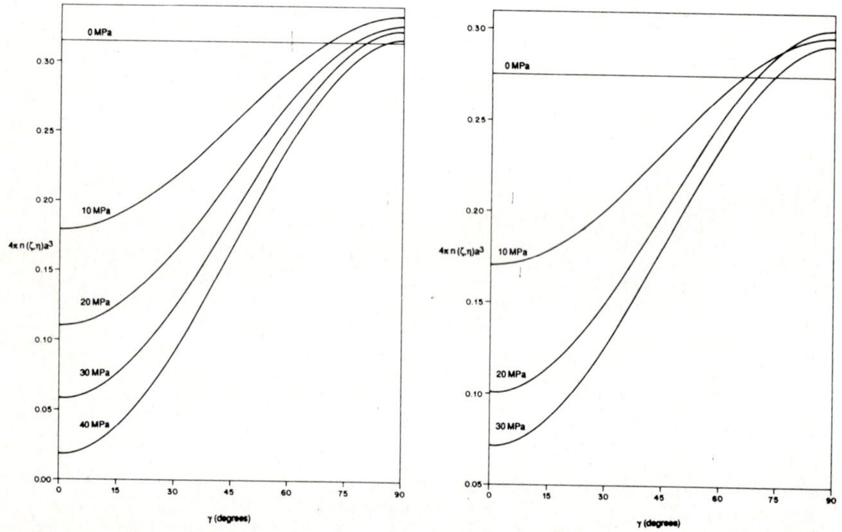

Figure 6
Distribution of microcrack normals $n(\zeta, \eta)$ for those cracks remaining open under an applied compressive stress of 0, 10, 20, 30 and 40 MPa along Ox_1 calculated from equation (7) using the values of W_{200} and W_{220} obtained from (a) the longitudinal and (b) the shear wave velocity measurements of Nur and Simmons [5]. γ is the angle between the crack normal $\underline{n}$ and the stress axis Ox_1. a is the crack radius.

ELASTIC WAVE PROPAGATION
M.F. McCarthy, M.A. Hayes, (Editors)

ELASTIC WAVE SCATTERING BY TWO PENNY-SHAPED CRACKS

Lars Peterson

Institute of Theoretical Physics
Chalmers University of Technology
S-412 96 Göteborg, Sweden

The scattering of waves by two penny-shaped cracks with zero surface traction in a homogeneous, isotropic linear elastic medium has been considered. We give numerical results in the form of back-scattered far-fields for pulsed incident waves. The examples are restricted to longitudinal waves of normal incidence on two cracks with a mutual axis of rotational symmetry, but the method is general in that transverse waves and waves of arbitrary incidence can also be handled. The method is based on integral equations relating displacements and stresses on the crack surface, and an integral representation of the scattered displacement field in terms of the crack-opening displacement. The integral equations are discretized by an expansion in trigonometric functions in the azimuthal direction and in orthogonal polynomials in the radial coordinate. The T matrix for a single crack is calculated, and then we construct the scattered displacement field for the combination of two cracks.

1. INTRODUCTION

The direct scattering problem for cracks has been considered by a number of authors. Ray methods, including the geometric theory of diffraction, are described by Achenbach et al [1]. Using integral equation methods the penny-shaped crack is treated in full generality by Martin and Wickham [2], who also give many further references, and by Krenk and Schmidt [3]. Shah et al [4] consider a Y-shaped crack near a free surface using the finite element method and Scandrett and Achenbach [5] study pulse scattering by a surface-breaking crack using finite differences. Boström and Olsson [6] use the null field approach to study nonplanar cracks.

The aim of the present paper is first to compute the T matrix for one penny-shaped crack. This T matrix can then be used as a building-block for more complicated scattering geometries [7,8] and this is demonstrated by considering two penny-shaped cracks, which in the numerical examples have coincident symmetry axies. The T matrix relates the expansion coefficients of the incoming field, when expanded in regular spherical vector wave functions, to the coefficients of the scattered field, expanded in terms of irregular spherical vector wave functions. It is possible to use the matrix equations of Krenk and Schmidt [3] to directly compute the crack opening displacement, given an incoming field in terms of regular vector wave functions. In contrast to Krenk and Schmidt, we then use an integral representation with the Green tensor expanded in spherical vector wave functions to compute the expansion of the scattered field and thereby the T matrix.

2. THE T MATRIX FOR ONE CRACK

Let the crack occupy the region $x^2+y^2 \leq a^2$, $z=0$, in an otherwise homogeneous elastic medium with Lamé parameters λ and μ and wavenumbers k_p and k_s (time harmonic conditions with the factor $\exp(-i\omega t)$ are assumed). We expand the incoming and scattered displacement fields in spherical vector wave functions (n is a multiple index) [8]:

$$\vec{u}^{in} = \sum_n a_n \text{Re}\vec{\Psi}_n \tag{1}$$

and

$$\vec{u}^{s} = \sum_n f_n \vec{\Psi}_n \tag{2}$$

$\text{Re}\vec{\Psi}_n$ denotes the regular part of $\vec{\Psi}_n$, or in other words that the spherical Hankel function in the expression for $\vec{\Psi}_n$ is changed to a spherical Bessel function. The linear relationship between the incoming and scattered fields is now expressed through:

$$f_n = \sum_{n'} T_{nn'} a_{n'} \tag{3}$$

where $T_{nn'}$ is the T matrix. It contains a complete discription of the scatterer. It is essential to note that it is a relationship between the expansion coefficients in terms of spherical vector wave functions.

Following Krenk and Schmidt [3] the scattered displacement field in the upper half-space is expressed in terms of cylindrical vector wave functions, which is natural in this case. This leads to integral equations involving both the displacements and stresses on the crack surface, which are then expanded in associated Legendre functions. These expansions give the correct variation for the displacements and stresses with distance from the crack edge [2,9], if we assume that this variation is the same as under static conditions (root singularity for stresses).

The symmetry with respect to the crack plane can be employed to divide the problem into a symmetric part and an antisymmetric part with regard to the displacement and stress components [3]. These symmetries hold for the total fields as well as for the scattered fields. The expansions mentioned above result in matrix equations relating the expansion coefficients of the stresses on the crack surface, due to an incoming field, to those of the displacements. These matrix equations can easily be solved. That is, given an incoming wave, we can compute the displacement field at the crack.

So far we have followed Krenk and Schmidt, but to obtain the T matrix we now proceed in the following way. First we use one of the partial waves $\text{Re}\vec{\Psi}_{n'}$ from the expansion Eq (1) as the incoming field and all the expansion coefficients are then dependent on the index n'. To obtain the scattered field, expanded in outgoing spherical waves, we start with the following integral representation, which holds for the case of zero stress on the crack surface, and which gives the scattered field $\vec{u}_s$ in terms of the crack opening displacement $\Delta\vec{u}$ [1,10]:

$$\vec{u}^{s} = \frac{k_s}{\mu} \iint_S \Delta\vec{u} \cdot (z \cdot \bar{\Sigma}) ds \tag{4}$$

where S is the upper surface of the crack and $z \cdot \bar{\Sigma}$ is the Green surface traction dyadic, see for example [8]. Thus we have the T matrix (compare Eq (3)):

$$T_{nn'} = \frac{ik_s}{\mu} \iint_S \Delta\vec{u}_{(n')} \cdot \vec{t}_z \{\mathrm{Re}\vec{\Psi}_n\} ds \tag{5}$$

where $\Delta\vec{u}_{(n')}$ indicates the crack-opening displacement caused by an incoming spherical partial wave $\mathrm{Re}\vec{\Psi}_{n'}$ and $\vec{t}_z$ is the surface traction operator. The three vector components of $\Delta\vec{u}$ are twice the expansion coefficients of the corresponding displacements on the crack surface. The surface integral in Eq (5) leads to expansion coefficients of the stresses of the same type as for the incoming waves and in an diagonality of the T matrix in the azimuthal index. Thus the T matrix is expressed in terms of matrices describing the relations between the stresses due to an incoming field and the displacement field on the crack surface and vectors consisting of the expansion coefficients of the stresses on the crack surface due to both incoming and outgoing spherical waves. The symmetric and antisymmetric cases are mutually exclusive and the incoming spherical vector wave functions are easily divided into the two groups.

3. SCATTERING BY TWO PENNY-SHAPED CRACKS

Once we know the T matrix for one penny-shaped crack it is possible to take into account the multiple scattering effects as Peterson and Ström [7] have done for acoustic waves and Boström [8] for elastic waves. If we are only interested in calculating the scattered field, we expand the incoming and scattered fields in spherical vector wave functions, relative to two origins, one within each scatterer. For the scattered field we then get:

$$\vec{u}^s = \sum_n f_n^1 \vec{\Psi}_n(\vec{r}_1) + \sum_n f_n^2 \vec{\Psi}_n(\vec{r}_2) \tag{6}$$

where:

$$f^1 = T^1[1-S(d)T^2S(-d)T^1]^{-1}[a^1+S(d)T^2a^2] \tag{7}$$

$$f^2 = T^2[1-S(-d)T^1S(d)T^2]^{-1}[a^2+S(-d)T^1a^1] \tag{8}$$

in compact matrix and vector notation. Here S is a matrix describing the translational properties of the spherical vector wave functions, d is the separation between the two origins and a^1 and a^2 are the expansion coefficients of the incoming wave with respect to the two origins.

4. NUMERICAL RESULTS

In the numerical examples we only present calculations for normal incidence (the rotationally symmetric case), and thus only azimuthal order $m = 0$ is needed. As a check we calculate the scattering cross-section for one penny-shaped crack and compare this with the results of Krenk and Schmidt. We can reproduce their results to within 3 significant figures for the case of incoming waves along the symmetry axis (due to programming errors the results in [3] are incorrect but the correct values are also available [11]).

The back-scattering amplitudes are the complex functions of the frequency multiplying the factor exp(ikr)/kr in the far-field expressions for the scattered field in the backward direction. These amplitudes are generally of nine types, corresponding to the combinations of the three possible modes (two transverse and one longitudinal) of the incoming fields and the three modes of the scattered fields.

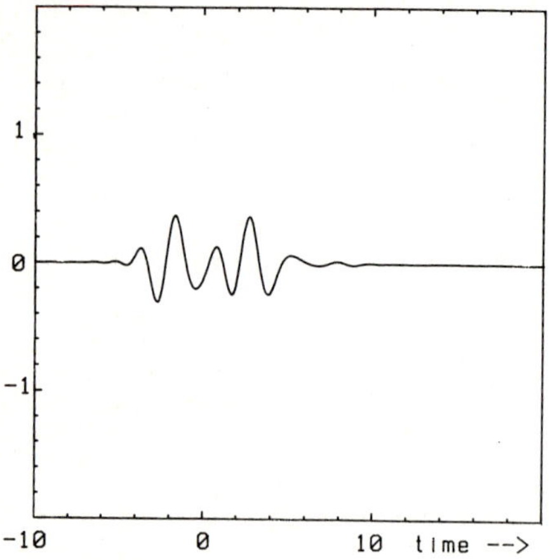

FIGURE 1
Reflected wave for $(k_s a)_{max} = 5$, $d = 4a$.

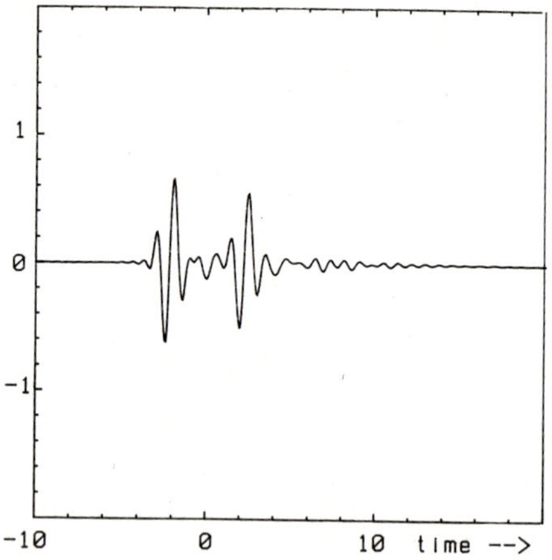

FIGURE 2
Same as Fig. 1 but for $(k_s a)_{max} = 10$.

Here we only consider normally incident plane longitudinal waves and present results for the longitudinal part of the scattered field.

If we consider an incoming pulse with a bounded frequency spectrum, it is possible to calculate the back-scattered pulse as a truncated fourier sum. Here we consider the case of a frequency spectrum for the incoming wave of the form $\sin^2(\pi k_s a/(k_s a)_{max})$ where $0 \le k_s a \le (k_s a)_{max}$. This gives a resonably distinct pulse, making it possible to distinguish between the reflections from the two cracks. This pulse is also normalized by setting its central peak to be of unit height. If we double the bandwidth of the frequency spectrum the pulse becomes half as wide in the time domain. In the figures time is given in terms of the dimensionless time $c_s t/a$ where c_s is the transverse wave speed.

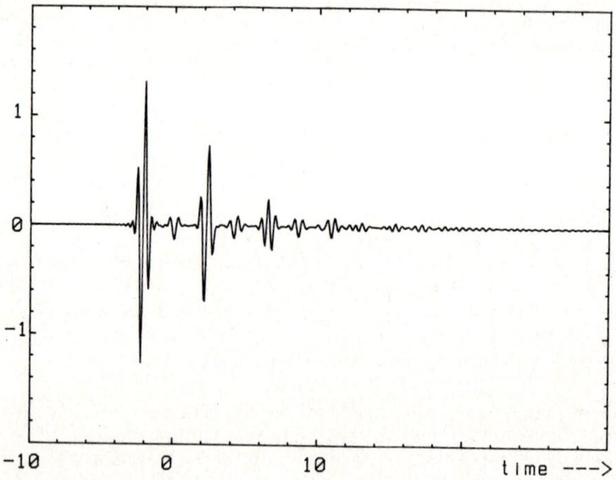

FIGURE 3
Same as Fig. 1 but with $(k_s a)_{max} = 20$.

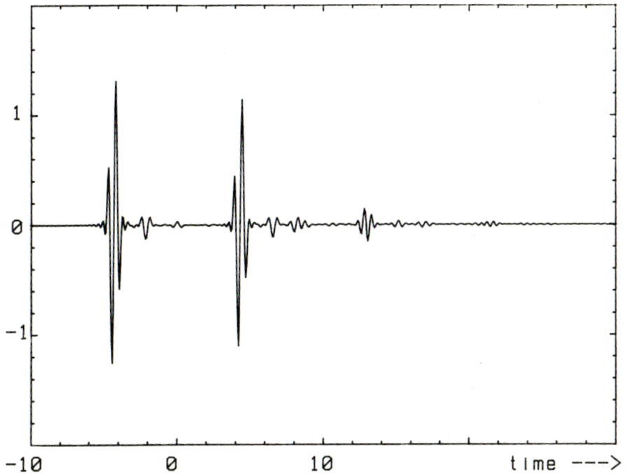

FIGURE 4
Same as Fig. 3 but for d = 8a.

In Fig. 1 we show the back-scattered pulse from two cracks with a separation four times the radius of the cracks for a frequency spectrum $0 < k_s a < 5$. We can see the corresponding results for $0 \leq k_s a \leq 10$ in Fig. 2 and for $0 \leq k_s a \leq 20$ in Fig. 3. In Fig. 1 it is just about possible to determine that there are two distinct scatterers, while in Fig. 2 and Fig. 3 it is possible to determine the separation quite accurately. We can also see that the second crack becomes more and more obscured by the first crack at higher frequences, as can be expected.

In the case of a separation between the cracks of eight times the crack radius, the shadowing effects are not so large, as can be seen in Fig. 4 ($0 \leq k_s a < 20$). Furthermore the wave has nearly forgotten the interaction with the first crack when it impinges on the second, thus the reflection from the second crack is nearly identical to that from the first crack. In the reflection from the two cracks it is possible to discern a peak which probably results from a Rayleigh wave which is excited at the crack edge,propagates over the crack, and then is re-emitted from the opposed crack edge. This peak arrives at about -2 time units, and it is also possible to see the same Rayleigh wave crossing the crack once more and then being re-emitted to arrive at 0 time units. In the reflection from the second crack we can see the first of these peaks (due to a Rayleigh wave on the second crack) at about 6.5 time units. At 13 time units we can see the result of a double internal reflection between the two cracks.

In our calculations we use a value of the Poisson number $\nu = 0.3$ and then the velocity of the Rayleigh wave is approximately half the velocity of the longitudinal wave. Measuring the time elapse between the different peaks in Fig. 4 it is thus possible to estimate the radius of the cracks relative to the separation and this agrees favourably with the true values.

Comparing the information that we get from a frequency spectrum with $0 < k_s a \leq 10$ (Fig. 2) with that from one with $0 < k_s a < 20$ (Fig. 3) we see that if we are only interested in the separation between the cracks the narrower bandwidth is sufficient. If we want to compare for example the separation with the diameter of the cracks we have to use the larger bandwidth since then, in Fig. 3 we can see a peak resulting from a Rayleigh wave propagating over the crack while this is impossible in Fig. 2.

5. CONCLUDING REMARKS

With the help of integral equations for displacements and stresses in a homogeneous, isotropic elastic medium and an integral representation of the scattered field we have calculated the T matrix for one penny-shaped crack. We have then used this T matrix to calculate the scattered field from two penny-shaped cracks. We have numerically calculated the back-scattering amplitude and we illustrate this with the reflection from two circular cracks of a plane longitudinal wave of normal incidence in the form of a pulse.

It is also possible to study a penny-shaped crack in a stratified elastic medium or in an elastic plate [12] where the T matrix calculated above can be used as a building-block. The method we use above to relate incoming and scattered displacement fields for one penny-shaped crack can also be used in calculations for a crack in the interface between two elastic media, but in this case it is impossible to divide the problem into a symmetric and an antisymmetric part and consequently this configuration presents a far more difficult problem to solve.

ACKNOWLEDGEMENTS

The author is indebted to Professor Anders Boström and Dr. Anthony Burden for many helpful discussions and for reading the manuscript.

REFERENCES

[1] Achenbach, J.D., Gautesen, A.K. and McMaken, H., Ray Methods for Waves in Elastic Solids (Pitman, Belfast, 1982)
[2] Martin, P.A. and Wickham, G.R., Proc. R. Soc. Lond. A 390 (1983) 91.
[3] Krenk, S. and Schmidt, H., Phil. Trans. R. Soc. Lond. A 308 (1982) 167.
[4] Shah, A.H., Wong, K.C. and Datta, S.K., Wave Motion 4 (1985) 319.
[5] Scandrett, C.L. and Achenbach, J.D., Wave Motion 2 (1987) 171.
[6] Boström, A. and Olsson, P., Wave Motion 1 (1987) 61.
[7] Peterson, B. and Ström, S., J. Acoust. Soc. Amer. 56 (1974) 771.
[8] Boström, A., J. Acoust. Soc. Amer. 67 (1980) 399.
[9] Krenk, S., J. Appl. Mech. 46 (1979) 821.
[10] Pao, Y.-H. and Varatharajulu, V., J. Acoust. Soc. Amer. 59 (1976) 1361.
[11] Krenk, S., private communication.
[12] Boström, A. and Karlsson, A., J. Appl. Mech. 52 (1985) 937.

ELASTIC WAVE PROPAGATION
M.F. McCarthy, M.A. Hayes, (Editors)
Elsevier Science Publishers B.V. (North-Holland), 1989

ON THE BOUNDARY INTEGRAL EQUATION FOR FLAT CRACKS

HA DUONG T.
CMAP, Ecole Polytechnique
91128 Palaiseau Cedex
France

The scattering problems of elastic waves from a penny-shaped crack have been treated by many researchers. A review of related works before 1983 can be found in the paper of Martin & Wickham [3]. However, the first kind boundary integral equation (BIE) that is satisfied by the crack opening displacement (COD) is in most cases transformed by more or less sophisticated techniques before any numerical calculation.

In this paper, we develop the advantages of working directly with the original BIE, using a variational technique to overcome the difficulties due to the first kind character of the equation, and to the hyper-singularity of the kernel. Our treatment is not restricted to the penny-shaped crack, and is valid for flat cracks of arbitrary shapes. Here we present the method for an anti-plane (thus scalar) harmonic problem, and refer to a forthcoming article for more details.

1. THE PROBLEM AND BIE.

If the incident wave is an anti-plane shear wave, so is the scattering wave, and the whole problem is a scalar wave one. And, with a convenient scaling of the measure units, we have the following problem (P) :

(1) $$(\omega^2 + \Delta)u = 0 \qquad \text{in } \Omega = \mathbb{R}^3 \setminus \overline{\Gamma},$$

(2) $$\frac{\partial u}{\partial n} = f \qquad \text{on } \Gamma,$$

(3) $$\frac{\partial u}{\partial n} - i\omega u = 0\left(\frac{1}{r}\right) \qquad (r = |x| \longrightarrow +\infty),$$

where Γ is a bounded sufficiently smooth domain of the plane $\Gamma_{12} = \{x_3 = 0\}$, representing the crack and Ω the surrounding elastic medium. The normal unit vector to Γ is chosen to be the same unit vector as that of the x_3-axis, and the upper face of Γ is denoted by Γ_+.

It is well-known that the solution of (P) is completely determined by its COD $\varphi = u_- - u_+$, where u_+ is the trace of u on the face Γ_+ :

(4) $$u(x) = -\int_\Gamma \frac{\partial}{\partial n_y}\left(\frac{e^{i\omega|x-y|}}{4\pi|x-y|}\right)\varphi(y)\,dy \qquad (x \in \Omega).$$

Taking the normal derivative of (4), and comparing to (2), we obtain the classical BIE for the COD of u :

(5) $$D\varphi(x) = f(x) , \qquad (x \in \Gamma) ,$$

where, unfortunately, the integral operator

$$\varphi \longrightarrow D\varphi = \left.\frac{\partial u}{\partial n}\right|_{\Gamma} , \quad \text{where } u \text{ is given by (4)} ,$$

can only be written with an hyper-singular kernel.

Our first proposition is then to replace this integral expression of D by an equivalent definition :

$$\begin{cases} D\varphi = \dfrac{\partial u}{\partial n} \quad \text{where } u \text{ is a solution of} \\ \bullet\ (1), (3), \text{ and the jump condition :} \\ \bullet\ (6) \qquad u_- - u_+ = \varphi \quad ; \quad \dfrac{\partial u_-}{\partial n} = \dfrac{\partial u_+}{\partial n} . \end{cases}$$

From this definition, we obtain easily the following

Theorem 1

The operator D is a continuous operator from the Sobolev space $H^{\frac{1}{2}}_{00}(\Gamma)$ into its dual $(H^{\frac{1}{2}}_{00}(\Gamma))'$, and is given by :

(7) $$D\varphi = R_{\Gamma} \circ T \circ P_{\Gamma}\, \varphi ,$$

where T is the pseudo-differential operator on $\mathbb{R}^2$ defined in terms of the Fourier variables :

(8) $$\widehat{T\psi}(\xi) = -\frac{i}{2} Z(\xi, \omega)\, \hat{\psi}(\xi) , \qquad \xi \in \mathbb{R}^2 .$$

The symbol Z is the following square root of $\omega^2 - |\xi|^2$:

(9) $$Z(\xi, \omega) = \begin{cases} \sqrt{\omega^2 - |\xi|^2} & \text{on } \{|\xi| < \omega\} , \\ i\sqrt{|\xi|^2 - \omega^2} & \text{on } \{|\xi| > \omega\} . \end{cases}$$

The operators R_{Γ} and P_{Γ} in formula (7) are respectively the restriction from Γ_{12} to Γ, and the extension from Γ to Γ_{12} by zero out of $\overline{\Gamma}$.

2. THE VARIATIONAL FORMULATION OF BIE (5).

This expression of D, joined to the following classical lemma 1, explains why it is natural that eq. (5) should be dealt with via a variational formulation.

Lemma 1

For $v \in (H_{00}^{\frac{1}{2}}(\Gamma))'$ and $w \in H_{00}^{\frac{1}{2}}(\Gamma)$, the bracket $< v , w >$ is equal to $< \tilde{v} , P_{\Gamma} w >$ where $\tilde{v}$ is any extension of v in $H^{-\frac{1}{2}}(\Gamma)$.

Thus, in the variational formulation of eq. (5):

To find $\varphi \in H_{00}^{\frac{1}{2}}(\Gamma)$ such that

$$b(\varphi, \psi) := < D\varphi , \psi > = < f , \psi > \qquad \forall \psi \in H_{00}^{\frac{1}{2}}(\Gamma), \tag{10}$$

the bilinear form b has the following expression in Fourier variables:

$$b(\varphi, \psi) = -\frac{i}{2} \int_{\mathbb{R}^2} Z(\xi, \omega)\, \hat{\varphi}_0(\xi)\, \overline{\hat{\psi}_0}(\xi)\, d\xi , \tag{11}$$

where $\varphi_0 = P_{\Gamma} \varphi$ is the extension of φ by zero out of $\overline{\Gamma}$.

We can then prove, by simple calculations, the main result that variational problem (10) is well posed:

Theorem 2

The bilinear form b is coercive on the space $H_{00}^{\frac{1}{2}}(\Gamma)$. More precisely, we have the following estimate:

$$| b(\varphi, \varphi) | \geqslant \frac{C}{(\omega^2 + 1)^{\frac{1}{2}}} \| \varphi \|^2_{\frac{1}{2}, \omega} , \tag{12}$$

where C is a constant independent of ω, and

$$\| \varphi \|^2_{\frac{1}{2}, \omega} := \int_{\mathbb{R}^2} (1 + \omega^2 + |\xi|^2)^{\frac{1}{2}} \, |\hat{\varphi}_0(\xi)|^2 \, d\xi . \tag{13}$$

The norm and the ω dependence in this coercivity estimate are set here for a future treatment of the time dependent problem. Finally, we have found in Bamberger [1] the simple and efficient idea of writing the decomposition

$$\frac{-\,iZ}{2} = \frac{\omega^2 - |\xi|^2}{2\,iZ} \quad ,$$

and using the Fourier correspondance

$$\frac{-1}{2\,iZ} \longleftrightarrow \frac{e^{i\omega|x|}}{4\pi|x|} \qquad (x \in \mathbb{R}^2) \quad ,$$

to obtain the following expression of b, where the integral kernel is now only weakly singular :

$$(14) \qquad b(\varphi,\psi) = \iint_{\Gamma\times\Gamma} \frac{e^{i\omega|x-y|}}{4\pi|x-y|} \{ \operatorname{curl} \varphi(x) \,.\, \operatorname{curl} \psi(y) - \omega^2 \varphi(x)\psi(y) \} \, dy\, dx \,.$$

This expression is known in the general case where Γ is a closed surface, see Hamdi [2] .

We recall briefly that the functions of the space $H_{00}^{\frac{1}{2}}(\Gamma)$ verify the "edge condition" that $\rho^{-\frac{1}{2}} \varphi \in L^2(\Gamma)$, where ρ is a smooth function equivalent to the distance to the boundary of Γ. Thus, if φ is smooth, it must vanish on the boundary $\partial\Gamma$. The discretisation process of eq. (5) is now clear : the best way to do it is by Galerkin methods, through the variational problem (10). The effective finite elements to be used must be at least of degree one, and null at the boundary $\partial\Gamma$.

REFERENCES.

[1] BAMBERGER A.
Approximation de la diffraction d'ondes élastiques. Une nouvelle approche (I) .
Rapport interne n° 91 (1983), CMAP, Ecole Polytechnique, France.

[2] HAMDI M.A.
Une formulation variationnelle par équations intégrales pour la résolution de l'équation de Helmholtz avec des conditions aux limites mixtes.
CRAS, série II, 292 (1981) p. 17-20 .

[3] MARTIN P.A. and WICKHAM G.R.
Diffraction of elastic waves by a penny-shaped crack : analytical and numerical results.
Proc. R. Soc. Lond. A 390 (1983) p. 91-129 .

ELASTIC WAVE PROPAGATION
M.F. McCarthy, M.A. Hayes, (Editors)

BORN SERIES APPROACH APPLIED TO THREE DIMENSIONAL ELASTODYNAMIC INCLUSION ANALYSIS BY BIE METHODS

M. KITAHARA* and K. NAKAGAWA**

* Faculty of Marine Science and Technology, Tokai University, Shimizu, Shizuoka 424 Japan
** Total System Institute, Shinjuku, Tokyo 162, Japan

The idea of Born series is applied to solve a system of boundary integral equations (BIE) for three dimensional elastodynamic multi-scattering problems by several inclusions in an elastic solid. Although the Born series strategy is explained for two inclusion problems, the method itself can be applicable for any numbers of inclusions.

1. INTRODUCTION

It is well known that the elastodynamic scattering problem is reduced to the singular integral equations over the surface of scatterers. To solve this singular integral equations, the boundary integral equation (BIE) method is adopted here. For the three dimensional elastodynamic inclusion problem, Kitahara, Nakagawa and Achenbach [1] have shown a detailed treatment of BIE for the single inclusion in an elastic solid. For a counterpart of infinite numbers of cavities which have periodicity, Achenbach and Kitahara[2] have treated this problem as the integral equation over a surface of scatterer in a reference cell. In this paper, we imagine several elastic inclusions in the three dimensional elastic solid, and consider the multi-scattering problem of these inclusions. When we apply the BIE method for this problem, the point is how to reduce the matrix dimension when we take the inverse of the matrix, because of the limitation of computer capacity. In a series of paper, Schuster and Smith [3]~[6] have shown an elegant method, hybrid BIE + Born series modeling scheme, to circumvent this matrix dimension problem for the two dimensional acoustic problem. We apply now Schuster and Smith's strategy to the three dimensinal elastodynamic multi-scattering problem and show the effectiveness of Born series approach for the present problem. The key-feature of the Born series strategy is to reduce the whole system-matrix to the algorithm of the sub-matrix system.

2. TREATMENT OF THREE DIMENSIONAL ELASTODYNAMIC INCLUSION PROBLEMS BY BIE

We show the treatment for the two inclusion problem, for simplicity, but the idea is true for any number of inclusions. The scattering problem is depicted in Fig.1. Surfaces of inclusions are denoted by S_1 and S_2, and $\mathbf{u}^I$ is the plane incident wave traveling in the surrounding elastic matrix. The time harmonic field is assumed now and the time factor $\exp(-i\omega t)$ will be omitted. Then the displacement field $\mathbf{u}$ in the exterior elastic matrix satisfies

$$\mu\Delta\mathbf{u} + (\lambda + \mu)\nabla\nabla \cdot \mathbf{u} + \rho\omega^2\mathbf{u} = \mathbf{0} \qquad (1)$$

where λ and μ are Lamé's constants in the matrix; ρ is the mass density and ω is the angular frequency. In the

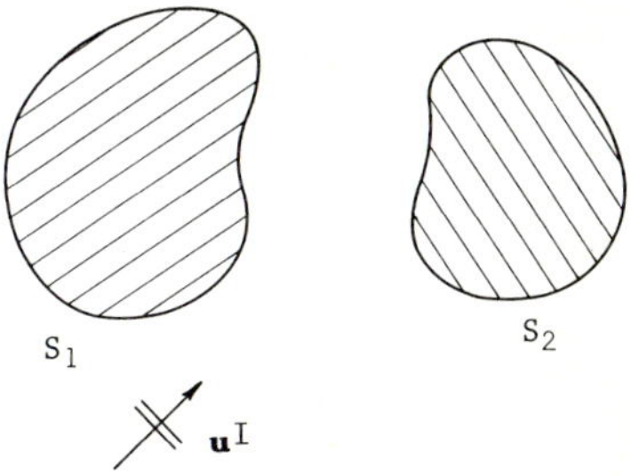

Fig.1 Scattering problems by two inclusions.

first inclusion enclosed by S_1, the displacement field $\bar{\mathbf{u}}$ satisfies

$$\bar{\mu}\Delta\bar{\mathbf{u}} + (\bar{\lambda} + \bar{\mu})\nabla\nabla\cdot\bar{\mathbf{u}} + \bar{\rho}\omega^2\bar{\mathbf{u}} = \mathbf{0} \tag{2}$$

where the physical quantities with single bar denote the quantities of the first inclusion. The displacement field $\bar{\bar{\mathbf{u}}}$ in the second inclusion enclosed by S_2 satisfies the same equation as eq. (2) which has the physical quantities with double bars.

The boundary integral equation for the surrounding matrix takes the following form (for example, see Ref. [1])

$$\begin{aligned} \mathbf{C}(\mathbf{x})\mathbf{u}(\mathbf{x}) = \int_{S_1+S_2} \mathbf{U}(\mathbf{x},\mathbf{y})\mathbf{t}(\mathbf{y})dS_y - \int_{S_1+S_2} \mathbf{T}(\mathbf{x},\mathbf{y})\mathbf{u}(\mathbf{y})dS_y \\ +\mathbf{u}^I(\mathbf{x}) , \qquad \mathbf{x} \in S_1 + S_2 \end{aligned} \tag{3}$$

where $\mathbf{U}(\mathbf{x},\mathbf{y})$ is the fundamental solusion, $\mathbf{C}(\mathbf{x})$ is the coefficient of the free term, and $\mathbf{t}$ is the traction vector. For the present 3D problem, the fundamental solusion has the following form

$$U_{ij}(\mathbf{x},\mathbf{y}) = \frac{1}{4\pi\mu}\left[\frac{e^{ik_T r}}{r}\delta_{ij} + \frac{1}{k_T^2}\frac{\partial}{\partial x_i}\frac{\partial}{\partial x_j}\left\{\frac{e^{ik_T r}}{r} - \frac{e^{ik_L r}}{r}\right\}\right] \tag{4}$$

where $r = |\mathbf{x}-\mathbf{y}|$, k_L and k_T are the longitudinal and transverse wave numbers, respectively. The traction kernel $\mathbf{T}(\mathbf{x},\mathbf{y})$ in eq. (3) can be expressed as

$$T_{ij}(\mathbf{x},\mathbf{y}) = -\left[\lambda U_{im,m}\delta_{jk} + \mu U_{ij,k} + \mu U_{ik,j}\right] n_k(\mathbf{y}) \tag{5}$$

where the differentiation is carried out at point $\mathbf{x}$ and $\mathbf{n}(\mathbf{y})$ is the unit outward normal vector at point $\mathbf{y}$. On the other hand, the BIE for the first inclusion can be written as

$$\bar{\mathbf{C}}(\mathbf{x})\bar{\mathbf{u}}(\mathbf{x}) = \int_{S_1} \bar{\mathbf{U}}(\mathbf{x},\mathbf{y})\bar{\mathbf{t}}(\mathbf{y})dS_y - \int_{S_1} \bar{\mathbf{T}}(\mathbf{x},\mathbf{y})\bar{\mathbf{u}}(\mathbf{y})dS_y, \quad \mathbf{x} \in S_1 . \tag{6}$$

The BIE for the second inclusion also has the following form

$$\bar{\bar{\mathbf{C}}}(\mathbf{x})\bar{\bar{\mathbf{u}}}(\mathbf{x}) = \int_{S_2} \bar{\bar{\mathbf{U}}}(\mathbf{x},\mathbf{y})\bar{\bar{\mathbf{t}}}(\mathbf{y})dS_y - \int_{S_2} \bar{\bar{\mathbf{T}}}(\mathbf{x},\mathbf{y})\bar{\bar{\mathbf{u}}}(\mathbf{y})dS_y, \quad \mathbf{x} \in S_2 . \tag{7}$$

After the discretization, we write eq. (3) for the surrounding elastic matrix as

$$\begin{bmatrix} \mathbf{H}_{11} & \mathbf{H}_{12} \\ \mathbf{H}_{21} & \mathbf{H}_{22} \end{bmatrix} \begin{Bmatrix} \mathbf{u}_1 \\ \mathbf{u}_2 \end{Bmatrix} = \begin{bmatrix} \mathbf{G}_{11} & \mathbf{G}_{12} \\ \mathbf{G}_{21} & \mathbf{G}_{22} \end{bmatrix} \begin{Bmatrix} \mathbf{t}_1 \\ \mathbf{t}_2 \end{Bmatrix} + \begin{Bmatrix} \mathbf{u}_1^I \\ \mathbf{u}_2^I \end{Bmatrix} . \tag{8}$$

In this expression, $\mathbf{u}_1(\mathbf{t}_1)$ is the displacement (traction) vector on the surface of S_1 and $\mathbf{u}_2(\mathbf{t}_2)$ is the displacement (traction) vector on the surface of S_2. The sub-matrices $\mathbf{H}_{12}$ and $\mathbf{G}_{12}$, for example, denote the double layer and single layer matrices which show the interaction effect from the second inclusion surface S_2 to the first inclusion surface S_1. The equation (6) for the first inclusion reduces to

$$\bar{\mathbf{H}}_{11}\bar{\mathbf{u}}_1 = \bar{\mathbf{G}}_{11}\bar{\mathbf{t}}_1 . \tag{9}$$

This expression can be rewritten as

$$\bar{\mathbf{K}}_{11}\bar{\mathbf{u}}_1 = \bar{\mathbf{t}}_1 \tag{10}$$

where

$$\bar{\mathbf{K}}_{11} = \bar{\mathbf{G}}_{11}^{-1}\bar{\mathbf{H}}_{11} \tag{11}$$

In the same way, equation (7) for the second inclusion finally reduces to

$$\bar{\bar{\mathbf{K}}}_{22}\bar{\bar{\mathbf{u}}}_2 = \bar{\bar{\mathbf{t}}}_2 \tag{12}$$

where

$$\bar{\bar{\mathbf{K}}}_{22} = \bar{\bar{\mathbf{G}}}_{22}^{-1}\bar{\bar{\mathbf{H}}}_{22} . \tag{13}$$

For the perfect bonding of inclusions, the continuity conditions on S_1 and S_2 can be written as

$$\mathbf{u}_1 = \bar{\mathbf{u}}_1\,,\ \mathbf{t}_1 + \bar{\mathbf{t}}_1 = \mathbf{0} \quad , \mathbf{x} \in S_1 \tag{14a}$$

$$\mathbf{u}_2 = \bar{\bar{\mathbf{u}}}_2\,,\ \mathbf{t}_2 + \bar{\bar{\mathbf{t}}}_2 = \mathbf{0} \quad , \mathbf{x} \in S_2 \tag{14b}$$

From the above continuity conditions (14a,b), eqs. (8), (10), and (12) can be combined in a system of equations

$$\begin{bmatrix} \mathbf{H}_{11} & \mathbf{H}_{12} \\ \mathbf{H}_{21} & \mathbf{H}_{22} \end{bmatrix} \begin{Bmatrix} \mathbf{u}_1 \\ \mathbf{u}_2 \end{Bmatrix} = \begin{bmatrix} \mathbf{G}_{11} & \mathbf{G}_{12} \\ \mathbf{G}_{21} & \mathbf{G}_{22} \end{bmatrix} \begin{bmatrix} -\bar{\mathbf{K}}_{11} & \mathbf{0} \\ \mathbf{0} & -\bar{\bar{\mathbf{K}}}_{22} \end{bmatrix} \begin{Bmatrix} \mathbf{u}_1 \\ \mathbf{u}_2 \end{Bmatrix} + \begin{Bmatrix} \mathbf{u}_1^I \\ \mathbf{u}_2^I \end{Bmatrix} . \tag{15}$$

This equation (15) is reduced to the final form

$$\begin{bmatrix} \mathbf{M}_{11} & \mathbf{M}_{12} \\ \mathbf{M}_{21} & \mathbf{M}_{22} \end{bmatrix} \begin{Bmatrix} \mathbf{u}_1 \\ \mathbf{u}_2 \end{Bmatrix} = \begin{Bmatrix} \mathbf{u}_1^I \\ \mathbf{u}_2^I \end{Bmatrix} \tag{16}$$

where

$$\begin{bmatrix} \mathbf{M} \end{bmatrix} = \begin{bmatrix} \mathbf{H} \end{bmatrix} + \begin{bmatrix} \mathbf{G} \end{bmatrix} \begin{bmatrix} \bar{\mathbf{K}}_{11} & \mathbf{0} \\ \mathbf{0} & \bar{\bar{\mathbf{K}}}_{22} \end{bmatrix} . \tag{17}$$

3. BORN SERIES APPROACH

Now the question is how to solve, effectively, the system of equations (16). First, we rearrange the system of eq. (16) in the following form

$$\begin{bmatrix} \mathbf{M}_{11} & \mathbf{0} \\ \mathbf{0} & \mathbf{M}_{22} \end{bmatrix} \begin{Bmatrix} \mathbf{u}_1 \\ \mathbf{u}_2 \end{Bmatrix} = \begin{Bmatrix} \mathbf{u}_1^I \\ \mathbf{u}_2^I \end{Bmatrix} + \begin{bmatrix} \mathbf{0} & -\mathbf{M}_{12} \\ -\mathbf{M}_{21} & \mathbf{0} \end{bmatrix} \begin{Bmatrix} \mathbf{u}_1 \\ \mathbf{u}_2 \end{Bmatrix} . \tag{18}$$

For the compact notation, we write the above equation (18) as

$$\mathbf{A}\mathbf{u} = \mathbf{u}^I + \mathbf{B}\mathbf{u} \tag{19}$$

where

$$\mathbf{A} = \begin{bmatrix} \mathbf{M}_{11} & \mathbf{0} \\ \mathbf{0} & \mathbf{M}_{22} \end{bmatrix} \tag{20a}$$

$$\mathbf{B} = \begin{bmatrix} \mathbf{0} & -\mathbf{M}_{12} \\ -\mathbf{M}_{21} & \mathbf{0} \end{bmatrix} \tag{20b}$$

and

$$\mathbf{u} = \{\mathbf{u}_1\,,\ \mathbf{u}_2\}^{\mathrm{T}} \quad , \quad \mathbf{u}^I = \{\mathbf{u}_1^I\,,\ \mathbf{u}_2^I\}^{\mathrm{T}} \ . \tag{21a,b}$$

Taking the inverse of $\mathbf{A}$ in eq. (20a) and operating $\mathbf{A}^{-1}$ on both hand sides of eq. (19), we have

$$\mathbf{u} = \mathbf{A}^{-1}\mathbf{u}^I + \mathbf{C}\mathbf{u} \tag{22}$$

where

$$\mathbf{A}^{-1} = \begin{bmatrix} \mathbf{M}_{11}^{-1} & \mathbf{0} \\ \mathbf{0} & \mathbf{M}_{22}^{-1} \end{bmatrix} \tag{23}$$

and

$$\mathbf{C} = \mathbf{A}^{-1}\mathbf{B} = \begin{bmatrix} \mathbf{M}_{11}^{-1} & \mathbf{0} \\ \mathbf{0} & \mathbf{M}_{22}^{-1} \end{bmatrix} \begin{bmatrix} \mathbf{0} & -\mathbf{M}_{12} \\ -\mathbf{M}_{21} & \mathbf{0} \end{bmatrix} . \tag{24}$$

Now we consider to solve eq. (22) iteratively. The first step is just to neglect the term $\mathbf{C}\mathbf{u}$ in the right hand side of eq. (22) and to put the zeroth order approximation as

$$\mathbf{u}_0 = \mathbf{A}^{-1}\mathbf{u}^I . \tag{25}$$

This expression (25) is the well-known Born approximation [7] which gives the good approximation for $\mathbf{u}$ in the low frequency range. Born series strategy works in the following way

$$\begin{aligned}
\mathbf{u}_1 &= \mathbf{u}_0 + \mathbf{C}\mathbf{u}_0 \\
\mathbf{u}_2 &= \mathbf{u}_0 + \mathbf{C}\mathbf{u}_1 = \mathbf{u}_0 + \mathbf{C}\mathbf{u}_0 + \mathbf{C}^2\mathbf{u}_0 \\
\mathbf{u}_3 &= \mathbf{u}_0 + \mathbf{C}\mathbf{u}_2 = \mathbf{u}_0 + \mathbf{C}\mathbf{u}_0 + \mathbf{C}^2\mathbf{u}_0 + \mathbf{C}^3\mathbf{u}_0 \\
&\vdots
\end{aligned} \tag{26}$$

For the nth order approximation $\mathbf{u}_n$, we can write

$$\mathbf{u}_n = \mathbf{u}_0 + \mathbf{C}\mathbf{u}_{n-1} = \sum_{N=0}^{n} \mathbf{C}^N \mathbf{u}_0 . \tag{27}$$

The question is the convergence of the series in eq. (27). When we see the detailed expression of C in eq. (24), we can expect

$$\|\mathbf{C}\| = \frac{\|\mathbf{B}\|}{\|\mathbf{A}\|} < 1 \tag{28}$$

from the physical meaning of the matrices $\mathbf{A}$ and $\mathbf{B}$ in eqs. (20a) and (20b). Namely, the matrix $\mathbf{A}$ expresses the self-influence matrix of each inclusion itself and the matrix $\mathbf{B}$ expresses the effect of interaction of two inclusions.

4. TWO SPHERICAL INCLUSION MODEL

In the numerical analysis, two inclusions will be assumed to be spherical inclusions with same radius a , as shown in Fig. 2. The distance b of two inclusions will be taken as $b = a/5$. Material properties of both inclusions are assumed here to be the same ($\bar{E} = \bar{\bar{E}}$, $\bar{\mu} = \bar{\bar{\mu}}$, $\bar{\rho} = \bar{\bar{\rho}}$) and we consider the following three types of inclusions:

Case A : two cavities (special case of inclusions)
Case B : $\bar{E}/E = 1/2$, $\mu = \bar{\mu} = 1/4$, $\bar{\rho}/\rho = 1$ (sofft inclusions)
Case C : $\bar{E}/E = 2$, $\mu = \bar{\mu} = 1/4$, $\bar{\rho}/\rho = 1$ (hard inclusions)

The Cartesian and spherical coordinate systems are designated in Fig. 3. The coordinate θ is measured from the north pole and the ϕ is measured from the x_1 axis in the $x_1 - x_2$ plane. The incident wave is assumed to be the following plane longitudinal wave

$$\mathbf{u}^I(\mathbf{x}) = \mathbf{e}_3 \exp(ik_L x_3) . \tag{29}$$

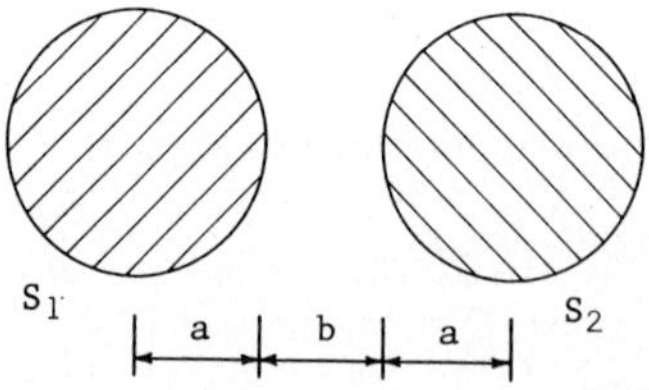

Fig.2 Two spherical inclusion model.

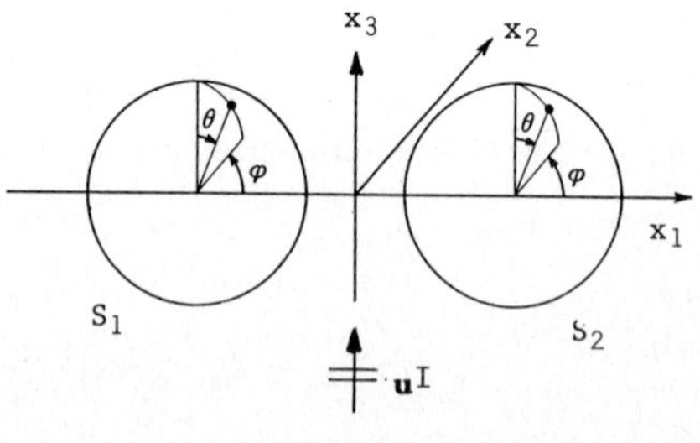

Fig.3 Coordinate systems.

5. PRELIMINARY BIE ANALYSIS FOR ONE INCLUSION

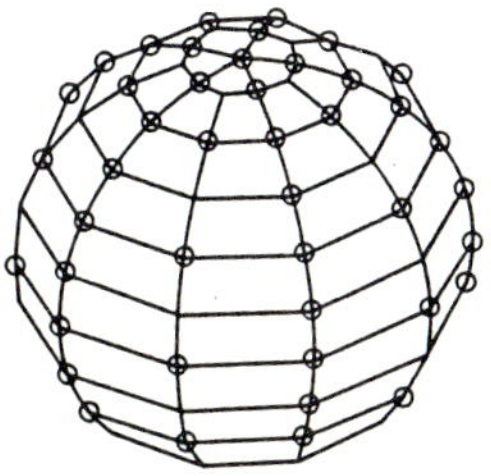

Fig.4 Boundary elements.

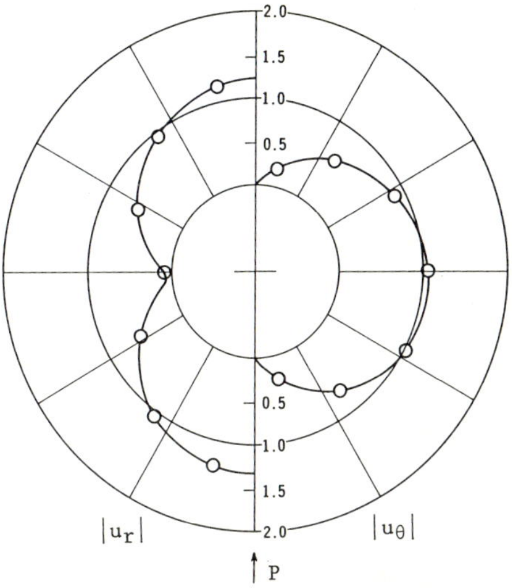

Fig.5 Boundary displacements $|u_r|$ and $|u_\theta|$ for the single inclusion (—:Pao and Mow[8], O:BIE, Soft inclusion of $\bar{E}/E$ =0.5, ak_L=1.0).

To clarify the accuracy of BIE methods for the present problem, first we consider the one spherical inclusion problem. For this problem, we can find the exact solution by Pao and Mow[8]. The boundary elements for our numerical BIE analysis are shown in Fig. 4. The number of boundary elements is M = 56 for this single inclusion and the constant approximation of unknown displacements over elements is adopted here. Fig. 5 shows the absolute values of the radial displacement $|u_r|$ and the circumferential component $|u_\theta|$ on the inclusion surface. The solid lines are Pao and Mow's solution and the circles are BIE solutions. Material properties of the inclusion have been chosen to be Case B (soft inclusion) in the last Section 4. The nondimensional longitudinal wave number ak_L of the incident wave is $ak_L = 1.0$ for this analysis.

6. BORN SERIES RESULTS FOR TWO INCLUSION MODEL

In order to confirm, numerically, the convergence of Born series in eq. (27), first we define the following quantities :

$$A_1 = \sum_{\alpha,i} |\mathbf{u}^S_{\alpha i}|/K_f \tag{30a}$$

$$A_2 = \sum_{\alpha,i} |\mathbf{u}^N_{\alpha i}|/K_f \tag{30b}$$

$$E = A_2/A_1 \ . \tag{30c}$$

In these expressions, $u^S_{\alpha i}$ is the ith $(i = 1, 2, 3)$ component of the scattered displacement field $(\mathbf{u}^S = \mathbf{u} - \mathbf{u}^I)$ on the αth $(\alpha = 1, 2, \cdots, M)$ element ; $u^N_{\alpha i}$ is the Nth term $(\mathbf{u}^N = \mathrm{C}^N \mathbf{u}_0)$ of the Born series in eq. (27), and K_f is the degree of freedom of the system ($K_f = M$, in this case). From the above definition of A_1 and A_2, the ratio $E = A_2/A_1$ means the contribution from the N th term of the series for the final scattered displacement amount. Fig. 6 shows this ratio of A_2/A_1 for each N in eq. (27). The cavity, soft and hard inclusions in this figure correspond to the case A, B, and C, in the inclusion model of Section 4.

To see the convergence of displacement field, the absolute values of the component in the θ-direction, $|u_\theta|$, are shown in Fig. 7 for the valus of the left surface of cavities in the section $\phi = 0°$ at $ak_L = 1.0$. The solid, dashed, and dashed-dotted lines are the converged value, 0th order, and 1st order approximations in the series of eq. (27), respectively. It is to be remarked that the 0th order approximation has no interaction effect, and so the 0th order approximation coincides with the results for the single cavity.

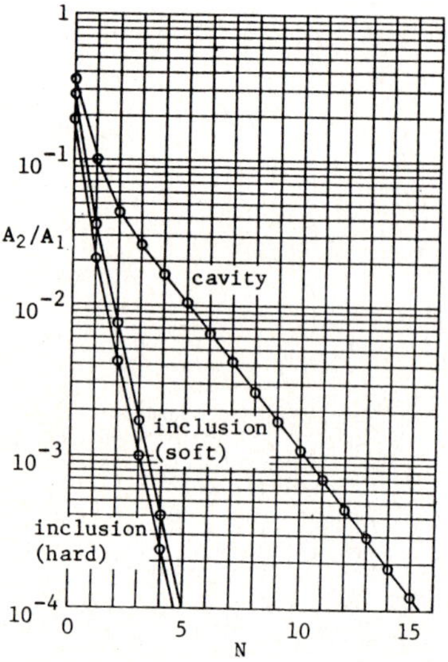

Fig.6 Convergence rate of Born series for ak_L=1.0 (N is the index of the series in eq.(27)).

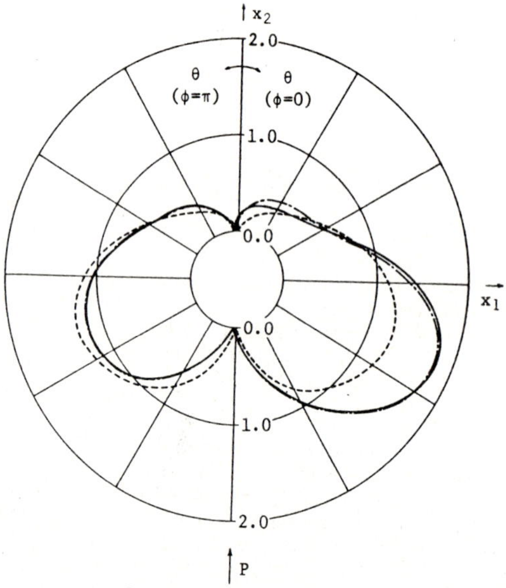

Fig.7 Displacement component $|u_\theta|$ on the left surface of cavities at ak_L=1.0 (Solid line: converged value, dashed line: 0th order approx., dashed-dotted line: 1st order approx.).

REFERENCES

[1] Kitahara, M., Nakagawa, K. and Achenbach, J. D.; Boundary integral equation method for elastodynamic scattering by a compact inhomogeneity, Computational Mechanics, in print.

[2] Achenbach, J. D. and Kitahara, M.; Reflection and transmission of an obliquely incident wave by an array of spherical cavities, J. Acoust. Soc. Am., Vol.80, pp. 1209-1214, 1986.

[3] Schuster, G. T. and Smith, L. C.; A comparison among four direct boundary integral methods, J. Acoust. Soc. Am., Vol.77, pp.850-864, 1985.

[4] Schuster, G. T.; A hybrid BIE+Born series modeling scheme : Generalized Born series, J. Acoust. Soc. Am., Vol.77, pp.865-879, 1985.

[5] Schuster, G. T.; Solution of the acoustic transmission problem by a perturbed Born series, J. Acoust. Soc. Am., Vol.77, pp.880-886, 1985.

[6] Schuster, G. T. and Smith, L. C.; Modeling scatterers embedded in plane-layered media by a hybrid Haskel-Thomson and boundary integral equation method, J. Acoust. Soc. Am., Vol.78, pp.1387-1394, 1985.

[7] Gubernatis, J. E., Domany, E., Krumhansl, J. A. and Huberman, M.; The Born approximation in the theory of the scattering of elastic waves by flaws, J. Acoust. Soc. Am., Vol.48, pp.2812-2819, 1977.

[8] Pao, Y. H. and Mow, C. C.; Diffraction of Elastic Waves and Dynamic Stress Concentrations, Crane Russak & Co. Inc., New York, 1973.

ELASTIC WAVE PROPAGATION
M.F. McCarthy, M.A. Hayes, (Editors)
Elsevier Science Publishers B.V. (North-Holland), 1989

ELASTIC AND THERMOELASTIC SCATTERING PROBLEMS AND NON-HOMOGENEOUS MEDIA

Marco CODEGONE

Dipartimento di Matematica, Politecnico di Torino
Corso Duca degli Abruzzi, 24
10129 TORINO (ITALY)

We study elastic and thermoelastic wave diffraction in R^3 through a heterogeneous medium of periodic structure, with period ε, which occupies a bounded domain. In the thermoelastic case a parameter η denotes the thermal conductivity of a bounded region of the medium. We study the following problems of perturbation. The first one is related to the limit behaviour when the period ε goes to zero. A homogenized problem is obtained with the convergence of the scattering frequencies. The other one considers the behaviour, as η goes to zero, of the scattering frequency in the origin, which has, for $\eta>0$, infinite multeplicity and splits, for $\eta=0$, into the set of an eigenvalue with infinite multiplicity and infinitely many scattering frequencies with finite multiplicity.

1. INTRODUCTION

The linear thermoelasticity system (see [3] for instance) reads

$$(1.1)\qquad \frac{\partial^2 u_i}{\partial t^2} - \frac{\partial \sigma^T_{ij}}{\partial x_j} = 0,$$

$$(1.2)\qquad \frac{\partial}{\partial t}(\theta + \beta \operatorname{div} \underline{u}) - \eta k \Delta \theta = 0,$$

where $\underline{u}$ and θ denote the displacement vector and the temperature, Δ is the Laplace operator, σ^T is the "total" stress tensor, which decomposes into two parts depending on $\underline{u}$ and θ according to

$$(1.3)\qquad \sigma^T = \sigma(\underline{u}) + \tilde{\sigma}(\theta),$$

$$(1.4)\qquad \sigma_{ij}(\underline{u}) = a_{ijlm} e_{lm}(\underline{u}) \; ; \quad e_{lm}(\underline{u}) = \frac{1}{2}\left(\frac{\partial u_l}{\partial x_m} + \frac{\partial u_m}{\partial x_l}\right),$$

$$(1.5)\qquad \tilde{\sigma}_{ij}(\theta) = -\beta\, \delta_{ij}\, \theta.$$

The part σ is the classical elasticity tensor (or isothermal elasticity tensor) which depends on the strain tensor $e(\underline{u})$ according to (1.4), where a_{ijlm} are the elasticity coefficients, which satisfy the symmetry and positivity conditions:

$$(1.6)\qquad a_{ijlm} = a_{ijml} = a_{mlij},$$

$$(1.7)\qquad a_{ijlm}e_{lm}e_{ij} \geqq C\, e_{ij}e_{ij}, \quad \forall\, e_{ij} \text{ symmetric},$$

for some $C \geqq 0$.
Of course, (1.1) should be considered in the distribution sense, and then it implies the transmission condition

$$(1.8)\qquad [\sigma^T_{ij}n_j] = 0,$$

on the eventual discontinuities of the medium, where the bracket denotes the jump across a surface with normal $\underline{n}$.
Let us recall that the strain-stress relation (1.4) in the isotropic case becomes

$$(1.9)\qquad \sigma_{ij}=\lambda e_{mm}\delta_{ij}+2\mu e_{ij} \longleftrightarrow a_{ijlm} = \lambda\delta_{ij}\delta_{lm}+\mu(\delta_{il}\delta_{jm}+\delta_{jl}\delta_{im}),$$

where λ, μ are the Lamé constants of the material.
In this paper we consider the specific case where the space R^3 is divided into two parts, a bounded one B_ε, with boundary Γ_ε, and the exterior region E_ε. The medium is supposed to have two different properties on the regions B_ε and E_ε. B_ε is enclosed in a bounded region B, which has a periodic structure. Each period εY is a homotopy of a fundamental period Y. The parallelopiped Y is divided in two parts the first one is strictly contained in the interior of Y, the second one encloses the boundary ∂Y of Y.
Then the coupling coefficient β, the conductivity ηk and the elasticity coefficients a_{ijlm} are functions of x of the form:

$$(1.10)\qquad \beta(x) = \begin{cases} \beta^B = \text{const.}>0 & \text{if } x\in B_\varepsilon, \\ \beta^E = \text{const.}>0 & \text{if } x\in E_\varepsilon, \end{cases}$$

$$(1.11)\qquad k(x) = \begin{cases} 1 & \text{if } x\in B_\varepsilon, \\ 0 & \text{if } x\in E_\varepsilon, \end{cases}$$

$$(1.12)\qquad a_{ijlm} = \begin{cases} a^B_{ijlm} = \text{const.} & \text{if } x\in B_\varepsilon, \\ a^E_{ijlm} = \text{isotropic const.}, & \text{if } x\in E_\varepsilon, \end{cases}$$

where it is understood that the a^E_{ijlm} are constants expressed in terms of Lamé constants λ^E, μ^E according to (1.9). The medium is not necessarily isotropic in B_ε. On the other hand, we note that (1.11) expresses that the thermal conductivity vanishes in the exterior region E_ε. More precisely the thermal conductivity will be $\eta k(x)$, with $k(x)$ given by (1.11), where η denotes a small parameter taking values $\geqq 0$. In fact, we shall also consider complex values of η in a neighbourhood of the origin, in order to use techniques of holomorphic functions.
On the interface Γ_ε we shall prescribe the boundary condition (1.8) and the continuity of the displacement vector; moreover, for $\eta\neq 0$, we shall prescribe a Neumann boundary condition for θ on the side B_ε, expressing the fact that the heat cannot pass across Γ_ε into the region E_ε where the conductivity vanishes:

$$(1.13)\qquad [\underline{u}] = 0; \qquad [\sigma^T_{ij}n_j] = 0 \qquad \text{on } \Gamma_\varepsilon,$$

$$(1.14)\qquad \frac{\partial\theta}{\partial n} = 0 \quad \text{on } \Gamma_\varepsilon, \text{ side } B_\varepsilon, \text{ for } \eta\neq 0, \text{ nothing for } \eta=0.$$

2. HOMOGENIZATION OF A THERMOELASTIC ADIABATIC SCATTERING PROBLEM

Let us seek for solutions of (1.1) and (1.2) depending on t by the factor $\exp(-\zeta t)$, i.e. of the form

$$\underline{u}(x,t) = e^{-\zeta t}\,\underline{u}(x)\,, \tag{2.1}$$

$$\theta(x,t) = e^{-\zeta t}\,\theta(x)\,, \tag{2.2}$$

We shall also denote $\zeta = i\omega$, and either ζ or ω will be called the corresponding frequency (in fact the genuine frequency is $\omega/(2\pi)$). Now we indicate the solution with $\underline{u}^\varepsilon$ and θ^ε. The system (1.1),(1.2) leads to the following problem:

$$-\frac{\partial\sigma_{ij}(\underline{u}^\varepsilon)}{\partial x_j} + \frac{\partial(\beta^\varepsilon\theta^\varepsilon)}{\partial x_i} + \zeta^2\,u_i^\varepsilon = 0\quad, \tag{2.3}$$

$$\zeta\,\beta^\varepsilon\,\mathrm{div}\,\underline{u}^\varepsilon + \zeta\,\theta^\varepsilon = 0\;, \tag{2.4}$$

which is equivalent to

$$-\frac{\partial}{\partial x_j}(\sigma_{ij}(\underline{u}^\varepsilon) + (\beta^\varepsilon)^2\mathrm{div}\,\underline{u}^\varepsilon\,\delta_{ij}) = -\,\zeta^2\,u_i^\varepsilon, \tag{2.5}$$

which is a modified elasticity (or adiabatic thermoelasticity) system. The behaviour at infinity of the solution $\underline{u}^\varepsilon$ is the same as for the elasticity system. In order to define it, we consider the Green function of the elasticity system:

$$G_j^\ell(x,\zeta) = \frac{1}{4\pi}\left(\frac{\delta_{j\ell}}{b^2}\,\frac{e^{\pm\zeta r/b}}{r} + \frac{1}{\zeta^2}\,\frac{\partial^2}{\partial x_j\partial x_\ell}\left(\frac{e^{\pm\zeta r/a}}{r} - \frac{e^{\pm\zeta r/b}}{r}\right)\right). \tag{2.6}$$

for $\zeta \neq 0$, with

$$a^2=\lambda^E+(\beta^E)^2+2\mu^E\;;\quad b^2=\mu^E\;;\quad r=|x|\,, \tag{2.7}$$

where the sign + or - are used in the so-called outgoing or incoming fundamental solution, respectively. This denomination is obvious on account of the dependence $\exp(-\zeta t)$ on time. The outgoing radiation condition at infinity is then:

(2.8) $\underline{u}^\varepsilon$ is superposition of foundamental solutions (2.6) with sign +

The limit problem, as $\varepsilon\searrow 0$, is obtained by the homogenization of the total strain tensor σ_{ij}^ε:

$$\sigma_{ij}^\varepsilon(x,\underline{u}^\varepsilon) = \begin{cases} \sigma_{ij}^B(x/\varepsilon,\underline{u}^\varepsilon)+(\beta^B(x/\varepsilon))^2\mathrm{div}\,\underline{u}^\varepsilon\delta_{ij} & \text{if } x\in B,\\ \sigma_{ij}^E(\underline{u}^\varepsilon)+(\beta^E)^2\mathrm{div}\,\underline{u}^\varepsilon\delta_{ij} & \text{if } x\in CB,\end{cases} \tag{2.9}$$

We recall that B is the bounded region with periodic structure, which encloses the B^ε region, while $CB\subset E^\varepsilon$. The tensor is then given by

$$\sigma_{ij}^\varepsilon(x,\underline{u}^\varepsilon) = a_{ijlm}^\varepsilon(x)\frac{\partial u_\ell^\varepsilon}{\partial x_m}\,,$$

where $a^{\varepsilon}_{ijlm}(x)$ takes two different constant values with periodic dependence for $x \in B$, and a constant value for $x \in CB$.
The homogenization limit gives:

$$\sigma^H_{ij}(x,\underline{u}^0) = a^H_{ijlm}(x)\frac{\partial u^0_\ell}{\partial x_m} ,$$

where

$$a^H_{ijlm}(x) = \begin{cases} a^h_{ijlm}=\text{const.} & \text{if } x\in B, \\ a^{CB}_{ijlm}=a^{\varepsilon}_{ijlm}=\text{same const. as for } \varepsilon>0 & \text{if } x\in CB, \end{cases}$$

a^h_{ijlm} is obtained by the solution of an elliptic problem in the fundamental period Y. The weak limit of $\underline{u}^{\varepsilon}$, as $\varepsilon \searrow 0$, in $(H^1_{loc}(R^3))^3$ solves the following homogenized problem:

(2.10) $$-\frac{\partial \sigma^H_{ij}(x,\underline{u}^0)}{\partial x_j} + \zeta^2 u^0_i = 0 ,$$

(2.11) $\underline{u}^0$ satisfies the outgoing radiation condition in the form (2.8)

The limit temperature θ^0 is given by the following expression:

(2.12) $$\theta^{\varepsilon} \longrightarrow \theta^0 \text{ in } L^2_{loc} \text{ weak} .$$

Remark 2.1: In the general case, a^H_{ijlm} is not isotropic, then the question, to obtain the relation between $\underline{u}^0$ and θ^0, requires the homogenization of the equation (2.4). ∎

Proposition 2.1: If ζ^{ε} is a sequence of scattering frequencies of the problems (2.5), (2.8), which converges to ζ^0, as $\varepsilon \searrow 0$, then ζ^0 is a scattering frequency of the homogenized problem. ∎

The proof is based on an a priori estimate for the scattering solutions corresponding to the scattering frequencies ζ^{ε}. On the other hand we have:

Proposition 2.2: If ζ^0 is a scattering frequency of the limit problem, we have that it exists at least a scattering frequency ζ^{ε} of the problem (2.5), (2.8) such that $\zeta^{\varepsilon} \longrightarrow \zeta^0$, as $\varepsilon \searrow 0$. ∎

This result is obtained by the meromorphic dependence of $\underline{u}^{\varepsilon}$ by the spectral parameter ζ.

3. PERTUBATION OF THE ZERO SCATTERING FREQUENCY IN THERMOELASTICITY.

In this section we fix $\varepsilon>0$ and consider the perturbation problem with respect to the thermal conductivity parameter $\eta k(x)$. In order to study the scattering frequencies near $\zeta=0$, we shall perform the dilatation

(3.1) $$\zeta = \eta\, z$$

where z is the new spectral parameter, which we consider in any bounded region of $\underline{C}$ (and then ζ of order $O(\eta)$). The equations (1.2), taking into account (2.1) and (2.2) and the Neumann boundary condition (1.14) for θ, becomes:

(3.2) $$-\Delta_N\theta = z\,(\theta + \beta^B \mathrm{div}\,\underline{u}) \quad \text{in } B_\varepsilon ,$$

where Δ_N denotes the Laplacian with Neumann boundary condition. Now we shall "solve" (1.1) with respect to $\underline{u}$ and substitute into (3.2) to obtain a functional equation in θ. In this manner we obtain some operator L, $A(\zeta)$ such that in a neighbourhood of $\zeta=0$ we have:

(3.3) $$\underline{u} = [A(\zeta) + \zeta^2]^{-1}\, L\, \theta .$$

In order to sustitute this into (3.2), we define the operator $K(\zeta)$ by:

(3.4) $$K(\zeta)\,\theta = \beta^B\, \mathrm{div}\, \{[A(\zeta) + \zeta^2]^{-1}\, L\, \theta\}.$$

Lemma 3.1: The operator $K(\zeta)$ is well defined for ζ in a neighbourhood of the origin. It is there a holomorphic function with values in $(L^2(B_\varepsilon))$. The operator $I+K(\zeta)$ is an isomorphism of $L^2(B_\varepsilon)$ for sufficiently small $|\zeta|$. Moreover, $K(0)$ is hermitian and

(3.5) $$(K(0)\theta,\theta)_{L^2(B_\varepsilon)} \geq 0 \qquad \forall\, \theta \in L^2(B_\varepsilon).$$ ∎

Now, (3.2) becomes

(3.6) $$-\Delta_N\theta = z[I+K(\eta z)]\theta,$$

which is an implicit eigenvalue problem in $L^2(B_\varepsilon)$. We shall write this under a more classical form by applying the isomorphism $[I+K(\zeta)]^{-1}$ (Lemma 3.1) to (3.6), which becomes:

(3.7) $$\Lambda(\zeta)\theta = z\theta \; ; \quad \zeta = \eta z , \quad \text{where}$$

(3.8) $$\Lambda(\zeta) \equiv [I+K(\zeta)]^{-1}(-\Delta_N) .$$

We then have:

Lemma 3.2: $\Lambda(\zeta)$ with sufficiently small $|\zeta|$ is a family of holomorphic unbounded operators of $L^2(B)$. For $\zeta=0$, $\Lambda(0)$ has eigenvalues, noted $z_i(0)$ which are real positive, tending to $+\infty$ as $i\to+\infty$ and with finite multiplicity. According to classical holomorphic perturbation theory, $\Lambda(\zeta)$ has the eigenvalues $z_i(\zeta)$ which are algebroid functions of ζ, i.e. they are holomorphic functions of some fractional power $\zeta^{1/p}$ of ζ, p integer >0. ∎

Coming back to (3.6), finding z as a function of η is equivalent to "solve" the implicit equation

(3.9) $$z = z_i(\eta z),$$

for each i, where z_i are the functions quoted in Lemma 3.2. We shall disregard the first eigenvalue, $z_i(\zeta)\equiv 0$, corresponding to the eigenfunction θ=const.. As we said in Lemma 3.2, $z_i(\zeta)$ has in general an algebraic singularity al $\zeta=0$, i.e. it is a p-valued function which is expressed as a holomorphic function f of $\zeta^{1/p}$ with the p values of it. In order to use the implicit function theorem for holomorphic functions, we use the same device as in [7]: we write

(3.10) $$\zeta^{1/p} = \eta^{1/p} z^{1/p} , \qquad z^{1/p} = \tilde{z} , \qquad \eta^{1/p} = \gamma ,$$

and (3.9) becomes

(3.11) $F(\tilde{z},\gamma) = 0, \qquad F(\tilde{z},\gamma) \equiv \tilde{z}^p - f_i(\gamma\tilde{z}).$

In order to solve in a neighbourhood of

$$\tilde{z} = z_i(0)^{1/p}, \quad \gamma = 0,$$

we check that at this point, $\partial F/\partial\tilde{z} \neq 0$. Then we obtain the implicit function $\tilde{z}(\gamma)$ and then

$$z = \tilde{z}^p(\eta^{1/p}),$$

which is a p-valued function of η, which we shall denote $f_i(\eta^{1/p})$. Finally, $\zeta = \eta f(\eta^{1/p})$, and we have proved the following:

Theorem 3.1: The considered thermoelasticity problem, with small η, has infinitely many scattering functions ζ near the origin, of the form

(3.12) $\zeta = \eta f_i(\eta^{1/p}),$

which have in general algebraic singularities (i.e. each one is a holomorphic function of some root $\eta^{1/p}$ of η). The values $f_i(0)$ are real and positive, and form a sequence tending to $+\infty$. It should be noticed that all the functions (3.12) are not necessarily defined simultaneously for sufficiently small η; but, taking a finite number of them, they are well defined for sufficiently small η. ∎

REFERENCES

[1] Codegone, M., "Scattering of Elastic Waves through a Heterogeneous Medium", Math.Meth.Appl.Sci. 2(1980),pp.271-287.
[2] Codegone, M. and Sanchez-Palencia, E., "Asymptotic of the scattering frequencies for a thermoelasticity problem with small thermal conductivity", Math.Modelling and Num.Analys., in print 1988.
[3] Dafermos, C.M., "On the existence and the asymptotic stability of solution to the equation of linear thermoelasticity", Arch.Rat.Mech.Anal. Vol.29(1968) pp.241-171.
[4] Kato, T., "Perturbation Theory for Linear Operators", Springer, New York (1966).
[5] Kupradze, V.D., "Potential Methods in the Theory of Elasticity", Israel Program for Sc. Translations, Jerusalem (1965).
[6] Leis, R., "Zur Theorie elastischer Schwingungen in inhomogenen Medien", Arch.Rat.Mech.Anal. 39(1970) pp.158-168.
[7] Ohayon, R. and Sanchez-Palencia, E., "On the vibration problem for an elastic body surrounded by slightly compressible fluid". R.A.I.R.O. Analyse Numerique, 17 (1983), pp. 311-326.
[8] Sanchez-Hubert, J. and Sanchez-Palencia, E., "Vibration and Coupling of Continuous Systems. Asymptotic Methods", Springer, Heidelberg (to be published, 1988-9).
[9] Sanchez-Palencia, E., "Perturbation of Eigenvalues in Thermoelasticity and Vibration of Systems with Concentrated Masses", in Trends on Application of Pure Mathematics to Mechanics, ed. Ciarlet and Roseau, Springer, Lecture Notes in Physics n.195, (1984).

ELASTIC WAVE PROPAGATION
M.F. McCarthy, M.A. Hayes, (Editors)

ADVANCED BOUNDARY ELEMENT CALCULATIONS FOR ELASTIC WAVE SCATTERING

William S. HALL and William H. ROBERTSON

Department of Mathematics and Statistics
Teesside Polytechnic
Middlesbrough, England

The scattering of time-harmonic elastic waves propagating in an infinite homogenous isotropic solid and incident on an arbitrarily shaped cavity may be formulated in terms of boundary integrals for the total displacement field. By considering the field on the boundary a system of boundary integral equations may be set up and solved numerically by the Boundary Element Method (BEM). The application of an advanced BEM scheme is described which incorporates quadratic boundary and unknown function representations and series expansion of singular kernel integrations. Results are given for a plane wave incident on a circular cylinder, with eigenfunction expansion solutions for comparison.

1. INTRODUCTION

One of the reasons for developing an accurate calculation procedure for elastic waves is to help in the analysis of the ultrasonic waves used in Non-Destructive Testing (NDT) to detect the presence, and hopefully the nature, of cracks in engineering structures. Further information on this application may be found in Coffey and Chapman [1].

2. INTEGRAL EQUATION FORMULATION

Following Pao and Varatharajulu [2] an integral equation formulation may be written for the amplitude of the time-harmonic wave displacement $\underline{u}(\underline{x})$ scattered by an obstacle in an elastic solid with Lamé constants λ and μ. For the two dimensional case of a traction-free cavity with boundary S this has been developed as

$$\int_S \underline{u}(\underline{x}').\underline{n}.\underline{\underline{\Sigma}}(\underline{x}|\underline{x}')ds' \quad + \quad \underline{u}^I(\underline{x}) \quad = \quad \epsilon\underline{u}(\underline{x})$$

where

$$\underline{\underline{\Sigma}} = \lambda \, \underline{\underline{I}}.\underline{\underline{G}} + \mu \, (\nabla \underline{\underline{G}} + \underline{\underline{G}}\nabla)$$

and

$$\underline{\underline{G}} = \frac{1}{4\pi\rho\omega^2}\left\{k_s \underline{\underline{I}} g_s(\underline{x}|\underline{x}') + \nabla\left[g_p(\underline{x}|\underline{x}') - g_s(\underline{x}|\underline{x}')\right]\nabla'\right\}$$

is the Green's displacement tensor in which

$$g_p(\underline{x}|\underline{x}') = i\pi H_0^{(1)}(k_p r), \qquad g_s(\underline{x}|\underline{x}') = i\pi H_0^{(1)}(k_s r)$$

are the free-space scalar Green's functions for compression and shear waves. k_p, k_s are compression and shear wavenumbers and

$$r = |\underline{x} - \underline{x}'|.$$

ϵ is given by Jones [3] as

$$\epsilon = \begin{cases} 1 & \underline{x}\epsilon V, \text{ external to the cavity} \\ \tfrac{1}{2} & \underline{x}\epsilon S, \text{ on the boundary} \\ 0 & \underline{x}\epsilon B, \text{ inside the cavity.} \end{cases}$$

Taking $\underline{x}\epsilon S$ and $\epsilon = \tfrac{1}{2}$ gives a boundary integral equation, which is treated by the Boundary Element Method in the following section. Once the surface field is known, taking $\underline{x}\epsilon V$ ($\epsilon = 1$) allows the far field displacements to be found.

3. BOUNDARY ELEMENT METHOD

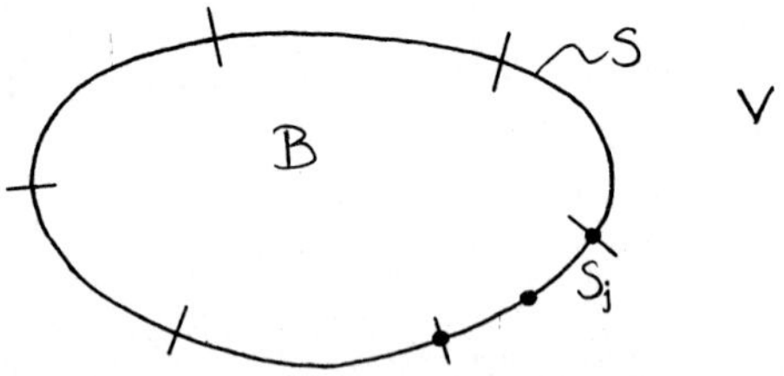

FIGURE 1

The boundary S is divided into N elements S_j so that

$$S = \bigcup_{j=1}^{N} S_j.$$

Each S_j is approximated by interpolating between given boundary points (or nodes) $\underline{x}_i$. Quadratic interpolation is used here, so that three nodes define an element, hence

$$\underline{x}(\xi) = \sum_{\alpha=1}^{3} M^{\alpha}(\xi)\underline{x}_{\alpha}, \qquad -1 \leqslant \xi \leqslant +1,$$

for each element, where

$$M^1(\xi) = \tfrac{1}{2}\xi(\xi-1), \qquad M^2(\xi) = (1-\xi)(1+\xi), \qquad M^3(\xi) = \tfrac{1}{2}\xi(\xi+1).$$

Unknown displacements are represented in the same way so that

$$\underline{u}(\underline{x}(\xi)) = \sum_{\alpha=1}^{3} M^{\alpha}(\xi)\underline{u}_{\alpha}.$$

where $\underline{u}_{\alpha}$ are displacements at nodes, ie $\underline{u}(\underline{x}_{\alpha})$.

The integral equation (1) then becomes

$$\epsilon\underline{u}(\underline{x}) - \sum_{j=1}^{N}\int_{-1}^{1}\left\{\sum_{\alpha=1}^{3}M^{\alpha}(\xi)\underline{u}_{\alpha}\right\}.\underline{n}.\underline{\underline{\Sigma}}(\underline{x}|\underline{x}(\xi))J(\xi)d\xi = \underline{u}^{I}(\underline{x}).$$

$J(\xi)$ is the Jacobian for the parametric representation. Using collocation and taking $\underline{x}$ to be each node in turn with $\epsilon=\frac{1}{2}$ leads to the following system of linear equations for the $\underline{u}_i$'s

$$\tfrac{1}{2}\underline{u}_i - \sum_{j=1}^{2N}\underline{\underline{A}}_j(\underline{x}_i).\underline{u}_j = \underline{u}^{I}(\underline{x}_i), \qquad i = 1,2,\ldots,2N.$$

This may be solved by standard Gaussian elimination techniques.

4. KERNEL INTEGRATION USING SINGULARITY SUBTRACTION AND SERIES EXPANSIONS

The coefficients $\underline{\underline{A}}_j(\underline{x}_i)$ of the linear equations contain integrals which will be singular if $\underline{x}_i \in S_j$, since $\underline{\underline{\Sigma}}(\underline{x}_i|\underline{x}(\xi)) \to 0$ as $\underline{x}(\xi) \to \underline{x}_i$, and the integrals thus need special treatment. Following Aliabadi, Hall and Phemister [4] and writing F for the singular integrand, a related function F^* may be used to subtract out the singularity, ie

$$\int_{-1}^{1}Fd\xi = \int_{-1}^{1}F^*d\xi + \int_{-1}^{1}(F-F^*)d\xi.$$

F^* is chosen to have the same singular behaviour as F but with a simpler form which is such that

(i) $(F-F^*)$ is regular

(ii) F^* can be integrated analytically.

A systematic way of finding F^* is to take the first few terms of the Taylor series expansion for F. This is used on the complicated expression given in Hall and Robertson [5] for F, which contains singular or badly-behaved terms involving

$$1/r, \quad \log r, \quad r_{i}, \quad r_{j}.$$

To obtain simplified forms for these, r is thus expanded in terms of the variable $\delta\xi = \xi-\xi_i$ about the point $\underline{x}_i = \underline{x}(\xi_i)$, giving

$$r = |\delta\xi|(d_o + d_1\delta\xi + d_2\delta\xi^2).$$

Also

$$M^{\alpha}(\xi) = m_0^{\alpha} + m_1^{\alpha}\delta\xi + m_2^{\alpha}\delta\xi^2$$

and

$$J(\xi)n_k = \begin{cases} x_2'+x_2''\delta\xi & k=1 \\ -x_1'-x_1''\delta\xi & k=2 \end{cases}.$$

where d_0, d_1, d_2, m_0^{α}, m_1^{α}, m_2^{α}, x_2', x_2'', x_1'. x_1'' are constants.

Using these expansions, the subtraction can be performed with an F^* function in which the singularities may be isolated in terms of the integration variable ξ through $\delta\xi$, ie

$$\int F d\xi = \int \left\{ F^*_{-1} + F^*_{\ell 1} + F^*_{\ell 2} \right\} d\xi + \int \left\{ F - F^*_{-1} - F^*_{\ell 1} - F^*_{\ell 2} \right\} d\xi,$$

where

$$F^*_{-1} = a_0/\delta\xi, \qquad F^*_{\ell 1} = a_1 \delta\xi \log|\delta\xi|, \qquad F^*_{\ell 2} = a_2 \delta\xi^2 \log|\delta\xi|.$$

a_0, a_1, a_2 are formed from constants in the expansions for $\underline{\underline{\Sigma}}$, $J\underline{n}$, and M^α. $1/\delta\xi$ is singular and $\delta\xi\log|\delta\xi|$, $\delta\xi^2\log|\delta\xi|$ are non-singular but not of polynomial form and do not integrate well by Gaussian methods. $(F - F^*_{-1} - F^*_{\ell 1} - F^*_{\ell 2})$ is a well-behaved remainder integrand to be integrated numerically, and

$$\int F^*_{\ell 1} d\xi, \quad \int F^*_{\ell 2} d\xi$$

are standard integrals.

If $\xi_i = 0$, which corresponds to $\underline{x}_i$ being the centre node of the element, then

$$\int F^*_{-1} d\xi = \int_{-1}^{1} \frac{a_0 d\xi}{\xi_i},$$

which has zero Cauchy principal value. A similar result occurs for the end nodes of the element, provided the adjoining element is considered at the same time. Hence, when the whole boundary is considered, the term $1/\delta\xi$ has no effect.

5. RESULTS

A circular cylinder of radius a is taken as a test case. A plane wave, specified to be either a compression (P-) or a shear (S-) wave, of unit amplitude and wavenumber k is incident upon this cylinder. The results given are for the real part of the x_1-component of the scattered displacement.

A sequence of results for the field on the surface of the cylinder, for increasing numbers of nodes, is given in Figure 2 for an incident compression wave such that $k_p a = 5$. For an accurate solution, 32 elements are required. Discretisations with fewer nodes show clearly that the quadratic behaviour within each element is insufficient to match the true solution. The far field at 100a for $k_p a = 5$ is calculated from the nodal values obtained with the 32-element discretisation, and is very close to the exact eigenfunction expansion solution [5] as shown in Figure 3.

Far field displacements for an incident shear wave such that $k_s a = 10$ are presented in Figure 4. A good solution is obtained in this case with 64 elements.

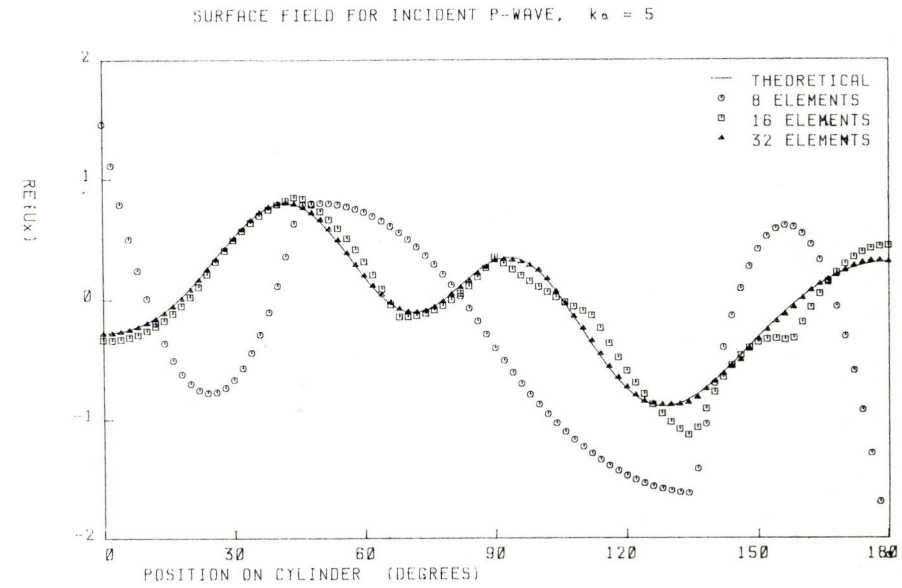

FIGURE 2

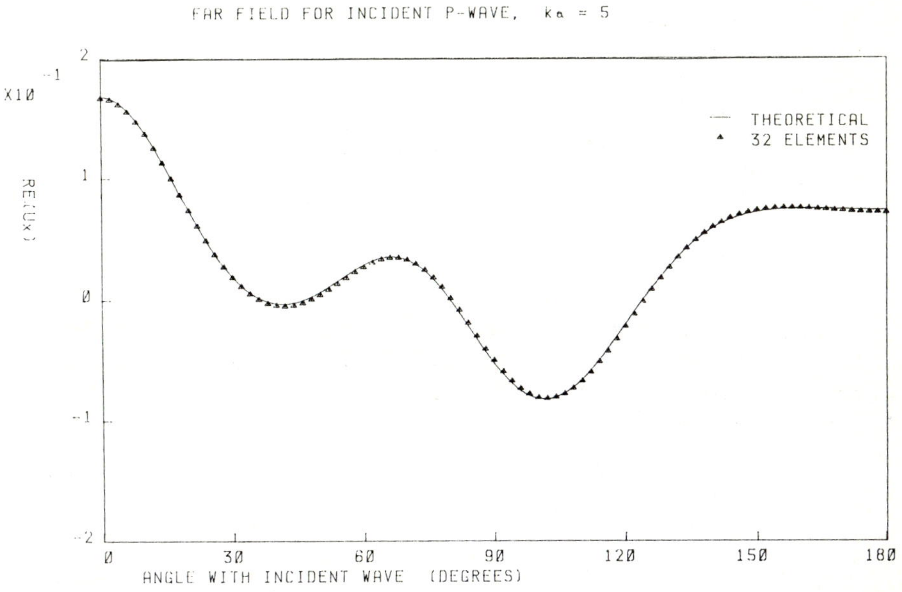

FIGURE 3

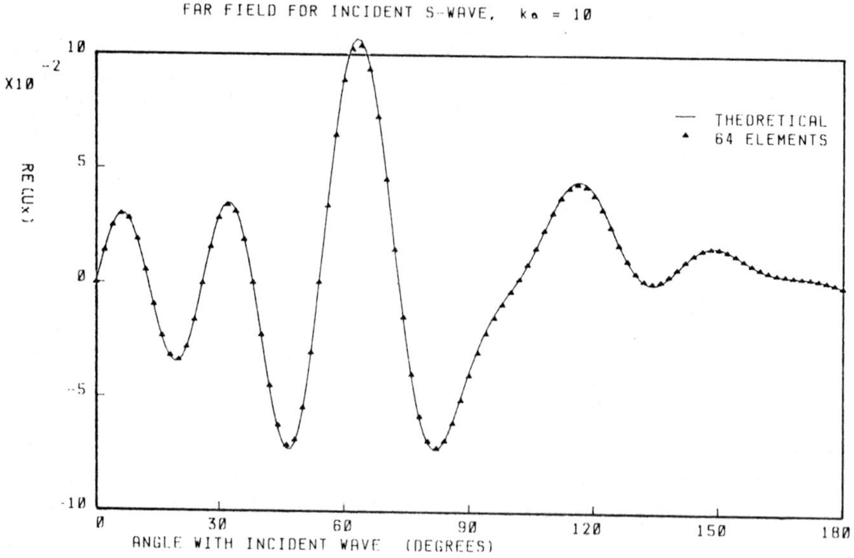

FIGURE 4

6. ACKNOWLEDGEMENTS

The authors wish to thank CEGB Scientific and Technical Branch, Manchester, and Dr R.K. Chapman for their collaboration and assistance, and Cleveland County Council for financial support (WHR).

7. REFERENCES

[1] Coffey,J.M. and Chapman,R.K., Nucl. Energy, 22, 319-333 (1983)

[2] Pao,Y.H. and Varatharajulu,V., J. Acoust. Soc. Am., 59, 1361-1371 (1976)

[3] Jones,D.S., IMA J. Appl. Math., 34, 83-97 (1985)

[4] Aliabadi,M.H., Hall,W.S. and Phemister,T.G., Int. j. numer. methods eng., 21, 2221-2236 (1985)

[5] Hall,W.S. and Robertson,W.H., Boundary Element Methods for 2D Elastic Wave Scattering (Teesside Polytechnic Maths Report, TPMR 88-1, 1988)

ELASTIC WAVE PROPAGATION
M.F. McCarthy, M.A. Hayes, (Editors)
© Elsevier Science Publishers B.V. (North-Holland), 1989

A PRACTICAL METHOD TO MEASURE THE DEPTH OF CRACKS IN CONCRETE STRUCTURES BY USING DIFFRACTED P WAVES

Masayoshi Nakano

Faculty of Engineering
Kansai University
Osaka, Japan

The model experiment suggests that a P pulse, diffracted at the edge of a crack and arriving at the surface of concrete structure, usually gives rise to a 180° shift of phase in the direction at a constant angle to the crack, and therefore, to the surface.

This phenomenon can be theoretically verified using analysis of diffraction by a semi-infinite crack (Achenbach et al.[1]), after some simplification to a two-dimensional case. It can be demonstrated that the constant angle depends solely on the Poisson ratio of the medium, and also that the phenomenon never takes place in any medium that has a Poisson ratio of more than 0.38.

Based on the result cited above, a simple and practical procedure is proposed in order to estimate the depth of cracks in concrete structures. Not only the simplicity but also the accuracy of this procedure suggest that it may well replace most previous methods.

1. INTRODUCTION

As a method of nondestructive testing of concrete structures the use of ultrasonic pulse has recently gained general popularity, on account of its wide applicability and practical degree of accuracy (BS4408 [2] and ASTM C597-83 [3]).

When two piezoelectric transducers are fixed as pulse source and receiver upon a concrete surface, crossing a crack, the depth of crack is obtained by measuring the arrival time of P pulse from source to receiver. (As to these techniques, see Jones [4].) Uneven contact between the transducers and the surface, or insufficiency in pulse energy, however, makes it difficult to identify from the record the moment of pulse arrival(Akashi and Amasaki[5]).

While rcently it has often been reported that situating the receiver farther from the crack inverts the initial motion of the pulse, the law governing the occurrence of this phenomenon has not so far been discussed. If we can make clear the law and obtain the depth of crack by finding the distance at which the inversion occurrs, the whole process of measurement will become very simple and accurate, for the inversion of initial motion is distinguishable at first glance of the record, and the process of measurement will no longer be affected so sensitively by unevenness of contact between transducer and surface or by energy of received pulse, as above. We decided therefore to endeavour to derive the relevant law from using theoretical analysis of elastic waves, and feel confident that we have succeeded in this endeavour.

Using several samples of concrete in each of which we cut a thin trench as a substitute for a crack, we then proceeded to confirm by model experiment that our theoretical result was plausible. Moreover, our theory shows that the boundary distance of inversion is determined solely by the Poisson ratio of the medium, and that this phenomenon never occurs in any medium the Poisson ratio of which has a value larger than 0.4. Similar experiments were carried out using duralumin (Poisson ratio:0.34), and poly-urethane resin (Poisson ratio:0.48); the results of both of these supported the theory.

2. THEORETICAL PROPOSITION

Figure 1(a) shows a schematic diagram of a section of concrete with a crack passing through it perpendicularly to its surface. If a dilatational pulse is radiated from a source (S), the receiver (R) located on the opposite side of the crack will record the diffracted P pulse.

Since the diffracted P pulse is the quickest of all to arrive at R, to consider only the P pulse is to ignore the effects of the free surface. Hereafter, the incident pulse is replaced by a sinusoidal plane wave, and we assume the width of crack is too small to affect significantly. The problem can thus be analysed by investigating a substitute model in which a P wave is diffracted by a half-plane slit ($y=0$, $x \geqq 0$) in an elastic space. Figure 1(b) shews this model.

The problem of diffraction by a slit is treated theoretically by Maue [6]. Achenbach et al. [1] have extended this problem to a three dimensional model, deriving a general solution in the case of a high frequency wave. We can easily accomplish our purpose only by simplifying their results to the two dimensional case and utilizing them for numerical calculations.

The exact representation in integral form of $\phi(x,y)$, the potential of a P wave diffracted by a slit, can be derived using the Wiener-Hopf technique. If it is assumed that

$$\rho/L(P) >> 1 \qquad (x=\rho\cos\psi,\ y=\rho\sin\psi) \tag{1}$$

in which $L(P)$ is the wave-length of P wave, then the integral expressing $\phi(x,y)$ is evaluated approximately through the method of steepest descent in the following form:

$$\phi(x,y) \sim A \frac{\exp\ i[k(\alpha)\rho-\pi/4]\cdot\sin\psi\ \sqrt{1-\cos\theta}}{\sqrt{\pi\rho}\ [2k(\alpha)]^{3/2}(k^2-1)\sqrt{1+\cos\psi}}$$
$$\times \frac{[(k^2-2\cos^2\theta)(k^2-2\cos^2\psi)-4\cos\theta\cos\psi\ \sqrt{1+\cos\theta}\ \sqrt{1+\cos\psi}\ \sqrt{k-\cos\theta}\ \sqrt{k-\cos\psi}\ ]}{(\cos\theta+\cos\psi)(k(r)-\cos\theta)(k(r)-\cos\psi)K(-\cos\theta)K(\cos\psi)} \tag{2}$$

Here $k=v(P)/V(S)$, $k(\alpha)=2\pi/L(P)$ and $k(r)=v(P)/v(R)$, $v(P)$, $v(S)$ and $v(R)$ denoting the velocities of propagation of dilatational, shear and Rayleigh waves, respectively. Function $K(X)$ is defined as follows:

$$K(X) = \exp[L(X)] \text{ and } L(X) = -\frac{1}{\pi}\int_1^k \tan^{-1}\left(\frac{4z^2\sqrt{z^2-1}\ \sqrt{k^2\ z^2}}{k^2-2z^2}\right)\left(\frac{dz}{z-X}\right) \tag{3}$$

The modulus of vertical displacement at R varies with ψ under fixed five values of θ as in Figure 2. The Poisson ratio of the medium is assumed to be 0.25. The change of sign of modulus means phase shift π, or, an inversion of the sense of displacement. The value of $\theta+\psi$ where the modulus vanishes will be expressed as 2ε.

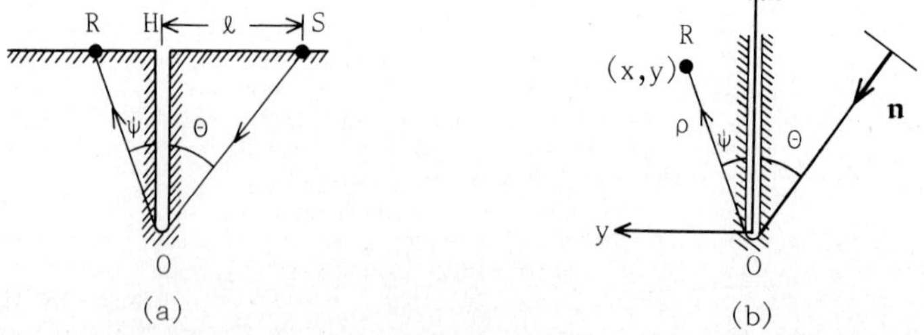

FIGURE 1
Schematic diagram of concrete with a crack (a), and the substituted model; an elastic space with a half-plane slit (b).

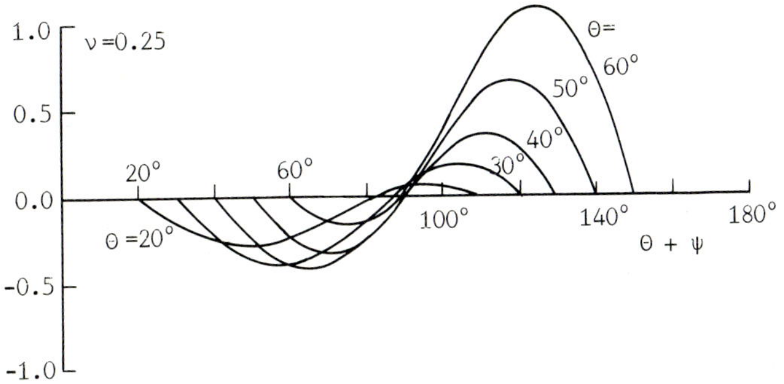

FIGURE 2
Variation of modulus of vertical displacement with receiving direction.

It may be observed from this figure that 2ε's take, roughly speaking, about 90°. More precise study of 2ε is shown in Figure 3. Figure 3 shows the variation of ψ with θ, such that θ+ψ coincides with 2ε. The value of ψ is obtained by using the condition that the numerator in formula (2) above vanishes. Three curves in the figure correspond to three cases that Poisson ratio are 0.2, 0.25 and 0.3, respectively. Let the maximal value of ε be ε(ν), then, ε = ε(ν) where θ = ψ. As ε(ν) depends only on ν, it can be obtained by solving the equation;

$$8x^4-4(k-1)x^3-4k(k+1)x^2+k^4 = 0, \quad x = \cos[\varepsilon(\nu)] \qquad (4),$$

which is easily derived from the numerator of (2), by the substitution in it θ = ψ = ε(ν). Table 1 denotes the value of ε(ν) for ν between 0.10 and 0.38. When ν becomes equal to or larger than 0.4, ε(ν) no longer exists. The variation of ε(ν) with ν is also shown as a solid curve in Figure 4.

If the depth of the crack is assumed to be d in concrete with a Poisson ratio ν, it is also assumed that the crack runs perpendicular to the surface. We set source S and receiver R, respectively, on either side of crack, and equidistant from it. This is shown as SH = HR = ℓ (say). Suppose that ε = ∠SOH = ∠HOR becomes equal to ε(ν) when ℓ takes a value ℓ^0. If ν is known in advance, ε(ν) can be determined from Table 1. Therefore, we derive the depth d as follows;

TABLE 1

ν	ε(ν)	ν	ε(ν)	ν	ε(ν)
0.01	54.38°	0.14	50.05°	0.27	40.86°
0.02	54.13°	0.15	49.59°	0.28	39.67°
0.03	53.87°	0.16	49.10°	0.29	38.36°
0.04	53.59°	0.17	48.58°	0.30	36.89°
0.05	53.31°	0.18	48.03°	0.31	35.27°
0.06	53.01°	0.19	47.44°	0.32	33.36°
0.07	52.70°	0.20	46.82°	0.33	31.18°
0.08	52.37°	0.21	46.15°	0.34	28.61°
0.09	52.03°	0.22	45.43°	0.35	25.49°
0.10	51.67°	0.23	44.66°	0.36	21.53°
0.11	51.30°	0.24	43.82°	0.37	16.01°
0.12	50.90°	0.25	42.92°	0.38	4.90°
0.13	50.49°	0.26	41.93°		

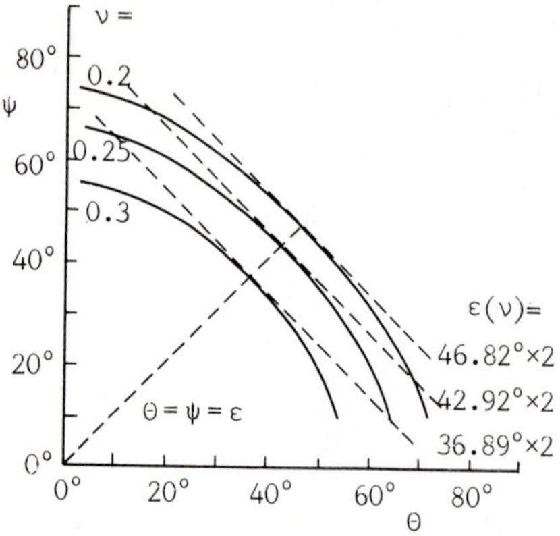

FIGURE 3
Variation of receiving direction (ψ) by incident direction (θ), where diffracted wave vanishes.

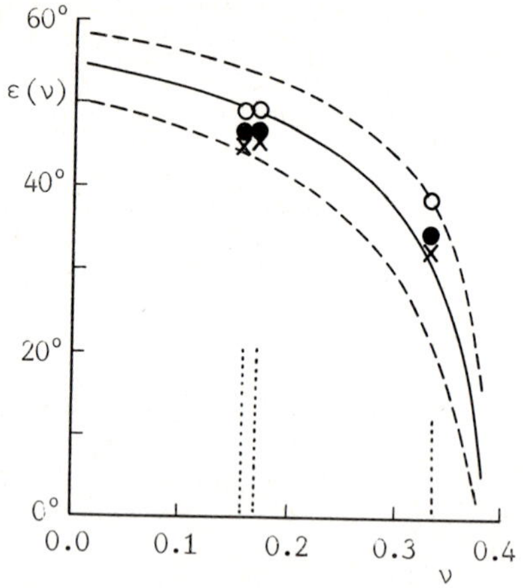

FIGURE 4
Variation of ε(ν) by ν (solid curve) and the region of error in model experiments described in Section 3 (domain enclosed by two broken lines).

$$d = \ell^0 \cot[\varepsilon(\nu)] \quad (5).$$

The Poisson ratio of concrete may be generally considered as to be 0.15 to 0.20, which implies that $\varepsilon(\nu)$ ranges from 47° to 50° and $\cot[\varepsilon(\nu)]$ from 0.84 to 0.93 (see Table 1). If we take account of accuracy of actual measurement, we shall be able to regard $\cot[\varepsilon(\nu)]$ as normally being 1, and hence more simply

$$d = \ell^0 \quad (6)$$

gives an approximate formula with which depth of crack in concrete may be estimated.

3. EXPERIMENTAL CONFIRMATION

Twelve samples of concrete in solid rectangular block form ($15x15x55cm^3$) are prepared, consisting of six with W/C=55% and others with W/C=95%. Into two of these blocks was cut perpendicular to one surface a trench of a depth of 10cm, one having a width of 0.3mm and the other one of 0.5mm; a trench of a depth of 5cm was cut into each of another pair, and one of a depth of 3cm into the last two blocks, the width varying within each of these pairs as in the first two. Poisson ratioes are 0.17 and 0.16 in samples with W/C=55% and 95%, respectively, and hence, $\varepsilon(\nu)$ in either case can be regarded as 49°.

At two points 1cm distant from the upper end of the crack, and on opposite sides of it, the source and the receiver are respectively fixed and the pulse reaching the receiver is photographed. Then, each transducer is shifted one centimeter farther from the end of the crack and fixed there and the pulse again photographed. This procedure is repeated until the initial sense of the pulse becomes inverted for the first time. Photo 1 shows a series of received pulses, representing those at ℓ=4, 6, 8, 10, 12, 14 and 16cm, respectively, in which the sample has a crack with depth 10cm and width 0.5mm.

We assume that at the first point at which the inversion of pulse occurs

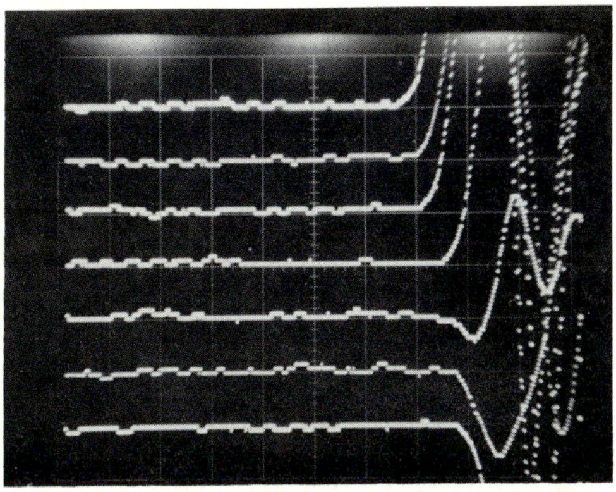

Photo 1
An example of a series of received pulses in case of a crack with 10cm depth in concrete.

SH = HR = ℓcm and that ℓ^0 in (5) should be the mean value of ℓ and ℓ-1. (In Photo 1, the inversion takes place first at ℓ=12cm. However, it will begin at ℓ =11cm if the transducer is shifted by 1cm.) Using ℓ^0 thus defined we immediately obtain the depth by (5) or (6).

The depth obtained in this way is derived as 0.5cm larger than its true value in each case. The error is clearly caused by the method of measurement in which the transducers are shifted every 1cm. Since the absolute value of error in ℓ^0 reaches 0.5cm the apparent value of $\varepsilon(\nu)$, determined from the ratio of depth (true value) to ℓ^0, $\varepsilon(\nu)$=arccot(d/ℓ^0), is also distributed within a certain domain. Naturally this domain becomes largest when d = 3cm, for the absolute error in ℓ^0 has more effect, the shallower the depth. The two broken lines in Figure 4 indicate the domain of apparent values of $\varepsilon(\nu)$ produced when d = 3cm. Three marks in the figure, ○, ●, × respectively show the apparent values of $\varepsilon(\nu)$ obtained by ℓ^0 in cases where d is 3cm, 5cm and 10cm. These values are all included in the domain constituting the region of error inevitable in this method of measurement.

In order to confirm that our theory also leads to plausible result in case of another Poisson ratio, solid duralmin is used in turn. Its Poisson ratio is 0.34, and hence, $\varepsilon(\nu)$ is about 30°. Three samples are prepared in each of which is cut a crack with depth of either 3, 5 or 10cm, and common width 1mm. The values of ℓ^0 are determined as 2.5, 3.5 and 6.5cm, respectively, from which the apparent values of $\varepsilon(\nu)$ are calculated as shown in Figure 4. These too are included in the domain of inevitable errors as in the case of concrete.

This theory further suggests that the phenomenon described above does not take place in media that have Poisson ratioes exceeding a certain limit; practically $\nu \geqq 0.38$. To exemplify this, a similar experiment is performed using as sample poly-urethane resin, whose Poisson ratio is 0.48. Cracks of the same six sets of dimensions as were inserted into our samples of concrete; inversion of the pulse was, however, not detectable in all of these samples.

4. CONCLUSIONS

We shall sum up as follows what is stated in the preceding sections.

1) Assume that the P wave is incident upon a slit of half-plane in an infinite elastic medium with Poisson ratio ν. Where θ is the angle between the slit and the direction of the propagation of the incident wave, the diffracted P wave caused by slit makes a π-shift of its phase in the fixed direction of angle ψ to

the plane of slit. ψ is determined by θ and ν.

2) Let $\theta + \psi = 2\varepsilon$, then, 2ε takes a fairly fixed value depending on ν. To express this more precisely, ε gets maximal when $\theta = \psi(=\varepsilon)$ and the maximal value $\varepsilon(\nu)$, a function of only ν, can be obtained by solving equation (4). (See Table 1 as to the values of $\varepsilon(\nu)$ in detail.)

The phase shift π in the diffracted wave does not occur in a medium with sufficiently large value of ν; practically $\nu \geqq 0.38$.

3) By utilizing the above phenomenon a practical and simple method of measuring the depth of a crack in concrete is proposed.

Assume that the crack in question runs perpendicularly to the surface of concrete and that the Poisson ratio of the concrete is already known. At two opposite points respectively on the surface of equal distances from the end of the crack, two transducers (as source and receiver) are fixed, and the pulse reaching the receiver is photographed. Then, two transducers are moved keeping equal distances from the end of crack. By repeating this procedure the positions of source and receiver at which the pulse motion is inversed can be discovered. Using the distance between these positions and $\varepsilon(\nu)$, we can eaily obtain the depth of crack.

4) The following problems remain for further investigation.

If the crack grows not perpendicular but largely inclined to the surface, the above method should be considerably modified. If the inclination of crack is sufficiently small, however, its approximate depth may be obtained by means of the above method.

Experiment suggests that the effect of the width of the crack can be considered negligible within a certain range of values, though the range of values which may be disregarded has not yet been accurately determined. Another problem to be tackled in the future is that of the measurement of the depth of two cracks existing close to each other.

The effect of inserted iron material is a further problem of great importance in the practical application of the method here presented.

ACKNOWLEDGEMENTS

The model experiments above cited were performed by researchers of Kajima Institute of Construction Technology. The writer should like to thank them for their enthusiastic cooperation.

The writer is also indebted to Prof. A.S. Gibbs, Faculty of Literature, Kansai University, for his kind advice and guidance concerning the English of this presentation.

REFERENCES

[1] Achenbach, J.D., Gautesen, A.K. and McMaken, H., Ray Methods for Waves in Elastic Solids (Pitman, London, 1982)
[2] British Standards, No.4408, Part 5, Recommendation for Nondestructive Methods of Pulse in Concrete (British Standard Institute, London, 1974)
[3] American Society for Testing Materials, Designation C597-83, Standard Test Method for Pulse Velocity through Concrete (Annual Book of ASTM Standards, 1985)
[4] Jones, R., Nondestructive Testing of Concrete (Cambridge Univ. Press, London, 1962)
[5] Akashi, T. and Amasaki, S., State-of-the-Art on Nondestructive Testing of Concrete (Concrete Journal, 23-12, pp.11-20, 1985, in Japanese)
[6] Maue, A.W., Die Beugung elastischer Wellen an der Halbebene (Z. angew. Math. Mech., 33-1/2, pp.1-10, 1953)

ELASTIC WAVE PROPAGATION
M.F. McCarthy, M.A. Hayes, (Editors)

TIME-DOMAIN BORN APPROXIMATION TO THE FAR-FIELD SCATTERING OF PLANE ELASTIC WAVES BY AN ELASTIC HETEROGENEITY.

Dirk Quak

Laboratory of Electromagnetic Research, Faculty of Electrical Engineering, Delft University of Technology, P.O. Box 5031, 2600 GA Delft, The Netherlands

The low-contrast or (first-order) Born approximation is applied to the time-domain scattering of a plane elastic wave by a non-homogeneous heterogeneity of bounded extent, embedded in an isotropic and homogeneous solid. Closed-form analytic expressions are obtained for the spherical-wave far-field scattering amplitude related to homogeneous heterogeneities. Relaxation effects and anisotropy of the heterogeneity are included. The results are, among others, useful as test cases for time-domain inverse scattering algorithms.

1. INTRODUCTION

In the present paper the Born approximation is applied to the time-domain scattering of a plane elastic wave by a non-homogeneous heterogeneity of bounded extent situated in a homogeneous, isotropic, perfectly elastic solid. The analysis is carried out entirely in the space-time domain. Anisotropy and relaxation effects of the heterogeneity are included. Closed-form analytic expressions are derived for the spherical-wave, far-field, scattering amplitudes in case the plane elastic wave is incident upon an homogeneous heterogeneity. Our analysis differs in several aspects from the work of previous authors. Gubernatis et.al. [1 - 2] present a frequency-domain analysis of the scattering problem of a homogeneous, isotropic inclusion, while the scattering analysis is based on the elastodynamic wave equation for the particle displacement of the wave mation. Due to the latter, contrasts in volume density of mass and in the stiffness (or compliance) parameters are handled differently. Wu and Richards [3] generalize this formulation to an elastic, inhomogeneous, isotropic heterogeneity. Rose and Richardson [4] give time-domain expressions for the particle velocity of the scattered elastic wavefield in the Born approximation.

In our analysis we start from the first order partial differential equations for the particle velocity and the stress. The contrast sources for the density of volume force and of strain rate in these equations for the scattered field are expressed in terms of the contrast in elastic parameters, the local particle velocity and the local stress. In recent papers [5 - 6], the equivalent problem of the time domain scattering of a plane acoustic wave by an acoustically penetrable object and of a plane electromagnetic wave by an electromagnetically penetrable object, respectively, is similarly dealt with.

2. FORMULATION OF THE SCATTERING PROBLEM

In a isotropic, homogeneous, unbounded perfectly elastic medium an arbitrary heterogeneity occupying the bounded domain D^S is present. The domain exterior to D^S is denoted by D. Domain D is mechanically characterized by its volume density of mass ρ and its Lame coefficients λ and μ. In any source-free subdomain of D, the elastic field quantities satisfy the elastic field equations

$$\partial_j\tau_{ij} - \rho\,\partial_t v_i = 0 \tag{1}$$

$$(\partial_i v_j+\partial_j v_i)/2 - s_{ijpq}\partial_t\tau_{pq} = 0, \tag{2}$$

in which s_{ijpq} denotes the compliance of the elastic solid. The heterogeneity D^S shows a contrast with respect to the surrounding medium. This contrast is accounted for by the contrast parameters $\delta\rho_{ik}(\mathbf{x},t)$ and $\delta s_{ijpq}(\mathbf{x},t)$, so the inertia relaxation function ρ^S_{ik} in the heterogeneity is given by $\rho^S_{ik}(\mathbf{x},t) = \rho\delta_{ik}\delta(t) + \delta\rho_{ik}(\mathbf{x},t)$ and the compliance relaxation function s^S_{ijpq} by $s^S_{ijpq}(\mathbf{x},t) = s_{ijpq}(\mathbf{x})\delta(t) + \delta s_{ijpq}(\mathbf{x},t)$, in which $\delta(t)$ denotes the unit impulse. Then the elastodynamic field quantities satisfy in D^S the equations

$$\partial_j\tau_{ij} - \rho\,\partial_t v_i = \partial_t \int_0^\infty \delta\rho_{ik}(\mathbf{x},t')\, v_k(\mathbf{x},t-t')\,dt' \tag{3}$$

$$(\partial_i v_j+\partial_j v_i)/2 - s_{ijpq}\partial_t\tau_{pq} = \partial_t \int_0^\infty \delta s_{ijpq}(\mathbf{x},t')\,\tau_{pq}(\mathbf{x},t-t')\,dt'. \tag{4}$$

In (3) and (4), the causality has been taken into account.

The heterogeneity is irradiated by either a uniform plane elastic P-wave or S-wave $\{\tau^i_{ij},v^i_j\}$ propagating in the direction of the unit vector $\boldsymbol{\alpha}$; it satisfies the sourcefree counterparts of (1)-(2). For an incident P-wave holds

$$\{\tau^i_{ij},v^i_j\} = \{TP^i_{ij},VP^i_j\}a(t-\alpha_p x_p/c_P), \tag{5}$$

where a(t) is the, somehow normalized, wave shape of the incident wave, and $\{TP^i_{ij},VP^i_j\}$ are the constant amplitude tensors of the P-wave.
For the case of an incident S-wave, we similarly have :

$$\{\tau^i_{ij},v^i_j\} = \{TS^i_{ij},VS^i_j\}a(t-\alpha_p x_p/c_S). \tag{6}$$

The scattered elastodynamic wave $\{\tau^S_{ij},v^S_j\}$ is defined as

$$\{\tau^S_{ij},v^S_i\} = \{\tau_{ij}-\tau^i_{ij}\ ,\ v_i-v^i_i\}. \tag{7}$$

With (1)-(4), it follows that τ^S_{ij} and v^S_i satisfy in R^3 the equations

$$\partial_j\tau^S_{ij} - \rho\,\partial_t v^S_i = \{-f^S_i(\mathbf{x},t),0\} \qquad \mathbf{x}\in\{D^S,D\} \tag{8}$$

$$(\partial_i v^S_j+\partial_j v^S_i)/2 - s_{ijpq}\partial_t\tau^S_{pq} = \{h^S_{ij}(\mathbf{x},t),0\} \qquad \mathbf{x}\in\{D^S,D\} \tag{9}$$

where

$$f^S_i(\mathbf{x},t) = -\partial_t \int_0^\infty \delta\rho_{ik}(\mathbf{x},t')v_k(\mathbf{x},t-t')\,dt' \tag{10}$$

is the contrast volume density of body force of the heterogeneity, and

$$h^{s}_{ij}(\mathbf{x},t) = \partial_t \int_0^{\infty} \delta s_{ijpq}(\mathbf{x},t')\ \tau_{pq}(\mathbf{x},t - t')\ dt' \tag{11}$$

is the contrast source density of strain rate of the heterogeneity.

3. REPRESENTATION OF THE SCATTERED FIELD

In our further analysis we shall use the time-domain volume source representation of the scattered wave field. For simplicity, we only give the expression for the particle velocity.

For the particle velocity of the scattered field we have

$$v^{s}_{j}(\mathbf{x},t) = \rho^{-1}\partial_t A_j - \rho^{-1} c_{ikpq}\partial_i \Psi_{jkpq} \quad \text{when } t \in \{t_0,\infty\}, \tag{12}$$

where t_0 is the instant at which the incident wave hits the heterogeneity. It is expressed in terms of the retarded elastodynamic vectorpotential

$$A_i(\mathbf{x},t) = \int_{t_0}^{\infty} dt' \int_{\mathbf{x}'\in D^s} G_{ij}(\mathbf{x} - \mathbf{x}',t - t') f^{s}_{j}(\mathbf{x}',t')d\mathbf{x}' \tag{13}$$

and the retarded elastodynamic tensorpotential

$$\Psi_{ijpq}(\mathbf{x},t) = \int_{t_0}^{\infty} dt' \int_{\mathbf{x}'\in D^s} G_{ij}(\mathbf{x} - \mathbf{x}',t - t') h^{s}_{pq}(\mathbf{x}',t')d\mathbf{x}', \tag{14}$$

in which

$$G_{ij}(\mathbf{x},t) = G_S(\mathbf{x},t)\delta_{ij} + \partial_i\partial_j \int_0^{\infty} [c_P^2 G_P(\mathbf{x},t-t'') - c_S^2 G_S(\mathbf{x},t-t'')]t''dt'' \tag{15}$$

with

$$G_{P,S}(\mathbf{x},t) = (4\pi c^2_{P,S}|\mathbf{x}|)^{-1}\delta(t - |\mathbf{x}|/c_{P,S}) \tag{16}$$

denotes the infinite medium Green's function of the three-dimensional elastodynamic wave equation

$$(c_P^2 - c_S^2)\partial_i\partial_k G_{kj} + c_S^2\partial_k\partial_k G_{ij} - \partial_t^2 G_{ij} = - \delta_{ij}\delta(\mathbf{x},t). \tag{17}$$

In (12)-(17) we have taken into account the condition of causality, i.e. we have assumed that the scattered wave field vanishes everywhere prior to t_0.

By letting $|\mathbf{x}| \to \infty$ in (12)-(14) we arrive at integral representations for the far-field amplitude radiation characteristic for the particle velocity of the scattered wave. To this end, we employ in the expression for $G_{ij}(\mathbf{x}-\mathbf{x}',t-t')$ that results from (15) the relation

$$|\mathbf{x} - \mathbf{x}'| = |\mathbf{x}| - \xi_k x'_k + \text{vanishing terms} \quad \text{as } |\mathbf{x}| \to \infty, \tag{18}$$

with $\boldsymbol{\xi} = \mathbf{x}/|\mathbf{x}|$ the unit vector in the direction of observation. This yields

$$G_{ij}(\mathbf{x}-\mathbf{x}',t-t') \sim (\delta_{ij}-\xi_i\xi_j)\frac{\delta(t-t'-|\mathbf{x}|/c_S+\xi_k x'_k/c_S)}{4\pi c_S^2|\mathbf{x}|} + \xi_i\xi_j\frac{\delta(t-t'-|\mathbf{x}|/c_P+\xi_k x'_k/c_P)}{4\pi c_P^2|\mathbf{x}|} \quad \text{as } |\mathbf{x}| \to \infty. \tag{19}$$

Using (19) in (13) we arrive at the asymptotic expression

$$A_i(\mathbf{x},t) \sim AP_i(\xi,t-|\mathbf{x}|/c_P)/4\pi c_P^2|\mathbf{x}|+AS_i(\xi,t-|\mathbf{x}|/c_S)/4\pi c_S^2|\mathbf{x}| \tag{20}$$

where the P-wave contribution is given by

$$AP_i(\xi,t) = \xi_i\xi_j \int_{\mathbf{x}'\in D^s} f_j^s(\mathbf{x}',t + \xi_k x_k'/c_P)d\mathbf{x}'; \tag{21}$$

and the S-wave contribution by

$$AS_i(\xi,t) = (\delta_{ij}- \xi_i\xi_j) \int_{\mathbf{x}'\in D^s} f_j^s(\mathbf{x}',t + \xi_k x_k'/c_S)d\mathbf{x}'. \tag{22}$$

Using (19) in (14) we obtain the asymptotic expression

$$\Psi_{ijpq}(\mathbf{x},t) \sim \Psi P_{ijpq}(\xi,t-|\mathbf{x}|/c_P)/4\pi c_P^2|\mathbf{x}|+ \Psi S_{ijpq}(\xi,t-|\mathbf{x}|/c_S)/4\pi c_S^2|\mathbf{x}| \tag{23}$$

where the P-wave contribution is given by

$$\Psi P_{ijpq}(\xi,t) = \xi_i\xi_j \int_{\mathbf{x}'\in D^s} h_{pq}^s(\mathbf{x}',t + \xi_m x_m'/c_P)d\mathbf{x}' \tag{24}$$

and the S-wave contribution by

$$\Psi S_{ijpq}(\xi,t) = (\delta_{ij}- \xi_i\xi_j) \int_{\mathbf{x}'\in D^s} h_{pq}^s(\mathbf{x}',t + \xi_m x_m'/c_S)d\mathbf{x}'. \tag{25}$$

The use of (20) and (23) in (12) leads to the asymptotic expression

$$v_j^s(\mathbf{x},t) \sim VP_j^s(\xi,t-|\mathbf{x}|/c_P)/4\pi c_P^2|\mathbf{x}| + VS_j^s(\xi,t-|\mathbf{x}|/c_S)/4\pi c_S^2|\mathbf{x}| \tag{26}$$

with

$$VP_j^s = \rho^{-1}\partial_t AP_j + \rho^{-1}c_{ikpq}(\xi_i/c_P)\partial_t\Psi P_{jkpq} \tag{27}$$

and

$$VS_j^s = \rho^{-1}\partial_t AS_j + \rho^{-1}c_{ikpq}(\xi_i/c_S)\partial_t\Psi S_{jkpq}. \tag{28}$$

For the stress a similar asymptotic expression holds i.e.

$$\tau_{ij}^s(\mathbf{x},t) \sim TP_{ij}^s(\xi,t-|\mathbf{x}|/c_P)/4\pi c_P^2|\mathbf{x}| + TS_{ij}^s(\xi,t-|\mathbf{x}|/c_S)/4\pi c_S^2|\mathbf{x}|. \tag{29}$$

So, in the far-field region the scattered field separates into a P-wave part and an S-wave part with the properties

$$VP_j^s = (\xi_p VP_p^s)\xi_j, \qquad \xi_p VS_p^s = 0 \tag{30}$$

respectively. Between the stress and the particle velocity far-field radiation characteristics the following relations exist:

$$TP_{ij}^s = - ZP_{ijq}VP_q^s, \qquad TS_{ij}^s = - ZS_{ijq}VS_q^s, \tag{31}$$

where $ZP_{ijq} = c_{ijpq}\xi_p/c_P$ is the plane P-wave impedance and $ZS_{ijq} = c_{ijpq}\xi_p/c_S$ is the plane S-wave impedance of the homogeneous medium surrounding the heterogeneity.

4. BORN APPROXIMATION

In the low-contrast, or (first-order) Born approximation the unknown values of the total elastodynamic wavefield in D^s are replaced by the known values of the incident wavefield. After this has been done, the volume densities of the contrast sources f_i^S and h_{ij}^S are explicitly known in D^S, namely (cf. (10)-(11))

$$f_i^S(\mathbf{x},t) = -\partial_t \int_0^\infty \delta\rho_{ik}(\mathbf{x},t')\, v_k^i(\mathbf{x},t-t')\, dt' \qquad \mathbf{x}\in D^S, \tag{32}$$

and

$$h_{ij}^S(\mathbf{x},t) = \partial_t \int_0^\infty \delta s_{ijpq}(\mathbf{x},t')\, \tau_{pq}^i(\mathbf{x},t-t')\, dt' \qquad \mathbf{x}\in D^S. \tag{33}$$

In the following, this approximation will be applied to homogeneous heterogeneties. In particular, we shall determine the far-field radiation characteristics of the scattered wavefield. Two cases are distinguished: (a) the incident wave is a P-wave and (b) the incident wave is an S-wave.

(a) <u>Incident P-wave</u>

From (5) and (32)-(33) we obtain in expression (27) for the scattered P-wave, after introducing the vector

$$u^{PP} = \xi/c_P - \alpha/c_P \tag{34}$$

through which the amplitude radiation characteristics of the scattered P-wave depend on the direction of observation ξ, the direction of propagation α of the incident wave, and the wavespeed c_P in the surrounding medium,

$$\rho^{-1}\partial_t AP_j(\xi,t) = -V\, \xi_i\xi_j\, VP_k^i \int_0^\infty \rho^{-1}\delta\rho_{ik}(t')\, T(u^{PP},t-t')dt', \tag{35}$$

and

$$\rho^{-1}c_{ikpq}(\xi_i/c_P)\partial_\tau \Psi P_{jkpq}(\xi,t) = V\, \xi_i\xi_j\xi_k(\rho c_P)^{-1}TP_{rs}^i \int_0^\infty c_{ikpq}\delta s_{pqrs}(t')\, T(u^{PP},t-t')dt' \tag{36}$$

in which

$$T(u,t) = V^{-1}\int_{\mathbf{x}\in D^S} \partial_t^2\, a(t + u_s x_s)\, dV \tag{37}$$

is a shape factor and V is the volume of the heterogeneity.

For the scattered S-wave the vector u^{SP} is introduced as

$$u^{SP} = \xi/c_S - \alpha/c_P. \tag{38}$$

Expression (28) then yields

$$\rho^{-1}\partial_t AS_j(\xi,t) = -V\,(\delta_{ij} - \xi_i\xi_j)\, VP_k^i \int_0^\infty \rho^{-1}\delta\rho_{ik}(t')\, T(u^{SP},t-t')dt' \tag{39}$$

and

$$\rho^{-1}c_{ikpq}(\xi_i/c_S)\partial_t \Psi S_{jkpq}(\xi,t) = V\,(\delta_{jk} - \xi_j\xi_k)\, \xi_i(\rho c_S)^{-1}TP_{rs}^i \int_0^\infty c_{ikpq}\delta s_{pqrs}(t')\, T(u^{SP},t-t')dt', \tag{40}$$

in which T(u,t) is again the shape factor as defined in (37).

(b) <u>Incident S-wave</u>

With (6) and (32)-(33) we obtain, after introducing the vector

$$u^{PS} = \xi/c_P - \alpha/c_S \tag{41}$$

in expression (27) for the scattered P-wave

$$\rho^{-1}\partial_t AP_j(\xi,t) = -V\, \xi_i\xi_j\, VS_k^i \int_0^\infty \rho^{-1}\delta\rho_{ik}(t')\, T(u^{PS},t-t')dt' \tag{42}$$

and

$$\rho^{-1}c_{ikpq}(\xi_i/c_P)\, \partial_\tau \Psi P_{jkpq}(\xi,t) = \tag{43}$$

$$V\ \xi_i \xi_j \xi_k (\rho c_P)^{-1} TS^i_{rs} \int_0^\infty c_{ikpq} \delta s_{pqrs}(t')\ T(u^{PS}, t - t')dt'$$

in which T(u,t) is again the shape factor as defined in (37).

For the scattered S-wave the vector u^{SS} is introduced correspondingly as

$$u^{SS} = \xi/c_S - \alpha/c_S. \qquad (44)$$

Expression (28) yields

$$\rho^{-1} \partial_t AS_j(\xi,t) = -V\ (\delta_{ij} - \xi_i \xi_j)\ VS^i_k \int_0^\infty \rho^{-1} \delta\rho_{ik}(t')\ T(u^{SS}, t - t')dt' \qquad (45)$$

and

$$\rho^{-1} c_{ikpq}(\xi_i/c_S) \partial_t \Psi S_{jkpq}(\xi,t) = \qquad (46)$$

$$V\ (\delta_{jk} - \xi_j \xi_k) \xi_i (\rho c_S)^{-1} TS^i_{rs} \int_0^\infty c_{ikpq} \delta s_{pqrs}(t')\ T(u^{SS}, t - t')dt'$$

in which T(u,t) is again the shape factor as defined in (37).

The shape factor T depends on the geometrical shape of the heterogeneity, on the normalized wave shape of the incident wave, and on one of the the vectors u^{PP}, u^{SP}, u^{PS} and u^{SS}. The shape factors for the ellipsoid, the elliptic cone of finite height, the elliptic cylinder of finite height and the tetrahedron can be derived and are presented in [5 - 6].

5. CONCLUSION

With the aid of analytical techniques, closed-form expressions for the far-field particle velocity in an elastic solid have been derived, when a plane elastic wave is incident upon a number of contrasting elastic heterogeneities having canonical geometry. The synthetic data thus obtained for the simple, though not trivial, geometries can be used as test cases for computational time-domain inversion algorithms.

REFERENCES

[1] Gubernatis, J.E., E. Domany and J.A. Krumhansl, "Formal aspects of the theory of the scattering of ultrasound by flaws in elastic materials," J. Appl. Phys. **48**, 2804-2811 (1977).

[2] Gubernatis, J.E., E. Domany, J.A. Krumhansl and M. Huberman, "The Born approximation in the theory of the scattering of elastic waves by flaws," J. Appl. Phys. **48**, 2812-2819 (1977).

[3] Wu, R., and K. Aki,"Scattering characteristics of elastic waves by an elastic heterogeneity", Geophysics **50**, 582-595 (1985).

[4] Rose, J.H. and J.M. Richardson, "Time Domain Born Approximation," J. Nondestructive Evaluation **3**, 45-53 (1982).

[5] Quak, D., A.T. de Hoop and H.J. Stam, "Time-domain farfield scattering of plane acoustic waves by a penetrable object in the Born approximation," J. Ac. Soc. Am. **80**, 1228-1234 (1986).

[6] Quak, D. and A.T. de Hoop, "Time domain Born approximation to the far-field scattering of plane electromagnetic waves by a penetrable object," Radio Science **21**, 815-821 (1986).

ELASTIC WAVE PROPAGATION
M.F. McCarthy, M.A. Hayes, (Editors)
Elsevier Science Publishers B.V. (North-Holland), 1989

THE PLANE PLATE MODEL APPLIED TO THE SCATTERING OF THE ULTRASONIC WAVES FROM CYLINDRICAL SHELLS

Gérard QUENTIN and Maryline TALMANT

Groupe de Physique des Solides de l'Ecole Normale Supérieure
Université Paris 7 - Tour 23 - 2 place Jussieu
75251 Paris Cedex 05 - France

Our experimental study of the acoustic scattering from very thin cylindrical shells immersed in a fluid and filled with air shows that the resonances occur at frequencies in very good agreement with the theoretical results of the Resonance Scattering Theory (R.S.T.)[1-4]. The agreement is also very good concerning the width of the resonances. At low frequencies two circumferential waves are observed: a "fast" one which is generally considered as a pseudo-Lamb wave S_o and a "slow" one about which there is no agreement between theoretical interpretations. We have tried to compare the physical parameters characterizing these waves with a very simple theoretical approach using the plate theory of Grabowska[5]. We present in this paper a comparison between this approach and the result of calculations issued from the Resonance Scattering Theory.

1. INTRODUCTION

At very low frequencies on very thin cylindrical shells ($b/a > 0.9$) where b and a are respectively the inner and outer radii, two circumferential waves propagate. The faster one has been related since many years to the generalized symmetric-Lamb wave S_o. There is no such agreement concerning the slow wave.

Junger[6] indicates that for these shells the eigenmodes of vibrations are modified by the loading from the fluid. Dragonette[7] calculates the form function for $b/a > 0.98$ at low frequencies ($k_1a < 90$, where k_1 is the wave number inside the fluid) ; he find two periodicities. The periodicity corresponding to the fast wave is related to the Lamb wave S_o and the other one appearing only in a limited frequency range to the Lamb wave A_o.

The results obtained lately by Veksler[8] in a wider range of frequencies, thickness and materials lead to the same conclusion. The Resonance Scattering Theory (R.S.T.) has been applied to this problem by Überall and his coworkers Breitenbach[9] shows a correspondance between the first whispering gallery mode and S_o and predicts at low frequencies resonances of a wave with a phase velocity smaller than the sound velocity in the external fluid. Subrahmanyan[11] predicts the existence of a Stoneley type wave in addition to the A_o wave. In France, the Resonance Scattering Theory has been developed by Derem and Rousselot. Maze et al[12] and Derem[13] have shown the correspondance between the $\ell = 2$ wave (first whispering gallery mode) and the S_o wave when the ratio b/a is larger than 0.7. More recently Izbicki et al[15,16] gave an interpretation of the slow wave observed at low frequencies in term of a pseudo-Stoneley wave.

In conclusion, the theoretical approach is in agreement concerning the fast wave : convergence of the whispering gallery mode G.E.1. ($\ell = 2$) towards symmetric Lamb mode S_o. Concerning the slow wave three different interpretations have been given ; either it is considered as an antisymmetric A_o Lamb wave or a

Rayleigh wave or a Scholte-Stoneley wave. The physical difference between these interpretations is of importance because in the two first cases the main part of the wave's energy is concentrated inside the material of the shell ; in the case of Scholte-Stoneley wave the acoustical energy is mainly located inside the fluid.

2. EXPERIMENTAL MEASUREMENTS USING PULSE EXCITATION AND RESONANCE SCATTERING THEORY

We have experimentally studied the scattering from shells with an internal radius b = 10 mm of various materials (duraluminum, stainless steel and copper) and various thicknesses ($0.96 << b/a < 0.99$) filled with air and immersed either in water or in ethanol[1-4].

From the ultrasonic spectra we have shown that we can deduce the group velocity of the waves circumnavigating the shell[3]. The measured attenuation of these waves is related to the resonance width. The experimental results have been compared with theoretical calculations of Rousselot[16] using the R.S.T. for shells of duraluminum and stainless steel with a thickness of 0.4 mm. The resonances of the fast wave are in very good agreement with those of the $\ell = 2$ wave. The same agreement is found between resonances of the slow wave and those of the $\ell = 0$ wave of the Resonance Scattering Theory. The attenuation of these waves and the domain of the observation of the slow wave agree also very well with theoretical predictions for low acoustic frequencies ($k_1 a < 140$). At higher frequencies a third wave is observed with resonances close to the $\ell = 3$ wave of the R.S.T.

In order to better understand the physical nature of the slow wave, we shall present now a comparison between the theoretical results of the R.S.T. and one very simple approach where we consider the shell as a thin plane plate. The limitation of this approach will also be discussed.

3. PLANE PLATE THEORY

The geometrical and physical parameters are given in Fig.1. The plane plate is made of an homogeneous isotropic solid immersed on the side $y > d$ in a fluid 1 and on the side $y < -d$ in a fluid 2. The following notations are used :

ρ_i = density of fluid i

C_i = sound velocity inside fluid i

ρ = density of the plate's material

C_L, C_T = velocity of the longitudinal and shear acoustic waves inside the plate

k = wave number of the wave propagation along $\vec{Ox}$

f = sound frequency

$k_i = 2\pi/C_i$

$R_i = (k^2 - k_i^2)^{1/2}$ $\quad i = 1,2,L,T$

The dispersion relationship for waves propagating along the plate can be written

$$\sum_{i=1,2} (S-C_{is})(A-C_{iA})-((C_{1s}C_{1A})-(C_{2s}C_{2A})^{1/2})=0 \qquad (1)$$

The various terms used in this equation are

$$S = 4k^2 R_T R_L \, th(R_L d)-(R_T^2+k^2)^2 th(R_T d)$$

$$A = 4k^2 R_T R_L \coth(R_L d) - (R_T^2 + k^2)^2 \coth(R_T d)$$

$$C_{is} = \frac{\rho_i}{\rho} \frac{R_L}{R_i} k_T^4 \operatorname{th}(R_L d) \operatorname{th}(R_T d)$$

$$C_{iA} = \frac{\rho_i}{\rho} \frac{R_L}{R_i} k_T^4 \coth(R_L d) \coth(R_T d)$$

The physical meaning of these terms will appear with specific examples. If fluids 1 and 2 are replaced by vacuum, equation (1) takes the very simple form

$$A\,S = 0 \tag{2}$$

which is the Lamb equation and can be separated into two equations :

$$S = 0 \quad \text{equation of the symmetric modes} \tag{3}$$

$$A = 0 \quad \text{equation of the antisymmetric modes} \tag{4}$$

At low frequencies two eigen modes of the plate exist. Mode S_o has a velocity corresponding to a wave number growing assymptotically towards $1/2\,k_T^2/(k_T^2-k_L^2)^{1/2}$ when the frequency goes to zero. In the same conditions the velocity of Lamb wave A_o goes to zero. Lamb equation has only real roots.

If the plate is immersed on both sides in the same fluid 1, we can follow Osborne and Hart's study[17]. The dispersion equation (1) can be simplified into :

$$(S - C_{1s})\,(A - C_{1A}) = 0 \tag{5}$$

and like in the preceding case separated into two equations

$$S - C_{1s} = 0 \tag{6}$$

$$A - C_{1A} = 0 \tag{7}$$

Osborne et al have shown that these equations have two different kinds of solution. The first kind of solutions are the generalized Lamb modes "S_i" and "A_i". These solutions correspond to complex values of the wave number k. The real part of k is very close of the solution of the Lamb equation (2) for a metal immersed in fluid. The imaginary part of k corresponds to the attenuation of the wave, during its propagation, related to reradiation inside the external fluid. In addition to these classical roots there exist two additional real roots for wave vector k. The first one, the S wave has a phase velocity very close to that of the surrounding fluid medium. The velocity of the second one, the A wave, grows monotonically from zero at zero frequency to the velocity of the fluid medium. These two waves corresponding to real roots of the wave equation and are not attenuated at all during their propagation. This property is similar to that observed for Scholte-Stoneley waves at the interface between a fluid and a semi-infinite solid medium. Consequently they are sometimes called pseudo-Stoneley waves.

In our experimental studies the external medium (1) is a fluid and the internal one (2) is air. The situation studied by Grabowska[5] corresponds very well to this experimental set-up. The dispersion equation (1) can also be simplified into:

$$(S - 1/2\,C_{1s})\,(A - 1/2\,C_{1A}) - 1/4\,C_{1s}\,C_{1A} = 0 \tag{8}$$

The two first factors of the first term are the same as in Osborne's equation except that $1/2\,C_{1s}$ (respectively $1/2\,C_{1A}$) replaces C_{1s} (respectively C_{1A}). That is they correspond to the same plane plate but immersed in a fluid with half the density of the real fluid. Equation (8) cannot be separated into two parts because of the presence of the coupling term $1/4\,C_{1s}\,C_{1A}$. If we calculate the solutions of this equation we obtain for the generalized Lamb waves values of the wave vector k with real part very close to solutions of Lamb's equation and imaginary parts very close to half the value obtained from Osborne's equation (5). This last observation is consistent with the reradiation of the wave only on one side. From the real solutions A and S of equation (5), only one remains : wave A.

4. COMPARISON BETWEEN THE PLANE PLATE THEORY AND THE RESONANCES SCATTERING THEORY

We present a comparison between these two theoretical approaches for two materials : duraluminum and stainless steel. In the case of the plane plate theory, we suppose a plate with a thickness of 0.4 mm immersed in water on one side and vacuum on the other. For the R.S.T. the parameters are the same but the inner radius of the shell is supposed to be 10 mm (b/a = 0.96).

In plate theory the dispersion curves are generally plotted against the parameters fe (e = plate thickness). In the R.S.T. they are plotted against the dimensionless frequency k_1a. These two scales are related through the relationship.

$$k_1 a = \frac{2\pi}{C_1(1-b/a)} fe$$

Our comparison is concerned with generalized Lamb waves A_o, S_o and A_1 and the wave A described previously.
In Fig. 2 we compare, for duraluminum, the generalized Lamb wave S_o and the $\ell = 2$ wave of the R.S.T. In the conditions used the agreement between the two dispersion curves is very good but at very low frequencies ($fe < 0.1$ MHZ x mm) the velocity of the $\ell = 2$ waves grows very rapidly with decreasing frequencies. At the opposite, the S_o Lamb waves goes toward a finite limit value at zero frequency. Concerning the attenuation coefficient of the $\ell = 2$ and S_o wave, they are close to one another. This coefficient has been deduced for the $\ell = 2$ wave from the resonance's width (Fig. 3).

In figure 4 we compare waves $\ell = 2$ and S_o and $\ell = 1$ and A_o for stainless steel. This comparison shows a very good agreement for values of the phase velocity. The same comparison on attenuation coefficients gives the same result.
A comparison between the phase velocity of wave A_1 and that of the $\ell = 3$ wave of the R.S.T. is shown in Fig. 5. This generalized Lamb wave can only be observed if

$$fe > C_T/2$$

Even if the agreement of the two dispersion curves is rather good, we have no explanation of the fact that the divergence is observed at a lower frequency for the $\ell = 3$ wave than for the A wave.

The most interesting comparison concerns the plane plate wave A and the $\ell = 0$ wave of the R.S.T. The comparison can only be made using the phase velocity because, in the case of the A wave, the attenuation coefficient is zero (k is a real root of the dispersion equation). The agreement between the A wave velocity and that of the $\ell = 0$ waves is very good (Fig. 6). We have plotted on the same curve the dispersion curve of wave A_o. When the velocity of the A_o wave approaches the velocity of sound in the fluid medium, the imaginary part of its wave vector becomes exceedingly large and this wave cannot be propagative any more. The frequency when this phenomenon appears is very close to the frequency fe_k where the resonances of the $\ell = 0$ wave disappears both theoretically and experimentally.

5. Conclusion

There is a very good agreement between the plane plate model and the Resonance Scattering Theory concerning most of the waves observed in very thin shells. The following correspondence have been established for the two materials studied with respect to the phase velocity Cp and the attenuation of the waves :
- The $\ell = 2$ wave of the R.S.T. and the pseudo-Lamb symmetric mode S_o
- The $\ell = 1$ wave of the R.S.T. and the pseudo-Lamb antisymmetric mode A_o for $fe > fe_k$
- The $\ell = 3$ wave of the R.S.T. and the pseudo-Lamb antisymmetric mode A_1.

The main new result of this study correspond to the similarity between the $\ell = 0$ wave of the R.S.T. and the pseudo-Stoneley wave A of the plane plate model.

ACKNOWLEDGEMENTS

This work has been supported by the French Direction des Recherches Etudes et Techniques under contract 86 - 120. We tank J.L. Rousselot for the calculations using the R.S.T.(16).

REFERENCES

[1] M. TALMANT and G. QUENTIN, Study of the pseudo-Lamb wave S_0 generated in thin cylindrical shells insonified by short ultrasonic pulses in water, Progress in Underwater Acoustics (H. Merklinger edit. Plenum Press). 1987, 137 - 144.

[2] M. TALMANT, Rétrodiffusion d'une impulsion brève par une coque cylindrique à paroi mince, Thèse de Doctorat, Université Paris 7, 1987.

[3] M. TALMANT and G. QUENTIN, Backscattering of a short ultrasonic pulse from thin shells to be published in J. Appl. Phys.

[4] M. TALMANT, G. QUENTIN, J.L. ROUSSELOT, J.V. SUBRAHMANYAN et H. ÜBERALL, Acoustic resonance of thin cylindrical shells and the Resonance Scattering Theory, submitted to J. Acoust. Soc. of Am.

[5] A. GRABOWSKA, Propagation of elastic wave in solid layer-liquid system, Archives of Acoustics, 1979, 4(1), 57-64.

[6] M.C. JUNGER, Radiation loading of cylindrical and spherical surfaces, J. Acoust. Soc. Am. 1952, 24(3), 288-289.

[7] L.R. DRAGONETTE, Evaluation of the relative importance of circumferential or creeping waves in the acoustic scattering from rigid and elastic solid cylinders and cylindrical shells, N.R.L. Report 8216, 1978.

[8] N.D. VEKSLER, Sound wave scattering by circular cylindrical shells, Wave Motion, 1986, 8, 525-536.

[9] E.D. BREITENBACH, Normal mode acoustic scattering from elastic cylindrical shells, Ph.D. Catholic University of America, Washington, 1979.

[10] E.D. BREITENBACH, H. ÜBERALL and K.B. YOO, Resonant acoustic scattering from elastic cylindrical shells, J. Acoust. Soc. Am., 1983, 74(4), 1267-1273.

[11] J.V. SUBRAHMANYAN, Creeping wave analysis through frequency plane for an obliquely incident plane wave scatterer, Ph.D. Catholic University of America, Washington, 1983.

[12] G. MAZE, J. RIPOCHE, A. DEREM et J.L. ROUSSELOT, Diffusion d'une onde ultrasonore par des tubes remplis d'air immergés dans l'eau, Acustica, 1984, 55, 69-85.

[13] A. DEREM, Diffusion des ondes acoustiques par des cylindres métalliques immergés : remarques concernant l'interprétation des phénomènes, Revue de Cethedec, 1982, 72.

[14] J.L IZBICKI, Diffusion acoustique par des cylindres et des tubes, Ondes guidées et résonances en basse-fréquence, Thèse d'Etat, Université du Havre, 1986.

[15] J.L. IZBICKI, G. MAZE et J. RIPOCHE, Diffusion acoustique par des tubes immergés dans l'eau : nouvelles fréquences observées en basse fréquence, Acustica, 1986, 61, 137-139.

[16] J.L. ROUSSELOT, Communication personnelle.

[17] M.F.M. OSBORNE et S.O. HART, Transmission, Reflection and Guiding of an Exponential Pulse by a Steel Plate in water. I. Theory, J. Acoust. Soc. of Am. 1945, 17(1), 1-18.

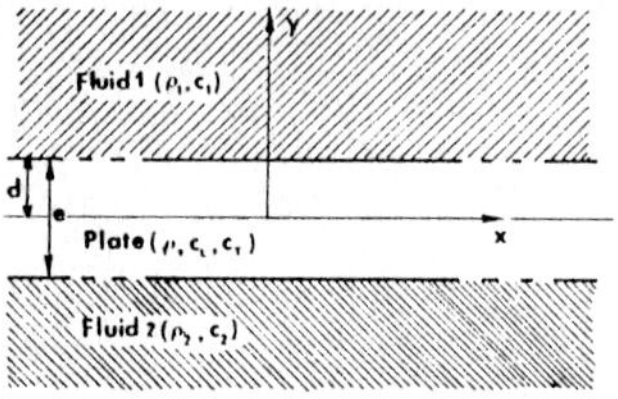

Fig. 1. Cross-section of the plane plate immersed into two different fluids.

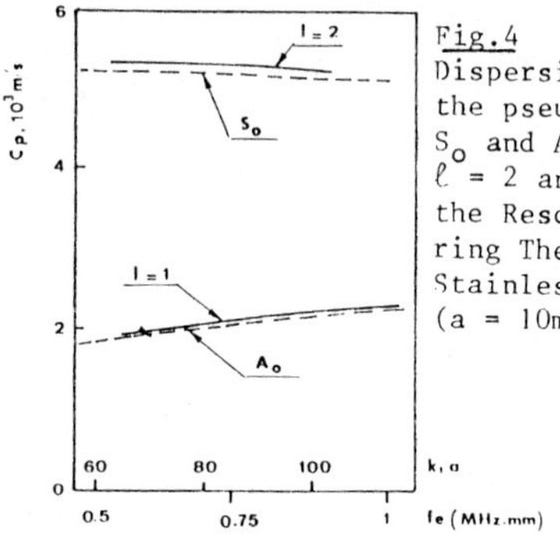

Fig. 4
Dispersion curves of the pseudo-Lamb waves S_o and A_o and of the $\ell = 2$ and $\ell = 1$ of the Resonance Scattering Theory (12). Stainless Steel shell (a = 10mm ; e = 0.4mm).

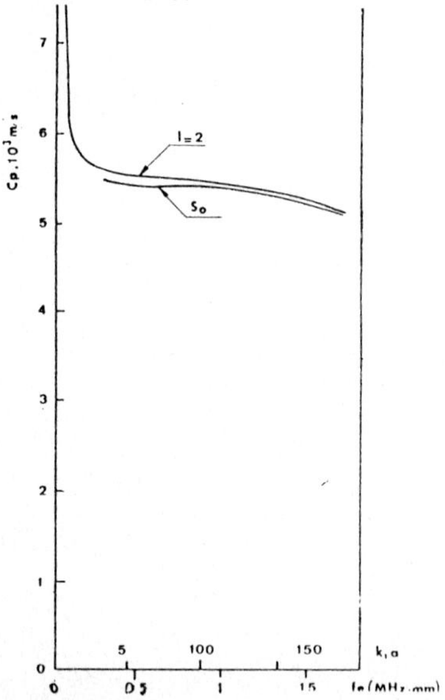

Fig. 2. Dipersion of the pseudo-Lamb wave S_0 and of the $\ell = 2$ wave of the Resonance Scattering Theory (12). Duraluminum shell (a=10mm ; e=0.4mm).

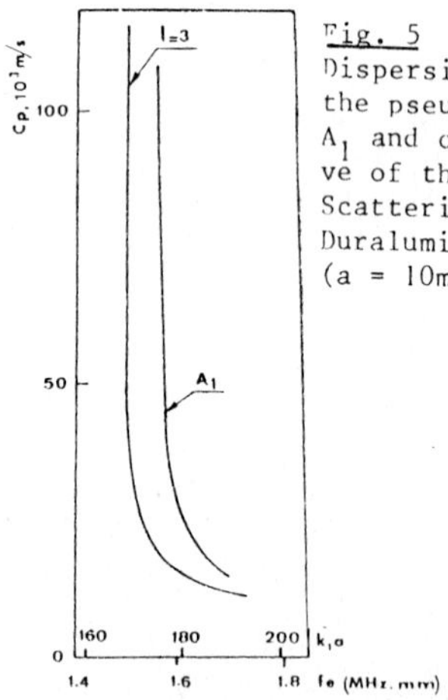

Fig. 5
Dispersion curve of the pseudo-Lamb wave A_1 and of the $\ell = 3$ wave of the Resonance Scattering Theory (12) Duraluminum shell (a = 10mm ; e = 0.4mm).

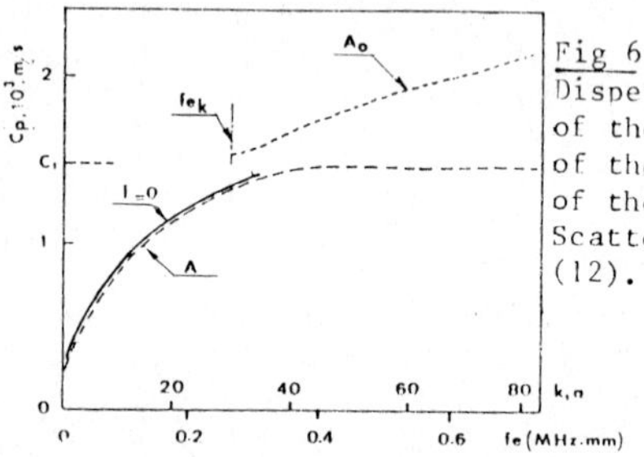

Fig 6
Dispersion curves of the A wave and of the $\ell = 0$ wave of the Resonance Scattering Theory (12). Duraluminum

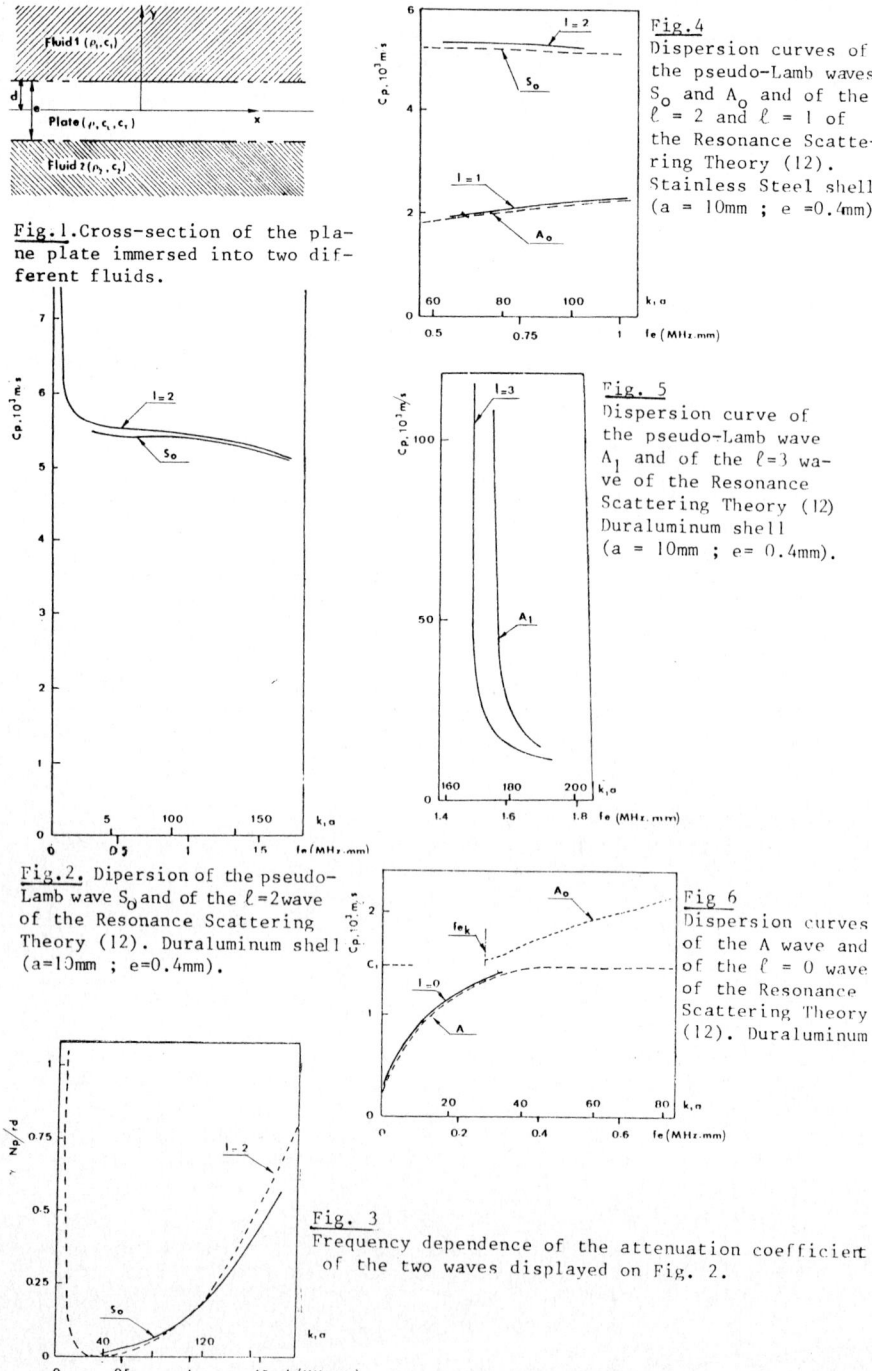

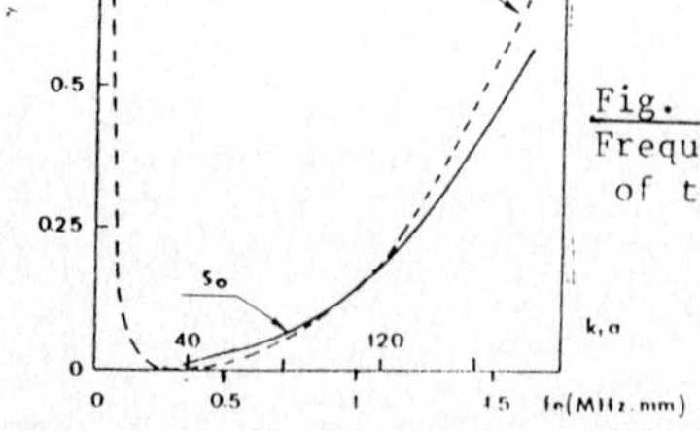

Fig. 3
Frequency dependence of the attenuation coefficient of the two waves displayed on Fig. 2.

ELASTIC WAVE PROPAGATION
M.F. McCarthy, M.A. Hayes, (Editors)

A SPACE-TIME FINITE-ELEMENT METHOD FOR THE COMPUTATION OF THREE-DIMENSIONAL ELASTODYNAMIC WAVE FIELDS (THEORY)

Hendrik J. Stam and Adrianus T. de Hoop

Laboratory of Electromagnetic Research, Faculty of Electrical Engineering, Delft University of Technology, P.O. Box 5031, 2600 GA Delft, The Netherlands

The theory of a space-time finite-element method for the numerical solution of elastodynamic wave problems in bounded time-invariant subdomains of three-dimensional space is developed. It is shown how the finite-element method can be regarded as to be based on a space-time elastodynamic reciprocity theorem of the time-correlation type. Particular local representations for the elastodynamic wave field are developed that can handle strongly inhomogeneous structures in which solid/solid interfaces are present.

1. INTRODUCTION

The theoretical essentials of a full space-time finite-element method, based on reciprocity, for the numerical solution of elastodynamic wave problems in bounded, time-invariant subdomains of three-dimensional space are developed. The elastodynamic wave fields are characterized by their particle velocity and stress, which are considered as their state variables. It is shown how the finite-element method can be regarded as to be based on a space-time elastodynamic reciprocity theorem of the time-correlation type [1]. In its turn this theorem is shown to be equivalent to a certain weighting procedure applied to the equation of motion and deformation rate equation. The medium in the configuration is taken to be linear, locally and instantaneously reacting, and time-invariant in its elastic behaviour. Arbitrary inhomogeneity and anisotropy are taken into account. Particular local representations for the elastodynamic wave field are developed that can handle strongly inhomogeneous structures in which solid/solid interfaces are present.

2. BASIC EQUATIONS OF ELASTODYNAMICS

The elastodynamic waves under consideration are small-amplitude disturbances. Position of observation in R^3 is specified by the Cartesian coordinates $\{x_1,x_2,x_3\}$. The subscript notation for Cartesian vectors and tensors in R^3 is employed and the summation convention applies. The corresponding lower-case Latin subscripts are to be assigned the values $\{1,2,3\}$. The time coordinate is denoted by t. Partial differentiation is denoted by ∂; ∂_p denotes differentiation with respect to x_p, ∂_t denotes the differentiation with respect to t. The geometrical configuration that we study is taken to be time-invariant. The medium in it is assumed to be linear, locally and instantaneously reacting, and time-invariant in its elastic behaviour. It may be arbitrarily inhomogeneous and anisotropic. The elastodynamic wave fields are characterized by their particle velocity v_r and stress τ_{pq}. The physical properties of the solid are characterized by its anisotropic volume density of mass ρ_{kr} and its compliance s_{ijpq}. In each subdomain of the configuration where the elastodynamic properties vary continuously with position, the elastodynamic field quantities are continuously differentiable and satisfy the (linearized) equation of motion

$$(2.1)\quad -\Delta_{kmpq}\partial_m\tau_{pq} + \rho_{kr}\partial_t v_r = f_k,$$

and the (linearized) deformation rate equation

$$(2.2)\quad \Delta_{ijmr}\partial_m v_r - s_{ijpq}\partial_t\tau_{pq} = h_{ij},$$

where f_k = volume source density of force, h_{ij} = volume density of strain rate and $\Delta_{kmpq} = (\delta_{kp}\delta_{mq} + \delta_{kq}\delta_{mp})/2$ is a unit tensor of rank four that specifically occurs in elastodynamics. It has the symmetry properties $\Delta_{kmpq} = \Delta_{mkpq} = \Delta_{mkqp} = \Delta_{kmqp}$. At interfaces between two different media the constitutive coefficients ρ_{kr} and s_{ijpq} in general jump by finite amounts. In all applications we shall assume that at a solid/solid interface the media are in rigid contact; then, the particle velocity v_r and the traction $t_k = \Delta_{kmpq}\nu_m\tau_{pq}$, where ν_m is the unit normal to the interface, are continuous across the interface. Finally, to have a unique solution of the wave problem in the bounded domain D (which later is identified with the domain of finite-element computation), we need initial conditions at some time $t = t_0$:

$$(2.3)\quad v_r(\mathbf{x},t_0) = v_r^I(\mathbf{x}),\quad \text{when}\quad \mathbf{x} \in D,\quad \tau_{pq}(\mathbf{x},t_0) = \tau_{pq}^I(\mathbf{x}),\quad \text{when}\quad \mathbf{x} \in D,$$

and boundary conditions on the boundary surface ∂D of D. The simplest of these are explicit ones that apply to either a prescribed particle velocity or a prescribed traction, i.e.,

$$(2.4)\quad v_r = v_r^B,\quad \text{when}\quad \mathbf{x} \in \partial D_1 \quad\text{and}\quad t_k = t_k^B,\quad \text{when}\quad \mathbf{x} \in \partial D_2$$

where v_r^B and τ_{pq}^B are the prescribed values of the particle velocity and the stress at the parts ∂D_1 and ∂D_2 of ∂D respectively. The intersection of ∂D_1 and ∂D_2 is empty and the union of ∂D_1 and ∂D_2 is ∂D.

3. THE SPACE-TIME FINITE-ELEMENT METHOD

As the starting point for the construction of a finite-element formulation of the elastodynamic wave problem we introduce the space-time elastodynamic reciprocity theorem of the time-correlation type [1]. Of the two states occurring in this theorem, State A is identified with the actual wave field that is to be approximated, while State B is considered as a computational one that remains to be chosen appropriately. As regards the space-time geometry in which the two states occur, the time-invariance implies that this geometry is the Cartesian product $D \times R$ of a time-invariant spatial domain $D \subset R^3$ and the real time axis R. The theorem is applied to the bounded domain D that consists of subdomains $\{D^{(n)};\ n = 1,..,N\}$ where the medium properties vary continuously with position. The boundary surface of $D^{(n)}$ is denoted by $\partial D^{(n)}$. The required reciprocity relation is then

$$(3.1)\quad \sum_{n=1}^{N} \int_{t\in R} dt \int_{x\in\partial D^{(n)}} \Delta_{mrpq}\nu_m[v_r^A(\mathbf{x},t)\tau_{pq}^B(\mathbf{x},t-\tau) + v_r^B(\mathbf{x},t-\tau)\tau_{pq}^A(\mathbf{x},t)]dA$$
$$= \partial_\tau \int_{t\in R} dt \int_{x\in D} \tau_{ij}^B(\mathbf{x},t-\tau)[s_{ijpq}^A(\mathbf{x})-s_{pqij}^B(\mathbf{x})]\tau_{pq}^A(\mathbf{x},t)dV$$
$$+ \partial_\tau \int_{t\in R} dt \int_{x\in D} v_k^B(\mathbf{x},t-\tau)[\rho_{kr}^A(\mathbf{x})-\rho_{rk}^B(\mathbf{x})]v_r^A(\mathbf{x},t)dV$$
$$+ \int_{t\in R} dt \int_{x\in D} \tau_{ij}^B(\mathbf{x},t-\tau)h_{ij}^A(\mathbf{x},t) - v_k^B(\mathbf{x},t-\tau)f_k^A(\mathbf{x},t)dV$$
$$+ \int_{t\in R} dt \int_{x\in D} \tau_{pq}^A(\mathbf{x},t)h_{pq}^B(\mathbf{x},t-\tau) - v_r^A(\mathbf{x},t)f_r^B(\mathbf{x},t-\tau)dV.$$

In this theorem, the state quantities have been assumed to be piecewise continuously differentiable in $\{D^{(n)},\ n=1,..,N\}$. First of all we shall show that Equation (3.1) can, from a particular point of view, also be regarded as a "weighted" form of the equation of motion (2.1) and the deformation rate equation (2.2) pertaining to the State A. To this end, we take the quantities $\{v_k^B,\tau_{ij}^B\}$ of State B to be continuously differentiable functions in the subdomains $\{D^{(n)},\ n=1,..,N\}$, subject to the choice $\rho_{rk}^B = 0$, $s_{ijpq}^B = 0$, $f_r^B = -\Delta_{rmij}\partial_m\tau_{ij}^B$ and $h_{pq}^B = \Delta_{pqmk}\partial_m v_k^B$. Substituting these quantities into (3.1) and using Gauss' divergence theorem in the subdomains $\{D^{(n)};n=1,..,N\}$, of D where both sides are continuously differentiable, we end up with

$$(3.2)\quad \int_{t\in R} dt \int_{x\in D} v_k^B(\mathbf{x},t-\tau)[-\Delta_{kmpq}\partial_m\tau_{pq}^A(\mathbf{x},t) + \rho_{kr}^A(\mathbf{x})\partial_t v_r^A(\mathbf{x},t) - f_k^A(\mathbf{x},t)]dV$$
$$+ \int_{t\in R} dt \int_{x\in D} \tau_{ij}^B(\mathbf{x},t-\tau)[\Delta_{ijmr}\partial_m v_r^A(\mathbf{x},t) - s_{ijpq}^A(\mathbf{x})\partial_t\tau_{pq}^A(\mathbf{x},t)$$
$$- h_{ij}^A(\mathbf{x},t)]dV = 0.$$

Upon taking the functions $\tau_{ij}^B = 0$ (and hence $f_r^B = 0$) throughout $D \times R$ and $v_k^B \neq 0$, Equation (3.2) represents the weighted form of the equation of motion (2.1) over the space-time domain $D \times R$ with the arbitrary weighting function v_k^B, while if we choose the functions $v_k^B = 0$ (and hence $h_{pq}^B = 0$) throughout $D \times R$ and $\tau_{ij}^B \neq 0$, Equation (3.2) is nothing but the weighted form of the equation of deformation rate (2.2) over the space-time domain $D \times R$ with the arbitrary weighting function τ^B_{ij}.

Formulation (3.2) is used to set up a space-time finite-element method. In this formulation the A-field is identified with the actual wave field and the B-field with the weighting field. In view of the time-invariance of D, the space-time domain over which the finite-element method is applied, is discretized into a union of elementary subdomains that are cylindrical in the time direction. In these subdomains local functions are defined that are the product of a function of the spatial variables and a function of time. The local functions are subsequently combined to global expansion functions that are used in the approximation of the wave field.

When taking a reciprocity relation as the point of departure of setting up a numerical scheme, it seems more or less natural to treat the States A and B in an equivalent manner, which implies that each specimen of the sequences of functions into which State A is expanded are also taken as a specimen of the State B. As far as (3.2) is concerned, this implies that the sequence of weighting functions is taken to be the same as the sequence of expansion functions.

In the application of the finite-element method we further take the elastodynamic properties of the medium and the known volume sources to be constant in each elementary subdomain. The surface sources are taken to be piecewise constant in the elementary subdomains of the discretized outer boundary of computation.

4. THE EXPANSION FUNCTIONS

Our expansion functions that are used to locally represent the elastodynamic wave field quantities in an elementary subdomain of the configuration are the product of a function of the spatial coordinates and a function of time. In our discussion we shall concentrate on the expansion functions of spatial coordinates because in the spatial direction strong inhomogeneities may occur. We take polynomial expansion functions of degree one which are the lowest-degree polynomials by which physically non-existing surface source distributions on interfaces of discontinuity can numerically be avoided in the representations.

To obtain these linear interpolation functions, the spatial domain is discretized into a union of tetrahedra, the vertices of which coincide with the nodes of the discretization. The position vectors of the four vertices $\{P(0), P(1), P(2), P(3)\}$ of T are denoted by $\{x_i(0), x_i(1), x_i(2), x_i(3)\}$. The vectorial areas $\{A_i(0), A_i(1), A_i(2), A_i(3)\}$ of the faces of T that are directed along the outward normals to the faces of T are given by

(4.1) $A_i(0) = \varepsilon_{ijk}[x_j(1)x_k(2) + x_j(2)x_k(3) + x_j(3)x_k(1)]/2$, etc.

where ε_{ijk} is the completely antisymmetric unit tensor of rank three (Levi-Civita tensor). The volume of T is given by

(4.2) $V = \varepsilon_{ijk}[- x_i(0)x_j(1)x_k(2) + x_i(1)x_j(2)x_k(3) - x_i(2)x_j(3)x_k(0)$

$+ x_i(3)x_j(0)x_k(1)]/6.$

Let the position vector b_i of the barycenter of T be introduced through

(4.3) $b_i = (1/4)\sum_{I=0}^{3} x_i(I)$,

then the linear scalar interpolation function that equals unity when $x_i = x_i(I)$ for I=0,1,2,3, and equals zero in the remaining three vertices of T can be written as

(4.4) $\phi_i(I;x_i) = 1/4 - (3V)^{-1}(x_j-b_j)A_j(I)$.

Since $x_i = \sum_{I=0}^{3}x_i(I)\phi_i(I;x_i)$, with $\sum_{I=0}^{3}\phi(I;x_i) = 1$, the functions $\{\phi(I;x_i);$ $I=0,1,2,3\}$ are nothing but the barycentric coordinates in T.

For the local spatial interpolation of the elastodynamic wave field, we let ourselves guide by the same arguments as that have been used in [2] for the representation of three-dimensional electromagnetic fields, i.e., we want functions that guarantee the continuity of all field-components that are continuous across an interface, while leaving the non-continuous components free to jump by finite amounts. In the realm of elastodynamic wave fields this implies that we construct functions that automatically guarantee the continuity of all components of the particle velocity and the continuity of the normal components of the stress (i.e., the traction) across solid/solid interfaces of discontinuity, while it leaves the tangential components of the stress free to jump by finite amounts.

The local functions $\phi(I;x_i)$ are employed to interpolate the particle velocity v_r and the stress τ_{pq} in a tetrahedron. Let $\{v_r(I;t),\tau_{pq}(I;t);$ $I = 0,1,2,3\}$ denote the values of $\{v_r,\tau_{pq}\}$ when $x_i = x_i(I)$ is approached via the interior of T, then the local representations of $\{v_r,\tau_{pq}\}$ are

(4.5) $\{v_r(x_i;t),\tau_{pq}(x_i;t)\} = \sum_{I=0}^{3} \{v_r(I;t),\tau_r(I;t)\}\phi(I;x_i)$ for $x_i \in T$.

In case the face $A_i(J)$ of T coincides with a solid/solid interface, face interpolation is used for $\tau_{pq}(I;t)$, with $I \neq J$. We represent $\tau_{pq}(I;t)$ with respect to the local basis $\{-(3V)^{-1}[x_i(J)-x_i(I)]\}$, with $J \neq I$ as follows:

(4.6) $\tau_{pq}(I;t) = -(3V)^{-1}\sum_{J=0}^{3} T_p(I,J;t)[x_q(J)-x_q(I)]$,

in which $T_p(I,J;t) = \tau_{pq}(I;t)A_q(J)$. Apart from a factor, $T_p(I,J;t)$ is the traction on $A_q(J)$, with $J \neq I$, in $x_i(I)$.

For the global representation of the particle velocity we use nodal interpolation throughout D, with vector components along the axes of the background reference frame. With this the continuity of all components of the particle velocity across solid/solid interfaces is automatically guaranteed. For the global representation of the stress nodal interpolation is used in the subdomains of D where the medium properties vary continuously with position; near solid/solid interfaces, face interpolation is employed. In this way the continuity of the traction across solid/solid interfaces is automatically guaranteed, while the tangential components of the stress are free to jump by finite amounts.

5. CONCLUSIONS

The theory of a space-time finite-element method for the numerical computation of three-dimensional elastodynamic wave motions in bounded, time-invariant configurations that may be anisotropic and strongly inhomogeneous, is presented. It is shown that the finite-element method can be considered to be based on a space-time reciprocity theorem of the time-correlation type. Linear local spatial expansion functions for the representation of the particle velocity and the stress in an elementary subdomain of a discretized geometry are constructed that automatically guarantee the continuity requirements at interfaces of discontinuity in material properties in an elastic configuration while leaving non-continuous elastodynamic field components free to jump by finite amounts.

REFERENCES

[1] De Hoop, A.T. and Stam, H.J., "Elastodynamic time-domain reciprocity theorems for solids with relaxation", Proceedings of the IUTAM Symposium on ELASTIC WAVE PROPAGATION, Galway, Ireland, 20-25 March 1988, Edited by: McCarthy/Hayes.

[2] Mur, G. and de Hoop, A.T., IEEE Transactions on Magnetics (1985), vol. MAG-21, no. 6, pp 2188-2191.

F

GENERAL THEORY OF ELASTIC WAVE PROPAGATION

ELASTIC WAVE PROPAGATION
M.F. McCarthy, M.A. Hayes, (Editors)

GAUSSIAN WAVE PACKETS IN LINEAR AND NONLINEAR ANISOTROPIC ELASTIC SOLIDS

Andrew N. Norris*

Department of Mechanics and Materials Science
Rutgers University
P.O. Box 909
Piscataway, NJ 08855, USA

Gaussian wave packets are high frequency, asymptotic solutions to the equations of elastodynamics. They can be used, for example, to model pulse propagation in complex materials with smoothly varying properties and sharp surfaces of material discontinuity. The fundamental departure from the usual geometrical optics development is that the phase function is assumed to be complex valued. This has important consequences for the behaviour of the solution in the neighbourhood of the unique central ray. For example, if the initial disturbance is in the shape of a gaussian envelope, the propagated pulse remains gaussian. Nonlinear effects are taken into account by assuming the strains remain small, so that weakly nonlinear wave theory can be used. A nonlinear phase modulation equation is derived; and solved for an initial disturbance corresponding to an acceleration wave. This example illustrates that one can obtain a much richer theory through the use of complex phase.

1. INTRODUCTION

A pulse of very short wavelength propagates in many respects like a particle whose centre follows the Fermat geodesics of the medium. If the disturbance is in the form of a travelling wavefront, then the amplitude is governed by the stretching or contraction of the transverse wavefront, which in turn depends upon the curvature. Problems arise at points in space where rays cross, such as caustics or foci. The curvature becomes unbounded and boundary layer corrections are required to make the solution uniform. However, if the initial disturbance is localized in the transverse direction, it may be modelled using a complex valued phase function. The corresponding radius of curvature is also complex, and never becomes zero. There are thus no caustics, obviating the need for boundary layer corrections.

This paper considers high frequency, localized solutions in smoothly varying elastic media. Most of the development, whether the medium is linear or non-linear, is concerned with the consequences of the eiconal equation for complex phase. The idea of complex phase is central to complex point source solutions [1,2] and gaussian beam summation methods [3,4]. Gaussian wave packets are a particular type of solution in which the initial pulse is in the form of gaussian envelope centered about one point, so that the phase is explicitly complex. Previous papers have considered GWPs of infinitesimal amplitude in acoustic [5] and isotropically elastic [6] media. Homogeneous anisotropic media were considered in [7]. In all these cases, the amplitude of the GWP

*This work was supported by the National Science Foundation, through grant number MDM 85-16256.

is nonsingular, so that no caustic corrections are required. Thus, GWPs are guaranteed to remain localized and bounded. References [5-7] also discuss the scattering of GWPs at smooth interfaces of material discontinuity. For a single incident GWP, several reflected and transmitted GWPs are generated.

It is also possible to consider nonlinear effects by appropriately scaling the amplitude of the displacement [8]. The strain is assumed small, although the acceleration can be of order unity. The nonlinearity causes a phase modulation in the same way that one would expect from weakly nonlinear geometrical optics [9,10]. It turns out that the modulation is always real, even though the phase is complex valued. Explicit results can be obtained for a particular type of initial condition, analogous to an acceleration wave, and the possibility of shock waves is discussed. We begin with the equations of motion.

2. EQUATIONS OF MOTION AND THE ASYMPTOTIC EXPANSION

Consider a smoothly varying material with strain energy function

$$W = \frac{1}{2} C_{ijk\ell}(\mathbf{X})E_{ij}E_{k\ell} + \frac{1}{6} C_{ijk\ell mn}(\mathbf{X}) E_{ij}E_{k\ell}E_{mn} + \cdots \tag{2.1}$$

Here $\mathbf{E}$ is the Cauchy-Green strain tensor, $\mathbf{X}$ is the material position vector, and the inhomogeneous nature of the medium is reflected in the dependence of the moduli upon $\mathbf{X}$. The equations of motion are

$$\rho u_{i,tt} = (C_{ijk\ell}\, u_{k,\ell})_{,j} + S_{ijk\ell mn}\, u_{m,n}\, u_{k,j\ell} + \cdots, \tag{2.2}$$

where $\mathbf{u}(\mathbf{X},t)$ is the displacement, and [8]

$$S_{ijk\ell mn} = C_{ijk\ell mn} + C_{mnj\ell}\delta_{ik} + C_{ijn\ell}\delta_{km} + C_{njk\ell}\delta_{im} \ . \tag{2.3}$$

Equation (2.2) reduces in the absence of the second term on the right hand side to the equations of linear elasticity in a smoothly varying medium.

Let L be some characteristic length associated with the problem, and $\varepsilon << 1$ a small positive number. The displacement $\mathbf{u}$ is assumed to be of order $\varepsilon^2 L$ and the strain of order ε. Let T be a characteristic time, for example $T = L/c$, where c is the wave propagation speed. Then the acceleration is assumed to be of order (L/T^2), i.e. of order unity with respect to the asymptotic parameter ε. This scaling is implicit if

$$\mathbf{u} = \varepsilon^2 L\, \mathbf{U}\, (\mathbf{X},t,\theta) + \text{c.c.} \ , \tag{2.4}$$

where $\mathbf{U}$ is a dimensionless complex vector function and c.c. denotes the complex conjugate. The fast phase function is assumed to be

$$\theta(\mathbf{X},t) = \frac{1}{\varepsilon T}\, \phi(\mathbf{X},t) \ , \tag{2.5}$$

and both θ and ϕ are complex valued. Note that $(\varepsilon T)^{-1}$ has the dimensions of frequency, and can be thought of as the centre frequency of the high frequency solution. Finally, we assume asymptotic expansions

$$\phi = \phi^{(0)}(\mathbf{X},t) + \varepsilon\, \phi^{(1)}(\mathbf{X},t) \ , \quad \mathbf{U} = \mathbf{U}^{(0)}(\mathbf{X},t,\theta) + \varepsilon\, \mathbf{U}^{(1)}(\mathbf{X},t,\theta) + \cdots \ . \tag{2.6}$$

3. THE EICONAL EQUATION AND ITS CONSEQUENCES

3.1. The Slowness Surface

The first in the sequence of equations obtained from eqs. (2.2) and (2.4)-(2.6), is of order unity,

$$(C_{ijk\ell}\phi_j^{(0)}\phi_\ell^{(0)} - \rho\phi_t^{(0)^2}\delta_{ik})U_{k,\theta\theta}^{(0)} = 0. \tag{3.1}$$

In order to simplify the order (ε) transport equation later, it is necessary to assume $\mathbf{U}^{(0)}$ is of the form

$$\mathbf{U}^{(0)}(\mathbf{X},t,\theta) = B(\mathbf{X},t,\theta)\ \mathbf{q}(\mathbf{X},t)\quad , \tag{3.2}$$

where $\mathbf{q}$ is the unit eigenvector of $C_{ijk\ell}\phi_j^{(0)}\phi_\ell^{(0)}(\mathbf{X},t)$. This corresponds to a single mode solution. Multiple mode solutions, and their interaction, are discussed in [8]. Thus, $\phi_t^{(0)}$ satisfies the secular equation

$$\det(C_{ijk\ell}\ p_j p_\ell - \rho\phi_t^{(0)^2}\delta_{ik}) = 0\ , \tag{3.3}$$

where

$$\mathbf{p} = \nabla\phi^{(0)}\ . \tag{3.4}$$

The roots of eq. (3.3) for a given $\phi_t^{(0)}$ describe the three sheets of the slowness surface. Let the one of interest be

$$\phi_t^{(0)} + f(\mathbf{p},\mathbf{X}) = 0\quad . \tag{3.5}$$

3.2. The Ray Equations

The eiconal equation (3.5) is a first order p.d.e. which may be solved along its characteristic curves, $t = \bar{t}(\tau)$, $\mathbf{X} = \bar{\mathbf{X}}(\tau)$, where τ is the ray parameter. Let an overdot denote the derivative along the ray, $d/d\tau$. Then $\dot{\bar{t}} = 1$, so that τ is the same as the elapsed time t on the ray. Also, $\phi_t^{(0)}$ remains constant on the ray, and is assumed to equal -1 for simplicity. Then

$$f(\mathbf{p},\mathbf{X}) = p\ v(\mathbf{X},\mathbf{n})\quad , \tag{3.6}$$

where $p = |\mathbf{p}|$, $\mathbf{n} = \mathbf{p}/p$ and v is the phase speed which solves $\det(C_{ijk\ell}n_j n_\ell - \rho v^2\delta_{ik}) = 0$. The vector $\mathbf{p}$ is the slowness vector for the ray, and the ray equations are

$$\dot{\bar{\mathbf{X}}}(t) = \frac{\partial f}{\partial \mathbf{p}} = \mathbf{c}(\bar{\mathbf{X}}(t),\mathbf{n}(t))\quad , \tag{3.7}$$

$$\dot{\mathbf{p}}(t) = -\frac{\partial f}{\partial \mathbf{X}} = -\frac{1}{v}\nabla v(\bar{\mathbf{X}}(t),\mathbf{n}(t))\ . \tag{3.8}$$

The vector $\mathbf{c}$ is the wave velocity $c_i = \rho^{-1}C_{ijk\ell}q_j q_k p_\ell$. It is constant along the ray in a homogeneous medium, so that the ray is a straight line in this particular case. Finally, the ray equation for $\phi^{(0)}$ implies that it remains constant in an arbitrarily inhomogeneous medium, i.e. $\phi^{(0)}(\bar{\mathbf{X}}(t),t) = \phi^{(0)}(\bar{\mathbf{X}}(0),0) \equiv \phi_o^{(0)}$.

3.3. The Derived Ray Equations

The ray equations permit a local expansion of the first order phase $\phi^{(0)}$ about a ray position $(\bar{\mathbf{X}}(t),t)$, specifically

$$\phi^{(0)}(\bar{\mathbf{X}}(t)+\Delta\mathbf{X},t+\Delta t) = \phi_0^{(0)} + (\mathbf{p}.\Delta\mathbf{X}-\Delta t) + O(\Delta^2) \ . \tag{3.9}$$

The phase behaves, to this order of approximation, like that of a plane progressing wave. Higher order corrections require the quantities $\nabla\nabla\phi^{(0)}$, $\nabla\phi_t^{(0)}$ and $\phi_{tt}^{(0)}$ on the ray. These follow from the variational equations corresponding to (3.7) and (3.8). The resulting equations, or derived ray equations after Hayes [11], have been discussed previously for an inhomogeneous isotropic medium [5,6], and for a homogeneous anisotropic medium [7].

Define the tensor $\mathbf{M}(t)$,

$$\mathbf{M}(t) = \nabla\nabla\phi^{(0)}(\bar{\mathbf{X}}(t),t) \ , \tag{3.10}$$

and its inverse

$$\mathbf{R}(t) = \mathbf{M}^{-1}(t) \ . \tag{3.11}$$

It can be deduced, using the methods of Section 4 below, that

$$\dot{\mathbf{R}}(t) = \mathbf{N} + \mathbf{RL} + \mathbf{L}^T\mathbf{R} + \mathbf{RHR} \ , \tag{3.12}$$

where

$$N_{ij} = \frac{\partial^2 f}{\partial p_i \partial p_j} \ , \quad L_{ij} = \frac{\partial c_j}{\partial X_i} \ , \quad H_{ij} = \frac{1}{v}\frac{\partial^2 v}{\partial X_i \partial X_j} \ . \tag{3.13}$$

The remaining second order derivates of $\phi^{(0)}$ follow by differentiating the eiconal eq.(3.5), using (3.6),

$$\nabla\phi_t^{(0)}(\bar{\mathbf{X}}(t),t) = -\mathbf{M}.\mathbf{c} - \frac{1}{v}\nabla v \ , \tag{3.14}$$

$$\phi_{tt}^{(0)}(\bar{\mathbf{X}}(t),t) = \mathbf{c}.\mathbf{M}.\mathbf{c} + \frac{1}{v}\,\mathbf{c}.\ \nabla v \ . \tag{3.15}$$

Combining these results, the local expansion of the phase $\phi^{(0)}$ about a ray position becomes

$$\phi^{(0)}(\bar{\mathbf{X}}(t)+\Delta\mathbf{X},t+\Delta t) = \phi_0^{(0)} + (\mathbf{p}.\Delta\mathbf{X} - \Delta t) + \frac{\Delta t}{v}\left(\frac{\Delta t}{2}\,\mathbf{c}-\Delta\mathbf{X}\right).\nabla v$$
$$+ \frac{1}{2}(\Delta\mathbf{X}-\mathbf{c}\Delta t).\mathbf{M}.(\Delta\mathbf{X}-\mathbf{c}\Delta t). \tag{3.16}$$

3.4. Discussion

The key ingredient in (3.16) is the tensor $\mathbf{M}(t)$, which is found by integrating the ordinary differential equation (3.12) in conjunction with (3.7) and (3.8), with initial conditions $\bar{\mathbf{X}}(0) = \mathbf{X}_0$, $\mathbf{p}(0) = \mathbf{p}_0$ and $\mathbf{M}(0) = \mathbf{M}_0$. The tensor $\mathbf{M}_0$ can in general be complex valued. In particular, if we specify that $\mathrm{Im}\mathbf{M}_0$ is positive definite, then it can be shown [5, Appendix B] that $\mathrm{Im}\mathbf{M}(t)$ is positive definite for all t. This in turn implies that the imaginary part of the variation of $\phi^{(0)}$ from its value on the ray is positive for all $(\Delta\mathbf{X},\Delta t)$.

The evolution of $\mathbf{M}(t)$ is particularly simple in a homogeneous medium, since then $\mathbf{L} = \mathbf{H} = 0$, and so

$$\mathbf{M}(t) = \mathbf{M}_0(I + t\mathbf{N}\mathbf{M}_0)^{-1} \tag{3.17}$$

The tensor $\mathbf{N}$ has been discussed in [7], where it is shown that $\mathbf{N}.\mathbf{p} = 0$. This latter property is true whether the medium is homogeneous or inhomogeneous. In an isotropic medium,

$$\mathbf{N} = c^2(I - \mathbf{n}\otimes\mathbf{n}) \ . \tag{3.18}$$

For example, an axially symmetric initial phase is defined by

$$\mathbf{M}_0 = a_1(\mathbf{I}-\mathbf{n}\otimes\mathbf{n}) + a_2\mathbf{n}\otimes\mathbf{n} \quad , \tag{3.19}$$

and therefore, in a homogeneous, isotropic medium,

$$\mathbf{M}(t) = \frac{a_1}{1+ta_1c^2}(\mathbf{I}-\mathbf{n}\otimes\mathbf{n}) + a_2\mathbf{n}\otimes\mathbf{n} \quad . \tag{3.20}$$

If a_1 is real and positive, corresponding to a locally spherical wavefront, the width of $\phi^{(0)}$ about its ray position increases with t. In general, we call **N** the spreading tensor since it governs the width of $\phi^{(0)}$ in the direction transverse to the phase direction in a homogeneous medium.

Note that **M**(t) in (3.20) becomes singular at some finite time if a_1 is real and negative. This corresponds to the focussing of a spherically converging wavefront. This type of geometrical singularity is removed by the initial condition that $\mathrm{Im}\mathbf{M}_0$ is positive definite. The initial phase $\phi^{(0)}(\mathbf{X},0)$ in the neighbourhood of $\mathbf{X}_0$ must then be of the form

$$\phi^{(0)}(\mathbf{X}_0+\Delta\mathbf{X},0) = \phi_0^{(0)} + \mathbf{p}_0.\Delta\mathbf{X} + \frac{1}{2}\Delta\mathbf{X}.\mathbf{M}_0.\Delta\mathbf{X} \quad . \tag{3.21}$$

Since $\mathbf{p}_0$ is real by assumption, the initial value of $\nabla\phi^{(0)}$ at points in the neighbourhood of $\mathbf{X}_0$ must be complex. The associated wave velocity vectors **c** are also complex, and so all rays emanating from the neighbourhood of $\mathbf{X}_0$ propagate into complex space. Ray congruences in real space, such as a focus, will not occur.

Finally, we note that the only known explicit solution of (3.12) for an inhomogeneous medium is for an isotropic material with constant velocity gradient [5].

4. THE GENERAL SOLUTION

4.1. The Transport Equation

So far we have only used the first in the sequence of asymptotic equations. The next one, of order ε, when multiplied by $U_{i,\theta}^{(0)}$ simplifies using (3.1), to

$$[C_{ijk\ell}\, U_{i,\theta}^{(0)}\, U_{k,\theta}^{(0)}\, p_j]_{,\ell} - \rho[U_{i,\theta}^{(0)}\, U_{i,\theta}^{(0)}\, \phi_t^{(0)}]_{,t}$$

$$+ \frac{1}{T}[C_{ijk\ell}p_j\phi_\ell^{(1)} - \rho\phi_t^{(0)}\phi_t^{(1)}\delta_{ik}][U_{i,\theta}^{(0)}\, U_{k,\theta}^{(0)}]_{,\theta}$$

$$+ \frac{L}{T^2} S_{ijk\ell mn}U_{i,\theta}^{(0)}\, U_{k,\theta\theta}^{(0)}\, p_j\, p_\ell\, p_n[U_{m,\theta}^{(0)} + \text{c.c.}] = 0 \quad . \tag{4.1}$$

This simplifies further along the ray $\bar{\mathbf{X}}(\tau)$ if we define

$$b(\tau,\theta) = B_{,\theta}(\bar{\mathbf{X}}(\tau),\tau,\theta) \quad . \tag{4.2}$$

Generalizing the procedure of [7,8], eq. (4.1) becomes

$$\frac{1}{\rho}\frac{\partial}{\partial\tau}(\rho b^2) + b^2\,\mathrm{Tr}[\mathbf{L}+\mathbf{NM}] + [\frac{\partial\phi^{(1)}}{\partial\tau} + \beta\,\mathrm{Re}\,b]\frac{\partial b^2}{\partial\theta} = 0 \quad , \tag{4.3}$$

where the real parameter β,

$$\beta = \frac{L}{T}\frac{1}{\rho} S_{ijk\ell mn}\, q_i q_k q_m p_j p_\ell p_n \quad , \tag{4.4}$$

is a dimensionless nonlinearity parameter.

The phase correction $\phi^{(1)}$ in eq. (4.3) is a function of the ray parameter and the total phase θ, i.e. $\phi^{(1)} = \phi^{(1)}(\tau,\theta)$ on the ray. The separate equation

$$\frac{\partial}{\partial\tau}(\rho b^2) + \rho b^2 \,\mathrm{Tr}[\mathbf{L}+\mathbf{NM}] = 0 \quad , \tag{4.5}$$

can be integrated as follows. Let **A** and **B** be tensors defined by

$$A_{ij}(t) = \frac{\partial X_i(t)}{\partial X_{oj}} \quad , \quad B_{ij}(t) = \frac{\partial p_i(t)}{\partial X_{oj}} \quad . \tag{4.6}$$

Evolution equations for **A** and **B** along a ray follow from (3.7), (3.8) and (3.13).

$$\dot{\mathbf{A}}(t) = \mathbf{L}^T\mathbf{A} + \mathbf{N}\,\mathbf{B} \quad , \tag{4.7}$$

$$\dot{\mathbf{B}}(t) = -\mathbf{H}\,\mathbf{A} - \mathbf{L}\,\mathbf{B} \ . \tag{4.8}$$

Equation (3.12) follows by noting that $\mathbf{R} = \mathbf{A}\,\mathbf{B}^{-1}$, and (4.5) simplifies using (4.7), to

$$\frac{\partial}{\partial\tau}(\rho b^2 \det \mathbf{A}) = 0 \quad . \tag{4.9}$$

This may be integrated directly to give b in terms of **A**. However, it is preferable to write everything in terms of the basic tensor **M**. Using det **A** = det **R** det **B**, and rewriting (4.8) as

$$\frac{d}{d\tau} \log \det \mathbf{B} = -\nabla.\mathbf{c} - \mathrm{Tr}(\mathbf{H}\,\mathbf{R}) \quad , \tag{4.10}$$

we obtain

$$b(\tau,\theta) = b(0,\theta)\, g(\tau) \quad , \tag{4.11}$$

where

$$g(t) = \left\{\frac{[c\rho^{-1}\det \mathbf{M}](t)}{[c\rho^{-1}\det \mathbf{M}](0)}\right\}^{1/2} \exp[\tfrac{1}{2}\int_0^t \mathrm{Tr}(\mathbf{H}\,\mathbf{R})\,dt] \ . \tag{4.12}$$

The remainder of (4.3) can then be integrated directly to give

$$\phi^{(1)}(\tau,\theta) = -\ \mathrm{Re}[b(0,\theta)\int_0^\tau \beta\, g(t)\, dt], \tag{4.13}$$

where $\phi^{(1)}(0,\theta) \equiv 0$. The value of θ at $(\bar{\mathbf{X}}(\tau), \tau)$ then follows from (2.5),(2.6) and (4.13) as the solution of the implicit equation

$$\theta = \theta_o - \ \mathrm{Re}[b(0,\theta)\int_0^\tau \beta g(t)\, \frac{dt}{T}] \ , \tag{4.14}$$

where $\theta_o = \frac{1}{\varepsilon T}\ \phi_o^{(0)}$.

The integral in (4.14) can be performed explicitly for a homogeneous medium. Then the integral in (4.12) is zero, and **M**(t) follows from (3.17). The function $\det(\mathbf{I}+t\mathbf{NM}_0)$ is quadratic in t since **N** is of rank two. Thus,

$$\det[\mathbf{I} + t\mathbf{NM}(0)] = 1 + \gamma t + \frac{\eta}{2}t^2 \quad , \tag{4.15}$$

where

$$\gamma = \mathrm{Tr}\,\mathbf{N}\,\mathbf{M}(0) \quad , \tag{4.16}$$

$$\eta = e_{ijk}\, e_{pqr}\, n_i n_p\, N_{j\ell}\, N_{km}\, M_{\ell q}(0)\, M_{mr}(0) \quad , \tag{4.17}$$

and e_{ijk} is the third order alternating tensor. Therefore,

$$\int_0^t g(t')dt' = \begin{cases} t & \eta = 0, \quad \delta = 0\ , \\ \frac{2}{\gamma}[(1+\gamma t)^{1/2}-1] & \eta = 0, \quad \delta \neq 0\ , \\ \frac{2}{\gamma}\log[1+\frac{\gamma t}{2}] & \eta \neq 0, \quad \delta = 0\ , \\ (\frac{2}{\eta})^{1/2}[\mathrm{ch}^{-1}(\frac{\gamma+\eta t}{\delta})-\mathrm{ch}^{-1}(\frac{\gamma}{\delta})] & \eta \neq 0, \quad \delta \neq 0\ , \end{cases} \tag{4.18}$$

where $\delta = (\gamma^2-2\eta)^{1/2}$. The case of $\eta = \delta = 0$ corresponds to a plane wavefront; $\eta = 0$, $\delta \neq 0$ when one principal wavefront curvature equals zero, i.e., a cylindrical wavefront; $\delta = 0$, $\eta \neq 0$ corresponds to equal principal curvatures or a spherical wavefront, and $\delta \neq 0$, $\eta \neq 0$ means the principal curvatures are distinct.

4.2. Discussion

The general solution now follows from (2.4), (3.2), with assumed initial amplitude of the form

$$B(\mathbf{X},0,\theta) = B^{(0)}(\mathbf{X},\theta)\ , \tag{4.19}$$

with $\theta(\mathbf{X},0)$ given by (2.5), (2.6.1) and (3.21), and $\phi^{(1)}(\mathbf{X},0) \equiv 0$. Using (4.2) and (4.11), we have

$$B(\bar{\mathbf{X}}(\tau),\tau,\theta) = B^{(0)}(\mathbf{X}_0,\theta)\ g(\tau)\ , \tag{4.20}$$

where θ must be determined from (4.14). Equations (4.20), with (3.2) where $\mathbf{q}(\bar{\mathbf{X}}(\tau),\tau) \equiv \mathbf{q}(\tau)$ is the eigenvector of $C_{ijk\ell}(\bar{\mathbf{X}}(\tau))p_j p_\ell(\tau)$, define the field right at the ray position $\bar{\mathbf{X}}(\tau)$ at time τ. The solution at neighbouring points follows by approximating the phase θ using (3.16). The correction to $\phi^{(1)}$ is assumed to be of higher order. This non-local ray solution is based upon the amplitude and displacement direction for the central ray $\bar{\mathbf{X}}(\tau)$, but uses a quadratic, or paraxial, approximation to the phase at neighbouring points. These simultaneous approximations are consistent in the sense that any further corrections to the amplitude or phase requires solving for higher order derivatives of the phase along the ray. Thus, Norris [6] discusses the procedure for (1) locally expanding the amplitude about the ray, (2) obtaining a higher order Taylor series for $\phi^{(0)}$, and (3) deriving higher order terms $\mathbf{U}^{(n)}$ in (2.6). All three problems are inter-related, and very quickly lead to exceedingly complicated systems of equations. It is not even clear, a priori, whether this would be a worthwhile endeavour to undertake. Thus, in the present theory, an initial $B^{(0)}$ in (4.19) of the form $B^{(0)} = A\exp(i\theta)$, with $\mathrm{Im}\mathbf{M}_0$ positive definite, is guaranteed to remain in the form of a localized solution, i.e. it decays exponentially away from the ray position. If we use a third order Taylor series for $\phi^{(0)}$, the quantity $\nabla\nabla\nabla\phi^{(0)}(\bar{\mathbf{X}}(t),t)$ will inevitably develop an imaginary component, so that $\exp(i\theta)$ will become unbounded in some directions. We have verified this by deriving and solving the necessary ray equations, analogous to (4.7) and (4.8), for a homogeneous, isotropic medium.

The present type of solution is therefore very well suited to gaussian wave packet disturbances of the form $B^{(0)} = A\exp(i\theta)$. It can also be shown [5], that these GWP solutions satisfy an energy conservation relation of the form

$$\int \rho(\mathbf{X}(\tau))\ |\mathbf{U}^{(0)}(\bar{\mathbf{X}}(\tau),\tau,\theta(\bar{\mathbf{X}}(\tau)+\mathbf{X},\tau))|^2 d\mathbf{X} = \text{constant}, \tag{4.21}$$

independent of τ.

4.3. Infinitesimal Gaussian Wave Packets

The nonlinear phase modulation equation (4.14) can be solved trivially if $\beta \equiv 0$, as is the case for a transverse wave in an isotropic solid [8]. This result for the transverse wave is in agreement with the known result that transverse acceleration waves do not develop shocks [12]. However, the parameter β is generally non-zero, even if the medium is linearly elastic, i.e. if $C_{ijk\ell mn} \equiv 0$ in (2.1). One way to simplify (4.14) is to assume a different scaling for the displacement, specifically $\mathbf{u} = o(\varepsilon^2 L)$, with the same form for the phase function (2.5). The analysis for the evolution of $\phi^{(0)}$ proceeds as before, but the transport equation differs from (4.3) in having no term with β. Therefore $\phi^{(1)} \equiv 0$, and so $\theta \equiv \theta_0$. In particular, if the initial disturbance is of GWP form with $\mathbf{X}_0 = 0$, and

$$\mathbf{u}(\mathbf{X},0) = U_0\mathbf{q}\exp[i\omega(\mathbf{p}_0.\mathbf{X} + \tfrac{1}{2}\mathbf{X}.\mathbf{M}_0.\mathbf{X})] + \text{c.c.} , \tag{4.22}$$

where $\omega \equiv 1/\varepsilon T$ and $U_0 = o(\varepsilon^2 L)$, then the subsequent solution can be approximated about the ray position $\bar{\mathbf{X}}(t)$. Thus, by (3.16) and (4.12),

$$\mathbf{u}(\mathbf{X},t) \sim U_0\mathbf{q}(t)\, g(t)\, \exp\{i\omega[\phi^{(0)}(\mathbf{X},t) - \phi_0^{(0)}]\} + \text{c.c.} , \tag{4.23}$$

where $\mathbf{q}(t)$ is the unit eigenvector at $\bar{\mathbf{X}}(t)$.

By assumption, $\text{Im}\mathbf{M}_0$ is positive definite. Therefore, the evolved GWP at time t is localized about $\bar{\mathbf{X}}(t)$. The GWP will be in the form of an oscillating signal of frequency ω, modulated by a gaussian envelope. The shape of the envelope, and hence the extent of the GWP, depends upon $\mathbf{M}(t)$. For example, in a homogeneous medium, $\mathbf{M}(t)$ is given explicitly in (3.17), and the GWP becomes simply

$$\mathbf{u}(\mathbf{X},t) = U_0\mathbf{q}\left[\frac{\det\mathbf{M}(t)}{\det\mathbf{M}_0}\right]^{\frac{1}{2}}\exp\{i\omega[\mathbf{p}_0.(\mathbf{X}-\bar{\mathbf{X}})+\tfrac{1}{2}(\mathbf{X}-\bar{\mathbf{X}}).\mathbf{M}(t).(\mathbf{X}-\bar{\mathbf{X}})]\} + \text{c.c.} \tag{4.24}$$

4.4. Gauss-Hermite Wave Packets

Infinitesimal initial disturbances of the following type can also be considered,

$$\mathbf{u}^H(\mathbf{X},0) = [1 + A^{(1)}_{ij}X_iX_j + A^{(2)}_{ijk\ell}X_iX_jX_kX_\ell + \ldots\ldots]\mathbf{u}^c(\mathbf{X},0) + \text{c.c.} , \tag{4.25}$$

where $\mathbf{A}^{(1)}$, $\mathbf{A}^{(2)}$ ... are constant and symmetric and $\mathbf{u}^c(\mathbf{X},0)$ is the complex valued first term on the right hand side of (4.22), i.e. $\mathbf{u} = 2\,\text{Re}\,\mathbf{u}^c$. One can think of (4.25) as being derived from (4.22) by differentiating with respect to the elements of $\mathbf{M}_0$. The evolved form of (4.25) at time t, then follows from (4.23) as

$$\mathbf{u}^H(\mathbf{X},t)=[1+\frac{1}{i\omega}A^{(1)}_{ij}\frac{\partial}{\partial M_{ij}(0)} + \frac{1}{(i\omega)^2}A^{(2)}_{ijk\ell}\frac{\partial}{\partial M_{ij}(0)\partial M_{k\ell}(0)} +..]\mathbf{u}^c(\mathbf{X},t)+\text{c.c.} \tag{4.26}$$

The major contribution to the derivatives of $\mathbf{u}^c(\mathbf{X},t)$ comes from the exponential term. Explicit differentiation gives

$$\mathbf{u}^H(\mathbf{X},t)\sim\mathbf{u}^c(\mathbf{X},t)\,\{1+y_iy_jV_{ik}V_{j\ell}[A^{(1)}_{k\ell}+ (y_py_qV_{pm}V_{qn}A^{(2)}_{k\ell mn} +\ldots)]\} + \text{c.c.} , \tag{4.27}$$

where $\mathbf{y} = \mathbf{X}-\bar{\mathbf{X}}(t)$, and in a homogeneous medium,

$$\mathbf{V}(t) = (\mathbf{I} + t\mathbf{N}\mathbf{M}_0)^{-1} . \tag{4.28}$$

5. APPLICATIONS OF INFINITESIMAL GWPs TO PIECEWISE HOMOGENEOUS MEDIA

Suppose that the initial GWP is axially symmetric, with $\mathbf{M}_0$ given by (3.19), and $\mathbf{n}$ in the X_3 direction. Choose the X_1 and X_2 directions as the principle directions of $\mathbf{N}$, so that $\mathbf{N}$ = diag $(S_1, S_2, 0)$ where S_1 and S_2 have dimensions of (speed)^2. In general, $S_1 \neq S_2$, although $S_1 = S_2 = c^2$ in an isotropic medium. It follows from (3.17), (3.19) and (4.24) that

$$\mathbf{u}(\mathbf{X},t) = \frac{\mathbf{q}\, U_0}{(1+a_1S_1t)^{\frac{1}{2}}(1+a_1S_2t)^{\frac{1}{2}}} \exp\{i\omega[(\frac{X_3}{v} - t) + \frac{1}{2}\frac{a_1}{1+a_1S_1t}(X_1-c_1t)^2 + \frac{1}{2}\frac{a_1}{1+a_1S_2t}(X_2-c_2t)^2 + \frac{1}{2}a_2(X_3-vt)^2]\} + \text{c.c.}, \quad (5.1)$$

where we have used the fact that $\mathbf{n.c} = v$ [7]. The centre of the GWP is at $\bar{\mathbf{X}}(t) = \mathbf{c}t$, but the lateral spreading is in the X_1 and X_2 directions, with rates depending upon S_1 and S_2, respectively.

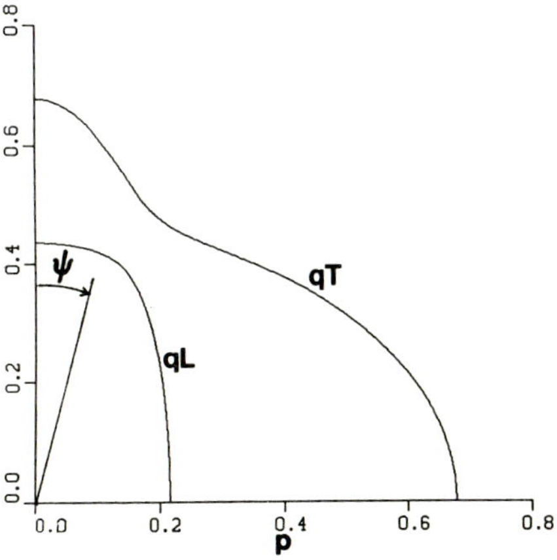

Fig. 1 Slowness surfaces for a transversely isotropic solid approximating a single ply, 60% glass fibre/epoxy composite. The fibre direction is horizontal, and only the in-plane modes are shown. The additional qTH surface is an ellipse. Units are μs/mm. From [7].

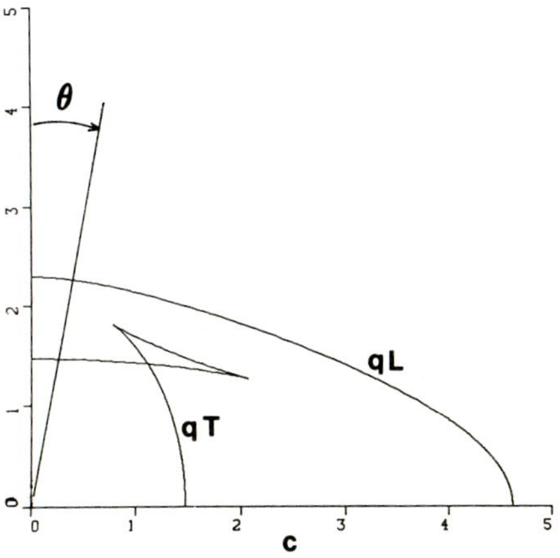

Fig. 2 The wave velocity surfaces associated with Figure 1. Units are mm/μs. From [7].

The factors S_1 and S_2 can vary considerably in an anisotropic medium. For example, a composite material composed of aligned glass fibres in an epoxy matrix can be considered as an effectively homogeneous transversely isotropic material. This effective medium approximation is valid if the fibre spacing is much shorter than the wavelength. Figures 1 and 2 show the section of the qL and qT slowness and wave surfaces in a plane containing the fibre axis. The associated spreading factor, say S_1, is shown in Figure 3 as a function of the wave velocity direction. Note that S_1 can be negative for the qT mode. This phenomenon occurs whenever the wave surface possesses cusps, and the present theory is still valid as long as $S_1 \neq 0$. When $S_1 < 0$, a pulse that would normally broaden will become narrower, and vice versa. This aspect is discussed further in [7]. The bulbous character of the qL curve in Figure 3 means that a GWP propagating in a direction about 45° to the fibre axis will spread much more rapidly than a pulse propagating parallel or perpendicular to the fibres. In fact, the spreading in the fibre direction is very small, roughly one tenth the value estimated from the wave speed using the isotropic spreading rate $S = c^2$. The low spreading rate indicates a channelling effect in the fibre direction. Some numerical examples are given in [7] of GWPs travelling in different directions, illustrating graphically the directional variation in the spreading.

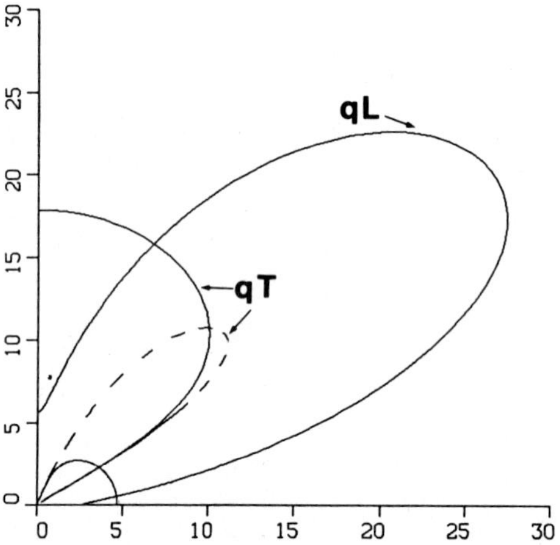

Fig. 3 The spreading factor S_1 for the slowness surfaces in Fig. 1, plotted as a function of the wave velocity direction. The dashed part of the qT curve is where S_1 is negative. Units are $mm^2/\mu s^2$. From [7].

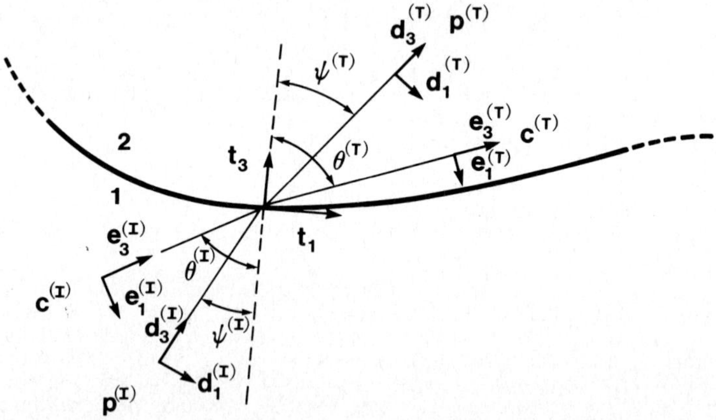

Fig. 4 The geometrical parameters for an incident and transmitted GWP at an interface between two dissimilar media.

The same reference [7], discusses in great detail the transmission and reflection of GWPs from smooth interfaces of material discontinuity. The continuity of traction and displacement at the interface produces new reflected and transmitted GWPs. Figure 4 illustrates the various geometrical parameters that enter into the problem for the incident (I) and a single transmitted (T) GWP. The **p** vector for each new GWP follows from Snell's law, and the initial amplitudes are determined by the usual reflection/transmission equations for a flat interface. The initial values of the new **M** tensors can be related to the incident **M** by locally matching the phases of each GWP near the point of incidence of the central ray. The equations for the new **M** tensors also include terms depending upon the interface curvature. Despite the complexity of these relations, it can be proved that the new **M** tensors are symmetric, and have positive definite imaginary part. Thus, the new GWPs are localized.

Figures 5 to 7 illustrate the type of solution obtained using the GWP theory plus the transmission/reflection relations. An axially symmetric GWP originates in water at a distance of 15mm from the smooth surface of the 60% fibre epoxy composite of Figures 1 to 3. The composite occupies $Y > 0$, and the X axis is the fibre direction. The incident direction is in the XY plane and is the only parameter varied in Figures 5 to 7, which show the envelope of both the qL and qT disturbances in the plane of XY. The envelope is defined as the sum of the two real quantities $\mathrm{Re}\{\mathbf{q}.\mathbf{u}\exp[-i\omega\mathbf{p}.(\mathbf{X}-\bar{\mathbf{X}}(t))]\}$ for each mode. This removes the high frequency oscillation in the **p** direction, making it easier to see the actual shapes.

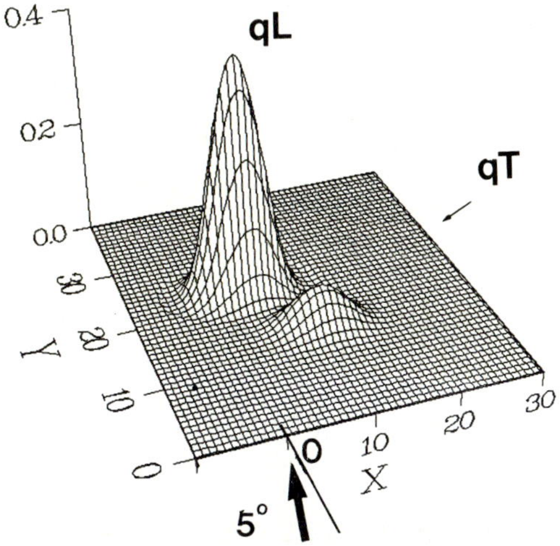

Fig. 5 The transmitted qL and qT GWPs for a GWP incident from water upon the fibre/epoxy composite. The interface is Y=0, the fibre axis is the X axis, the unit of length is mm, angle of incidence 5°, at t=10 μsec after the instant the incident GWP is at (0,0).

Note from Figure 5 that the qT GWP is further away from the normal direction, i.e. the Y axis. Such a state of affairs is impossible in an isotropic medium, but correlates with the data in Figures 1 and 2. Similarly, for 10° incident angle in Figure 6, the qT direction of propagation is further away from the normal than the qL direction. Also, the different rates of spreading are evident by comparison with Figure 5. Considerably greater spreading rates are apparent in Figure 7, where the incident angle is closer to the critical angle of 19°. Note, however, that there is a change in **M** due to transmission across the interface Y=0, so that Figure 3 alone is not sufficient to predict the transmitted spreading. Also, the direction of spreading is evident from the projected contours in Figures 6 and 7. Note in particular, that the qT GWP in Figure 6 does not spread in a direction orthogonal to its propagation direction.

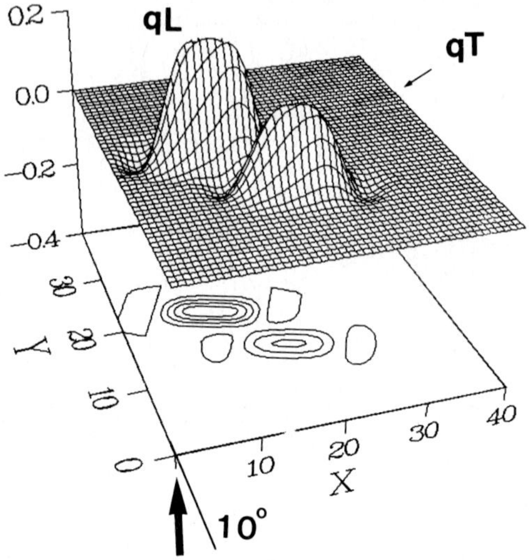

Fig. 6 The transmitted GWPs for incidence of 10°.

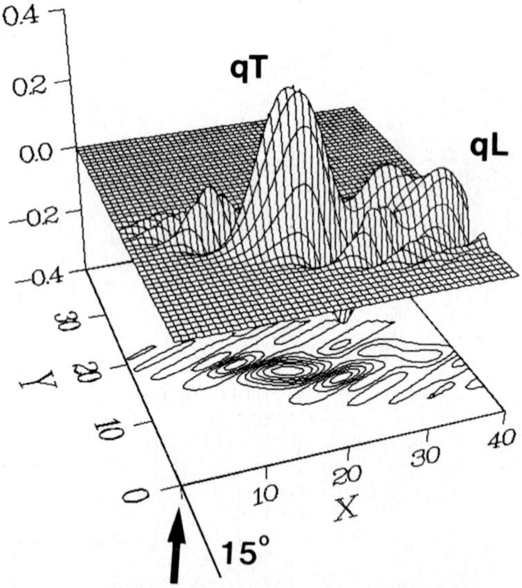

Fig. 7 The transmitted GWPs for 15° incident angle. Note that part of the qL GWP is not shown.

6. FINITE AMPLITUDE SOLUTIONS

6.1. Acceleration Waves

The phase modulation equation (4.14) cannot be solved explicitly for an initial disturbance of gaussian form, i.e. $b(0,\theta) = A\, e^{i\theta}$, A complex. However, we note that the modulation $\theta-\theta_0$ is always real, independent of $b(0,\theta)$. Following [8], consider the initial profile defined by

$$b(0,\theta) = A\,\theta\, h_{-1}(-\theta) \quad , \tag{6.1}$$

where $A=A_1+iA_2$ and $h_{-1}(z)=0$ if Re $z<-1$, $h_{-1}(z)=1$ if Re $z>0$, and is smoothly varying for $-1\leq \text{Re } z\leq 0$.

The initial wavefront associated with (6.1) is the strip $0 \leq \text{Re}\,\theta \leq 1$. The region behind the wavefront is $\text{Re}\theta < 0$. Let the initial position $\mathbf{X}_0 = 0$ at $t=0$ be behind the wavefront, i.e., $\theta_0 < 0$. Then (4.14) can be solved to give

$$\theta = \theta_0\{1 + \text{Re}[A \int_0^\tau \beta\, g(t)\, \frac{dt}{T}]\}^{-1} \tag{6.2}$$

The full solution then follows from (3.2), (4.2), (4.11), (4.12), (6.1) and (6.2), and is valid for times before the first possible time when the numerator in (6.2) vanishes, indicating the onset of shocks [10]. In order to gain further insight into the possibility of a shock developing, consider an axially symmetric initial phase with $\mathbf{M}_0$ given by (3.19), and assume $\mathbf{N}$ is of isotropic form, i.e. (3.18). Then $g(t) = 1/(1+a_1c^2t)$, and the shock condition in a homogeneous medium is that there is a $t_S > 0$ for which

$$\beta\, \text{Re}[\frac{A}{a_1c^2T} \quad \log\,(1+a_1c^2t_S)] = -1 \quad . \tag{6.3}$$

Before considering (6.3), note that the acceleration associated with (6.2) is

$$\mathbf{a} = \frac{L}{T^2}\, \mathbf{q}\; 2\{1 + \text{Re}[A \int_0^\tau \beta\, g(t)\, \frac{dt}{T}]\}^{-1}\, \text{Re}[\frac{A_1+iA_2}{1+a_1c^2t}] + O(\varepsilon) \quad . \tag{6.4}$$

The sign of $\mathbf{a}.\mathbf{q}$ is initially determined by A_1, however, if $a_1 = i\alpha$, $\alpha > 0$, the sign ultimately depends upon A_2. For this choice of a_1, it follows [8] from (6.3) that a shock will always occur if $\beta A_2 < 0$. We therefore have the strange result that the condition for shock development is independent of the initial acceleration.

The explanation is that the acceleration can change sign. Thus, in a medium for which only compressive waves develop shocks, an initially expansive wave can become compressive, and thence form a shock. An alternative way of looking at it is that the initial strain depends upon $\text{Re}(A\theta)$. This becomes $(A_1\theta_0 - \alpha A_2r^2/2\varepsilon T)$ in a neighbourhood of $\mathbf{X}_0$ transverse to the direction $\mathbf{n}$. The strain at $\mathbf{X}_0$ depends upon A_1, but nearby, on the order of $\varepsilon^{1/2}L$ away, it depends upon A_2. Thus, the strain could be compressive at $\mathbf{X}_0$, but the surrounding area is expansive. This type of initial situation cannot be defined by a real phase function θ (i.e. a_1 real). The only way the acceleration could change sign for a real phase function is if the initial wavefront is converging ($a_1 < 0$). Then a focus causes a 180° phase change. The solution (6.1) with $a_1 = i\alpha$, $\alpha > 0$ corresponds to an initially plane wavefront, and subsequently a diverging wavefront, i.e. no focussing occurs. Thus, in the context of nonlinear geometrical optics, by allowing the phase to take on complex values, solutions can be obtained that are outside the purview of geometrical optics with real phase.

REFERENCES

[1] Deschamps, G.A., Electron. Lett. 7 (1971) 684.
[2] Norris, A.N., J. Opt. Soc. Am. A3 (1986) 2005.
[3] Cerveny , V., Popov, M.M. and Psencik, I., Geophys. J.R. Astr. Soc. 70 (1982) 109.
[4] White, B.S., Norris, A.N., Bayliss, A. and Burridge, R., Geophys. J.R. Astr. Soc. 89 (1987) 579.
[5] Norris, A.N., White, B.S. and Schrieffer, J.R., Proc. Roy. Soc. London A412 (1987) 93.
[6] Norris, A.N., Acta Mechanica, in print.
[7] Norris, A.N., Wave Motion 9 (1987) 509.
[8] Norris, A.N., submitted.
[9] Seymour, B.R. and Mortell, M.P., Nonlinear Acoustics, in: Nemat-Nasser, S., (ed), Mechanics Today, Vol. 2 (Pergamon Press, London, 1975) pp. 251-312.
[10] Hunter, J.K. and Keller, J.B., Comm. Pure Appl. Math. 36 (1983) 547.
[11] Hayes, W.D., Proc. Roy. Soc. London A320 (1970) 209.
[12] McCarthy, M.F., Singular Surfaces and Waves, in: Eringen, A.C., (ed), Continuum Physics, Vol. 2 (Academic, New York, 1975) pp. 450-523.

ELASTIC WAVE PROPAGATION
M.F. McCarthy, M.A. Hayes, (Editors)

AN ASYMPTOTIC DESCRIPTION OF AN ELASTODYNAMIC BEAM

John G. Harris

216 Talbot Laboratory
104 South Wright Street
Urbana, Illinois 61801, U.S.A.

INTRODUCTION

The purpose of this paper is to describe and illustrate the mathematical ways of representing highly directional wavefields or beams that are radiated by microwave frequency transducers. The complex source-point method and the parabolic wave approximation are discussed, the former being used, in part, to motivate the latter. The main conclusion is that the parabolic wave approximation is perhaps the most useful method of description.

This work has been motivated by an interest in transducers used for active nondestructive testing. Such transducers typically operate from 10 to 100 MHz. Therefore they have large apertures, when measured in wavelengths, and radiate highly directional wavefields.

1. A GAUSSIAN BEAM AND THE COMPLEX SOURCE-POINT METHOD

Deschamps [2] noted that if complex values are assigned to the coordinates of a monopole source (for a scalar wavefield) the resulting disturbance represented a Gaussian beam. This idea has been used by Pott and Harris [3,4] to investigate the reflection, transmission and anomalous scattering of a Gaussian beam at a fluid-solid interface.

Consider the wavefield excited by an acoustic monopole; namely,

$$\Phi = C_1 \exp(ikQ)/(4\pi Q) \tag{1.1}$$

where

$$Q = [(x_1 - q_1)^2 + (x_2 - q_2)^2 + (x_3 - q_3)^2]^{1/2} \tag{1.2}$$

k is the wave number and C_1 is a constant. The term Φ is the velocity potential. If we let the source coordinates (q_1,q_2,q_3) take the values $(0,0,ia)$, where a is a positive constant, choose the square root so that $\mathrm{Re}(Q) \geqslant 0$ and then approximate Q for small τ, Q becomes

$$Q = x_3(1 + \tau^2) - ia(1 - \tau^2) . \tag{1.3}$$

The parameter

$$\tau = r/[2(x_3^2 + a^2)]^{1/2} \tag{1.4}$$

where $r = (x_1^2 + x_2^2)^{1/2}$, and is called the paraxial parameter. The approximation that τ be small is called the paraxial approximation. Hence in the paraxial approximation, Eq. (1.1) becomes

$$\Phi = C_1[4\pi(x_3 - ia)]^{-1} \exp[ikx_3(1 + \tau^2) + ka(1 - \tau^2)] . \tag{1.5}$$

Equation (1.5) represents a wave that propagates mainly in the x_3 direction but spreads slightly as it does so. Moreover, it represents a wave whose geometrical decay is x_3^{-1} and whose cross-section is described by $\exp(-ka\tau^2)$.

Suppose that in Eq. (1.5), we set $x_3 = 0$. Then

$$\Phi = iC_1(4\pi a)^{-1}e^{ka}\exp(-kr^2/2a) \; . \tag{1.6}$$

This suggests that the beam is radiated from an aperture of radius $b = (2a/k)^{1/2}$. Assuming b is fixed and known,

$$a = kb^2/2 \; . \tag{1.7}$$

But this is the Rayleigh distance; that is, it is the distance over which an acoustic beam, excited by a rigid piston vibrating in an infinite, rigid baffle, remains well-collimated, and it is the distance at which it first begins to evolve into a spherical wave [1]. This strongly suggests that b is the underlying radial scale, while a is the underlying propagation scale.

Introduce the scaled coordinates $\alpha = r/b$ and $\beta = x_3/a$, where a and b are connected by Eq. (1.7), into the reduced wave equation, set $\Phi = \exp(ikx_3)k^{-2}$ $C(\alpha,\beta)$ and then set

$$C(\alpha,\beta) = \sum_0^\infty C^n(\alpha,\beta)\,(kb)^{-2n} \; . \tag{1.8}$$

This leads to a recursive system of partial differential equations; namely,

$$\frac{1}{\alpha}\frac{\partial}{\partial\alpha}\left(\alpha\frac{\partial C^n}{\partial\alpha}\right) + 4i\frac{\partial C^n}{\partial\beta} = -4\frac{\partial^2 C^{n-1}}{\partial\beta^2} \tag{1.9}$$

where C^{-1} is defined to be zero. The left-hand side is known as the parabolic approximation to the wave equation. Suppose that at $x_3 = 0$ we impose the particle velocity

$$\partial\Phi/\partial x_3 = U\exp[-(r/b)^2] \; . \tag{1.10}$$

Then the solution of Eq. (1.9), subject to Eq. (1.10), for n = 0, is

$$C^o = U[k(\beta - i)]^{-1}\exp[i\alpha^2/(\beta - i)] \; . \tag{1.11}$$

Apart from a multiplicative constant the Φ calculated from C^o is identical to that given by Eq. (1.5).

Much of the motivation for introducing the scaling used above and attempting the expansion Eq. (1.8) came from a study of a paper by Tjøtta and Tjøtta [5]. They used a similar approach to approximate the pressure field radiated by a baffled piston source. We call this wavefield a uniform beam and consider it next.

2. A UNIFORM BEAM

The Gaussian beam is rather special. It retains its Gaussian cross-section into the farfield because there are no edge-diffracted waves to distort it. The uniform beam, mentioned in the previous section, is more typical. It starts out as two distinct waves, a generalized plane wave emitted from the face of the aperture and an edge-diffracted wave, but at approximately the Rayleigh distance the two waves begin to merge so that in the farfield the beam has evolved into a spherical wave with a radiation pattern. If we examine the uniform beam further, we soon realize that there is a boundary layer in the neighborhood of the edge, there is a second one surrounding $r = b$ from approximately a wavelength in front of the aperture out to approximately the Rayleigh distance, and finally the axis of the beam is a caustic formed by the rays from the edge. Buchal and Keller [6] have studied this problem in considerable detail using the method of matched asymptotic expansions. However, the resulting expansions are somewhat involved when contrasted with that used for the Gaussian beam in the previous section. Here we shall explore the possibility of using this simpler expansion for the uniform beam.

Consider a problem in which the velocity potential Φ satisfies the reduced wave equation and at $x_3 = 0$ the boundary condition is

$$\partial\Phi/\partial x_3 = UH(1 - r/b) \tag{2.1}$$

where $H(x)$ is the Heaviside function. Using arguments similar to those explained in Born and Wolf [7] to describe the boundary diffraction wave, Φ can be represented as follows:

$$\Phi(\underset{\sim}{x}) = \Phi_g(\underset{\sim}{x}) + \frac{U}{2\pi ik}\int_0^{2\pi} \frac{b}{R}\,\frac{\hat{n}\cdot\hat{R}}{1 - (\hat{x}_3\cdot\hat{R})^2}\, e^{ikR}\, d\phi' \tag{2.2}$$

and

$$\Phi_g(\underset{\sim}{x}) = UH(1 - r/b)e^{ikx_3}/ik\ . \tag{2.3}$$

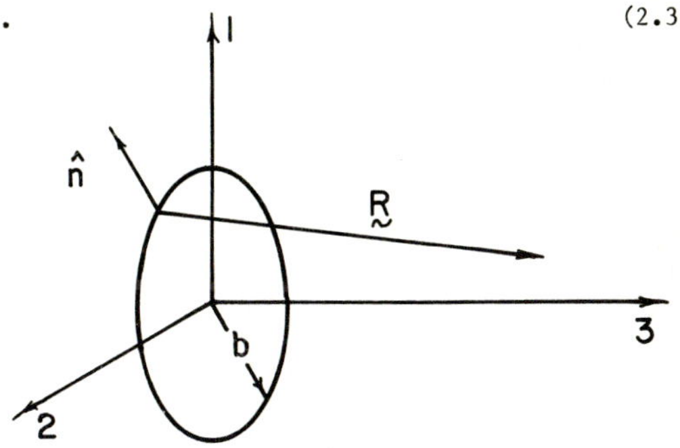

Figure 1. The uniform beam is excited by an axisymmetric source distribution of radius b lying in the $x_3 = 0$ plane. The vector $\underset{\sim}{R}$ joins $(b,\phi',0)$ to (r,ϕ,x_3) and $\hat{n}$ is normal to the circle $r = b$.

Here the unit vector $\hat{R} = R/R$, and $\underset{\sim}{R}$ and $\hat{n}$ are shown in Fig. 1. The term Φ_g represents the wave emitted from the face of the aperture, while the integral represents the edge-diffracted wave. If we introduce the scaled coordinates α and β, take β to be order one and $\alpha < 1$, then for kb large

$$\begin{aligned}\Phi(\underset{\sim}{x}) = \Phi_g - \Phi_g \exp[i(1 + \alpha^2)/\beta]\ \{&[J_0(2\alpha/\beta) - i\alpha J_1(2\alpha/\beta)] \\ &- i(kb)^{-2}\beta^{-3}[J_0(2\alpha/\beta) + i\alpha J_1(2\alpha/\beta)]\}\ .\end{aligned} \tag{2.4}$$

If β is allowed to become large while $\alpha < 1$, then Eq. (2.4) becomes equal to the well-known farfield approximation [1]. However, if we let β become small we do not recover the boundary condition. If we return to Eq. (2.2), introduce the scaled coordinates α and $\gamma = kx_3$, and take both α and γ less than 1, then for kb large we obtain an asymptotic expansion in inverse powers of (kb) rather than $(kb)^2$. Moreover, we can recover the boundary condition from this expansion. It is therefore more suitable to think of Eq. (2.4) as describing the intermediate field.

Note that with this sort of problem we can usually calculate the farfield from an integral representation. Therefore we need not match our asymptotic expansion to one describing the very nearfield, but can connect our expansion to the boundary conditions by matching it to the farfield expansion. If we were to solve the present problem using the same scheme as described in Eqs. (1.8) and (1.9), and determined the unknown constants as just indicated, then the answer, for $\alpha < 1$, would be identical to Eq. (2.4). We shall not do this here because we do much the same thing in elastodynamics in the next section.

3. AN ELASTODYNAMIC BEAM

Here we describe the propagation of a uniform elastodynamic beam. We could also consider a Gaussian elastodynamic beam, but the work would add nothing to what has already been said in the previous sections. The work here is an abridgement of a longer paper [8].

Imagine that a compressional wave transducer is coupled to a homogeneous, isotropic, linearly elastic halfspace by a thin layer of ideal fluid. The halfspace occupies the region $x_3 \geqslant 0$ (Fig. 1). To model this a distribution of normal traction

$$t_3(r/b) = (\lambda + 2\mu)T\, H(1 - r/b) \qquad (3.1)$$

is imposed upon the surface $x_3 = 0$. The normal traction $t_3 = -\tau_{33}$, λ and μ are Lamé's elastic constants, and T is a constant. The tangential tractions are zero.

The equations of elastodynamics are summarized in Achenbach [9]. It suffices to note that, for an axisymmetric problem, the particle displacement $\underset{\sim}{u}$ is given by

$$\underset{\sim}{u} = \nabla\Phi + \nabla \times \Psi\hat{\phi} \qquad (3.2)$$

where $\hat{\phi}$ is a unit vector pointing in the direction of increasing ϕ. Note that here Φ is a displacement potential, whereas in the previous section it was a velocity potential. Integral representations of the waves excited by a normally directed point load acting on a halfspace are constructed in Achenbach [9]. Those for a more general distribution of normal traction are constructed in the same way. These representations can be asymptotically approximated for positions in the farfield. These approximations are given by Eqs. (17) and (18) of [8].

We now proceed to construct an asymptotic solution, having the form of Eq. (2.4), to the intermediate wavefield. The potential Φ is calculated exactly as was done in Eqs. (1.8)-(1.11). The potential Ψ is calculated in a similar way. Firstly Ψ is set equal to $\exp(ik\kappa x_3)\, k^{-2}\, S(\alpha,\beta)$, then $S(\alpha,\beta)$ is expanded as

$$S(\alpha,\beta) = \sum_0^\infty S^n(\alpha,\beta)\,(kb)^{-2n-1} \qquad (3.3)$$

and finally the S^n are found to satisfy

$$\frac{1}{\alpha}\frac{\partial}{\partial\alpha}\left(\alpha\frac{\partial S^n}{\partial\alpha}\right) - \frac{1}{\alpha^2}S^n + 4i\kappa\frac{\partial S^n}{\partial\beta} = -4\frac{\partial^2 S^{n-1}}{\partial\beta^2} \qquad (3.4)$$

where S^{-1} is assumed to be zero. Here k is the compressional wave number and κ is the ratio of the compressional wave speed to the shear wave speed.

Because we only know the farfield expressions to leading order, we can only match the C^o and S^o terms. Solving Eqs. (1.9) and (3.4) for n = 0 and matching to the farfield approximations gives

$$C^o = T[1 - e^{i/\beta} + 2e^{i/\beta} \int_0^{\alpha/\beta} e^{i\beta t^2} J_1(2t)dt] \tag{3.5}$$

and

$$S^o = -\frac{4iT}{\kappa\beta} \exp[i\kappa(1 + \alpha^2)/\beta]\, J_1(\frac{2\alpha}{\beta}) \; . \tag{3.6}$$

The C^o can also be written as

$$C^o = T - T \exp[i(1 + \alpha^2)/\beta] \sum_0^\infty (-i\alpha)^m J_m(\frac{2\alpha}{\beta}) \; . \tag{3.7}$$

Note that Eq. (3.7) is in agreement with Eq. (2.4).

Knowing the potentials we can now calculate the compressional and shear particle displacements

$$\underset{\sim}{u}^L = e^{ikx_3} \underset{\sim}{U}^L \tag{3.8}$$

and

$$\underset{\sim}{u}^T = e^{ik\kappa x_3} \underset{\sim}{U}^T \; . \tag{3.9}$$

The $\underset{\sim}{U}^L$ and $\underset{\sim}{U}^T$ are described by

$$U_r^L = \frac{2T}{k(kb)\beta} \exp[i(1 + \alpha^2)/\beta] J_1(\frac{2\alpha}{\beta}) \tag{3.10}$$

$$U_3^L = i\, C^o/k \tag{3.11}$$

and

$$U_r^T = -\, i\kappa\, S^o/[k(kb)] \; . \tag{3.12}$$

Note that not only have we captured the compressional wave radiated by the face of the transducer but also the edge-diffracted waves that are of order $(kb)^{-1}$.

In Figs. 2 and 3 we have plotted the normalized magnitude of the leading order displacement term A, where A = (kU_3/T). Examining the two figures shows that beyond β = 0.2 the oscillations in the direction of propagation broaden considerably until, near β = 1, the beam starts to evolve into a spherical wave. Further, though the beam starts with considerable structure in its cross-section, by the time it has propagated outward to a region near β = 1 it starts to resemble a Gaussian beam. However, the presence of the small sidelobes indicate the continued influence of the edge waves.

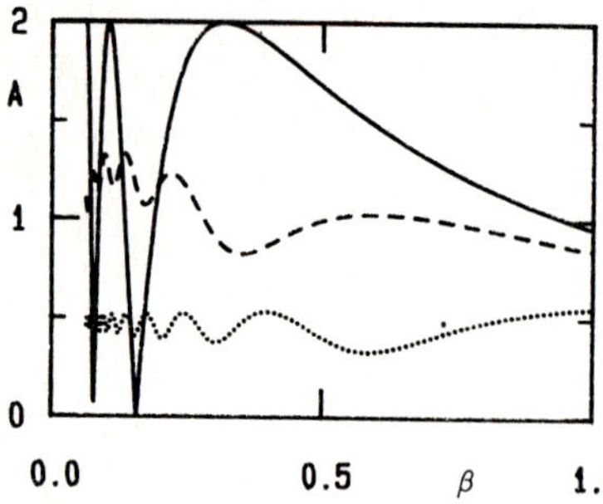

Figure 2 Solid line, α = 0; dashed line α = 0.5; dotted line α = 1.

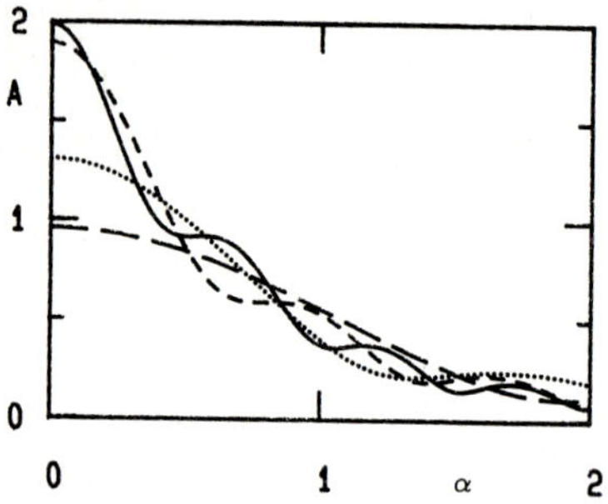

Figure 3 Solid line, β = 0.3, short dashed line, β = 0.5; dotted line, β = 0.7; long-dashed line, β = 1.

CONCLUSIONS

(i) An asymptotic technique that can be used to approximate the wavefield of an ultrasonic transducer of large aperture in a region where the wavefield still retains its highly collimated character has been developed and applied to two model beam profiles, a Gaussian beam and a uniform beam.

(ii) Examining Eqs. (3.10)-(3.12) suggests that near $\beta = 1$ the uniform elastodynamic beam behaves somewhat as if it were a plane compressional wave. However, the edges generate both a transversely polarized compressional wave and a transversely polarized shear wave. The relative simplicity of these expressions is noteworthy.

ACKNOWLEDGEMENTS

I thank Hyung-Chul Choi for carrying out the numerical calculations. This work was supported by the National Science Foundation through Grant No. MSM 85-13928.

REFERENCES

[1] Pierce, A. D., Acoustics (McGraw-Hill, New York, 1981) pp. 231-245.
[2] Deschamps, G. A., Electron. Lett. 7 (1971) 684.
[3] Pott, J. and Harris, J. G., J. Acoust. Soc. Am. 76 (1984) 1829.
[4] Harris, J. G. and Pott, J., J. Acoust. Soc. Am. 78 (1985) 1072.
[5] Tjøtta, J. N. and Tjøtta, S., J. Acoust Soc. Am. 68 (1980) 334.
[6] Buchal, R. N. and Keller, J. B., Commun. Pure Appl. Math. 13 (1960) 85.
[7] Born, M. and Wolf, E., Principles of Optics, 4th ed. (Pergamon, Oxford, 1970) pp. 449-453.
[8] Harris, J. G., ASME J. Appl. Mech. 55 (1988) in print.
[9] Achenbach, J. D., Wave Propagation in Elastic Solids (North-Holland, Amsterdam, 1980) pp. 66-78, 310-321.

ELASTIC WAVE PROPAGATION
M.F. McCarthy, M.A. Hayes, (Editors)

IDENTIFICATION OF ELASTIC MATERIAL SYMMETRY BY ACOUSTIC MEASUREMENT

Stephen C. Cowin

Department of Biomedical Engineering, Tulane University, New Orleans, LA, 70118, USA*

The problem considered is that of identifying the type of elastic symmetry of a material from a small specimen of the material using acoustic wave measurements. The method described also determines the elastic constants of the material.

1. INTRODUCTION

This work addresses the question of identification of the particular type of elastic symmetry - isotropic, transversely isotropic, orthorhombic or othotropic, cubic, hexagonal, tetragonal, monoclinic or triclinic - possessed by a material considered to be a linearly elastic solid. The only symmetry element or operation employed in classifying a material with regard to its elastic symmetry is the plane of reflective or <u>mirror symmetry</u>. The classification of all the material symmetries of linear elastic solid on the basis of the number and orientation of its planes of mirror symmetry was accomplished by Cowin and Mehrabadi [1]. This classification scheme simplifies the classical scheme devised by Voigt. Classical crystallography requires a number of macroscopic symmetry elements or operations in addition to planes of mirror symmetry. These include centers of symmetry and one-, two-, three-, four- or six-fold rotation and inversion axes.

The new classification scheme, which is restricted to linear elastic solids, employs only planes of symmetry and is described in section 4, following two sections that introduce the generalized Hooke's law (section 2) and the concepts of acoustic tensor, specific directions and specific axes (section 3). The measurement of the elasticity coefficients is described in section 5, and section 6 is a summary and integration of the results presented.

2. Anisotropic Elasticity

The constitutive relation for linear anisotropic elasticity is the generalized Hooke's law,

$$T_{ij} = C_{ijkm}E_{km}, \tag{1}$$

* This investigation was supported by USPHS, Research Grant DE06859 from the National Institutes of Health, Bethesda, MD 20205

which is the most general linear relation between the stress tensor **T** whose components are T_{ij} and the strain tensor **E** whose components are E_{ij}, where the strain has been assumed to be measured from an unstressed reference state. The coefficients of linearity, namely C_{ijkm}, are the components of the fourth-rank elasticity tensor. As is customary in the discussion of linear anisotropic elasticity, we introduce a 6x6 matrix denoted by **c** to represent the components of C_{ijkm}. The shift from an index system with a range of three $(i,j = 1,2,3)$ to one with a range of $6(\alpha,\beta = 1,2,3,4,5,6)$ is accomplished by the following rules for replacing a pair of indices ij by a single lower case Greek index: change 11 to 1, 22 to 2, 33 to 3, 23 to 4, 13 to 5 and 12 to 6. The generalized Hooke's law (1) can then be rewritten in the form

$$\begin{bmatrix} T_{11} \\ T_{22} \\ T_{33} \\ T_{23} \\ T_{13} \\ T_{12} \end{bmatrix} = \begin{bmatrix} c_{11} & c_{12} & c_{13} & c_{14} & c_{15} & c_{16} \\ c_{12} & c_{22} & c_{23} & c_{24} & c_{25} & c_{26} \\ c_{13} & c_{23} & c_{33} & c_{34} & c_{35} & c_{36} \\ c_{14} & c_{24} & c_{34} & c_{44} & c_{45} & c_{46} \\ c_{15} & c_{25} & c_{35} & c_{45} & c_{55} & c_{56} \\ c_{16} & c_{26} & c_{36} & c_{46} & c_{56} & c_{66} \end{bmatrix} \begin{bmatrix} E_{11} \\ E_{22} \\ E_{33} \\ 2E_{23} \\ 2E_{13} \\ 2E_{12} \end{bmatrix} \tag{2}$$

where the 6x6 matrix in (2) is the matrix **c**. Two second rank tensors formed from C_{ijkm} are needed in subsequent developments. These two tensors are defined by

$$A_{ij} = C_{ijkk}\ ,\quad B_{ij} = C_{ikjk}, \tag{3}$$

and, in terms of the components of **c**, they have the following representatives:

$$\mathbf{A} = \begin{bmatrix} c_{11}+c_{12}+c_{13} & c_{16}+c_{26}+c_{36} & c_{15}+c_{25}+c_{35} \\ c_{16}+c_{26}+c_{36} & c_{12}+c_{22}+c_{23} & c_{14}+c_{24}+c_{24} \\ c_{15}+c_{25}+c_{35} & c_{14}+c_{24}+c_{34} & c_{13}+c_{23}+c_{33} \end{bmatrix} \tag{4}$$

$$\mathbf{B} = \begin{bmatrix} c_{11}+c_{55}+c_{66} & c_{16}+c_{26}+c_{45} & c_{15}+c_{46}+c_{35} \\ c_{16}+c_{26}+c_{45} & c_{22}+c_{44}+c_{66} & c_{24}+c_{34}+c_{56} \\ c_{15}+c_{46}+c_{35} & c_{24}+c_{34}+c_{56} & c_{33}+c_{44}+c_{55} \end{bmatrix} \tag{5}$$

3. SPECIFIC DIRECTIONS AND SPECIFIC AXES

Let **n** denote the direction in which the wave is propagating and let **q** denote a unit vector in the direction of the amplitude vector. The value of **q** for a wave in the direction **n** is the solution to the eigenvalue problem,

$$Q_{ik}(\mathbf{n})q_k = \rho v^2 q_i, \tag{6}$$

where ρ is the density of the anisotropic material, v is the phase velocity of the wave and $\mathbf{Q}(\mathbf{n})$ is the acoustic tensor in the direction $\mathbf{n}$ defined in terms of C_{ijkm} by

$$Q_{ik}(\mathbf{n}) = C_{ijkm}n_j n_m \tag{7}$$

A wave is said to be longitudinal if $\mathbf{q} = \mathbf{n}$ and to be transverse if $\mathbf{q}\cdot\mathbf{n}$ is zero.

Borgnis [2] introduced the concept of specific directions for longitudinal waves in anisotropic elastic materials. He noted that there are only specific directions in which a pure longitudinal wave can propagate in an anisotropic elastic material. For a longitudinal wave, the **n** and **q** of (6) must coincide, so the specific directions **a** are the solutions to the non-linear eigenvalue problem

$$Q_{ik}(\mathbf{a})a_k = \rho v_\ell^2 a_i \tag{8}$$

where v_ℓ is the phase velocity of the longitudinal wave. For any linear elastic material the existence of at least one specific direction is assured by an application of the fixed point theorem. This application of the fixed point theorem is due to Stippes and was generalized by Truesdell [3].

The concept of a specific axis related to pure transverse wave propagation will now be introduced. Let **a** denote a unit vector and let **b** denote a direction perpendicular to **a**. If, for all **b** such that $\mathbf{a}\cdot\mathbf{b} = 0$, it is possible to propagate a pure transverse wave with amplitude in the **a** direction, the axis **a** is said to be a pure shear amplitude specific axis or, for brevity, a specific axis. The condition for a specific axis **a** to exist is that **a** be an eigenvector of all $\mathbf{Q}(\mathbf{b})$ for which $\mathbf{a}\cdot\mathbf{b}$ is zero, thus

$$Q_{ik}(\mathbf{b})a_k = \rho v_s^2 a_i, \text{ for all } \mathbf{b} \text{ such that } \mathbf{a}\cdot\mathbf{b} = 0, \tag{9}$$

where v_s is the shear wave velocity in the direction **b**.

4. DETERMINATION OF MATERIAL SYMMETRY

The problem of determining the type of elastic symmetry possessed by a material given the numerical values of the fourth rank elasticity tensor relative to an arbitrary coordinate system was considered by Cowin and Mehrabadi [1]. They solved the problem by constructing the number and orientation of the planes of symmetry characterizing each of the ten traditional and distinct elastic symmetries and by proving a theorem which provides the machinery for determining the number and orientation of the planes of symmetry of a given **c** relative to an arbitrary coordinate system. The main result of Cowin and Mehrabadi [1] is that a set of necessary and sufficient conditions for the vector **a** to be a normal to a plane of symmetry of a material of given elasticity is that **a** be an eigenvector of **A** and **B**, given by (4) and (5), respectively, a specific direction (8) and a specific axis (9).

It can be shown, Cowin [4], that if **a** is a specific direction and a specific axis, then it is also an eigenvector of **A** and **B**.

The combination of this result with the theorem of Cowin and Mehrabadi [1] described above proves the following theorem: A necessary and sufficient condition that **a** be a normal to a plane of symmetry in a linear elastic material is that it be a specific direction and a specific axis.

Cowin and Mehrabadi [1] showed that each of the ten distinct elastic symmetries can be characterized uniquely by the number and orientation of the planes of mirror symmetry it possesses. Triclinic symmetry has no planes of symmetry and monoclinic has just one. Orthorhombic or orthotropic symmetry has three mutually orthogonal planes of symmetry. Hexagonal [7] and hexagonal [6] symmetries each have three planes of symmetry whose normals all lie in one plane and make angles that are multiples of 120° with each other. Hexagonal [7] is distinguished from hexagonal [6] in that one of the normals to a plane of symmetry is a base vector for the symmetry coordinate system for hexagonal [6], but not for hexagonal [7]. Similarly, tetragonal [7] and tetragonal [6] symmetries have five planes of symmetry, four of whose normals lie in the fifth plane of symmetry and make angles of 45° with respect to one another. Tetragonal [7] is distinguished from tetragonal [6] in that three of the normals to planes of symmetry are the base vectors for a symmetry coordinate system for tetragonal [6], while only one normal to a plane of symmetry is a base vector for a symmetry coordinate system for tetragonal [7]. Cubic symmetry has nine planes of symmetry, all intersecting at multiples of 45°. Hexagonal [5] or transversely isotropic symmetry has one plane of isotropy, which is a plane of symmetry in which each vector is a normal to a plane of symmetry. Isotropic symmetry is characterized by the fact that every plane passing through a point is a plane of isotropy.

5. MEASUREMENT OF THE ELASTICITY COEFFICIENTS

In this section a method for determination of all the components of **c** from a single specimen using elastic wave propagation data is outlined. A specimen in the form of a cube is cut from the material. A Cartesian coordinate system with axes x_1, x_2 and x_3 parallel to the edges of the cube is designated. Consider first wave propagation in the **n**-direction. We do not know *a priori* the directions of particle oscillation for the three modes of propagation in that direction. Thus, it is necessary to generate a variety of waves with dominant particle oscillations in different directions. The arrival of the wave at the other face perpendicular to the **n**-axis is sensed by another transducer or set of transducers. Transducers are available which are sensitive to longitudinal oscillations and shear oscillations. A longitudinal transducer would pick up the longitudinal component of the wave and a shear transducer could be used to find the transverse components. Shear transducers are sensitive to oscillations in a particular direction. Thus, it is necessary to be able to rotate the shear transducer through 180° to find the dominant transverse components. Once the three modes of particle oscillation for a propagating wave have been determined, waves are generated in which each of these is the dominant one. Wave speeds are determined using these "dominant mode" waves. The following measurements are made: the density of the material, the velocities of the three modes of propagation in a particular direction and the particle displacement associated with each of those modes. For the direction $\mathbf{n}$ the three measured velocities are denoted by $v_i(\mathbf{n})$ $(i = 1,2,3)$, and the notation $a_{1j}(\mathbf{n})$, $a_{2j}(\mathbf{n})$, $a_{3j}(\mathbf{n})$ $(j = 1,2,3)$ is used for the components of a unit vector parallel to the amplitude vector for each $v_i(\mathbf{n})$. From these numbers the components of $\mathbf{Q}(\mathbf{n})$ can be calculated as follows:

$$
\begin{aligned}
Q_{11}(\mathbf{n}) &= a_{11}^2\rho v_1 + a_{12}^2\rho v_2 + a_{13}^2\rho v_3 \\
Q_{22}(\mathbf{n}) &= a_{21}^2\rho v_1 + a_{22}^2\rho v_2 + a_{23}^2\rho v_3 \\
Q_{33}(\mathbf{n}) &= a_{31}^2\rho v_1 + a_{22}^2\rho v_2 + a_{33}^2\rho v_3 \\
Q_{12}(\mathbf{n}) &= a_{11}a_{21}\rho v_1^2 + a_{12}a_{22}\rho v_2^2 + a_{13}a_{23}\rho v_3^2 \\
Q_{13}(\mathbf{n}) &= a_{11}a_{31}\rho v_1^2 + a_{12}a_{32}\rho v_2^2 + a_{13}a_{33}\rho v_3^2 \\
Q_{23}(\mathbf{n}) &= a_{21}a_{32}\rho v_1^2 + a_{22}a_{32}\rho v_2^2 + a_{23}a_{33}\rho v_3^2
\end{aligned}
\qquad (10)
$$

This method is used to determine the acoustic tensors Q(1), Q(2) and Q(3) where 1,2, and 3 indicate the following three unit normals

$$\mathbf{n} = \mathbf{e}_1,\ \mathbf{n} = \mathbf{e}_2,\ \mathbf{n} = \mathbf{e}_3, \qquad (11)$$

respectively.

The specimen is then subjected to six bevel cuts so that opposing surfaces perpendicular to the following three unit normals are exposed:

$$\sqrt{2}\mathbf{n} = \mathbf{e}_2 + \mathbf{e}_3,\ \ \sqrt{2}\mathbf{n} = \mathbf{e}_1 + \mathbf{e}_3,\ \ \sqrt{2}\mathbf{n} = \mathbf{e}_1 + \mathbf{e}_2\ . \qquad (12)$$

Measurements identical to those described above are performed and the value of the acoustic tensor $\mathbf{Q}(\mathbf{n})$ is determined for each direction. These acoustic tensors are denoted by $\mathbf{Q}(4)$, $\mathbf{Q}(5)$ and $\mathbf{Q}(6)$ corresponding to the ordering of the $\mathbf{n}$'s in equation (11) above. From these six acoustic tensors the values of the matrix $\mathbf{c}$ are determined from the following formulas:

$$
\begin{aligned}
c_{11} &= Q_{11}(1),\ c_{12} = 2Q_{12}(6) - c_{16} - c_{26} - c_{66}, \\
c_{22} &= Q_{22}(2),\ c_{13} = 2Q_{13}(5) - c_{15} - c_{35} - c_{55}, \\
c_{33} &= Q_{33}(3),\ c_{23} = 2Q_{23}(4) - c_{24} - c_{34} - c_{44},
\end{aligned}
\qquad (13)
$$

$$
\begin{aligned}
c_{44} &= Q_{33}(2) = Q_{22}(3),\ c_{15} = Q_{13}(1),\ c_{16} = Q_{12}(1),\ c_{56} = Q_{23}(1), \\
c_{55} &= Q_{33}(1) = Q_{11}(3),\ c_{24} = Q_{23}(2),\ c_{26} = Q_{12}(2),\ c_{46} = Q_{13}(2), \\
c_{66} &= Q_{22}(1) = Q_{11}(2),\ c_{34} = Q_{23}(3),\ c_{35} = Q_{13}(3),\ c_{45} = Q_{12}(3),
\end{aligned}
\qquad (14)
$$

$$
\begin{aligned}
c_{14} &= 2Q_{13}(6) - c_{15} - c_{46} - c_{56} = 2Q_{12}(5) - c_{16} - c_{45} - c_{56}, \\
c_{25} &= 2Q_{23}(6) - c_{24} - c_{46} - c_{56} = 2Q_{12}(4) - c_{26} - c_{46} - c_{45}, \\
c_{36} &= 2Q_{23}(5) - c_{34} - c_{45} - c_{56} = 2Q_{13}(4) - c_{35} - c_{45} - c_{46}.
\end{aligned}
\qquad (15)
$$

These results were obtained by Van Buskirk et al. [5] and they relate the components of **c** to measured quantities.

6. SUMMARY AND INTEGRATION OF RESULTS

In the previous section an acoustic wave approach was described for the measurement of the twenty-one independent elastic constants in the matrix **c** of the most general linearly elastic anisotropic solid. The method required that one be able to measure the density of the material, the velocities of the three modes of wave propagation in each of six directions, and the particle displacements associated with each of those modes. In section 4 it was shown that, given **c**, one could determine all the normals to the planes of symmetry of the material. In order for **a** to be a normal to a plane of symmetry of a material it had to be an eigenvector of the tensors **A** and **B** given by (4) and (5), respectively, a specific direction (8) and a specific axis (9). Once all the normals to the planes of symmetry for the material have been determined, the particular type of elastic symmetry enjoyed by the material can be determined from the information in the last paragraph of section 4.

REFERENCES

[1] Cowin, S.C. and Mehrabadi, M.M., Quart. J. Mech. Appl. Math. 40(1987) 451.
[2] Borgis, F.E., Phys. Rev. 98(1955) 1000.
[3] Truesdell, C., J. Acoust. Soc. Am. 40(1966) 729.
[4] Cowin, S.C., in preparation.
[5] Van Buskirk, W.C., Cowin, S.C. and Carter, R., J. Mat. Sci 21(1986) 2759

ELASTIC WAVE PROPAGATION
M.F. McCarthy, M.A. Hayes, (Editors)

SOLUTION TO MULTI-DIMENSIONAL WAVE PROPAGATION PROBLEMS IN SOLIDS BY THE METHOD OF CHARACTERISTICS

M. ZIV

Holon Technological Institute, P.O. Box 305, Holon 58102, Israel

A presentation is given of solutions to multi-dimensional wave propagation in solids. These solutions are obtained by a method based on the theory of characteristics. Dynamic integration must follow the solution on wave surfaces propagating in space and time from sources of excitation; this satisfies the actual geometric representation of hyperbolic partial differential equations, Huygen's principle, and Fermat's principle of least time. While for one-dimensional analysis a characteristics formulation suffices, it must be realized that multi-dimensional analysis necessitates two additional major steps. First, the characteristics formulation must be extended to accommodate strong discontinuities. Secondly, the characteristics numerical integration must be confined to each region of influence encountered in the solution domain. Only then is adequate resolution reached for multi-dimensional wave propagation problems; the numerical results are then intrinsically stable and computer usage is minimal. Since in the resultant deformation the various wave fronts are explicitly revealed, physical interpretation is ensured. The characteristics approach, well-posed, readily lends itself in a compatible manner to nonlinear transient deformation where strong shock waves occur.

1. INTRODUCTION

The phenomenon of transient wave propagation in solids has created increased interest in recent years. Solutions have been sought to wave problems, particularly those which arise in fields of geophysics, space technology, and testing of materials. Some of the geophysical problems of current interest include protection from nuclear explosions, detection of underground structures and minerals, and underground testing of nuclear bombs. In space technology scientists encounter complex wave problems associated with projectiles (and penetration) impact on space vehicles, the impulsive ignition of solid propellant motors, impact landings of space vehicles, external or chemical energy deposition in space vehicles, and an abrupt separation of structure components from space vehicles. Tests of new materials are being performed now by emitting wave signals through the tested material and are known as nondestructive evaluation and acoustic emission.

These problems generally require the treatment of finite bodies subjected to concentrated impulsive loads. Waves generated under these conditions are inherently multi-dimensional and of large deformation. However, the required solution to many of these problems is confined to domains influenced only by some of the boundaries of the body in question and not by all of them. Therefore, a material response is sought rather than a structural response. It follows then that a wave propagation analysis is mandatory rather than that of standing waves.

Of the available computational techniques the characteristics approach is the most appropriate for solving wave propagation problems since it is the only one compatible with the actual geometric representation of hyperbolic partial differential equations. The objective of this article is to point out the necessary analytical steps in the multi-dimensional computational method which is based on the characteristics theory.

2. THEORY OF CHARACTERISTICS

Basic field equations that govern the dynamic deformation of a medium are partial differential equations of the hyperbolic type. Since the basic equations are defined, however, exclusively for a dynamically deformed element just excited, they are not numerically integrable as they stand for the subsequent motion in the medium. The ensuing material motion is led by wave surfaces propagating in space and time from the source of excitation. Under these circumstances, it is clear that the basic governing equations must be transformed to hold along and within the generated wave surfaces so that the material motion can be integrated. The procedure of rewriting the governing system of partial differential equations of the hyperbolic type as interior differential equations on wave surfaces is known as the wave characteristics formulation. In conjunction with these interior equations (wave characteristics equations), the formulation yields the speeds of the wave surfaces and their space-time location (bicharacteristics curves) in the medium.

The well-known conventional procedure of obtaining the wave characteristics formulation for one-dimensional problems of wave propagation in solids cannot be applied to multi-dimensional cases. Moreover, while for the integration process to take place in the well-known one-dimensional cases a characteristics formulation suffices, multi-dimensional analysis necessitates two additional major steps before integration. Sections 3, 4, and 5 are devoted to the major analytical steps needed in the multi-dimensional computational method based on the characteristics theory, and Section 6 describes the all-events process of integration needed for nonlinear transient deformation. Solutions to representative wave propagation problems are summarized in Section 7.

3. CHARACTERISTICS FORMULATION

The well-known directional derivative approach for obtaining the characteristics formulation is directly applicable only when the unknown dependent variables are scalar and vector quantities, as in cases of multi-dimensional hydrodynamics, e.g. [1], and one-dimensional motion in solids, e.g. [2]. In multi-dimensional transient deformation of solids, however, where inseparable tensor quantities are involved, the number of the indeterminate first partial derivatives in the dynamical field equations exceeds the number of available equations, see Eqs.(6)-(10) of [3]. Therefore, rather than a direct treatment of these equations, which implies a direct dynamical analysis, it is expedient to apply the kinematical approach of [3] based on Hadamard's "weak" discontinuity relations which exist across a wave surface where the dependent variables are continuous while their partial derivatives are discontinuous.

With the kinematical approach of [3], a multi-dimensional characteristics formulation has been derived in [4] for hyperelastodynamics, Eqs.(17)-(22) or Eqs. (35)-(39) which may be written in terms of Eq.(25) or Eq.(42). In [5], a multi-dimensional characteristics formulation has been derived for hypoelastodynamics exhibiting compressible rate-sensitive material flow, Eqs.(17)-(21). These equations with Eq.(70) of [4] are applicable to elastic-perfectly plastic and soil materials.

4. STRONG DISCONTINUITY CHARACTERISTICS

In one-dimensional wave propagation problems, the characteristics formulation is applicable directly, irrespective of the existence of strong discontinuities in the solution domain. This is so because the conventional one-dimensional characteristics formulation consists of ordinary differential equations having the same form as Rankine-Hugoniot relations, which are suitable both to weak and strong motions.

By contrast, the multi-dimensional formulation of Sec.3 yields, along the path of integration, characteristics equations containing partial derivatives which are compatible with weak motion only. In a multi-dimensional region, however, the motion (irrespective of the nature of the applied load) involves strong discontinuities as a result of wave reflection from the boundaries of the solid. Consequently, strong discontinuity relations must be used in extending the multi-dimensional characteristics equations of Sec.3 to make them compatible with strong discontinuous motion phenomena. In [6], geometrical strong discontinuity formulas, Eqs.(32′)-(37′), are given by which the wave characteristics equations of Sec.3 can be extended in order to make them compatible with strong material motion, where the dependent variables themselves are discontinuous across the wave surface.

5. INTEGRATION SCHEME FOR EACH REGION OF INFLUENCE

In the well-known one-dimensional wave propagation problems, the characteristics formulation is applicable directly to the entire solution domain. This is a result of the reference spatial coordinate being normal to all existing wave fronts at all times. For this reason and for the one discussed in Sec.4, the characteristics formulation is all that is needed for integration to take place in one-dimensional problems.

The picture changes completely in the case of multi-dimensional motion. Here the solution domain must be divided into regions of influence, each of which calls for a separate computational algorithm. A geometric construction is, therefore, needed describing each of the existing regions of influence encountered in the solution domain of a given boundary-value problem. Such a construction is given in [7].

Sections 3,4 and 5 specify the three major steps that must be followed prior to the integration of the characteristics equations along their bicharacteristics curves in a multi-dimensional wave motion problem. In the next section, an integration method is described for waves of large deformation.

6. ALL-EVENTS METHOD OF INTEGRATION ALONG WITH THE EXISTING WAVES

In the well-known one-dimensional method of characteristics when applied to the numerical solution of elastic-plastic transient deformation of solids, [2], [8]-[11], two far-reaching idealizations are imposed on both the rate-independent and the rate-dependent theories of anelasticity: the leading wave was assumed to propagate with the constant elastic speed and the evaluation of the unknowns at a space-time point was based solely on past information. Because of the first idealization, wave propagation problems with a concave upward stress-strain curve could not be solved quantitatively since the convergence of the bicharacteristics curves formed a noncharacteristics shock front. Because of the second idealization, in problems with a downward concavity of a stress-strain curve, an elaborate numerical analysis, including iteration procedure, was needed in order to obtain adequate resolution for the problem at hand.

An all-events analytical method has been presented in [12] for the step-by-step integration of the wave characteristics equations governing, without detriment to generality, large-amplitude uniaxial strain motion of inelastic materials having a stress-strain curve of upward concavity. In particular, inherent in this method is the capability of determining deformation associated with converging shock waves. Converging shock waves arise when fast unloading waves overtake the slower leading wave and/or when the applied load has a gradual rise time. Then wavelets of high stresses propagate faster than those of low stresses because of the upward bend of the stress-strain curve. When an unloading process exists during the deformation, this method has also the capability of locating in the space-time solution domain the permanent strain boundary which separates the motion of deformation from that of the rigid body. Significantly, functions at a grid point are numerically evaluated by taking account of all events, past and present, which surround that point in a total linear manner, in the sense that the determination of the variable coefficients of the incremental terms in the characteristics equations is uncoupled. First, the functions in the incremental terms are obtained and then the dependent coefficients are determined separately. The nonlinearity of the deformation, however, is exhibited by actual bicharacteristics curves that vary from point to point inside the variable reference grid. The reference grid is formed by the varying bicharacteristics curves of the leading wave propagating into the medium with the varying chord-shock speed. Based on these features, the all-events method of integration along with the existing waves is essentially analytical where in its final stage of development a straightforward numerical integration is required which is free of iteration procedure, stability criteria, and of any other numerical analysis.

7. RESULTS AND CONCLUSIONS

Based on the theory of characteristics, a computational method has been developed for the evaluation of multi-dimensional transient deformation of solids in which waves emanate from impulsive sources and their reflections from boundaries map a strong discontinuity region into the solid. This method is essentially analytical where in its final stages of development a straightforward coarse step-by-step numerical integration is required which is free of iteration procedure, stability criteria, and of any other numerical analysis. While for one-dimensional problems a wave characteristics formulation suffices, the multi-dimensional analysis necessitates two additional major analytical steps compatible with the initial-boundary value problem in question. First, the characteristics formulation must be extended to accommodate strong discontinuities. Second, the characteristics integration must be confined to each region of influence encountered in the solution domain. Only then is adequate resolution reached in multi-dimensional wave propagation problems; the numerically converging results are then intrinsically stable and computer usage is minimal. Since in the resultant deformation the various wave fronts are explicitly revealed, physical interpretation is ensured, [13].

In the course of pursuing the computational method, the following contributions have been made: a. The development of a general technique for extracting the wave characteristics formulation from a given system of hyperbolic partial differential equations governing dynamic deformation of a solid medium, [3]-[5]; b. The development of geometrical strong discontinuity formulas for the characteristics equations in order to make them compatible with strong material motions, [6]; c. The development of a geometric construction, for a given boundary-value problem, of the regions of influence and their recurring points, [7]; d. The development of an all-events technique of integration of the characteristics equations along with the existing waves for shock waves in nonlinear solids, [12] (The resultant deformation is given in [14]); e. The obtainment of a computational technique for the initial deformation along a guided impulsive precursor wave that interacts with interfaces of a layered matrix, [15].

REFERENCES

[1] Richardson, D.J., The Solution of Two-Dimensional Hydro-dynamic Equations by the Method of Characteristics, in: Adler, B., (ed.), Methods in Computational Physics (Academic Press, Vol. 3, 1984).

[2] Cristescu, N., Dynamic Plasticity (North-Holland, Amsterdam, 1067).

[3] Ziv, M., Int. J, Solids Structures (1969) 1135.

[4] Ziv, M., J. Acoust. Soc. Am. (1981) 218.

[5] Ziv, M., J. Appl. Mech. Trans. ASME (1977) 419.

[6] Ziv, M., J. Acoust. Soc. Am. (1987) 22.

[7] Ziv, M., Bull. Seismol. Soc. Am. (1975) 1359.

[8] Zukas, J.A., Nicholas, T., Greszczuk, L.B., and Curran, D.R., Impact Dynamics (Wiley, New York, 1982) pp. 112-114.

[9] Phillips, A. and Zabinski, M.P., Ing. Arch. (1972) 367.

[10] Lee, E.H. and Liu, D.T., An Example of the Influence of Yield on High Pressure Wave Propagation, in: Kolsky, H. and Prager, W. (eds.), Stress Waves in Anelastic Solids (Springer-Verlag, Berlin, 1964).

[11] Nowacki, W.K., Stress Waves in Non-Elastic Solids (Pergamon, New York, 1978).

[12] Ziv, M., J. Acoust. Soc. Am. (1984) 566.

[13] Ziv, M., J. Acoust. Soc. Am. (1976) 9.

[14] Ziv, M., J. Acoust. Soc. Am. (1984) 574.

[15] Ziv, M., J. Acoust. Soc. Am. (1988) March.

ELASTIC WAVE PROPAGATION
M.F. McCarthy, M.A. Hayes, (Editors)

COUPLING EFFECTS IN STRUCTURAL WAVE PROPAGATION

Mahir SAYIR

Institute of Mechanics
Swiss Federal Institute of Technology
Zurich, Switzerland

Three problems of elastic wave propagation in structures will be discussed. Their common feature is dynamic coupling resulting from both inertial and elastic interaction between different modes of motion.

1. REFLEXIONS AND TRANSMISSION OF FLEXURAL WAVES AT THE RIGHT ANGLE CORNER OF A UNIDIRECTIONALLY FIBER-REINFORCED FRAME

Consider a frame with two beam elements connected at a right angle (Fig. 1) and send a *flexural* wave in one of the beam elements towards the corner. An analysis similar to the one in [1] will reveal that two flexural and two longitudinal waves are generated at the corner. These waves are complemented by stationary motion on both sides of the corner, not carrying energy and decaying rapidly with distance. In metallic *isotropic* frames, pulses in the *low* and *middle frequency range,* generate very weak longitudinal waves even though the transmitted longitudinal motion carries *five* times more energy than the reflected one. The incident energy is delivered mainly to the reflected and transmitted flexural waves in approximately equal parts. The middle frequency range for which this situation remains valid may reach for a 10 mm thick square cross-section frequencies as high as 4-5 kHz. Shear coupling which reinforces the transmitted longitudinal motion and weakens the transmission of flexural waves, appears at much higher values of frequency in isotropic structures. The characteristic "small" parameters which determine the nature of the dynamic behaviour in isotropic frames are the thickness-flexural wavelength ratio $\varepsilon_F := \pi h/\lambda_F$ and the flexural-longitudinal wavelength ratio $\varepsilon_L := \lambda_F/\lambda_L$.

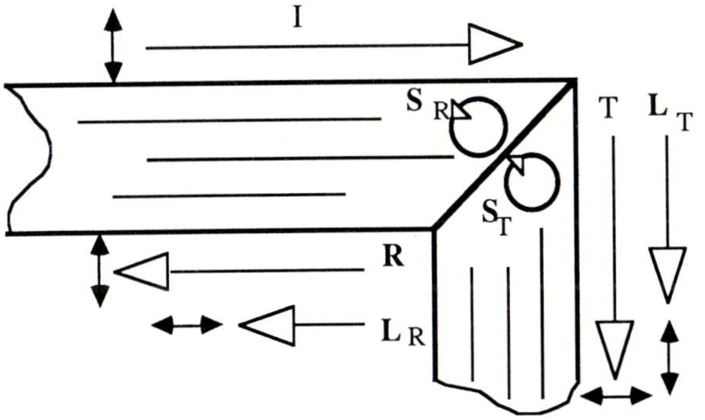

FIGURE 1
Incident flexural wave I, reflected flexural wave R, reflected longitudinal wave.L_R, transmitted flexural wave T, transmitted longitudinal wave L_T, stationary motions S_R and S_T

In a *composite* frame built out of a soft matrix reinforced by stiff fibers in the axial direction of the individual beam elements, a third parameter becomes extremely important, namely the shear-longitudinal moduli ratio $\varepsilon_M := G/E$. The theory developed in [2] and confirmed in experiments as reported in [3] and [4] shows that the magnitude of a dimensionless "number"

$$p := \varepsilon_F/\sqrt{\varepsilon_M}$$

essentially influences the flexural behaviour of a composite beam. If this number reaches values of order 1 due to the fact that ε_M may be quite small (for example $1/10^2$) in composite beams, *shear coupling* becomes very strong and the phase speed of flexural waves reaches values not far from the shear wave speed c_2 even in the middle frequency range mentioned above. Cross sections do not remain plane and, even in a first order theory, shearing deformation must be taken into account since it influences essentially normal stress distributions in the cross sections. The frequency-wavelength relation was derived in [2] and reads

$$h^2\omega^2/4 = c_2^2\, \varepsilon_M\, (p^2\text{-}pThp)$$

for a rectangular cross section, where h is the thickness and ω is the circular frequency of the sinusoidal pulse. Using this relation and formulating boundary conditions at the corner, one obtains by straightforward analysis the amplitudes of the two reflected and two transmitted waves along with those of the stationary motions as a function of frequency. With the help of these amplitudes the energy-flux ratios with respect to the energy content of the incident waves can be determined. The result is represented in Fig. 2 for two different E/G ratios. For the lower ratio, shear coupling reinforces transmission of longitudinal modes and weakens the transmitted flexural mode. For the higher ratio, the impedance of transmission becomes very high for both modes; most of the incoming energy is reflected in flexural mode. Thus, if the matrix is sufficiently soft with respect to the fibers, transmission of flexural energy is limited to low frequencies. This is mainly due to the fact that rotations at the corner are "neutralized" by strong shear coupling. Such features can be conveniently used in vibration isolation design.

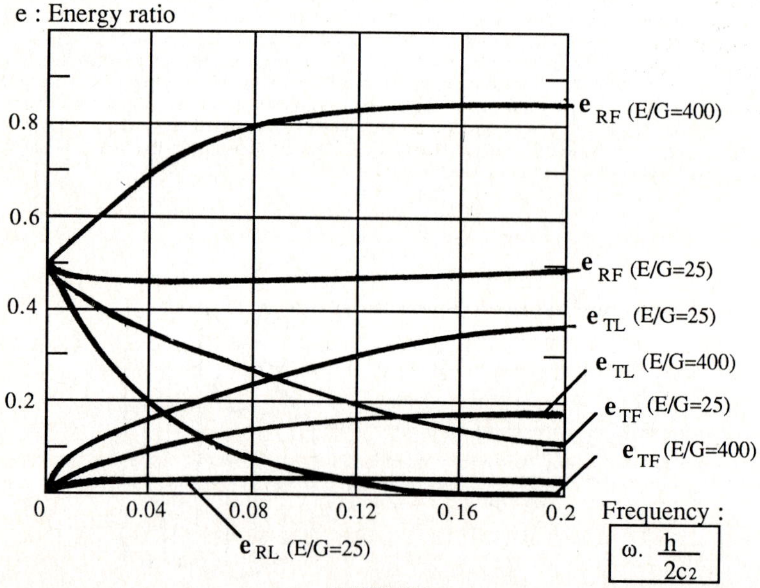

FIGURE 2
Reflected and transmitted energy fluxes: label F for flexural, label L for longitudinal waves

2. DYNAMIC ELASTIC-PLASTIC BEHAVIOUR OF A FRAME IN COUPLED BENDING AND TORSION

A rectangular metalic space frame clamped at its two ends is subjected to an impact in the middle of the transversal beam element, perpendicular to the plane of the frame. A .22 lead bullet is used, loading time is about 40-60 μs and the impact momentum is varied between 0.5 to 1 Ns in the case of steel (Fig. 3.).

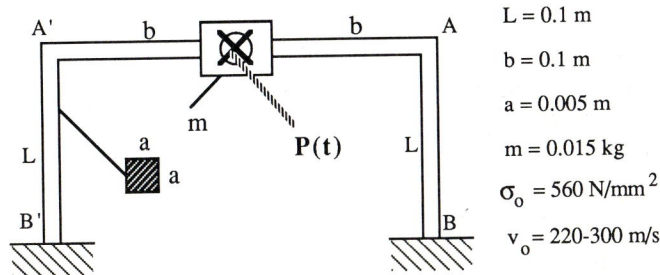

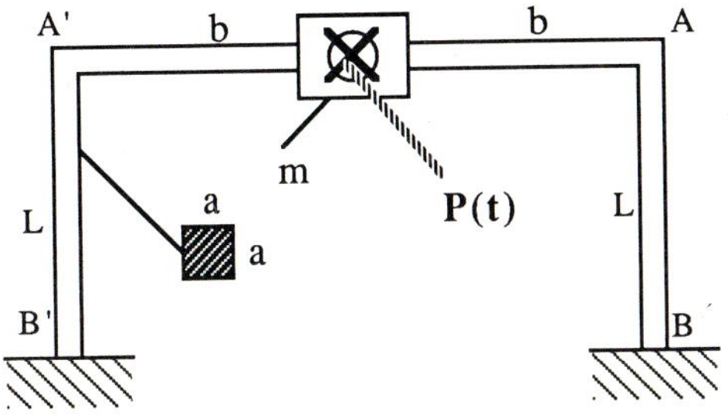

FIGURE 3
Rectangular frame dynamically loaded in the middle of transversal beam element

The purpose of the investigation is to study and understand the development of localized plastic deformation in the various stages of the motion. In approximately 20 μs plastic deformation develops on both sides of the mass in the middle which is hit by the bullet. Since the axial extension of the plastic zone remains very small with respect to beam dimensions, hinge-like behaviour may be assumed and the dynamic behaviour can further be studied by elastic analysis. Plastic deformation simply fixes the value of the bending moment at the hinges to the yield value.

When the flexural wave reaches the corner A, it finds a situation very similar to the one of the previous examle of section 1 for the case of isotropic behaviour. Two reflected and two transmitted waves are generated along with stationary motion localized to the corner. Reflected and transmitted trosional waves in the present example behave very much like the reflected and transmitted longitudinal waves of the previous one.The corresponding formal expressions for the various amplitudes are also very similar. Since the dynamic impedance of torsional waves is much larger than the one for flexural waves in the low and middle frequency-range, little energy is reflected or transmitted in the torsional mode and the incoming energy is delivered in approximately equal parts mainly to the reflected and transmitted flexural waves. Thus, the amplitude of the torque at the corner remains small and, since torque in one beam element is equivalent to bending moment in the other, bending moments remain also small. Thus, *no plastic deformation can develop at the corners in this phase of motion.* This is very different from equivalent static loading where one might expect plastic hinges at the corners in a collapse mode. Fig. 4 shows the bending moment as a function of time at 10 mm distance from the corner for the first 300 μs.

In addition to the plastic hinges in the middle, plastic deformation in the transversal beam AA´ may occur after 120-150 μs at 25-35 mm distance from the corners A, A´, when the reflected part of the flexural wave meets with the incident wave *in phase* and reinforces amplitudes up to the yield limit (Fig. 5).

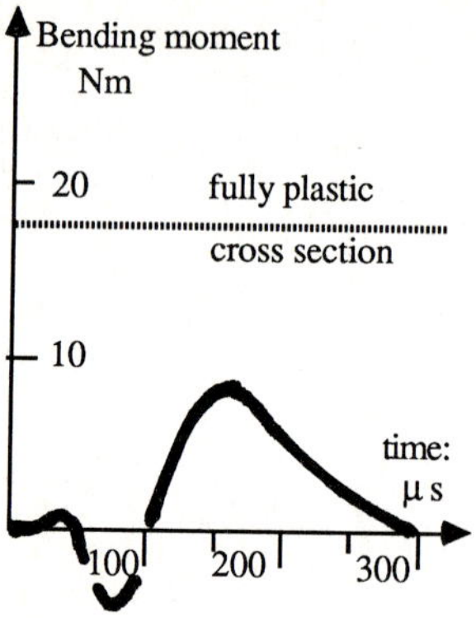

FIGURE 4
Bending moment at 10 mm distance from the corner

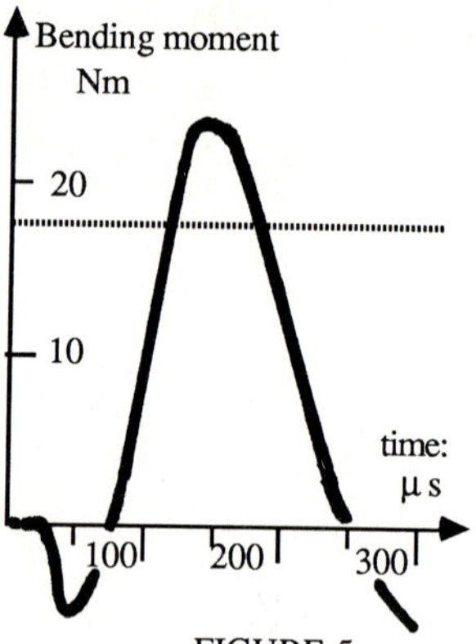

FIGURE 5
Bending moment at 30 mm distance from the corner

Due to the dispersive nature of flexural waves, flexural deformation develops rather slowly in the clamped beam elements AB, A´B´ (Fig. 3). Even though only small parts of energy are transmitted in torsional modes around the corner, this energy cumulates fairly quicly because of another interesting property of transmission and reflection at the corner A: If an incident torsional wave reaches it, approximately 80 % of its energy is *reflected back!* Thus the clamped beams act as *"torsional traps"*. Any incoming energy in a trosional mode is largely conserved and cumulated! After about 300 μs, the torque at the clamping reaches values which remain fairly constant in the subsequent phase and comparable to bending moments which take longer to develop. Eventually, about 1 ms after impact, localized plastic deformation occurs at the clampings by combined torsional and bending stresses; a dynamic collapse-like mechanism appears, involving fairly large "plastic" rotations coupled with elastic vibrations of moderate amplitude. In the last phase, plastic deformation ceases completely by dissipation and only elastic vibrations remain active for some time.

3. ELASTIC IMPACT OF SPHERICAL OBJECTS ON SHELLS

If a steel ball of radius r_b hits an *elastic plate* (for example a glass plate) of thickness h comparable to r_b, the kinetic energy of the ball is largely transformed in flexural energy of the plate by waves traveling a few times across the thickness of the plate during the time of contact. This flexural energy is carried away by corresponding waves radiating from the impact region, and the ball hardly bounces back (see [5], [6]). Fig. 6 shows the coefficient of restitution $e := v_1/v_o$ as a function of the ratio $2r_b/h$, where v_o is the velocity of the ball before and v_1 after impact.

If a similar experiment is carried out with a thin *elastic shell* instead of a plate [7], the coefficient of restitution reaches a *minimum* for a certain radius and *increases* for larger balls (Fig. 7). A careful analysis shows that this is due to the fact that in a shell, flexural waves are stongly coupled with waves in the *membrane mode*. Whereas flexural waves carry all energy away from the region of impact (displacement-time behaviour : Heaviside-function-like), membrane waves *store* potential energy which flows back and generates reverse motion (displacement-time behaviour : oscillatory, Bessel-function-like). As the contact time increases with the ball radius, the reverse motion of the shell at the point of impact causes more and more energy to flow back to the ball and increases the coefficient of restitution with increasing ball radius.

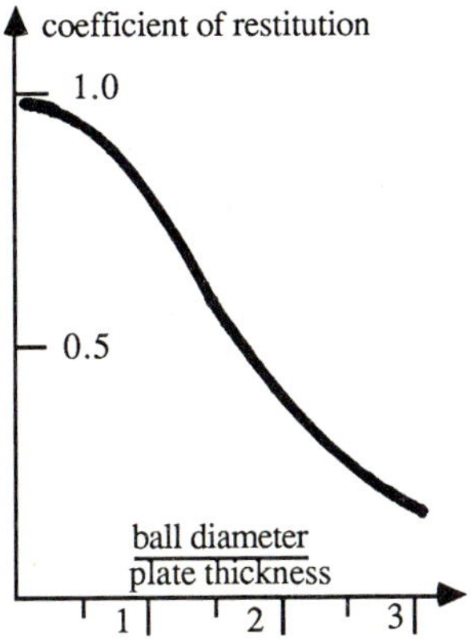

FIGURE 6
Coefficient of restitution for a steel ball of radius r_b on a glass plate of thickness h, impact velocity $v_o = 3.4$ m/s

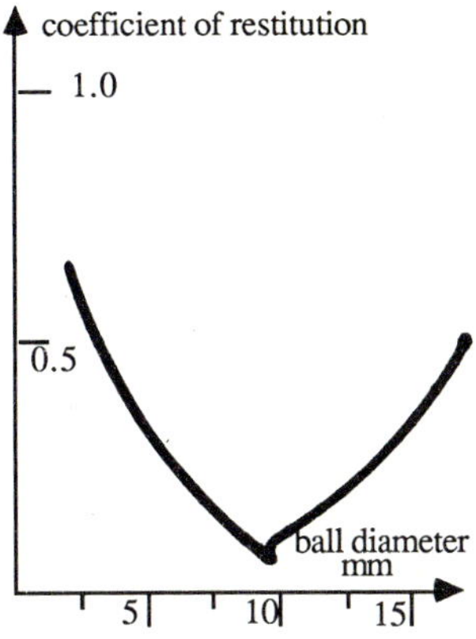

FIGURE 7
Coefficient of restitution for a glass shell with radius of curvature 0.55 m, thickness 1.8 mm, impact velocity $v_o = 2,.9$ m/s

Membrane waves coupled with flexural modes play an important role in the dynamics of *wavy roof structures* subjected to accidental impact (Fig. 8). One of the failure modes of such structures is a longitudinal crack starting at the boundary of the wavy plate. This is mainly due to membrane waves generated together with flexural waves at the point of impact. Such membrane waves travel along the axis of the cylindrical individual elements of the wavy plate, reach the boundary and generate by reflection strong tensile hoop stresses which induce failure in the brittle fiber-reinforced cement mostly used in the design of wavy roof structures.

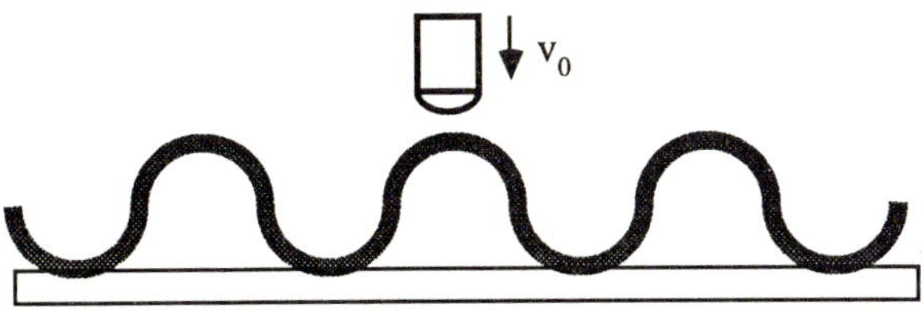

FIGURE 8
A wavy roof structure subjected to impact

4. CONCLUSION

Coupling between different modes of elastic wave propagation may cause particular patterns of energy transfer. A thorough understanding of such patterns and of the parameters influencing them is quite often a prerequisite for successful dynamical design. The three examples shortly described above illustrate some interesting physical features of dynamic mode coupling accessible to both analytical and experimental analysis.

REFERENCES

[1] Lee, J.P. and Kolsky, H., The Generation of Stress Pulses at the Junction of Two Non-collinear Rods, J.Appl.Mech., 39, (1972), 809-813.

[2] Sayir, M., Flexural Vibrations of Strongly Anisotropic Beams, Ing.Arch., 49, (1980), 309-330.

[3] Kolsky, H. and Mosquera, J.M., Dynamic Loading of Fiber-Reinforced Beams, Mech. of Mat. Behaviour, Elsevier Science Pub. (1984), 201-218.

[4] Sayir, M., Theoretical and Experimental Results on the Dynamic Behaviour of Composite Beams, Plates and Shells, Lecture Notes in Engineering, Springer Verlag, 28 (1986), 72-88.

[5] Zener, C., The Intrinsic Elasticity of Large Plates, Phys. Rev., 59, (1941), 669-673.

[6] Koller, M.G., Elastischer Stoss von Kugeln auf dicke Platten, Ph.D. Thesis ETH Zurich No. 7299, (1983).

[7] Koller, M.G. and Busenhart, M., Elastic Impact of Spheres on Thin Shallow Spherical Shells, Int. J. Impact Engng. 4, (1986), 11-21.

ELASTIC WAVE PROPAGATION
M.F. McCarthy, M.A. Hayes, (Editors)

EXACT SOLUTIONS FOR DIFFRACTION-INDUCED CRACK PROPAGATION AND DISLOCATION EMISSION

Louis M. BROCK

Engineering Mechanics
University of Kentucky
Lexington, Kentucky, United States of America

The importance of fracture-dislocation interaction has been demonstrated, but generally by quasi-static analyses. Here, exact dynamic solution development is outlined, and some results are utilized to study this interaction in order to gain insight into some important aspects of fracture under dynamic loading.

1. INTRODUCTION

Several analyses[1-3] have demonstrated the importance of dislocation emission from crack edges in the characterization of fracture. Other work[4,5] has shown that dislocation arrays can lower the effective stress intensity factor of a crack. These studies are, however, quasi-static, so that information on rates and the ordering of events is difficult to obtain, and the possibility of dynamic overshoot common in fracture problems[6] must be neglected. This article, therefore, outlines some efforts to obtain and then utilize exact dynamic solutions for fracture-dislocation interaction in 2D unbounded solids containing semi-infinite cracks.

2. TWO CANONICAL WAVE PROPAGATION PROBLEMS

2.1. Definition

Burridge and Knopoff[7] have shown that classical Volterra[8] dislocations can be represented as body forces. The body forces are linear combinations of concentrated forces and force doublets, which suggests the following two canonical wave propagation problems: The first canonical problem considers a concentrated in-plane force near a semi-infinite crack; in the second problem, the concentrated force is anti-plane. Clearly, differentiation, convolution, and superposition of the first problem solution would yield solutions for Mode I and Mode II fracture interaction with edge and climb dislocations; similarly, the second problem solution could be used to generate solutions for Mode III fracture-screw dislocation interaction. Indeed, the methodologies developed for the canonical problems might well prove to be applicable to the solution of the interaction problems themselves.

2.2. Present Solution Status

If there are no characteristic lengths, then the first canonical problem can be solved easily; indeed, the corresponding interaction problem itself may be solveable exactly, e.g. Burgers and Freund[9]. Generally, however, the interaction problems of interest in [1-5] exhibit characteristic lengths. Freund[10] has, nevertheless, presented a solution for equal and opposite normal concentrated forces applied to opposite surfaces of a stationary crack at a fixed distance from the crack edge. Brock has generalized the results to the case of normal and tangential concentrated forces applied to only one of the crack surfaces[11] and, by superposition, to the first canonical problem itself[12] in the case where both the force and crack remain stationary. Unfortunately, the solution is in the form of multiple integrals which were, for purposes of analysis, unwieldy.

For the second canonical problem, however, the situation is more favorable: Achenbach[13] has treated Mode III crack growth under largely arbitrary anti-plane loading by adopting methods for supersonic flow problems[14], and the results have been used to study stress intensity factor response. Nevertheless, the complete solution expressions are, in general, multiple integrations. While not nearly as formidable as those mentioned above, they do make analysis inconvenient.

Brock, therefore, considered the canonical problem instead, and was able to generate rather simple complete solutions[15]. More recently, the same solution techniques were applied to the interaction problem itself, and complete solutions for the both arbitrary screw dislocation motion near a growing Mode III crack, and for plane SH-wave diffraction obtained[16,17]. The solutions are either first integrals or in closed form. Thus, a rather complete study of the types of problems considered in [1-5] can be made for Mode III interaction, and some results are presented here. These results involve idealized situations, but allow some useful physical insights to be gained. Therefore, the wave propagation solutions employed do not make full use of the generality of the basic solution. In particular, the situations treated involve constant speeds of dislocation and crack propagation.

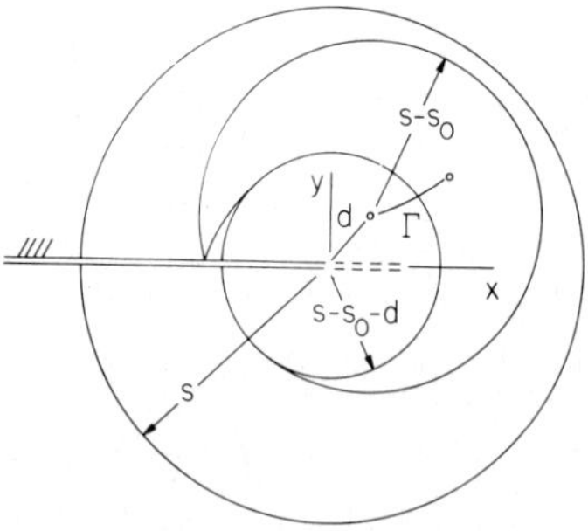

FIGURE 1
Basic Solution wave pattern

3. SCREW DISLOCATION MOTION NEAR A GROWING CRACK

3.1. Basic Solution

Consider the situation illustrated in Figure 1. For $s<0$, where $s=(\text{time})\times(\text{rotational wave speed})$, a screw dislocation is in equilibrium with a stationary semi-infinite crack a distance d from the edge. For $s>0$, the crack begins to extend with an arbitrary, sub-critical speed. The signal from this extension reaches the dislocation at $s=d$ in the form of a cylindrical SH-wave and, at some $s=s_0>d$, the dislocation begins to move along an arbitrary contour Γ at an arbitrary, sub-critical speed. The resulting wave pattern shown in Figure 1 shows clearly that (d,s_0) are characteristic lengths. As indicated above, the exact solution is available in [16]. Although hypothetical, it provides the apparatus to treat the following problems:

3.2. Dislocation Emission due to SH-wave Diffraction

Consider a step-stress pulse in the form of a plane SH-wave approaching a stationary crack along a path normal to the crack plane. At s=0, diffraction occurs, and at some $s=s_o>0$, a screw dislocation is emitted from the crack edge and moves along the crack plane with a constant, subcritical speed. The exact solution for this problem can be obtained by superposing the appropriate special case of the solution to the basic problem with the solution to the SH-wave diffraction problem. The latter problem has been treated in [13], and the complete solution is available in [16]. From this superposition, the Peach-Koehler force[18] acting on the dislocation can be derived. This force is completely in the crack plane, and is given by

$$\frac{2}{\pi}\tau b\left[\sqrt{\left(\frac{s-cT_o}{cT_o}\right)} + \tan^{-1}\sqrt{\left(\frac{cT_o}{s-cT_o}\right)}\right] - \frac{\mu b^2}{2\pi}\frac{1}{2cT_o}\sqrt{\left(\frac{1+c}{1-c}\right)}, \quad T_o = s-s_o \qquad (1a,b)$$

where $s-cT_o>0$ and $\tau>0$ is the step-stress pulse magnitude, while $b>0$ is the Burger's vector magnitude, and the screw dislocation is right-handed. The parameter $c(0<c<1)$ is the dislocation speed non-dimensionalized with respect to the rotational wave speed. The quantity μ is the shear modulus.

Theory[1-3,19] suggests that this process can, in fact, occur only when the stress corresponding to the Peach-Koehler force exceeds the lattice friction stress, which itself can be identified with the material yield stress, σ. Thus, it is easily shown that (1a) must exceed σb. By setting (1a) equal to σb, and eliminating s, we obtain an equation for the instantaneous crack edge-dislocation separation cT_o, parameterized by $(\tau,\sigma,\mu,b,c,s_o)$. Application of the principle of the argument[20] to this equation shows that two real roots can exist. The smaller root defines the region at the crack edge in which the dislocation was created, and so must be less than the core radius h of the dislocation[21]; the larger root is the distance d^* at which the emitted dislocation arrests. By this interpretation, we can obtain from the roots of the equation the approximations

$$\beta\frac{d^*}{h} = \left[\frac{1}{4\pi}\frac{\mu b}{\sigma h}\sqrt{\left(\frac{1+c}{1-c}\right)}\right]^2, \quad s^* = \frac{d^*}{c} + s_o, \quad s_o = \beta\left(\frac{\pi\sigma}{2\tau}\right)^2\left[1+\sqrt{\left(\frac{d^*}{\beta h}\right)}\right]^2 h \qquad (2a\text{-}c)$$

which are valid for $d^*/s_o \ll 1$. Here s^* corresponds to the instant of dislocation arrest, and $\beta h(0<\beta<1)$ is the distance from the crack edge at which the dislocation fully coalesced upon its emission.

Two observations are in order: First, the arrest distance d^* is independent of the loading on the crack, i.e. SH-wave diffraction. More importantly, it should be noted that quasi-static counterparts to (2a)[1-3] are similar in form, except that there is no speed parameter c. Therefore, d^* demonstrates a distinctive dynamic overshoot effect.

The question can also be asked: does dislocation emission precede or lag behind fracture at the crack edge? It is easily shown[16] that fracture initiation will precede this emission process only if

$$s_o > \frac{\pi}{2}\left(\frac{\mu}{2\tau}\right)^2 hk_c^2, \quad k_c = \frac{1}{\mu\sqrt{h}}K_3^c \qquad (3a,b)$$

where K_3^c is the critical value of the dynamic stress intensity factor K_3. Indeed, fracture can precede dislocation arrest only if

$$\lambda_1\sqrt{(1+c)^3} - \lambda_2\sqrt{c}\sqrt{(1-c^2)} - \sqrt{c}(1-c) \geq 0 \qquad (4a)$$

$$\lambda_1 = \frac{1}{4\pi^2}\frac{\tau b}{\mu h}\left(\frac{\mu}{\sigma}\right)^2, \quad \lambda_2 = \frac{1}{2\sqrt{(2\pi)}}\frac{\mu}{\sigma}k_c \tag{4b,c}$$

It can be shown[16] that for an aluminum-like material, for example, inequality (3a) can only be satisfied if c~1, while inequality (4a) is satisfied only when c~0 or c~1. Thus, the dynamic analysis presented here predicts, appropriately, that, even in a linear elastic model, fracture under dynamic loading in this type of material cannot be purely brittle. The analysis also confirms the dislocation-induced lowering of the stress intensity factor noted in [4,5] since it can be shown that, once emission occurs

$$K_3 = 2\tau\sqrt{\left(\frac{2}{\pi}T_o\right)} - \frac{\mu b}{\sqrt{(2\pi)}}\sqrt{\left(1+\frac{1}{c}\right)}\frac{1}{\sqrt{T_o}} \tag{5}$$

Again, the presence of c indicates a dynamic effect: the lowering, or shielding[5] due to the dislocation is larger.

3.3. Effects of Crack Motion on Dislocation Trajectories

Using approximations of the basic problem solution described, Brock and Jolles[22] examined screw dislocation trajectories resulting from crack motion. They assumed, however, that the dislocations always migrate toward the crack edge. In [17], the results of [16] were used to develop exact solutions for a stationary array of screw dislocations near a growing semi-infinite crack. In particular, the anti-plane displacement w due to a single dislocation was found to be

$$w = w_o + W \tag{6}$$

$$\frac{2\pi}{b}w_o = \text{Im}\,\ell n\left(\frac{\sqrt{z}+\sqrt{D}}{\sqrt{z}-\sqrt{D}}\right), \quad z = re^{i\theta}, \quad D = de^{i\psi} \tag{7a-c}$$

$$\frac{4\pi}{b}W = \text{Im}\,\ell n\left(\frac{\sqrt{z}+\sqrt{D}}{\sqrt{z}-\sqrt{D}}\right)\left(\frac{\sqrt{\bar{D}}+\sqrt{z}}{\sqrt{\bar{D}}-\sqrt{z}}\right)\left(\frac{\sqrt{z'}-\sqrt{D'}}{\sqrt{z'}+\sqrt{D'}}\right)\left(\frac{\sqrt{\bar{D}'}-\sqrt{z'}}{\sqrt{\bar{D}'}+\sqrt{z'}}\right), \quad s \geq r \tag{7d}$$

where (r,θ) and (d,ψ) are, as seen in Figure 2, the polar coordinates of, respectively, the observation point and dislocation positions. The parameter C defines the crack extension, which begins at s=0. In eq. (7), $(\bar{\ })$ denotes the complex conjugate, and

$$z' = r'e^{i\theta'} = z - C', \quad C' = C(s'), \quad s' - C' = s - x \tag{8a-d}$$

The presence of the retarded time variable s' shows that W is the displacement component generated by the wave motion, while w_o is induced by the equilibrium state of the crack and dislocation. This observation can be confirmed by the work of [4]. Finally, in eq. (7) we also have the similar formulas

$$D' = d'e^{i\psi'} = D - C' \tag{9}$$

By appropriate superposition of eq. (6), the Peach-Koehler force f on the dislocation, indicated schematically in Figure 2, is easily calculated for an array of dislocations. For the purposes of illustration, consider 7

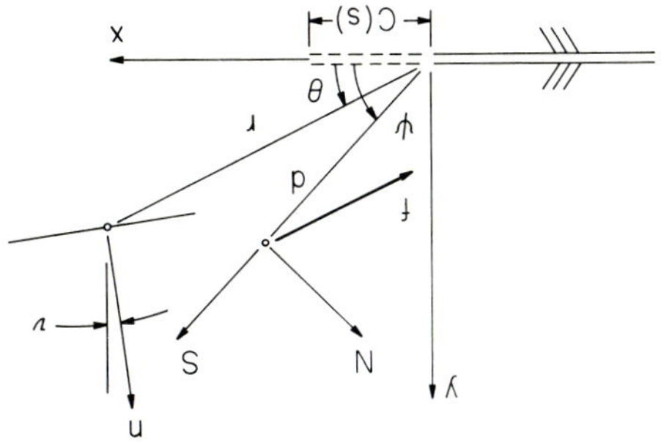

FIGURE 2
Geometry for single screw dislocation

pairs of dislocations arranged in a circle of radius d around (x,y)=0. Figure 3 shows the total f on various array members at different instants. The x-axis intercepts denote the crack edge, which moves at a constant speed which is 20% of the rotational wave speed, i.e. $\dot{C} = 0.2$. Figure 3 shows that the motion signal causes a prominent and instantaneous jump in both the magnitude and direction of f. The direction behavior indicates that dislocations could very well migrate toward the crack edge.

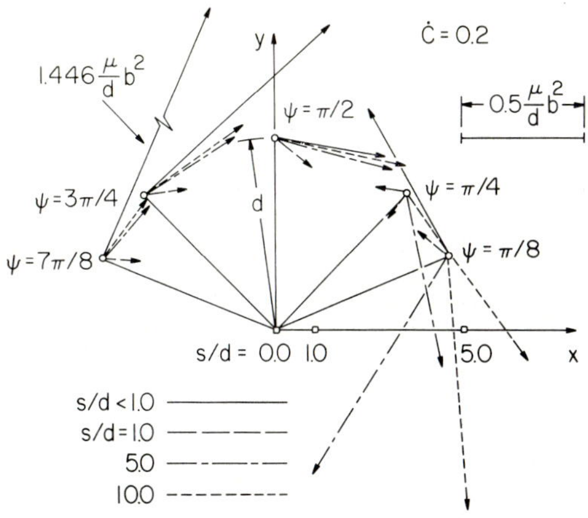

FIGURE 3
Total force on members of array

4. BRIEF DISCUSSION

The results given here illustrate that, by first studying two canonical wave propagation problems which may exhibit characteristic lengths, exact solutions to some dynamic problems of fracture-dislocation interaction can be obtained. Study of these solutions, in turn, yields important information about the characterization of fracture. In particular, the relative brittleness of fracture under dynamic loading can be calibrated in terms of definite time spans. More generally, the solutions indicate clearly the importance of including dynamic/wave propagation effects in studies of fracture.

ACKNOWLEDGEMENTS

This work was supported by NSF Grant MEA 8319605.

REFERENCES

[1] Rice, J.R. and Thomson, R., Philosophical Magazine 29 (1974) 73.
[2] Li, J.C.M., Dislocation Sources, in: Dislocation Modelling of Physical Systems (Pergamon, New York, 1981) pp. 498-518.
[3] Ohr, S.M., Materials Science and Engineering 72 (1985) 1.
[4] Majumdar, B.S. and Burns, S.J., Acta Metallurgica 29 (1981) 579.
[5] Thomson, R.M. and Sinclair, J.E., Acta Metallurgica 30 (1982) 1325.
[6] Achenbach, J.D. and Brock, L.M., On Quasistatic and Dynamic Fracture, in: Sih, G.,(ed.), Dynamic Crack Propagation (Noordhoff, Leyden, 1973) pp. 529-541.
[7] Burridge, R. and Knopoff, L., Bulletin of the Seismological Society of America 54 (1964) 1875.
[8] Love, A.E.H., A Treatise on The Mathematical Theory of Elasticity (Dover, New York, 1944).
[9] Burgers, P. and Freund, L.B., ASME Journal of Applied Mechanics 49 (1982).
[10] Freund, L.B., International Journal of Engineering Science 12 (1974) 179.
[11] Brock, L.M., International Journal of Solids and Structures 18 (1982) 467.
[12] Brock, L.M., Quarterly of Applied Mathematics 43 (1985) 201.
[13] Achenbach, J.D., Zeitschrift fur angewandte Mathematik und Physik 21 (1970) 887.
[14] Evvard, J.C., Distribution of Wave Drag and Lift in the Vicinity of Wingtips at Supersonic Speeds, NACA Technical Note No. 1382 (July 1947).
[15] Brock, L.M., Quarterly of Applied Mathematics 44 (1986) 265.
[16] Brock, L.M., An Exact Dynamic Analysis of Dislocation Emission and Fracture, University of Kentucky Department of Engineering Mechanics Technical Report (January 1988).
[17] Brock, L.M., Transient Generalized Forces due to Dislocation Array -Growing Crack Interaction, University of Kentucky Department of Engineering Mechanics Technical Report (February 1988).
[18] Mura, T., Micromechanics of Defects in Solids (Martinus Nijhoff, The Hague, 1982).
[19] Bilby, B.A., Cottrell, A. and Swinden, K.H., Proceedings of the Royal Society of London A272 (1963) 304.
[20] Hille, E., Analytic Function Theory, Volume 1 (Ginn, Waltham, 1959).
[21] Hirth, J.P. and Lothe, J., Theory of Dislocations (Wiley, New York, 1982).
[22] Brock, L.M. and Jolles, M., International Journal of Solids and Structures 23 (1987) 607.

ELASTIC WAVE PROPAGATION
M.F. McCarthy, M.A. Hayes, (Editors)

ON INDUCED DISCONTINUITIES IN A CLASS OF LINEAR MATERIALS WITH INTERNAL STATE PARAMETERS

Peter J. CHEN

Sandia National Laboratories
Albuquerque, New Mexico 87185
USA

Angelo C. MORRO

DIBE - University of Genova
Viale Causa 13, 16145 Genova
Italy

The behavior of induced discontinuities is considered for a class of materials with internal state parameters. The constitutive relation provides the Cauchy stress tensor in terms of the infinitesimal strain tensor and a symmetric tensor which represents internal state parameters. The differential equations are derived which govern the evolution of induced discontinuities behind longitudinal and transverse shock waves. The influence of constitutive coefficients on the growth of induced discontinuities is discussed.

1. INTRODUCTION

Knowledge of the way in which waves propagate in material bodies is of great importance in modelling material behavior. Wave speed and attenuation are determined by material properties which then become measurable through experiments on wave features. Depending on the type of experiments we may consider different descriptions of waves. In particular time harmonic waves are considered in the case of periodic perturbations and propagating singular surfaces when discontinuities are of relevance. Here the attention is focussed on singular surfaces and specifically on induced discontinuities behind shock waves. It is our purpose to show how the behavior of induced discontinuities may give remarkable information on material coefficients.

The materials to which our investigation applies belong to a class of solids with internal parameters. The constitutive relation for the stress and the rate law for the internal state parameters are taken to be linear. Then we determine exact differential equations for the induced discontinuities behind longitudinal and transverse shock waves. The equations indicate that the behavior of the induced discontinuities depends on the shock amplitudes and, non-linearly, on the wave surface geometries. Specific solutions are obtained that give the global behavior of induced discontinuities for the case of plane waves with initially flat profiles.

The internal state parameters are sufficiently flexible to permit interpretations of material responses in various contexts. For example, we can model relaxation processes as exhibited by viscoelastic materials or relaxation in shear and stiffening in extension (compression) as exhibited by some chemically reacting mixtures [1]. In this sense our results on induced discontinuities will prove effective in ascertaining certain material behavior.

2. THE MODEL OF SOLID WITH INTERNAL STATE PARAMETERS

We consider a solid whose states are pairs $(\mathbf{E}, \boldsymbol{\xi})$ of the infinitesimal strain tensor $\mathbf{E}$ and a symmetric tensor $\boldsymbol{\xi}$. The tensor $\boldsymbol{\xi}$ represents the internal state parameters and depends on time according to the evolution equation

$$\dot{\xi} = \alpha\,\xi + \lambda_\xi(\mathrm{tr}\,\mathbf{E})\mathbf{1} + 2\,\mu_\xi\,\mathbf{E}.$$

Here $\alpha, \lambda_\xi, \mu_\xi$ are scalar quantities with the only restriction that $\alpha < 0$, which guarantees asymptotic stability. The stress is specified by letting the Cauchy stress tensor $\mathbf{T}$ be given by

$$\mathbf{T} = \lambda(\mathrm{tr}\,\mathbf{E})\,\mathbf{1} + 2\,\mu\,\mathbf{E} + \xi.$$

The quantities λ, μ may be regarded as the instantaneous Lamé coefficients and are required to satisfy the inequalities

$$\mu > 0, \qquad 3\,\lambda + 2\,\mu > 0.$$

3. SHOCK WAVES

A shock wave is defined to be a propagating singular surface S across which $\mathbf{u}$ and ξ are continuous while the derivatives of $\mathbf{u}$ and ξ suffer jump discontinuities across S but are continuous everywhere else. The surface S is represented by the parametric equations

$$x_i = Y_i(V^\Gamma, t)$$

where latin subscripts denote Cartesian components and V^Γ, $\Gamma = 1,2$, is a pair of surface parameters. We denote by $a_{\Gamma\Delta}$ and $b_{\Gamma\Delta}$, respectively, the components of the surface metric and the second fundamental form of the surface.
Let U_N be the speed of propagation and $\mathbf{N}$ the unit normal of the wave. The global form of balance of linear momentum implies that, across S,

$$[\mathbf{T}]\mathbf{N} = -\rho_0 U_N[\dot{\mathbf{u}}]. \tag{3.1}$$

We let $a_i = [\dot{u}_i]$ represent the components of the amplitude vector of the shock. It follows from (3.1) that every shock wave is either longitudinal or transverse. The speed of the longitudinal shock, U_{NL}, is given by

$$\rho_0 U_{NL}^2 = \lambda + 2\,\mu$$

and that of the transverse shock, U_{NT}, is given by

$$\rho_0 U_{NT}^2 = \mu.$$

At a longitudinal shock $a_i = aN_i$. At a transverse shock a_i is a surface vector, $a_i N_i = 0$, and the relation between the spatial components a_i and the surface components a^Γ is

$$a_i = Y_{i;\Gamma}a^\Gamma, \qquad a^\Gamma = a_i Y^\Gamma_{i;}.$$

As shown in [2] (cf. also [3]), the amplitude a of a longitudinal shock evolves in time according to the equation

$$2\,\rho_0\,U_{NL}\frac{da}{dt} = (\lambda + 2\mu)b^\Gamma_\Gamma a + \frac{\lambda_\xi + 2\mu_\xi}{U_{NL}}a. \tag{3.2}$$

For the transverse shock wave there are three possibilities of representing the evolution equation depending on whether the spatial components, the surface components or the amplitude itself are considered. For example,

$$2\,\rho_0 U_{NT}\frac{da_i}{dt} = \mu\, b^{\Gamma}_{\Gamma} a_i + \frac{\mu_\xi}{U_{NT}} a_i. \tag{3.3}$$

These results give the explicit dependence of the wave amplitudes on the wave surface geometries and the properties of the material. If as is the case of some chemically reacting mixtures, the material exhibits relaxation in shear and stiffening in extension then the amplitude of a plane shear wave will always decay exponentially, but that of a plane longitudinal wave may grow or decay exponentially depending on whether $\lambda_\xi + 2\mu_\xi$ is positive or negative. The case when λ_ξ is nearly equal to $-2\mu_\xi$ is of interest; the amplitude of a plane longitudinal shock evolves slowly as compared to the decay of the amplitude of a plane transverse shock.

4. INDUCED DISCONTINUITIES

Roughly speaking, an induced discontinuity is the slope of the wave profile immediately behind a shock. The procedure for the derivation of the induced discontinuities hinges essentially on the use of

$$[\operatorname{div} \dot{\mathbf{T}}] = \rho_0[\ddot{\mathbf{u}}].$$

Here we summarise the main results (cf. [2]).

Longitudinal shock wave. Starting from the assumption that the amplitude of the shock is uniform over the entire shock surface, viz. $a_{;\Gamma} = 0$, it follows that

$$[Y^{\Gamma}_{i;} N_j \dot{u}_{i,j}] = 0,$$

whereby the surface components of the normal derivative of the velocity are continuous across the shock. Then the induced discontinuity is the normal component of the normal derivative of the velocity, viz. $[N_i N_j \dot{u}_{i,j}]$. Its evolution equation turns out to be

$$2\,\rho_0\, U_{NL}\,\frac{d}{dt}[N_i N_j \dot{u}_{i,j}] = (\lambda + 2\mu) b^{\Gamma}_{\Gamma}[N_i N_j \dot{u}_{i,j}] + \frac{\lambda_\xi + 2\mu_\xi}{U_{NL}}[N_i N_j \dot{u}_{i,j}]$$
$$+\frac{3(\lambda + 2\mu)}{2} b^{\Gamma}_{\Delta}\, b^{\Delta}_{\Gamma}\, a + \frac{\lambda + 2\mu}{4}(b^{\Gamma}_{\Gamma})^2 a + \frac{3(\lambda_\xi + 2\mu_\xi)^2}{4\,\rho_0\, U^4_{NL}} a - \frac{\alpha(\lambda_\xi + 2\mu_\xi)}{U^2_{NL}} a. \tag{4.1}$$

The solution of (4.1) depends on the solution of (3.2). By way of example, consider the case of a plane wave, preceded by a plane shock, with an initially flat profile. Accordingly we take the initial conditions as

$$a(0) = a_0, \qquad [N_i N_j \dot{u}_{i,j}](0) = 0.$$

Integration of (4.1) yields

$$[N_i N_j \dot{u}_{i,j}](t) = a_0\Big(\frac{3(\lambda_\xi + 2\mu_\xi)^2}{8\,\rho_0^2\, U^5_{NL}} - \frac{\alpha(\lambda_\xi + 2\mu_\xi)}{2\,\rho_0\, U^3_{NL}}\Big)t\, \exp\Big(\frac{\lambda_\xi + 2\mu_\xi}{2\,\rho_0\, U^2_{NL}}\, t\Big).$$

As a consequence

$$\operatorname{sgn}[N_i N_j \dot{u}_{i,j}] = +(-)\operatorname{sgn} a_0 \quad \text{if} \quad \frac{3(\lambda_\xi + 2\mu_\xi)^2}{4\,\rho_0\, U^2_{NL}} - \alpha(\lambda_\xi + 2\mu_\xi) > (<) 0.$$

Since $\alpha < 0$, either of the preceding possibilities may occur depending on the value of α when

$$\lambda_\xi + 2\mu_\xi < 0.$$

In this case the magnitude of the induced discontinuity increases monotonically with t for small t, attains a maximal value, and decreases monotonically to zero as $t \to \infty$. Otherwise the magnitude of the induced discontinuity is monotonically increasing in time and tends to infinity as $t \to \infty$.

Transverse shock wave. Assume that the surface variation of the amplitude vector, $a_{i;\Delta}$, is proportional to the normal of the wave surface. It follows that $a^{\Gamma}_{;\Delta} = 0$ and

$$[N_i N_j \dot{u}_{i,j}] = 0$$

whereby the normal component of the normal derivative of the velocity is continuous across a transverse shock. Then the induced discontinuity consists of the surface components of the normal derivative of the velocity, viz. $[Y^{\Sigma}_{i;} N_j \dot{u}_{i;j}]$. The evolution equation can be written as

$$\begin{aligned} 2\,\rho_0\,U_{NT}\,\frac{d}{dt}[Y^{\Sigma}_{i;} N_j \dot{u}_{i;j}] &= \mu\, b^{\Gamma}_{\Gamma}[Y^{\Sigma}_{i;} N_j \dot{u}_{i,j}] + 2\,\mu\, b^{\Gamma}_{\Delta}[Y^{\Delta}_{i;} N_j \dot{u}_{i,j}] + \frac{\mu_\xi}{U_{NT}}[Y^{\Sigma}_{i;} N_j \dot{u}_{i,j}] \\ &+ \mu\, b^{\Gamma}_{\Gamma}\, b^{\Sigma}_{\Delta}\, a^{\Delta} + \frac{\mu}{2} b^{\Gamma}_{\Delta}\, b^{\Delta}_{\Gamma}\, a^{\Sigma} + \frac{\mu}{4}(b^{\Gamma}_{\Gamma})^2 a^{\Sigma} + \frac{3\,\mu_\xi^2}{4\rho_0 U^4_{NT}} a^{\Sigma} - \frac{\alpha\mu_\xi}{U^2_{NT}} a^{\Sigma}. \end{aligned} \tag{4.2}$$

By analogy with the previous case consider a plane transverse wave with initial conditions

$$a^{\Sigma}(0) = a^{\Sigma}_0, \qquad [Y^{\Sigma}_{i;} N_j \dot{u}_{i,j}](0) = 0.$$

Then the solution of (4.2) is given by

$$[Y^{\Sigma}_{i;} N_j \dot{u}_{i;j}](t) = a^{\Sigma}_0 \Big(\frac{3\,\mu_\xi^2}{8\,\rho_0^2\,U^5_{NT}} - \frac{\alpha\,\mu_\xi}{2\,\rho_0\,U^3_{NT}}\Big) t \exp\Big(\frac{\mu_\xi}{2\,\rho_0\,U^2_{NT}} t\Big).$$

Hence

$$\operatorname{sgn}[Y^{\Sigma}_{i;} N_j \dot{u}_{i,j}] = +(-)\operatorname{sgn} a^{\Sigma}_0 \quad \text{if} \quad \frac{3\mu_\xi^2}{4\,\rho_0\,U^2_{NT}} - \alpha\,\mu_\xi > (<) 0.$$

Depending on the value of α either of the preceding possibilities may occur when $\mu_\xi < 0$. For both longitudinal and transverse waves with curved surfaces the system of equations which must be solved simultaneously is the equation of the components of the shock amplitude vector (3.2) or (3.3), the evolution equation (4.1) or (4.2), and the equation

$$\frac{d}{dt} b^{\Gamma}_{\Delta} = U_N\, b^{\Gamma}_{\Lambda}\, b^{\Lambda}_{\Delta}, \qquad U_N = U_{NL},\ U_{NT},$$

giving the evolution of the mixed components of the second fundamental form.

REFERENCES

[1] Gruschka, H.D. and Wecken, F., Gasdynamic Theory of Detonation (Gordon & Breach, New York, 1971).
[2] Chen, P.J. and Morro, A., Meccanica 22 (1987), 14.
[3] Chen, P.J., Nuovo Cimento B 81 (1984), 113.

ELASTIC WAVE PROPAGATION
M.F. McCarthy, M.A. Hayes, (Editors)

ESTIMATES FOR DYNAMICAL SOLUTIONS IN FINITE THERMOELASTODYNAMICS

M. ARON[1] and R.E. CRAINE[2]

[1]Department of Mathematics, Plymouth Polytechnic, Plymouth
PL4 8AA, UK
[2]Faculty of Mathematical Studies, The University, Southampton
SO9 5NH, UK

1. INTRODUCTION

In recent years growth properties of solutions to the dynamical equations in the theory of elasticity have been extensively investigated for both the linear and nonlinear cases. Most of these results have been obtained using either the logarithmic convexity method or the concavity technique. A discussion of the known results and a comprehensive list of references up to 1981 can be found in [1]. In the context of nonlinear elasticity authors have concentrated their efforts on finding restrictions on the material and on the initial data that would imply the non-existence of solutions to the dynamical problem, globally in time. As corollaries to their main results the authors have deduced important conclusions regarding the stability of solutions to the equilibrium equations and the behaviour, in time, of solutions to the dynamical equations, in the time-interval of their existence. Generalizations of some of these results to nonlinear thermoelasticity have been made by Hills and Knops [2,3], Dafermos [4] and Chirita [5]. In [4] and [5] continuous dependence results are obtained, firstly, under convexity assumptions and, secondly, when a strong ellipticity condition is satisfied.

In this paper we prove that the null solution to the equations of finite thermoelastodynamics depends continuously on the data provided that the Helmholtz free energy satisfies a certain growth inequality, that the specific heat is bounded uniformly from below by a positive constant and that the data is restricted in a manner set out in section 3 of the paper. The restriction that is placed upon the Helmholtz free energy neither implies nor (in general) is implied by the assumptions considered in [4] and [5]. In the proof of our continuous dependence result we obtain an estimate for the growth in time of a certain energy functional that, in turn, implies estimates for the displacement, strain, velocity and temperature corresponding to a smooth (classical)* solution of the underlying equations. These estimates have the character of a priori estimates when either the external heat supply is zero or when an a priori lower bound for the absolute temperature can be found from the physics of the problem.

The governing equations for finite thermoelastodynamics are stated in section 2 and the particular initial-boundary-value problem considered in this paper is then presented. An energy functional is defined in section 3 and manipulations involving this functional, the governing differential equations and the constitutive assumptions enable an integral equation for the functional to be derived. The solution to the integral equation is then found using a Gronwall-type inequality due to Brezis [6].

2. PRELIMINARIES

An elastic body in its undeformed reference configuration C occupies the domain Ω. We assume that Ω is an open, bounded and connected subset of R^3 that is properly regular (see Fichera [7]), so that the integration on the boundary Σ of Ω is well defined and integration by parts over Ω is permissible. The material form of the equation governing the balance of linear momentum in a continuous material is

$$\text{Div}\,\underline{T} + \rho_0\underline{b} = \rho_0\ddot{\underline{u}}\ , \tag{2.1}$$

(the reader can consult Eringen and Suhubi [8], for example, for a derivation of the basic equations stated in this section). In equation (2.1) ρ_0 denotes the density of the body in C, $\underline{b}$ is the body force per unit mass of C, $\underline{u}$ represents the displacement of a point in Ω at time t and $\underline{T}$ is the first Piola-Kirchhoff stress tensor. The material form of the balance of energy equation is

$$\rho_0\dot{\varepsilon} = \underline{T}\cdot\dot{\underline{F}} - \text{Div}\,\underline{Q} + \rho_0 r\ , \tag{2.2}^{\dagger}$$

where ε is the internal energy and r the external heat supply (both measured per unit mass in C), $\underline{Q}$ is the referential heat flux and $\underline{F}$ (= $\underline{I}$+Grad $\underline{u}$, $\underline{I}$ being the identity tensor) denotes the deformation gradient tensor. Throughout this paper Div and Grad represent the divergence and gradient operators with respect to the position $\underline{X}$ in C and a superposed dot denotes partial differentiation with respect to t keeping $\underline{X}$ fixed.

Let η denote the entropy per unit mass in C and Θ represent the (strictly positive) absolute temperature. Then, the second law of thermodynamics can be written

$$\rho_0\dot{\eta} + \text{Div}(\underline{Q}\Theta^{-1}) - \rho_0 r\Theta^{-1} \geq 0\ . \tag{2.3}$$

After defining the free energy function ψ by

$$\psi = \varepsilon - \Theta\eta\ , \tag{2.4}$$

a standard development of the constitutive theory for a thermoelastic material leads to

$$\psi = \bar{\psi}(\underline{F},\Theta),\quad \underline{T} = \rho_0\partial\bar{\psi}/\partial\underline{F},\quad \eta = -\partial\bar{\psi}/\partial\Theta,\quad \underline{Q} = \bar{\underline{Q}}(\underline{F},\Theta,\text{Grad}\Theta)\ . \tag{2.5}$$

Using equations (2.4) and (2.5) the energy equation (2.2) becomes

$$\rho_0\Theta\dot{\eta} = -\text{Div}\,\underline{Q} + \rho_0 r\ , \tag{2.6}$$

and hence inequality (2.3) simplifies to

$$-\underline{Q}\cdot\text{Grad}\,\Theta \geq 0\ . \tag{2.7}$$

Next, denote the constant reference temperature of the system by Θ_R and define the temperature difference θ through

$$\theta = \Theta - \Theta_R\ . \tag{2.8}$$

In this paper the initial and boundary conditions considered are

$$\underline{u} = \underline{u}_0,\quad \dot{\underline{u}} = \underline{v}_0\ \text{ and }\ \theta = \theta_0\ \text{ all on } \Omega \times (-\infty,0]\ , \tag{2.9}$$

$$\underline{u} = \underline{0}\ \text{ on } \Sigma_1 \times (-\infty,+\infty),\quad \underline{T}\,\underline{N} = \underline{t}_s\ \text{ on } \Sigma_2 \times (-\infty,+\infty)\ , \tag{2.10}$$

$$\theta = 0\ \text{ on } \Sigma_3 \times (-\infty,+\infty),\quad \underline{Q}\cdot\underline{N} = Q_s\ \text{ on } \Sigma_4 \times (-\infty,+\infty)\ . \tag{2.11}$$

In (2.9) - (2.11) $\underline{N}$ is the unit outward normal to Σ and Σ_i (i=1,...,4) are subsets of the boundary Σ such that $\Sigma_1 \cup \Sigma_2 = \Sigma$, $\Sigma_1 \cap \Sigma_2 = \varphi$, $\Sigma_1 \neq \varphi$, $\Sigma_3 \cup \Sigma_4 = \Sigma$ and $\Sigma_3 \cap \Sigma_4 = \varphi$, where φ denotes the empty set. Attention is restricted to problems with dead mechanical loading only, i.e. situations where $\underline{b}$ and $\underline{t}_s$ do not depend on the deformation.

In the following section we consider dynamical solutions to equations (2.1) and (2.6) which are consistent with equations (2.5) and inequality (2.7) and satisfy the initial and boundary conditions (2.9) to (2.11).

3. DERIVATION OF ESTIMATES

Define the functional E(t) by

$$E(t) = \int_\Omega \rho_0(\bar{\varepsilon} - \Theta_R\eta + \tfrac{1}{2}\dot{\underline{u}}\cdot\dot{\underline{u}})d\Omega \ . \qquad (3.1)^{\S}$$

On differentiating (3.1) with respect to t, substituting for $\dot{\varepsilon}$ and $\dot{\eta}$ from equations (2.2) and (2.6) respectively and using the boundary conditions (2.10) and (2.11) we obtain the result

$$\dot{E}(t) = \int_{\Sigma_2} \underline{t}_S\cdot\dot{\underline{u}}d\Sigma + \int_\Omega \rho_0\underline{b}\cdot\dot{\underline{u}}d\Omega + \int_\Omega \Theta_R\Theta^{-2}\underline{Q}\cdot\text{Grad}\Theta d\Omega - \int_{\Sigma_4} \theta\Theta^{-1}Q_S d\Sigma + \int_\Omega \rho_0\theta\Theta^{-1}r d\Omega \ . \qquad (3.2)$$

With the aid of inequality (2.7), equation (3.2) implies that

$$\dot{E}(t) \leq \int_{\Sigma_2} \underline{t}_S\cdot\dot{\underline{u}}d\Sigma + \int_\Omega \rho_0\underline{b}\cdot\dot{\underline{u}}d\Omega - \int_{\Sigma_4} \theta\Theta^{-1}Q_S d\Sigma + \int_\Omega \rho_0\theta\Theta^{-1}r d\Omega \ . \qquad (3.3)$$

Two ways of developing inequality (3.3) are now presented.

(i) Gurtin [9] has defined processes that are consistent with a thermally passive environment to be those that satisfy an inequality which, for our problem, can be written

$$\int_{\Sigma_4} \theta\Theta^{-1}Q_S d\Sigma - \int_\Omega \rho_0\theta\Theta^{-1}r d\Omega \geq 0 \ . \qquad (3.4)$$

Combining the inequalities (3.3) and (3.4) yields

$$\dot{E}(t) \leq \int_{\Sigma_2} \underline{t}_S\cdot\dot{\underline{u}}d\Sigma + \int_\Omega \rho_0\underline{b}\cdot\dot{\underline{u}}d\Omega \qquad (3.5)$$

(cf. [9, inequality (9.13)]) and, after the integration of (3.5) and use of the Cauchy-Schwartz inequality, we obtain

$$E(t) \leq E(0) + \int_{\Sigma_2}\left[\int_0^t \underline{t}_S\cdot\dot{\underline{u}}d\tau\right] d\Sigma + \rho_0\int_0^t \|\underline{b}\|_\Omega \|\dot{\underline{u}}\|_\Omega \, d\tau \ . \qquad (3.6)$$

In writing (3.6) we adopt the notation, used throughout this paper, that for any vector function $\underline{f}$

$$\|\underline{f}\|^2_\Omega \doteq \int_\Omega \underline{f}\cdot\underline{f} \, d\Omega \ . \qquad (3.7)$$

(ii) In the second approach it is assumed that

$$\Theta \geq K > 0 \text{ in } \Omega, \quad \theta Q_S \geq 0 \text{ on } \Sigma_4 \ . \qquad (3.8)^+$$

It should be noted that the constant K in $(3.8)_1$ can possibly be obtained from the underlying physics of the problem. Observe also that the postulate $(3.8)_2$ includes the commonly used boundary condition $\underline{Q}\cdot\underline{N} = \bar{h}\,(\Theta-\Theta_R) = \bar{h}\theta$ on Σ_4, where $\bar{h}$ (>0) is the heat transfer coefficient. Integration of inequality (3.3) with respect to t and the introduction of the assumptions (3.8) yields, after two uses of the Cauchy-Schwartz inequality, the result

$$E(t)\leq E(0) + \int_{\Sigma_2}\left\{\int_0^t \underline{t}_S\cdot\dot{\underline{u}}\,d\tau\right\}d\Sigma + \rho_0\int_0^t \|\underline{b}\|_\Omega\,\|\dot{\underline{u}}\|_\Omega\,d\tau + \rho_0 K^{-1}\int_0^t \|r\|_\Omega\,\|\theta\|_\Omega\,d\tau, \quad (3.9)$$

where for a scalar function g we define

$$\|g\|^2 \doteq \int_\Omega g^2 d\Omega . \quad (3.10)$$

Since inequality (3.6) can be obtained formally from (3.9) by letting $K\to\infty$ (keeping all other quantities fixed) we shall subsequently develop in detail only the more general inequality (3.9).

With the further assumption that

$$\underline{t}_S = \underline{0} \text{ on } \Sigma_2 \times (-\infty,0] , \quad (3.11)$$

integration by parts of the first integral in (3.9) and further uses of the Cauchy-Schwartz inequality leads to

$$\int_{\Sigma_2}\left\{\int_0^t \underline{t}_S\cdot\dot{\underline{u}}\,d\tau\right\}d\Sigma \leq \|\underline{t}_S\|_{\Sigma_2}\,\|\underline{u}\|_{\Sigma_2} + \int_0^t \|\dot{\underline{t}}_S\|_{\Sigma_2}\,\|\underline{u}\|_{\Sigma_2}\,d\tau , \quad (3.12)$$

where we have defined

$$\|\underline{h}\|^2_{\Sigma_2} \doteq \int_{\Sigma_2} \underline{h}\cdot\underline{h}\,d\Sigma \quad (3.13)$$

for any vector function $\underline{h}$. Next, μ is defined to be the solution of the eigenvalue problem

$$\Delta\underline{u} = \underline{0} \text{ in } \Omega, \quad \underline{u} = \underline{0} \text{ on } \Sigma_1, \quad (\underline{N}\cdot\nabla)\underline{u} = \mu\underline{u} \text{ on } \Sigma_2, \quad (3.14)$$

where Δ denotes the Laplacean, and we then obtain the Stekloff inequality

$$\mu = \min_{\underline{u} = \underline{0} \text{ on } \Sigma_1} \left\{ \frac{\|\text{Grad}\,\underline{u}\|^2_\Omega}{\|\underline{u}\|^2_{\Sigma_2}} \right\} \quad (3.15)$$

(see [10]). With the aid of (3.12) and (3.15) the inequality (3.9) can be written

$$\begin{aligned} E(t) \leq E(0) &+ \mu^{-\frac{1}{2}} \|\underline{t}_S\|_{\Sigma_2}\,\|\text{Grad}\,\underline{u}\|_\Omega + \mu^{-\frac{1}{2}}\int_0^t \|\dot{\underline{t}}_S\|_{\Sigma_2}\,\|\text{Grad}\,\underline{u}\|_\Omega\,d\tau \\ &+ \rho_0\int_0^t \|\underline{b}\|_\Omega\,\|\dot{\underline{u}}\|_\Omega\,d\tau + \rho_0 K^{-1}\int_0^t \|r\|_\Omega\,\|\theta\|_\Omega\,d\tau . \end{aligned} \quad (3.16)$$

A Taylor series expansion in temperature (with remainder) of the free energy function ψ, together with the result $(2.5)_3$ and the definition of the specific heat c leads to

$$\bar{\varepsilon} - \Theta_R\eta = \bar{\psi}_R + c^*\theta^2/\Theta^* , \qquad (3.17)$$

where c^* is the value of $\frac{1}{2}c$ at temperature Θ^* (Θ^* lying between Θ_R and Θ) and $\bar{\psi}_R = \bar{\psi}(\Theta = \Theta_R)$. The formula (3.17), first obtained by Ericksen [11], coupled with the assumptions

$$c^*/\Theta^* \geq p , \quad \int_\Omega \bar{\psi}_R d\Omega \geq k\|\mathrm{Grad}\ \underline{u}\|_\Omega^2 \qquad (3.18)$$

for some positive constants p and k, then yields

$$\int_\Omega \rho_0(\bar{\varepsilon}-\Theta_R\eta)d\Omega \geq \rho_0 k\|\mathrm{Grad}\ \underline{u}\|_\Omega^2 + \rho_0 p\|\theta\|_\Omega^2 . \qquad (3.19)$$

Both of the postulates (3.18) have been used by other authors. Inequality $(3.18)_1$, which implies that the specific heat is strictly positive, appears in [12] whilst $(3.18)_2$ is identical to the stability condition stated by Knops and Wilkes [10] and used extensively elsewhere.

It now follows directly from (3.1), (3.7) and (3.19) that

$$\|\dot{\underline{u}}\|_\Omega^2 \leq 2\rho_0^{-1}\ E(t), \quad \|\mathrm{Grad}\ \underline{u}\|_\Omega^2 \leq (\rho_0 k)^{-1}\ E(t), \quad \|\theta\|_\Omega^2 \leq (\rho_0 p)^{-1}\ E(t), \qquad (3.20)$$

and hence inequality (3.16) can be written

$$E(t) \leq E(0) + \|\underline{t}_S\|_{\Sigma_2}\ (\rho_0\mu k)^{-\frac{1}{2}}\{E(t)\}^{\frac{1}{2}} + \int_0^t H(\tau)\ \{E(\tau)\}^{\frac{1}{2}}\ d\tau , \qquad (3.21)$$

where $H(\tau)$ is defined by

$$H(\tau) = (2\rho_0)^{\frac{1}{2}}\ \|\underline{b}\|_\Omega + \rho_0^{\frac{1}{2}}K^{-1}p^{-\frac{1}{2}}\ \|r\|_\Omega + (\rho_0 k\mu)^{-\frac{1}{2}}\ \|\dot{\underline{t}}_S\|_{\Sigma_2} . \qquad (3.22)$$

With the result

$$\|\underline{t}_S\|_{\Sigma_2}\ (\rho_0\mu k)^{-\frac{1}{2}}\ \{E(t)\}^{\frac{1}{2}} \leq \tfrac{1}{2}\alpha\|\underline{t}_S\|_{\Sigma_2}^2 + (2\alpha\rho_0\mu k)^{-1}\ E(t) , \quad \alpha > 0, \qquad (3.23)$$

and the further postulate

$$\|\underline{t}_S\|_{\Sigma_2} \leq T , \quad T = \text{const.}, \quad T \geq 0 , \qquad (3.24)$$

inequality (3.21) becomes

$$\{1-(2\alpha\rho_0\mu k)^{-1}\}\ E(t) \leq E(0) + \tfrac{1}{2}\alpha T^2 + \int_0^t H(\tau)\ \{E(\tau)\}^{\frac{1}{2}}\ d\tau . \qquad (3.25)$$

Provided that $\alpha > (2\rho_0\mu k)^{-1}$, $H(\cdot) \in L_1[0,t^*]$ and $E^{\frac{1}{2}}(\cdot) \in C[0,t^*]$ the Gronwall-type inequality derived by Brezis [6] and used recently by Aron [13] allows us to deduce from (3.25) that

$$\{E(t)\}^{\frac{1}{2}} \leq \mathcal{E}(\alpha,t) \doteq \{\alpha\rho_0\mu k(2\alpha\rho_0\mu k-1)^{-1}\ (2E(0) + \alpha T^2)\}^{\frac{1}{2}}$$
$$+ \alpha\rho_0\mu k\ (2\alpha\rho_0\mu k-1)^{-1}\int^t H(\tau)d\tau , \quad t \in [0,t^*] . \qquad (3.26)$$

If at time t the minimum value of $E(\alpha,t)$ occurs at $\alpha = \alpha^*(t)$ then it follows from (3.20) and (3.26) that

$$\|\dot{\underline{u}}\|_\Omega \leq 2^{\frac{1}{2}}\rho_0^{-\frac{1}{2}}E(\alpha^*,t), \qquad \|\mathrm{Grad}\ \underline{u}\|_\Omega \leq (\rho_0 k)^{-\frac{1}{2}}E(\alpha^*,t), \qquad (3.27)$$
$$\|\theta\|_\Omega \leq (\rho_0 p)^{-\frac{1}{2}}E(\alpha^*,t) \quad \text{all for } t \in [0,t^*] .$$

Moreover, with the Poincaré inequality

$$\|\underline{u}\|_\Omega \leq \ell\|\mathrm{Grad}\ \underline{u}\|_\Omega, \quad \ell = \text{const.}, \quad \ell > 0, \qquad (3.28)$$

we obtain from $(3.27)_2$ the estimate

$$\|\underline{u}\|_\Omega \leq \ell(\rho_0 k)^{-\frac{1}{2}}E(\alpha^*,t) \quad \text{for } t \in [0,t^*] . \qquad (3.29)$$

Finally, we note from (3.22) and (3.26) that in the formal limit $K \to \infty$ the expression for $H(\tau)$ is simplified and consequential changes occur in E.

FOOTNOTES AND REFERENCES

* Since the existence of solutions under the hypotheses adopted in this paper has not yet been established our results are purely formal. Moreover, it is known that when a solution does exist singularities may develop in the solution after some finite time. Given smooth initial data, however, one would expect that there will always be an interval of time in which a smooth solution does exist [1].

† In the space of tensors T we define the inner product of $\underline{A}$, $\underline{B} \in T$ by $\underline{A}\cdot\underline{B}$ = trace ($\underline{A}\ \underline{B}^T$), where $\underline{B}^T$ denotes the transpose of $\underline{B}$.

§ The quantity $\bar{\varepsilon}$ is defined from $\bar{\psi}$ in an obvious way.

\+ When the external heat supply r is zero assumption $(3.8)_2$ implies that inequality (3.4) is satisfied.

[1] Knops, R.J., Instability and the Ill-posed Cauchy Problem in Elasticity, in: Hopkins, H.G. and Sewell, M.J., (eds.), Mechanics of Solids (Pergamon, New York-Oxford, 1981) pp. 357-382.

[2] Hills, R.N. and Knops, R.J., Proc. Roy. Soc. Edin. (A) 72 (1974) 239.

[3] Hills, R.N. and Knops, R.J., SIAM J. Appl. Math. 30 (1976) 424.

[4] Dafermos, C.M., Arch. Rat. Mech. Anal. 70 (1979) 167.

[5] Chirita, C., Arch. Mech. 35 (1983) 117.

[6] Brezis, H., Operateurs Maximaux Monotones et Semi-Groups de Contractions dans les Espaces de Hilbert (North-Holland, Amsterdam, 1973) p.157.

[7] Fichera, G., Lectures on Elliptic Differential Systems and Eigenvalue Problems, (Springer-Verlag, Berlin, 1965) p. 11.

[8] Eringen, A.C. and Suhubi, E.S., Elastodynamics, vol. 1, (Academic Press, New York-London, 1974) ch. 1.

[9] Gurtin, M.E., Modern Continuum Thermodynamics, in: Mechanics Today, vol. 1 (Pergamon, New York, 1972) ch. IV,

[10] Knops, R.J. and Wilkes, E.W., Theory of Elastic Stability, in: Truesdell, C., (ed.), Handbuch der Physik, vol. VIa/3 (Springer-Verlag, Berlin-Heidelberg-New York, 1973), p. 265.

[11] Ericksen, J.L., Int. J. Solids Structures 2 (1966) 573.

[12] Naghdi, P.M. and Trapp, J.A., Arch. Rat. Mech. Anal. 51 (1973) 165.

[13] Aron, M., Z. Angew. Math. Phys. 35 (1984) 424.

ELASTIC WAVE PROPAGATION
M.F. McCarthy, M.A. Hayes, (Editors)

EXTERIOR BOUNDARY VALUE PROBLEMS IN ELASTODYNAMICS*

George C. HSIAO

Department of Math Sciences
University of Delaware
Newark, Delaware 19716, U.S.A.

Wolfgang L. WENDLAND

Mathematisches Institut A
Universität Stuttgart
7000 Stuttgart 80, FRG

In this paper, we present a boundary integral equation method for treating exterior boundary value problems in elastodynamics. The essence of this method is to reduce the boundary value problems to a system of strongly elliptic Fredholm integral equations of the first kind. Existence and uniqueness of solutions are established in suitable Sobolev spaces. In particular, it is shown that this approach is most desirable for deriving the asymptotic expansions for the solutions of low frequency scattering in the two-dimensional case.

1. INTRODUCTION

This paper summarizes some recent results obtained by the authors [5], [6] with respect to the low frequency behaviors of the solutions of exterior two-dimensional fundamental boundary value problems for the harmonic oscillations in elastodynamics. Our approach is based on the reduction of boundary value problems via the direct method to boundary integral equations of the first kind with weakly singular and hypersingular kernels. The corresponding integral operators here are strongly elliptic pseudodifferential operators of orders -1 and $+1$, respectively on the compact boundary manifold. In particular, the following results are established. For the Dirichlet problem, the solution converges on compact subdomain to the stationary solution with the order of $|\ln \omega|^{-1}$, as frequency $\omega \to 0$. This corresponds to the uniform convergence on every compact subdomain of the time dependent solution, for increasing time, to the stationary solution. On the other hand, for the Neumann problem, the solution converges on compact subdomain with the order of $|\omega^2 \ln \omega|$ to its stationary solution provided the given traction satisfies the equilibrium condition (see, Section 2 below).

We remark that our results are similar to those for the Helmholtz equation. The low frequency asymptotics for the Helmholtz equation has been analyzed by using classical Fredholm boundary integral equations of the second kind (see, e.g. [8]). For the Neumann problem, the corresponding integral operator is a small perturbation of the limiting operator

*This work was supported in part by the Center for Advanced Study of the University of Delaware, the Naval Research Laboratory under contract N00014-87-C-2121 and the Stiftung Volkswagenwerk.

with $\omega = 0$, while for the Dirichlet problem, the classical integral operator admits an eigenvalue for $\omega = 0$. This has been handled by a singular perturbation procedure as in [2] and [8] or by a regular perturbation procedure of a modified stationary integral equation of the second kind as in [10] and [11]. Our approach recovers the results of [8] for the Helmholtz equation as indicated in [9], and provides an alternative means other than the ones given by [1] for constructing approximate solutions of the Neumann problem in the case of elastodynamics.

2. STATEMENT OF PROBLEMS

Let Γ be a sufficiently smooth simple closed curve in $\mathbb{R}^2$ and Ω be the exterior domain. Assume that Ω is filled with a homogeneous isotropic elastic material with Lamé constants μ and λ such that $\lambda > -\mu$ and $\mu > 0$. Then the governing equation for the displacement field $\vec{u}$ in elastodynamics is given by

$$\mu \Delta \vec{u} + (\lambda + \mu) \operatorname{grad} \operatorname{div} \vec{u} + \omega^2 \vec{u} = \vec{0} \quad \text{in} \quad \Omega \,. \tag{1}$$

We consider here two fundamental boundary value problems for (1), the *Dirichlet problem* (D) and the *Neumann problem* (N) which consist of (1) together with boundary conditions

$$\vec{u}|_\Gamma = \vec{\phi} \tag{2}$$

and

$$T[\vec{u}]|_\Gamma := \lambda(\operatorname{div} \vec{u})\vec{\nu} + 2\mu \mathcal{E}(\vec{u}) \cdot \vec{\nu}|_\Gamma = \vec{\psi}, \tag{3}$$

respectively, in addition to the usual radiation conditions [1], [7] given in terms of the potential wave $\vec{u}_p$ and the solenoidal wave $\vec{u}_s$ associated with the wave numbers $k_s, k_p \geq 0$,

$$k_p^2 := \omega^2/(\lambda + 2\mu) \quad \text{and} \quad k_s^2 := \omega^2/\mu. \tag{4}$$

Here $\vec{\nu}$ denotes the exterior normal vector to Γ with respect to $\mathbb{R}^2 \backslash \overline{\Omega}$, and $\mathcal{E}(\vec{u}) := (\nabla \vec{u} + \nabla \vec{u}')/2$ is the strain tensor with $\nabla \vec{u}'$ designating the transpose of $\nabla \vec{u}$.

The amplitude $\vec{u}(x;\omega)$ for $\omega > 0$ will be compared with the stationary displacement field $\vec{u}_0$ of the corresponding boundary value problems for the elastostatic equation,

$$\mu \Delta \vec{u}_0 + (\lambda + \mu) \operatorname{grad} \operatorname{div} \vec{u}_0 = \vec{0} \quad \text{in} \quad \Omega \tag{5}$$

subject to the growth conditions

$$\vec{u}_0 = O(1) \quad \text{as} \quad |x| \to \infty \tag{6}$$

and

$$\vec{u}_0 = \frac{\lambda + 3\mu}{4\pi(\lambda + 2\mu)} \left\{ -\ln|x| + \frac{\lambda + \mu}{\lambda + 3\mu} \frac{x\,x}{|x|^2} \right\} \int_\Gamma \vec{\psi}\, ds + o(1) \quad \text{as} \quad |x| \to \infty, \tag{7}$$

respectively for the *Dirichlet problem* (D_0) and for the *Neumann problem* (N_0). From (6), it is shown in [4], [5] that the solution $\vec{u}_0$ of (D_0) has the asymptotic development

$$\vec{u}_0 = \vec{\eta}_0 + O\left(\frac{1}{|x|+1}\right) \quad \text{as} \quad |x| \to \infty,$$

where $\vec{\eta}_0$ denotes the rigid translation. The growth condition (7) for the problem (N_0) will become transparent, after we introduce the fundamental solution for (5). As will be seen, if the given traction $\vec{\psi}$ does not satisfy the equilibrium condition, namely $\int_\Gamma \vec{\psi}\, ds \neq \vec{0}$, not only the stationary solution $\vec{u}_0$ will not be bounded, but also the corresponding time dependent solution of the vibration problem will have *logarithmic blowup* with time as in [10] from the term of $O(\ln \omega)$ in (17) below for the low frequency.

3. BOUNDARY INTEGRAL EQUATIONS

We begin with the Betti representation formula for the unknown amplitude $\vec{u}(x)$ of (1):

$$\vec{u}(x) = \int_\Gamma \{\mathbf{G}_\omega(x,y)\vec{v}(y) - \mathbf{F}_\omega(x,y)\vec{t}(y)\}\, ds_y,\ x \in \Omega\,. \tag{8}$$

Here

$$\mathbf{F}_\omega(x,y) = \frac{i}{4\mu} H_0^{(1)}(k_s|x-y|)\mathbf{I} + \frac{i}{4\omega^2}\nabla_y\nabla_y \{H_0^{(1)}(k_s|x-y|) - H_0^{(1)}(k_p|x-y|)\} \tag{9}$$

is the fundamental displacement tensor in terms of the Hankel function $H_0^{(1)}$, and $\mathbf{G}_\omega(x,y)$ is the fundamental stress tensor defined by

$$\mathbf{G}_\omega(x,y) = (T_y[\mathbf{F}_\omega(x,y)])'. \tag{10}$$

We note that solutions of (1) can be represented by (8) where the right hand side is completely determined by the Cauchy data $\vec{u}|_\Gamma =: \vec{v}$ and $T[\vec{u}]|_\Gamma =: \vec{t}$. Consequently, if we let x approach Γ, we find, with the jump relations on Γ, the equations between the Cauchy data $\vec{v}$ and $\vec{t}$. Thus, for the *Dirichlet problem* (D), we arrive at the boundary integral equation for $\vec{t}$:

$$\mathbf{V}_\omega\vec{t}(x) := \int_\Gamma \mathbf{F}_\omega(x,y)\vec{t}(y)\, ds_y = \vec{\phi}_\omega(x) := -\frac{1}{2}\vec{\phi}(x) + \int_\Gamma \mathbf{G}_\omega(x,y)\vec{\phi}(y)\, ds_y, \tag{11}$$

and for the *Neumann problem* (N), the boundary integral equation for $\vec{v}$:

$$\mathbf{D}_\omega\vec{v}(x) := -T_x\int_\Gamma \mathbf{G}_\omega(x,y)\vec{v}(y)\, ds_y = \vec{\psi}_\omega(x) := -\frac{1}{2}\vec{\psi}(x) - \int_\Gamma T_x[\mathbf{F}_\omega(x,y)]\vec{\psi}(y)\, ds_y. \tag{12}$$

Both (11) and (12) are boundary integral equations of the first kind with weakly singular and hypersingular kernels, respectively. It can be shown that the solution of (11) is unique provided ω^2 is not an eigenvalue of the interior elastodynamic Dirichlet problem, and so is the solution of (12) when ω^2 is not an eigenvalue of the interior elastodynamic Neumann problem (see, e.g. [2]). In what follows, we shall call ω an exceptional frequency, if ω^2 is an eigenvalue of the interior elastodynamic Dirichlet or Neumann problem. Then there is an interval $0 < \omega < \omega_1$ without exceptional frequencies. For these ω, (11) and (12) may be resolved for the unknowns $\vec{t}$ and $\vec{v}$ respectively. In fact, based on corresponding results in elastostatics [3], [4], (11) and (12) can be easily solved for low frequency. For this purpose, we need the asymptotic series development of the fundamental tensor $\mathbf{F}_\omega(x,y)$.

4. LOW FREQUENCY APPROXIMATIONS

Let us first introduce the boundary operators $\mathbf{V}_0$ and $\mathbf{D}_0$ for the elastostatic equation (5):

$$\mathbf{V}_0\vec{t}_0(x) := \int_\Gamma \mathbf{F}_0(x,y)\vec{t}_0(y)\,ds_y, \quad \mathbf{D}_0\vec{v}_0(x) := \int_\Gamma \mathbf{G}_0(x,y)\vec{v}_0(y)\,ds_y, \quad x \in \Gamma, \qquad (13)$$

where $\mathbf{F}_0$ and $\mathbf{G}_0$ are respectively the fundamental displacement and stress tensors for (5) given by

$$\mathbf{F}_0(x,y) = \frac{\lambda+3\mu}{4\pi\mu(\lambda+2\mu)}\left\{-\ln|x-y|\mathbf{I} + \frac{\lambda+\mu}{\lambda+3\mu}\,\frac{(y-x)(y-x)}{|x-y|^2}\right\} \qquad (14)$$

$$\mathbf{G}_0(x,y) = (T_y[\mathbf{F}_0(x,y)])'. \qquad (15)$$

The fundamental tensor $\mathbf{F}_\omega$ has the series expansion [1]:

$$\begin{aligned}\mathbf{F}_\omega(x,y) &= \mathbf{F}_0(x,y) - \frac{\lambda+3\mu}{4\pi\mu(\lambda+2\mu)}\{\ln\omega - \gamma_0\}\mathbf{I} \\ &\quad + \sum_{m=1}^{\infty}(\mathbf{K}_m^1(x,y) + \ln\omega\,\mathbf{K}_m^2(x,y))\omega^{2m},\end{aligned} \qquad (16)$$

where

$$\begin{aligned}\gamma_0 = {} & \frac{\lambda+2\mu}{2(\lambda+3\mu)}\ln\mu + \frac{\mu}{2(\lambda+3\mu)}\ln(\lambda+2\mu) - \frac{\lambda+\mu}{\lambda+3\mu} \\ & + \ln 2 - 0.5772 + i\,\pi/2,\end{aligned}$$

$$\mathbf{K}_m^2(x,y) = O(|x-y|^{2m}) \quad \text{and} \quad \mathbf{K}_m^1(x,y) = O(|x-y|^{2m}\ln|x-y|).$$

This expansion converges absolutely and uniformly for bounded $|x-y|$. Moreover the boundary integral operators defined by the kernels $\mathbf{K}_m^l$, $l = 1,2$ are classical pseudodifferential operators of order $-2m-1$. Applying the traction operator T on Γ to (16) one finds a corresponding expansion for $\mathbf{G}_\omega$ (see [1]). This leads to the following decompositions for $\mathbf{V}_\omega$ and $\mathbf{D}_\omega$, namely

$$\mathbf{V}_\omega\vec{t} = \mathbf{V}_0\vec{t} + \frac{\lambda+3\mu}{4\pi\mu(\lambda+2\mu)}\{\ln\omega - \gamma_0\}\int_\Gamma \vec{t}\,ds + O(\omega^2\ln\omega), \qquad (17)$$

$$\mathbf{D}_\omega\vec{v} = \mathbf{D}_0\vec{v} + \mathbf{B}_\omega\vec{v}, \qquad (18)$$

where

$$\mathbf{B}_\omega \vec{v}(x) = -\sum_{m=1}^{\infty} T_x \left\{ \int_\Gamma (T_y[\mathbf{K}^1_m(x,y) + \ln \omega \, \mathbf{K}^2_m(x,y)])' \vec{v}(y) \, ds_y \right\} \omega^{2m}$$
$$= O(\omega^2 \ln \omega).$$

We remark that from (14), the growth condition (7) can be easily derived from the Betti representation formula for the solutions of (5). Furthermore, expansions (17) and (18) suggest that as a good approximation, for small frequency ω, one may solve (11) and (12) by neglecting terms of $O(\omega^2 \ln \omega)$. This leads to the following results [5], [6].

Theorem 1. Let $\vec{\phi} \in (H^{s+1}(\Gamma))^2$, $s \in I\!R$ and ω_1 be the first exceptional frequency. Then for $0 \le \omega \ll \omega_1$, the solution $\vec{t}$ of (11) admits the representation:

$$\vec{t} = \vec{t}_0 + \beta_1(\omega)\vec{t}_1 + \beta_2(\omega)\vec{t}_2 + O(\omega^2 \ln \omega)$$

with

$$\vec{\eta} := \frac{1}{\alpha} \int_\Gamma \vec{t} \, ds = \vec{\eta}_0 + \beta_1(\omega)\vec{\eta}_1 + \beta_2(\omega)\vec{\eta}_2 + O(\omega^2 \ln \omega)$$

where $(\vec{t}_j, \vec{\eta}_j) \in (H^s(\Gamma))^2 \times I\!R^2$, $j = 0,1,2$, are the unique solution pairs of

$$\mathbf{V}_0 \vec{t}_j - \vec{\eta}_j = \vec{f}_j, \quad \int_\Gamma \vec{t}_j \, ds = \vec{b}_j. \tag{19}$$

Here $\vec{f}_0 = \vec{\phi}_0$, $\vec{f}_j = \vec{0}$, $j = 1,2$, and $\vec{b}_0 = \vec{0}$, $\vec{b}_j = \vec{e}_j$, $j = 1,2$. The coefficients β_1 and β_2 are the unique solutions of the algebraic system:

$$\beta_1(\vec{e}_1 - \alpha\vec{\eta}_1) + \beta_2(\vec{e}_2 - \alpha\vec{\eta}_2) = \alpha\vec{\eta}_0,$$

with

$$\frac{1}{\alpha} = \frac{\lambda + 3\mu}{4\pi\mu(\lambda + 2\mu)} \{\ln \omega - \gamma_0\}.$$

Theorem 2. Let $\vec{\psi} \in (H^s(\Gamma))^2$, $s \in I\!R$ and ω_1 be the first exceptional frequency. Then for $0 \le \omega \ll \omega_1$, the solution $\vec{v}$ of (12) admits the representation:

$$\vec{v} = \vec{v}_0 + \mathbf{M}(x)\vec{\alpha} + O(\omega^2 \ln \omega).$$

Here $\vec{v}_0 \in (H^{s+1}(\Gamma))^2$ is the unique solution of the system

$$\mathbf{D}_0 \vec{v}_0 = \vec{\psi}_0, \quad \int_\Gamma \vec{v}_0 \cdot \vec{m}^\star_k \, ds = 0, \quad k = 1,2,3, \tag{20}$$

where $\vec{m}^\star_k$, $k = 1,2,3$, together with $\eta_{jk} \in I\!R$ are the unique solution pairs of

$$\mathbf{V}_0 \vec{m}^\star_k - \sum_{j=1}^{3} \eta_{jk} \vec{m}_j(x) = 0, \quad \int_\Gamma \vec{m}_j(y) \cdot \vec{m}^\star_k \, ds = \delta_{jk}, \quad j,k = 1,2,3, \tag{21}$$

and $\mathbf{M}(x) := (\vec{m}_1, \vec{m}_2, \vec{m}_3)$ *is the* 2×3 *matrix with* $\vec{m}_j = \vec{e}_j,\ \ j = 1, 2,$ *and* $\vec{m}_3 = (-x_2, x_1)'$. *The vector* $\vec{\alpha} = (\alpha_1, \alpha_2, \alpha_3) \in I\!R^3$ *is explicitly given by*

$$\alpha_k = -\sum_{j=1}^{3} \eta_{jk} \int_\Gamma \vec{m}_j(X) \cdot \vec{\psi}(x)\, ds_x + \frac{1}{\alpha} \sum_{i=1}^{2} \int_\Gamma \vec{\psi} \cdot \vec{e}_i \delta_{ik}\, ds, \quad k = 1, 2, 3.$$

Asymptotic expansions for the corresponding potential fields are easily obtained from the results in Theorems 1 and 2. In particular, we see that for the problem (N), we have

$$\vec{u} = \vec{u}_0 + \frac{\lambda + 3\mu}{4\pi\mu(\lambda + 2\mu)} \{\ln \omega - \gamma_0\} \int_\Gamma \vec{\psi}\, ds + \mathbf{B}_\omega \vec{v} + O(\omega^2 \ln \omega) \tag{22}$$

from which asymptotic behaviors for $\vec{u}$ may be deduced.

REFERENCES

[1] Ahner, J. E., Hsiao, G. C., On the two-dimensional exterior boundary–value problems of elasticity. *SIAM J. Appl. Math* 31 (1976), 677–685.

[2] Colton, D., Kress, R., *Integral Equation Methods in Scattering Theory* (John Wiley & Sons, New York, 1983).

[3] Hsiao, G. C., Kopp, P., Wendland, W. L., Some applications of a Galerkin collocation method for integral equations of the first kind. *Math. Meth. Appl. Sci.* 6 (1984) 280–325.

[4] Hsiao, G. C., Wendland, W. L., On a boundary integral equation method for some exterior problems in elasticity, *Proceeding of Tbilisi University*, Tbilisi University Press, Tbilisi, (1985), 31–60

[5] Hsiao, G. C., Wendland, W. L., On the low frequency asymptotics of the exterior 2–D Dirichlet problem in dynamic elasticity, in: *Inverse and Ill–Posed Problems*, Engl and Groetsch (eds.), (Academic Press, 1987) pp. 461–482.

[6] Hsiao, G. C., Wendland, W. L., Boundary integral equation methods for exterior problems in elastostatics and elastodynamics, in preparation.

[7] Kupradze, V. D., *Potential Methods in the Theory of Elasticity* (Israel Programs for Scientific Translations, Jerusalem, 1965).

[8] MacCamy, R. C., Low frequency acoustic oscillations. *Quart. Appl. Math.* 23 (1965), 247–256.

[9] Wendland, W. L., Boundary element methods and their asymptotic convergence, in: *Theoretical Acoustics and Numerical Techniques*, Filippi (ed.), (Springer–Verlag, 1983) pp. 135–216.

[10] Werner, P., Zur Asymptotik der Wellengleichung und der Wärmeleitungsgleichung in zweidimensionalen Aussenräumen, *Math. Meth. Appl. Sci.* 7 (1985), 170–201.

[11] Werner, P., Low frequency asymptotics for the reduced wave equation in two–dimensional exterior spaces, *Math. Meth. Appl. Sci.* 8 (1986), 134–156.

ELASTIC WAVE PROPAGATION
M.F. McCarthy, M.A. Hayes, (Editors)

CIRCULARLY POLARISED PLANE WAVES IN TRANSVERSELY-ISOTROPIC ELASTIC MATERIALS

D. RYAN (Regional Technical College, Sligo)

and

M. HAYES (University College, Dublin)

1. INTRODUCTION

The purpose of this note is to obtain all the circularly polarised plane wave solutions for transversely-isotropic, homogeneous, linearly elastic media. Both homogeneous and inhomogeneous plane waves are considered. Whereas homogeneous circularly polarised waves may propagate only in certain directions, namely those directions of propagation for which the sheets of the slowness surface touch or intersect, the circles of polarisation of inhomogeneous circularly polarised waves cover the entire sphere.

We seek solutions of the displacement equations of motion for which the displacement $\underline{u}$ has the form $\underline{u} = \underline{A} \exp iw(\underline{S}.\underline{x} - t)$ where $\underline{A}$ and $\underline{S}$ are complex vectors (bivectors) and w is real. For circularly polarised waves $\underline{A}$ is isotropic : the scalar product $\underline{A}.\underline{A} = 0$. It is always possible to write the slowness bivector $\underline{S}$ in the form

$$\underline{S} = T e^{ip} \underline{C} , \quad \underline{C} = m\hat{\underline{m}} + i\hat{\underline{n}} , \tag{1}$$

where T, p and m are real scalars, and $\hat{\underline{m}}$ and $\hat{\underline{n}}$ are orthogonal unit vectors. On prescribing $\underline{C}$, T and P may be determined from the secular equation. Prescribing $\underline{C}$ is equivalent to prescribing an ellipse, the directional ellipse, whose principal axes are along $\hat{\underline{m}}$ and $\hat{\underline{n}}$. As shown by Gibbs [1] the bivector exp(ip) $\underline{C}$ has real and imaginary parts which form a pair of conjugate semi-diameters of this ellipse. This pair gives the directions of the planes of constant phase and the planes of constant amplitude. The phase velocity and attentuation are then determined when T is known. Thus for homogeneous plane waves a direction is specified whereas for inhomogeneous plane waves an ellipse is prescribed.

It has been shown [2] that a necessary and sufficient condition for the propagation of circularly polarised waves is that the secular equation has a double root - for prescribed $\underline{C}$ in the case of inhomogeneous waves, or for prescribed $\hat{\underline{n}}$ in the case of homogeneous waves.

2. BASIC EQUATIONS

The constitutive equations for the material may be written

$$\begin{aligned}
t_{11} &= d_{11} u_{1,1} + d_{12} u_{2,2} + d_{13} u_{3,3} , \quad t_{12} = (d_{11} - d_{12})/2 \; (u_{1,2} + u_{2,1}) , \\
t_{22} &= d_{12} u_{1,1} + d_{11} u_{2,2} + d_{13} u_{3,3} , \quad t_{13} = d_{44} (u_{1,3} + u_{3,1}) , \\
t_{33} &= d_{13}(u_{1,1} + u_{2,2}) + d_{33} u_{3,3} , \quad t_{23} = d_{44}(u_{2,3} + u_{3,2}) ,
\end{aligned} \tag{2}$$

where t_{ij} are the stresses, and $u_{i,j} \equiv \partial u_i / \partial x_j$, the displacement gradients.

The equations of motion in the absence of body forces are

$$d_{11}u_{1,11} + \tfrac{1}{2}(d_{11} - d_{12})u_{1,22} + d_{44}u_{1,33} + \tfrac{1}{2}(d_{11} + d_{12})u_{2,12} + (d_{13} + d_{44})u_{3,13} = \rho\partial^2 u_1/\partial t^2 ,$$

$$\tfrac{1}{2}(d_{11} + d_{12})u_{1,12} + \tfrac{1}{2}(d_{11} - d_{12})u_{2,11} + d_{11}u_{2,22} + d_{44}u_{2,33} + (d_{13} + d_{44})u_{3.23} = \rho\partial^2 u_2/\partial t^2 , \quad (3)$$

$$(d_{13} + d_{44})(u_{1,13} + u_{2,23}) + d_{44}(u_{3,11} + u_{3,22}) + d_{33}u_{3,33} = \rho\partial^2 u_3/\partial t^2 ,$$

where ρ is the density .

Let the displacement be

$$\underline{u} = \underline{A} \exp iw(\underline{S}.\underline{x} - t) . \quad (4)$$

Its real part is

$$\underline{u}^+ = [\underline{A}^+ \cos w(\underline{S}^+.\underline{x} - t) - \underline{A}^- \sin w(\underline{S}^+.\underline{x} - t)] \exp -w\underline{S}^-.\underline{x} . \quad (5)$$

The planes of constant phase are $\underline{S}^+.\underline{x}$ = constant and the planes of constant amplitude are $\underline{S}^-.\underline{x}$ = constant. The wave train is elliptically polarised in the plane of $\underline{A}^+$ and $\underline{A}^-$, a pair of conjugate radii at $\underline{x} = \underline{x}^*$ being

$$\underline{A}^+ \exp - w\,\underline{S}^-.\underline{x}^* \quad \text{and} \quad \underline{A}^- \exp - w\,\underline{S}^-.\underline{x}^* . \quad (6)$$

The sense of description of the waves is from $\underline{A}^+$ to $\underline{A}^-$.

Inserting (1) and (4) into the equations of motion (3) leads to the propagation condition and secular equation :

$$(\underline{Q}(\underline{C}) - \rho T^{-2} e^{-2ip}\,\underline{1})\,\underline{A} = 0 , \quad (7)$$

$$\det(\underline{Q} - \rho T^{-2} e^{-2ip}\,\underline{1}) = 0 , \quad (8)$$

where the elements of $\underline{Q}$ are

$$\begin{pmatrix} d_{11}C_1^2 + \tfrac{1}{2}(d_{11} - d_{12})C_2^2 + d_{44}C_3^2 ; & \tfrac{1}{2}(d_{11} + d_{12})C_1C_2 ; & (d_{13} + d_{44})C_1C_3 \\ \tfrac{1}{2}(d_{11} + d_{12})C_1C_2 ; & \tfrac{1}{2}(d_{11} - d_{12})C_1^2 + d_{11}C_2^2 + d_{44}C_3^2 ; & (d_{13} + d_{44})C_2C_3 \\ (d_{13} + d_{44})C_1C_3 ; & (d_{13} + d_{44})C_2C_3 ; & d_{44}(C_1^2 + C_2^2) + d_{33}C_3^2 \end{pmatrix} . \quad (9)$$

One root of the secular equation (8) is

$$\rho T^{-2} e^{-2ip} = \tfrac{1}{2}(d_{11} - d_{12})(C_1^2 + C_2^2) + d_{44}C_3^2 , \quad (10)$$

the two other roots being solutions of the biquadratic

$$(\rho T^{-2} e^{-2ip})^2 - [(d_{11} + d_{44})(C_1^2 + C_2^2) + (d_{33} + d_{44})C_3^2](\rho T^{-2} e^{-2ip}) + d_{33}d_{44}C_3^4 + [d_{11}d_{33} + d_{44}^2 - (d_{13} + d_{44})^2](C_1^2 + C_2^2)C_3^2 + d_{11}d_{44}(C_1^2 + C_2^2)^2 = 0 . \quad (11)$$

3. CIRCULARLY POLARISED WAVES

In order that circularly polarised waves may propagate for given $\underline{C}$, the secular equation must have a double root. This is possible if either (a) the root (10) is also a root of the biquadratic (11) , or (b) the biquadratic has a double root.

Case (a)

On inserting (10)into equation (11), we find the two possibilities

(i) $C_1^2 + C_2^2 = 0$, and (ii) $(C_1^2 + C_2^2)R = C_3^2$,

where R is given by

$$2[(d_{11} + d_{12})(d_{33} - d_{44}) - 2(d_{13} + d_{44})^2]R = (d_{11} + d_{12})(d_{11} - d_{12} - 2d_{44}) . \quad (12)$$

For Possibility (i)

$$S_1^2 + S_2^2 = 0 , \quad d_{44} S_3^2 = \rho , \quad \underline{A} = (S_2, - S_1, 0) = S_2(1, \mp i, 0) , \quad (13)$$

$$\underline{A}.\underline{S} = 0 , \quad \underline{u} = (1, + i, 0) \exp[iw(\pm i S_2 x + S_2 y + (\rho/d_{44})^{\frac{1}{2}} z - t)] .$$

The circle of polarisation lies in the x - y plane. Since $\underline{A}.\underline{S} = 0$, the projection of the ellipse of $\underline{S}$ upon the plane of $\underline{A}$, the x - y plane , is also a circle. If S_2, which is arbitrary, is chosen to be real, then the planes of constant phase are $S_2 y + (\rho/d_{44})^{\frac{1}{2}} z$ = constant, the planes of constant amplitude being x = constant.

For Possibility (ii)

$$(S_1^2 + S_2^2)R = S_3^2 , \quad [\tfrac{1}{2}(d_{11} - d_{12}) + Rd_{44}] (S_1^2 + S_2^2) = \rho, \quad (14)$$

$$\underline{A} = (C_2 + iN C_1 , - C_1 + iNC_2 , iqNC_3) ,$$

where

$$[4(d_{13}+d_{44})^2 (d_{12}+d_{44}) - (d_{11}+d_{12})^2 (d_{33}-d_{44})]N^2 = -2(d_{13}+d_{44})^2 (d_{11}-d_{12}-2d_{44}), \quad (15)$$

$$(1 + q^2R) N^2 = 1 , \quad 2(d_{13} + d_{44})qR = -(d_{11} + d_{12}) .$$

We note from $(14)_2$, that $S_1^+ S_1^- + S_2^+ S_2^- = 0$, and hence from $(14)_1$, $S_3^+ S_3^- = 0$. Thus the planes of constant phase are orthogonal to the planes of constant amplitude. The polarisation circle may not be in the x - y plane.

Example 1

The polarisation circle may be chosen to lie in the x - z plane. This may be attained as follows. Choose

$$C_1^+ = 0 , \quad C_2^- = 0 , \quad C_3^- = 0 , \quad C_1^- = NC_2^+ , \quad \underline{C} = (iNC_2^+, C_2^+, C_3^+) . \quad (16)$$

Then from $(14)_1$,

$$(1-N^2)R(C_2^+)^2 = (C_3^+)^2 , \quad (17)$$

$$\underline{A}^+ = [(1-N^2) C_2^+, 0, 0] , \quad \underline{A}^- = [0, 0, qNC_3^+] .$$

Using $(15)_2$, it is seen that $|\underline{A}^+| = |\underline{A}^-|$. Also

$$[\tfrac{1}{2}(d_{11} - d_{12}) + d_{44} R] (C_3^+)^2 T^2 = \rho R . \quad (18)$$

Both C_2^+ and T are determined by C_3^+ .

Case (b) Double Roots of Biquadratic

The biquadratic has the double root

$$2\rho\, T^{-2}\, e^{-2ip} = (d_{11} + d_{44})\,(C_1^2 + C_2^2) + (d_{33} + d_{44})C_3^2 \;, \tag{19}$$

provided C_i satisfy

$$(d_{11} - d_{44})(C_1^2 + C_2^2) + (d_{44} - d_{33})C_3^2 = \pm\, 2i(d_{13} + d_{44})(C_1^2 + C_2^2)^{\frac{1}{2}}C_3 . \tag{20}$$

Corresponding to the upper (+) sign

$$(C_1^2 + C_2^2)^{\frac{1}{2}} = ik^{\pm}\, C_3 \;, \tag{21}$$

where $k^{\pm}$ are given by

$$(d_{11} - d_{44})k^{\pm} = (d_{13} + d_{44}) \pm [(d_{13} + d_{44})^2 + (d_{11}-d_{44})(d_{44}-d_{33})]^{\frac{1}{2}} . \tag{22}$$

The corresponding amplitude bivector is

$$\underline{A} = (C_1, C_2, \; - i(C_1{}^2 + C_2{}^2)^{\frac{1}{2}}) = (C_1, C_2, k^{\pm}\, C_3) \;. \tag{23}$$

In general the planes of constant phase are not orthogonal to the planes of constant amplitude. Indeed $\underline{S}^+$ and $\underline{S}^-$ are parallel to a pair of conjugate diameters of the spheroid $x^2 + y^2 + (k^{\pm})^2 z^2 = 1$.
Also $\underline{A}.\underline{S} \neq 0$. The circle of polarisation may not lie in the x - y plane.

Example 2

By choice of $\underline{C}$, the circle of polarisation may lie in the xz plane. Choose

$$C_1^- = 0 \;, \quad C_2 = 0 \;, \quad C_3^+ = 0 \;. \tag{24}$$

Then, from (21) and (23) ,

$$C_1^+ = -k^{\pm}\, C_3^- \;, \quad \underline{C} = (-k^{\pm}\, C_3^- \;, \; 0, \; i\, C_3^-) \;,$$

$$\underline{A} = C_1^+(1, 0, -i) \;, \tag{25}$$

$$2\,\rho\, T^{-2} = [(d_{11} + d_{44})\,(k^{\pm})^2 - (d_{33} + d_{44})]\,(C_3^-)^2 \;.$$

This and Example 1 show that two circularly polarised waves with identical amplitude bivectors in the xz plane may propagate with different directional ellipses.
Finally, corresponding to the lower (-) sign in (20) ,

$$(C_1{}^2 + C_2{}^2)^{\frac{1}{2}} = -ik^{\pm}\, C_3 \;, \tag{26}$$

and the corresponding amplitude bivector is

$$\underline{A} = (C_1, C_2, \; - k^{\pm}\, C_3) \;. \tag{27}$$

References

[1] Gibbs, J.W. Elements of Vector Analysis, 1881, 1884 (privately printed) = pp 17 - 90, Vol. 2, Part 2 Scientific Papers, Dover Publications, New York 1961.

[2] Hayes, M. Inhomogeneous Plane Waves, Arch. Rational Mech. Anal., 85, 41(1984) .

G
MAGNETO-THERMO-ELECTROMAGNETIC ELASTIC WAVE PROPAGATION

ELASTIC WAVE PROPAGATION
M.F. McCarthy, M.A. Hayes, (Editors)

ELECTROELASTIC NONLINEARITIES, BIASING DEFORMATIONS AND PIEZOELECTRIC VIBRATIONS

H.F. Tiersten

Department of Mechanical Engineering,
Aeronautical Engineering & Mechanics
Rensselaer Polytechnic Institute
Troy, New York 12180-3590

1. INTRODUCTION

In the interaction of the quasi-static electric field with deformable insulators the condition of rotational invariance causes a combination of the electric field and the deformation gradients to occur in the constitutive equations along with the finite strain[1-4]. In addition, the Maxwell electrostatic stress tensor is present as well as the usual mechanical stress tensor. Since the resulting system is intrinsically nonlinear, the linear dynamic equations are more general than those of linear piezoelectricity when a bias is present and reduce to them only in the absence of a bias[5,6]. Even in the simplest case of stress-free thermal deformation, which is just about always present, the more general equations arise when the fixed reference coordinates at the reference temperature are employed[6].

The advantage of the use of reference coordinates, which cannot be employed within the usual linear theory, in the accurate calculation of the temperature sensitivity of high precision quartz resonators is shown. In the treatment the equation for the perturbation in eigenfrequency of the piezoelectric solution due to a bias[7], which is obtained from the more general linear equations, is employed. However, both the biasing state and the vibrational solution are obtained by solving systems of unbiased linear equations. The change in frequency resulting from any bias may readily be calculated from the perturbation equation when the linear piezoelectric solution and biasing state are known.

The very accurate asymptotic differential equation describing the transversely varying essentially thickness modes[8] used in the technologically important high precision contoured quartz crystal resonators is presented and the approximations made in obtaining the equation are briefly described. The importance of the phenomenon of energy trapping[9-11], which is employed in the aforementioned contoured crystal resonators, is discussed. The continuous representation of the acoustic surface wave mode shape along the transmission path in acoustic surface wave resonators is presented[12] along with calculations of the normal acceleration sensitivity for certain specific rectangular support configurations[13-15]. The significance of these calculations is discussed.

2. ELECTROELASTIC EQUATIONS AND BIASING DEFORMATIONS

In an earlier paper[6] finite deformation theory was discussed in some detail in order to give some indication of the reasons for the nonlinear electroelastic equations being fundamentally different from the linear piezoelectric equations. Sections I - III of Ref.6 are considered to be part of this work and we, of course, employ the notation of Ref.6 here. Nevertheless, certain portions of Ref.6 are included here for emphasis. As in Ref.6 the present position y_i of a material point is a function of the reference position X_L and time t, which enables us to write $y_i = y_i(X_L, t)$. A schematic diagram of a deformation is shown in Fig.1 of Ref.6, in which the unit normal to a material point on the reference position of the surface is denoted N_L and the unit normal to the same material point on the present position of the surface is denoted n_i, which, of course, differs from N_L. The mechanical displacement of a point is denoted by u_K in the same figure. We now consider Sec.I of Ref.6 with the exception of the last paragraph to be part of this work for the sake of brevity.

As in Ref.6 we now note that under a static bias the material points move from the reference coordinates X_L to the intermediate coordinates ξ_α, and we may write $\xi_\alpha = \xi_\alpha(X_L)$. Then in the superposed small dynamic motion the material points move from the intermediate coordinates ξ_α to the present coordinates y_i, and we have $y_i = y_i(\xi_\alpha, t) = \hat{y}_i(X_L, t)$.

A schematic diagram showing the intermediate biasing state as well as the present and reference configurations is shown in Fig.2 of Ref.6. As already noted ξ_α denotes the coordinates of the intermediate position of the material point which was at the reference position X_L. The unit normal to a material point on the intermediate position of the surface is denoted ν_α, which, of course, differs from the normal N_L to the reference position of the same point of the surface. The static displacement of a material point from its reference position to its intermediate position is denoted w_K and the dynamic displacement of the same point from the intermediate position to the present position is denoted u_β. In this nonlinear description the small-field dynamic equations may use either the intermediate coordinates ξ_α or the reference coordinates X_L as independent variables. Since in the typical situation it is undesirable to measure the geometry each time the bias is varied, it is advantageous to use the X_L as independent variables. Moreover, in the applications envisaged although the static biasing mechanical displacement w_L of a material point is large compared with u_β, it is nevertheless small. For the sake of brevity we now consider all of Sec.II of Ref.6 not written here to be part of this work.

As in Ref.6 we now note that the equation for the first perturbation of the eigenvalue of the piezoelectric solution due to a bias mentioned in the Introduction may be written in the form[7]

$$\Delta_\mu = H_\mu / 2\omega_\mu, \quad \omega = \omega_\mu - \Delta_\mu, \tag{2.1}$$

where ω_μ and ω are the unperturbed and perturbed eigenfrequencies, respectively, and in the general case H_μ is given in Eq.(56) of Ref.6. Again, for brevity we now consider Sec.III of Ref.6 to be part of this work.

We now note that purely linear equations must be referred to the intermediate coordinates ξ_α which change with temperature. In earlier work[16] on the temperature derivatives of the effective elastic constants of quartz from measurements of the thickness vibrations of rotated quartz plates using the linear equations the changes in density and thickness of the plate were properly taken into account, but the change in the normal to the major surfaces of the anisotropic plate with respect to the principal axes of the quartz with temperature was omitted. This same omission was made in all subsequent applications[17,18] of those temperature derivatives. Once it is realized that this change in normal with temperature should be included, it is clear that nonlinear deformation theory must be employed. Furthermore, when the nonlinear theory is employed it is also clear that it is advantageous to refer the description to the reference coordinates X_L, which are defined at the reference temperature T_o at which all geometric measurements are made, because the mass density and geometry of the fixed reference state at T_o do not change. In more recent work using the nonlinear description referred to the X_L coordinates the first temperature derivatives of the fundamental elastic constants of quartz were obtained[19]. Numerous applications[20-24] of the proper nonlinearly based formalism have shown that it consistently yields results that are considerably more accurate than the incomplete linearly based description.

3. VIBRATIONS OF PLATES

In order for the solution for straight-crested waves in an isotropic elastic plate with traction-free boundary conditions to satisfy the differential equations and boundary conditions certain relations between the circular frequency ω and propagation wavenumber ξ must be satisfied. These relations take the form of what are called dispersion curves. For the case of elastic waves having displacement components in the direction of propagation and normal to the surfaces of the plate an example of such curves is shown in Fig.1, which was obtained by Mindlin[25] approximately 30 years ago. In the figure Ω denotes the frequency normalized with respect to the lowest thickness shear frequency $\omega_o = (\pi/2h)\sqrt{\mu/\rho}$, where 2h is the thickness of the plate and μ and ρ are the shear modulus and density, respectively, and $\gamma = 2\xi h/\pi$ denotes the dimensionless propagation wavenumber. It is clear from the figure that γ can be real or imaginary (γ can also be complex[26,27] but that is not discussed here). Real γ corresponds to propagating waves and trigonometric standing waves and imaginary (and complex) γ corresponds to decaying waves. Clearly then, imaginary (and complex) γ cannot occur in an infinite plate, but they can occur and, indeed, are important in bounded plates.

The lowest dispersion curve emanating from the origin as a parabola is for the flexural wave in the plate. The curvature of the parabola at the origin is correctly predicted by the elementary theory of the flexure of thin plates[28,29], but it very quickly, i.e., at very low frequencies, becomes incorrect, as is clear from the figure. The straight line emanating from the origin is for the lowest extensional wave in the plate. The slope of this line at the origin is correctly predicted by the elementary theory of the extension of thin plates[28,29], which predicts that the straight line continues forever. The figure shows that this is clearly incorrect.

In the electronics industry high precision resonators and filters employ plates operating in the vicinity of the pure thickness modes, for which $\gamma = 0$ in Fig.1, such as, e.g., the fundamental thickness-shear mode and the third and fifth harmonics. We wish to obtain asymptotic dispersion relations and the associated differential equations in the plane of the plate which very accurately describe the modes in the vicinity of the pure thickness frequencies, in much the same way as the elementary theories of the extension and flexure of thin plates describe the extensional and flexural behavior of the plate in the appropriate limits. However, before we do this we should introduce and briefly explain the concept of energy trapping[9-11] which is so important for the use of the aforementioned transversely varying essentially thickness modes in the electronics industry. To this end we have shown typical dispersion curves in the vicinity of the thickness frequency in Fig.2. The upper curve is for the completely unelectroded plate and the lower curve is for the completely electroded plate, both of which are also shown in the figure. If the partially electroded plate shown in the upper left of Fig.2 is operated at a resonant frequency in between the two thickness frequencies so that the dispersion curves shown dominate the mode, then it is clear from the curves that in the electroded region the mode shape along the plate is trigonometric and in the unelectroded region it decays exponentially. This is called trapping and it enables the edge of the resonator to be held essentially without losing energy from the mode.

It has been shown[8] that in the general anisotropic case for an odd ($n = 1, 3, 5, \ldots$) mode through the thickness in the vicinity of that thickness frequency to lowest order in the small decay numbers ξ and ν along the unelectroded plate the dispersion equation can be written in the form

$$M_n \xi^2 + Q_n \xi\nu + P_n \nu^2 - \frac{n^2 \pi^2 \bar{c}^{(1)}}{4h^2} + \rho\omega^2 = 0\,, \tag{3.1}$$

where M_n, P_n and Q_n are defined in Eqs.(74) of Ref.8. Since quartz has small piezoelectric coupling and the decay (or wave) numbers along the plate are small and the product of two small quantities is negligible, only the thickness dependence of all electrical variables has been taken into account in obtaining (3.1). Aside from this restriction, the solution yielding (3.1) satisfies the differential equations and boundary conditions on the major surfaces of the plate systematically to second order in the small decay numbers along the plate. In obtaining the solution[8] the components of the mechanical displacement vector were transformed to the orthogonal eigenvector triad of the pure thickness solution while continuing to refer the independent variables to the original Cartesian coordinate system. This operation, which is essential to the solution, was performed because in the vicinity of a thickness mode one eigenvector (here $\underset{\sim}{u}_n^{(1)}$ with piezoelectrically stiffened eigenvalue $\bar{c}^{(1)}$) is large while the other two are small. The analysis holds unless there is a coincidence (or near coincidence) of two or more thickness-frequencies. In the treatment Eq.(3.1) is ultimately obtained as the coefficient of the $\underset{\sim}{u}_n^{(1)}$ (denoted u_1^n) equation of motion. It has further been shown[8] that for small wavenumbers $\bar{\xi}$ and $\bar{\nu}$, i.e., trigonometric behavior, along the electroded plate the dispersion equation takes the form

$$M_n \bar{\xi}^2 + Q_n \bar{\xi}\bar{\nu} + P_n \bar{\nu}^2 + \frac{n^2 \pi^2}{4h^2} \hat{c}^{(1)} - \rho\omega^2 = 0\,, \tag{3.2}$$

where

$$\hat{c}^{(1)} = \bar{c}^{(1)}\left(1 - \frac{8k_1^2}{n^2\pi^2} - 2\hat{R}\right), \quad k_1^2 = \frac{e_{26}^2}{\bar{c}^{(1)}\varepsilon_{22}}, \quad \hat{R} = \frac{2\rho' h'}{\rho h}, \tag{3.3}$$

and $\hat{c}^{(1)}$ differs from $\bar{c}^{(1)}$ as shown because of the mass loading and shorting effect of the electrodes.

It has been shown[8] by virtue of (3.2) and the manner of derivation of (3.1) that for each odd harmonic overtone, to second order in the wavenumbers $\bar{\xi}$ and $\bar{\nu}$ for the electroded plate we have

$$\left(M_n\bar{\xi}^2 + Q_n\bar{\xi}\bar{\nu} + P_n\bar{\nu}^2 + \frac{n^2\pi^2\hat{c}^{(1)}}{4h^2} - \rho\omega^2\right) u_1^n = 0, \tag{3.4}$$

where u_1^n is the asymptotic solution function for the nth odd harmonic. We now postulate that the homogeneous differential equation for u_1^n governing the mode shapes along the surface of the electroded plate may be written in the form

$$M_n\frac{\partial^2 u_1^n}{\partial X_1^2} + Q_n\frac{\partial^2 u_1^n}{\partial X_1\partial X_3} + P_n\frac{\partial^2 u_1^n}{\partial X_3^2} + \frac{n^2\pi^2\hat{c}^{(1)}}{4h^2}u_1^n - \rho\ddot{u}_1^n = 0, \tag{3.5}$$

since the substitution of trigonometric dependence on X_1, X_3 and t in (3.5) yields (3.4). It has further been shown[8] that in the inhomogeneous case in which a voltage is applied across the surface electrodes, the nth odd harmonic family of modes is governed by the inhomogeneous equation

$$M_n\frac{\partial^2 \tilde{u}_1^n}{\partial X_1^2} + Q_n\frac{\partial^2 \tilde{u}_1^n}{\partial X_1\partial X_3} + P_n\frac{\partial^2 \tilde{u}_1^n}{\partial X_3^2} - \frac{n^2\pi^2\hat{c}^{(1)}}{4h^2}\tilde{u}_1^n - \rho\ddot{\tilde{u}}_1^n = \rho\omega^2(-1)^{\frac{n-1}{2}}\frac{e_{26}}{c^{(1)}}\frac{4Ve^{i\omega t}}{n^2\pi^2}, \tag{3.6}$$

where

$$c^{(1)} = \bar{c}^{(1)}(1 - k_1^2), \quad u_1^n = \tilde{u}_1^n(X_1,X_3,t)\sin\frac{n\pi X_2}{2h}. \tag{3.7}$$

Clearly, Eq.(3.6) can be transformed to a coordinate system X_1', X_3' in the plane of the plate, in which the mixed derivative term does not appear. When this is done we obtain

$$M_n'\frac{\partial^2 \tilde{u}_1^n}{\partial X_1'^2} + P_n'\frac{\partial^2 \tilde{u}_1^n}{\partial X_3'^2} + \frac{n^2\pi^2\hat{c}^{(1)}}{4h^2}\tilde{u}_1^n - \rho\ddot{\tilde{u}}_1^n = \rho\omega^2(-1)^{\frac{n-1}{2}}\frac{e_{26}}{c^{(1)}}\frac{4Ve^{i\omega t}}{n^2\pi^2}, \tag{3.8}$$

where M_n' and P_n' are expressed in terms of M_n, P_n and Q_n, in accordance with Eqs.(100) of Ref.8 and X_1' and X_3' are expressed in terms of X_1 and X_3, in accordance with Eqs.(97) of Ref.8.

A schematic diagram of a plano-convex resonator is shown in Fig.3. It has been shown that the slowly varying thickness of a spherically contoured resonator can be represented in the form[30,8]

$$2h = 2h_o[1 - (X_1'^2 + X_3'^2)/4Rh_o], \tag{3.9}$$

the substitution of which in (3.8) and expansion to first order in $X_1^2 + X_3^2$ yields

$$M_n' \frac{\partial^2 u^n}{\partial X_1'^2} + P_n' \frac{\partial^2 u^n}{\partial X_3'^2} - \frac{n^2\pi^2 \hat{c}^{(1)}}{4h_o^2}\left(1 + \frac{(X_1'^2 + X_3'^2)}{2Rh_o}\right) u^n + \rho\omega^2 u^n =$$

$$= \rho\omega^2 (-1)^{(n-1)/2} \frac{e_{26}}{c^{(1)}} \frac{4V}{n^2\pi^2}, \tag{3.10}$$

where $\hat{u}_1^n = u^n(X_1', X_3')e^{i\omega t}$. It has been shown that the eigensolutions of the associated homogeneous problem, i.e., of (3.10) with $V = 0$, can be written in the form[30,8]

$$u_{nmp} = e^{-\alpha_n (X_1'^2/2)} H_m(\sqrt{\alpha_n} X_1') e^{-\beta_n (X_3'^2/2)} H_p(\sqrt{\beta_n} X_3'), \tag{3.11}$$

where H_p and H_m are Hermite polynomials and

$$\alpha_n^2 = \frac{n^2\pi^2 \hat{c}^{(1)}}{8Rh_o^3 M_n'}, \quad \beta_n^2 = \frac{n^2\pi^2 \hat{c}^{(1)}}{8Rh_o^3 P_n'}. \tag{3.12}$$

The eigenfrequencies ω_{nmp} obtained from the solution (3.11) are given by

$$\omega_{nmp}^2 = \frac{n^2\pi^2 \hat{c}^{(1)}}{4h_o^2 \rho}\left[1 + \frac{1}{n\pi}\left(\frac{2h_o}{R}\right)^{1/2}\left(\sqrt{\frac{M_n'}{\hat{c}^{(1)}}}\,(2m+1) + \sqrt{\frac{P_n'}{\hat{c}^{(1)}}}\,(2p+1)\right)\right] \tag{3.13}$$

where $n = 1,3,5, \ldots$, $m,p = 0,2,4, \ldots$. We now note that since the mode is sharply confined to the vicinity of the center of the plate, the edge of the electrode and, of course, the edge of the plate were both ignored in the treatment used in obtaining the solution.

The only modes of interest are the harmonics for which $m = p = 0$ and $H_m = H_p = 1$. Contoured AT-cut quartz crystal resonators[31] are used as high precision oscillators because the pure thickness frequency used in the flat AT-cut is thermally compensated. The plate is contoured to confine the mode to the vicinity of the center of the plate in accordance with (3.11) so that the edges can be supported without damaging the quality factor (Q). The contouring causes a small change in the orientation of the zero temperature cut, which is very important technologically. Figure 4 shows a comparison[32,8] of loglog plots of the calculated shift in rotation angle for the zero temperature cut for the fundamental mode for a few center thicknesses and different electrode sizes with the known measured design curves[33] for the fundamental mode of the plano-convex resonator. It can be seen from the figure that the calculated results are not straight lines, but curves and, in fact, a different curve for each different center thickness and each electrode size. Although all the calculated curves tend to follow the general trend of the single measured design line, they deviate from it at large R differently for different center thicknesses. However, the calculation becomes invalid for some large R along some ill-defined curve because the edge of the electrode, which was ignored in the analysis, becomes important for larger values of R. This occurs at lower values of R for larger $2h_o$. Nevertheless, in the really practical regions the calculated curves deviate from the measured line by not more than 1.5'. For n = 3 and n = 5 for the AT-cut there is no change in angle. More recently, similar calculations for the new improved doubly-rotated SC-cut[34] were performed[35] for n = 1,3,5 and the results are plotted in Fig.5, in which the circle on the curve n = 3 was determined from measurements because the precise φ angle cannot be calculated sufficiently accurately because of inaccuracies in the measured temperature derivatives of the elastic constants of quartz[16,19]. The curves shown in Fig.5 were calculated before measurements were made. They were subsequently verified experimentally[36].

4. NORMAL ACCELERATION SENSITIVITY OF ACOUSTIC SURFACE WAVE RESONATORS

A plan view and cross-section of the resonator with the substrate extending beyond rectangular supports is shown in Fig.6. We now briefly present the proper continuous representation of the acoustic surface wave mode shape in resonators with grooved reflectors, which was obtained in recent work[12] that neglected the very small scattering into bulk waves. The straight-crested surface wave displacement field may be written in the known form,

$$u_j = \alpha_j(X_2)e^{i\xi(X_1-Vt)}, \quad \alpha_j = \sum_{m=1}^{4} C^{(m)}A_j^{(m)}e^{i\beta_m\xi X_2} . \tag{4.1}$$

It has been shown[12] that the variable-crested resonant surface wave mode shape with variable amplitude along the transmission path is very accurately approximated by

$$u_j = \cos\frac{\pi X_3}{2w}\,\mathrm{Re}\left[\alpha_j(X_2)\hat{C}^R(X_1)e^{i\xi(X_1-s)} + \alpha_j^*(X_2)\hat{C}^L(X_1)e^{-i\xi(X_1-s)}\right]e^{-i\omega t}, \tag{4.2}$$

where the variations along the transmission path are given in Eqs.(45) of Ref.12. The amplitude of the standing surface wave mode, i.e., the part multiplying cos ωt, along the transmission path is plotted in Fig.7.

We now know g_γ^μ and from an appropriate static analysis $\hat{c}_{L\gamma M\alpha}$ for normal acceleration can be found[15]. Hence, H_μ in Eq.(2.1) can be evaluated. Such calculations have been performed using the known values of the second order[37] and third order[38] elastic constants of quartz. The calculated acceleration sensitivity is plotted as the solid curve in Fig.8 as a function of the planar aspect ratio a/b for b = 5.0 mm and given overhangs ℓ_a and ℓ_b, which are denoted in the figure. In addition, the normal acceleration sensitivities that were calculated in earlier work for the cases of simple[13] and rigid[14] edge supports are plotted in Fig.8 as a function of the planar aspect ratio a/b for the same value of b. It can be seen from the figure that all the sensitivity curves go through zero near the same value of a/b. The sensitivity vanishes for certain planar aspect ratios even though the influence of cylindrical flexure on surface waves is very great, since the wave is located exactly where the flexural biasing deformation is largest, because cylindrical flexure in the two orthogonal spanning directions produces changes in frequency of opposite sign. It is quite clear from the figure that in the vicinity of the zero-crossings the slope of the sensitivity curve is much shallower for the configuration with the overhangs than for the other cases. This means that for the configuration considered here very low normal acceleration sensitivities can be obtained (parts in 10^{11} per g) over a reasonably large and, indeed, practical range of aspect ratio. Such sensitivities have recently been observed at Raytheon with this support configuration.

REFERENCES

1. R.A. Toupin, J. Rational Mech. Anal. 5, 849 (1956).
2. H.F. Tiersten, Int. J. Eng. Sci. 9, 587 (1971).
3. R.N. Thurston, "Waves in Solids," in Encyclopedia of Physics, edited by C. Truesdell (Springer, Berlin, 1974), Vol. VIa/4.
4. D.F. Nelson, Electric, Optic and Acoustic Interactions in Dielectrics (Wiley, New York, 1979).
5. J.C. Baumhauer and H.F. Tiersten, J. Acoust. Soc. Am. 54, 1017 (1973).
6. H.F. Tiersten, J. Acoust. Soc. Am. 70, 1567 (1981).
7. H.F. Tiersten, J. Acoust. Soc. Am. 64, 832 (1978).
8. D.S. Stevens and H.F. Tiersten, J. Acoust. Soc. Am. 79, 1811 (1986).
9. W. Shockley, D.R. Curran and D.J. Koneval, J. Acoust. Soc. Am. 41, 981 (1967).
10. M. Onoe and H. Jumonji, Electronics and Comm. Eng. (Japan), 84 (1965).
11. R.D. Mindlin and P.C.Y. Lee, Int. J. Solids Struct. 2, 125 (1966).
12. H.F. Tiersten, J.T. Song and D.V. Shick, J. Appl. Phys. 62, 1154 (1987).
13. D.V. Shick and H.F. Tiersten, Proceedings of the 40th Annual Symposium on Frequency Control, U.S. Army Electronics Technology and Devices Laboratory, Fort Monmouth, New Jersey and Institute of Electrical and Electronics Engineers, New York, IEEE Cat. No. 86CH2330-9, 262 (1986).
14. H.F. Tiersten and D.V. Shick, Proceedings of the 41st Annual Symposium on Frequency Control, U.S. Army Electronics Technology and Devices Laboratory, Fort Monmouth, New Jersey and Institute of Electrical and Electronics Engineers, New York, IEEE Cat. No. 87CH2427-3, 282 (1987).
15. H.F. Tiersten and D.V. Shick, to be published in the Proceedings of the 2nd European Frequency and Time Forum.
16. R. Bechmann, A.D. Ballato, and T.J. Lukaszek, Proc. IRE 50, 1812 (1962).
17. M.B. Schulz, B.J. Matsinger and M.G. Holland, J. Appl. Phys. 41, 2755 (1970).
18. M.F. Lewis, G. Bell and E. Patterson, J. Appl. Phys. 42, 476 (1971).
19. B.K. Sinha and H.F. Tiersten, J. Appl. Phys. 50, 2732 (1979).
20. B.K. Sinha and H.F. Tiersten, J. Appl. Phys. 51, 4659 (1980).
21. D.F. Williams and F.Y. Cho, Proceedings of the 34th Annual Symposium on Frequency Control, U.S. Army Electronics Research and Development Command, Fort Monmouth, New Jersey, 302 (1980).
22. G. Theobald, G. Marianneau, R. Prétot and J.J. Gagnepain, Proceedings of the 33rd Annual Symposium on Frequency Control, U.S. Army Electronics Research and Development Command, Fort Monmouth, New Jersey, 238 (1979).
23. P. Levesque, M. Valdois, D. Hauden, J.J. Gagnepain, P. Hartemann and J. Uebersfeld, 1979 Ultrasonics Symposium Proceedings, IEEE Cat. No. 79CH1482-9, Institute of Electrical and Electronics Engineers, New York, 896 (1979).
24. B.K. Sinha, W.J. Tanski, T.J. Lukaszek and A. Ballato, J. Appl. Phys. 57, 767 (1985).
25. R.D. Mindlin, Structural Mechanics (Pergamon Press, Oxford, 1960), p.199.
26. R.D. Mindlin and M.A. Medick, J. Appl. Mech., Trans. ASME, 1 (1958).
27. M. Onoe, J. Acoust. Soc. Am. 30, 1159 (1958).
28. R.D. Mindlin, "An Introduction to the Mathematical Theory of the Vibration of Elastic Plates," U.S. Army Signal Corps Eng. Lab., Fort Monmouth, New Jersey (1955). Signal Corps Contract DA-36-03956-56772.
29. H.F. Tiersten, Linear Piezoelectric Plate Vibrations (Plenum, New York, 1969), Chap.10.

30. C.J. Wilson, J. Phys. D7, 2449 (1974).
31. A. Ballato, "Doubly-Rotated Thickness Mode Plate Vibrators," in Physical Acoustics, edited by W.P. Mason and R.N. Thurston (Academic, New York, 1977), Vol.XIII, Sec.IV.
32. D.S. Stevens, H.F. Tiersten and B.K. Sinha, J. Appl. Phys. 54, 1704 (1983).
33. L.A. Tyler, prepared for the Department of the Army under Contract No. DA-039-SC-71061 by Union Thermoelectric Division of Comptometer Corporation, Niles, Illinois, 1961 AD274031, Fig.16.
34. A. Ballato, E.P. Eer Nisse and T.J. Lukaszek, 1978 Ultrasonics Symposium Proceedings (Institute of Electrical and Electronics Engineers, New York, 1978), IEEE Cat. No. CH1344-1SU, p.144.
35. D.S. Stevens and H.F. Tiersten, Proceedings of the 38th Annual Symposium on Frequency Control, U.S. Army Electronics Research and Development Command, Fort Monmouth, New Jersey and Institute of Electrical and Electronics Engineers, New York, IEEE Cat. No. 84CH2062-8, 176 (1984).
36. J.A. Kosinski, Proceedings of the 39th Annual Symposium on Frequency Control, U.S. Army Electronics Research and Development Command, Fort Monmouth, New Jersey and Institute of Electrical and Electronics Engineers, New York, IEEE Cat. No. 85CH2186-5, 400 (1985).
37. R. Bechmann, Phys. Rev. 110, 1060 (1958).
38. R.N. Thurston, H.J. McSkimin and P. Andreatch, Jr., J. Appl. Phys. 37, 267 (1966).

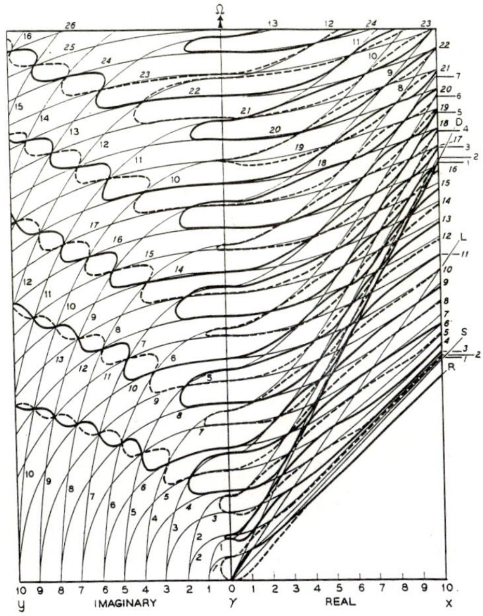

Figure 1 Frequency Spectrum of an Infinite Isotropic Elastic Plate with Traction-Free Surfaces for Real and Imaginary Wavenumbers for a Poisson's Ratio of 0.31 (after Mindlin, Ref.25)

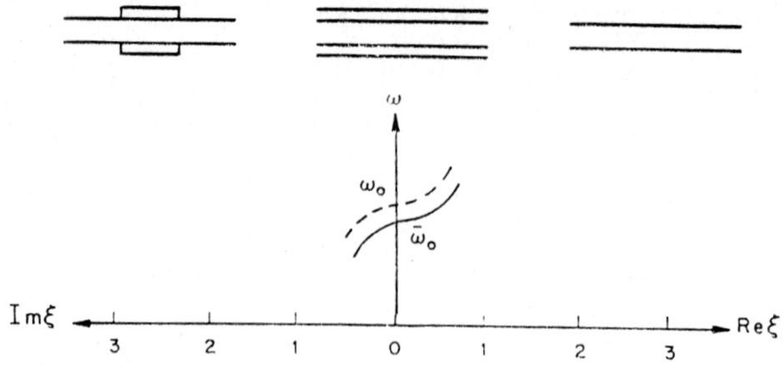

Figure 2 Typical Dispersion Curves in the Vicinity of a Cutoff Frequency for the Unelectroded and Fully Electroded Plate for the Explanation of Trapping in the Partially Electroded Plate

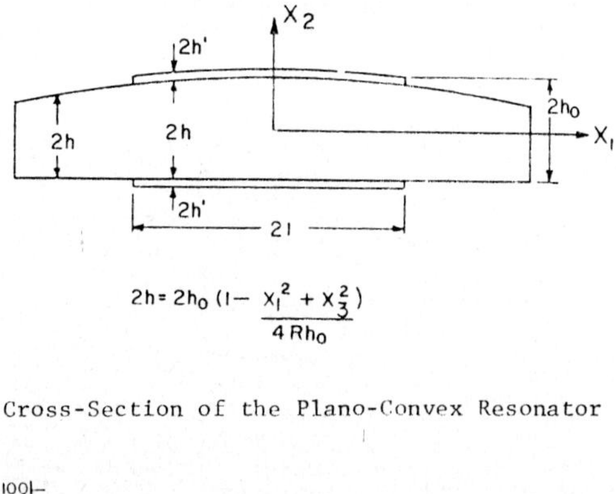

$$2h = 2h_o\left(1 - \frac{x_1^2 + x_3^2}{4Rh_o}\right)$$

Figure 3 Cross-Section of the Plano-Convex Resonator

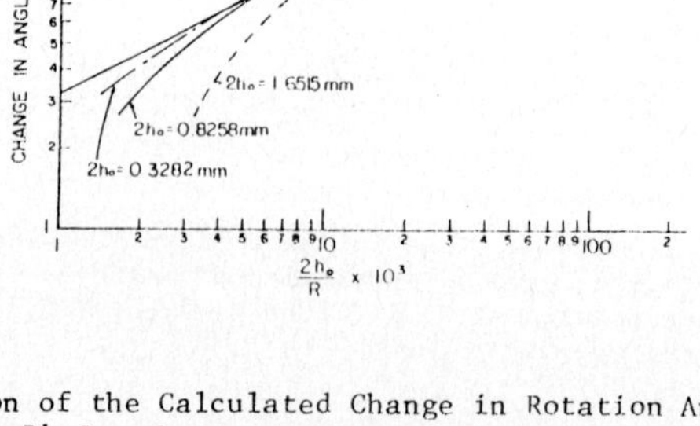

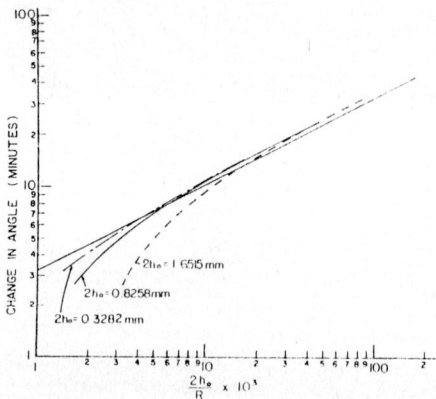

Figure 4 Comparison of the Calculated Change in Rotation Angle $\Delta\theta$ from $\theta = -35^{\circ}8.5'$ for the Zero Temperature Cut for the Fundamental Mode of the Contoured Resonator as a Function of the Center Thickness to Radius of Contour Ratio with the Measured Design Curve

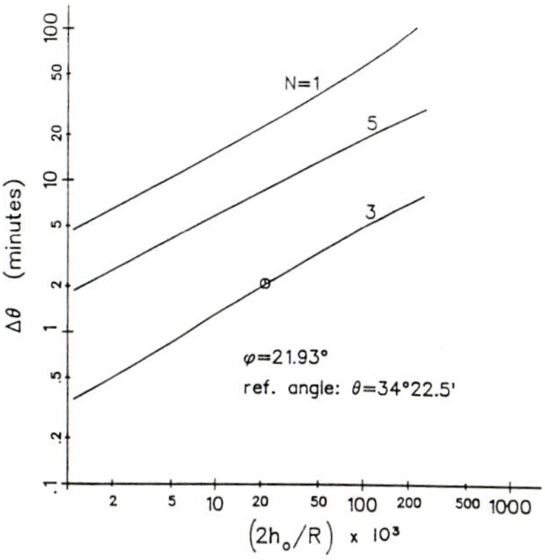

Figure 5 Change in Rotation Angle $\Delta\theta$ from $\theta = -34^\circ 22.5'$ for the Zero Temperature Coefficient of Frequency SC-Cut at 25°C for an Actual Fixed Value of $\varphi = 21.93^\circ$ as a Function of the Ratio of the Center Thickness-to-Radius of Contour for the First, Third and Fifth Harmonics

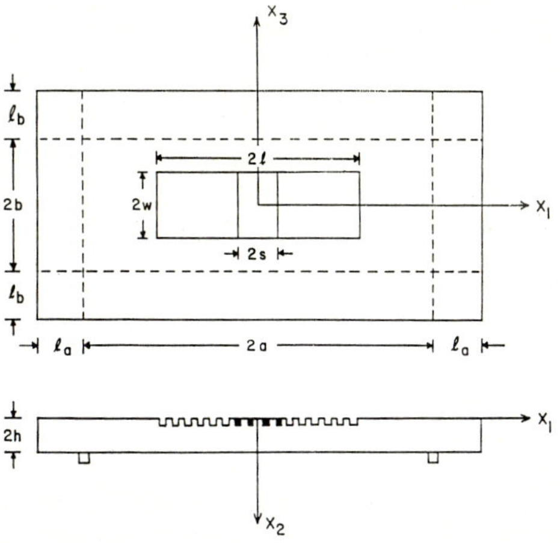

Figure 6 Plan View and Cross Section of Rectangular ST-Cut Quartz Plate with the Substrate Extending Beyond Rectangular Supports

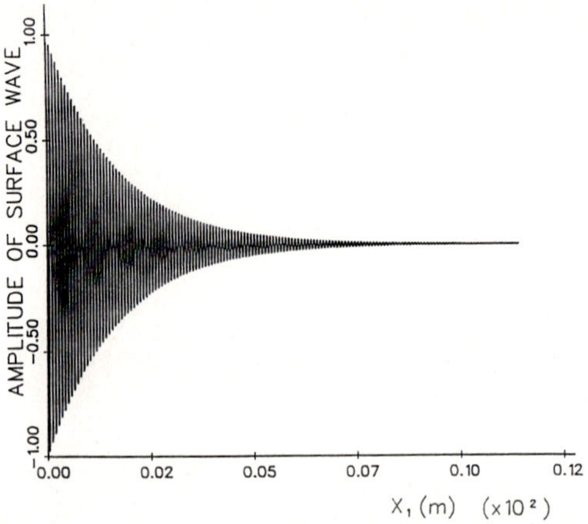

Figure 7 Standing Surface Wave Mode Shape Along Transmission Path at Resonance

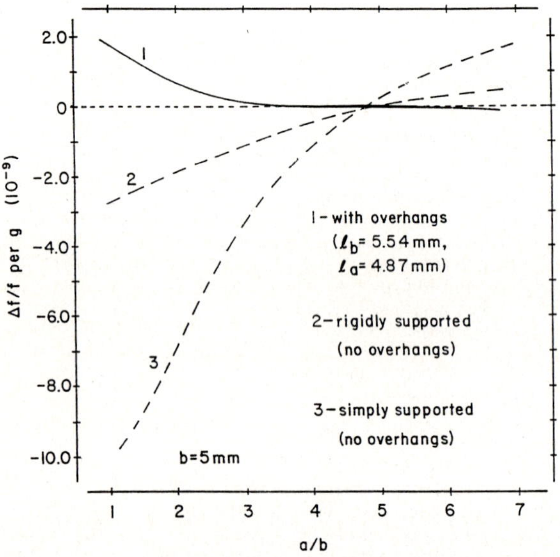

Figure 8 Calculated Normal Acceleration Sensitivity Versus Planar Aspect Ratio a/b for the Support Configuration Shown in Fig.6 (Solid Line) and the Configurations Treated in Two Earlier Works (Dotted Lines)

ELASTIC WAVE PROPAGATION
M.F. McCarthy, M.A. Hayes, (Editors)

NONLINEAR WAVE PHENOMENA IN MAGNETOSTRICTIVE ELASTIC BODIES

Gérard A. MAUGIN

Université Pierre-et-Marie Curie
Laboratoire de Modélisation en Mécanique,associé au C.N.R.S.,
Tour 66, 4 place Jussieu, 75252 Paris Cédex 05 , France.

This contribution concerns an applied-mathematics approach to the essential nonlinear dynamical characteristic features of some devices that use magnetostrictive elastic elements . A special emphasis is placed on the propagation of nonlinear surface waves on a half-space and on the nonlinear vibrations of a resonating slab, both problems using a multiple-scale technique and being reduced to very similar systems of nonlinear ordinary differential equations.

1. INTRODUCTION

Although less used than piezoelectric devices , magnetostrictive elements are also promising and their nonlinear dynamical behaviour is of special interest since it exhibits specific features . These are the object of the present contribution. In particular , the very nonlinearity of magnetostrictive phenomena (magnetostriction is the production of a stress *even* in the applied magnetic field and thus is, *per se* , a nonlinear coupling effect in contradistinction with piezoelectricity or piezomagnetism [1]-[2]) and the peculiarity of magnetic fields as compared to electric ones , make that (i) the wave problems always involve a *bias* magnetic field ,(ii) there is a discrepancy of one order between mechanical and magnetic entities as soon as one considers some iterative procedure and a resulting hierarchy of equations , and (iii) when a boundary is present, there always is a matching of interior and exterior solutions for the magnetoacoustic problem. With this in mind we first recall the nonlinear , rotationally invariant equations of the magnetoelasticity of insulators in the quasi-magnetostatic framework [3]. Having commented upon the linear and nonlinear *bulk-wave* characteristics , we then examine in greater detail the far-field solution for surface magnetoacoustic waves and the nonlinear characteristics of magnetostrictive resonators. The defect of *anisochronism* of these two devices is determined as well as the spatial evolution of amplitudes from a source in the former case or the time evolution of modes in the latter case. Whenever possible, comparisonsare drawn with the perfect elastic case (e.g.,[4]-[7]) and the electroelastic case (e.g., [8]-[9]). Previous results pertaining to the linear and linearized cases may be found in various works [10]-[12].

2. BASIC EQUATIONS

All equations are necessarily written in the material description so that all all boundaries are fixed. These equations consist of the equations of motion, Maxwell's equations and the associated boundary (jump) conditions, together with the nonlinear , coupled, stress and magnetic-induction constitutive equations. For a material domain D_0 of regular boundary ∂D_0 equipped with unit outward normal of components N_K^0 we have [3] :

$$0 = \varrho_0 \frac{\partial^2 u_i}{\partial t^2} - \frac{\partial T^t_{Ki}}{\partial X_K} \quad , 0 = \partial B_K / \partial X_K \quad , 0 = H_K + \partial \emptyset / \partial X_K \quad \text{in } D_0 \qquad ,(2.1)$$

$$0 = \partial^2 \emptyset / \partial X_K \partial X_K \qquad \text{outside } D_0 \qquad ,(2.2)$$

$$0 = N_K [\![T^t_{Ki}]\!] \quad , \quad 0 = N_K [\![B_K]\!] \quad , \quad 0 = [\![\emptyset]\!] \quad \text{at} \quad D_0 \qquad (2.3)$$

with

$$T^t_{Ki} = \delta_{Ri}(C_{KRMN} u_{M,N} + \gamma_{KRMNPQ} u_{M,N} u_{P,Q} + \delta_{KRMNPQAB} u_{M,N} u_{P,Q} u_{A,B} + B_{KRMN} H_M H_N + \text{h.o.t}) \qquad ,(2.4)$$

$$B_K = \mu_{KL} H_L - \bar{B}_{KLMN} H_L u_{M,N} + \frac{1}{6} \chi_{KLMN} H_L H_M H_N + \text{h.o.t} \qquad ,(2.5)$$

where upper and lower case Latin indices refer to material and current Cartesian coordinates , respectively. The symbols ϱ_0 , $u_i = x_i - X_K \delta_{Ki}$, x_i , B_K , H_K , $\emptyset$ and T^t_{Ki} denote, respectively, the matter density at the reference configuration , the elastic displacement , the nonlinear motion, the Lagrangian magnetic field and induction , the magnetostatic scalar potential, andthe total (nonsymmetric) Piola-Kirchhoff stress tensor. C_{KLMN} and μ_{KL} are tensors of elasticity coefficients and of magnetic permeabilities of the second order ; χ_{KLMN} is a tensor of magnetic susceptibilities of the fourth order ; γ_{KRMNPQ} and $\delta_{KRMNPQAB}$ are *effective* tensors of elasticity coefficients of the third and fourth orders; B_{KLMN} and $\bar{B}_{KLMN}$ are *effective* tensors of magnetostriction coefficients of the first order . Effective magnetostrictions of higher orders have been discarded. The material is assumed to be centrosymmetric (e.g., cubic).

In the *linearized* theory with a bias magnetic field of magnitude H^0 , a typical *magnetoelastic coupling coefficient* (usually a small parameter [10]) is defined by $\epsilon_m = \beta \bar{\beta} (H^0)^2$, where β and $\bar{\beta}$ are typical components of the tensors B_{KLMN} and $\bar{B}_{KLMN}$. This causes a slight *decrease* in the acoustic speed [11] whereas piezoelectricity expressed in terms of electric fields causes an *increase* in the acoustic speed [1],[2] . We also note that the boundary conditions (2.3) are *always* in the form of *jump* conditions since there always exists an exterior magnetic field , magnetism presenting no analogue to the earthing condition of electricity .

In the case of surface waves propagating in the *linearized* theory with a bias field $\underline{H}^0$ of any direction relative to the real wave vector, one obtains a *four-mode* solution of the generalized Rayleigh-wave type in the form ($\partial D_0 = \{ X_2 = 0 \}$, $D_0 = \{ X_2 > 0 \}$:

$$(u_K, \emptyset) = \sum_{\alpha=1}^{4} (G^\alpha_K, G^\alpha) \exp (-k_0 \chi_\alpha X_2) \exp [i(\omega t - k_0 X_1)] \qquad (2.6)$$

for $X_2 > 0$. However, for centrosymmetric cubic crystals and $\underline{H}^0$ set orthogonal to the sagittal plane $P_S = \{ X_3 = 0 \}$, a split of the general solution (2.6) is achieved , yielding a pure elastic Rayleigh wave (with displacement polarized parallel to P_S) on the one hand and a shear horizontal (SH) magnetoelastic wave [13] of components $(u_3, \emptyset)$ akin to the Bleustein-Gulyaev mode of piezoelectricity [2], [8] on the other hand .

For an *elastic resonator* , $D_0 = \{ |X_1| < h \}$, the *linear* version of eqns. (2.1)-(2.5) provides *stationary-wave* solutions between the two faces $|X_1| = h$, of a purely elastic nature as (frequency normalized to one, u may be any component u_K which uncouples from the other two)

$$u = U_0 \sin t \sin x \quad , \quad x = X_1 \quad , \quad h = (2q+1)\frac{\pi}{2} = \frac{\pi}{2} , \frac{3\pi}{2} , \frac{5\pi}{2}, \ldots \qquad (2.7)$$

The following two sections reconsider the *nonlinear* magnetoelastic version of (2.6) and (2.7) for small amplitudes, nonlinearities and coupling coefficients.

3. FAR-FIELD SOLUTION FOR SURFACE MAGNETOELASTIC WAVES

Let ϵ_1 and ϵ_2 be the two small parameters related to nonlinear (e.g., third order) elastic effects and magnetostriction, respectively. We assume that $\epsilon_2 = O(\epsilon_1)$ [12] . Then , if one tries a straightforward expansion , *à la* Poincaré, of the *full multimode* surface-wave problem governed by eqns.(2.1)-(2.5) , one is led to a zeroth-order solution which, even in the presence of the bias magnetic field, is purely elastic and of the form

$$u_K^{(o)} = U_1 \sum_{\alpha=1}^{3} C_K^\alpha \exp(-k_o \chi_\alpha X_2) \exp[i(\omega t - k_o X_1)] \qquad ,(3.1)$$

while, at the first order of approximation, both magnetostriction and nonlinear elasticity cause a production of second harmonic (that grows with travelled distance from the source) and only magnetostriction causes an anisochronism (alteration in the propagation velocity of the ω-component) at that order [12], nonlinear elasticity requiring one order higher (i.e., fourth-order elasticity) to cause a similar effect on bulk waves [8]. Since surfac waves can travel over long distances on a substrate and can be efficiently used in a number of nonlinear operations [8] , it is salient to envisage a uniformly valid far-field solution . To that purpose a two-space technique is used with slow variables $\xi = \epsilon X_1$, $\eta = \epsilon X_2$, so that

$$u_K(X_1,X_2,t) = \tilde{u}_K(X_1,X_2,\xi,\eta,t) = \epsilon_1 \tilde{u}_K^{(o)} + \ldots \quad ,$$
$$\phi_{,K}(X_1,X_2,t) = -\tilde{H}_K(X_1,X_2,\xi,\eta,t) = -H_K^o + \epsilon_2 \tilde{\phi}_{,K}^{(1)} + \ldots \quad . \qquad (3.2)$$

The zeroth-order equations are identically satisfied by taking $\tilde{u}_K^{(o)}$ as a linear combination of harmonics of (3.1) with spatially slowly varying amplitudes in the form (c.c. = complex conjugate)

$$\tilde{u}_K^{(o)} = \frac{1}{2} U_1 \sum_{n=1}^{\infty} \sum_{\alpha=1}^{3} [C_K^\alpha A_n^\alpha(\zeta) \exp(in\Psi^\alpha) + c.c.] \quad ,$$
$$\Psi^\alpha = \omega t - k_o X_1 - i k_o \chi_\alpha X_2 \quad , \quad \zeta = \xi + \mu_n^\alpha \eta \quad , \qquad (3.3)$$

some freedom being left in this solution through the functional dependence of the A_n^α's , where μ_n^α is some kind of depth parameter for the n-th harmonic of the α-th mode [4]. Following along the same line as in the purely elastic case [4], [8] , it is then shown that the *secularity condition* imposed at the next order of approximation provides a system of complex *coupled-amplitude equations* in the manner of nonlinear optics in absorbing transparent media [14] . For instance , for H^o set orthogonal to P_S and a centrosymmetric cubic crystal , we may envisage an SH elastic mode coupled to ϕ (compare [15]) . Limiting the analysis to interactions between the fundamental and the second harmonic, there remain two complex coupled amplitudes A_1 and A_2 for $\tilde{u}_3^{(o)}$ and these satisfy the nonlinear complex ODE's ($^+$ denotes the complex conjugate)

$$\frac{dA_1}{d\zeta} = - a\, A_1^+ A_2 - b\, A_1 \quad , \quad \frac{dA_2}{d\zeta} = c\, A_1^2 - d\, A_2 \qquad ,(3.4)$$

where all coefficients are complex but b and d vanish with magnetostriction and H^0. If $b = d = 0$, the system (3.4) admits *exact* solutions with a behaviour $|A_2| \propto \tanh(A\zeta)$, $|A_1| \propto \operatorname{sech}(A\zeta)$ as in index-matching media [14]. For b and d nonzero a numerical solution has to be sought which, in fact, exhibits a slight deviation from the abovementioned exact solution only for very large travelled distances (i.e., distances much larger than the shock-formation distance L_B computed from the nonlinear, purely elastic case [8]). (see Figure 1). Notice that the method used here differs from that of Lardner recently used in pure elasticity [5] and electroelasticity [9].

4. MAGNETOELASTIC RESONATOR

We consider a one-mode elastic process coupled to magnetic properties and assume that the thickness 2h, much smaller than L_B, of the resonator, is normalized to the spatial interval $(0, \pi)$. We examine the nonlinear case by using a *Galerkin representation* followed by a multiple-*time* scale technique in the manner of Googerdy and Peddleson [7]. After appropriate nondimensionalization eqns. (2.1)-(2.5) give (fourth-order elasticity and susceptibility are discarded) $(x = X_1)$

$$\begin{aligned}
&u_{tt} - u_{xx}\,(1 + 2\,\nu u_x) = \beta(\phi_x^2)_x \quad , \\
&\phi_{xx} - \bar{\beta}(\phi_x u_x)_x = 0 \quad , \qquad 0 < x < \pi \quad , \\
&\phi_{xx} = 0 \quad , \quad x < 0 \;,\; x > \pi \;, \qquad\qquad (4.1) \\
&[\![\phi_x - \bar{\beta}\phi_x u_x]\!] = 0 \;, \; [\![H^0]\!] = 0 \;, \; u(x,t) = 0 \text{ at } x=0,\; x = \pi \;, \\
&u(x,0) = \sin x \;, \; u_t(x,0) = 0 \qquad \text{(initial conditions)} \;,
\end{aligned}$$

where u is the excited scalar elastic mode (of any polarization). The functions sin nx provide a natural basis for elastic stationary modes and the sought nonlinear magnetoelastic solution can be projected on it. The Galerkin representation reads

$$u(x,t) = \sum_{n=1}^{\infty} f_n(t)\sin nx \;, \quad \phi_x(x,t) = -H^0 + \sum_{n=1}^{\infty} \Phi_n(t)\cos nx \;. \qquad (4.2)$$

Substituting from (4.2) into (4.1) and accounting for the orthonormality of basis functions yield an infinite set of nonlinear second-order ODE's for the f_n's coupled to the Φ_n's. Then these are expanded using a two-time scale technique with $\xi = t$, $\eta = \epsilon t$. For a two-mode interaction we write $f_n = \epsilon_1 f_{n0} + \ldots$, $f_{n0} = C_n(\eta)\cos[\alpha_n(\eta) - n\xi]$ and the *secularity* condition yields

$$\begin{aligned}
&\frac{dC_1}{d\eta} = -\frac{1}{4} C_1 C_2 \;, \qquad \frac{dC_2}{d\eta} = \frac{1}{16} C_1^2 \;, \\
&\frac{d\alpha_1}{d\eta} = \frac{1}{4} C_2 - \epsilon_m \;, \qquad \frac{d\alpha_2}{d\eta} = -\frac{1}{16}\frac{C_1^2}{C_2} - 4\,\epsilon_m
\end{aligned} \qquad (4.3)$$

Equations $(4.3)_{1-2}$ integrate immediately to (compare the solution of (3.4) whenever b=d=0 ; $\zeta = \eta$, A = 1/8 , $C_1(0) = 1$, $C_2(0) = 0$)

$$C_1 = \mathrm{sech}(\eta/8) \quad , \quad C_2 = \frac{1}{2} \tanh(\eta/8) \qquad .(4.4)$$

This is the same as in the purely elastic case [7] . Going back to the displacement field, one finds, using the integrals of eqns.$(4.3)_{3-4}$:

$$u_{10} = \mathrm{sech}(\eta/8) \cos(\xi + \eta\epsilon_m - \ln(\cosh(\eta/8))) \sin x \qquad ,(4.5)$$

so that the *time-dependent* anisochronism of the resonator reads ($\omega_0 = 1$)

$$\frac{\Delta\omega}{\omega_0} \simeq \epsilon_m + \epsilon\frac{t}{128} + O(\epsilon^4 t^3) \qquad .(4.6)$$

We see that the two problems examined are, in some way, dual to one another and they in fact reduce to solving very similar systems of first-order ODE's. The magnetoelastic couplings in both cases, clearly cause a very slight deviation from the results pertaining to the pure nonlinear elastic analysis. This is like in electroelasticity: the essential nonlinearities for coupled acoustic waves remain those related to the mechanical behavior.

REFERENCES

[1] Ristic , V.M., *Principles of Acoustic Devices* (J.Wiley-Interscience , New York , 1982) .

[2] Maugin, G.A., *Continuum Mechanics of Electromagnetic Solids* (North-Holland, Amsterdam, 1988).

[3] Abd-Alla, A.N. and Maugin, G.A., *J.Acoust.Soc.Amer. 82* (1987) 1746.

[4] Planat M., *J.Appl.Phys. 57* (1985) 4911.

[5] Parker, D.F., *Int.J.Engng.Sci. 26* (1988) 59.

[6] Lardner, R.W., *J.Elasticity 16* (1986) 63.

[7] Googerdy, A. and Peddleson, J. Jr., Paper presented at *1984 S.E.S. Meeting,* (Blaksburg, Virginia, October 1984).

[8] Maugin, G.A., *Nonlinear Electromechanical Effects and Applications - A Series of Lectures* (World Scientific, Singapore, 1985).

[9] Tupholme, G.E. and Harvey, A.P., *Int.J.Engng.Sci. 26* (in the press, 1988) .

[10] Maugin, G.A., *Int.J.Engng.Sci. 19* (1981) 321.

[11] Maugin, G.A., and Hakmi, A., *J.Acoust.Soc.Amer. 76* (1984) 826.

[12] Abd-Alla, A.N. and Maugin, G.A., Linear and Nonlinear Surface Waves on Magnetostrictive Crystals in a Bias Magnetic Field, in : Parker, D.F. and Maugin, G.A.(eds) , *Recent Advances in Surface Acoustic Waves* , Springer Series on Wave Phenomena (Springer-Verlag, Berlin , 1988).

[13] Maugin, G.A., Surface Waves Waves with Transverse Horizontal Polarization , in: Hutchinson, J.W.(ed), *Advances in Applied Mechanics* , Vol.23 (Academic Press, New York, 1983), pp.373-434.

[14] Baldwin, G.C., *An Introduction to Nonlinear Optics* (Plenum Press, New York, 1959).

[15] Kalyanasundaram, N., *J.Sound Vibrations 96* (1984) 411.

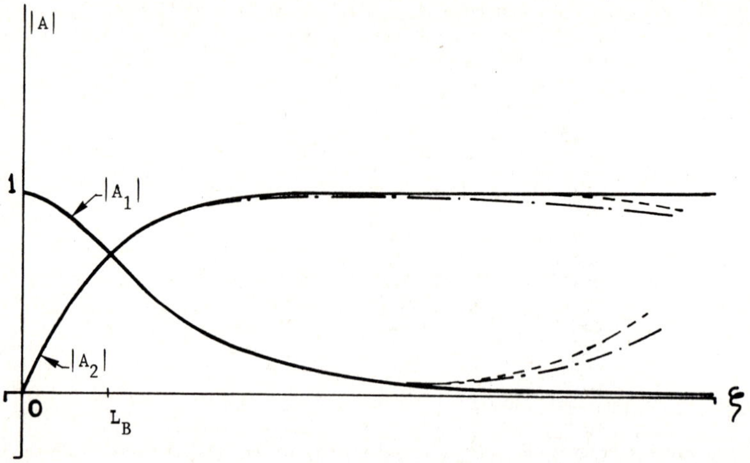

Figure 1 - Evolution of amplitude of the first two harmonics of surface magnetoacoustic waves with distance (—— : purely elastic case; -·-·- positive magnetostriction; ---- negative magnetostriction; We are indebted to M.Sabir for this computation)

ELASTIC WAVE PROPAGATION
M.F. McCarthy, M.A. Hayes, (Editors)

ACCELERATION WAVES IN ELASTIC DIELECTRICS WITH POLARIZATION INERTIA[1]

Sadık Dost

Department of Mechanical Engineering, The University of Calgary, Calgary, Alberta, Canada, T2N IN4

The propagation of acceleration waves of arbitrary form in deformable elastic dielectrics with polarization inertia effect is investigated by adopting the singular surface approach. For isotropic materials, assuming no deformation and polarization in the medium ahead of the wave front, the wave speeds for principal acceleration and polarization waves are obtained explicitly and the conditions of existance for real wave speeds are investigated. Numerical values of transverse wave speeds for $KTaO_3$ are evaluated as an example.

1. INTRODUCTION

Inclusion of polarization inertia effects in deformable elastic dielectrics has been the subject of some recent publications [1-6]. This made it possible to obtain a purely polarization wave as would be the case in a rigid dielectric [7]. Some publications have accomodated such an inertia term and even attributed to it a numerical value for some dielectrics [5 , 6] . In this paper the propagation of acceleration waves in such a medium is studied.

The propagation condition is derived by adopting the singular surface approach. Because of the presence of the second time derivative of the polarization vector on the righthand side of the intermolecular force balance equation, the jumps of balance equations yield two coupled acceleration waves associated with displacement and polarization. For isotropic materials [6 , 8] , assuming no polarization and deformation in the medium before the passage of the wave front, speeds of principal waves are obtained. Existence of real wave speeds for principal waves is investigated. In general these two principal waves are coupled. It is observed that if both mechanical and polarization waves are transverse and normal to each other, these two waves are decoupled and the speeds are always real. If these two transverse waves are travelling in the same direction, the waves are coupled and the corresponding speeds are real if only a condition interms of material coefficients is satisfied.

The results are then specialized for $KTaO_3$, for which some linear material constants and polarization inertia are available [6]. It is shown that in this special case two real transverse wave speeds exist and their numerical values are obtained.

2. BASIC EQUATIONS

The basic equations of elastic dielectrics with polarization inertia effects are given below in the reference state [6 - 10] :

$$T_{Kk,K} + \rho_o f_k = \rho_o \ddot{x}_k \quad , \quad C + D_{K,K} + D_K X_{L,k} x_{k,KL} = 0$$

$$E^{(L)}_{Kk,K} - \frac{\rho_o}{\rho} \varphi_{,k} - \frac{\partial \Sigma}{\partial p_k} - \frac{\rho_o}{\rho} \frac{\partial \varepsilon}{\partial p_k} = \frac{\rho_o}{\rho} m \ddot{p}_k \tag{2.1}$$

where x_k and X_K $(k, K = 1,2,3)$ are spatial and material cartesian coordinates respectively; a comma and a dot are used to denote partial and material differentiation with respect to coordinates and time, respectively; ρ_o , ρ , m and C denote the density in the reference state, the density in the deformed state, the polarization inertia and charge density per unit volume while f_k , p_k and φ represent body forces, polarization and the Maxwell electric potential, respectively; $x_{k,K}$ and $X_{K,k}$ are the deformation and inverse deformation gradients; and $\Sigma(x_{k,K}, p_{k,K}, p_k)$ and $\varepsilon(p_k, \varphi_{,k})$ represent the strain and electrical energy functions, where $p_{k,K}$ is the polarization gradient.

In Eqs. (2.1) T_{Kk}, $E^{(L)}_{Kk}$ and D_K are the first Piola-Kirchhoff stress tensor, the two-point electric tensor and the electric displacement vector defined as

$$T_{Kl} = \frac{\rho_o}{\rho} t_{kl} X_{K,k} \quad , \quad E^{(L)}_{Kl} = \frac{\rho_o}{\rho} E^{(L)}_{kl} X_{K,k} \quad , \quad D_K = d_k X_{K,k} \tag{2.2}$$

where t_{kl} , $E^{(L)}_{kl}$ and d_k are the Cauchy stress tensor, the electric tensor and the electric displacement vector, respectively, defined as

$$t_{kl} = t^{(L)}_{kl} + t^{(M)}_{kl} \quad , \quad E^{(L)}_{kl} = \frac{\rho}{\rho_o} \frac{\partial \Sigma}{\partial p_{l,K}} x_{k,K} \quad , \quad d_k = \frac{\partial \varepsilon}{\partial \varphi_{,k}} \tag{2.3}$$

and where

$$t^{(L)}_{kl} = \frac{\rho}{\rho_o} \frac{\partial \Sigma}{\partial x_{l,K}} x_{k,K} \quad , \quad t^{(M)}_{kl} = \varepsilon \delta_{kl} - \frac{\partial \varepsilon}{\partial \varphi_{,l}} \varphi_{,k} \tag{2.4}$$

The explicit forms of $t^{(L)}_{kl}$ and $E^{(L)}_{kl}$ for isotropic materials were given in [8].

3. PROPAGATION OF ACCELERATION WAVES

An acceleration wave in an elastic dielectric is defined as a propagating surface of discontinuity σ on which $\mathbf{x}$, $\mathbf{p}$ and φ and their first order derivatives are continuous but their second and higher order derivatives suffer jumps. Compatibility conditions [8] require that

$$\begin{aligned} [x_{k,KL}] &= a_k x_{m,K} x_{n,L} n_m n_n , \quad [\dot{x}_{k,K}] = - U a_k x_{l,K} n_l \quad , \quad [\ddot{x}_k] = U^2 a_k \\ [p_{k,KL}] &= b_k x_{m,K} x_{n,L} n_m n_n , \quad [\dot{p}_{k,K}] = - U b_k x_{l,K} n_l \quad , \quad [\ddot{p}_k] = U^2 b_k \\ [\varphi_{,KL}] &= \omega x_{m,K} x_{n,L} n_m n_n \quad , \quad [\dot{\varphi}_{,K}] = - U \omega x_{l,K} n_l \quad , \quad [\ddot{\varphi}] = U^2 \omega \end{aligned} \tag{3.1}$$

where $\mathbf{n}$ is the unit normal to σ and U represents the speed of propagation, and **a** , **b** and ω are called the amplitudes of acceleration jump, polarization jump and maxwell potential jump, respectively.

The jump of Eq. (2.1) yields that $[\varphi_{,KL}] = 0$, which implies that φ and its any order derivatives will be continuous on σ. Using Eqs. (3.1), the jumps of Eqs. $(2.1)_{1,3}$ lead to the following equations

$$\begin{bmatrix} \mathbf{Q} - \rho U^2 \mathbf{I} & \mathbf{P} \\ \mathbf{P} & \mathbf{R} - mU^2 \mathbf{I} \end{bmatrix} \begin{bmatrix} \mathbf{a} \\ \mathbf{b} \end{bmatrix} = 0, \tag{3.2}$$

where, for isotropic materials, the components of tensors **Q**, **P** and **R** are given by

$$Q_{ij} = 2 \frac{\partial t^{(L)}_{ki}}{\partial c^{-1}_{jl}} c^{-1}_{lm} n_k n_m + \frac{\partial t^{(L)}_{ki}}{\partial \pi^{-1}_{jl}} \pi^{-1}_{lm} n_k n_m - \frac{\partial t^{(L)}_{ki}}{\partial p_j} p_m n_k n_m,$$

$$P_{ij} = \frac{\partial t^{(L)}_{ki}}{\partial \pi^{-1}_{lj}} c^{-1}_{kl} n_l n_m, \quad R_{ij} = \frac{\partial E^{(L)}_{kl}}{\partial \pi^{-1}_{lj}} c^{-1}_{ml} n_k n_m, \tag{3.3}$$

where $c^{-1}_{kl} = x_{k,K} x_{l,K}$ and $\pi^{-1}_{kl} = x_{k,K} p_{l,K}$. For a nontrivial solution of **a** and **b**, the determinant of the coefficients matrix in Eqs. (3.2) must vanish, which will determine the speed of propagation, U. Since this characteristic equation is a polynomial of twelfth degree and an implicit function in U, it seems very difficult to find the wave speeds from it.

We now determine the speeds of propagation corresponding to a nonzero set of (**a b**) for principal waves propagating in isotropic dielectrics. Assuming there is no deformation and no polarization in the medium before the passage of the wave front, and defining $\mathbf{a} = a\boldsymbol{\nu}$ $(a = |\mathbf{a}| \neq o)$ and $\mathbf{b} = b\boldsymbol{\mu}$ $(b = |\mathbf{b}| \neq o)$, the propagation condition corresponding to a non - zero set (**a**, **b**) yields

$$(Q - \rho U^2)(R - mU^2) - P^2 = 0, \tag{3.4}$$

where

$$Q = Q_1 + Q_2 (\boldsymbol{\nu} \cdot \mathbf{n})^2, \quad R = R_1 + R_2 (\boldsymbol{\mu} \cdot \mathbf{n})^2$$
$$P = P_1\, \boldsymbol{\nu} \cdot \boldsymbol{\mu} + P_2 (\mathbf{n} \cdot \boldsymbol{\nu})(\mathbf{n} \cdot \boldsymbol{\mu}) \tag{3.5}$$

The coefficients Q_1, $Q_2, \ldots$, P_2 in Eqs. (3.5) are given as follows in terms of material constants $h_1, h_2, \ldots, h_{15}$, $\Phi^{(a)}$, $\Psi^{(a)}$ (see [8]):

$$Q_1 = 2(h_8 + h_9), \quad P_1 = h_1 + 2h_{13} + 3h_{15}, \quad R_1 = h_4,$$

$$Q_2 = h_1 + 2h_7 + 2h_8 + 4h_9 + h_{13} + 3h_{15} + \Phi^{(7)} + \Phi^{(8)} + \Phi^{(9)} + \Psi^{(7)} + \Psi^{(8)} + \Psi^{(9)},$$

$$P_2 = \Psi^{(7)} + \Psi^{(8)} + \Psi^{(9)}, \quad R_2 = h_2 + 2h_{14} + \Psi^{(1)} + \Psi^{(13)} + \Psi^{(15)}. \tag{3.6}$$

In general there will be two principal waves traveling at the speeds $\mp U$ and $\mp V$, which are the roots of Eq. (3.4). Eq. (3.4) may be written as

$$U^4 - (Q/\rho + R/m)U^2 + (QR - P^2)/\rho m = 0 . \tag{3.7}$$

First of all, since $(Q/\rho + R/m)^2 - 4(QR - P^2)/\rho m$ is always positive, the roots U^2 and V^2 are real :

$$\left.\begin{matrix}U^2\\V^2\end{matrix}\right\} = \frac{1}{2}(Q/\rho + R/m) \pm \frac{1}{2}\{(Q/\rho - R/m)^2 + 4P^2/\rho m\}^{1/2} . \tag{3.8}$$

Secondly, as seen from Eqs. (3.8), if $QR-P^2 > 0$, these roots will be positive. Therefore if this condition is satisfied, there will be two pair of waves propagating at the speeds $\mp U$ and $\mp V$.

Special Cases : Let the principal direction $\mathbf{n_1}$ be the direction of propagation $\mathbf{n}$.

1. Now if $\boldsymbol{\nu} \,//\, \mathbf{n}$ and $\boldsymbol{\mu} \,//\, \mathbf{n}$, both waves are longitudinal. In this case

$$Q = Q_L = Q_1 + Q_2 , \; R = R_L = R_1 + R_2 \text{ and } P = P_L = P_1 + P_2 . \tag{3.9}$$

The corresponding wave speeds U_L and V_L will be obtained from Eqs. (3.8) by replacing Q, R and P with Q_L, R_L and P_L, respectively.

2. If $\boldsymbol{\nu} \,//\, \mathbf{n}$ and $\boldsymbol{\mu} \perp \mathbf{n}$ (therefore $\boldsymbol{\mu} \perp \boldsymbol{\nu}$), the mechanical wave is longitudinal but the polarization wave will be transverse. In this case

$$Q = Q_L = Q_1 + Q_2 , \quad R = R_T = R_1 , \quad P = 0 \tag{3.10}$$

Therefore the waves are **decoupled** and the speeds are

$$U = U_{AL} = \mp\sqrt{Q_L/\rho} , \text{ and } V = V_{PT} = \mp\sqrt{R_1/m} . \tag{3.11}$$

Note that if an initial polarization exists in the medium ahead of the wave front, in this case P will not be zero.

3. If $\boldsymbol{\nu} \perp \mathbf{n}$, $\boldsymbol{\mu} \,//\, \mathbf{n}$ (i.e. $\boldsymbol{\mu} \perp \boldsymbol{\nu}$), we will have mechanical transverse and polarization longitudinal waves respectively.

For this case

$$Q = Q_T = Q_1 , \quad R = R_L = R_1 + R_2 , \quad P = 0 . \tag{3.12}$$

The waves are **decoupled** again and the speeds are

$$U = U_{AT} = \mp \sqrt{Q_1/\rho} \quad \text{and} \quad V = V_{PL} = \mp \sqrt{R_L/m} \,. \tag{3.13}$$

4. If $\boldsymbol{\nu} \perp \mathbf{n}$ and $\boldsymbol{\mu} \perp \mathbf{n}$. In this case we have two possiblities either i) $\boldsymbol{\nu} \perp \boldsymbol{\mu}$ or ii) $\boldsymbol{\nu} \,//\, \boldsymbol{\mu}$

i) In this case

$$Q = Q_T = Q_1 \,, \quad R = R_T = R_1 \quad \text{and} \quad P = 0 \,, \tag{3.14}$$

and the decoupled transverse wave speeds are

$$U = U_{AT} = \mp \sqrt{Q_1/\rho} \quad \text{and} \quad V = V_{PT} = \mp \sqrt{R_1/m} \,. \tag{3.15}$$

ii) For this case both transverse waves are propagating in the same direction and

$$Q = Q_T = Q_1 \,, \quad R = R_T = R_1 \quad \text{and} \quad P = P_T = P_1 \,. \tag{3.16}$$

The corresponding speeds will be obtained from Eqs. (3.5) by replacing Q, R, and P with Q_1, R_1 and P_1, respectively, i.e.

$$\begin{matrix} U_T^2 \\ V_T^2 \end{matrix} = \frac{1}{2}\,(U_{AT}^2 + V_{PT}^2) \mp \frac{1}{2}\,\{ (U_{AT}^2 - V_{PT}^2)^2 + 4\, P_1^2/\rho m\}^{1/2} \,. \tag{3.17}$$

Note that the terms Q_1 and R_1 involve only linear constitutive coefficients while P_1 contains non-linear coefficients too. In the case of decoupled transverse waves (i.e. case 4(i), the speeds are known only in terms of Q_1 and R_1. Therefore the speed $U_{AT} = \sqrt{Q_1/\rho}$ may be considered as the limiting speed of a plane acoustic transverse wave propagating in an elastic material while $V_{PT} = \sqrt{R_1/m}$ as the speed of polarization wave propagating in a rigid dielectric. Some numerical values for these speeds will be given for $KTaO_3$ later. The speeds of coupled transverse waves in case 4 ii) will be determined in terms of U_{AT}, V_{PT} and P_1 only. In the linear case, P_1 will contain only the linear term h_1, and the corresponding speeds from Eqs. (3.17), U_T and V_T will differ from U_{AT} and V_{PT} due to the interaction term h_1.

4. AN APPLICATION TO $KTaO_3$:

For $KTaO_3$, numerical values of some linear constitutive coefficients and the polarization inertia were obtained in [6] by using the experimental data given at 15°K in [11]. Using these numerical values we obtain

$$\begin{aligned} Q_1 &= 2(h_8 + h_9) \cong 1.20 \times 10^{11}\ \text{kgm}^5/\text{s}^2 \,, \\ R_1 &= h_4 \cong 4.47 \times 10^{-10}\ \text{kgm}^5/\text{C}^2\,\text{s}^2 \,, \\ P_1 &= h_1 \cong 7.10\ \text{kgm}^2/\text{Cs}^2 \,, \end{aligned} \tag{4.1}$$

$$\rho = 6970 \text{ kg/m}^3 \quad \text{and} \quad m = 1.54 \times 10^{-17} \text{ kgm}^3/\text{C}^2 \,. \tag{4.2}$$

Using the numerical values given in Eqs. (4.1) and (4.2), the speeds in Eqs. (3.15) are obtained respectively as

$$U_{AT} = 4160 \text{ m/s} \quad \text{and} \quad V_{PT} = 5387 \text{ m/s} \,. \tag{4.3}$$

The speed $U_{AT} = 4160$ m/s is in agreement with the acoustic phonon velocity of transverse wave obtained as a temperature independent (limiting) velocity (or velocity obtained at or above 300°K) in [11]. Since U_{AT} is the speed of a plane principal acceleration wave propagating in a linear undisturbed medium this agreement was expected.

Now, using these speeds, given in Eqs. (4.3), in Eqs. (3.17), we obtain

$$\begin{matrix} U^2 \\ V^2 \end{matrix} = 2.315 \times 10^7 \mp 0.999 \times 10^7 \tag{4.4}$$

from which two real transverse wave speeds corresponding to mechanical and polarization acceleration waves are respectively obtained as

$$U_T = 3627.5 \text{ m/s} \quad \text{and} \quad V_T = 5756.8 \text{ m/s} \tag{4.5}$$

Due to the presence of the term P_1, which is characterizing the interaction between strain and polarization gradient, the speeds U_T and V_T are different than those corresponding to the decoupled case given in Eqs. (4.3). While one of the waves is travelling at a lower speed, the other one is travelling faster.

REFERENCES

[1] Maugin, G.A.; **Arch. Mech., 28,** 679-692 (1976).

[2] Maugin, G.A.; **Arch. Mech., 29 (1),** 143-151 (1977).

[3] Maugin, G.A. and J.Pouget : **J. Acoust. Soc. I. Am., 68 (2),** 575-587 (1980).

[4] Askar, A., J.Pouget and G.A. Maugin; "The Mechanical Behaviour of Electromagnetic Solid Continua" (Ed. G.A. Maugin), Elsevier Science Publishers B.V. (North-Holland), IUTAM-IUPAP, 151-156 (1984).

[5] Dvorak, V. : **Physical Review, 167 (2),** 525-528 (1967).

[6] **Şahin, E. and** S. Dost; **Int. J. Engng. Sci.** (in press).

[7] Dost, S. and E.Şahin; **Int. J. Engng. Sci., 24 (8),** 1445-1451 (1986).

[8] Dost, S.; **Int. J. Engng. Sci., 21 (11),** 1305-1311 (1983).

[9] Chowdhury, K.L., M.Epstein and P.G.Glockner, **Int. J. Non-Linear Mech., 13,** 311-322 (1979).

[10] Chowdhury, K.L. and P.G. Glockner; **Int. J. Nonlinear Mech., 15,** 263-269 (1980).

[11] Axe, J.D., J.Harada and G.Shirane; **Physical Review B, 1 (3),** 1227-1233 (1970).

[12] Barrett, H.H.; **Phys. Letters 26 A,** 217 (1968).

[13] Şuhubi, E.S.; **Int. J. Engng. Sci. 8,** 699 (1970).

---

1 The results presented here were obtained in the course of research sponsored by the Natural Science and Engineering Council of Canada, Grant No A-1628.

ELASTIC WAVE PROPAGATION
M.F. McCarthy, M.A. Hayes, (Editors)

ELECTROMAGNETIC WAVE STUDY OF ELASTIC WAVE PROPAGATION

Robert E. Green, Jr.

Materials Science & Engineering Department
The Johns Hopkins University
Baltimore, MD 21218 U.S.A.

Experimental measurements are described where electromagnetic waves of different frequencies have been used to study elastic wave propagation in solids. Included are optical probing of linear elastic ultrasonic waves in transparent solids, optical detection of acoustic emission signals (stress waves), thermal imaging of the heat generated by high-power ultrasonic waves propagating in metals, synchrotron x-ray imaging of plastic deformation caused by high-power ultrasonic waves, pulsed laser generation of elastic waves, optical interferometric detection of elastic waves, and full-field heterodyne holographic imaging of elastic wave fields. Specific areas where theoretical analyses would be welcome are mentioned.

1. INTRODUCTION

Elastic wave propagation studies afford very useful and versatile measurement techniques for characterization of solid materials. However, the use of elastic wave measurements for such purposes has as a prerequisite the careful documentation of the behavior of the elastic waves themselves. Visible, infrared, and x-ray electro-magnetic waves provide powerful tools for this documentation. In addition recent applications of pulsed lasers and advances in interferometric and holographic techniques present unique opportunities to obtain new insights into elastic wave propagation in solid materials.

2. OPTICAL PROBING OF TRANSPARENT SOLIDS

Liu [1] used a narrow laser beam to probe the transverse elastic wave fields in an isotropic fused quartz block as generated by an AC-cut quartz crystal. As with all finite sized transducers the ultrasonic beam exhibited diffraction spread. Figure 1 shows the experimental diffraction spread data recorded as a function of distance from the transverse wave generating transducer for different elastic wave frequencies. The experimental results are similar to those observed with longitudinal waves in that the diffraction spread decreases with increasing frequency for a constant diameter transducer, i.e. the more nearly the elastic wave approximates to a plane wave. To the best of the present authors knowledge, unlike the case for longitudinal wave generating transducers where a theory exists based on a vibrating piston model, no theory has ever been developed to explain the observed diffraction spread from a transverse wave transducer.

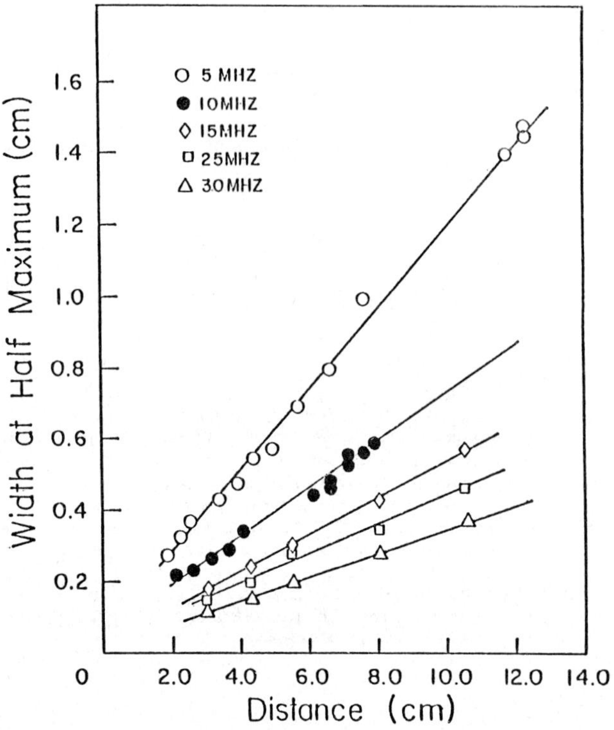

FIGURE 1
Laser beam measured diffraction spread of transverse ultrasonic waves in isotropic solid.

3. OPTICAL DETECTION OF ACOUSTIC EMISSION

Figure 2 compares 200 microsec waveforms from fracture of a glass capillary against an aluminum test block as detected (a) with a non-contact laser interferometer and (b) with a conventional contact piezoelectric transducer [2]. The portion of the optically detected signal to the left of the vertical dashed line agrees within experimental accuracy with the theoretically computed vertical surface displacement versus time record for a step-function time dependence point source load applied to the surface of an isotropic linear elastic half-space. The arrow indicates an enlarged view of this portion of the same waveform. As can be seen, the commercial piezoelectric transducer introduces a multitude of oscillations and does not faithfully reproduce the initial portion of the waveform characteristic of the source [2]. Although the optically detected initial portion of the acoustic emission waveform is in reasonable agreement with theory, no adequate theoretical treatment has been developed to explain the features of the recorded signal following this initial portion, of which undoubtedly the geometrical shape of the test structure plays a major role.

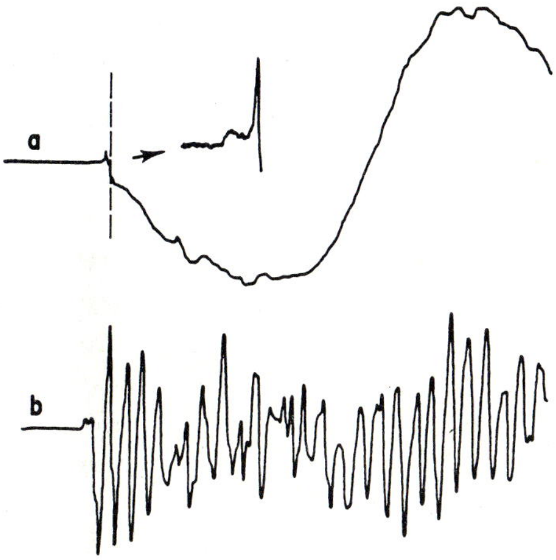

FIGURE 2
Same acoustic emission waveform detected with (a) laser interferometer and (b) conventional piezoelectric transducer.

4. HIGH-POWER ULTRASONIC WAVES

Since controversy existed concerning the heat generated in solids by high-power ultrasonic waves [3], Mignogna et al. [4] used an infrared television system to monitor the surface temperature distribution in solids subjected to high-power ultrasound. Half-wavelength specimens of resonant length exhibited heating at displacement nodes, while single crystal and fine-grained polycrystalline specimens of non-resonant length were observed to heat only at the horn/specimen interface. These observations are in agreement with elasticity theory. Hot zones were also detected at saw-cuts, drilled holes, fatigue cracks, and grain boundaries. Measurement of low-amplitude (8MHz) ultrasonic velocity and attenuation simultaneously with insonation of metal test specimens with high-power (20 kHz) ultrasound provided strong indirect evidence that the high-power ultrasound caused dislocation motion. In order to obtain more direct evidence, an aluminum single crystal of resonant length was insonated so that a displacement node and hence strong plastic deformation would be expected to occur at the mid-section. Subsequently, white beam topographic images of the mid-section were obtained at the National Synchrotron Light Source at Brookhaven National Laboratory. Figure 3 shows an optically enlarged image of one of the synchrotron white beam x-ray diffraction "spots". The extreme asterism and inhomogeniety in this image clearly indicate that high-power ultrasound caused severe plastic deformation and lattice bending throughout the bulk of the mid-section. Subsequent to this result, a large-grained zinc specimen subjected to high-power ultrasound was observed to heat excessively at one large grain, Fig. 4, and to plastically bend into a curved shape as a result interaction of the high-power ultrasound with the grain microstructure. Although many practical applications of high-power ultrasound for mechanical deformation and stress relieving of metals have been reported [5], no satisfactory theoretical treatment explaining the interaction of high-power ultrasound with metals has appeared.

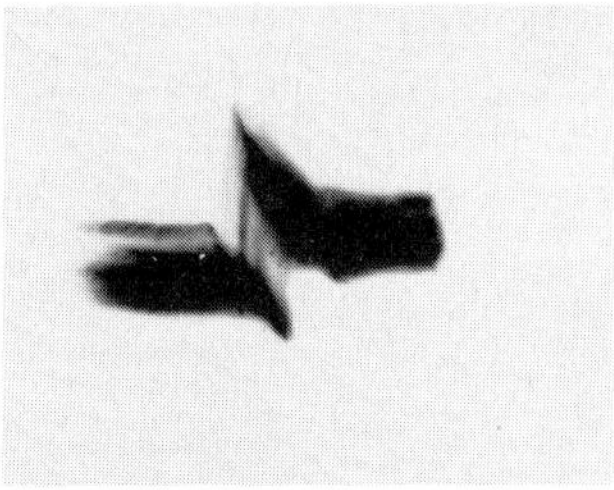

FIGURE 3
Synchrotron white beam x-ray diffraction "spot" showing plastic deformation in an aluminum crystal caused by high-power ultrasound.

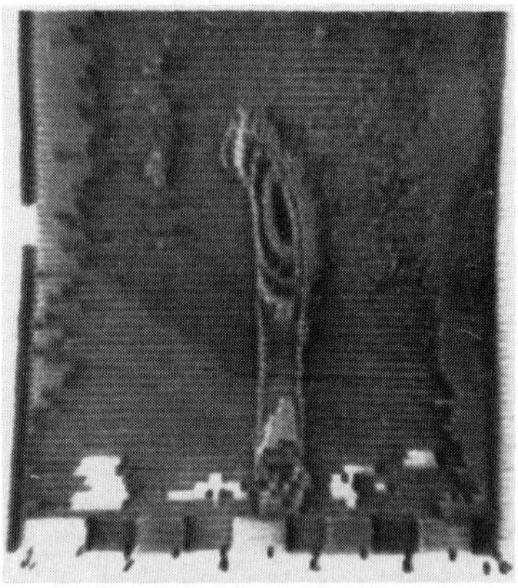

FIGURE 4
Infrared image of heat generated at large grain in zinc specimen caused by high-power ultrasound.

5. OPTICAL GENERATION AND DETECTION OF ELASTIC WAVES

Laser beam ultrasound generation and optical interferometric detection affords the opportunity to make truly non-contact ultrasonic measurements. Incorporation of scanning or full-field imaging techniques would greatly increase the capability of testing large structures without the present necessity of either immersing the test object in a water tank or using water squirter acoustic coupling. In addition measurements can be made on hot bodies, in hostile environments, and in outer space. Figure 5, taken from the work of Rosen [6], shows a schematic diagram of such a non-contact system which used a Nd:YAG pulsed laser to generate Rayleigh waves at the surface of a variety of metal specimens. The surfaces tested had been microstructurally modified by electron beam or high-power laser irradiation to either form an amorphous layer on the crystalline bulk, to produce different layers of transformed phases, or to induce case hardening of steel components. The surface waves were detected in a non-contact manner using a dual laser beam interferometer coupled to a spectrum analyzer, which permitted absolute measurement of ultrasonic wave velocity and attenuation as a function of frequency.

Recent developments in optical elastic wave detection involve incorporation of fiber-optics in the internal design of the interferometer resulting in a much more compact and rugged instrument [7]. Figure 6 shows the fidelity of the waveform detected with such an instrument. This elastic wave pulse was generated by a 5MHz transducer acoustically coupled to one side of a 3 inch thick aluminum block and detected with a fiber-optic interferometer on the opposite side. Another recent development is the ability to obtain full-field images of elastic wavefield displacements on the surface of solid materials using heterodyne holographic double-pulse laser techniques [8]. Figure 7 shows such a full-field image obtained by incidence of a laser pulse on the back surface of a graphite/epoxy plate and holographic interferometric imaging of the elastic wave displacement field on the opposite face after a selected delay time. The anisotropy inherent in the construction of the composite plate is evident in the wavefield. In addition, the effect of a delamination on the wavefield is visible on the left side of the image.

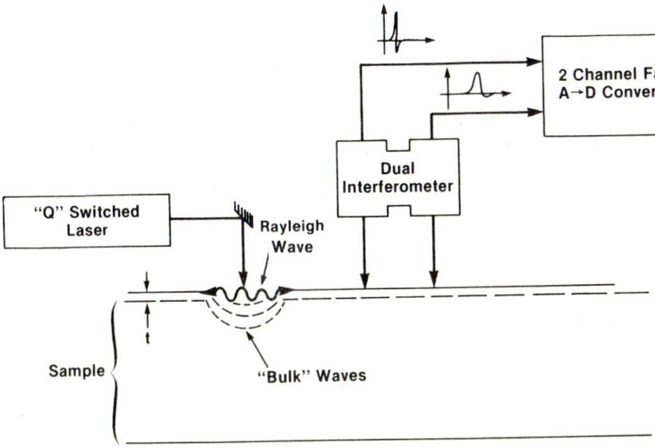

FIGURE 5
Schematic of laser beam generation and dual laser beam interferometric detection of Rayleigh waves at the surface of metals.

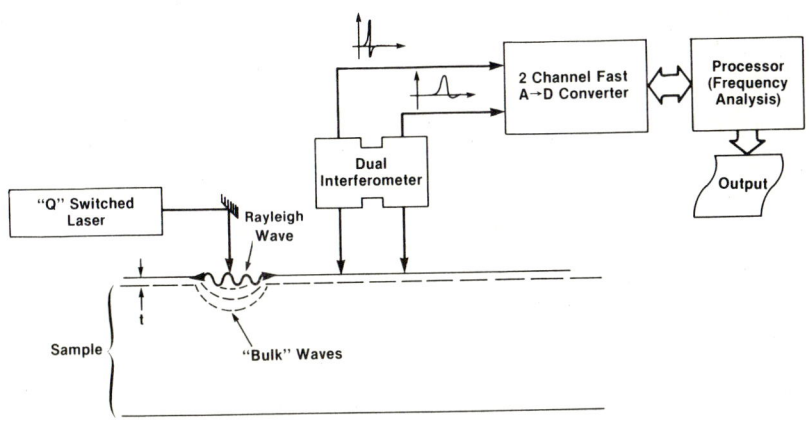

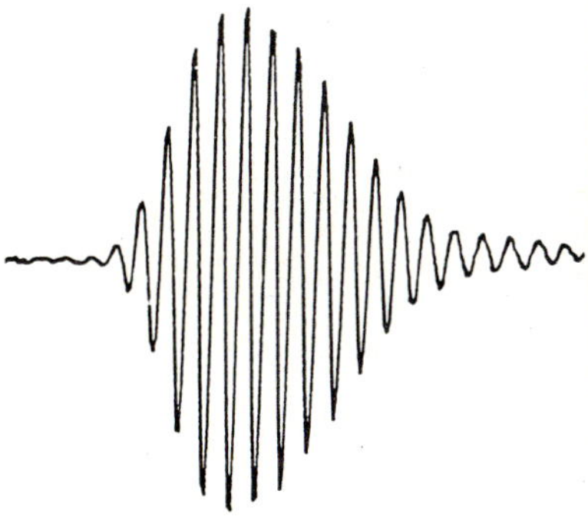

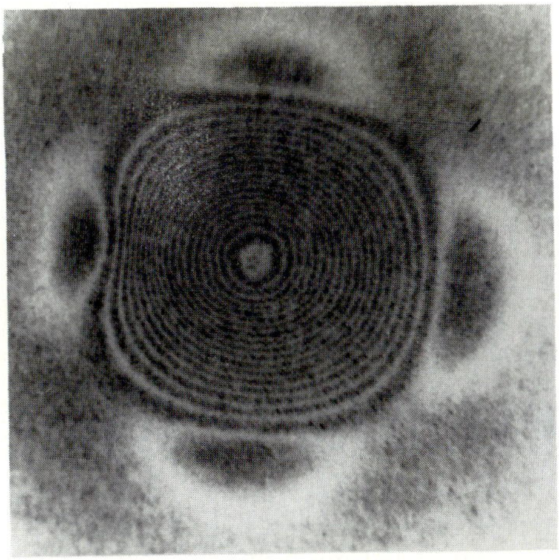

FIGURE 6
40 microsec time record of 5 MHz elastic wave pulse in aluminum block detected by fiber-optic interferometer.

FIGURE 7
Holographically recorded surface displacements caused by laser generated elastic waves in graphite/epoxy composite plate.

6. CONCLUSIONS

The results of several experiments have been presented where optical, infrared, and x-ray electromagnetic waves have been used to obtain information about elastic waves inside or on the surface of solid materials. Suggestions are made where additional theoretical work would be desirable.

7. ACKNOWLEDGEMENTS

The research described in this paper was partially supported by the Army Research Office, the National Science Foundation, the Naval Sea Systems Command, the Department of Energy, the Office of Naval Research and the National Aeronautics and Space Administration.

8. REFERENCES

[1] J.M. Liu, Ph.D. Dissertation, Mechanics Department, The Johns Hopkins University (198).

[2] R.E. Green, Jr., in Mechanics of Nondestructive Testing, W.W. Stinchcomb (ed.), (Plenum, NY 1980), pp. 55-76.

[3] R.E. Green, Jr., Ultrasonics 13, 117-127 (1975).

[4] R.B. Mignogna, R.E. Green, Jr., J.C. Duke, Jr., E.G. Henneke II, and K.L. Reifsnider, Ultrasonics 19, 159-163 (1981).

[5] A. Puskar, The Use of High-Intensity Ultrasonics, Elsevier Scientific Publishing Co., N.Y. (1982).

[6] M. Rosen, Center for Nondestructive Evaluation, The Johns Hopkins University, Baltimore, MD 21218, U.S.A. (Private Communication).

[7] J.B. Spicer and J.W. Wagner, Center for Nondestructive Evaluation, The Johns Hopkins University, Baltimore, MD 21218, U.S.A. (Private Communication).

[8] M. Ehrlich and J.W. Wagner, Center for Nondestructive Evaluation, The Johns Hopkins University, Baltimore, MD 21218, U.S.A. (Private Communication).

ELASTIC WAVE PROPAGATION
M.F. McCarthy, M.A. Hayes, (Editors)

ELECTROMAGNETICALLY COUPLED ELASTIC WAVE PROPAGATION PROBLEMS IN EDDY CURRENT NONDESTRUCTIVE TESTING

R.K.T. HSIEH

Department of Mechanics
Royal Institute of Technology
S-100 44 Stockholm
Sweden

One important application of eddy current testing methods is the inspection of metallic objects for flaws or defects such as cracks, voids, or foreign inclusions. The presence of the flaw modifies the low frequency electromagnetic field of an induction coil. The change in the field is recognized as a flaw-induced e.m.f. either in the energizing coil or in a separate pick-up coil.
This paper presents a theory for electromagnetically coupled elastic wave propagation and its application to an eddy current flaw detection which includes elastic deformation. The theoretical description of the defect is based on the concept of material multipoles. This paper also provides results which can facilitate practical applications of the theory to nondestructive testings for the characterization of the material defect, such as its moduli, size effect and shape effects.

1. INTRODUCTION

Eddy current testing (ECT) is the name given to an industrial inspection method which uses low frequency electromagnetic fields. The field is generated by an induction coil carrying an electric current, and the effect of the test object on the field is detected by the change in the impedance of the coil or by a change in th e.m.f. induced in a second coil. To produce the change the test object must be conducting or ferromagnetic. It is from the currents induced in the conducting matter that the name for the method is derived. The field configuration is sensitive to the shape, position, conductivity and permeability of the test body. These are the basic properties of the test object and in principle, an arbitrary small variation of any one of them is detected by observing the corresponding change in the e.m.f. induced in the coil. As cracks and foreign inclusions in the test object cause changes in its shape, conductivity and permeability, ECT can also be used to their detection. This particular application of eddy current testing is known as eddy current flaw detection (ECFT).

In conventional ECFT, the effect of a flaw in a test body is represented by a suitable electromagnetic dipole. Such approximation implies that the dimension of the flaw is smaller than the skin depth in the conducting matter which corresponds to the wavelength in free space, and that only "infinitesimal" defects can be dealt with. If in addition, the dimension of the flaw is smaller than the wavelength of the incident field, the electromagnetic fields are governed by the uncoupled (quasi static) zero frequency form of the Maxwell equations, also called the small flaw scattering equations. It is understood here that the effect of the flaw is to scatter the incident field. Though eddy current testing was first subject of systematic study during World War II, it has not further developed essentially [1]-[2]. Recently because of the development of increasingly powerful computers, more attention has however been paid to further develop ECT [3]. In this study, the accent is laid on the development of an ECTF for an electromagnetically coupled elastic test object with a finite flaw. The flaw is being modelled using the concept of material multipole [4]-[6].

Throughout this paper, time-harmonic motions are considered but the term $\exp(-i\omega t)$ is generally omitted.

2. MATERIAL MULTIPOLE DEFECT

Let us consider a linear homogeneous elastic solid, the constitutive behavior of which is given, as usual, by

$$t_{ij}(\underline{x}) = C_{ijmn}\, u_{m,n}(\underline{x}), \tag{1}$$

where $\underset{\sim}{t}$ is its stress, $\underline{u}$ its elastic displacement, $\nabla\underline{u}$ its symmetric elastic strain and $\underset{\sim}{C}$ its elastic moduli which, for the isotropic case, may be written as

$$C_{ijkl} = \lambda\delta_{ij}\delta_{kl} + \bar{\mu}(\delta_{ik}\delta_{jl} + \delta_{il}\delta_{jk}), \tag{2}$$

where λ and $\bar{\mu}$ are the Lamé's coefficients.

Let us further consider an array of discrete point body forces which is in the elastic material. The array is respectively centered at the point-array location $\underline{x}'$ and its corresponding point body forces are located at $\underline{x}'+\underline{d}^{\alpha}$ (α = 1, ... , K).

The elastic solid is now under the exertion of the body forces with density

$$\underline{f}(\underline{x}) = \sum_{\alpha=1}^{K} \underline{f}^{\alpha}\, \delta(\underline{x}-\underline{x}'-\underline{d}^{\alpha}) \tag{3}$$

and the elastic displacement has to verify the equation

$$C_{ijmn} u_{m,ni}(\underline{x}) + \rho\, \omega^2 u_j + f_j(\underline{x}) = 0. \tag{4}$$

Using the concept of the elastic Green function, the displacement field due to the array of point forces relative to its respective centre $\underline{x}'$ in the presence of another elastic source $\underline{u}^{o}$ can be written as

$$u_j(\underline{x}) = u_j^{o} + \sum_{n=0}^{\infty} \frac{(-1)^n}{n!} P_{is_1\ldots s_n} G_{ij,s_1 \ldots s_n}(\underline{x},\underline{x}'), \tag{5}$$

where the tensor $P_{is_1\ldots s_n}$ defined by

$$P_{is_1\ldots s_n} = \sum_{\alpha=1}^{K} f_i^{\alpha} d_{s_1}^{\alpha} \ldots d_{s_n}^{\alpha}, \tag{6}$$

is called the elastic multipole of order n and the elastic Green function $\underset{\sim}{G}$ is defined through the relationship

$$C_{ijmn}\, G_{mp,nj}(\underline{x},\underline{x}') + \rho\, \omega^2 G_{ip}(\underline{x},\underline{x}') + \delta_{ip}\, \delta(\underline{x}' - \underline{x}) = 0, \tag{7}$$

the symbols δ_{ij} and $\delta(\underline{x}' - \underline{x})$ are respectively the Kronecker delta and the Dirac function.
For an isotropic solid of infinite extent, the Green function $\underset{\sim}{G}$ has been found, see e.g. [7].

Eq (5) is valid for both permanent and induced material multipoles. Eqs (1)-(7) describe the mechanics of multipole force arrays in an elastic solid.

In the following sections, uses will be made of the concept of induced material multipoles of type inhomogeneity to model volume defects, such as foreign inclusions, cracks and voids, in coupled elastic materials.

An inhomogeneity can be respectively defined by the material moduli and the mass density

$$C_{ijkl}(\underline{x}) = C_{ijkl} + \Delta C_{ijkl}\, \gamma(\underline{x}), \tag{8}$$

$$\rho(\underline{x}) = \rho + \Delta\rho\, \gamma(\underline{x}), \tag{9}$$

Here $\Delta C_{ijkl} = C^*_{ijkl} - C_{ijkl}$, $\Delta\rho = \rho^* - \rho$, where $\underset{\sim}{C}^*$ is the material constant in the inhomogeneity, ρ^* is its mass density and $\gamma(\underline{x})$ is the indicative function characterizing its position, size and shape.

For physical reasons, we have that

$$dP_i = - \Delta\rho\, \omega^2\, u_i\, dV. \tag{10}$$

The other "inhomogeneous" material multipoles can then be defined as

$$dP_{ij} = - \Delta C_{ijmn}\, u_m\, dS_n \quad \text{and} \quad dP_{ijl} = 2\, \Delta C_{ijkl}\, u_k\, dV. \tag{11}$$

These various material multipoles characterize an elastic inhomogeneity. A fictitious body force $\underline{f}^{def}$ can instead be determined therefrom and

$$f_i^{def} = \Delta\rho\, \omega^2 \gamma\, u_i + \Delta C_{ijkl}\, (\gamma\, u_{k,l})_{,j} \quad . \tag{12}$$

For a void, we would have $\Delta\underset{\sim}{C} = -\underset{\sim}{C}$. For a crack, in addition, the displacement in eq (10) would be the discontinuity of the displacement across the crack surfaces $[\underline{u}]$. As can be verified, one of the advantages of the present approach is that though an inhomogeneity is contained in the elastic solid, only the Green function of the matrix is needed for the determination of the elastic fields.
Material multipoles are of course different for different constitutive laws, see e.g. [8]. However, it has been shown that the concept of induced material multipole for any chosen material, here for elastic solids, can even be applied to coupled materials, see [5].

3. AN ECTF MODEL FOR ELECTROMAGNETICALLY COUPLED ELASTIC SOLID

It is known that if an electrically conducting magnetoelastic solid is subjected to electromagnetic forces while immersed in a magnetic field, the electromagnetic field is still governed by Maxwell's equations with a modified Ohm's law, while the elastic deformation field is determined by a modified Hooke's law, see e.g. [9].

Under the assumptions of low excitation frequency (less than 10^7 Hz), the basic field equations of eddy current problems for isotropic, electromagnetically coupled elastic solids in the absence of external source current, with the neglection of displacement current and free electric charge density, may be summarized as follows.

- The quasi-static form of the Maxwell's equations :

$$\nabla \times \underline{E} - i\omega\, \underline{B} = 0, \qquad \nabla . \underline{B} = 0, \qquad \nabla \times \underline{H} = \underline{J}, \qquad \nabla . \underline{J} = 0 . \tag{13}$$

The modified Ohm's law and the magnetic constitutive relation are respectively

$$\underline{J} = \sigma(\underline{E} - i\omega\, \underline{u} \times \underline{B}) \qquad \text{and} \qquad \underline{B} = \mu \underline{H} , \tag{14}$$

where σ is the conductivity, μ the magnetic permeability, $\underline{E}$, $\underline{B}$, $\underline{J}$ and $\underline{u}$ are respectively the electric field, the magnetic induction field, the electric current density and the elastic displacement.

- The equation of linear momentum :

$$\nabla . \underset{\sim}{t} = - \rho\omega^2 \underline{u} , \tag{15}$$

in which for simplicity, the linearized constitutive relation for the elastic stress, eq (1), is assumed and the ponderomotive electromagnetic forces are neglected.

- The interface conditions :

$$\underline{n} \times [\underline{E}]=0, \quad \underline{n}.[\underline{J}]=0, \quad \underline{n} \times [\underline{H}]=\underline{K}, \quad \underline{n}.[\underline{B}]=0, \quad [\underline{u}]=0, \quad [\underset{\sim}{t}].\underline{n}=0, \tag{16}$$

where $[.]$ denotes the jump of a certain quantity on the two sides of the interface, $\underline{K}$ is the surface current density.

An electromagnetically coupled elastic inhomogeneity can be defined in a similar way as in the previous section. Together with eqs (8)-(9), its characterizations are then given by

$$\sigma(\underline{x}) = \sigma + \Delta\sigma\ \Upsilon(\underline{x}),$$

$$\mu(\underline{x}) = \mu + \Delta\mu\ \Upsilon(\underline{x}), \tag{17}$$

where $\Delta\sigma=\sigma^*-\sigma$, $\Delta\mu=\mu^*-\mu$ are respectively the perturbation values of the material properties between the inhomogeneity and the matrix solid. σ^* and μ^* are respectively the electric conductivity and the magnetic permeability in the inhomogeneity.

Similarly also, the effect of an electromagnetically coupled elastic inhomogeneity can be characterized by the fictitious body force, eq (12), the fictitious current $\underline{J}^{def}$ and the fictitious electric charge ρ_e^{def}

$$\rho_e^{def} = \frac{\Delta\sigma}{i\omega} \nabla.(\gamma\,(i\omega\,\underline{u}x\underline{B} - \underline{E}))\ ,$$

$$\underline{J}^{def} = \nabla x\left(\frac{\Delta\mu\ \gamma}{\mu} \left(\frac{1}{\mu(\underline{x})}\right) \underline{B}\right) \ . \tag{18}$$

The ECFT problems can thus be modelled by a set of coupled differential equations given by eq (4) and eq (13) respectively augmented by eq (12) and eq (18).

It can be seen that both the displacement $\underline{u}$ and the magnetic flux $\underline{B}$ can be determined consecutively. After some algebra, the electric field $\underline{E}$ can be then obtained from

$$\nabla^2\underline{E} + i\omega\mu\sigma\ \underline{E} = i\omega\mu\ \underline{J}^{ind}\ , \tag{19}$$

where we have defined

$$\underline{J}^{ind} = \frac{1}{\mu}\ \nabla\left[\underline{B}.(\nabla x\underline{u}) - \underline{u}.(\nabla x\underline{B})\right] + \frac{1}{\mu\sigma}\ \nabla\rho^{def} - (\underline{J}^{def} - i\omega\sigma\ \underline{u}x\underline{B}).$$

The quantity $(\omega\mu\sigma)^{\frac{1}{2}}$ is the reciprocal of the skin depth.

It can be seen that eq (19) admits the solution

$$E_i = \int (G^e\ \nabla E_i - E_i\ \nabla G^e)\ dS - i\omega\mu \int J_i^{ind}\ G^e\ dV\ , \tag{20}$$

where the coupled electric Green function G^e is defined by

$$\nabla^2\ G^e(\underline{x},\underline{x}') + i\omega\mu\sigma\ G^e(\underline{x},\underline{x}') = -\ \delta(\underline{x}'-\underline{x})\ ,$$

and the Green function has been found, [10].

4. REFERENCES

[1] Libby, H.L., Introduction to Electromagnetic Nondestructive Test Methods (John Wiley & Sons, New York, 1971).

[2] Burrows, M., PhD dissertation, University of Michigan, Ann Arbor (1964).

[3] Yamamoto, Y. and Miya, K. (eds.), Electromagnetomechanical Interactions in Deformable Solids and Structures (North Holland, Amsterdam, 1987).

[4] Hsieh, R.K.T., Int. J. Engng. Sci. 20 (1982) 201.

[5] Zhou, S.A., PhD dissertation, Royal Institute of Technology, Stockholm, (1987).

[6] Zhou, S.A. and Hsieh, R.K.T., Int. J. Engng. Sci. 26 (1988) 13.

[7] Achenbach, J.D., Wave Propagation in Elastic Solids (North Holland, Amsterdam, 1976).

[8] Hsieh, R.K.T., Vörös, G. and Kovàcs, I., Physica 101B (1980) 201.

[9] Moon, F.C., Magneto-Solid Mechanics (John Wiley & Sons, New York, 1984).

[10] Panorsky, W.K.H. and Melba, P., Classical Electricity and Magnetism (Addison-Wesley, Reading, Mass., 1955).

ELASTIC WAVE PROPAGATION
M.F. McCarthy, M.A. Hayes, (Editors)

Lower Symmetry of the Elastic and Piezoelectric Tensors from Rotational Coupling

D. F. Nelson

Department of Physics
Worcester Polytechnic Institute
Worcester, MA 01609-2280

A long wavelength lattice dynamical calculation of the elastic and piezoelectric tensors yields linear coupling to rotation as well as to strain in *dynamic* interactions. This leads in the general anisotropic case to 45 independent elastic components (rather than 21 as believed for over a century) and to 27 independent piezoelectric components (rather than 18). The derivation differs from traditional ones by including optic modes in characterizing the crystal, by not using the adiabatic approximation, and by recognizing the existence of spontaneous or constant parts of the optic mode coordinates. The new coupling should be observable at temperatures close to a second order phase transition where a zone-center soft mode has fallen into the hypersonic region.

From the beginning of the formulation of the modern theory of elasticity by Cauchy in the 1820s there was controversy over whether there were 15 or 21 independent components of the stiffness tensor c_{ijkl} in a general anisotropic (triclinic) crystal until the experiments of Voigt in the late 1880s gave strong support to theories of 21 independent components[1]. Except for one challenge by Laval[2] in the 1950s, which was refuted by Lax[3], 21 has been the accepted number ever since.

Here we derive that there can be 45 independent stiffness components in a general anisotropic crystal. The derivation is based on lattice dynamics in the long wavelength limit but which does not invoke the adiabatic approximation by which the inertial effects of the optic modes are usually ignored. We show that the optic mode inertial effects cause a linear elastic coupling to rotations, as well as to strains. We believe these inertial effects make measurable contributions to stiffness components of crystals when the frequency of a zone-center soft mode driving a second order phase transition drops into the hypersonic region.

Some time ago we examined the total stress tensor of a dielectric crystal possessing all long wavelength modes (optic and acoustic) of mechanical motion in interaction with the electromagnetic field including questions of uniqueness[4,5]. We showed that the total spatial frame stress tensor t_{ij}^T is uniquely defined if (a) it appears as the (negative) flow of momentum in the momentum conservation law,

$$\frac{\partial}{\partial t}[\rho \dot{x}_i + \epsilon_o(\mathbf{E}\times\mathbf{B})_i] - \frac{\partial t^T_{ij}}{\partial z_j} = 0\,, \tag{1}$$

(b) it possesses a continuous scalar product with the unit normal $\mathbf{n}$ across any surface fixed in the spatial frame,

$$(t^T_{kl})^o n_l - (t^T_{kl})^i n_l = 0\,, \tag{2}$$

and (c) it reduces in a vacuum to the Maxwell stress tensor

$$m_{ij} = \epsilon_o E_i E_j + \frac{1}{\mu_o} B_i B_j - \frac{1}{2}(\epsilon_o E_k E_k + \frac{1}{\mu_o} B_k B_k)\delta_{ij}\,, \tag{3}$$

where $\mathbf{E}$ is the electric field and $\mathbf{B}$ is the magnetic induction.

We showed that for a dielectric crystal having arbitrary symmetry, structural complexity, and nonlinearity, possessing all long-wavelength modes of mechanical motion (both optic and acoustic), and interacting with the electromagnetic field, the total stress tensor satisfying the three conditions is given by

$$t^T_{ij} = t^E_{ij} + m_{ij} - \rho \dot{x}_i \dot{x}_j\,. \tag{4}$$

Here $-\rho \dot{x}_i \dot{x}_j$ is the (negative) flow of material momentum, ρ being the deformed mass density, $\mathbf{x}$ being the spatial (center-of-mass) position vector, and the dot representing the (total) material time derivative d/dt. Both this term and the Maxwell stress tensor are symmetric upon interchange of i and j. The term t^E_{ij} does *not*, in general, possess that symmetry.

We call t^E_{ij} the elastic stress tensor for two reasons. First, it is the stress tensor that appears in the spatial frame dynamical equation for the center-of-mass variable[6] (which is the generalized elasticity equation) *if* the body forces are expressed in the electric plus Lorentz force form,

$$\rho \ddot{x}_i = t^E_{ij,j} + q^D E_i + (\mathbf{j}^D \times \mathbf{B})_i\,. \tag{5}$$

Here $q^D = -\nabla\cdot\mathbf{P}$ and $\mathbf{j}^D = \frac{\partial \mathbf{P}}{\partial t} + \nabla\times(\mathbf{P}\times\dot{\mathbf{x}})$ are, respectively, the dielectric charge and dielectric current expressed in terms of the polarization. Second, the elastic stress tensor can be expressed entirely in terms of mechanical variables of the medium:

$$t^E_{ij} = t^\Pi_{ij} + \frac{1}{J}\sum_{\nu=1}^{N-1} m^\nu \ddot{y}^{T\nu}_i y^{T\nu}_j \tag{6}$$

with t^Π_{ij} given by

$$t^\Pi_{ij} = \frac{1}{J}\frac{\partial(\rho^0\Sigma)}{\partial E_{BC}}\bigg|_{\Pi^\nu_A} x_{i,B} x_{j,C} = t^\Pi_{ji}\,. \tag{7}$$

Here $\mathbf{y}^{T\nu}$ is one of N-1 vector internal coordinates[9] that are closely related to optic mode coordinates, m^ν is the material frame mass density associated with the ν-internal coordinate, $\rho^0\Sigma$ is the stored energy per unit material frame (undeformed) volume, $E_{BC} = \frac{1}{2}(x_{k,B}x_{k,C} - \delta_{BC})$ is the Green finite strain tensor, $\Pi^\nu_A = X_{A,i} y^{T\nu}_i - Y^\nu_A$ is a rotationally invariant measure of the internal coordinate, and $x_{i,B} = \partial x_i/\partial X_B$. The (total) internal coordinate $\mathbf{y}^{T\nu} = \mathbf{Y}^\nu + \mathbf{y}^\nu$ consists of a constant or spontaneous part $\mathbf{Y}^\nu$ present in the natural (unperturbed) state of the crystal and a part that varies from external influences (e.g. deformations or electric fields).

Equation (6) clearly shows that the elastic stress tensor is asymmetric. It also shows that the antisymmetric part of t^E_{ij} is not only an internal coordinate or optic mode effect, but also that it is only a dynamic effect. Thus, the static stress tensor is perfectly symmetric. In other words, the antisymmetric part is always dispersive. Furthermore, it can be surmised from the time derivatives in Eq. (6) that the antisymmetric part of t^E_{ij} will be measurably large only for acoustic wave frequencies near an optic mode resonant frequency. To bring such a resonant frequency near to the accessible acoustic region will probably require the use of a soft mode whose frequency approaches zero at a second order phase transition.

We wish to emphasize that our conclusion that the elastic stress tensor is asymmetric is a strong statement. It is *not* conditional on some hypothetical constitutive assumption. It is, on the contrary, a derived property for *any* real dielectric crystal.

The antisymmetric part of the elastic stress tensor, and hence of the total stress tensor, creates a torque that is balanced[7] by a change in the internal angular momentum density $\mathbf{l} = \Sigma_\nu \rho^\nu \mathbf{y}^{T\nu} \times \dot{\mathbf{y}}^{T\nu}$ of the optic modes:

$$\frac{1}{J}\frac{d}{dt}(Jl_i) = \epsilon_{ijk} t^T_{kj} = \epsilon_{ijk} t^E_{kj} \tag{8}$$

where J is the Jacobian of the transformation from material $\mathbf{X}$ to spatial $\mathbf{x}$ coordinates and ϵ_{ijk} is the permutation symbol. Thus, the usual requirement that the antisymmetric part of the stress tensor vanish to insure angular momentum conservation is not relevant to a medium represented by a manifold of N vector matter continua (N=number of particles per primitive unit cell), N-1 of which lead to 3N-3 optic modes, and one of which leads to the 3 acoustic modes.

In view of the above results it is worthwhile linearizing Eq. (5) to see if the general, nonlinear, asymmetric elastic stress tensor remains asymmetric in the linear limit and thus causes a lowering of the symmetry of the stiffness tensor and the piezoelectric stress tensor. Accomplishing this linearization requires (a) use of the stored energy written to bilinear order,

$$\rho^0\Sigma = \sum_{\nu\mu} {}^{20}M^{\nu\mu}_{AB}\Pi^\nu_A\Pi^\mu_B + \sum_\nu {}^{11}M^\nu_{ABC}\Pi^\nu_A E_{BC} + {}^{02}M_{ABCD}E_{AB}E_{CD}, \tag{9}$$

where the ${}^{mn}M$ are expansion constants having tensor indices as subscripts, internal coordinate designations as postsuperscripts, and mnemonic notation (mn = 20, 11, 02) indicating the numbers of the Π^ν_A and E_{BC} fields to which they couple as presuperscripts, (b) solving the internal motion equations,

$$m^\nu \ddot{y}^{T\nu}_i = q^\nu E_i - \frac{\partial \rho^0\Sigma}{\partial \Pi^\nu_C} X_{C,i}, \tag{10}$$

for a harmonic excitation (angular frequency ω), where q^ν is the charge density associated with the ν-internal motion, and (c) linearizing Eq. (6)[10]. For displaying explicitly the frequency dependence of the result it is necessary to transform from the vector internal coordinates to the scalar normal (optic) mode coordinates[11] by

$$\eta^k = \sum_\nu (m^\nu)^{\frac{1}{2}} n^{k\nu}_j y^\nu_j, \tag{11}$$

a similar relation between N^k, the spontaneous part of the normal coordinate, and $\mathbf{Y}^\nu$, and

$$2\sum_\nu {}^{20}M^{\nu\mu}_{jl} n^{k\nu}_j/(m^\nu m^\mu)^{\frac{1}{2}} = \Omega^2_k n^{k\mu}_l\,. \tag{12}$$

Here $\mathbf{n}^{k\nu}(\nu = 1, 2, \ldots N-1; k = 1, 2, \ldots 3N-3)$ are the optic mode orthonormal eigenvectors, and Ω_k are the transverse optic mode frequencies.

The result of this calculation is

$$t^E_{ij} = c_{ijab}(\omega)u_{a,b} - e_{hij}(\omega)E_h \tag{13}$$

where

$$c_{ijab}(\omega) = 2\,{}^{02}M_{ijab} - \sum_k \frac{{}^{11}N^k_{ij}\,{}^{11}N^k_{ab}}{\Omega^2_k - \omega^2}$$

$$+\omega^2 \sum_{km}\sum_{\nu} \frac{{}^{11}N^k_{ij} n^{k\nu}_a n^{m\nu}_b N^m}{\Omega^2_k - \omega^2} + \omega^2 \sum_{km}\sum_{\nu} \frac{{}^{11}N^k_{ab} n^{k\nu}_i n^{m\nu}_j N^m}{\Omega^2_k - \omega^2}$$

$$-\,\omega^2 \sum_{kmn}\sum_{\nu\mu} \frac{\Omega^2_k}{\Omega^2_k - \omega^2} N^n n^{k\nu}_i n^{n\nu}_j n^{k\mu}_a n^{m\mu}_b N^m \tag{14}$$

$$e_{hij}(\omega) = -\sum_k \frac{c^k_h\,{}^{11}N^k_{ij}}{\Omega^2_k - \omega^2} + \omega^2 \sum_{km}\sum_{\nu} \frac{c^k_h n^{k\nu}_i n^{m\nu}_j N^m}{\Omega^2_k - \omega^2} \tag{15}$$

and for compactness we use

$${}^{11}N^k_{ab} = \sum_\nu n^{k\nu}_j\,{}^{11}M^\nu_{jab}/(m^\nu)^{\frac{1}{2}}, \tag{16}$$

$$c^k_j = \sum_\nu q^\nu n^{k\nu}_j/(m^\nu)^{\frac{1}{2}}\,. \tag{17}$$

Note that for $\omega \neq 0$

$$c_{jiab}(\omega) \neq c_{ijab}(\omega) = c_{abij}(\omega) \neq c_{abji}(\omega), \tag{18}$$

$$e_{hij}(\omega) \neq e_{hji}(\omega) \tag{19}$$

which leads in the most general anisotropic case to 45 independent components of the stiffness tensor[12] $c_{ijab}(\omega)$ and 27 independent components of the piezoelectric stress tensor $e_{hij}(\omega)$. Because of this generalized symmetry the displacement gradient $u_{a,b} = \partial u_a/\partial x_b$, not the infinitesimal strain, is the correct independent variable. The displacement gradient is a sum of infinitesimal strain, $u_{(a,b)} = (u_{a,b} + u_{b,a})/2$, plus infinitesimal rotation, $u_{[a,b]} = (u_{a,b} - u_{b,a})/2$. Note that terms in $c_{ijab}(\omega)$ and $e_{hij}(\omega)$ that give new antisymmetric contributions are proportional to ω^2 as expected from the derivatives appearing in Eq. (6). It should be emphasized that, though the lowered symmetry expressed in Eqs. (18) and (19) is true for any $\omega \neq 0$, the size of the antisymmetric contribution is negligible whenever the acoustic wave frequency is much less than all optic mode frequencies. Thus, for common, stable materials the traditional higher symmetry of the stiffness tensor and the piezoelectric stress tensor remains an adequate approximation for frequencies of even 10 GHz.

It should also be noted that the terms in $c_{ijab}(\omega)$ and $e_{hij}(\omega)$ that give new antisymmetric contributions depend directly on N^m, the spontaneous or constant parts of the optic mode coordinates. That such quantities exist is readily shown by calculating the spontaneous polarization of a pyroelectric or ferroelectric crystal. The result,

$$P^s_j = \sum_\nu q^\nu Y^\nu_j = \sum_k c^k_j N^k, \tag{20}$$

would not exist without the existence of N^k, or equivalently of the spontaneous part $\mathbf{Y}^\nu$ of the internal coordinates.

We wish to emphasize that the above derivation of lowered symmetry of the elastic and piezoelectric tensors is based on accepted long-wavelength lattice dynamics of real crystals. It obtains a new result through (a) the inclusion of optic modes in the formulation, (b) the lack of use of the adiabatic approximation, and (c) the recognition that optic mode coordinates have, in general, spontaneous parts. In contrast, the Laval proposal[2], while having some predictions in common with this work, was based on hypothetical non-central forces created by the charge density of valence and conduction electrons. The existence of the non-central forces was believed unrelated to the use or lack of use of the adiabatic approximation. For the latter reason the Lax refutation[3] of the Laval proposal invoked the adiabatic approximation. Clearly the purely electronic motions envisaged in the Laval model are appropriately handled by that approximation. For these reasons we regard the present derivation as quite distinct from the Laval speculation. Furthermore, we note that the use of the adiabatic approximation in the Lax refutation makes it inapplicable to the present work.

Lastly, we should note that an underdamped zone-center soft mode has been observed in chloranil at a frequency of about 2 cm^{-1} = 60 GHz[13] at a temperature slightly below the second order phase transition. Since acoustic waves at frequencies of order of 30 GHz can be studied by Brillouin scattering, the terms proportional to $\omega^2/(\Omega_s^2-\omega^2)$, where s denotes the soft mode, should be measurably large. On the other hand, we do not believe the acoustic velocity anomalies observed very near or below the commensurate-incommensurate phase transitions in $BaMnF_4$ by Fritz[14] and in $RbH_3(Se0_3)_2$ by Esayan et al.[15] result from this mechanism. For one thing, the frequencies of measurement, 4 to 30 MHz in Ref. 15 and 15 to 90 MHz in Ref. 16, are very low. Secondly, the incommensurate phase possesses a helical structure modulation that may produce a measurable acoustic activity effect[16,17] which modifies the acoustic velocity through a fifth rank tensor term. However, an explanation of the acoustic velocity anomalies based on wavevector dispersion[18] has not proven successful[19], nor has a model based on phasons[20].

I wish to acknowledge useful conversations with J. F. Scott concerning the anomalies near commensurate-incommensurate phase transitions.

1. For a historical review, see A.E.H. Love, *A Treatise on the Mathematical Theory of Elasticity*, Fourth Edition (University Press, Cambridge, 1927), reprinted (Dover, New York, 1944), pp.1-31.

2. J. Laval, C. R. Acad. Sci., Paris 232, 1947 (1951); in *L'etat solide*, LXe Congres de Physique Solvay, (Stoops, Bruxelles, 1951), p. 273; C.R. Acad. Sci., Paris 238, 1773 (1954).

3. M. Lax, in *Lattice Dynamics, Proceedings of the International Conference, Copenhagan, 1963* (Pergamon, Oxford, 1964), p. 583. This paper gives quite complete referencing of the controversy.

4. D. F. Nelson and M. Lax, Phys. Rev. B 13, 1770 (1976).

5. D. F. Nelson, *Electric, Optic, and Acoustic Interactions in Dielectrics* (Wiley, New York, 1979), pp. 134-141.

6. Ref. 5, pp. 129-131.

7. Ref. 5, pp. 146-149.

8. Ref. 5, pp. 131-132.

9. Ref. 5, pp. 77-80.

10. The electric and Lorentz body forces give no linear terms to the center-of-mass or elasticity equation.

11. Ref. 5, pp. 204-205.

12. If the new stiffness tensor were to be expressed in contracted (matrix) notation, it would have to be expressed as a symmetric 9×9 matrix, not the traditional symmetric 6×6 matrix.

13. D. M. Hanson, J. Chem. Phys. 63, 5046 (1975).

14. I. J. Fritz, Phys. Lett. 51A, 219 (1975).

15. S. Kh. Esayan, V.V. Lemanov, N. Manatkulov, and L. A. Shuvalov, Kristallografiya 26, 1086 (1981) [Sov. Phys.- Crystallogr. 26, 619 (1981)].

16. A. S. Pine, Phys. Rev. B2, 2049 (1970).

17. J. Joffrin and A. Levelut, Solid State Commun. 8, 1573 (1970).

18. V. Dvorak and S. K. Esayan, Solid State Commun. 44, 901 (1982).

19. J. F. Scott, in *Nonlinearity in Condensed Matter*, A. R. Bishop, et al., editors (Springer, Berlin, 1987), p. 320.

20. R. J. Gooding and M. B. Walker, Phys. Rev. B35, 6831 (1987).

ELASTIC WAVE PROPAGATION
M.F. McCarthy, M.A. Hayes, (Editors)

INHOMOGENEOUS MAGNETOELASTIC PLANE WAVES

Philippe BOULANGER

Département de Mathématique
Université Libre de Bruxelles
Campus Plaine, C.P.218/1, Boulevard du Triomphe
1050 Bruxelles, Belgium.

In this paper non magnetizable isotropic elastic solids of infinite electrical conductivity are considered. The field and constitutive equations are linearized about an undeformed equilibrium configuration characterized by a static magnetic field, and the propagation of time-harmonic inhomogeneous plane waves is studied. In particular, special inhomogeneous waves without purely mechanical counterpart are obtained.

1. INTRODUCTION

Waves propagating in non magnetizable isotropic elastic solids of infinite or finite conductivity have been studied by several authors (see [1] and the references quoted in this paper). They are called magnetoelastic waves. In one of the first papers on this subject, due to L.Knopoff [2], the equations are linearized about an undeformed equilibrium configuration characterized by a static magnetic field. Later, J.Bazer [3], in the case of infinite conductivity, presented an exhaustive study of homogeneous plane waves and of wave fronts based upon the linearized equations. He later also considered one-dimensional non linear waves, and shock waves [1][4]. Here we return to the linearized equations and present a study of inhomogeneous time-harmonic plane waves (planes of equal amplitude different from planes of equal phase) using the method proposed by M.A.Hayes [5] in 1984. The electrical conductivity is assumed to be infinite.

2. BASIC EQUATIONS

The basic equations of magnetoelasticity are the equation of motion, the Maxwell equations, the usual constitutive equation of non linear isotropic elasticity, and, for the case of infinite conductivity considered here, the assumption that the electric field in the comoving frame is zero. The electromechanical interaction is due only to the Lorentz force entering the equation of motion.
After elimination of the electric field, and linearization about an undeformed equilibrium configuration characterized by the static uniform magnetic field $\mathring{\underset{\sim}{B}}$ these equations reduce to

$$\rho_o \partial_t^2 \underset{\sim}{u} = \text{div}\, \underset{\sim}{\tau} + (\text{rot}\, \underset{\sim}{b}) \times \mathring{\underset{\sim}{B}}, \tag{1}$$

$$\partial_t \underset{\sim}{b} = \text{rot}(\partial_t \underset{\sim}{u} \times \mathring{\underset{\sim}{B}}) \quad , \quad \text{div}\, \underset{\sim}{b} = 0, \tag{2}$$

$$\underset{\sim}{\tau} = \lambda(\text{div}\, \underset{\sim}{u})\underset{\sim}{1} + \mu\{\text{grad}\, \underset{\sim}{u} + (\text{grad}\, \underset{\sim}{u})^T\}, \tag{3}$$

where ρ_o is the mass density in the equilibrium configuration, $\underset{\sim}{u}$ the displacement from this configuration, $\underset{\sim}{\tau}$ the stress tensor (the undeformed configuration being stress-free), $\underset{\sim}{b}$ the additionnal magnetic field. The contribution of the displacement current in the equation of motion is relativistic compared

to the inertia term and has consequently been neglected.
The equation (3) is Hooke's law, λ and μ being the Lamé constants.

3. THE PROPAGATION CONDITION AND THE SECULAR EQUATION

We now consider inhomogeneous time-harmonic plane wave solutions of the basic equations. Thus

$$(\underset{\sim}{u},\underset{\sim}{b}) = \mathcal{R}e\{(\underset{\sim}{U},\underset{\sim}{B})\exp i\omega(t - \underset{\sim}{S}.\underset{\sim}{x})\}, \tag{5}$$

where the bivectors (complex vectors) $\underset{\sim}{U}$, $\underset{\sim}{B}$ are the complex amplitudes of $\underset{\sim}{u}$ and $\underset{\sim}{b}$, ω is the angular frequency (real), and $\underset{\sim}{S} = \underset{\sim}{S}^+ + i\underset{\sim}{S}^-$ is the slowness bivector. This bivector may be written [5]

$$\underset{\sim}{S} = N\underset{\sim}{C} \quad , \quad \underset{\sim}{C} = m\underset{\sim}{\hat{m}} + i\underset{\sim}{\hat{n}}, \tag{6}$$

where N is a complex number, $\underset{\sim}{\hat{m}}$ and $\underset{\sim}{\hat{n}}$ are orthogonal real unit vectors, and m is a real number. In order to determine all the possible waves the complex number N and the amplitude bivectors $\underset{\sim}{U}$, $\underset{\sim}{B}$ have to be found for every choice of the bivector $\underset{\sim}{C}$ [5]. We here assume $m \neq 0$, since $m = 0$ corresponds to the case of homogeneous waves.
Inserting (5) into (1) (2) (3) and eliminating $\underset{\sim}{\tau}$ and $\underset{\sim}{B}$ gives the following propagation condition for the determination of the amplitude bivector $\underset{\sim}{U}$:

$$\{\underset{\sim}{Q}(\underset{\sim}{S}) - \underset{\sim}{1}\}\underset{\sim}{U} = \underset{\sim}{0}, \tag{7}$$

where the tensor $\underset{\sim}{Q}(\underset{\sim}{S})$ called magnetoacoustic tensor is given by

$$\underset{\sim}{Q}(\underset{\sim}{S}) = (v_L^2 - v_T^2 + \overset{\circ}{a}{}^2)\underset{\sim}{S} \otimes \underset{\sim}{S} + \{v_T^2\underset{\sim}{S}.\underset{\sim}{S} + (\overset{\circ}{\underset{\sim}{a}}.\underset{\sim}{S})^2\}\underset{\sim}{1} - (\overset{\circ}{\underset{\sim}{a}}.\underset{\sim}{S})(\underset{\sim}{S} \otimes \overset{\circ}{\underset{\sim}{a}} + \overset{\circ}{\underset{\sim}{a}} \otimes \underset{\sim}{S}), \tag{8}$$

with v_L^2, v_T^2 and $\overset{\circ}{\underset{\sim}{a}}$ defined by

$$v_L^2 = (\lambda + 2\mu)/\rho_\circ \quad , \quad v_T^2 = \mu/\rho_\circ \quad , \quad \overset{\circ}{\underset{\sim}{a}} = \overset{\circ}{\underset{\sim}{B}}/\sqrt{\rho_\circ}. \tag{9}$$

The speeds v_L, v_T are the speeds of longitudinal and transverse elastic waves in the absence of magnetic field. For perfect fluids (magnetohydrodymanic), it may be shown that the magnetoacoustic tensor is still given by (8) but with $v_L = 0$ and $v_T^2 = p'_\circ$ (derivative of the pressure with respect to the mass density evaluated at $\rho_\circ$).
When the amplitude $\underset{\sim}{U}$ of the displacement is known, the amplitude $\underset{\sim}{B}$ of the additional magnetic field is obtained from (2) :

$$\underset{\sim}{B} = - i\omega\sqrt{\rho_\circ}\,\underset{\sim}{S} \times (\underset{\sim}{U} \times \overset{\circ}{\underset{\sim}{a}}). \tag{10}$$

Inserting $(6)_1$ into (7) gives the eigenvalue problem

$$\{\underset{\sim}{Q}(\underset{\sim}{C}) - N^{-2}\underset{\sim}{1}\}\underset{\sim}{U} = \underset{\sim}{0} \tag{11}$$

for the determination of N^{-2} and $\underset{\sim}{U}$. This problem has to be solved for every choice of the bivector $\underset{\sim}{C} = m\underset{\sim}{\hat{m}} + i\underset{\sim}{\hat{n}}$. The properties of eigenvalues and eigenbi-

vectors of a complex symmetric matrix are given in [5].
The equation for the determination of the eigenvalues N^{-2} is called the secular equation. It is a cubic equation which is factored so that one root is N_T^{-2} given by

$$N_T^{-2} = v_T^2 \underset{\sim}{C}.\underset{\sim}{C} + \overset{\circ}{a}{}^2 C_{\|}^2, \tag{12}$$

and the two other ones are the roots of the quadratic equation

$$N^{-4} - N^{-2}(v_L^2 + v_T^2 + \overset{\circ}{a}{}^2)\underset{\sim}{C}.\underset{\sim}{C} + v_L^2 v_T^2(\underset{\sim}{C}.\underset{\sim}{C})^2 + \overset{\circ}{a}{}^2 \underset{\sim}{C}.\underset{\sim}{C}(v_L^2 C_{\|}^2 + v_T^2 \underset{\sim}{C}_{\perp}.\underset{\sim}{C}_{\perp}) = 0, \tag{13}$$

where $C_{\|}$ and $\underset{\sim}{C}_{\perp}$ are defined by

$$C_{\|} = (\overset{\circ}{\underset{\sim}{a}}.\underset{\sim}{C})/\overset{\circ}{a} \quad , \quad \underset{\sim}{C}_{\perp} = \underset{\sim}{C} - C_{\|}(\overset{\circ}{\underset{\sim}{a}}/\overset{\circ}{a}). \tag{14}$$

Assuming $\overset{\circ}{\underset{\sim}{a}} \neq \underset{\sim}{0}$, it is easily seen that the root (12) is also a root of (13) if and only if $C_{\|} = 0$ or $\underset{\sim}{C}_{\perp}.\underset{\sim}{C}_{\perp} = 0$. We defer the analysis of these cases to sections 6 and 7, and it is now assumed that $C_{\|} \neq 0$ and $\underset{\sim}{C}_{\perp}.\underset{\sim}{C}_{\perp} \neq 0$. Then the bivectors or vectors $\underset{\sim}{C}$, $\overset{\circ}{\underset{\sim}{a}}$, $\underset{\sim}{C} \times \overset{\circ}{\underset{\sim}{a}}$ are linearly independent and taking the dot product of (11) with these bivectors gives the system

$$\begin{aligned}
&\{N^{-2} - (v_L^2 + \overset{\circ}{a}{}^2)\underset{\sim}{C}.\underset{\sim}{C}\}\underset{\sim}{C}.\underset{\sim}{U} + (\overset{\circ}{\underset{\sim}{a}}.\underset{\sim}{C})\overset{\circ}{\underset{\sim}{a}}.\underset{\sim}{U} = 0,\\
&-(v_L^2 - v_T^2)(\overset{\circ}{\underset{\sim}{a}}.\underset{\sim}{C})\underset{\sim}{C}.\underset{\sim}{U} + (N^{-2} - v_T^2\underset{\sim}{C}.\underset{\sim}{C})\overset{\circ}{\underset{\sim}{a}}.\underset{\sim}{U} = 0, \qquad (15)\\
&(N^{-2} - v_T^2\underset{\sim}{C}.\underset{\sim}{C} - \overset{\circ}{a}{}^2 C_{\|}^2)(\underset{\sim}{C} \times \overset{\circ}{\underset{\sim}{a}}).\underset{\sim}{U} = 0,
\end{aligned}$$

for the determination of the eigenbivectors $\underset{\sim}{U}$.
In what follows cartesian coordinate axes x_1, x_2, x_3 are chosen along the unit vectors $\underset{\sim}{\hat{n}}$, $\underset{\sim}{\hat{m}}$, $\underset{\sim}{\hat{n}} \times \underset{\sim}{\hat{m}}$, so that $\underset{\sim}{C} = (i, m, 0)$ and $\overset{\circ}{a}C_{\|} = i\overset{\circ}{a}_1 + m\overset{\circ}{a}_2$.

4. THE TRANSVERSE WAVE

The root (12) of the secular equation, and the corresponding amplitudes $\underset{\sim}{U}$, $\underset{\sim}{B}$ obtained from (15) and (10), may be written

$$N_T^{-2} = v_T^2(m^2 - 1) + (i\overset{\circ}{a}_1 + m\overset{\circ}{a}_2)^2 \;,\; \underset{\sim}{U} = \alpha \underset{\sim}{C} \times \overset{\circ}{\underset{\sim}{a}} \;,\; \underset{\sim}{B} = - i\omega\alpha\sqrt{\rho_\circ}\,\overset{\circ}{a}C_{\|}\underset{\sim}{U}, \tag{16}$$

where α is an arbitrary complex factor. Since $\underset{\sim}{C}.\underset{\sim}{U} = 0$, this wave is transverse. Using the geometrical interpretations given in [5], it turns out that, when $\overset{\circ}{a}_3 \neq 0$, the displacement $\underset{\sim}{u}$ and the magnetic field $\underset{\sim}{b}$ are both elliptically polarized in the plane orthogonal to $\overset{\circ}{\underset{\sim}{a}}$ and the polarization ellipses are similar and similarly situated. When $\overset{\circ}{a}_3 = 0$, that is when the static magnetic field is in the plane of the slowness bivector, $\underset{\sim}{u}$ and $\underset{\sim}{b}$ are both linearly polarized along the x_3-axis.
For $m = \pm 1$, we obtain an "universal solution" in the sense that it does not depend on the elastic coefficients but only on the static magnetic field. In particular the same solution is valid for perfect fluids. The slowness bivector of this wave is given, up to a ± sign, by

$$(\overset{\circ}{a}{}_1^2 + \overset{\circ}{a}{}_2^2)\underset{\sim}{S} = (\overset{\circ}{a}_1, \overset{\circ}{a}_2, 0) \pm i(\overset{\circ}{a}_2, -\overset{\circ}{a}_1, 0). \tag{17}$$

It is isotropic ($\underset{\sim}{S}.\underset{\sim}{S} = 0$) since $\underset{\sim}{S}^+.\underset{\sim}{S}^- = 0$ and $|\underset{\sim}{S}^+| = |\underset{\sim}{S}^-|$. The phase speed is the magnitude of the projection of $\overset{\circ}{\underset{\sim}{a}}$ onto the direction of $\underset{\sim}{S}^+$ (Alfvèn speed), and $\underset{\sim}{S}^-$ is orthogonal to $\overset{\circ}{\underset{\sim}{a}}$. When $\overset{\circ}{a}_3 \neq 0$, $\underset{\sim}{S}^+$ is not along $\overset{\circ}{\underset{\sim}{a}}$, $\underset{\sim}{u}$ and $\underset{\sim}{b}$ are elliptically polarized in the plane orthogonal to $\overset{\circ}{\underset{\sim}{a}}$ and the projections of the polarization ellipses onto the plane of the bivector $\underset{\sim}{S}$ are circles. When $\overset{\circ}{a}_3 = 0$, $\underset{\sim}{S}^+$ is along $\overset{\circ}{\underset{\sim}{a}}$, $\underset{\sim}{u}$ and $\underset{\sim}{b}$ are linearly polarized in the direction orthogonal to the plane of $\underset{\sim}{S}$.

5. THE TWO OTHER WAVES

The roots N_+^{-2}, N_-^2 of the quadratic equation (13), and the corresponding amplitudes $\underset{\sim}{U}$, $\underset{\sim}{B}$ obtained from (15) and (10) are given by

$$N_\pm^{-2} = \frac{1}{2}(m^2-1)(v_L^2 + v_T^2 + \overset{\circ}{a}{}^2) \pm \frac{1}{2}\{(m^2-1)^2(v_L^2 - v_T^2 + \overset{\circ}{a}{}^2)^2 - 4(m^2-1)(v_L^2 - v_T^2)(i\overset{\circ}{a}_1 + m\overset{\circ}{a}_2)^2\}^{1/2},$$

$$\underset{\sim}{U} = \alpha\{\underset{\sim}{C} - \frac{\overset{\circ}{a}C_{||}}{N_\pm^{-2} - N_T^{-2}}\underset{\sim}{C} \times (\overset{\circ}{\underset{\sim}{a}} \times \underset{\sim}{C})\}, \tag{18}$$

$$\underset{\sim}{B} = i\omega\alpha\sqrt{\rho_\circ}\, N_\pm \frac{N_\pm^{-2} - v_T^2(m^2-1)}{N_\pm^{-2} - N_T^{-2}}\underset{\sim}{C} \times (\overset{\circ}{\underset{\sim}{a}} \times \underset{\sim}{C}).$$

Here $m^2 \neq 1$, and since $\underset{\sim}{C}.\underset{\sim}{U} \neq 0$ and $\underset{\sim}{C} \times \underset{\sim}{U} \neq \underset{\sim}{0}$, these waves are neither transverse nor longitudinal. The displacement $\underset{\sim}{u}$ and the magnetic field $\underset{\sim}{b}$ are both elliptically polarized. When $\overset{\circ}{a}_3 = 0$, the polarization ellipses are both in the plane of the slowness bivector.

It is known [5] that a necessary and sufficient condition that the complex symmetric matrix $\underset{\sim}{Q}(\underset{\sim}{C})$ have an isotropic bivector $\underset{\sim}{U}$ is that this matrix have a double eigenvalue N^{-2}. This occurs when the quadratic equation (13) has a double root, that is either when

$$\overset{\circ}{a}_2 = 0 \text{ and } m^2 = \{(v_L^2 - v_T^2 + \overset{\circ}{a}{}^2)^2 - 4(v_L^2 - v_T^2)\overset{\circ}{a}{}_1^2\}(v_L^2 - v_T^2 + \overset{\circ}{a}{}^2)^{-2}, \tag{19}$$

or when

$$\overset{\circ}{a}_1 = 0 \text{ and } m^2 = \{(v_L^2 - v_T^2 + \overset{\circ}{a}{}^2)^2 - 4(v_L^2 - v_T^2)\overset{\circ}{a}{}_2^2\}^{-1}(v_L^2 - v_T^2 + \overset{\circ}{a}{}^2)^2. \tag{20}$$

Let us consider the case when (19) holds, so that the double root is

$$N^{-2} = -2\overset{\circ}{a}{}_1^2(v_L^2 - v_T^2)(v_L^2 + v_T^2 + \overset{\circ}{a}{}^2)(v_L^2 - v_T^2 + \overset{\circ}{a}{}^2)^{-2}. \tag{21}$$

Then $\underset{\sim}{S}^+$ is along the x_1-axis, and $\underset{\sim}{S}^-$ is along the x_2-axis so that it is orthogonal to $\overset{\circ}{\underset{\sim}{a}}$ and $\underset{\sim}{S}^+$. The corresponding displacement is circularly polarized in a plane parallel to $\underset{\sim}{S}^-$, and the magnetic field $\underset{\sim}{b}$ is elliptically polarized also in a plane parallel to $\underset{\sim}{S}^-$. The possibility of such a wave is due to the static magnetic field, since when $\overset{\circ}{\underset{\sim}{a}}$ vanishes its slowness becomes infinite ($N^{-2} = 0$). When $\overset{\circ}{a}_3 = 0$, $\underset{\sim}{S}^+$ is along $\overset{\circ}{\underset{\sim}{a}}$, $\underset{\sim}{u}$ is circularly polarized in the plane of the slowness bivector, $\underset{\sim}{b}$ is elliptically polarized in this plane, and since $\underset{\sim}{C}.\underset{\sim}{B} = 0$, its polarization ellipse is similar and similarly situated to the ellipse associated to $\underset{\sim}{C}$ when rotated through a quadrant [5]. The case when (20) holds does not yield any further result, since it leads to the same geometrical configuration of $\overset{\circ}{\underset{\sim}{a}}$, $\underset{\sim}{S}$, $\underset{\sim}{U}$, $\underset{\sim}{B}$ but rotated in space through a quadrant around the x_3-axis.

6. ORTHOGONAL SETTING OF THE STATIC MAGNETIC FIELD : $C_{\|} = 0$

When $C_{\|} = 0$, that is when $\overset{\circ}{\underset{\sim}{a}}.\underset{\sim}{S} = 0$, (8) shows that the magnetoacoustic tensor $\underset{\sim}{Q}(\underset{\sim}{S})$ is the same as the acoustical tensor of the purely elastic case, except that v_L^2 is replaced by $v_L^2 + \overset{\circ}{a}^2$. Thus the results for isotropic elastic bodies [5] may be used with this substitution.

7. ISOTROPIC $\underset{\sim}{C}_\perp$: $\underset{\sim}{C}_\perp.\underset{\sim}{C}_\perp = 0$

The bivector $\underset{\sim}{C}_\perp$ is isotropic when

$$\overset{\circ}{a}_2 = 0 \text{ and } m^2 = \overset{\circ}{a}{}_3^2/\overset{\circ}{a}{}^2, \text{ or } \overset{\circ}{a}_1 = 0 \text{ and } m^2 = \overset{\circ}{a}{}^2/\overset{\circ}{a}{}_3^2. \tag{22}$$

Let us assume that the first eventuality holds. Then the root (12) of the secular equation is also a root of (13) so that the secular equation has the double root N_T^{-2} and the simple root N_L^{-2} :

$$N_T^{-2} = -(v_T^2 + \overset{\circ}{a}{}^2)(\overset{\circ}{a}{}_1^2/\overset{\circ}{a}{}^2) \quad , \quad N_L^{-2} = -v_L^2(\overset{\circ}{a}{}_1^2/\overset{\circ}{a}{}^2). \tag{23}$$

For each of the roots (23), $\underset{\sim}{S}^+$ is along the x_1-axis, and $\underset{\sim}{S}^-$ is along the x_2-axis so that it is orthogonal to $\overset{\circ}{\underset{\sim}{a}}$ ans $\underset{\sim}{S}^+$. Also the magnitude of the projection of $\underset{\sim}{S}^+$ onto $\overset{\circ}{\underset{\sim}{a}}$ is respectively $1/(v_T^2 + \overset{\circ}{a}{}^2)^{1/2}$, $1/v_L$.
When $\underset{\sim}{C}_\perp$ is isotropic, the bivectors or vectors $\underset{\sim}{C}$, $\overset{\circ}{\underset{\sim}{a}}$, $\underset{\sim}{C} \times \overset{\circ}{\underset{\sim}{a}}$ are not linearly independent. Then in order to obtain the eigenbivectors $\underset{\sim}{U}$ corresponding to the eigenvalues (23) we take the dot product of (11) with the three linearly independent bivectors or vectors $\underset{\sim}{C}_\perp$, $\bar{\underset{\sim}{C}}_\perp$ (complex conjugate of $\underset{\sim}{C}_\perp$), $\overset{\circ}{\underset{\sim}{a}}$. This yields the system

$$\begin{gathered}
\{N^{-2} + (v_T^2 + \overset{\circ}{a}{}^2)(\overset{\circ}{a}{}_1^2/\overset{\circ}{a}{}^2)\}\underset{\sim}{C}_\perp.\underset{\sim}{U} = 0, \\
2(v_L^2 - v_T^2 + \overset{\circ}{a}{}^2)\overset{\circ}{a}{}_3^2\underset{\sim}{C}_\perp.\underset{\sim}{U} - \overset{\circ}{a}{}^2\{N^{-2} + (v_T^2 + \overset{\circ}{a}{}^2)(\overset{\circ}{a}{}_1^2/\overset{\circ}{a}{}^2)\}\bar{\underset{\sim}{C}}_\perp.\underset{\sim}{U} \\
+ 2i\overset{\circ}{a}_1(v_L^2 - v_T^2)(\overset{\circ}{a}{}_3^2/\overset{\circ}{a}{}^2)\overset{\circ}{\underset{\sim}{a}}.\underset{\sim}{U} = 0, \\
i\overset{\circ}{a}_1(v_L^2 - v_T^2)\underset{\sim}{C}_\perp.\underset{\sim}{U} - \{N^{-2} + v_L^2(\overset{\circ}{a}{}_1^2/\overset{\circ}{a}{}^2)\}\overset{\circ}{\underset{\sim}{a}}.\underset{\sim}{U} = 0.
\end{gathered} \tag{24}$$

Corresponding to the eigenvalue N_T^{-2}, we obtain from (24) and (10) the amplitudes

$$\underset{\sim}{U} = \alpha \underset{\sim}{C}_{\perp} \quad , \quad \underset{\sim}{B} = \alpha\omega \sqrt{\rho_\circ}\, N_T \overset{\circ}{a}_1 \underset{\sim}{U}, \tag{25}$$

so that $\underset{\sim}{u}$ and $\underset{\sim}{b}$ are circularly polarized in the plane orthogonal to $\overset{\circ}{\underset{\sim}{a}}$. Corresponding to the eigenvalue N_L^{-2}, we obtain from (24) and (10) the amplitudes

$$\underset{\sim}{U} = \alpha\{(v_L^2 - v_T^2)\underset{\sim}{C} - (\overset{\circ}{\underset{\sim}{a}}.\underset{\sim}{C})\overset{\circ}{\underset{\sim}{a}}\} \quad , \quad \underset{\sim}{B} = \alpha\omega\sqrt{\rho_\circ}\, N_L (v_L^2 - v_T^2)\overset{\circ}{a}_1 \underset{\sim}{C}_{\perp}, \tag{26}$$

so that $\underset{\sim}{u}$ is elliptically polarized in a plane parallel to $\underset{\sim}{S}^-$, while $\underset{\sim}{B}$ is circularly polarized in the plane orthogonal to $\overset{\circ}{\underset{\sim}{a}}$. Also $\underset{\sim}{C}_{\perp}.\underset{\sim}{U} = 0$ which means [5] that the projection of the polarization ellipse of $\underset{\sim}{u}$ onto the plane orthogonal to $\overset{\circ}{\underset{\sim}{a}}$ is a circle.

When the second eventuality (22) holds, no further result is obtained, since it leads to the same geometrical configuration of $\overset{\circ}{\underset{\sim}{a}}$, $\underset{\sim}{S}$, $\underset{\sim}{U}$, $\underset{\sim}{B}$ but rotated in space through a quadrant around the x_3-axis.

Finally, if $v_L^2 - v_T^2 = \overset{\circ}{a}{}^2$, N_T^{-2} and N_L^{-2} given in (23) are equal, and the secular equation has a triple root. The corresponding amplitudes $\underset{\sim}{U}$, $\underset{\sim}{B}$ are then given by (25).

REFERENCES

[1] Bazer, J. and Ericson, W.B., Arch.Rat.Mech.Anal., 55 (1974) 124.
[2] Knopoff, L., J.Geophys.Research, 60 (1955) 441.
[3] Bazer, J., Geophys.J.R.Astr.Soc., 25 (1971) 207.
[4] Bazer, J. and Karal, F., Geophys.J.R.Astr.Soc., 25 (1971) 127.
[5] Hayes, M.A., Arch.Rat.Mech.Anal., 85 (1984) 41.

ELECTROMECHANICAL WAVES IN ELASTIC DIELECTRICS

Jerzy Pawel NOWACKI

Institute of Fundamental Technological Research
Polish Academy of Sciences
Warsaw, Poland

The paper is devoted to the investigation of the propagation of waves in dielectrics with polarization gradient. First, we consider the two-dimensional waves assuming that the solution depends on the variables x_1, x_2 and t only. The equations of piezoelectricity with this functional dependance lead to two independent systems. Next we confine ourselves to the problems of wave propagation in a half-space. The Rayleigh type waves and the Bleustein-Gulyaev waves are discussed. Finally, we consider the propagation of waves in a dielectric layer. The nature of the vibration is determined by the corresponding characteristic equation.

1. INTRODUCTION

Elastic dielectric materials, in which occurs the linear piezoelectric effect, have become important in modern technology. The classical linear theory of piezoelectricity created by Voigt [1] and developed by Toupin [2] is concerned with the interaction between mechanical strain and the electric field and the electric polarization. This theory was extended by Mindlin [3] by assuming the dependence of the stored energy function also on the polarization gradient. This extension accommodates observed phenomena, such as electromechanical interactions in symmetric and non-symmetric equations proposed by Mindlin which are derivable as a long wave limit of the equation of the modern dynamical theory of crystal lattice of electronicaly polarizable atoms [4]. It also includes a surface energy of deformation and polarization, which is absent from the previous theories but which has been observed in the laboratory and calculated from atomic considerations [5,6].

In recent years this theory has received considerable attention from researchers in the field of continuum mechanics. A number of papers have been devoted to the dynamical problems. For example, by including magnetic effects, plane monochromatic waves in quartz have been investigated in [7]. Glockner and his coworkers considered the propagation of waves in thermoelastic centrosymmetric isotropic dielectrics [8, 9,10]. The Lamb problem has been investigated in [11].

In the present paper we will consider two problems. The propagation of waves in a homogeneous dielectric half-space and in a plate bounded by two parallel planes

$$c_{44}\ \nabla^2 \mathbf{u} + (c_{12} + c_{44})\nabla\nabla.\mathbf{u} + d_{44}\ \nabla^2 \mathbf{P} + (d_{12} + d_{44})\nabla\nabla.\mathbf{P} + \rho\mathbf{f} = \rho\ \frac{\partial^2 \mathbf{u}}{\partial t^2},$$

$$d_{44}\ \nabla^2 \mathbf{u} + (d_{12} + d_{44})\nabla\nabla.\mathbf{u} + (b_{44} + b_{77})\nabla^2 \mathbf{P} + (b_{12} + b_{44} - b_{77})\nabla\nabla.\mathbf{P} - a\mathbf{P} - \nabla\varphi + E^0 = 0,$$

$$\nabla.\mathbf{P} - \varepsilon_0\ \nabla^2 \varphi = 0. \qquad (2.5)$$

Assume that all effects depend only on the variables x_1, x_2 and t. With this functional dependence for **u**, **P** and ϕ, eqns (2.5) lead to the following two independent systems

$$c_{44}\ \nabla_1^2 u + (c_{12} + c_{44})\partial_1 (\partial_1 u_1 + \partial_2 u_2) + d_{44}\ \nabla_1 P_1^2 + (d_{12} + d_{44})\partial_1 (\partial_1 P_1 + \partial_2 P_2) = \rho \ddot{u}_1,$$

$$c_{44}\ \nabla_1^2 u_2 + (c_{12} + c_{44})\partial_2 (\partial_1 u_1 + \partial_2 u_2) + d_{44}\nabla_1^2 P_2 + (d_{12} + d_{44})\partial_2 (\partial_1 P_1 + \partial_2 P_2) = \rho \ddot{u}_2, \qquad (2.6)$$

$$d_{44}\ \nabla_1^2 u_1 + (d_{12} + d_{44})\partial_1 (\partial_1 u_1 + \partial_2 u_2) + (b_{44} + b_{77})\nabla_1^2 P_1 + (b_{12} + b_{44} - b_{77})\partial_1 (\partial_1 P_1 + \partial_2 P_2) - aP_1 - \partial_1 \phi = 0,$$

$$d_{44}\ \nabla_1^2 u_2 + (d_{12} + d_{44})\partial_2(\partial_1 u_1 + \partial_2 u_2) + (b_{44} + b_{77})\nabla_1^2 P_2 + (b_{12} + b_{44} - b_{77})\partial_1 (\partial_1 P_1 + \partial_2 P_2) - aP_2 - \partial_2 \phi = 0,$$

$$\nabla.P - E_0\ \nabla_1^2 \phi = 0,$$

and

$$c_{44}\ \nabla_1^2 u_3 + d_{44}\ \nabla_1^2 P_3 = \rho \ddot{u}_3,$$

$$d_{44}\ \nabla_1^2 u_3 + (b_{44} + b_{77})\nabla_1^2 P_3 - aP_3 = 0. \qquad (2.7)$$

The system (2.6), in which u_1, u_2, P_1, P_2 and ϕ are unknown fields, is rather complex. Thus, to find the solution, we introduce the functions ψ, X, Γ and Λ connected to the displacement and polarization fields by relations

$$u_1 = \partial_1 \psi + \partial_2 \Gamma, \quad u_2 = \partial_2 \psi - \partial_1 \Gamma,$$

$$P_1 = \partial_1 X + \partial_2 \Lambda, \quad P_2 = \partial_2 X - \partial_1 \Lambda. \qquad (2.8)$$

After some algebra and when we confine ourselves to monochromatic waves propagating along the x_2 axis with the wave number k and frequency ω, eqns.(2.6) - (2.7) can be recast in the forms

$$(\partial_1^2 - k^2)[\, l_1^2 (\partial_1^2 - k^2)^2 - (\partial_1^2 - k^2)\left(1 - \frac{b_{11}\omega^2}{\hat{a} c_1}\right) - \frac{\omega^2}{c_1}\,]\ [\psi^*, X^*, \phi^*] = 0, \qquad (2.9)$$

2. GOVERNING EQUATIONS

The system of basic linear equation for an elastic dielectric including polarization gradient effects occupying the domain V and bounded by the surface S is given by

(i) The field equations

$$\sigma_{ij,i} + \rho f_j = \rho \frac{\partial^2 u_j}{\partial t^2}, \quad \sigma_{ij} = \sigma_{ji},$$

$$E_{ij,i} + {}_L E_j + E_j^{MS} = E_j^0,$$

$$-e_0 \varphi_{,ii} + P_{i,i} = 0, \qquad \text{in V},$$

$$\varphi_{,ii} = 0. \qquad \text{in V'}. \tag{2.1}$$

(ii) The boundary conditions

$$\sigma_{ij} n_i = k_j ,$$

$$E_{ij} n_i = S_j$$

$$(P_i - \varepsilon_0 \| \varphi_{,i} \|) = 0 \qquad \text{on S}. \tag{2.2}$$

(iii) The constitutive equations

$$\sigma_{ij} = d_{12} \delta_{ij} P_{nn} + d_{44} (P_{ji} + P_{ij}) + c_{12} \delta_{ij} S_{kk} + 2c_{44} S_{ij},$$

$$- {}_L E_k = a P_k ,$$

$$E_{ij} = b_{12} \delta_{ij} P_{nn} + b_{44} (P_{ij} + P_{ji}) + b_{77} (P_{ij} - P_{ji}) + 2d_{44} S_{ij} + b_0 \delta_{ij} \tag{2.3}$$

(iv) The kinematic relations

$$E_i^{MS} = - \varphi_{,j} , \quad S_{ij} = \frac{1}{2}(u_{i,j} + u_{j,i}), \quad P_{ij} = P_{j,i} \tag{2.4}$$

where σ_{ij} , E_{ij} and S_{ij} designate the components of the stress tensor, the electric tensor and the strain tensor, respectively: u_i , P_i , E_i , E_i^{MS}, E_i^0 , f_i and n_i are respectively the components of displacement vector, polarization vector, effective local electric force vector, Maxwell self-field vector, external electric field vector, body force vector and the exterior unit normal vector: φ, $\| \psi_{,i} \|$, denote Maxwell potential, jump in $\varphi_{,i}$ across S: k_i and S_i are surface traction and electric force, while V' stands for the external vacuum and ε_0 its permittivity. Substituting (2.3) and (2.4) into (2.1) yields the Navier's equations of elastic dielectrics as

$$[l_1^2 (\partial_1^2 - k^2)^2 - (\partial_1^2 - k^2)\left(1 - \frac{\hat{b}_{44}\omega^2}{ac_2^2}\right) - \frac{\omega^2}{c_2^2}] (\Lambda^*, \Gamma^*) = 0,$$

$$[l_2^2 (\partial_1^2 - k^2)^2 - (\partial_1^2 - k^2)(1 - \eta_2) - \sigma_2^2) \; (u_3^*, P_3^*) = 0, \tag{2.10}$$

where

$$\hat{a} = a + \varepsilon_0^{-1} \;, \hat{b}_{44} = b_{44} + b_{77} \;, c_1^2 = \frac{c_{11}}{\rho} \;, c_2^2 = \frac{c_{44}}{\rho},$$

$$l_1^2 = \frac{c_{11}b_{11} - d_{11}^2}{\hat{a}c_{11}} > 0, \; l_2^2 = \frac{c_{44}\hat{b}_{44} - d_{44}^2}{ac_{44}} > 0,$$

$$\eta_1 = \frac{b_{11}\,\omega^2}{\hat{a}c_1^2} \;, \quad \eta_2 = \frac{\hat{b}_{44}\,\omega^2}{ac_2^2}, \quad \rho_1 = \frac{\omega}{c_1} \;, \; \rho_2 = \frac{\omega}{c_2} \;.$$

The general form of solution is given by

$$\phi = (A_1 e^{-\beta_1 x_1} + A_2 e^{-\beta_2 x_2})\, e^{(ikx_2 - i\omega t)},$$

$$X = (x_1 A_1 e^{-\beta_1 x_1} + x_2 A_2 e^{-\beta_2 x_2}) e^{(ikx_2 - i\omega t)},$$

$$\phi = \varepsilon_0^{-1} \; x + A_5\, e^{-kx_1}\, e^{(ikx_2 - i\omega t)} \;, \tag{2.11}$$

$$\Lambda = (B_1\, e^{-\gamma_1 x_1} + B_2\, e^{-\gamma_2 x_2}) e^{(ikx_2 - i\omega t)},$$

$$\Gamma = (\tau_1 B_1\, e^{-\gamma_1 x_1} + \tau_2 B_2\, e^{-\gamma_2 x_2}) e^{(ikx_2 - i\omega t)},$$

and

$$u_3 = C_1 e^{-\gamma_1 x_1} + C_2 e^{-\gamma_2 x_2},$$

$$P_3 = \tau_1 C_1 e^{-\gamma_1 x_1} + \tau_2 C_2 e^{\gamma_2 x_2}, \tag{2.12}$$

where

$$\beta_{1,2} = \left[k^2 + \frac{1}{l_1^2}\{ (1-\eta_1) \pm [(1 - \eta_1)^2 + 4\sigma_1^2 \; l_1^2]^{1/2}\}\right]^{1/2},$$

$$\gamma_{1,2} = \left[k^2 + \frac{1}{l_1^2}\{ (1-\eta_1) \pm [(1 - \eta_2)^2 + 4\sigma_1^2 \; l_1^2]^{1/2}\}\right]^{1/2},$$

$$x_\alpha = -\frac{c_{11}}{d_{11}} \frac{\beta_\alpha^2 - k^2 + \sigma_1^2}{\beta_\alpha^2 - k^2} \;, \qquad \tau_\alpha = -\frac{c_{44}}{d_{44}} \frac{\gamma_\alpha^2 - k^2 + \sigma_2^2}{\gamma_\alpha^2 - k^2} \;.$$

3. THE HALF-SPACE

In the infinite dielectric domain the above waves propagate separately and independently. But in the case considered here these waves are linked by the boundary conditions and as well, as we must satisfy the Laplace equation for the electric potential in vacuum:

$$\nabla^2 \phi^* = 0. \tag{3.1}$$

The solution of (3.1) has the form

$$\phi^* = A_0\, \phi^* (x_1) e^{(ikx_2 - i\,\omega t)} \tag{3.2}$$

3.1. Rayleigh waves.

The system of eqns.(2.6)-(3-1) should be accompanied by the boundary conditions. We assume that the surface $x_1 = 0$ is free of stresses i.e.

$$\sigma_{11}(0,x_2,t) = \sigma_{12}(0,x_2,t) = 0,$$

$$E_{11}(0,x_2,t) = E_{12}(0,x_2,t) = 0, \tag{3.3}$$

$$P_1 - \varepsilon_0 \,\|\, \sigma,\phi \,\|_{x_1=0} = 0, \quad (\phi^+ - \phi^-)\,|_{\,x_1=0} = 0$$

The exploration of these boundary conditions yields a system of six homogeneous algebraic equations, the coefficients of which are the functions of k . On equating to zero the determinant of the system we obtain the characteristic equation for k^2. The smallest of ratios of this equation yields the phase velocity $v = \omega/k$ of the propagation of surface Rayleigh waves in dielectric isotropic half-space.

3.2. Bleustein-Gulyaev waves.

The system of equations (2.7) should now be accompanied by the following boundary conditions:

$$\sigma_{13}\ (0,x_2,t) = E_{13}\ (0,x_2,t) = \theta \tag{3.4}$$

Substituting the (2.12) into (3.2) we obtain the system of equations for constants C_1 and C_2 . Next, we find that the characteristic equation has only a trivial solution. This proves the non-existance of such a waves in isotropic dielectric materials.

4. THE ELASTIC LAYER

Now, we assume that both edges are free of stresses.

4.1. Plane waves.

The boundary conditions (3.1) must be satisfied for $x_1 = \pm h$. The general problem of propagation of waves may be reduced to the solution of two simple problems, i.e. to the consideration of the symmetric and antisymmetric vibrations only. These vibrations are characterized by the symmetry of the displacement u_2 and polarization P_2 with respect to the plane $x = \theta$ The displacement u_1 and the polarization P_1 are antysymmetric with respect to the same plane. This, as before leads to the system of algebraic equations for constants A_i and B_i .

4.2. SH-waves.

Using the trigonometric functions we are able to rewrite (2.12) for monochromatic waves in the form

$$\begin{aligned} u_3^*(x_1) &= C_1 \operatorname{ch} \gamma_1 x_1 + C_2 \operatorname{ch} \gamma_1 x_1 + C_3 \cos i\gamma_2 x_1 + C_4 \cos i\gamma_2 x_1, \\ P_3^*(x_1) &= \tau_1 C_1 \operatorname{ch} \gamma_1 x_1 + \tau_1 C_2 \operatorname{ch} \gamma_1 x_1 + \tau_2 C_3 \cos i\gamma_2 x_1 + \tau_2 C_4 \cos i\gamma_2 x_1, \end{aligned} \tag{4.1}$$

In the symmetric case the boundary conditions $\sigma_{13} = \mathcal{E}_{13} = 0$ for $x_1 = \pm h$ have the form

$$\begin{aligned} &c_{44}\,(C_2 \gamma_1 \operatorname{ch} \gamma_1 h + i\gamma_2 C_4 \cos i\gamma_2 h) + d_{44}\,(\gamma_1 \tau_1 C_2 \gamma_1 \operatorname{ch} \gamma_1 h + \\ &+ i\gamma_2 \tau_2 C_4 \cos i\gamma_2 h) = 0, \\ &d_{44}\,(C_2 \gamma_1 \operatorname{ch} \gamma_1 h + i\gamma_1 C_4 \cos i\gamma_2 h) + \hat{b}_{44}(\gamma_1 \tau_1 C_2 \gamma_1 \operatorname{ch} \gamma_1 h + \\ &+ i\gamma_2 \tau_2 C_4 \cos i\gamma_2 h) = 0, \end{aligned} \tag{4.2}$$

which leads to transcendental characteristic equation

$$(\tau_1 - \tau_2)\gamma_1 \gamma_2 \cos i\gamma_2 h = 0, \tag{4.3}$$

which can be reduced, after some elementary manipulation, to

$$\gamma_1 = 0\,,\ \gamma_2 = \frac{n\Pi}{2h} \qquad n = 1,3,5.... \,. \tag{4.4}$$

Analogously, for antysymmetric case we obtain

$$\gamma_1 = 0\,,\ \gamma_2 = \frac{n\Pi}{2h} \qquad n = 2,4,6... \,. \tag{4.5}$$

REFERENCES

[1] Voigt, W, Lehrbuch der Kristallphysik, Teubner, Leipzig, 1910.
[2] Toupin, R.A., The Elastic Dielectric, J.Ration.Mech.Anal.,vol.5, pp.849-915, 1956.
[3] Mindlin, R.D., Polarization Gradient in Elastic Dielectrics, Int.J.Solids Struct., vol.4, pp.637-642, 1968.
[4] Cochran, W. Phonons in Perfect Lattices, Plenum Press, New York (1966).
[5] Tosi, M.P., Solid State Physics, vol.16 Academic Press, New York (1964).
[6] Benson, G.C. and Yun, K.S., The Solid-Gas Interface Edited by E.A.Flood Dekker, New york (1967).
[7] Mindlin, R.D. and Toupin, R.A., Int.J.Solids Structures 7,1219 (1971).
[8] Chowdhury, K.L. and Glockner, P.G., On Thermoelastic Dielectrics, Int.J.Solids Struct. 13, 1173-1182 (1977).
[9] Nowacki, J.P. and Glockner, P.G. Some Dynamical Problems of Thermoelastic Dielectrics, Int.J.Sol.Structures 15, 183 (1979).
[10] Nowacki, J.P. and Glockner, P.G. Propagation of Waves in the Interior of a Thermoelastic Dielectric Half-space, Int.J.Engng.Sci. vol.19 pp.603-613
[11] Nowacki, J.P. and Trimarco, C., A Dynamic Problem for a Piezoelectric Medium, Proceedings of Euromech 226 (in print).

ELASTIC WAVE PROPAGATION
M.F. McCarthy, M.A. Hayes, (Editors)

FINITE AMPLITUDE WAVE PROPAGATION IN MAGNETIZED PERFECTLY ELECTRICALLY CONDUCTING ELASTIC MATERIALS

M. M. CARROLL

University of California
Berkeley, California 94720 USA

M. F. McCARTHY

University College
Galway, Ireland

Plane progressive and standing wave solutions are obtained for homogeneous, isotropic nonlinear elastic solids which are perfect electrical conductors and are subjected to a uniform magnetic induction field which acts in the direction of propagation. The motions are plane and circularly polarized, and have sinusoidal time variation and periodic spatial variation which depends on the shear response of the material and the magnitude of the magnetic induction field. A time dependent transverse magnetic induction field is induced by the mechanical disturbance. The problem of finding the spatial structure of the standing wave reduces to a problem of motion of a particle in a central force field.

1. INTRODUCTION

In earlier papers, closed form solutions were obtained for plane circularly polarized progressive waves and standing waves in isotropic nonlinear elastic solids [1-4] and in isotropic nonlinear dielectrics [5,6]. A feature of these solutions is that the waves are monochromatic, i.e., the circular polarization inhibits harmonic generation.

In the present paper, similar solutions are obtained for homogeneous, isotropic nonlinear elastic solids which are perfect electrical conductors and are subjected to a uniform magnetic induction field acting in the direction of propagation. It is assumed that the rates of the processes involved are such that the electromagnetic field may be regarded as quasi-static. With this assumption, the solutions are again sinusoidal in time and the effects of the biasing magnetic field are made explicit.

2. BASIC EQUATIONS

In describing the response of an isotropic perfectly electrically conducting elastic material, we can choose the deformation tensor $\underline{\gamma}$ and the magnetic induction field $\underline{b}$ as independent variables* and the Cauchy stress $\underline{T}$ and the magnetic field $\underline{h}$ as dependent variables. The deformation tensor is defined by $\underline{\gamma} = \underline{F}\,\underline{F}^T$, where $\underline{F}$ is the deformation gradient tensor. For nonmagnetic materials the constitutive equations then have the form

$$\underline{T} = \hat{\underline{T}}(\underline{\gamma}) \quad , \quad \underline{h} = \underline{b}/\xi \quad , \qquad (2.1)$$

*The tensor $\underline{\gamma}$ is not an appropriate independent variable for anisotropic media.

where $\hat{\underline{T}}$ is an isotropic tensor function, i.e.,

$$\hat{\underline{T}}(\underline{Q}\,\underline{\gamma}\,\underline{Q}^T) = \underline{Q}\,\hat{\underline{T}}(\underline{\gamma})\underline{Q}^T \tag{2.2}$$

for all orthogonal tensors $\underline{Q}$ and ξ is the constant magnetic permeability of the material. The constitutive equations must be modified for incompressible materials. In this case all admissible deformations are isochoric so that $\det \underline{\gamma} = 1$. The Cauchy stress now has an undetermined hydrostatic pressure, so that

$$\underline{T} = -p\underline{1} + \tilde{\underline{T}}(\underline{\gamma}) \quad , \tag{2.3}$$

where $\underline{1}$ denotes the unit tensor, p is an arbitrary scalar field, and the domain of $\tilde{\underline{T}}$ is restricted to isochoric deformations.

The representation theorem for isotropic tensor functions leads to the following representation of $\underline{T}$:

$$\underline{T} = \alpha_0\underline{1} + \alpha_1\underline{\gamma} + \alpha_{-1}\underline{\gamma}^{-1} \quad , \tag{2.4}$$

where α_0, α_1 and α_{-1} are scalar functions of the invariants

$$I_1 = \mathrm{tr}\,\underline{\gamma} \quad , \quad I_2 = \mathrm{tr}\,\underline{\gamma}^* \quad , \quad I_3 = \det \underline{\gamma} \tag{2.5}$$

and $\underline{\gamma}^*$ is the adjugate of $\underline{\gamma}$. The scalar response functions must meet certain conditions to ensure the existence of an energy density, but we do not make these conditions explicit here. The constitutive equation for isotropic incompressible media has the form (2.5) except that α_0 in (2.5) is now an arbitrary scalar field ($\alpha_0 = -p$) and α_1 and α_{-1} are functions of the invariants I_1 and I_2.

We assume that there is no applied body force field and that the rates of the processes considered are such that the electromagnetic field may be regarded as quasi-static. The motion of the material is then governed by the equations [7]

$$\mathrm{div}\,\underline{\Sigma} = \rho\ddot{\underline{x}} \quad , \quad \dot{\underline{B}} = \underline{0} \quad , \quad \mathrm{div}\,\underline{b} = 0 \tag{2.6}$$

where

$$\underline{\Sigma} = \underline{T} - \frac{1}{2\xi}B^2\underline{1} + \frac{1}{\xi}\underline{F}\,\underline{B} \otimes \underline{F}\,\underline{B} \quad , \quad \underline{B} = (\det \underline{F})^{-1}\underline{F}\underline{b} \quad , \tag{2.7}$$

ρ is the mass density and $\ddot{\underline{x}}$ is the acceleration. It follows from $(2.6)_2$ that, if the motion of the body is described by the mapping $\underline{x} = \underline{x}(\underline{X},t)$, then $\underline{B} = \underline{B}(\underline{X})$ is independent of t. In particular, if $\underline{B}$ is independent of $\underline{X}$ at any instant, i.e., if the magnetic induction field as measured in the reference configuration is uniform, then $\underline{B}$ will be a constant vector at all subsequent times. It is with such situations that we are concerned in this paper.

3. SIMPLE SHEARING AND MAGNETIZATION

Consider an isotropic perfectly electrically conducting elastic material which is magnetized so that the particle which is at (X,Y,Z) in a rectangular Cartesian system is subjected to the uniform magnetic induction field

$$B_x = 0 \quad , \quad B_y = 0 \quad , \quad B_z = B \quad . \tag{3.1}$$

Let the material also be subjected to a simple shear, so that the particle initially at (X,Y,Z) is now at (x,y,z), given by

$$x = X + \kappa Z \quad , \quad y = Y \quad , \quad z = Z \quad . \tag{3.2}$$

Here κ (> 0) is the amount of shear and the vector $\underline{B}$ is applied in the plane of shear and normal to the direction of shear.

An elementary computation, using (3.2), gives the components of the deformation tensor $\underline{\gamma}$ and its inverse $\underline{\gamma}^{-1}$ as

$$\underline{\gamma} = \begin{bmatrix} 1+\kappa^2 & 0 & \kappa \\ 0 & 1 & 0 \\ \kappa & 0 & 1 \end{bmatrix} \quad , \quad \underset{\sim}{\gamma}^{-1} = \begin{bmatrix} 1 & 0 & -\kappa \\ 0 & 1 & 0 \\ -\kappa & 0 & 1+\kappa^2 \end{bmatrix} \quad . \tag{3.3}$$

The invariants I_i are given by equations (2.5) and (3.2) as

$$I_1 = I_2 = 3 + \kappa^2 \quad ; \quad I_3 = 1 \quad . \tag{3.4}$$

The stress is obtained by substituting from (3.1)-(3.3) into (2.7). Thus, we have

$$\underline{\Sigma} = \begin{Bmatrix} \sigma_1 & 0 & \mu\kappa \\ 0 & \sigma_2 & 0 \\ \mu\kappa & 0 & \sigma_3 \end{Bmatrix} \quad , \tag{3.5}$$

where

$$\begin{aligned} \sigma_1 &= \alpha_o + \alpha_1(1+\kappa^2) + \alpha_{-1} - \frac{1}{2\xi} B^2 + \frac{1}{\xi} \kappa^2 B^2 \quad , \\ \sigma_2 &= \alpha_o + \alpha_1 + \alpha_{-1} - \frac{1}{2\xi} B^2 \quad , \\ \sigma_3 &= \alpha_o + \alpha_1 + \alpha_{-1}(1+\kappa^2) + \frac{1}{2\xi} B^2 \quad , \\ \mu &= \alpha_1 - \alpha_{-1} + \frac{1}{\xi} B^2 \quad . \end{aligned} \tag{3.6}$$

The function μ is called the generalized shear modulus and we note that its value is an increasing function of the applied magnetic induction field $\underline{B}$.

It follows from (3.2) and $(2.7)_2$ that the magnetic induction $\underline{b}$ at the point (x,y,z) has components

$$b_x = \kappa B \quad , \quad b_y = 0 \quad , \quad b_z = B \tag{3.7}$$

so that the magnetic induction has a component in the direction of shear.

Finally, we rewrite the shear response equation $(3.5)_5$ in the form

$$\sigma = \mu(\kappa^2, B^2)\kappa \tag{3.8}$$

and we assume that this relation can also be inverted, for given B, in the form

$$\kappa = \nu(\sigma^2, B^2)\sigma \quad , \tag{3.9}$$

where ν is the generalized shear compliance.

4. CIRCULARLY POLARIZED PROGRESSIVE WAVES

We now consider a circularly polarized harmonic progressive wave of the form

$$x = X + A\sin(kZ-\omega t) \quad , \quad y = Y + A\cos(kZ-\omega t) \quad , \quad z = Z \tag{4.1}$$

propagating in perfectly electrically conducting isotropic elastic material in the direction of a biasing magnetic induction field in the z direction,

$$B_x = 0 \quad , \quad B_y = 0 \quad , \quad B_z = B \ . \tag{4.2}$$

We investigate the possibility of such waves in the presence of a biasing longitudinal magnetic induction field.

An elementary calculation gives the components of the deformation tensor $\underline{\gamma}$ as

$$\underline{\gamma} = \begin{bmatrix} 1+\kappa^2\cos^2\zeta & \kappa^2\cos\zeta\sin\zeta & \kappa\cos\zeta \\ \kappa^2\cos\zeta\sin\zeta & 1+\kappa^2\sin^2\zeta & \kappa\sin\zeta \\ \kappa\cos\zeta & \kappa\sin\zeta & 1 \end{bmatrix} , \tag{4.3}$$

with

$$\kappa = kA \quad , \quad \zeta = \omega t - kz \ . \tag{4.4}$$

The deformation tensor $\underset{\sim}{\gamma}$ in (4.3) is related to that in (3.3) by an orthogonal transformation

$$\underline{\gamma}(\kappa^2,\zeta) = \underline{Q}(\zeta)\underline{\gamma}(\kappa^2,0)\underline{Q}(\zeta)^T \ , \tag{4.5}$$

with

$$\underline{Q}(\zeta) = \begin{bmatrix} \cos\zeta & -\sin\zeta & 0 \\ \sin\zeta & \cos\zeta & 0 \\ 0 & 0 & 1 \end{bmatrix} \tag{4.6}$$

Thus, $\underline{Q}(\zeta)$ describes a rotation through an angle ζ about the z-axis and the local deformation in the circularly polarized wave is a shear of amount kA, such that the direction and plane of shear at any location z rotate about the z-axis with angular frequency ω. Since the magnetic induction field is invariant under the rotation $\underline{Q}(\zeta)$, it follows that the stress $\underline{\Sigma}$ and magnetic induction field associated with the fields (4.1) and (4.2) are obtained from those associated with (3.1) and (3.2) by applying the rotation $\underline{Q}(\zeta)$. Thus, we have

$$\underline{\Sigma} = \begin{bmatrix} \sigma_1\cos^2\zeta + \sigma_2\sin^2\zeta & (\sigma_1-\sigma_2)\cos\zeta\sin\zeta & \mu\kappa\cos\zeta \\ (\sigma_1-\sigma_2)\cos\zeta\sin\zeta & \sigma_1\sin^2\zeta + \sigma_2\cos^2\zeta & \mu\kappa\sin\zeta \\ \mu\kappa\cos\zeta & \mu\kappa\sin\zeta & \sigma_3 \end{bmatrix} , \tag{4.7}$$

and

$$b_x = \kappa B\cos\zeta \quad , \quad b_y = \kappa B\sin\zeta \quad , \quad b_z = B \ , \tag{4.8}$$

where the response functions in (4.7) are defined in (3.6), with $\kappa = kA$. Clearly, the magnetic induction field $\underline{b}$ at any location z makes an angle arc tan κ with the direction in which the wave propagates and rotates about the z axis with angular velocity ω.

The equations of motion (2.6) reduce to

$$(\mu\kappa \cos \zeta)' = \rho\omega^2 A \sin \zeta \quad ,$$
$$(\mu\kappa \sin \zeta)' = -\rho\omega^2 A \cos \zeta \quad , \tag{4.9}$$
$$\sigma_3' = 0 \quad ,$$

where the prime denotes differentiation with respect to z. The first two equations are compatible and reduce to the algebraic equation

$$\omega^2/k^2 = \mu(k^2A^2,B^2)/\rho \quad , \tag{4.10}$$

which determines the phase velocity of the wave. Since κ and B are constants, equation $(4.9)_3$ is satisfied identically for both compressible and incompressible solids (choosing p = constant). The magnetic induction field equations $(2.6)_1$ and $(2.6)_2$ are also satisfied identically since B is a constant. Thus, the circularly polarized progressive wave (4.1) can propagate in the presence of a biasing magnetic induction field (4.2). The interaction effects are (i) the influence of the magnetic induction field in tending to increase the phase velocity in (4.10), (ii) the influence of the magnetic induction field on the normal stresses in (4.7), (iii) the appearance of a propagating transverse magnetic induction field in (4.8), which is in phase with the direction of shearing.

5. CIRCULARLY POLARIZED STANDING WAVES

The sinusoidal finite amplitude standing waves of the form

$$x = X + \phi(Z)\cos \omega t + \psi(Z)\sin \omega t \quad ,$$
$$y = Y + \phi(Z)\sin \omega t - \psi(Z)\cos \omega t \quad , \tag{5.1}$$
$$z = Z$$

can exist in all <u>incompressible</u> isotropic elastic solids [2,4] as well as in all incompressible elastic dielectrics [8]. We now investigate the effect on this motion of a uniform biasing magnetic induction field

$$B_x = 0 \quad , \quad B_y = 0 \quad , \quad B_z = B \tag{5.2}$$

in an incompressible isotropic perfectly electrically conducting elastic material.

We introduce the functions $\kappa(Z)$ and $\theta(Z)$, defined by

$$\kappa = \sqrt{\phi'^2 + \psi'^2} \quad , \quad \theta = \text{arc tan}(\psi'/\phi') \tag{5.3}$$

so that

$$\phi' = \kappa \cos \theta \quad , \quad \psi' = \kappa \sin \theta \quad . \tag{5.4}$$

An elementary calculation shows that the deformation tensor $\underline{\gamma}$ associated with the motion (5.1) has the form (4.3) except that κ is now a function of z and ζ is given by

$$\zeta = \omega t - \theta \quad . \tag{5.5}$$

It follows that the stress and magnetic induction are again given by (4.7) and (4.8), with the understanding that σ_1, σ_2 and σ_3 involve the arbitrary pressure $\alpha_0 = -p$. If we assume that p does not depend on x or y, the equations of motion $(2.6)_1$ reduce to

$$(\mu\kappa \cos\zeta)' = -\rho\omega^2(\phi \cos\omega t + \psi \sin\omega t) \quad ,$$
$$(\mu\kappa \sin\zeta)' = -\rho\omega^2(\phi \sin\omega t - \psi \cos\omega t) \quad , \tag{5.6}$$
$$\sigma_3' = 0 \quad ,$$

while equations $(2.6)_2$ and $(2.6)_3$ are satisfied identically.

Substitution for κ and ζ from (5.4) and (5.5) in $(5.6)_{1,2}$ shows that these equations are compatible and reduce to the pair of coupled nonlinear ordinary differential equations

$$(\mu\phi')' = -\rho\omega^2\phi \quad , \quad (\mu\psi')' = -\rho\omega^2\psi \quad . \tag{5.7}$$

The arbitrary pressure p can then be chosen so that $(5.6)_3$ is satisfied.

As discussed in [2,4], the change of variables

$$\Phi = \mu\phi' \quad , \quad \Psi = \mu\psi' \tag{5.8}$$

reduces equations (5.7) to the form

$$\Phi'' = -\rho\omega^2\nu(\sigma^2,B^2)\Phi \quad , \quad \Psi'' = -\rho\omega^2\nu(\sigma^2,B^2)\Psi \quad , \tag{5.9}$$

with the generalized shear compliance defined in (3.9) and

$$\sigma^2 = \Phi^2 + \Psi^2 \quad . \tag{5.10}$$

Since B is constant, these are the equations of motion for a particle in an attractive central force field, with (Φ,Ψ) as Cartesian coordinates of the particle and z as time. These equations may also be written in polar form

$$\sigma'' - \sigma\theta'^2 = -\rho\omega^2\nu\sigma \quad , \quad \sigma\theta'' + 2\sigma'\theta' = 0 \quad , \tag{5.11}$$

with

$$\Phi = \sigma\cos\theta \quad , \quad \Psi = \sigma\sin\theta \tag{5.12}$$

The first integrals of equations (5.11) are

$$\frac{1}{2}(\sigma'^2 + h^2/\sigma^2) + \rho\omega^2\int_0^\sigma \nu(s^2,B^2)s\,ds = T \tag{5.13}$$

and

$$\sigma^2\theta' = H \quad , \tag{5.14}$$

where the constants T and H are the energy and angular momentum in the central force orbit. Equations (5.13) and (5.14) are first order ordinary differentials in σ and θ, so that the problem is reduced to quadratures [2,4].

The magnetic induction field $\underline{b}$ is once again given by (4.8) except that κ is a function of z and ζ is given by (5.5). Thus, as before, the effect of the mechanical wave is to make the magnetic induction field nonuniform through the introduction of a transverse component.

ACKNOWLEDGMENTS

This work was supported by a contribution from the Shell Companies Foundation, in support of a Shell Distinguished Chair at the University of California, Berkeley.

REFERENCES

[1] Carroll, M. M., Acta Mechanics 3 (1967) 167.
[2] Carroll, M. M., Journal of Elasticity 7 (1977) 411.
[3] Carroll, M. M., Journal of Elasticity 8 (1978) 323.
[4] Carroll, M. M., Journal of Applied Mechanics 46 (1979) 867.
[5] Carroll, M. M., Quarterly of Applied Mathematics 25 (1967) 319.
[6] Carroll, M. M., Applied Physics 19 (1979) 321.
[7] McCarthy, M. F., Proc. Vib. Probs. 4 (1967) 337.
[8] McCarthy, M. F., and Carroll, M. M., "Circularly Polarized Waves of Finite Amplitude in Nonlinear Dielectrics," in print.

ELASTIC WAVE PROPAGATION
M.F. McCarthy, M.A. Hayes, (Editors)

The Stability of Plane Waves in Generalised Thermoelasticity

N.H. Scott[1]

Abstract

Equations governing infinitesimal sinusoidal wave propagation in an homogeneously prestrained anisotropic thermoelastic solid have been extended by previous authors to the case of generalised thermoelasticity, which includes the effect of a small relaxation time τ. This generalisation changes the nature of the system of equations from parabolic to hyperbolic. All four low frequency squared wave speeds have the same value in each theory, whilst three of the squared wave speeds at high frequency (corresponding to elastic wave behaviour) differ by only $O(\tau)$ from those of the classical theory. The fourth is radically different; it becomes wave like in character under the new theory with a large but finite real and positive squared wave speed which is $O(\tau^{-1})$. All these waves are shown to be stable for all frequencies.

1 Introduction

By generalised thermoelasticity we mean a theory of heat conduction in which the heat flux vector $q_i(\mathbf{x},t)$ satisfies the equation

$$q_i + \tau_0 \dot{q}_i = -K_{ij}\theta_{,j} \qquad (1)$$

instead of the usual Fourier law of heat conduction. Here $\theta(\mathbf{x},t)$ is the temperature excess, K_{ij} is the thermal conductivity tensor and τ_0 a positive relaxation time whose vanishing reduces (1) to the usual Fourier law. Commas denote space derivatives and superposed dots time derivatives. If $\tau_0 = 0$ the temperature is governed by the usual diffusion equation which is, of course, parabolic in nature. If $\tau_0 > 0$ the governing equation is of hyperbolic type and is a wave equation predicting a thermal wave with finite speed as opposed to the infinite speed of temperature diffusion in the case $\tau_0 = 0$. This thermal wave is often termed *second sound* and Chandrasekhariah [1] has written a recent review of the subject to which the reader is referred for further details and references.

Sharma and Sidhu [2] considered plane wave propagation in anisotropic generalised thermoelasticity basing their analysis upon that of Chadwick [3] for anisotropic classical thermoelasticity ($\tau_0 = 0$). The purpose of the present paper is to extend to generalised thermoelasticity a result of the author [4] that under certain conditions all homogeneous plane waves in classical thermoelasticity are stable.

2 Basic equations

We consider a homogeneous anisotropic thermally conducting elastic body initially at a uniform temperature T resting in a state of equilibrium after a finite, homogeneous prestrain. A small particle displacement $\mathbf{u}(\mathbf{x},t)$ and temperature excess $\theta(\mathbf{x},t)$ are then superposed on

[1]School of Mathematics, University of East Anglia, Norwich, NR4 7TJ, U.K.

the uniform state. The basic equations of linear generalised thermoelasticity are taken in the form [2]

$$d_{ijkl}u_{k,jl} - \beta_{ij}\theta_{,j} = \rho\ddot{u}_i,$$

$$\kappa_{ij}\theta_{,ij} - T\beta_{ij}\dot{u}_{i,j} = \rho c\dot{\theta} + \tau_0(\rho c\ddot{\theta} + T\beta_{ij}\ddot{u}_{i,j}). \tag{2}$$

All other quantities occurring in equations (2) are constants evaluated in the uniform state, namely, the isothermal elasticity tensor d_{ijkl}, the temperature coefficient of stress β_{ij}, the density ρ, the symmetrised thermal conductivity tensor κ_{ij}, the specific heat at constant deformation c and the thermal relaxation time τ_0. If τ_0 vanishes Chadwick's [3] equations of classical linear thermoelasticity are recovered.

In the sequel we shall make the following three assumptions. (1) The isothermal elasticity tensor satisfies the strong ellipticity condition: $d_{ijkl}a_ib_ja_kb_l > 0$ for all nonzero real vectors $\mathbf{a}$ and $\mathbf{b}$. (2) The specific heat at constant deformation is positive: $c > 0$. (3) The symmetrised conductivity tensor is positive definite: $\kappa_{ij}a_ia_j > 0$ for all nonzero real vectors $\mathbf{a}$. No assumption need be made concerning the temperature coefficient of stress β_{ij}.

We consider now only those solutions of (2) which represent a plane wave of the form

$$(\mathbf{u}, \theta) = (\mathbf{U}, \Theta)exp[i\omega(s\mathbf{x}.\mathbf{n} - t)] \tag{3}$$

in which $\mathbf{U}$ and Θ are the constant complex displacement and temperature amplitudes, ω is the real angular frequency,

$$s = s_+ + is_-$$

is the complex slowness and $\mathbf{n}$ is a real unit vector defining the direction of propagation of the wave. On substituting the waveforms (3) into the field equations (2) we arrive at

$$(1 - i\omega\tau_0)\rho v^2 det(\rho v^2\mathbf{I} - \hat{\mathbf{Q}}) + i\omega\kappa c^{-1} det(\rho v^2\mathbf{I} - \tilde{\mathbf{Q}}) = 0, \tag{4}$$

the secular equation of Sharma and Sidhu [2], which if $\tau_0 = 0$ reduces to the secular equation derived by Chadwick [3] in classical thermoelasticity. In (4), $v = s^{-1}$ is the complex wave speed and $\kappa = \kappa_{ij}n_in_j$ is the conductivity in the direction $\mathbf{n}$, whilst $\tilde{\mathbf{Q}}$ and $\hat{\mathbf{Q}}$ are the isothermal and isentropic acoustic tensors defined by

$$\tilde{Q}_{ik} = d_{ijkl}n_jn_l, \;\; \hat{Q}_{ik} = \tilde{Q}_{ik} + (T/\rho c)\beta_{ij}\beta_{kl}n_jn_l,$$

respectively, each of which is symmetric because of the well-known symmetry property

$$d_{ijkl} = d_{klij}.$$

This symmetry property together with assumptions (1) and (2) is sufficient to ensure that each tensor has three real, positive eigenvalues.

We may non-dimensionalise the secular equation (4) in terms of the typical isothermal elasticity $\gamma = \tilde{Q}_{ik}n_in_k > 0$ by defining

$$\xi = \gamma^{-1}\rho v^2, \;\; \omega^* = \gamma c\kappa^{-1}, \;\; \tau = \tau_0\omega^*, \;\; \omega' = \omega/\omega^*$$

and arrive at

$$(1 - i\omega\tau)f(\xi) + i\omega g(\xi) = 0, \tag{5}$$

where

$$f(\xi) = \xi(\xi - \hat{q}_1)(\xi - \hat{q}_2)(\xi - \hat{q}_3), \tag{6}$$

$$g(\xi) = (\xi - \tilde{q}_1)(\xi - \tilde{q}_2)(\xi - \tilde{q}_3). \tag{7}$$

The notation ω' has been replaced by ω for convenience and $\tilde{q}_i$ and $\hat{q}_i$ $(i = 1, 2, 3)$ denote the real, positive eigenvalues of the non-dimensional tensors $\gamma^{-1}\tilde{\mathbf{Q}}$ and $\gamma^{-1}\hat{\mathbf{Q}}$, respectively. The quantity τ is a non-dimensional measure of the relaxation time τ_0 and for most materials possessing such a relaxation time seems to lie between 10^{-3} and 1.

Equation (5) is to be regarded as a quartic equation to be solved for ξ as a function of ω once the eigenvalues $\tilde{q}_i$, $\hat{q}_i$, and the relaxation time τ have been specified. The four roots $\xi_i(\omega)$ $(i = 1, 2, 3, 4)$ trace out four branches in the complex ξ plane for positive ω.

3 Stability in classical thermoelasticity: $\tau = 0$

A plane wave is said to be *stable* if its amplitude remains bounded at all distances in the direction of travel. For $\omega > 0$ this requires $s_+ s_- \geq 0$, which when put in terms of ξ becomes

$$\Im\, \xi \leq 0. \tag{8}$$

Thus, in order to prove the stability of all plane waves with positive frequency in generalised thermoelasticity we must show that each branch $\xi_i(\omega)$ remains entirely within the lower half of the complex ξ plane for $\omega \geq 0$ (though a branch may touch the positive real ξ axis).

Such stability has already been demonstrated by the author [4] in the case of classical thermoelasticity but since the results have not yet appeared in the literature they will be recapitulated here. A key result is the following deduction from assumptions (1) and (2):

Theorem 1 *If the eigenvalues of the two acoustic tensors are a set of six distinct numbers then they interlace according to*

$$\tilde{q}_1 < \hat{q}_1 < \tilde{q}_2 < \hat{q}_2 < \tilde{q}_3 < \hat{q}_3, \tag{9}$$

where $\tilde{q}_1 > 0$ *and the orderings* $\tilde{q}_1 < \tilde{q}_2 < \tilde{q}_3$ *and* $\hat{q}_1 < \hat{q}_2 < \hat{q}_3$ *have been adopted.*

If the six numbers are not distinct then common factors may be removed from (5) and the definitions (6) and (7) of $f(\xi)$ and $g(\xi)$ and it is found that the remaining (distinct) eigenvalues satisfy inequalities of the form (9). Each such factor removed from (5) corresponds to a non-dispersive, unattenuated, purely elastic wave that is decoupled entirely from thermal effects.

In the low frequency limit of the classical theory we find that the four branches may be approximated by their Taylor expansions:

$$\xi_i(\omega) = \hat{q}_i - i\omega g(\hat{q}_i)/f'(\hat{q}_i) + O(\omega^2), \quad (i = 1, 2, 3) \tag{10}$$

$$\xi_4(\omega) = -i\omega g(0)/f'(0) + O(\omega^2), \tag{11}$$

where prime denotes differentiation, whilst in the high frequency limit we have

$$\xi_i(\omega) = \tilde{q}_i + i\omega^{-1} f(\tilde{q}_i)/g'(\tilde{q}_i) + O(\omega^{-2}), \quad (i = 1, 2, 3) \tag{12}$$

$$\xi_4(\omega) = -i\omega + \hat{q}_1 + \hat{q}_2 + \hat{q}_3 - \tilde{q}_1 - \tilde{q}_2 - \tilde{q}_3 + O(\omega^{-1}). \tag{13}$$

Theorem 1 and assumption (3) may be used to show that each of these branches satisfies $\Im\, \xi(\omega) \leq 0$ in each limit of ω and so is stable. It is possible further to deduce [4]:

Theorem 2 *A consequence of Theorem 1, together with assumption (3), is that small amplitude sinusoidal vibrations in a classical homogeneous thermoelastic solid are necessarily stable for all frequencies.*

4 Stability in generalised thermoelasticity: $\tau > 0$

The stability results of the last section may be extended to generalised thermoelasticity by essentially similar arguments.

First, we note that for low frequencies the branches of (5) are given by (10) and (11) even for non-zero τ since τ does not appear until the $O(\omega^2)$ terms [2]. Therefore, each wave is certainly stable for small enough frequencies.

However, with $\tau > 0$ the high frequency limit differs from (12) and (13) obtained previously. As $\omega \to \infty$ we see that the four limiting values of ξ are given by the four roots of the real quartic equation

$$F(\xi) \equiv -\tau f(\xi) + g(\xi) = 0. \tag{14}$$

We now prove

Theorem 3 *For each $\tau > 0$, equation (14) has four real roots $\xi = \bar{q}_i$, say, $(i = 1,2,3,4)$, which interlace according to*

$$\bar{q}_1 < \tilde{q}_1 < \hat{q}_1 < \bar{q}_2 < \tilde{q}_2 < \hat{q}_2 < \bar{q}_3 < \tilde{q}_3 < \hat{q}_3 < \bar{q}_4, \tag{15}$$

with $\bar{q}_1 > 0$, provided that the the six acoustic eigenvalues are distinct.

Proof We are able to deduce from Theorem 1 the following strict inequalities

$$\begin{aligned}
F(0) &= -\tilde{q}_1\tilde{q}_2\tilde{q}_3 < 0,\\
F(\tilde{q}_1) &= -\tau\tilde{q}_1(\tilde{q}_1 - \hat{q}_1)(\tilde{q}_1 - \hat{q}_2)(\tilde{q}_1 - \hat{q}_3) > 0,\\
F(\hat{q}_1) &= (\hat{q}_1 - \tilde{q}_1)(\hat{q}_1 - \tilde{q}_2)(\hat{q}_1 - \tilde{q}_3) > 0,\\
F(\tilde{q}_2) &= -\tau\tilde{q}_2(\tilde{q}_2 - \hat{q}_1)(\tilde{q}_2 - \hat{q}_2)(\tilde{q}_2 - \hat{q}_3) < 0,\\
F(\hat{q}_2) &= (\hat{q}_2 - \tilde{q}_1)(\hat{q}_2 - \tilde{q}_2)(\hat{q}_2 - \tilde{q}_3) < 0,\\
F(\tilde{q}_3) &= -\tau\tilde{q}_3(\tilde{q}_3 - \hat{q}_1)(\tilde{q}_3 - \hat{q}_2)(\tilde{q}_3 - \hat{q}_3) > 0,\\
F(\hat{q}_3) &= (\hat{q}_3 - \tilde{q}_1)(\hat{q}_3 - \tilde{q}_2)(\hat{q}_3 - \tilde{q}_3) > 0,\\
F(+\infty) &= -\infty < 0.
\end{aligned} \tag{16}$$

Since $F(\xi)$ is a quartic polynomial continuous in ξ it is an immediate consequence of these inequalities that equation (14) has four real roots that satisfy the inequalities (15).

If the six eigenvalues are not distinct then factors may be removed from (5),(6) and (7) as before. Such factors occur also in (14) showing that if there is a coincidence between one of the values $\tilde{q}_i$ and one of the values $\hat{q}_i$ then one of the $\bar{q}_i$ is also equal to this common value. When all such coincident eigenvalues have been removed it is found that the remaining (distinct) eigenvalues satisfy strict inequalities of the form (15). □

On defining the real quartic polynomial

$$h(\xi) \equiv (\xi - \bar{q}_1)(\xi - \bar{q}_2)(\xi - \bar{q}_3)(\xi - \bar{q}_4), \tag{17}$$

we find that for high frequencies the four branches of $\xi(\omega)$ are given by

$$\xi_i(\omega) = \bar{q}_i - i\omega^{-1}\tau^{-1}f(\bar{q}_i)/h'(\bar{q}_i) + O(\omega^{-2}), \quad (i = 1,2,3,4) \tag{18}$$

for $\tau > 0$. The high frequency branch (13) in the case $\tau = 0$ which terminates at $-i\infty$ has been replaced for all $\tau > 0$ by the branch of (18) that terminates at the finite point $\bar{q}_4$. This is the chief distinguishing feature between the cases $\tau = 0$ and $\tau > 0$. Of course, their is no claim that the low and high frequency branches of $\xi(\omega)$ are connected together in the order in which they have been numbered and in general they are not (see FIG 2).

The inequalities (15) of Theorem 3 may be used to verify that each of the branches (18) satisfies $\Im\, \xi(\omega) < 0$ for ω large enough and is therefore stable. Thus we see that each of the four generalised thermoelastic waves is stable for low frequencies and for high frequencies. This may be extended to

Theorem 4 *For all $\tau > 0$ and for all positive frequencies each of the four branches $\xi(\omega)$ of (5) satisfies $\Im\, \xi(\omega) < 0$ and so gives a stable wave.*

Proof We have seen that the branches of $\xi(\omega)$ touch the real ξ axis at the points $0, \hat{q}_1, \hat{q}_2, \hat{q}_3$ (corresponding to $\omega = 0$) or $\bar{q}_1, \bar{q}_2, \bar{q}_3, \bar{q}_4$ (corresponding to $\omega = \infty$) and lie in $\Im\, \xi(\omega) < 0$ close to these points. But no branch may cut the real axis in any other point, as (5) may be rewritten as

$$f(\xi)/g(\xi) = -i\omega/(1 - i\omega\tau), \tag{19}$$

indicating that for finite, non-zero ω the ratio $f(\xi)/g(\xi)$ is complex so that ξ may not be real. Thus each branch must remain within the region $\Im\, \xi(\omega) < 0$, proving stability. □

For most materials that exhibit non-zero τ we find that

$$\tau \ll 1$$

and then the roots $\bar{q}_i$ of (14) are approximately

$$\bar{q}_i = \tilde{q}_i + \tau f(\tilde{q}_i)/g'(\tilde{q}_i) + O(\tau^2), \quad (i = 1, 2, 3) \tag{20}$$

$$\bar{q}_4 = \frac{1}{\tau} + \hat{q}_1 + \hat{q}_2 + \hat{q}_3 - \tilde{q}_1 - \tilde{q}_2 - \tilde{q}_3 + O(\tau). \tag{21}$$

Thus the first three roots $\bar{q}_i$ differ little from $\tilde{q}_i$ whilst the fourth is radically different from its counterpart in classical thermoelasticity.

The numerical results presented in FIGS 1 and 2 do not correspond to any actual material, but physically allowable values of the dimensionless eigenvalues have been chosen. These may be read off from the scales if desired. In each FIGURE the broken curves correspond to classical and the full curves to generalised thermoelasticity for the value of τ given. FIG 1 illustrates the situation of the previous paragraph where only one branch is much different in the two theories. In FIG 2, however, the branches join together quite differently in the two theories.

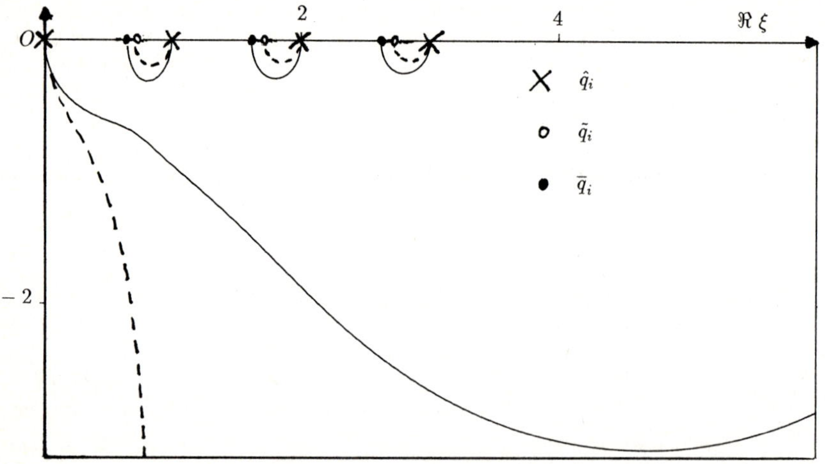

FIG 1 THE FOUR BRANCHES OF $\xi(\omega)$, $(\tau = 0.15)$

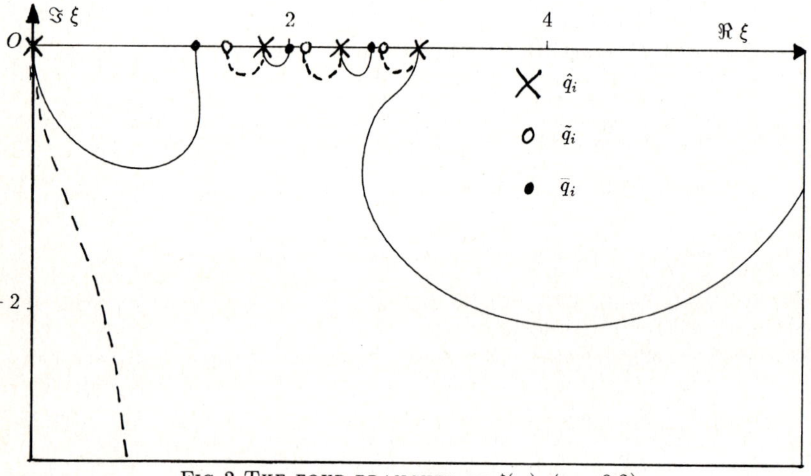

FIG 2 THE FOUR BRANCHES OF $\xi(\omega)$, $(\tau = 0.2)$

References

[1] D.S.Chandrasekharaiah, Thermoelasticity with second sound:A review, *Appl. Mech. Reviews*, **39**, 1986, 355–376.

[2] J.N.Sharma and R.S.Sidhu, On the propagation of plane harmonic waves in anisotropic generalised thermoelasticity, *Int. J. Engng Sci.*, **24**, 1986, 1511–1516.

[3] P.Chadwick, Basic properties of plane harmonic waves in a prestressed heat-conducting elastic material, *J. Thermal Stresses*, **2**, 1979, 193–214.

[4] N.H.Scott, Stability in thermoelasticity, Irish Mechanics Group, July 1985, Cork, Ireland.

ELASTIC WAVE PROPAGATION
M.F. McCarthy, M.A. Hayes, (Editors)
Elsevier Science Publishers B.V. (North-Holland), 1989

ELASIC WAVES IN THE PRESENCE OF A NEW TEMPERATURE SCALE

Witold KOSIŃSKI

Polish Academy of Sciences
Institute of Fundamental Technological Research
Warsaw, Poland

The paper is devoted in the derivation of a set of constitutive laws for a thermoelastic material with a new temperature scale, which allow to investigate the propagation of thermal waves. An essential property of the new temperature is its continuity even in the case of shock waves in contrast to the discontinuity in the absolute temperature. Moreover, a non-linear version of the so-called-Vernotte-Cattaneo law for heat conduction appears in the present set of equations in the form of the prolongeted equation of the kinetic equation for the new temperature scale. The velocity and the corresponding amplitudes of weak thermal and waves are derived additional to the shock jump condition for the heat flux vector.

1. INTRODUCTION

The classical theory of the conduction of heat rests upon Fourier's law which expresses the hypothesis that the flux of heat is proportional to the gradient of temperatute distribution. As a concequence of this hypothesis, the thermal disturbance in a finite region instantaneously affects all points ot the body. This behaviour, which implies an *infinite* speed of propagation of thermal disturbances has motivated proposals to modify Fourier's law to include the time derivative of the heat flux accompanied by *a thermal relaxatiqn time* [1-5] or by introducing the temperature rate as an additional state variable in the description [6-7].

In the present paper we keep the original proportionality relation between the heat flux vector and the temperature gradient with one small modification: in that relation the gradient operates on *a new temperature scale*, which is related to the absolute temperature by a kinetic equation. Consequently the new temperature scale is a functional of the absolute one and its continuous behaviour in case of shock waves is essential in the theory.

2. GOVERNING EQUATIONS

In the the rational mechanics approach a non-linear thermoelastic material body, with its reference mass density ρ_0, is characterized by constitutive eqations for the internal energy ε, the absolute temperature ϑ, the Piola-Kirchhoff stress S and the heat flux vector q in terms of the deformation (displacement) gradient F, the specific entropy η and the temperature gradient $\nabla\vartheta$. If v and $\dot{v}$ denotes the particle velocity and the acceleration vector, respectively, and Div is the divergence operator in the reference placement of the body, then in the presence of the body force b and the body heat supply r, the balance laws for the linear momentum

$$\rho_0 \mathrm{v} - \mathrm{Div}\,\mathrm{S} = \rho_0 \mathrm{b}$$

and the energy

$$\rho_0 \overline{(\varepsilon + 0.5\, \mathrm{v}^2)}^{\cdot} - \mathrm{Div}(\mathrm{vS} - \mathrm{q}) = \rho_0 (\mathrm{b}\cdot \mathrm{v} + r),$$

together with the unilateral differential constraint

$$\rho_0 \dot{\eta} + \mathrm{Div}\,(\mathrm{q}\,\vartheta^{-1}) \geq \rho_0\, r\, \vartheta^{-1} \tag{2.1}$$

which represents the second law of thermodynamics, lead to the well-known consequences: 1. the independence of the energy function of the temperature gradient, 2. the potential relations

$$\vartheta = \partial_\eta \hat{\varepsilon}(\mathrm{F},\eta), \quad \mathrm{S} = \rho_0 \partial_{\mathrm{F}} \hat{\varepsilon}(\mathrm{F},\eta), \tag{2.2}$$

3. the residual inequality

$$-\vartheta^{-1}\mathrm{q}\cdot\nabla\vartheta \geq 0. \tag{2.3}$$

The last inequality places restriction on the constitutive equation for the heat flux vector $\mathrm{q} = \hat{\mathrm{q}}(\mathrm{F},\eta,\nabla\vartheta)$. If the equation takes the form of the phenomenological Fourier law

$$\mathrm{q} = -k\nabla\vartheta, \tag{2.4}$$

where k is the thermal conductivity, then (2.3) results in $k \geq 0$, since $\vartheta > 0$. It is known that no pure thermal waves may propagate in the medium described by the above system of equations.

Now we are giving a possible simple way of improvement of the equations that does not change the essence of the proportionality law (2.4) and at the same time overcome the difficulty.

Let $\beta \in (0,\infty)$ represent a new temperature scale which is related to the previous one by the solution of the following initial value problem (the kinetic equation)

$$\dot{\beta} = f(\vartheta, \beta), \quad \beta(t_0) = \beta_0. \tag{2.5}$$

Here the function f will be determined latter. The initial instant t_0 is the beginning of the process considered. In the present approach the aforementioned proportionality law for the heat flux is following

$$\mathbf{q} = b \nabla \beta, \tag{2.6}$$

where the coefficient b may, in general, be a function of F and η. In the final derivation the simplest case of constant b will be assumed. The constitutive equation for the specific energy ε, the temperature ϑ and the stress S are assumed in the form of smooth function relations with F, η and $\nabla\beta$ as independent variables. Now restrictions imposed on that relations by the second law of thermodynamics (2.1) are to be derived.
Using the both balance laws and the chain rule property in differentiating the energy function $\hat{\varepsilon}$, we obtain

$$\rho_0 (\vartheta - \partial_\eta \hat{\varepsilon})\dot{\eta} + (\mathrm{S} - \rho_0 \partial_{\mathrm{F}} \hat{\varepsilon}) \cdot \dot{\mathrm{F}} - \rho_0 \partial_{\nabla\beta} \hat{\varepsilon} \cdot \overline{\nabla \beta}^{\cdot} - \frac{1}{\vartheta} \mathbf{q} \cdot \nabla \vartheta \geq 0. \tag{2.7}$$

To find the second mixed derivative of β the prolongation of the kinetic equation (2.5) is to be considered. It takes the form

$$\overline{\nabla \beta}^{\cdot} = \partial_\vartheta f \nabla \vartheta + \partial_\beta f \nabla \beta. \tag{2.8}$$

Hence we can see that the both time derivatives $\dot{\eta}$ and $\dot{\mathrm{F}}$ are independent of other variables and their derivatives. Consequently the inequality (2.7) will be satisfied provided the following relationships

$$\vartheta - \partial_\eta \hat{\varepsilon} = 0 \quad \text{and} \quad \mathrm{S} - \rho_0 \partial_{\mathrm{F}} \hat{\varepsilon} = 0, \tag{2.9}$$

and the inequality

$$- \rho_0 \partial_{\nabla\beta} \hat{\varepsilon} \cdot \overline{\nabla \beta}^{\cdot} - \frac{1}{\vartheta} \mathbf{q} \cdot \nabla \vartheta \geq 0 \tag{2.10}$$

hold. The former gives the classical potential relations while the latter needs further analysis. Due to the prolongated equation (2.8) we can write

$$-(\rho_0 \partial_{\nabla\beta} \hat{\varepsilon}\, \partial_\vartheta f + \frac{1}{\vartheta} \mathbf{q}) \cdot \nabla \vartheta - \rho_0 \partial_{\nabla\beta} \hat{\varepsilon}\, \partial_\beta f \cdot \nabla \beta \geq 0. \tag{2.11}$$

Now we are going to solve the last inequality under particular assumption concerning the form of the energy function $\hat{\varepsilon}$. We assume, namely that $\hat{\varepsilon}$ can be written as the sum

$$\hat{\varepsilon}(\mathrm{F}, \eta, \nabla \beta) = \hat{\varepsilon}_1(\mathrm{F}, \eta) + 0.5\, \hat{\varepsilon}_2 (\nabla \beta)^2, \tag{2.12}$$

where we could assume that $\hat{\varepsilon}_2$ depends on F and η as well. For further derivation, however, we consider the simplest case of constant $\hat{\varepsilon}_2$. Then due to (2.6) we obtain

$$-(\rho_0 \hat{\varepsilon}_2 \partial_\vartheta f + \frac{b}{\vartheta}) \nabla \beta \cdot \nabla \vartheta - \rho_0 \hat{\varepsilon}_2 \partial_\beta f (\nabla \beta)^2 \geq 0. \tag{2.13}$$

The case of constant b leads to two relations

$$\rho_0 \hat{\varepsilon}_2 \partial_\vartheta f + \frac{b}{\vartheta} = 0 \text{ and } \rho_0 \hat{\varepsilon}_2 \partial_\beta f \leq 0. \tag{2.14}$$

Integrating the former we get the general form of the function f

$$f(\vartheta,\beta) = (\rho_0 \hat{\varepsilon}_2)^{-1} b \log\{\vartheta_0 (\vartheta)^{-1}\} + f(\vartheta_0,\beta), \tag{2.15}$$

where ϑ_0 is a reference temperature. Since $\hat{\varepsilon}$ must be positive, from $(2.14)_2$ follows, that the partial derivative $\partial_\beta f$ cannot be positive, and consequently the solution of the kinetic equation is stable.

To get the explicit representation of f and to show a link between the present approach and that basing on the Maxwell-Vernotte-Cattaneo (M-V-C) type equations let us make the following assumption:

the prolongated kinematic equation (2.8) *together with the constitutive equation for the heat flux* (2.6) *lead to the rate type equation*

$$\mathrm{T}\,\dot{\mathbf{q}} + \mathbf{q} = -k \nabla \vartheta, \tag{2.16}$$

where T *is a thermal relaxation time-valued function.*

The immediate consequences of this are the following relations

$$k = -\mathrm{T}\, b\, \partial_\vartheta f \quad \text{and} \quad 1 = -\mathrm{T}\, \partial_\beta f. \tag{2.17}$$

which show that *f cannot be independent of β*. The integrability (compatibility) condition of (2.17) is

$$\partial_\beta (k\, \mathrm{T}^{-1}) - b\, \partial_\vartheta (\mathrm{T}^{-1}) = 0. \tag{2.18}$$

Let τ_0 and k_0 be a characteristic material constant representing the relaxation time (cf.[[5]) and the coefficient of thermal conductivity at ϑ_0, respectively, and assume that $b = k_0$. Then the dimensional analysis together with the requirement $\hat{\varepsilon}_2 = 0$ when $\tau_0 = 0$, lead to the following representation

$$\hat{\varepsilon}_2 = k\, \tau_0 (\rho_0 \vartheta_0)^{-1} \tag{2.19}$$

Due to (2.17) and (2.18), a particular representation of T will imply forms of f and k. Now we list three forms of T and the corresponding of f and T.

A. If $\mathrm{T}(\beta) = \tau_0$, then $k(\vartheta,\beta) = k_0 \vartheta_0 \vartheta^{-1}$ and

$$f(\vartheta,\beta) = \tau_0^{-1} \vartheta_0 \log(\vartheta_0 \vartheta^{-1}) - (\beta - \beta_0)\tau_0^{-1}. \tag{2.20}$$

B. If $\mathrm{T}(\beta) = \tau_0 \beta \vartheta_0^{-1}$, then $k(\vartheta,\beta) = k_0 \beta \vartheta^{-1}$ and

$$f(\vartheta,\beta) = \tau_0^{-1}\vartheta_0 \log(\vartheta_0\vartheta^{-1}) + \tau_0^{-1}\vartheta_0 \log(\beta_0\beta^{-1}). \tag{2.21}$$

C. If $T(\beta) = \tau_0\vartheta_0\beta^{-1}$, then $k(\vartheta,\beta) = k_0\vartheta_0^2(\beta\vartheta)^{-1}$ and

$$f(\vartheta,\beta) = \tau_0^{-1}\vartheta_0 \log(\vartheta_0\vartheta^{-1}) - 0.5(\tau_0\ \vartheta_0)^{-1}(\beta^2 - \beta_0^2). \tag{2.22}$$

The direct inspection of the listed constitutive functions shows that the temperature ϑ_0 plays the role of a critical value. If we check the experimental results concerning the dependence of the thermal conductivity on the temperature, then for many materials one observes two regions of that dependence. At rather low temperatures (below 20 or 30°K) k is an increasing function of ϑ, while for higher temperatures k decreases with ϑ. Moreover the form of f in all three cases secures that β is decreasing if $\vartheta > \vartheta$.

3. WAVE VELOCITIES

Before the case A of the constitutive equations will be considered, let us notice that in the model proposed in the previous section the sef of field and constitutive equations in the case of shock waves has the following property, as far as the jumps of discontinuity of state variables are concerned, independently of the particular forms of the functions $f(\vartheta,\beta)$, $T(\beta)$ and $k(\vartheta,\beta)$. Namely

$$V[\![\beta]\!] = 0, \quad V[\![q]\!] + [\![bf]\!]N = 0, \tag{3.1}$$

where V and N is the velocity and the normal versor of the shock wave, respectively. The continuity of the new temperature scale is an essential property of the model proposed in Sec.2. Moreover, in the case A the relation $(3.1)_2$ takes the form

$$[\![\ q\ ]\!] = k_0(V\tau_0)^{-1}\vartheta_0 \log\frac{\vartheta^-}{\vartheta^+}\ N. \tag{3.2}$$

Comparing the present result with that obtained by Achenbach in [1] we notice the difference resulting from the linearization made in [1]; there instead of $\log\frac{\vartheta^-}{\vartheta^+}$ the jump $[\![\vartheta]\!]$ appares. Coefficients are the same in both relations.

For the case A under consideration, neglecting the mechanical coupling, from the field equations written as a first order system in ϑ and $z = \nabla\beta$, we can derive the equation for the velocity of weak discontinuity wave, i.e. for the characteristic velocity λ. It reads

$$\lambda^2(\rho_0 c_V\lambda^2 + \lambda k_0\vartheta^{-1} z\cdot n - k_0\vartheta_0\ (\tau_0\ \vartheta\ )^{-1} = 0 \tag{3.3}$$

where n is normal to the wave front and $c_V = \vartheta\ \partial_\vartheta\hat{\eta}$ is the specific heat at constant volume. Two vanishing velocities correspond to two amplitude vectors (i.e.right eigenvectors) which are perpendicular to **n**. Other two nonvanishing velocities are related to the amplitude vectors $\mathbf{r}_i = \vartheta_0 (\lambda_i \tau_0 \vartheta)^{-1} \mathbf{n}$, $i = 1,2$, which are parallel; they are equal but opposite directed when the waves are symmetric , i.e.

$$\lambda^2 = k_0 \vartheta_0 \, (\tau_0 \rho_0 c_v \vartheta)^{-1}. \tag{3.4}$$

It can happen only if $\mathbf{z} = \nabla\beta$ is perpendicular to **n**, i.e. when $\mathbf{q} \cdot \mathbf{n} = 0$. This is rather obvious in view of the physical process of heat conduction , not mentioning the second law of thermodynamics. The requirement the thermal waves be symmetric in any case is nonrealistic [4]. Note that the wave velocity is temperature dependent.

The case of thermomechanical waves with the explicit representation of the free energy function ψ of the form, suitable for the description of a rock-like medium under very high confining pressure,

$$\psi(F,\vartheta,\nabla\beta) = a_1(\vartheta)(\det F - 1) + a_2(\vartheta)(\mathrm{tr} FF^T - 3) + a_3(\vartheta)\log\det F + 0.5\,\hat{\varepsilon}_2 (\nabla\beta)^2$$

will be considered in the next paper.

REFERENCES

[1] Achenbach, J.D., The influence of heat conduction on propagating stress jumps, J.Mech.Phys.Solids,vol.16,(1968) 273-282.

[2] Gurtin, M.E.and Pipkin A.C., A general theory of heat conduction with finite wave speeds, Arch.Ration.Mech.Anal.,vol.31,(1968) 113-126.

[3] Kosiński, W., Thermal waves in inelastic bodies, Arch. Mechanics (AMS),vol.27,(1975) 733-748.

[4] Mihàilescu, M. and Suliciu, I.,Finite and symmetric thermomechanical waves in materials with internal state variables, Int.J.Solids Struct., vol.12,(1976) 559-575.

[5] Chester, M., Second sound in solids, Phys.Rev.,vol.131,(1963) 2013-2015.

[6] Bogy, D.B. and Naghdi, P.M., On heat conduction and wave propagation in rigid solids, J.Math.Physics,vol.11,(1970) 917-923.

[7] McCarthy, M.F., Thermodynamics of materials of differential type, Proc.Vibr.Probl.(now J.Techn. Phys.),vol.13,(1972) 137-143.

AUTHOR INDEX

Achenbach, J.D. 419
Alkemade, J.A. 293
Aron, M. 541
Bamberger, A. 241
Barnett, D. M. 33
Basta, S. 95
Beltzer, A. I. 427
Bogy, D.B. 279,307
Borejko, P. 39
Bostrom, A 101,383,401
Boulanger, Ph. 603
Braga, A.M.B. 301
Brazier-Smith, P. 247
Brock, L.M. 531
Burridge,R. 229
Carroll, M. M. 617
Chadwick, P. 3
Chang,H.- W. 229
Chapman,C. H. 413
Chen, P. J. 537
Chen, P.L. 59
Chin, L. C. 287
Christiansen, P. 167
Coates, R.T. 413
Codegone, M. 455
Corones, J. 349
Cowin, S.C. 513
Cox, E. A. 173
Craine, R.E. 541
Daher, N. 147
Dassios, G. 407
Datta, S. K. 101,383
David, E. A. 125
de Billy, M. 89
de Hoop, A. 369,485
de Maag, J.W. 293
de Vries, S.M. 315
Dermenjian,Y. 241
Dieulesaint, E. 17
Dost, S. 577
Downey, H.A. 279
Dunn, J.E. 133
Dunwoody, J. 107
Felsen, L.B. 195
Fosdick, R.L. 133
Fu, Y.B. 141
Fukuoka, H. 75
Gabay, C. C. 195
Gaunaurd, G. C. 335,395
Ghaleb, M.J. 309
Gracewski,S. M. 307
Green, R.E. 583
Green, W.A. 189
Ha-Duong, T. 445
Haddow, J. B. 161
Hall. W. S. 461
Harris, J. G. 507
Hayes, M. 553
Herrman, G. 301
Hirao, M. 75
Hosten, B. 95
Hsiao, G. C. 321,547
Hsieh, R.K.T. 589
Itabashi, M. 223
Jensen, F. B. 211
Johnson. G.C. 45
Joly, P. 241
Kawata, K. 223
Kitahara, M. 449
Kleinman, R. E. 321
Klosner, J. M. 195
Konczak, Z. 281
Kooij, B. J. 217
Kosinski,W. 631
Kuo,M.K. 59,419
Lee, S. Y. 287
Lenz, J. 81
Lothe, J.33
Lu, I.T. 195
Majorkowska-Knap, K. 81
Mal, A. K. 67
Martin, P. A. 389
Maugin,G. A. 147,571
McCarthy, M.F. 617
McCoy, J. J. 359
Meister, E. 375
Molina, C. 253
Morro, A. 537
Mortell, M.P. 173
Muir, F. 259
Nakagawa, K. 449
Nakano, M. 467
Nelson, D.F. 597
Norris, A. N. 493
Nowacki , J. P. 609
Olsson, P. 101,383,401
Pao, Y. H. 235
Parker, D.F. 125
Pastrone, F. 253
Peterson, L. 439
Piau, M. 309
Pierce, A. D. 205
Planat, M. 113
Pouget, J. 179
Quak, D. 473
Quentin, G. 89,479
Rao, Govinda,
Robertson W. H. 461
Rossmanith, H. P. 51
Royer, D. 17
Ryan, D. 553
Sabina, F. J. 327
Sayers, C. M. 433
Sayir, M. 525
Schmidt, H. 211
Schoenberg, M. 259
Schuetz. L. 321
Scott, N.H. 141,625
Soerensen, M. 167
Speck,F. - O. 375
Stam, H.J. 369, 485
Sugimoto, N. 265
Tait, R. J. 161
Talmant, M. 479
Teymur, M. 155
Tiersten, H.F. 559
Wang, T.K. 59
Wegner, J. 161
Wendland, W.L. 547
Werby, M. F. 395
Wickham, G.
Willis, J. R. 327
Xu,P.C. 67
Yang, H. J. 273 287, 419
Yeh, C.S. 59
Zhang, Y. 113
Ziegler, F. 39, 235
Ziv, M. 519